Lecture Notes in Computer Science 16587

Founding Editors

Gerhard Goos
Juris Hartmanis

Editorial Board Members

Elisa Bertino, *Purdue University, West Lafayette, IN, USA*
Wen Gao, *Peking University, Beijing, China*
Bernhard Steffen, *TU Dortmund University, Dortmund, Germany*
Moti Yung, *Columbia University, New York, NY, USA*

Florent Foucaud · Aline Parreau
Editors

Combinatorial Algorithms

37th International Workshop, IWOCA 2026
Clermont-Ferrand, France, June 8–11, 2026
Proceedings

Editors
Florent Foucaud [iD]
Université Clermont Auvergne
Aubière, France

Aline Parreau [iD]
CNRS and Université Lyon 1
Villeurbanne, France

ISSN 0302-9743 ISSN 1611-3349 (electronic)
Lecture Notes in Computer Science
ISBN 978-3-032-27731-2 ISBN 978-3-032-27732-9 (eBook)
https://doi.org/10.1007/978-3-032-27732-9

Preface

We are happy to write this preface for the proceedings of the 37th International Workshop on Combinatorial Algorithms (IWOCA 2026), now that the main work of the Programme Committee (PC) of selecting the papers for presentation is completed. The PC of IWOCA 2026 consisted of 38 members from diverse geographic areas, genders, and scientific expertise (related to combinatorial algorithms). We are happy to report that the collaboration was successful, and we enjoyed working together towards a successful conference!

IWOCA 2026 was organized at the Université Clermont Auvergne in Clermont-Ferrand, in the volcanic region of Auvergne, in the heart of France. See https://iwoca2 026.limos.fr for the event's website. In order to accommodate difficult personal situations or travel-related issues, and to lower the carbon footprint of the conference, it was decided that authors of accepted papers could present them online. The conference took place on June 8–12, 2026, including a collaborative research workshop on June 11–12.

IWOCA is an annual conference series that started in 1989 as AWOCA (Australasian Workshop on Combinatorial Algorithms), and became an international conference in 2007, having been held (in alphabetical order) in Australia, Canada, Czech Republic, Finland, France, Germany, India, Indonesia, Italy, Japan, Singapore, South Korea, Taiwan, UK, and USA. In 2026, IWOCA returned to France, for the first time since 2013, where it was held in Rouen (IWOCA 2020 was also planned in Bordeaux, France but had to be held online due to the COVID-19 pandemic). The conference brought together researchers on diverse topics related to combinatorial algorithms, as specified in the Call for Papers.

Three invited talks were presented at IWOCA 2026, by Sergio Cabello (University of Ljubljana, Slovenia), Pierre Fraigniaud (CNRS and Université Paris Cité, France), and Neeldhara Mishra (Indian Institute of Technology Gandhinagar, India). Their abstracts are included in the frontmatter below.

One usual focus of IWOCA that we particularly enjoy is its collaborative aspect. Thus, IWOCA 2026 featured a traditional open problem session, and a 1.5-day collaborative workshop at the end of the conference, to hopefully create new collaborations and fostered even more scientific exchange. Moreover, following the model of many recent conferences in France and elsewhere, the organizers also implemented an "Extended Stay Support Scheme (ESSS)" for participants who wished to extend their stay in France. The intention was to lower the carbon footprint of the conference by also encouraging collaborations. Typically, the ESSS funds a 1-week research visit of a conference participant with a colleague in a French institution, just before or after the conference. The funding is provided by the host institution. As the time of writing, already a dozen participants expressed their interest in the ESSS, and we remove hope IWOCA 2026 and its ESSS will foster many fruitful collaborations!

The PC received 110 abstract submissions, out of which 94 were fully refereed (the others were either retracted before the full paper submission deadline, or desk-rejected). Each submission was reviewed by at least three PC members with the help of trusted external referees, and evaluated on its quality, originality, and relevance to the conference. The PC selected 37 papers for presentation at the conference and inclusion in the proceedings.

For the first time in the history of IWOCA, in order to increase the fairness of the selection process (including papers co-authored by PC members), we used a double-blind submission system (facilitated via the EasyChair submission platform). We believe it was successful and came with only minor constraints, and hope that double-blind reviewing can be generalized in our research community.

The PC also selected two papers to receive the Best Paper Award and the Best Student Paper Award, respectively. These awards were sponsored by Springer. The awardees were:

- Best Paper Award: *(Even hole, triangle)-free graphs revisited* by Beatriz Martins and Nicolas Trotignon;
- Best Student Paper Award: *Minimum Clique Bicoloring* by Shunsuke Hamada, Yuto Okada, Hirotaka Ono, and Yota Otachi.

We also organized two special issues for full versions of the selected papers: one in the *Journal of Computer and System Sciences* (published by *Elsevier*), for 5-6 selected papers, and one in the online diamond open-access journal *Discrete Mathematics & Theoretical Computer Science* (published by the institutional publisher *Episciences*), where all full versions of papers can be submitted.

Finally, we would like to thank all authors who submitted their papers to IWOCA 2026 for their interest in IWOCA, the invited speakers and authors of accepted papers for their talks and participation to the conference, the PC members for their time and energy, and the 156 external reviewers for their expertise. We thank Springer for publishing the proceedings of IWOCA 2026 in the ARCoSS subline of their LNCS series and for their financial support towards the best paper and the best student paper awards, and the editors-in-chief of the *Journal of Computer and System Sciences* and *Discrete Mathematics & Theoretical Computer Science* for agreeing to edit the two IWOCA 2026 special issues. We thank the Steering Committee for giving us the opportunity to serve as program chairs of IWOCA 2026 and for their continuous support, and the Local Organizing Committee for organizing the conference. Last but not least, we also thank all the institutional sponsors that helped to keep the registration fee of IWOCA 2026 as low as possible.

April 2026   Florent Foucaud
                                                                Aline Parreau

Organization

Program Committee Chairs

Florent Foucaud	Université Clermont Auvergne, France
Aline Parreau	CNRS and Université Lyon 1, France

Steering Committee

Maria Chudnovsky	Princeton University, USA
Henning Fernau	Universität Trier, Germany
Ralf Klasing	CNRS and Université de Bordeaux, France
Tomasz Radzik	King's College London, UK
Bill Smyth	McMaster University, Canada
Wing-Kin (Ken) Sung	Chinese University of Hong Kong, China

Program Committee

Cristina Bazgan	Université Paris Dauphine-PSL, France
Petra Berenbrink	Universität Hamburg, Germany
Davide Bilò	Università dell'Aquila, Italy
Hans L. Bodlaender	Utrecht University, Netherlands
Hans-Joachim Böckenhauer	Eidgenössische Technische Hochschule Zürich, Switzerland
Tiziana Calamoneri	Sapienza Università di Roma, Italy
Dibyayan Chakraborty	University of Leeds, UK
Mónika Csikós	Université Paris Cité, France
Antoine Dailly	INRAE and Université Clermont Auvergne, France
Peter Dankelmann	University of Johannesburg, South Africa
Jessica Enright	University of Glasgow, UK
Henning Fernau	Universität Trier, Germany
Florent Foucaud	Université Clermont Auvergne, France
Serge Gaspers	University of New South Wales, Australia
Pavol Hell	Simon Fraser University, Canada
Ling-Ju Hung	National Taipei University of Business, Taiwan
Ralf Klasing	CNRS and Université de Bordeaux, France

Thierry Lecroq	Université de Rouen Normandie, France
Daniel Lokshtanov	University of California, Santa Barbara, USA
Arnaud Mary	Université Lyon 1, France
Martin Milanič	University of Primorska, Slovenia
Valia Mitsou	Université Paris Cité, France
Pedro Montealegre	Universidad Adolfo Ibáñez, Chile
Amer Mouawad	American University of Beirut, Lebanon
Lucia Moura	University of Ottawa, Canada
Alantha Newman	CNRS and ENS de Lyon, France
Prajakta Nimbhorkar	Chennai Mathematical Institute, India
Aline Parreau	CNRS and Université Lyon 1, France
Rajiv Raman	Indraprastha Institute of Information Technology Delhi, India
Adele Rescigno	University of Salerno, Italy
Matthieu Rosenfeld	Université de Montpellier, France
Sagnik Sen	Indian Institute of Technology Dharwad, India
Maria Serna	Universitat Politècnica de Catalunya, Spain
Ana Silva	Universidade Federal do Ceará, Brazil
Uéverton Souza	Instituto de Matemática Pura e Aplicada, Brazil
Prafullkumar Tale	Indian Institute of Science Education and Research Pune, India
Kunihiro Wasa	Hosei University, Japan
Binhai Zhu	State University of Montana, USA

Additional Reviewers

Adak, Rajat	Burjons, Elisabet
Adamson, Duncan	Buzzega, Giovanni
An, Hyung-Chan	Cabello, Sergio
Antony, Dhanyamol	Cameron, Kathie
Apollonio, Nicola	Carvalho Sandes, Nelson
Arana, Carmen	Casel, Katrin
Aubian, Guillaume	Chu, Huairui
Bailey, Robert	Cichacz, Sylwia
Banik, Aritra	Colli, Giordano
Bellitto, Thomas	Concha-Vega, Pablo
Bergé, Pierre	Conte, Alessio
Berthe, Gaétan	Cordasco, Gennaro
Blömer, Johannes	Corò, Federico
Bousquet, Nicolas	Coste, Louann
Breitkopf, Tom-Lukas	D'Arco, Paolo
Bruno, Roberto	Darmon, Sasha

Das, Arun Kumar
Das, Syamantak
David, Julien
Davot, Tom
De, Minati
de Colnet, Alexis
Delepine, Thomas
de Paula, Samuel P.
Dey, Sanjana
Di Fonso, Alessia
Di Stefano, Gabriele
Dreyer, Simon
E S, Ajaykrishnan
El Sabeh, Remy
Elbassioni, Khaled
Erlebach, Thomas
Fernandes, Tatiane
Fleischmann, Pamela
Froncek, Dalibor
Gahlawat, Harmender
Gargano, Luisa
Gehnen, Matthias
Gollin, Pascal
Gonçalves, Daniel
Groenland, Carla
Gupta, Shiwali
Götte, Thorsten
Habib, Michel
Herrmann, Anton
Hintze, Lukas
Huynh, Tony
Inamdar, Tanmay
Ingels, Florian
Isenmann, Lucas
Jabal Ameli, Afrouz
Jacob, Dalu
Jacob, Hugo
Jauregui, Benjamin
Johnson, Laura
Jooken, Jorik
Kamata, Tonan
Klein, Kim-Manuel
Komm, Dennis
Koutris, Paris
Kralovic, Richard

Kurita, Kazuhiro
Lamme, Jeroen
Lauria, Massimo
Lehtilä, Tuomo
Leucci, Stefano
Liedloff, Mathieu
Limouzy, Vincent
Lintzmayer, Carla Negri
Lorieau, Lucas
Lucke, Felicia
M, Komathi
Maack, Marten
Majumdar, Diptapriyo
Mallem, Maher
Mann, Kevin
Mao, Yaping
Molter, Hendrik
Morini, Fiorenza
Muehlenthaler, Moritz
Mömke, Tobias
Najem, Sara
Nanoti, Saraswati
Narayanaswamy, N.S.
Nemery, Edouard
Nichterlein, André
Ochem, Pascal
Ogrin, Laura
Olver, Neil
Ong, Ryan
P D, Pavan
Paesani, Giacomo
Pallathumadam, Sreejith K
Panolan, Fahad
Paul, Kaustav
Pedrosa, Lehilton L. C.
Pellegrini, Marco Antonio
Perez, Anthony
Pierron, Théo
Rabie, Mikaël
Rai, Ashutosh
Raso, Mario
Richer, Camille
Rolvien, Maurice
Romashchenko, Andrei
Röglin, Heiko

Sakai, Yoshifumi
Saladi, Rahul
Salo, Ville
Salvo, Ivano
Sampaio, Rudini
Sandeep, R.B.
Santos Morais, Cicero Samuel
Sawada, Joe
Schlöter, Jens
Seetharaman, Sanjay
Sharma, Vikram
Shin, Yongho
Siebertz, Sebastian
Silveira, Rodrigo
Sinaimeri, Blerina
Skodinis, Emanuel
Stocker, Moritz

Takaoka, Asahi
Tiwary, Hans Raj
Vaccaro, Ugo
van Renssen, André
Vaz, Daniel
Verbeek, Kevin
Vetrik, Tomas
Viana, Luiz Alberto
Vigny, Alexandre
Vušković, Kristina
Whittington, Philip
Wong, Dennis
Xu, Xiaoyang
Yuditsky, Yelena
Zamaraev, Viktor
Zeh, Norbert
Žerovnik, Janez

Local Organizing Committee

Samuel Avril	ENS de Lyon
Gaétan Berthe	CNRS and Université Clermont Auvergne
Béatrice Bourdieu	CNRS and Université Clermont Auvergne
Martine Caccioppoli	CNRS and Université Clermont Auvergne
Antoine Dailly (Co-chair)	INRAE and Université Clermont Auvergne
Jona Dirks	Université Clermont Auvergne
Solène Drouet	Université Clermont Auvergne
Florent Foucaud (Co-chair)	Université Clermont Auvergne
Yan Gerard	Université Clermont Auvergne
Claire Hilaire	Université Clermont Auvergne
Sophie Huiberts	CNRS and Université Clermont Auvergne
Victoria Kaial	Université Clermont Auvergne
Adrien Leduque	INRAE and Université Clermont Auvergne
Pham Le vu	Université Clermont Auvergne
Lucas Lorieau	CNRS and Université Clermont Auvergne
Anirudh Rachuri	Université Clermont Auvergne
Nicolas Schivre	Université Clermont Auvergne
Alexandre Vigny	Université Clermont Auvergne
Xiaofeng Wang	Université Clermont Auvergne
Loïc Yon	Université Clermont Auvergne

Abstracts of Invited Talks

Computational Geometry with Predictions

Sergio Cabello

University of Ljubljana, Slovenia

Algorithms with predictions have received considerable attention in recent years. Given a problem instance and a prediction for the solution, we would like to devise a procedure that solves the problem in such a way that the algorithm performs "better" when the prediction is "close" to the correct information. "Better" may be, for example, in terms of running time, approximation factor, or competitive ratio. Different measures of "closeness" may be used depending on the problem, or alternatively we may assume that predictions obey some probabilistic models.

I will explain some recent research in computational geometry that follows this paradigm of algorithms with predictions. In particular, we will look at the problem of searching for a target point at some unknown location when we get information about the approximate distance to the target, and at the problem of computing the Delaunay triangulation in the plane when we are given a triangulation that is close to be the Delaunay triangulation.

Most of the material will be based on joint work with Timothy M. Chan and Panos Giannopoulos.

Algorithmic Meta-Theorems for Distributed Computing

Pierre Fraignaud

CNRS, Université Paris Cité, France

Algorithmic meta-theorems can be viewed as theorems applying to large classes of algorithmic problems at once. An archetypal example of algorithmic meta-theorems is Courcelle's theorem (1990), which states that every graph property definable in the monadic second-order logic of graphs can be decided in linear time on graphs of bounded treewidth. Such a theorem is remarkable for many reasons, in particular because it establishes tight connections between sequential algorithms complexity, graph theory, and logic. It is only recently that meta-theorems have been formulated for distributed computing. This talk will survey some of the recent contributions in this field. In particular, it will address distributed decision for graph properties definable in first-order logic in graphs of bounded expansion, and distributed certification for graph properties definable in monadic second-order logic in graphs of bounded treewidth, as well as in graphs of bounded clique-width.

On Fairness with Graphical Valuations

Neeldhara Misra

Indian Institute of Technology Gandhinagar, India

The existence of envy-free up to any item (EFX) allocations is widely regarded as a central open problem in discrete fair division, and remains open beyond three agents. A recent line of work has carved out a structurally rich and surprisingly tractable corner of the landscape: the class of graphical valuations, introduced by Christodoulou, Fiat, Koutsoupias, and Sgouritsa (EC 2023), in which agents are vertices of a graph, items are edges, and each item is valued only by its two endpoint-agents. Their remarkable observation that EFX always exists on such instances has catalyzed a fast-moving body of follow-up work. This talk will survey this trajectory, leading to various intriguing questions this young area has surfaced. We hope to demonstrate that graphical valuations have become a productive meeting point for structural graph theory, parameterized complexity, and fair division.

Contents

On the Complexity of Vertex-Splitting into an Interval Graph 1
*Faisal N. Abu-Khzam, Dipayan Chakraborty, Lucas Isenmann,
and Nacim Oijid*

Bounds on Linear Turán Number for Trees 16
Rajat Adak and Pragya Verma

Exact Algorithms for Edge Deletion to Cactus 32
Sheikh Shakil Akhtar and Geevarghese Philip

Minimizing the Weighted Makespan with Restarts on a Single Machine 45
*Aflatoun Amouzandeh, Klaus Jansen, Lis Pirotton, Rob van Stee,
and Corinna Wambsganz*

Parameterized Algorithms for k-Inversion 60
Dhanyamol Antony, L. Sunil Chandran, Dalu Jacob, and R. B. Sandeep

On the Rank and the General Position Number in Cycle Convexity 73
*Júlio Aráujo, Samuel N. Aráujo, Pedro P. Medeiros, Nicolas Nisse,
and Caroline Silva*

Hardness Results on Bondage and Reinforcement Problems in Chordal
Graphs 88
Deepak M. Bakal and Y. M. Borse

Degree Realization with Maximum Matching 103
Amotz Bar-Noy, Igor Kalinichev, David Peleg, and Dror Rawitz

Vertex-Critical Graphs in Subfamilies of $(P_4 + \ell P_1)$-Free Graphs 118
Iain Beaton and Ben Cameron

Breadth-First Search Trees with Many or Few Leaves 131
Jesse Beisegel, Ekkehard Köhler, Robert Scheffler, and Martin Strehler

The Parameterized Complexity of Scheduling with Precedence Delays:
Shuffle Product and Directed Bandwidth 146
Hans L. Bodlaender and Maher Mallem

Parameterized Algorithms for Computing MAD Trees 161
  Tom-Lukas Breitkopf, Vincent Froese, Anton Herrmann,
  André Nichterlein, and Camille Richer

Dominating Set with Quotas: Balancing Coverage and Constraints 175
  Sobyasachi Chatterjee, Sushmita Gupta, Saket Saurabh,
  Sanjay Seetharaman, and Anannya Upasana

Reachability in Graphs with Polynomially Many Surface Non-separating
Cycles is in UL 190
  Neelabjo Shubhashis Choudhury, Chetan Gupta, and Raghunath Tewari

Enumerating Spanners in Directed Temporal Graphs 204
  Lapo Cioni, Andrea Marino, Jason Schoeters, and Takeaki Uno

Beer Path Problems in Temporal Graphs 220
  Andrea D'Ascenzo, Giuseppe F. Italiano, Sotiris Kanellopoulos,
  Anna Mpanti, Aris Pagourtzis, and Christos Pergaminelis

On $(1, \leq l)$-Locating-Dominating Codes in Infinite Triangular Grid 236
  Soura Sena Das, Tuomo Lehtilä, and Sagnik Sen

Realizing Planar Linkages in Polygonal Domains 251
  Thomas Depian, Carolina Haase, Martin Nöllenburg, and André Schulz

Online Drone Coverage of Targets on a Line 266
  Stefan Dobrev, Konstantinos Georgiou, Evangelos Kranakis,
  Danny Krizanc, Lata Narayanan, Jaroslav Opatrny, Denis Pankratov,
  and Sunil Shende

Set Selection with Uncertain Weights: Non-Adaptive Queries
and Thresholds 281
  Christoph Dürr, Arturo Merino, José A. Soto, and José Verschae

Removable Online Knapsack: Exploiting Recourse and Bounded Item Sizes ... 296
  Dimitris Fotakis, Laurent Gourvès, Aris Pagourtzis,
  and Panagiotis Patsilinakos

Layer-Based Width for PAFP on DAGs: A BFS-Width-2 Normal Form
and Exact-Length Width-2 Tractability 311
  Samuel German

Solid-Resolving Sets on Directed Graphs 327
  Anni Hakanen and P. D. Pavan

Minimum Clique Bicoloring 341
    Shunsuke Hamada, Yuto Okada, Hirotaka Ono, and Yota Otachi

On the ($\leq p$)-Inversion Diameter of Oriented Graphs 356
    Frédéric Havet, Clément Rambaud, and Caroline Silva

Hardness of SetCover Reoptimization 370
    Klaus Jansen, Tobias Mömke, and Björn Schumacher

Conflict-Free Cuts in Planar and 3-Degenerate Graphs with 1-Regular
Conflicts 385
    Subrahmanyam Kalyanasundaram and Subodh Kumar

Tight Upper Bounds on Color Reversal by Local Inversions 400
    Hitendra Kumar, Kumud Singh Porte, and R. B. Sandeep

Domination and Coverage Problems Under Vulnerability Constraints 415
    Ioannis Lamprou, Nikolaos Lazaropoulos, Ioannis Sigalas,
    Ioannis Vaxevanakis, and Vassilis Zissimopoulos

Fast Order Statistics with Group Inequality Testing 430
    Adiesha Liyanage, Brendan Mumey, and Braeden Sopp

An Algorithm for Monitoring Edge-Geodetic Sets in Chordal Graphs 442
    Clara Marcille and Nacim Oijid

(Even Hole, Triangle)-Free Graphs Revisited 456
    Beatriz Martins and Nicolas Trotignon

One Sequence to Rule Them All: $\mathcal{O}(1)$-Time Parallel Generation
of Mixed-Radix Gray Codes 470
    Lucia Moura, Prangya Parida, Brett Stevens, and Aaron Williams

On the Complexity of Signed Domination 486
    Sangam Balchandar Reddy

Improved Bounds on Proper Conflict-Free Coloring of Graphs 500
    Ankit Sharma, Kaustav Paul, and Arti Pandey

Cryptographic Applications of Combinatorial Ranking for Integer
Compositions 515
    Gustavo Zambonin, Larissa Gremelmaier Rosa, Ricardo Custódio,
    and Daniel Panario

Minimizing Makespan in Sublinear Time via Weighted Random Sampling 531
Bin Fu, Yumei Huo, and Hairong Zhao

Author Index 547

On the Complexity of Vertex-Splitting into an Interval Graph

Faisal N. Abu-Khzam[1][(✉)], Dipayan Chakraborty[2], Lucas Isenmann[3],
and Nacim Oijid[4]

[1] Lebanese American University, Beirut, Lebanon
`faisal.abukhzam@lau.edu.lb`
[2] LIS, Centrale Méditerranée, Aix-Marseille Université, Marseille, France
[3] Université de Strasbourg, Strasbourg, France
[4] Umeå University, Umeå, Sweden

Abstract. Vertex splitting is a graph modification operation in which a vertex is replaced by multiple vertices such that the union of their neighborhoods equals the neighborhood of the original vertex. We introduce and study vertex splitting as a graph modification operation for transforming graphs into interval graphs. Given a graph G and an integer k, we consider the problem of deciding whether G can be transformed into an interval graph using at most k vertex splits. We prove that this problem is NP-hard, even when the input is restricted to subcubic planar bipartite graphs. We further observe that vertex splitting differs fundamentally from vertex and edge deletions as graph modification operations when the objective is to obtain a chordal graph, even for graphs with maximum independent set size at most two. On the positive side, we give a polynomial-time algorithm for transforming, via a minimum number of vertex splits, a given graph into a disjoint union of paths, and that splitting triangle free graphs into unit interval graphs is also solvable in polynomial time.

1 Introduction

Interval graphs form a fundamental subclass of intersection graphs and arise naturally in a wide range of applications, including scheduling, genetics, and computational biology [5,6,14,15,20,25]. An interval graph is the intersection graph of intervals on the real line. This class admits strong structural characterizations and supports efficient algorithms: recognizing interval graphs and constructing interval representations can be done in linear time [6].

From an algorithmic standpoint, interval graphs are appealing because of their one-dimensional structure, which enables efficient solutions to many graph

This research project was supported by the Lebanese American University under the President's Intramural Research Fund PIRF0056.
The fourth author was partly supported by the Kempe Foundation Grant No. JCSMK24-515 (Sweden).

problems that are computationally intractable on general graphs. Once an interval representation is available, problems such as graph coloring, maximum clique, maximum independent set, and minimum vertex cover admit linear-time algorithms [14], and efficient solutions are also known for Hamiltonian path, domination variants, and a variety of scheduling and resource allocation problems [6,20]. At the same time, the very same rigid structure makes interval graphs difficult to obtain from arbitrary graphs. Classical graph modification problems such as Vertex Deletion or Edge Deletion into interval graphs, unit interval graphs, or chordal graphs are known to be NP-complete, even under strong restrictions on the input graph [16,17,21].

In this work, we investigate an alternative modification mechanism based on *vertex splitting*. Given a graph $G = (V, E)$, a vertex split replaces a vertex $v \in V$ by two new vertices, each inheriting a (possibly disjoint) subset of the neighbors of v. Rather than removing vertices or edges, vertex splitting redistributes adjacency, allowing a single vertex to be represented by multiple intervals. This operation preserves information while relaxing local structural constraints.

We introduce and study the following problem: given a graph G and an integer k, can G be transformed into an interval graph using at most k vertex splits? We view this problem as a finer-grained notion of distance to the class of interval graphs, distinct from deletion-based modifications. In particular, we show that there exist graphs for which the minimum number of vertex splits required to obtain a chordal graph is strictly larger than the minimum number of vertex deletions required to achieve the same goal, even when the input graph has independence number 2. This separation shows that known results for vertex deletion into chordal or interval graphs do not directly transfer to vertex splitting, and that splitting-based modification requires separate analysis.

We should mention that transforming a graph into an interval graph via a sequence of k vertex splitting operations can be viewed as a transformation/modification into a multi-intersection graph of intervals: each vertex is represented by one or more intervals. This differs slightly from the notion of a k-Interval Graph [19] and that of an Internal-k-graph [7]. In these (latter) older notions, the objective is to represent each vertex by k intervals.

Vertex splitting has been studied in several other contexts, including planarization and graph thickness [10,18], rigidity theory [23], flow and decomposition problems [24], and clustering and graph editing, where it enables the modeling of overlapping structures [2–4,11,12]. However, its role as a mechanism for enforcing classical intersection graph structure has remained largely unexplored.

Our contributions can be summarized as follows. We initiate a systematic study of vertex splitting as a graph modification operation toward interval graphs. We prove that transforming a graph into an interval graph using a bounded number of vertex splits is NP-hard, even under strong restrictions on the input. We show that vertex splitting exhibits complexity behavior that differs fundamentally from vertex deletion, including a strict separation between chordal vertex deletion and chordal vertex splitting. At the same time, we identify tractable cases: we show that splitting any graph into a disjoint union of paths can be decided in polynomial time. In fact, we show that splitting a given subcubic triangle free graphs into an interval graph is NP-hard but splitting triangle free graphs into unit interval graphs is solvable in polynomial time.

Due to space restrictions, proofs of results marked with (∗) are omitted, and can be found in the full version of the paper [1].

2   Preliminaries

Let $G = (V, E)$ be a graph and let v be a vertex of G. Throughout this article, we use standard graph-theoretic notation and assume all considered graphs are simple, unweighted and undirected. The sets $V(= V(G))$ and $E(= E(G))$ denote the vertex set and the edge set of G, respectively. We denote by $N(v)$ the *open neighborhood* of a vertex v. The chromatic number of G is denoted by $\chi(G)$, and $\overline{G}$ denotes the complement of G, i.e., $V(\overline{G}) = V(G)$ and $E(\overline{G}) = \{xy \in V(\overline{G})^2 | x \neq y$ and $xy \notin E(G)\}$. When no ambiguity arises, we identify a vertex subset $S \subseteq V(G)$ with the subgraph induced by S, and write $\chi(S)$ instead of $\chi(G[S])$. We denote by $\alpha(G)$ the size of a maximum independent set of G.

Chordal, Interval and Unit Interval graphs. A graph is said to be *chordal* if it contains no induced cycle of length at least 4. A graph is an *interval graph* if it is the intersection graph of intervals on the real line. A graph is a *unit interval graph* if it is the intersection graph of unit-length intervals; such graphs are also known as *indifference graphs*. It is well-known that an interval graph is chordal. Furthermore, a classical result states that a graph is a unit interval graph if and only if it is chordal and contains no induced claw, net, or tent [22] (see Fig. 1).

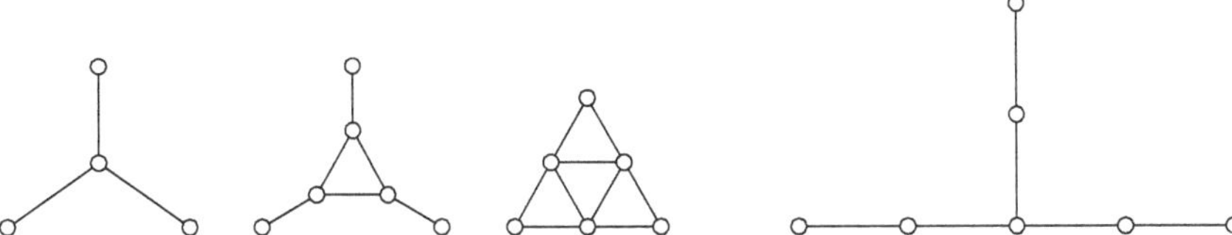

Fig. 1. The claw, net and tent graphs.   Fig. 2. The graph T_2.

We say that a vertex v is *simplicial* if $G[N(v)]$ is a clique. A classical characterization, due to Dirac, states that a graph is chordal if and only if it admits a *perfect elimination ordering*, that is, an ordering $v_1, \ldots, v_n$ of the vertices such that v_i is simplicial in $G[\{v_i, \ldots, v_n\}]$ for every i [9,14].

A (t, d)-*star* is a graph consisting of a central vertex v, the star center, and t vertex-disjoint paths of length d each that have v as endpoint (each). Figure 2 shows a $(3, 2)$-star, which we refer to hereafter (for simplicity) as T_2, also known as a subdivided claw.

A *tree* is a connected cycle-free graph. A *caterpillar* is a tree such that removing all its degree-one vertices (i.e. leaves) yields a path, called the spine. Let $\mathcal{C}at$ be the class of caterpillars. For k an integer, let $\mathcal{C}at_k$ be the set of caterpillar graphs of maximum degree k. The following characterization will be used in the proof of our main result.

Lemma 1 ($*$). *Let T be a tree. Then T is T_2-free if and only if T is a caterpillar.*

Vertex Splitting. We define a *(vertex) splitting* of a vertex $v \in V(G)$ as a graph modification operation in which v is replaced by two new vertices v_1 and v_2, such that every neighbor of v becomes adjacent to either v_1 or v_2, or both. Formally, splitting $v \in V(G)$ yields a graph G' defined as follows. Let $A, B \subseteq N_G(v)$ with $A \cup B = N_G(v)$. Then

$$V(G') = \big(V(G) \setminus \{v\}\big) \cup \{v_1, v_2\},$$
$$E(G') = \big(E(G) \setminus \{vx : x \in N_G(v)\}\big) \cup \{v_1x : x \in A\} \cup \{v_2x : x \in B\}.$$

Each vertex splitting increases the number of vertices by one. A *split* on a graph G is formally defined as the tuple $s = (v, A, B)$, where v is a vertex of G and $A, B \subseteq N_G(v)$ with $A \cup B = N_G(v)$. Moreover, an entity $\sigma = (s_1, s_2, \ldots, s_k)$ is defined to be a *sequence* of splits on a graph G_1 if s_i is a split on G_i resulting in the graph G_{i+1}, for all $i \in \{1, 2, \ldots, k\}$. Given a split (v, A, B), the particular case where $A \cap B = \emptyset$ is of interest, at least from a practical standpoint where splitting is supposed to result in clusters (see [3] for an example). This will be referred to as *exclusive* vertex splitting in the sequel. On the other hand, when the condition $A \cap B = \emptyset$ is not mandatory, the vertex splitting is referred to as *inclusive*. Therefore, an exclusive vertex splitting is also inclusive.

Chordal, Interval and Unit Interval Vertex Splitting. Let G be a graph. We denote by $ChVS(G)$ the minimum length of a split sequence which turns G into a chordal graph. Similarly we define $ChVXS(G)$ as the minimum length of an exclusive split sequence which turns G into a chordal graph. Similarly we define $IVS(G)$, $IVXS(G)$, $UIVS(G)$ and $UIVXS(G)$ for interval and unit interval graphs.

3   Complexity of Vertex Splitting Into Interval Graphs

We prove the NP-completeness of deciding whether a graph can be transformed into an interval graph using at most a given number of vertex splits. We first focus on the exclusive variant of vertex splitting and show its NP-hardness.

Since inclusive and exclusive splitting are not equivalent in general, hardness for one variant does not automatically transfer to the other. We next show that, on triangle-free graphs, any sequence of inclusive splits leading to an interval (or unit interval, or chordal) graph can be converted into a sequence of exclusive splits of no greater length. This assertion, formalized in Lemma 2, allows us later to deduce hardness results for the inclusive variant.

Lemma 2. *Let G be a triangle free graph and let σ be a sequence of l splits resulting in an interval graph (respectively, a unit interval graph, or a chordal graph). Then there exists a sequence of at most l exclusive splits resulting in a graph of the same class.*

Proof. Let us proceed by induction on the number of non exclusive splits. If all splits of σ are exclusive then we are done. Otherwise let k be the number of non exclusive splits of σ and suppose that the statement is true for sequences with less than $k - 1$ non exclusive splits.

Consider the last non exclusive split $s_i = (v, A, B)$ of the sequence $\sigma = (s_1, ..., s_l)$ where $A \subseteq N(v)$ and $B \subseteq N(v)$ such that $A \cup B = N(v)$ and $A \cap B \neq \emptyset$. Let v_1 and v_2 be the two created vertices such that $N(v_1) = A$ and $N(v_2) = B$.

For a set X of vertices of G during the split sequence σ, $D(X)$ denotes the set X and the the descendants of each of the vertices of X, i.e. the set of vertices resulting from the split sequence of the vertices in X. We consider the split $s_i' = (v, A, B \setminus A)$. For every split $s_j = (v_j, A_j, B_j)$ for $j > i$, we define $s_j' = (v_j, A_j', B_j')$ as follows: if $v_j \in D(A \cap B)$, $A_j' = A_j \setminus D(v_2)$ and $B_j' = B_j \setminus D(v_2)$, if $v_j \in D(v_2)$, $A_j' = A_j \setminus D(A \cap B)$ and $B_j' = B_j \setminus D(A \cap B)$, otherwise $A_j' = A_j$ and $B_j' = B_j$. We define the sequence $\sigma' = (s_1, ...s_{i-1}, s_i', ..., s_l')$. The splits $s_i', ..., s_l'$ are exclusive. Therefore the sequence σ' has $k - 1$ exclusive splits.

The resulting graph $G(\sigma')$ is a subgraph of $G(\sigma)$ where we have removed the edges between $D(v_2)$ and $D(A \cap B)$.

Finally, since triangle-free interval graphs are forests and interval graphs are T_2-free, by Lemma 1, they are precisely disjoint unions of caterpillars; and since any subgraph of a union of caterpillars is also a union of caterpillars, $G(\sigma')$ is an interval graph and σ' requires one less non-exclusive split than σ. We conclude by induction.

For unit interval graphs, the conclusion is the same: the class of triangle free unit interval graphs is monotone (closed under taking arbitrary subgraph) because it is the class of union of paths. Furthermore, the conclusion is the same for chordal graphs: the class of triangle-free chordal graphs is monotone because it is nothing but the class of forests. $\qquad\square$

Corollary 1 ($*$)**.** *For a triangle-free graph G,*

- $ChVS(G) = ChVXS(G)$.
- $IVS(G) = IVXS(G)$.
- $UIVS(G) = UIVXS(G)$.

Theorem 1. *Let G be a graph and $k \geq 0$ be an integer. Determining if $IVXS(G) \leq k$ is* NP-*complete, even if G is a planar bipartite subcubic graph.*

Proof. The proof is a reduction from the HAMILTONIAN PATH problem on cubic planar graphs, which is known to be NP-complete [13]. Let G be a cubic planar graph on n vertices $v_1, \ldots, v_n$, and let G' be the graph obtained by subdividing every edge of G exactly once. Formally, build G' as follows:

- For each vertex $v_i \in V(G)$, we add a vertex u_i in $V(G')$.
- for each edge $(v_i v_j) \in E(G)$, with $i < j$, we add a vertex u_{ij} in $V(G')$ and the two edges $(u_i u_{ij})$ and $(u_{ij} u_j)$.

The construction is depicted in Fig. 3.

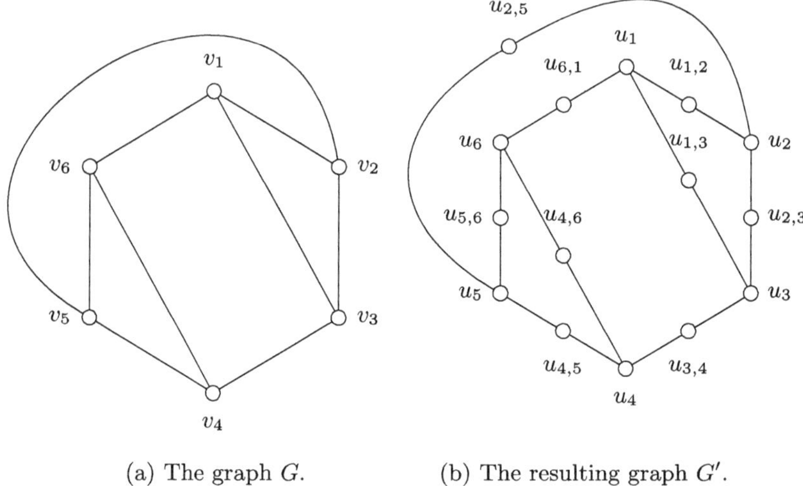

(a) The graph G.   (b) The resulting graph G'.

Fig. 3. A planar cubic graph G and the resulting graph G' of the reduction.

Since G is cubic, $|E(G)| = \frac{3}{2}n$, and thus $|V(G')| = \frac{5}{2}n$, and $|E(G')| = 3n$. Note that the vertices of G' can be partitioned into two independent sets $A = \{u_i\}_{1 \leq i \leq n}$ and $B = \{u_{ij} | (v_i v_j) \in E(G)\}$. Thus, G' is planar, bipartite and subcubic.

Let $k = \frac{n}{2} + 1$. We prove that an Hamiltonian path of G exists if, and only if, G' can be transformed into an interval graphs in k exclusive splits.

Suppose first that there exists a Hamiltonian path in G. Up to a relabeling of the vertices of G and G', suppose that this path is $(v_1, v_2, \ldots, v_n)$. Consider the following k exclusive splits of G': For each vertex $u_{i,j}$ with $i \neq j+1$, we split the vertex $u_{i,j}$ into v_i and v_j with v_i adjacent to u_i and v_j adjacent to u_j. Since $n - 1$ vertices $u_{i,j}$ satisfying $j = i + 1$ exist, and there is in total $\frac{3}{2}n$ vertices $u_{i,j}$, we performed $\frac{3}{2}n - (n - 1) = \frac{n}{2} + 1$ exclusive splits. The resulting graph is depicted in Fig. 4

By construction, for each $1 < i < n$, exactly one neighbor of u_i has been split, and two neighbors of each of u_1 and u_n have been split. Each of these exclusive splits have transformed vertices of degree 2 into two leaves. Moreover, for $1 \leq i \leq n$, the vertex u_i has not been split. Thus, $\{u_1, u_{1,2}, u_2, \ldots, u_{n-1}, u_{n-1,n}, u_n\}$ is a "dominating" path of the resulting graph (in the sense that any other vertex has a neighbor in the path). Finally, the resulting graph is a caterpillar, which is an interval graph.

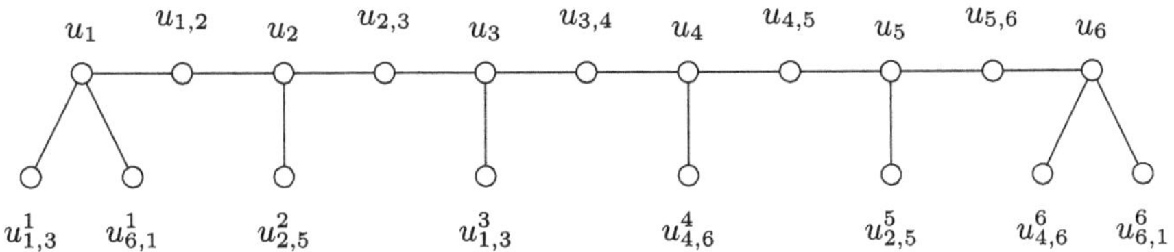

Fig. 4. The resulting graph after the flipping sequence.

Conversely, suppose that G' can be turned into an interval graph T after the application of k exclusive splits. First, note that, since G' is bipartite, all of its cycles have even length. And observing that any cycle resulting from a split must be an induced cycle of the original graph, we deduce that T has to be a forest. Moreover, since the exclusive splitting of a vertex does not change the number of edges, but increases the number of vertices by exactly one, and since $|V(G')| + k = |E(G')| + 1$, which implies that $|E(T)| = |V(T)| - 1$, we conclude that T must be a tree as it contains no cycles. Moreover, since T is an interval graph, it is T_2-free. Hence, using Lemma 1, T is a caterpillar.

Now let $P = (w_1, \ldots, w_m)$ be a dominating path of the caterpillar. First, we note that no leaf attached to an internal node of P can be a vertex u_i. Indeed, all the neighbors of u_i have degree 2, and thus, if u_i is a leaf, its neighbor in P can only have one other neighbor. So it is not an internal node of P. Up to a change of P into P' by adding the (at most two) u_i's that are leaves attached to the extremities of P, we can assume that all the u_is are in P. Now, by contracting all the vertices $u_{i,j}$ that are internal vertices of P, we obtain a path containing all the u_is, which corresponds to a Hamiltonian path of G.   □

Corollary 2 (∗). *Splitting into Interval Graphs is NP-complete even when restricted to planar bipartite subcubic graphs.*

4   Splitting into a Disjoint Union of Paths

Since splitting into a caterpillar graph is already NP-hard, it seems natural to look at splitting into a disjoint union of paths. In the sequel, we denote by $\mathcal{P}$ the class of graphs consisting of disjoint union of paths (each).

A trail of length k in a graph is a set of k different edges $\{e_1, \ldots e_k\}$, such that there exist $k+1$ (not necessarily different) vertices $\{v_1, \ldots, v_{k+1}\}$ such that for $1 \le i \le k$, we have $e_i = v_i v_{i+1} \in E$. For a graph G, we denote by $tp(G)$ the minimum number of trails required to partition the edge set of G into disjoint trails. The following lemma, relating $tp(G)$ to the number of odd-degree vertices in G could be of interest by itself.

Lemma 3. *Let G be a connected graph with $2p$ vertices of odd degree. Then $tp(G) = p$ if $p \ge 1$ and $tp(G) = 1$ if $p = 0$. If G is not connected, then we have $tp(G) = \sum\limits_{C \in cc(G)} tp(C)$, where $cc(G)$ is the set of connected component of G.*

Proof. The fact that $tp(G) = \sum\limits_{C \in cc(G)} tp(C)$, where $cc(G)$ is the set of connected component of G is straightforward. The proof is by induction on p. If $p = 0$ ($p = 1$ resp.), G is Eulerian (semi-Eulerian resp.). Thus, it contains an Eulerian cycle (trail resp.) which is a trail that covers all its edges. Thus, its edges can be covered by a single trail.

If $p \geq 2$, let u, v be two vertices of odd degree in the same connected component C of G. If C contains only two vertices of odd degree, we can cover all the edges of C with an eulerian trail P. Otherwise, consider any trail P from u to v. If removing of the edges forming P disconnects C, and we have an Eulerian connected component C' of $C \setminus P$, then we take an Eulerian tour of C' and insert it in P simply as follows: let x be a vertex of P which is in C'. We combine the tour starting and ending at x with P to obtain a longer trail. We do the same for every (possible) Eulerian componet of $C \setminus P$. When this process ends, we add P in the partition of the edges. After that, $G \setminus P$ has 2 vertices less of odd degree (the endpoints of P, since each loses exactly one edge), and no new connected component with only even degree vertices, which provides the result by induction. $\qquad\square$

Theorem 2. *Let $G = (V, E)$ be a graph. The minimum number of splits needed to transform G into a disjoint union of paths is*

$$\sum_{v \in V} \left(\left\lceil \frac{d(v)}{2} \right\rceil - 1 \right) + r,$$

where r is the number of connected components of G that contain no vertex of odd degree.

Proof. Since vertex splitting applies independently to connected components, and $\mathcal{P}$ is closed under disjoint union, it suffices to prove the statement for a connected graph, and then sum over components. So assume G is connected. Then $r \in \{0, 1\}$.

Let σ be a split sequence transforming G into $H \in \mathcal{P}$. Consider an original vertex $v \in V(G)$ and let $D(v)$ be the set of vertices of H that originate from v (the descendants of v after applying σ). Every edge incident to v in G remains present in H and is incident to at least one vertex of $D(v)$; therefore

$$\sum_{x \in D(v)} d_H(x) \geq d_G(v).$$

However, as union of paths is closed under edge deletion, we can without loss of generality suppose that all the splits are exclusive, and that each edge incident to v is not duplicated, which leads to

$$\sum_{x \in D(v)} d_H(x) = d_G(v).$$

Since $H \in \mathcal{P}$, every vertex of H has degree at most 2, so $\sum_{x \in D(v)} d_H(x) \leq 2|D(v)|$. Hence $d_G(v) \leq 2|D(v)|$, i.e. $|D(v)| \geq \lceil d_G(v)/2 \rceil$. Each split increases the total number of vertices by 1, and $|D(v)| = 1$ before any split on descendants of v. Thus at least $|D(v)| - 1 \geq \lceil d_G(v)/2 \rceil - 1$ splits are necessary "because of v". Summing over all $v \in V(G)$ yields the unconditional lower bound

$$|\sigma| \geq \sum_{v \in V(G)} \left(\left\lceil \frac{d_G(v)}{2} \right\rceil - 1 \right). \tag{$*$}$$

If $r = 0$, we are done with the lower bound. Assume now $r = 1$, i.e. G has no odd-degree vertices (so G is Eulerian). If equality holds in $(*)$, then for every v we must have $|D(v)| = \lceil d_G(v)/2 \rceil$ and every descendant has degree exactly 2 (because the degree sum over $D(v)$ is exactly $d_G(v)$ and the maximum is 2). Consequently, every vertex of H has degree exactly 2, so each connected component of H is a cycle. In particular, $H \notin \mathcal{P}$ unless $E = \emptyset$. Therefore, in the Eulerian connected case, at least one additional split is necessary beyond $(*)$, giving the lower bound

$$|\sigma| \geq \sum_{v \in V(G)} \left(\left\lceil \frac{d_G(v)}{2} \right\rceil - 1 \right) + 1.$$

To show the upper bound, let $T = \{P_1, \ldots, P_k\}$ be a partition of $E(G)$ into trails (as in the definition of $tp(G)$). We build a split sequence that realizes these trails as vertex-disjoint paths.

Fix a vertex $v \in V(G)$ and let the trails of T use v a total of $t(v)$ times, counting occurrences (so an internal occurrence contributes 2 incident edges of the trail at v, and an endpoint occurrence contributes 1 incident edge). Then the $d_G(v)$ incident edges of v are grouped into $t(v)$ blocks, where each block contains the edges incident to v and belonging to the same trail. Since the partition $T = \{P_1, P_2, \ldots, P_k\}$ is assumed to be minimum, each block has size 2, except possibly for one block of size 1 if v is an endpoint of a trail. Hence $t(v) \geq \lceil d_G(v)/2 \rceil$ for all v.

Now perform splits so that v is replaced by exactly $t(v)$ descendants, one descendant for each block, adjacent precisely to the edges of that block. After this, each descendant has degree 2 (for internal blocks) or degree 1 (for endpoint blocks). Doing this simultaneously for all vertices makes each trail P_i become an actual path component in the resulting graph, and different trails become vertex-disjoint. Therefore the resulting graph lies in $\mathcal{P}$.

The number of splits used "at v" is exactly $t(v) - 1$. Therefore the total number of splits is

$$\sum_{v \in V(G)} (t(v) - 1).$$

If $r = 0$ (there are odd-degree vertices), choose T so that each odd-degree vertex appears as an endpoint exactly once overall; then each $t(v)$ can be taken equal to $\lceil d_G(v)/2 \rceil$, yielding exactly $\sum_v (\lceil d_G(v)/2 \rceil - 1)$ splits.

If $r = 1$ (Eulerian connected), any trail partition must contain at least one trail, hence it must have two endpoints. This forces exactly one vertex to contribute an endpoint block (size 1), which increases $\sum_v t(v)$ by exactly 1 compared to the ideal pairing into blocks of size 2 everywhere. Equivalently, we can realize all vertices with $t(v) = d_G(v)/2$ except at one chosen vertex where we take $t(v) = d_G(v)/2 + 1$. This yields exactly $\sum_v (\lceil d_G(v)/2 \rceil - 1) + 1$ splits.

Thus, in both cases, the stated number of splits is achievable. Combined with the lower bounds, this proves optimality for the case where G is a connected graph. Summing over all connected components gives the general formula with the additional $+r$. $\qquad\square$

Note that the construction in the proof of Theorem 2 yields a polynomial-time algorithm that outputs an explicit splitting sequence transforming any graph into a disjoint union of paths.

Corollary 3 ($*$). *The minimum number of splits required to transform a graph into a disjoint union of paths, and the corresponding split sequence, can be computed in time $O(|E(G)||V(G)|)$.*

Corollary 4 ($*$). *If G is a triangle-free graph, the minimum number of (exclusive or inclusive) splits required to transform G into a unit interval graph can be computed in polynomial time.*

5   Splitting Versus Deletion

We now consider the (more general) class of chordal graphs. Our proofs have focused on the (sub)class of interval graphs, so one may suspect that the problem is also NP-hard in this case. However, this remains a conjecture and we therefore pose it as an open question. Note that the proof of Theorem 1 cannot be applied as triangle free chordal graphs correspond exactly to forests, and splitting into forests is already known to be solvable in polynomial time [11].

We note that modification into a chordal graph is well studied when the editing operation is edge or vertex deletion. Of course, vertex splitting can present additional challenges despite its close relationship with deletion. For example, if splitting k vertices results in a graph belonging to a hereditary class Π, then deleting the same set of vertices has the same effect, but the converse is not necessarily true. In this section, we study the relationship between vertex splitting and the two deletion operations when the goal is to obtain a chordal graph. We denote by $ChED(G)$ and $ChVD(G)$ the minimum number of edge deletions and vertex deletions, respectively, required to transform G into a chordal graph.

We prove that for any graph G, $ChVD(G) \leq ChVS(G) \leq ChED(G)$ and that these bounds are not tight.

We also focused our attention on graphs having $\alpha(G) \leq 2$ because they do not contain any induced cycle of length at least 6. Therefore to turn such a graph to a chordal graph, we only have to get rid of cycles of length 4 or 5.

5.1   Vertex Splitting Versus Edge Deletion into Chordal Graphs

We first consider the relationship with Chordal Edge Deletion which is known to be NP-complete [17].

Theorem 3. *For every graph G, $ChVS(G) \leq ChED(G)$. Moreover, $ChVS(G)$ is not lower-bounded by $ChED(G)$ even on graphs having $\alpha(G) \leq 2$.*

Proof. Let G be a graph. Suppose we have a sequence of edge deletions turning the G into a chordal graph. For every edge deletion xy, replace this operation by splitting x into two copies: one which is only connected to y (and becomes therefore a leaf) and the other which is connected to the other neighbors of x. This sequence of splits on G has the same length as the edge-deletion sequence, and it turns the graph into a chordal one. In fact, each leaf copy is a simplicial vertex and can be removed, or ignored (no effect on the graph's chordality). The remaining graph is the same as the graph obtained after the deletion of the edges. Then $ChVS(G) \leq ChED(G)$.

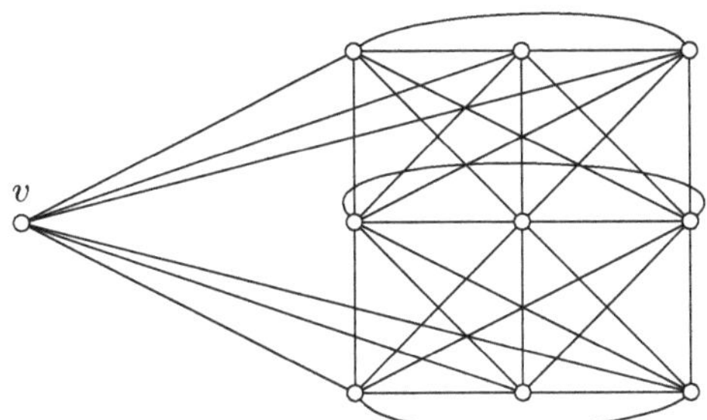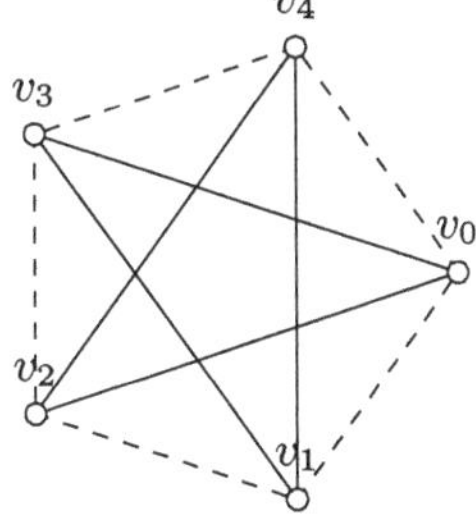

Fig. 5. The graph G_k for $k = 3$. This graph satisfies $ChVS(G_k) = 1$ and $ChED(G_k) \geq k$ and $\alpha(G_k) = 2$.

Fig. 6. In the case where G has 5 vertices, we show that G is C_5.

For any integer k, we consider the graph G_k defined as follows: consider the vertices $v, a_1, \ldots, a_k, b_1, \ldots, b_k, c_1, \ldots, c_k$. Make $\{a_1, \ldots, a_k\}$, $\{b_1, \ldots, b_k\}$ and $\{c_1, \ldots, c_k\}$ cliques and add the edges va_i, $a_i b_j$, $b_i c_j$ and $c_j v$ for every i and $j \in [k]$ (Fig. 5).

With only one split, we can turn this graph into a chordal graph by splitting the vertex v into two copies v_1 and v_2 such that v_1 is adjacent to the vertices $a_1, \ldots, a_k$ and v_2 is adjacent to the vertices $c_1, \ldots, c_k$. The vertex elimination order is $v_1, v_2, a_1, \ldots, a_k, b_1, \ldots, b_k, c_1, \ldots, c_k$. This graph needs at least k edge deletions because there are k edge independent induced C_4's: v, a_i, b_i, c_i for each $i \in [k]$. Thus $ChVS(G_k) = 1$ and $ChED(G_k) \geq k$.   □

5.2   Vertex Splitting Versus Vertex Deletion into Chordal Graphs

We now turn to the relationship with Vertex Deletion. The problem $ChVD$ is known to be NP-complete [16].

Theorem 4 (∗). *Let Π be a hereditary graph class. For any graph G, $\Pi VD(G) \leq \Pi VS(G)$. In particular, for any graph G, $ChVD(G) \leq ChVS(G)$.*

Theorem 5. *The parameters $ChVD$ and $ChVS$ are not equivalent, i.e. for any $k \geq 1$, there exists a graph G_k such that $ChVD(G_k) = 1$ and $ChVS(G_k) \geq k$.*

Proof. We prove that, for any $k \geq 1$, there exists a graph G_k such that $ChVD(G_k) = 1$ and $ChVS(G_k) \geq k$. Let G_k be the graph obtained from a star on $k + 1$ vertices by transforming each of its edges into a cycle of order 4. We have $ChVD(G_k) = 1$, as removing the center of the star transforms the graph into a union of P_3.

We now prove that at least k splits are required to turn G_k into a chordal graph. This graph has $3k + 1$ vertices and $4k$ edges. Let σ be a sequence of at most $k - 1$ splits on G_k. We denote by G_σ the obtained graph. This graph has at most $3k + 1 + k - 1 = 4k$ vertices. The number of edges has not decreased (more precisely it is the same if the splits are exclusive). Thus G_σ has at least $4k$ edges. Thus G_σ is not acyclic and contains therefore a cycle. Consider an induced cycle of G_σ. This cycle cannot be of size 3 because G_k is triangle-free and splits do not decrease the girth of the graph. We conclude that $ChVS(G_k) \geq k$.   □

We now focus on graphs such that their complement is triangle-free: in other words graphs G such that $\alpha(G) \leq 2$. In this case, we can prove that Chordal Vertex Deletion is"nearer" to Chordal Vertex Splitting than in the general case. The next Lemma is used to prove Theorem 6.

Lemma 4. *For any graph G, if $\alpha(G) \leq 2$ and $\chi(\overline{G}) \geq 3$, then G contains an induced C_4 or an induced C_5.*

Proof. By contradiction, let G be a graph with the smallest number of vertices that does not satisfy Lemma 4. Since $\chi(\overline{G}) \geq 3$, there exists $k \geq 1$ and vertices $v_1, \ldots v_{2k+1}$, such that $\{v_1, \ldots, v_{2k+1}\}$ is a cycle in $\overline{G}$. Moreover since $\alpha(G) \leq 2$, then $k \geq 2$. If $|G| = 5$, then necessarily $k = 2$. Since $\alpha(G) \leq 2$, it is not possible that there exists i such that $v_i v_{i+2}$ is not an edge, otherwise there would be a non triangle v_i, v_{i+1}, v_{i+2}. We deduce that G is C_5 (see Fig. 6), a contradiction.

Otherwise $|G| \geq 6$ (see Fig. 7. Consider the subgraph of G induced by the vertices $\{v_1, v_4, v_2, v_5\}$. By hypothesis, it cannot be a C_4. Thus, as we already know that $v_1 v_2$ and $v_4 v_5$ are non-edges in G and that $v_2 v_4$ is an edge in G (otherwise (v_2, v_3, v_4) is a non-triangle), there must be one other non-edge among $\{v_1 v_4, v_2 v_5, v_1 v_5\}$. If $v_1 v_5$ is a non-edge, then $(v_1, v_2, v_3, v_4, v_5)$ is an induced subgraph which satisfies the hypothesis of the lemma, which is not possible by the case $|G| = 5$. Therefore $v_1 v_4$ ($v_2 v_5$ resp.) is a non-edge, and thus the subgraph induced by $\{v_1, v_4, v_5, \ldots v_{2k+1}\}$ ($\{v_1, v_2, v_5, v_6, \ldots v_{2k+1}\}$ resp.) is a smaller counter example, which contradicts the minimality of G. A contradiction. We conclude that no smallest counter-example exists.   □

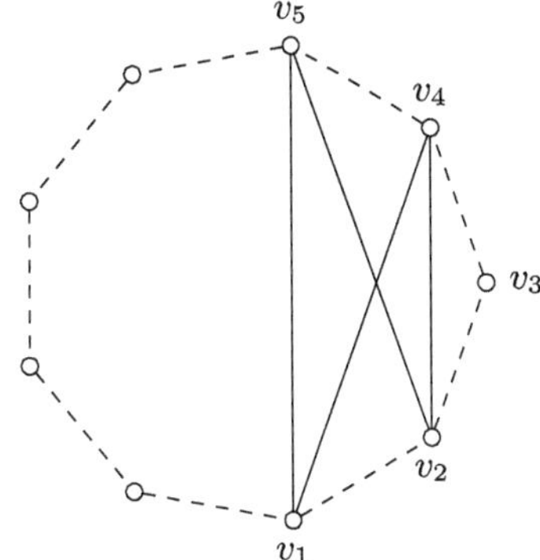
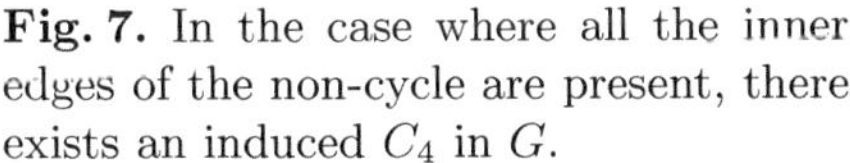

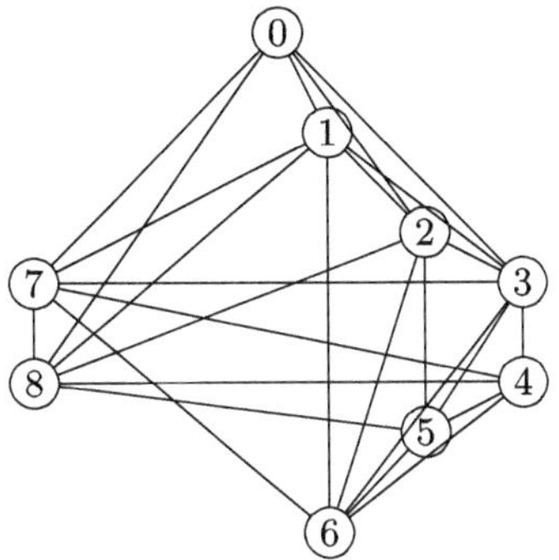

Fig. 7. In the case where all the inner edges of the non-cycle are present, there exists an induced C_4 in G.

Fig. 8. Example of a graph G which has $ChVD(G) = 2$ and $ChVS(G) > 2$ with $\alpha(G) = 2$. Vertices $0, 1, 2, 3$ and $4, 5, 6$ are forming cliques.

Theorem 6. *Let G be a graph such that $\alpha(G) \leq 2$. Then $ChVD(G) = 1 \iff ChVS(G) = 1$.*

Proof. If $ChVS(G) = 1$, then $ChVD(G) = 1$ by deleting the split vertex. Now assume that $ChVD(G) = 1$. There exists a vertex v, such that $G \setminus \{v\}$ is chordal. As $\alpha(G \setminus \{v\}) \leq 2$, we deduce by Lemma 4 that $\chi(\overline{G \setminus \{v\}}) \leq 2$. Thus $G \setminus \{v\}$ can be partitioned into two complete subgraphs A and B.

We define G' by splitting v into v_1 and v_2 such that v_1 is connected to $N(v) \cap A$ and v_2 is connected to $N(v) \cap v_2$, the vertices v_1 and v_2 are simplicial. We start an elimination order of G' by deleting these two vertices. Then, as $G \setminus \{v\}$ is chordal, we complete this elimination order of G' by concatenating the elimination order of $G \setminus \{v\}$. Thus $ChVS(G) = 1$.   □

We now prove that the generalization of the previous equivalence is false.

Theorem 7. *The parameters $ChVD(G)$ and $ChVS(G)$ are different among graphs G having $\alpha(G) \leq 2$. More precisely, there exists a graph G such that $ChVD(G) < ChVS(G)$ such that $\alpha(G) = 2$.*

Proof. We consider the graph G on 9 vertices described on Fig. 8. We can check that $\alpha(G) = 2$ as its complement is triangle-free. Remark that the cycles $(7, 0, 2, 6)$ and $(8, 1, 3, 4)$ are induced. As these cycles are disjoint, then $ChVD(G) \geq 2$. The induced subgraph $G \setminus \{7, 8\}$ is chordal as there exists an elimination order which is $0, 1, 2, 3, 4, 5, 6$.

We checked with a computer, by brute force, that $ChVS(G) \geq 3$ by trying all possible splits on one vertex of the cycle $(7, 0, 2, 6)$ and on one vertex of the cycle $(8, 1, 3, 4)$.   □

6   Concluding Remarks

We initiated a study of vertex splitting as a graph modification operation toward interval graphs. We proved that deciding whether a graph can be transformed into an interval graph using at most k vertex splits is NP-complete, even when the input is restricted to subcubic planar bipartite graphs. We also showed that, on triangle-free graphs, inclusive and exclusive splitting coincide for chordal, interval, and unit interval (target) graphs.

On the positive side, we presented a polynomial-time algorithm that computes the minimum number of splits needed to transform a given graph into a disjoint union of paths, together with a corresponding split sequence. As a consequence splitting triangle-free graphs into unit interval graphs is solvable in polynomial time.

We further compared vertex splitting with edge and vertex deletion when the target graph is chordal. We showed that chordal vertex splitting is always upper-bounded by chordal edge deletion, and that chordal vertex deletion is always upper-bounded by chordal vertex splitting. At the same time, we presented strong separations between these parameters, even on graphs with independence number at most two. These results confirm that splitting-based distance to chordal and interval graph classes behaves differently from deletion-based distance and requires separate analysis.

We conclude with several open questions. What is the complexity of chordal vertex splitting and unit interval vertex splitting in general? Are there fixed-parameter tractable or approximation algorithms for interval vertex splitting under natural parameters? As for auxiliary parameters, we note that splitting into interval graphs is FPT when parameterized by the treewidth, and the number of splits of the input graph. This follows immediately from Courcelle's Theorem [8] since the problem is expressible via monadic second order logic [10]. However, the problem remains interesting when parameterized by other auxiliary parameters such as twin-width and modular width, to name a few.

References

1. Abu-Khzam, F.N., Chakraborty, D., Isenmann, L., Oijid, N.: On the complexity of vertex-splitting into an interval graph (2026)
2. Abu-Khzam, F.N., Davot, T., Isenmann, L., Thoumi, S.: On the complexity of 2-club cluster editing with vertex splitting. In: Fomin, F.V., Xiao, M. (eds.) International Computing and Combinatorics Conference, pp. 3–14. Springer, Cham (2025). https://doi.org/10.1007/978-981-95-0218-9_1
3. Abu-Khzam, F.N., Egan, J., Gaspers, S., Shaw, A., Shaw, P.: Cluster Editing with Vertex Splitting. In: Lee, J., Rinaldi, G., Mahjoub, A.R. (eds.) ISCO 2018. LNCS, vol. 10856, pp. 1–13. Springer, Cham (2018). https://doi.org/10.1007/978-3-319-96151-4_1
4. Abu-Khzam, F.N., Isenmann, L., Merchad, Z.: Bicluster editing with overlaps: a vertex splitting approach. In: Fernau, H., Zhu, B., (eds.) Combinatorial Algorithms - 36th International Workshop, IWOCA 2025, Bozeman, MT, USA, July 21–24,

2025, Proceedings, vol. 15885 of Lecture Notes in Computer Science, pp. 146–159. Springer, Cham(2025)

5. Benzer, S.: On the topology of the genetic fine structure. Proc. Natl. Acad. Sci. **45**(11), 1607–1620 (1959). Introduced interval graphs in genetics

6. Booth, K.S., Lueker, G.S.: Testing for the consecutive ones property, interval graphs, and graph planarity using pq-tree algorithms. J. Comput. Syst. Sci. **13**(3), 335–379 (1976)

7. Brown, D.E., Flesch, B., Langley, L.J.: Interval k-graphs and orders. Order **35**(3), 495–514 (2018)

8. Courcelle, B.: The monadic second-order logic of graphs. i. recognizable sets of finite graphs. Inf. Comp. **85**(1), 12–75 (1990)

9. Dirac, G.A.: On rigid circuit graphs. Abh. Math. Semin. Univ. Hambg. **25**, 71–76 (1961)

10. Eppstein, D., et al.: On the planar split thickness of graphs. Algorithmica **80**(3), 977–994 (2018)

11. Firbas, A.: Establishing hereditary graph properties via vertex splitting, Technische Universität Wien, PhD thesis (2023)

12. Firbas, A., et al.: The complexity of cluster vertex splitting and company. Discret. Appl. Math. **365**, 190–207 (2025)

13. Garey, M.R., Johnson, D.S., Tarjan, R.E.: The planar Hamiltonian circuit problem is np-complete. SIAM J. Comput. **5**(4), 704–714 (1976)

14. Golumbic, M.C.: See Chapter 5 for interval graphs and scheduling applications. Algorithmic graph theory and perfect graphs, Academic Press (1980)

15. Iwata, S., Yamada, T.: Interval graph-based formulation for a scheduling problem with resource constraints and deadlines. Comput. Oper. Res. **31**(10), 1675–1687 (2004)

16. Lewis, J.M., Yannakakis, M.: The node-deletion problem for hereditary properties is np-complete. J. Comput. Syst. Sci. **20**(2), 219–230 (1980)

17. Natanzon, A., Shamir, R., Sharan, R.: Complexity classification of some edge modification problems. Discret. Appl. Math. **113**(1), 109–128 (2001)

18. Nöllenburg, M., Sorge, M., Terziadis, S., Villedieu, A., Wu, H.-Y., Wulms, J.: Planarizing graphs and their drawings by vertex splitting. J. Comput. Geometry **16**(1), 333–372 (2025)

19. Trotter, W.T., Jr., Harary, F.: On double and multiple interval graphs. J. Graph Theory **3**(3), 205–211 (1979)

20. Tucker, A.C.: Structure theorems for some circular-arc graphs. Discret. Math. **7**(2), 167–195 (1974)

21. van Bevern, R., Komusiewicz, C., Moser, H., Niedermeier, R.: Measuring Indifference: Unit Interval Vertex Deletion. In: Thilikos, D.M. (ed.) WG 2010. LNCS, vol. 6410, pp. 232–243. Springer, Heidelberg (2010). https://doi.org/10.1007/978-3-642-16926-7_22

22. Wegner, G.: Eigenschaften der Nerven homologisch-einfacher Familien im Rn (1967)

23. Whiteley, W.: Vertex splitting in isostatic frameworks. Structutal Topology, 1990, núm. (1990)

24. Zhang, C.Q.: Circular flows of nearly Eulerian graphs and vertex-splitting. J. Graph Theory **40**(3), 147–161 (2002)

25. Zhang, J., Pevzner, P.A.: Using interval graphs for genome mapping and rearrangement. Proc. Nat. Acad. Sci. **102**, 17373–17378 (2005)

Bounds on Linear Turán Number for Trees

Rajat Adak[1]($\boxtimes$) (iD) and Pragya Verma[2]($\boxtimes$) (iD)

[1] Department of Computer Science and Automation, Indian Institute of Science
Bangalore, Bangalore, India
`rajatadak@iisc.ac.in`
[2] Department of Computer Science and Engineering, Indian Institute of Technology
Bombay, Bombay, India
`24d0376@iitb.ac.in`

Abstract. A hypergraph H is said to be *linear* if every pair of vertices lies in at most one hyperedge. Given a family $\mathcal{F}$ of r-uniform hypergraphs, an r-uniform hypergraph H is said to be *$\mathcal{F}$-free* if it contains no member of $\mathcal{F}$ as a subhypergraph. The *linear Turán number* $ex_r^{\text{lin}}(n, \mathcal{F})$ denotes the maximum number of hyperedges in an $\mathcal{F}$-free linear r-uniform hypergraph on n vertices.

Gyárfás, Ruszinkó, and Sárközy [*Linear Turán numbers of acyclic triple systems*, European J. Combin. (2022)] initiated the study of bounds on the linear Turán number for acyclic 3-uniform linear hypergraphs.

In this paper, we extend the study of linear Turán numbers for acyclic systems to higher uniformity. We first give a construction for linear r-uniform trees with k edges that yields the lower bound $ex_r^{\text{lin}}(n, T_k^r) \geq n(k-1)/r$, under mild divisibility and existence assumptions. Next, we study hypertrees with four edges. We prove the exact bound $ex_r^{\text{lin}}(n, B_4^r) \leq (r+1)n/r$ and characterize the extremal hypergraph class, where B_4^r is formed from S_3^r by appending a hyperedge incident to a degree-one vertex. We also prove the bound $ex_r^{\text{lin}}(n, E_4^r) \leq (2r-1)n/r$ for the crown E_4^r. Finally, we give a construction showing $ex_r^{\text{lin}}(n, P_4^r) \geq (r+1)n/r$ under suitable assumptions and conclude with a conjecture on sharp upper bound for P_4^r.

Keywords: Linear Turán Number · Uniform Linear Hypergraph · Steiner System · Expansion

1 Introduction

Extremal graph theory is largely concerned with Turán-type problems, which ask for the maximum number of edges in a graph or hypergraph that avoids a prescribed forbidden substructure. In this work, we focus on a more recent line of research, namely the study of linear Turán numbers. The Turán problem for hypergraphs has been investigated extensively; see, for instance, the surveys [10, 12, 18] and the book [13].

The study of Turán numbers has a long history, beginning with the classical result of Turán (1941) for cliques. Let $ex(n, K_{r+1})$ denote the maximum number

© The Author(s), under exclusive license to Springer Nature Switzerland AG 2026
F. Foucaud and A. Parreau (Eds.): IWOCA 2026, LNCS 16587, pp. 16–31, 2026.
https://doi.org/10.1007/978-3-032-27732-9_2

of edges in a graph on n vertices that does not contain a copy of K_{r+1} as a subgraph. The Turán graph, denoted by $T(n, r)$, is a complete r-partite graph on n vertices whose partite sets are as nearly equal in cardinality as possible (balanced complete r-partite).

Theorem 1. (Turán [25]) *For $n \geq r \geq 1$, $ex(n, K_{r+1}) \leq |E(T(n, r))|$. Equality holds if and only if $G \cong T(n, r)$.*

In contrast to graphs, where subgraphs are defined in a unique way, hypergraphs admit several natural notions of containment for a given structure. In particular, the same object (such as a path or cycle) can be defined under different interpretations, including Berge, Minimal and Linear notions. This additional flexibility makes Turán-type problems in hypergraphs significantly more complex. The extension of the Turán problem to hypergraphs has been widely studied. Turán-type results for minimal paths and cycles were obtained by Mubayi and Verstraëte [21], while exact results for linear paths (in a uniform hypergraph) were established by Füredi, Jiang, and Seiver [11]. For Berge paths, tight results were given by Győri, Katona, and Lemons [17], with a remaining case resolved in [6]. In this paper, we focus exclusively on the linear framework.

1.1   Preliminaries

In a hypergraph $H = (V(H), E(H))$, $V(H)$ denotes the set of vertices, and $E(H)$ is the set of hyperedges in H. For $v \in V(H)$, $d(v)$ denotes the degree of the vertex in H, that is, the number of hyperedges incident on v. Let $\delta(H)$ denote the minimum vertex degree in H, similarly let $\Delta(H)$ be the maximum vertex degree.

H is called r-uniform if each $e \in E(H)$ contains exactly r vertices. A hypergraph is *linear* when any two distinct hyperedges intersect in at most one vertex.

1.2   Linear Turán Number

For a family $\mathcal{F}$ of linear r-uniform hypergraphs, the *linear Turán number*, $ex_r^{\lin}(n, \mathcal{F})$, denotes the maximum number of hyperedges in an n-vertex linear r-uniform hypergraph that contains no member of $\mathcal{F}$ as a subhypergraph. When $\mathcal{F} = \{F\}$, we abbreviate this notation to $ex_r^{\lin}(n, F)$.

A key construction in this framework is the *expansion* of a graph. Given a graph G, its expansion G^r is obtained by inflating each graph edge into an r-element hyperedge using $r - 2$ new vertices (kept disjoint across edges). The result is automatically a linear r-uniform hypergraph.

The term *linear Turán number* was introduced by Collier-Cartaino, Graber, and Jiang [5], who established foundational bounds for linear cycles. Yet the phenomenon predates the terminology: the celebrated work of Ruzsa and Szemerédi [22] can be stated as

$$n^{2 - \frac{c}{\sqrt{\log n}}} \leq ex_3^{\lin}(n, C_3^3) = o(n^2),$$

18   R. Adak and P. Verma

where $c > 0$ and C_3^3 denotes the 3-uniform expansion of the C_3 (3-edge cycle).

The systematic study of acyclic structures in this context was initiated by Gyárfás, Ruszinkó, and Sárközy [14], who investigated acyclic triple systems and obtained several bounds for 3-uniform linear expansions. Zhang and Wang [28] extended this direction to the 4-uniform setting and proved corresponding bounds for a variety of forbidden configurations.

The current picture continues to rapidly develop. Zhou and Yuan [29] derived general r-uniform bounds, including for P_k^r, showing $ex_r^{\text{lin}}(n, P_k^r) \leq \frac{(2r-3)kn}{2}$ for $r \geq 3, k \geq 4$. Complementing this line, Khormali and Palmer [20] and Zhang et al. [27] investigated star-forests, establishing bounds for sufficiently large n.

For linear hypertrees on four hyperedges, the setting already becomes rich. Expansions of trees yield several distinct configurations. For $k = 2$, the only possibility is P_2^r (where P_k denotes a path on k edges). It is easy to check that $ex_r^{\text{lin}}(n, P_2^r) = \lfloor \frac{n}{r} \rfloor$. For $k = 3$, we obtain precisely P_3^r and S_3^r (where S_k denotes a star on k edges). In [28], it was shown that $ex_r^{\text{lin}}(n, P_3^r) \leq n$. When $k = 4$, there are three expansions: P_4^r , S_4^r, and B_4^r, where B_4 is the tree obtained by attaching an extra edge to a leaf of S_3.

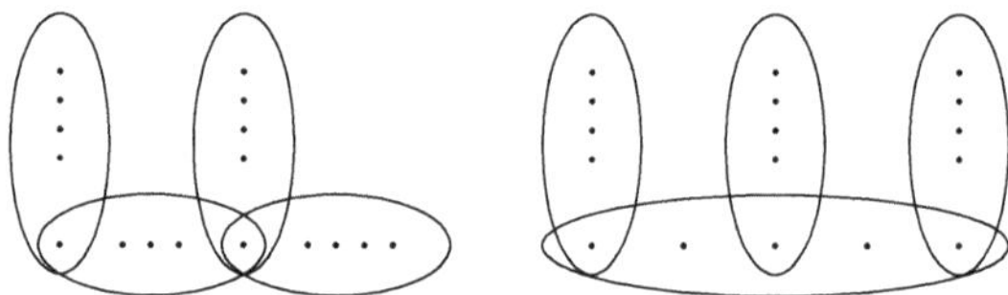

Fig. 1. Configurations of B_4^r and E_4^r (in the figure $r = 5$).

Alongside these expansions, there is a notable non-expansion configuration known as the *crown* E_4^r, consisting of three pairwise disjoint hyperedges together with a fourth hyperedge, called the *base*, that intersects each of the other three. Crowns have been studied in a variety of settings in [2,3,24,26]. In a related direction, Fletcher [9] obtained the bound $ex_3^{\text{lin}}(n, E_4^3) < \frac{5n}{3}$ for the 3-uniform case. More generally, let T_k^r denote a linear r-uniform hypertree on k hyperedges, that is, a connected, *acyclic* r-uniform hypergraph in which any two hyperedges intersect in at most one vertex. Here, *acyclic* means that there is no sequence of distinct hyperedges $e_1, \ldots, e_t$ ($t \geq 3$) with $e_i \cap e_{i+1} \neq \emptyset$ for all i (indices modulo t) and all these intersections are distinct. Similar to E_4^r, such a hypertree need not arise as an expansion configuration, depending on k.

Finding upper bounds on the Turán number of trees is already very difficult even in the graph setting (see the Erdős–Sós conjecture [8], 1963). Therefore, we focus on small hypertrees.

1.3   Steiner Systems

A Steiner system $S(t, r, n)$ consists of an n-vertex set together with blocks of size r such that every t-subset lies in exactly one block. In particular, $S(2, r, n)$ repeatedly emerges in the study of linear Turán extremal problems, as its block count realizes the maximum possible edge density of a linear r-uniform hypergraph on n vertices. This connection is not new: Steiner triple systems $S(2,3,n)$ (also denoted by $STS(n)$) have already appeared in the study of linear Turán numbers for triple systems in [14–16,23], while Steiner quadruple systems $S(2,4,n)$ (denoted by $SQS(n)$) arise naturally in the 4-uniform setting, for instance in [28]. These systems serve as extremal or near-extremal constructions in several linear Turán-type problems.

Note that $S(2,r,n)$ may not exist for all values of r and n; we return to this point after the following lemma.

Lemma 1. *Let H be a Steiner system $S(2,r,n)$ (assuming it exists). Then*

$$|E(H)| = \frac{\binom{n}{2}}{\binom{r}{2}} = \frac{n(n-1)}{r(r-1)}, \qquad and \qquad d(v) = \frac{n-1}{r-1} \ \textit{for every } v \in V(H).$$

Proof. Each edge covers $\binom{r}{2}$ distinct vertex-pairs, and by the definition, we get that every pair among the $\binom{n}{2}$ pairs is covered exactly once. Hence $|E(H)| = \binom{n}{2}/\binom{r}{2}$.

For the degree: fix $v \in V(H)$. The $(n-1)$ pairs $\{v,u\}$ $(u \neq v)$ must each be covered exactly once. Each edge containing v covers exactly $(r-1)$ such pairs, so $d(v) = \frac{(n-1)}{(r-1)}$. $\qquad\square$

Note that the divisibility conditions implicit in Lemma 1 are necessary for the existence of a Steiner system $S(2,r,n)$. Indeed, the degree condition $d(v) = \frac{n-1}{r-1}$ requires that $(r-1) \mid (n-1)$, and the edge count $|E(H)| = \frac{\binom{n}{2}}{\binom{r}{2}}$ requires that $\binom{r}{2} \mid \binom{n}{2}$. It is known that these necessary conditions are also sufficient for all sufficiently large n [19].

For small parameters, existence is characterized in several cases: $S(2,3,n)$ exists if and only if $n \equiv 1,3 \pmod 6$, and $S(2,4,n)$ exists if and only if $n \equiv 1,4 \pmod{12}$. Moreover, if r is a prime power, Steiner systems of the form $S(2,r,r^2)$ (corresponding to affine planes of order r) are known to exist. In general, however, existence of $S(2,r,r^2)$ is not guaranteed. The Bruck–Ryser–Chowla theorem [1,4] provides necessary conditions (when $r \equiv 1$ or $2 \pmod 4$), ruling out certain values of r (for example, $r = 6$).

However, a complete characterization of the existence of $S(2,r,n)$ for arbitrary parameters (r,n) is not known.

2   Our Results

We begin by determining the exact extremal value for linear stars.

Proposition 1. *For $r \geq 3$, $ex_r^{\mathrm{lin}}(n, S_k^r) \leq \dfrac{n(k-1)}{r}$. Equality holds if and only if the linear r-uniform hypergraph is $(k-1)$-regular, given it exists.*

To complement this, we obtain a matching lower bound for general linear hypertrees under mild divisibility and existence assumptions, via block constructions arising from Steiner systems.

Theorem 2. *Let $r \geq 3$ and $k \geq 2$. Let $t = (r-1)(k-1)+1$. Assume that $t \mid n$ and that the Steiner system $S(2, r, t)$ exists. Then, $ex_r^{\mathrm{lin}}(n, T_k^r) \geq \frac{n(k-1)}{r}$. This bound is sharp when T_k^r is S_k^r.*

Next, we turn to hypertrees on four hyperedges. For the configuration B_4^r, we obtain a sharp extremal characterization.

Theorem 3. *For $r \geq 3$, $ex_r^{\mathrm{lin}}(n, B_4^r) \leq \dfrac{(r+1)n}{r}$. Equality holds if and only if the linear r-uniform hypergraph is the union of disjoint Steiner systems $S(2, r, r^2)$, given that the Steiner system $S(2, r, r^2)$ exists.*

For the crown hypergraph E_4^r, we obtain the following upper bound.

Theorem 4. *For $r \geq 3$, $ex_r^{\mathrm{lin}}(n, E_4^r) \leq \dfrac{(2r-1)n}{r}$.*

The linear Turán number for S_4^r follows directly from Proposition 1. The upper bound for P_4^r follows from the bound on P_k^r in [29], and we obtain a lower bound for P_4^r under certain divisibility and existence assumptions.

Theorem 5. *Assume $r^2 \mid n$ and that the Steiner system $S(2, r, r^2)$ exists. Then, $ex_r^{\mathrm{lin}}(n, P_4^r) \geq \dfrac{(r+1)n}{r}$.*

3   Proof of Proposition 1 and Theorem 2

Proof of Proposition 1. If H is S_k^r-free, then $d(v) \leq k-1$ for all $v \in V(H)$. Since each edge consists of r vertices, we get an analogue of the handshaking lemma as $\sum_{v \in V(H)} d(v) = r|E(H)|$. Thus, $r|E(H)| \leq n(k-1) \implies |E(H)| \leq \dfrac{n(k-1)}{r}$. Therefore we get $ex_r^{\mathrm{lin}}(n, S_k^r) \leq \dfrac{n(k-1)}{r}$ and clearly for equality we need $d(v) = k-1$ for all $v \in V(H)$. Note that a linear r-uniform $(k-1)$-regular hypergraph may not exist. In such cases, equality is not feasible. Clearly, $r \mid n(k-1)$ is a necessary condition. Also it is necessary to have,

$$|E(H)| \binom{r}{2} \leq \binom{n}{2} \implies (k-1)(r-1) \leq (n-1)$$

These conditions are not sufficient. For example, when $(n, k, r) = (10, 4, 3)$, both the above conditions are satisfied, yet no such hypergraph exists. In fact, there is no known general characterization for the existence of regular linear r-uniform hypergraphs.

Now we give a constructive proof for Theorem 2 under the divisibility and existence assumptions.

Proof of Theorem 2. Let $n = qt$ and partition the n-vertex set into q disjoint parts $V(H) = V_1 \sqcup V_2 \sqcup \cdots \sqcup V_q$, such that $|V_i| = t$. Suppose each V_i induces H_i, a component of H, and each H_i is a copy of $S(2, r, t)$.

Since each H_i is linear and r-uniform by definition, and edges of disjoint components are disjoint, we have H to be linear and r-uniform.

By Lemma 1, we have

$$|E(H)| = \sum_{i=1}^{q} |E(H_i)| = \frac{n}{t} \cdot \frac{t(t-1)}{r(r-1)} = \frac{n(t-1)}{r(r-1)} = \frac{n(r-1)(k-1)}{r(r-1)} = \frac{n(k-1)}{r}$$

Claim. Every linear r-uniform hypertree with k edges has exactly $(r-1)k+1$ vertices.

Proof. Starting with one edge, clearly it contains r vertices. When we add a new edge in a linear r-uniform hypertree, it intersects the existing vertex set in exactly one vertex, hence contributes precisely $r-1$ new vertices. After adding $k-1$ further edges, the total vertex count is $r + (k-1)(r-1) = (r-1)k+1$. $\square$

Suppose there exists a T_k^r in H, clearly it must be in some H_i. From the above claim $|V(H_i)| \geq (r-1)k+1 = \big((r-1)(k-1)+1\big) + (r-1) = t + (r-1)$. Thus we get a contradiction.

Therefore, H is a T_k^r-free graph and $|E(H)| = \frac{n(k-1)}{r}$. Thus, under the given divisibility and existence assumptions we have $ex_r^{\mathrm{lin}}(n, T_k^r) \geq \frac{n(k-1)}{r}$. From proposition 1 it is clear that the bound is sharp when $T_k^r \cong S_k^r$.

4   Proof of Theorem 3

4.1   Proof of Inequality

Suppose the statement in Theorem 3 is false. Assume H to be a minimal counterexample. That is, H is a B_4^r-free linear r-uniform hypergraph on n vertices and

$$|E(H)| > \frac{(r+1)n}{r}$$

and for any subhypergraph H' of H, Theorem 3 holds for H'.

Claim. $\delta(H) \geq 2$.

Proof. If H contains an isolated vertex, then we can drop that vertex, resulting in a smaller counterexample. Thus, suppose there exists $v \in V(H)$, such that $d(v) = 1$. Let $H' = H \setminus \{v\}$. Clearly,

$$|E(H')| = |E(H)| - 1 > \frac{(r+1)n}{r} - 1 > \frac{(r+1)n}{r} - \frac{(r+1)}{r} = \frac{(r+1)(n-1)}{r}$$

Thus H' is a counterexample for Theorem 3, contradicting the minimality of H. $\square$

H must be connected, otherwise, we will get at least one connected component of H as a counterexample, contradicting minimality of H.

Claim. $\Delta(H) \geq (r+2)$.

Proof. Suppose $\Delta(H) \leq (r+1)$, then $d(v) \leq (r+1)$ for all $v \in V(H)$. Therefore, $\sum_{v \in V(H)} d(v) \leq n(r+1)$. Since H is r-uniform, we have

$$\sum_{v \in V(H)} d(v) = r|E(H)| > (r+1)n$$

Thus we get a contradiction.   □

Let $v \in V(H)$ be such that $d(v) = \Delta(H) = k$. Thus we get an expanded star $S = S_k^r$ centered at v. Let the edges of S be $E(S) = \{e_1, e_2 \ldots, e_k\}$. Let $u \in e_1 \setminus \{v\}$. Since $\delta(H) \geq 2$, there exists $f \in E(H)$ containing u such that $f \neq e_1$. Suppose f intersects t many hyperedges from $E(S)$, without loss of generality let these edges be $\{e_1, e_2, \ldots, e_t\}$. Since H is r-uniform, $t \leq r$.

Note that, $k \geq r+2$, therefore $t \leq k-2$ and thus $e_{t+1}, e_{t+2} \in E(S)$. It is easy to see that the edges f, e_1, e_{t+1}, e_{t+2} form a B_4^r. Thus we get a contradiction. Therefore, no such counterexample exists.

4.2   Characterizing Extremal Hypergraphs

For characterizing the extremal hypergraphs, we will only focus on the cases when $S(2, r, r^2)$ exists. If $S(2, r, r^2)$ does not exist, then the bound in Theorem 3 is not tight.

• If Part

Assume H to be a disjoint union of copies of a Steiner system $S(2, r, r^2)$. Fix one component $C \cong S(2, r, r^2)$ with $|V(C)| = r^2$.

Claim. $|E(C)| = r(r+1)$.

Proof. From Lemma 1, we have

$$|E(C)| = \frac{\binom{r^2}{2}}{\binom{r}{2}} = \frac{r^2(r^2-1)}{r(r-1)} = r(r+1)$$

□

If H has t components, then $n = tr^2$ and $|E(H)| = t \cdot r(r+1) = \frac{(r+1)n}{r}$.

Lemma 2. *If $C \cong S(2, r, r^2)$, and $e \in E(C)$, then for any $v \in V(C)$ such that $v \notin e$, there exists a unique $f \in E(H)$ containing v such that $f \cap e = \emptyset$.*

Proof. Fix an edge $e \in E(C)$ and a vertex $v \in V(C) \setminus e$. From Lemma 1 we have $d(v) = r + 1$. Hence there are exactly $r + 1$ edges of C containing v.

For each vertex $u \in e$, the pair $\{u, v\}$ is contained in a unique edge, which we denote by e_u. Clearly, if $u \neq u'$, then $e_u \neq e_{u'}$. Hence, the r vertices of e give rise to r distinct edges containing v, each intersecting e in exactly one vertex.

Since v lies in exactly $r + 1$ edges in total, there exists precisely one further edge f containing v that is different from all the e_u. This edge f cannot intersect e; otherwise it would coincide with e_u for some $u \in e$. Therefore, $f \cap e = \emptyset$.   □

Claim. C is B_4^r-free.

Proof. Suppose there exists a B_4^r in C and assume it consists of edges e_1, e_2, e_3 and e_4, where $e_1 \cap e_2 \cap e_3 = \{v\}$ and $e_3 \cap e_4 \neq \emptyset$. Clearly $v \notin e_4$ and e_1, e_2 contain v and are disjoint from e_4. Thus we get a contradiction to Lemma 2. $\qquad\square$

Thus, $|E(H)| = \dfrac{(r+1)n}{r}$ and H is B_4^r-free.

• Only If Part

Now let H be a B_4^r-free linear r-uniform hypergraph on n vertices satisfying $|E(H)| = \dfrac{(r+1)n}{r}$. Without loss of generality, assume H is connected. We need to show that H is $S(2, r, r^2)$.

Claim. H is $(r+1)$-regular

Proof. Since H is r-uniform, $\sum_{v \in V(H)} d(v) = r|E(H)| = (r+1)n$. Suppose there exists a vertex $v \in V(H)$ such that $d(v) = k \geq r + 2$. Now consider the S_k^r centered at v. It is easy to see that there exists a vertex in S_k^r, other than v, with degree at least 2, otherwise $H \cong S_k^r$, which will result, $n = k(r-1) + 1$ and $|E(H)| = k = \frac{n-1}{r-1} < \frac{(r+1)n}{r}$, a contradiction.

Thus, using the arguments as in the proof of inequality in Sect. 4.1, we will get a B_4^r in H.

Thus every vertex has degree at most $r + 1$ and the average is $r + 1$, forcing $d(v) = r + 1$ for all $v \in V(H)$, i.e. H is $(r+1)$-regular. $\qquad\square$

Let $v \in V(H)$ and the $r + 1$ edges containing v be $e_1, \ldots, e_{r+1}$. By linearity, these edges are pairwise disjoint outside v, so the set

$$X := \{v\} \cup \bigcup_{i=1}^{r+1} (e_i \setminus \{v\})$$

has size, $|X| = 1 + (r+1)(r-1) = r^2$.

Claim. $V(H) = X$.

Proof. Suppose for contradiction $V(H) \setminus X \neq \emptyset$. Since H is connected, there is an edge $f \in E(H)$ meeting X and containing at least one vertex outside X. Choose such an f. Clearly, $v \notin f$. Let $u \in f \cap X$ (necessarily $u \neq v$). Without loss of generality assume $u \in e_1$.

Now f has size r, and contains at least one vertex outside X. So among the remaining vertices of f, at most $r - 1$ lie in X. Each vertex of $f \cap \{X \setminus \{v\}\}$ lies in *exactly one* of the star edges $e_1, e_2, \ldots, e_{r+1}$, so f can intersect at most $r - 1$ of these $r + 1$ edges (including e_1). Therefore there exist *two* distinct edges, say e_a and e_b among $\{e_2, \ldots, e_{r+1}\}$, that are disjoint from f.

But then the hyperedges, f, e_1, e_a, e_b form a copy of B_4^r, contradicting that H is B_4^r-free. $\qquad\square$

Finally, we show H is a Steiner system $S(2, r, r^2)$. Because H is linear, any unordered pair of vertices is contained in *at most one* edge. Therefore, counting vertex-pairs inside edges gives

$$|E(H)| \binom{r}{2} \leq \binom{r^2}{2}.$$

But substituting $|E(H)| = \dfrac{(r+1)n}{r} = \dfrac{(r+1)r^2}{r} = r(r+1)$ yields equality:

$$r(r+1) \binom{r}{2} = r(r+1) \cdot \frac{r(r-1)}{2} = \frac{r^2(r^2-1)}{2} = \binom{r^2}{2}.$$

Hence every pair of vertices in H lies in *exactly one* edge, i.e. H is $S(2, r, r^2)$.

5   Proof of Theorem 4

We begin by outlining the main idea of the proof, and then present the argument in a sequence of steps.

Proof idea. We argue by contradiction. Assuming the hypergraph has more than $\frac{(2r-1)n}{r}$ edges, we show that the degree distribution forces the existence of an edge whose vertices have large total degree. We then show that such an edge must contain vertices with sufficiently large degrees, which yields a copy of E_4^r with that edge as its base, a contradiction.

Step 1: *Minimal counterexample and basic properties.*
Suppose the statement in Theorem 4 is false. Let H be a minimal counterexample. Thus, H is an E_4^r-free linear r-uniform hypergraph on n vertices with $|E(H)| > \frac{(2r-1)n}{r}$, and every proper subhypergraph satisfies the bound.
    Since H is minimal, we may assume that $\delta(H) \geq 2$. Indeed, if there exists an isolated vertex $v \in V(H)$, we may remove it without affecting the number of edges. Similarly, if there exists a vertex $v \in V(H)$ with $d(v) = 1$, then for $H' = H \setminus \{v\}$,

$$|E(H')| = |E(H)| - 1 > \frac{n(2r-1)}{r} - 1 > \frac{(n-1)(2r-1)}{r},$$

contradicting minimality. Hence $\delta(H) \geq 2$.

Step 2: *A local condition forcing a crown.*
We will now identify a configuration that guarantees the presence of E_4^r.

Lemma 3. *Let H be a linear r-uniform hypergraph. If there exists an edge $e \in E(H)$ containing three vertices a, b, c with $d(a) \geq 2r$, $d(b) \geq r+1$, and $d(c) \geq 2$, then H contains a copy of E_4^r with base e.*

Proof. Since $d(c) \geq 2$, pick an edge $f \neq e$ containing c. By linearity, any edge through b intersects f in at most one of the $r-1$ vertices in $f \setminus \{c\}$. Thus at most $r-1$ such edges intersect f, and since $d(b) - 1 \geq r$, there exists $g \neq e$ through b disjoint from f.

Similarly, an edge h through a can intersect $f \cup g$ in at most $2(r-1)$ vertices, but $d(a) - 1 \geq 2r - 1 > 2(r-1)$, so there exists $h \neq e$ disjoint from both f and g.

Thus, f, g, h are pairwise disjoint edges intersecting e, giving a copy of E_4^r with base e. $\qquad\square$

Strategy. We will show that such an edge e must exist by a global counting argument.

Step 3: *Measuring edge density via degrees.*
For each edge $e \in E(H)$, define $s(e) = \sum_{v \in e} d(v)$.

Idea. The quantity $s(e)$ measures how *dense* the neighbourhood of e is. Our goal is to show that some edge has very large $s(e)$.

It is easy to observe that

$$\sum_{e \in E(H)} s(e) = \sum_{e \in E(H)} \left(\sum_{v \in e} d(v) \right) = \sum_{v \in V(H)} \left(\sum_{e \ni v} d(v) \right) = \sum_{v \in V(H)} d(v)^2.$$

Step 4: *Controlling large-degree vertices.*
Set $A = 4r - 3$ and $B = r^2 + 3r - 3$. The vertices with degree at least A, are called *large vertices*. Let $L(H) = \{v \in V(H) : d(v) \geq A\}$ be the set of large vertices and define

$$s^*(e) = \begin{cases} \min\{s(e), B\}, & \text{if } e \text{ contains a large vertex,} \\ s(e), & \text{otherwise.} \end{cases}$$

Why truncation? Edges containing very large-degree vertices can artificially inflate $s(e)$. The truncation $s^*(e)$ controls this effect.

Lemma 4. *If $v \in L(H)$ and e contains v, then for all $u \in e \setminus \{v\}$, $d(u) < r+1$.*

Proof. Suppose there exists $u \in e \setminus \{v\}$ such that $d(u) \geq r + 1$. Note that $d(v) \geq A \geq 2r$. Let $c \in e \setminus \{u, v\}$. Clearly $d(c) \geq 2$, since $\delta(H) \geq 2$. Thus from Lemma 3 we have a E_4^r in H. $\qquad\square$

Step 5: *Bounding truncation error.*
Let $E_v(H)$ denote the set of hyperedges in $E(H)$ containing the vertex v.

Lemma 5. *If $v \in L(H)$ and $d(v) = d$, then $\sum_{e \in E_v(H)}(s(e) - s^*(e)) \leq d^2 - Ad$.*

Proof. From Lemma 4, for each $e \in E_v(H)$, $s(e) \leq d + r(r - 1)$. Clearly, for $e \in E_v(H)$, $s^*(e) = \min\{s(e), B\}$. Thus,

$$s(e) - s^*(e) \leq \max\{0,\ s(e) - B\} \leq \max\{0, (d + r(r - 1)) - B\} = \max\{0, d - A\}.$$

Since $d \geq A$, the maximum equals $d - A$. There are $|E_v(H)| = d$ edges containing v, so $\sum_{e \in E_v(H)}(s(e) - s^*(e)) \leq d(d - A) = d^2 - Ad$. $\qquad\square$

This bounds how much truncation reduces the contribution from edges containing large vertices.

Step 6: *Balancing the degree sequence.*
Since $\sum_v d(v) = r|E(H)| > r\frac{n(2r-1)}{r} = n(2r - 1)$, and H is minimal, we have $r|E(H)| = n(2r - 1) + l$ for some $l \in [r]$, we compare the degree sequence with a near-uniform baseline.

Lemma 6. *There exists a sequence of functions $f_0, f_1, \ldots, f_k : V(H) \to \mathbb{N}$ and a partition $V(H) = I \sqcup D$ such that:*

1. *For all $v \in V(H)$, $2r - 1 \leq f_0(v) \leq 3r - 1$ and $\sum_v f_0(v) = (2r - 1)n + l$;*
2. *$f_k(v) = d(v)$ for all $v \in V(H)$;*
3. *For $i \in [k - 1]$, there exist $x_i \in I$, $y_i \in D$ with $f_i(x_i) = f_{i-1}(x_i) + 1$, $f_i(y_i) = f_{i-1}(y_i) - 1$, and all other values unchanged;*
4. *if $v \in D$, then $f_0(v) = 2r - 1$*

Idea. We start with an almost regular degree sequence and gradually *unbalance* it to match the actual sequence. Because a sum of squares is minimized when its components are as equal as possible, tracking the exact increase during this process provides a rigorous baseline to lower bound the total edge density via $\sum_v d(v)^2$.

Proof. Label the vertices in $V(H)$ as v_1 through v_n so that $(d(v_i))_{i=1}^n$ is in non-decreasing order. We follow the following algorithm to define f_0.

1. First set $f_0(v) = 2r - 1$ for all $v \in V(H)$.
2. Set $R = l$ and $i = n$.
3. While $R > 0$,
   - Update $f_0(v_i) = \min\{(2r - 1 + R), d(v_i)\}$ and $R = R - (f_0(v_i) - 2r + 1)$ and $i = i - 1$. (Note that update to $f_0(v_i)$ happens only when $d(v_i) > f_0(v_i)$)

Clearly we get $2r - 1 \leq f_0(v) \leq 2r - 1 + l \leq 3r - 1$ and $\sum_v f_0(v) = (2r - 1)n + l$.

Now we iteratively transform f_i into the degree function d while preserving the total sum. We define f_i for $i > 0$, assuming f_{i-1} is already defined. Take the minimal $a \in [n]$ such that $f_{i-1}(v_a) > d(v_a)$ and the maximal $b \in [n]$ such that $f_{i-1}(v_b) < d(v_b)$. If no such a and b exist, then $f_{i-1} = d$, thus take $i - 1 = k$. Otherwise, define $f_i(v_a) = f_{i-1}(v_a) - 1, f_i(v_b) = f_{i-1}(v_b) + 1$, and $f_i(v_c) =$

$f_{i-1}(v_c)$ for $c \notin \{a, b\}$. This strictly decreases $\sum_v |f(v) - d(v)|$, so the process terminates after finitely many steps at $f_k = d$.

Let D be the set of vertices that are ever decreased and I the rest. Any vertex in D cannot have been among the initially increased vertices (while defining f_0), since it ends up requiring a decrease. Hence $f_0(v) = 2r - 1$ for all $v \in D$. □

Definition 1. *Consider a sequence $\{f_i\}_{i=0}^k$ of functions as in Lemma 6.*

- *For $0 \le i \le k$, define $T_i = \sum_v f_i(v)^2$.*
- *For $i \in [k]$, define $\Delta_i = T_i - T_{i-1}$.*
- *For $v \in V(H)$, let $I_v = \{i \in [k] : f_i(v) \ne f_{i-1}(v)\}$.*
- *For $v \in V(H)$, define $\Delta_v = \sum_{i \in I_v} \Delta_i$.*

Interpretation. The quantity T_i captures how concentrated the degree sequence is at stage i. The increments Δ_i track how this concentration changes, and Δ_v measures the total contribution of a vertex v to this change.

Lemma 7. *Let $v \in L(H)$ with $d(v) = d$. Then $\Delta_v \ge d^2 - Ad + (r-2)(3r-1)$.*

Proof. If $v \in L(H)$, then $v \in I$, so v is never decreased. Let $x = f_0(v)$, thus $2r - 1 \le x \le 3r - 1$. Each time $f_{i-1}(v)$ increases from $t - 1$ to t (in $f_i(v)$), the square increases by $t^2 - (t-1)^2 = 2t - 1$. Summing over the $(d - x)$ increments from x to d gives $\sum_{t=x+1}^d (2t - 1) = d^2 - x^2$.

Each such increment step is paired with a decrease of some $y \in D$ (since the sum remains unchanged). Vertices in D start at $2r - 1$ and are never increased, so the maximum possible square-loss in any single decrease is $(2r - 1)^2 - (2r - 2)^2 = 4r - 3 = A$. Hence, the total paired loss over the $(d - x)$ increments is at most $A(d - x)$. Therefore,

$$\Delta_v \ge (d^2 - x^2) - A(d - x) = d^2 - Ad + (Ax - x^2) \tag{1}$$

It is easy to check that the term $Ax - x^2 = x(A - x)$ is minimized on $[2r-1, 3r-1]$ at $x = 3r - 1$, giving

$$Ax - x^2 \ge (3r - 1)\big((4r - 3) - (3r - 1)\big) = (3r - 1)(r - 2) \tag{2}$$

Thus, from Eqs. (1) and (2) we get the required bound. □

Step 7: *Global bound.*
Let $T^*(H) - \sum_{e \in E(H)} s^*(e)$.

Lemma 8. *If H is a minimal counterexample on n vertices, then*

$$T^*(H) \ge n(2r - 1)^2 + l(4r - 1) + (r - 2)(3r - 1)|L(H)|$$

Proof. From Lemma 6, $T_k = \sum_v f_k(v)^2 = \sum_v d(v)^2 = \sum_e s(e)$. Thus,

$$T^*(H) = \sum_{e \in E(H)} s(e) - \sum_{e \in E(H)} (s(e) - s^*(e)) = T_k - \sum_{e \in E(H)} (s(e) - s^*(e)) \quad (3)$$

If an edge e does not contain a large vertex, then by definition $s^*(e) = s(e)$. Thus the sum can be restricted to the edges containing at least one large vertex, but from Lemma 4 we know that no edge can contain two large vertices. Thus,

$$\sum_{e \in E(H)} (s(e) - s^*(e)) = \sum_{v \in L(H)} \sum_{e \in E_v(H)} (s(e) - s^*(e)) \quad (4)$$

Since, for $i \in [k]$, $\Delta_i = T_i - T_{i-1}$, by telescoping we get, $T_k = T_0 + \sum_{i=1}^{k} \Delta_i = T_0 + \sum_{v \in I} \Delta_v \geq T_0 + \sum_{v \in L(H)} \Delta_v$. Thus from Eqs. (3) (4) we get,

$$T^*(H) \geq T_0 + \sum_{v \in L(H)} \Delta_v - \sum_{v \in L(H)} \sum_{e \in E_v(H)} (s(e) - s^*(e)) \quad (5)$$

Note that $\sum_v f_0(v) = (2r - 1)n + l$. Clearly, T_0 is minimized when for all $v \in V(H)$, $f_0(v)$ is as equal as possible. Therefore,

$$T_0 \geq (n - l)(2r - 1)^2 + l(2r)^2 = n(2r - 1)^2 + l(4r - 1)$$

Let $d(v) = d$, thus applying Lemmas 5 to 7 to Eq. (5) we get,

$$T^*(H) \geq T_0 + |L(H)|(d^2 - Ad + (r - 2)(3r - 1)) - |L(H)|(d^2 - Ad)$$
$$\geq n(2r - 1)^2 + l(4r - 1) + (r - 2)(3r - 1)|L(H)|.$$

$\square$

This combines all previous estimates to obtain a strong lower bound on the total truncated edge contribution.

Step 8: *Extracting a dense edge and resulting a contradiction.*
Since $|E(H)| = \frac{n(2r-1)+l}{r}$ where $l > 0$. Thus from Lemma 8,

$$\frac{T^*(H)}{|E(H)|} \geq \frac{(2r - 1)^2 n + l(4r - 1) + (r - 2)(3r - 1)|L(H)|}{\frac{n(2r-1)+l}{r}} \quad (6)$$

From Eq. (6) we get,

$$\frac{T^*(H)}{|E(H)|} > \frac{(2r - 1)^2 n + l(2r - 1)}{\frac{n(2r-1)+l}{r}} = r(2r - 1)$$

Since $T^*(H) = \sum_{e \in E(H)} s^*(e)$, we get that there exists an $e \in E(H)$ such that $s^*(e) \geq r(2r-1)+1$. Note that for $r \geq 3$, $(2r^2 - r + 1) - B = (2r^2 - r + 1) - (r^2 +$

$3r - 3) = (r - 2)^2 > 0$. Thus, $s^*(e) > B$ and therefore, $s^*(e) = s(e) = \sum_{v \in e} d(v)$. Since, $s(e) > B$, e does not contain any large vertex.

By averaging argument, there exists $x \in e$ such that $d(x) \geq 2r$. Suppose for all $y \in e \setminus \{x\}$, $d(y) \leq r$. Since e has no large vertex, $d(x) \leq 4r - 4$. Thus, $s(e) \leq (4r - 4) + (r - 1)r = r^2 + 3r - 4$. But $s(e) = s^*(e) \geq 2r^2 - r + 1$. Thus,

$$r^2 + 3r - 4 \geq 2r^2 - r + 1 \implies 0 \geq r^2 - 4r + 5 = (r - 2)^2 + 1$$

Thus, we get a contradiction. Therefore, there exists $y \in e \setminus \{x\}$ such that $d(y) \geq r + 1$. Now since $\delta(H) \geq 2$, $d(z) \geq 2$ for $z \in e \setminus \{x, y\}$. Thus, by Lemma 3, H contains a crown, which is a contradiction.

6   Proof of Theorem 5

Let $n = qr^2$ and partition the n-vertex set into q disjoint parts $V(H) = V_1 \sqcup V_2 \sqcup \cdots \sqcup V_q$, such that $|V_i| = r^2$. Suppose each V_i induces H_i, a component of H, and each H_i is a copy of $S(2, r, r^2)$.

Since each H_i is linear and r-uniform by definition, and edges of disjoint components are disjoint, we have H to be linear and r-uniform.

By Lemma 1, we have

$$|E(H)| = \sum_{i=1}^{q} |E(H_i)| = \frac{n}{r^2} \cdot \frac{r^2(r^2 - 1)}{r(r - 1)} = \frac{(r + 1)n}{r}.$$

Suppose there exists a P_4^r in H, clearly it must be in some H_i. Let e_1, e_2, e_3 and e_4 be the edges of P_4^r, such that $e_1 \cap e_3 = \emptyset$, $e_1 \cap e_4 = \emptyset$ and $v \in e_3 \cap e_4$. Note that $v \notin e_1$ and we get two edges e_3 and e_4 containing the vertex v that are disjoint from e_1. This contradicts Lemma 2.

Therefore, H is a P_4^r-free graph with $|E(H)| = \frac{(r+1)n}{r}$. Thus, we have $ex_r^{\mathrm{lin}}(n, P_4^r) \geq \frac{(r+1)n}{r}$. It is easy to check that for $q = 1$, the bound is strict, since $S(2, r, r^2)$ maximizes the number of edges.

Conclusion and Future Directions

In this paper, we studied linear Turán numbers for r-uniform linear hypertrees. We provided a general construction giving the lower bound $ex_r^{\mathrm{lin}}(n, T_k^r) \geq \frac{n(k-1)}{r}$ under mild assumptions, and showed that this bound is sharp for stars. For hypertrees on four edges, we determined the exact extremal value for B_4^r, proved an upper bound for the crown E_4^r, and established a lower bound for P_4^r.

We conclude by proposing a conjectural sharp upper bound for P_4^r, matching the lower bound we provided in Theorem 5.

Conjecture 1. $ex_r^{\mathrm{lin}}(n, P_4^r) \leq \dfrac{(r + 1)n}{r}$. Equality holds if and only if the linear r-uniform hypergraph is the union of disjoint Steiner systems $S(2, r, r^2)$, given that the Steiner system $S(2, r, r^2)$ exists.

A natural direction for future work is to determine a sharp upper bound for E_4^r.

Another direction that can be of interest is to obtain sharp bounds for P_k^r, as the known bound, $ex_r^{\mathrm{lin}}(n, P_k^r) \leq \frac{(2r-3)kn}{2}$ from [29], can be far from optimal, and we already have sharp bound for paths in the graph setting (check [7]).

References

1. Bruck, R.H., Ryser, H.J.: The nonexistence of certain finite projective planes. Can. J. Math. **1**(1), 88–93 (1949)
2. Carbonero, A., Fletcher, W., Guo, J., Gyárfás, A., Wang, R., Yan, S.: Crowns in linear 3-graphs, arXiv preprint (2021)
3. Carbonero, A., Fletcher, W., Guo, J., Gyárfás, A., Wang, R., Yan, S.: Crowns in linear 3-graphs of minimum degree 4. Electron. J. Comb. P4–17 (2022)
4. Chowla, S., Ryser, H.J.: Combinatorial problems. Can. J. Math. **2**, 93–99 (1950)
5. Collier-Cartaino, C., Graber, N., Jiang, T.: Linear Turán numbers of linear cycles and cycle-complete Ramsey numbers. Comb. Probab. Comput. **27**(3), 358–386 (2018)
6. Davoodi, A., Győri, E., Methuku, A., Tompkins, C.: An Erdős-Gallai type theorem for uniform hypergraphs. Eur. J. Comb. **69**, 159–162 (2018)
7. Erdős, P., Gallai, T.: On maximal paths and circuits of graphs. Acta Math. Acad. Sci. Hungar. **10**, 337–356 (1959)
8. Erdos, P.: Extremal problems in graph theory. In: Theory of Graphs and its Applications (Proc. Sympos. Smolenice, 1963), pp. 29–36 (1964)
9. Fletcher, W.: Improved Upper Bound on the Linear Turán Number of the Crown, arXiv preprint (2021)
10. Füredi, Z.: Turán type problems. Surveys in combinatorics **166**, 253–300 (1991)
11. Füredi, Z., Jiang, T., Seiver, R.: Exact solution of the hypergraph Turán problem for k-uniform linear paths. Combinatorica **34**(3), 299–322 (2014)
12. Füredi, Z., Simonovits, M.: The history of degenerate (bipartite) extremal graph problems. In: Erdős Centennial, pp. 169–264. Springer, Cham (2013)
13. Gerbner, D., Patkós, B.: Extremal finite set theory, Chapman and Hall/CRC (2018)
14. Gyárfás, A., Ruszinkó, M., Sárközy, G.N.: Linear Turán numbers of acyclic triple systems. Eur. J. Comb. **99**, 103435 (2022)
15. Gyárfás, A., Sárközy, G.N.: Turán and Ramsey numbers in linear triple systems. Discret. Math. **344**(3), 112258 (2021)
16. Gyárfás, A., Sárközy, G.N.: The linear Turán number of small triple systems or why is the wicket interesting? Discrete Math. **345**(11), 113025 (2022)
17. Győri, E., Katona, G.Y., Lemons, N.: hypergraph extensions of the Erdős-Gallai theorem. Eur. J. Comb. **58**, 238–246 (2016)
18. Keevash, P.: Hypergraph Turán Problems. Surv. Comb. **392**, 83–140 (2011)
19. Keevash, P.: The existence of designs. arXiv preprint (2014)
20. Khormali, O., Palmer, C.: Turán numbers for hypergraph star forests. Eur. J. Comb. **102**, 103506 (2022)
21. Mubayi, D., Verstraëte, J.: Minimal paths and cycles in set systems. Eur. J. Comb. **28**(6), 1681–1693 (2007)
22. Ruzsa, I.Z., Szemerédi, E.: Triple systems with no six points carrying three triangles. Combinatorics (Keszthely, 1976), Coll. Math. Soc. J. Bolyai **18**(939–945), 2 (1978)

23. Sárközy, G.N.: Turán and Ramsey numbers in linear triple systems II. Discret. Math. **346**(1), 113182 (2023)
24. Tang, C., Wu, H., Zhang, S., Zheng, Z.: On the Turán Number of the Linear 3-Graph C_{13}. Electron. J. Comb. P3–46 (2022)
25. Turán, P.: Egy gráfelméleti szélsoértékfeladatról. Mat. Fiz. Lapok **48**(3), 436 (1941)
26. Zhang, L.-P., Broersma, H., Wang, L.: Generalized crowns in linear r-graphs. Electron. J. Comb. pp. P1–29 (2025)
27. Zhang, L.-P., Broersma, H., Wang, L.: Turán numbers of general star forests in hypergraphs. Discret. Math. **348**(1), 114219 (2025)
28. Zhang, L., Wang, L.: The Linear Turán Numbers of Acyclic Linear 4-graphs, pp. 1–13. Acta Mathematicae Applicatae Sinica, English Series (2025)
29. Zhou, J., Yuan, X.: Turán problems for star-path forests in hypergraphs. Discret. Math. **348**(11), 114592 (2025)

Exact Algorithms for Edge Deletion to Cactus

Sheikh Shakil Akhtar[(✉)] and Geevarghese Philip

Chennai Mathematical Institute, Chennai, India
shakil@cmi.ac.in

Abstract. We study two related problems on simple, undirected graphs: Edge Deletion to Cactus and Spanning Tree to Cactus.

Edge Deletion to Cactus has been known to be NP-hard on general graphs at least since 1988. In their recent work where they showed—*inter alia*—that Edge Deletion to Cactus is NP-hard even when restricted to bipartite graphs, Koch et al. (Disc. Appl. Math., 2024) posed the complexity of Spanning Tree to Cactus as an open problem. Our first main result is that Spanning Tree to Cactus can be solved in polynomial time.

We use the polynomial-time algorithm for Spanning Tree to Cactus as a black box to obtain an exact algorithm that solves Edge Deletion to Cactus in $\mathcal{O}^\star(n^{n-2})$ time, where n is the number of vertices in the input graph and the $\mathcal{O}^\star()$ notation hides polynomial factors. This is a significant improvement over the brute-force algorithm for Edge Deletion to Cactus that enumerates all subsets of the edge set.

We further improve the running time for Edge Deletion to Cactus to $\mathcal{O}^\star(3^n)$ using dynamic programming; this is our second main result. To the best of our knowledge, this is the fastest known exact-exponential algorithm for Edge Deletion to Cactus.

1 Introduction

A *cactus graph* is a connected graph in which every edge belongs to at most one cycle. Equivalently, any two simple cycles in a cactus share at most one vertex. Another equivalent characterization is that every block (biconnected component) of the graph is either a single edge or a simple cycle. This characterization implies that cactus graphs can be recognized in polynomial time using a block decomposition algorithm, such as the classical linear-time algorithm for computing biconnected components by Hopcroft and Tarjan [9]. Cactus graphs form a natural generalisation of trees and arise in several areas of graph theory, network design, and algorithmic graph modification problems [3,8].

In this paper, we study two closely related edge modification problems whose objective is to transform a given connected graph into a cactus by deleting or selecting edges.

F. Foucaud and A. Parreau (Eds.): IWOCA 2026, LNCS 16587, pp. 32–44, 2026.
https://doi.org/10.1007/978-3-032-27732-9_3

SPANNING TREE TO CACTUS
Input: A simple, undirected connected graph $G = (V, E)$ and a spanning tree T of G.
Question: Find the maximum size of a set $F \subseteq E \backslash E(T)$ such that $H = (V, E(T) \cup F)$ is a spanning cactus subgraph of G.

EDGE DELETION TO CACTUS
Input: A simple, undirected connected graph $G = (V, E)$.
Question: Find the size of a minimum sized set $F \subseteq E$ such that $H = (V, E \setminus F)$ is a connected cactus subgraph of G.

The second problem, EDGE DELETION TO CACTUS, belongs to the broad class of graph modification problems, in which the goal is to transform a given graph into a graph belonging to a specified class using a minimum number of edits. Such problems have been extensively studied for graph classes including forests, interval graphs, chordal graphs, and planar graphs. A closely related problem is VERTEX DELETION TO CACTUS, which asks for a minimum-sized set of vertices whose removal transforms the input graph into a cactus. Several results in this direction are known [1, 2, 11, 13].

The computational complexity of EDGE DELETION TO CACTUS was first investigated by El-Mallah and Coulborn [5], who proved that the problem is NP-complete for general graphs. More recently, Koch et al. [10] strengthened this result by showing that the problem remains NP-complete even when restricted to bipartite graphs. In the same work, the authors explicitly posed the complexity of the first problem, SPANNING TREE TO CACTUS, as an open question.

To the best of our knowledge, no polynomial-time algorithm was previously known for SPANNING TREE TO CACTUS, and no exact exponential-time algorithm was known for EDGE DELETION TO CACTUS with a running time substantially better than brute-force enumeration over all subsets of edges.

Our Contributions. We resolve both problems algorithmically.

First, we design a polynomial-time algorithm for SPANNING TREE TO CACTUS.

Theorem 1. *There exists a polynomial-time algorithm that solves* SPANNING TREE TO CACTUS.

This algorithm is based on a reduction to the Maximum Independent Set problem on Edge Path Tree (EPT) graphs [6, 7, 12], obtained by constructing an auxiliary intersection graph of the unique paths between the endpoints of the non-tree edges. By combining this polynomial-time routine with enumeration over all spanning trees of G, we obtain our first exact algorithm for EDGE DELETION TO CACTUS. (In the following theorem the notation $\mathcal{O}^\star(.)$ hides the polynomial factors on n.)

Theorem 2. *There exists an algorithm that solves* EDGE DELETION TO CACTUS *in time* $\mathcal{O}^\star(n^{n-2})$.

Although this algorithm is less efficient than the one described later, it still improves substantially over the brute-force method of enumerating all subsets of the edge set, which would require $\mathcal{O}^{\star}(2^{n^2})$ time in the worst case.

Next, we solve the EDGE DELETION TO CACTUS problem independently in order to obtain a more efficient exact algorithm.

Theorem 3. *There exists an algorithm that solves* EDGE DELETION TO CACTUS *in time* $\mathcal{O}^{\star}(3^n)$.

Organisation of the Paper. We first establish structural properties of cactus graphs and edge-minimal non-cactus graphs. Using these properties together with results on EPT graphs, we prove Theorem 1. We then use this algorithm as a subroutine to obtain Theorem 2.

Finally, we develop additional structural observations on cactus graphs and use them to design a dynamic programming algorithm that proves Theorem 3.

2   Preliminaries

We use standard notations from graph theory. All graphs considered in this paper are simple and undirected.

Let $G = (V, E)$ be a graph. Throughout the paper, we write $n := |V|$ to denote the number of vertices of the input graph. The *size* of a graph is defined as $|V| + |E|$.

For any graph H, we denote its vertex set by $V(H)$ and its edge set by $E(H)$. For two graphs G_1 and G_2, their union, denoted by $G_1 \cup G_2$, is the graph with vertex set $V(G_1) \cup V(G_2)$ and edge set $E(G_1) \cup E(G_2)$.

For a graph G and a vertex $x \in V(G)$, we denote by $\deg_G(x)$ the degree of x in G. For any subset $X \subseteq V(G)$, we denote by $G[X]$ the subgraph of G induced by X. For any subset $X \subseteq V(G)$, we denote by $G - X$ the induced subgraph $G[V(G) \setminus X]$. If $X = \{v\}$ for some $v \in V(G)$, we simply write $G - v$. For any subset $F \subseteq E(G)$, we denote by $G - F$ the graph $(V(G), E(G) \setminus F)$. If $F = \{e\}$ for some $e \in E(G)$, we simply write $G - e$.

A *block* of a graph is a maximal biconnected subgraph.

We now define Edge Path Tree (EPT) graphs following Golumbic and Jamison [6,7]. Let T be a tree, and let $\mathcal{P}$ be a collection of nontrivial simple paths in T; that is, each path in $\mathcal{P}$ contains at least one edge. We define the *edge intersection graph* of $\mathcal{P}$ in T, denoted by $\Gamma(\mathcal{P}, T)$, as the graph whose vertex set corresponds to the elements of $\mathcal{P}$, with two vertices joined by an edge if and only if their corresponding paths share at least one edge in T. A graph G is called an *EPT graph* if there exist a tree T and a collection of paths $\mathcal{P}$ in T such that $G = \Gamma(\mathcal{P}, T)$.

3   Spanning Tree to Cactus

In this section, we first prove Theorem 1 by presenting a polynomial-time algorithm for the problem SPANNING TREE TO CACTUS. We then use this result to establish Theorem 2, which yields an algorithm for EDGE DELETION TO CACTUS with running time $\mathcal{O}^{\star}(n^{n-2})$.

We require the following theorem for the design of our algorithm. A pictorial illustration of the statement is given in Fig. 1.

Theorem 4. *Let $G = (V, E)$ be an edge-minimal non-cactus connected graph; that is, for every $e \in E$, the graph $G - e$ is a cactus. Then, for any two edges $e_1 = a_1b_1$ and $e_2 = a_2b_2$, there exists a cycle C in G that contains both e_1 and e_2.*

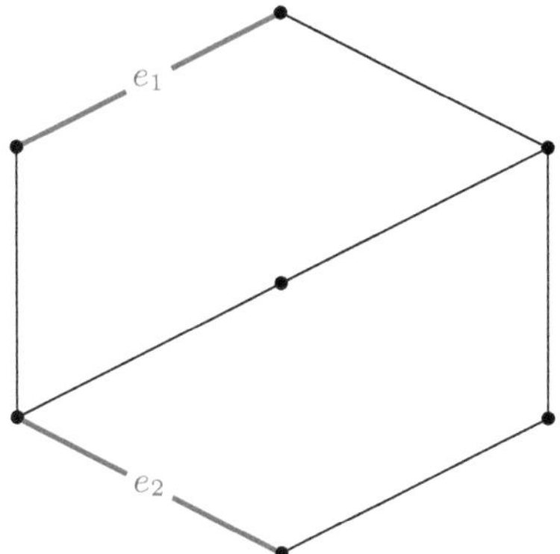

Fig. 1. An edge-minimal non-cactus graph with two marked edges e_1 and e_2, such that there exists a cycle which contains both e_1 and e_2, as claimed in Theorem 4.

The following theorem establishes a key structural property that is central to the design of our algorithm.

Theorem 5. *Let $T = (V, E)$ be a tree, and let $\hat{E}$ be a set of edges with endpoints in V such that $\hat{E} \cap E = \emptyset$. Define a new graph $H := (V, E \cup \hat{E})$. Then H is not a cactus if and only if there exist two edges in $\hat{E}$ such that the unique paths between their respective endpoints in T share at least one edge.*

Proof. Suppose first that there exist two edges in $\hat{E}$, say e'_1 and e'_2, such that the unique paths between their respective endpoints in T, denoted by P'_1 and P'_2, share at least one edge. Then H contains two distinct cycles, namely $P'_1 \cup e'_1$ and $P'_2 \cup e'_2$, that share an edge, and hence H is not a cactus.

Conversely, suppose that H is not a cactus. Let $E^{\star} \subseteq \hat{E}$ be a minimal set such that the graph

$$H^{\star} := (V, E \cup E^{\star})$$

is not a cactus; that is, for every $e \in E^\star$, the graph $H^\star - e$ is a cactus. Clearly, $H^\star$ is a subgraph of H, and we must have $|E^\star| \geq 2$.

Assume, for a contradiction, that for every pair of edges in $\hat{E}$, and hence also in $E^\star$, the unique paths between their respective endpoints in T are edge-disjoint. Let H' be an edge-minimal non-cactus subgraph of $H^\star$; that is, for every $e \in E(H')$, the graph $H' - e$ is a cactus. Then H' must contain all edges of $E^\star$, and therefore it contains at least two edges from $\hat{E}$; denote them by e_1 and e_2.

Let P_1 be the unique path in T between the endpoints of e_1, say a_1 and b_1, and let P_2 be the unique path in T between the endpoints of e_2, say a_2 and b_2. By Theorem 4, there exists a cycle, denoted by $\tilde{C}$, in H' that contains both e_1 and e_2. Define

$$\tilde{P} := \tilde{C} - e_1.$$

Then $\tilde{P}$ is a path between a_1 and b_1 and contains the edge e_2.

Note that $\tilde{P}$ is distinct from P_1, since $\tilde{P}$ contains the edge e_2, whereas P_1 does not. Consider the graph

$$\hat{C} := \tilde{P} \cup P_1.$$

This graph contains the edge e_2 but not e_1. Moreover, it is connected and every edge belongs either to $\tilde{P}$ or to P_1, which are two distinct paths between a_1 and b_1. Therefore, every edge in $E(\tilde{P}) \Delta E(P_1)$ lies in a cycle.

In particular, the edge e_2 is contained in a cycle of $\tilde{C}$; denote this cycle by $C^\star$. This cycle is distinct from $P_2 \cup \{e_2\}$, since it contains edges from P_1, whereas $P_2 \cup \{e_2\}$ does not, as $E(P_1) \cap E(P_2) = \emptyset$ by assumption.

Observe that the cycle $C^\star$ is present in the graph

$$H'' := (V, (E \cup E^\star) \setminus \{e_1\}).$$

Thus, the graph H'' contains two distinct cycles, $C^\star$ and $P_2 \cup \{e_2\}$, that share the edge e_2. This contradicts the minimality of $E^\star$, as $H'' = H^\star - e_1$. Therefore, there must exist two distinct edges in $E^\star$ and hence in $\hat{E}$, say e_i and e_j, such that the unique paths between their endpoints share at least one edge.

3.1   The Algorithm

Using Theorem 5, we will design an algorithm for the problem SPANNING TREE TO CACTUS.

Let $G = (V, E)$ be a connected graph with T as its spanning tree, which is given as input. The elements of $E \setminus E(T)$ are precisely the non-tree edges.

Construct an auxiliary graph $\mathcal{A}_T$ as follows:

- The vertices of $\mathcal{A}_T$ correspond to the unique paths in T associated with the non-tree edges. For example, for a non-tree edge $e = ab$, there is a vertex in $\mathcal{A}_T$ corresponding to the unique path in T between a and b.
- Two vertices in $\mathcal{A}_T$ are adjacent if and only if their corresponding tree paths share at least one edge.

Observe that, for any set of non-tree edges $F(\subseteq E \setminus E(T))$, the graph $H :=$ $(V, E(T) \cup F)$ is a cactus if and only if the vertices of $\mathcal{A}_T$ corresponding to the set F forms an independent set. This is due to Theorem 5.

Lemma 1. *Let $G = (V, E)$ be a connected graph and let T be a spanning tree of G. Let $F \subseteq E \setminus E(T)$. We construct an auxiliary graph $\mathcal{A}_T$ as shown above. Then the graph $H := (V, E(T) \cup F)$ is a cactus if and only if the set of vertices of $\mathcal{A}_T$ corresponding to F forms an independent set in $\mathcal{A}_T$.*

Proof. By Theorem 5, H is a cactus if and only if, there exists no two edges from F such that the unique paths between their endpoints intersect. Thus, H is a cactus if and only if the set of vertices of $\mathcal{A}_T$ corresponding to F forms an independent set in $\mathcal{A}_T$.

As stated earlier the graph $\mathcal{A}_T$ is an EPT graph. Moreover, Tarjan [12] showed that the size of a maximum independent set in an EPT graph can be computed in time polynomial in the size of the graph.

This yields a polynomial-time algorithm for our problem: the auxiliary graph $\mathcal{A}_T$ can be constructed in polynomial time, and the maximum independent set of $\mathcal{A}_T$ can then be computed in polynomial time.

Algorithm 1. SPANNINGTREETOCACTUS(G, T)

Require: A connected graph $G = (V, E)$ and a spanning tree T of G
Ensure: The maximum number of non-tree edges that can be added to T such that the resulting graph remains a cactus
1: Construct the auxiliary graph $\mathcal{A}_T$ as defined earlier (each vertex corresponds to a non-tree edge of G, and two vertices are adjacent if and only if the corresponding tree paths in T share at least one edge)
2: Compute a maximum independent set of $\mathcal{A}_T$ using the algorithm of Tarjan [12]
3: **return** the size of this independent set

Lemma 2. *Algorithm 1 is correct and runs in polynomial time.*

Proof. The correctness of Algorithm 1 follows directly from Lemma 1.

We now analyse its running time. Let T be a spanning tree of the input graph G. The number of non-tree edges in G is at most $\mathcal{O}(n^2)$. For each such edge, the algorithm constructs a vertex in the auxiliary graph $\mathcal{A}_T$. Hence, the size of $\mathcal{A}_T$ is bounded by $\mathcal{O}(n^2)$.

The edges of $\mathcal{A}_T$ are determined by testing pairwise intersections of the corresponding tree paths, which can be carried out in polynomial time. Therefore, the construction of $\mathcal{A}_T$ takes polynomial time in n.

As $\mathcal{A}_T$ is an EPT graph, the size of a maximum independent set in it can be computed in time polynomial in the size of the graph [12]. Since $\mathcal{A}_T$ has size polynomial in n, this step also runs in polynomial time.

Consequently, Algorithm 1 runs in time polynomial in n.

Observe that the above algorithm can be used to design another algorithm for EDGE DELETION TO CACTUS, although not as efficient as the one from Theorem 3. Using any existing algorithm to enumerate all possible spanning tree of the graph G (input to EDGE DELETION TO CACTUS), we call the above algorithm for each spanning tree of G and return the overall maximum value reported. Winter [14], provides an $\mathcal{O}^\star(\tau(G))$ time algorithm for enumerating all spanning tree of a graph G, where $\tau(G)$ is number of spanning trees of a graph G. By the well-known Cayley's formula, $\tau(G)$ can be at most n^{n-2}. Thus, we obtain an $\mathcal{O}^\star(n^{n-2})$-time algorithm for EDGE DELETION TO CACTUS, which improves upon the brute-force approach of enumerating all subsets of the edge set of G, an approach that would require $\mathcal{O}^\star(2^{n^2})$ time.

Thus, we have established the following theorem.

Theorem 2. *There exists an algorithm that solves* EDGE DELETION TO CACTUS *in time* $\mathcal{O}^\star(n^{n-2})$.

Proof. Let $F \subseteq E$ be a minimum-size set of edges whose deletion transforms $G = (V, E)$ into a cactus, and let

$$H := (V, E \setminus F).$$

Then H is a connected cactus subgraph of G. In particular, H is a maximum-sized spanning cactus subgraph of G.

Since H is connected, it contains a spanning tree T'. The tree T' is also a spanning tree of G.

Starting from T', the graph H can be obtained by adding to T' a set of non-tree edges such that the resulting graph remains a cactus. By Theorem 1, Algorithm 1 computes, for a given spanning tree T, the maximum number of non-tree edges that can be added to T while preserving the cactus property. For the particular tree T', the algorithm therefore returns $|E(H)| - |E(T')|$.

Consequently, if we execute Algorithm 1 for every spanning tree of G and select the largest value obtained, we recover the size of a maximum spanning cactus subgraph of G, namely $|E(H)|$.

The number of spanning trees of an n-vertex graph is at most n^{n-2}. Since Algorithm 1 runs in polynomial time for each spanning tree, the overall running time is $\mathcal{O}^\star(n^{n-2})$.

This yields an algorithm for EDGE DELETION TO CACTUS with running time $\mathcal{O}^\star(n^{n-2})$.

4   A Faster Exact Algorithm

In this section, we prove Theorem 3 and thereby obtain, using dynamic programming, an improved algorithm for EDGE DELETION TO CACTUS with running time $\mathcal{O}^\star(3^n)$.

We observe that solving the above problem is equivalent to finding the size of a maximum-sized spanning cactus subgraph of G. Indeed, the complement of the edge set of such a spanning cactus is a minimum-size set of edges whose removal transforms G into a cactus. Thus, we will try to solve the problem of finding the size of the largest spanning cactus of the input graph.

For $X \subseteq V$, where $G[X]$ is connected, we define $I(X)$ to be the number of edges in a maximum-sized spanning cactus subgraph of $G[X]$.

Let $x \in X$ be such that $\deg_{G[X]}(x) \geq 2$, and let $A \subseteq X \setminus \{x\}$ be a non-empty set. Set $B := X \setminus (A \cup \{x\})$.

Suppose that B is also nonempty and that both $G[A \cup \{x\}]$ and $G[B \cup \{x\}]$ are connected. We then define $J_X[x, A]$ by

$$J_X[x, A] := \max\{ \, |E(\mathcal{C})| \mid \mathcal{C} \text{ is a spanning connected subgraph of } G[X],$$
$$x \text{ is a cut vertex of } \mathcal{C},$$
$$\mathcal{C}[A \cup \{x\}] \text{ and } \mathcal{C}[B \cup \{x\}] \text{ are cacti} \, \}.$$

Intuitively, $J_X[x, A]$ is the maximum number of edges in a spanning cactus subgraph of $G[X]$ that is forced to decompose into two cactus subgraphs on $A \cup \{x\}$ and $B \cup \{x\}$, joined together only at the vertex x.

Before proceeding further, we must ensure that $J_X[x, A]$ is well defined for all $x \in X$ with $\deg_{G[X]}(x) \geq 2$ and for all subsets $A \subseteq X \setminus \{x\}$ such that both A and B are nonempty, where $B := X \setminus (A \cup \{x\})$, and such that both $G[A \cup \{x\}]$ and $G[B \cup \{x\}]$ are connected. The idea is to construct separate spanning trees of the two connected subgraphs $G[A \cup \{x\}]$ and $G[B \cup \{x\}]$, each rooted at x, and then to glue them together at x. Their union is a connected spanning subgraph of $G[X]$ in which the removal of x separates the graph into two forests, so that x is a cut vertex.

We now proceed with the design of the algorithm to find the maximum sized spanning cactus subgraph of a connected graph. For every set X such that $G[X]$ is connected and $|X| \leq 5$, we compute $I(X)$ and all values $J_X[x, A]$ (whenever they are defined) by brute force. If $|X| > 5$ and $G[X]$ is a connected cactus subgraph, then we set

$$I(X) := |E(G[X])|.$$

We now consider the remaining case in which $|X| > 5$ and $G[X]$ is connected but not a cactus. Before describing the algorithm for this case, we need to establish some structural properties of cactus graphs that will be used in the design of the algorithm.

Any connected non-cactus graph, with at least five vertices, must have a maximum sized spanning cactus subgraph with a cut vertex. If a maximum spanning cactus already has more than one block, then it necessarily has a cut vertex. Otherwise it is a simple cycle, and since the original graph is not a cactus, we can swap in a non-cycle edge and remove a cycle edge to obtain another spanning cactus of the same size that now has a cut vertex.

Lemma 3. *Let $H = (V, E)$ be a connected graph that is not a cactus, and assume that $|V| \geq 5$. Then there exists a maximum-sized spanning cactus subgraph of H that has a cut vertex. (See Fig. 2 for an example.)*

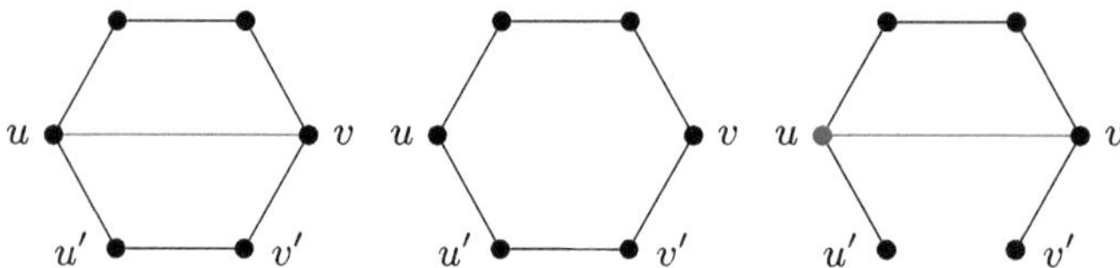

Fig. 2. From left to right: the graph H, a spanning cycle $\mathcal{C}$ of H, and a spanning cactus subgraph $\mathcal{C}'$ of H with cut vertex u, respectively.

Proof. Let $\mathcal{C}$ be a maximum-sized spanning cactus subgraph of H.

If $\mathcal{C}$ has more than one block, then it already has a cut vertex and there is nothing to prove. Otherwise, $\mathcal{C}$ consists of a single block and is therefore a simple cycle.

Since H is not a cactus, there exists an edge $e = uv \in E(H)$ that is not contained in $\mathcal{C}$. Let $e' = u'v' \in E(\mathcal{C})$. Without loss of generality, we may assume that $u \neq u'$ and $v \neq v'$, since $|V| \geq 5$.

Define
$$\mathcal{C}' := \big(V,\ (E(\mathcal{C}) \cup \{e\}) \setminus \{e'\}\big).$$

Then $\mathcal{C}'$ is a spanning cactus subgraph of H with the same number of edges as $\mathcal{C}$.

We now show that u is a cut vertex of $\mathcal{C}'$. Removing the edge e' from the cycle $\mathcal{C}$ turns it into a path. Adding the edge $e = uv$ reconnects this path at the vertex u but does not create a second internally disjoint path between the vertices u and u'. Consequently, in the graph $\mathcal{C}' - u$, the vertices that lie on the former path between u and u' are separated from the rest of the graph.

Thus, deleting u disconnects $\mathcal{C}'$, and therefore u is a cut vertex of $\mathcal{C}'$.

In the following lemma, we show that gluing two cactus graphs together at a single common vertex preserves the cactus property. Intuitively, every cycle in the resulting graph remains entirely within one of the original cacti, and therefore no two cycles can share an edge.

Lemma 4. *Let $\mathcal{C}$ and $\mathcal{C}'$ be two edge-disjoint cacti, such that $V(\mathcal{C}) \cap V(\mathcal{C}') = \{v\}$. Then $\mathcal{C} \cup \mathcal{C}'$ is also a cactus.*

Proof. Any cycle in $\mathcal{C} \cup \mathcal{C}'$, has all its edges either in $\mathcal{C}$ or in $\mathcal{C}'$. Thus, if two cycles in $\mathcal{C} \cup \mathcal{C}'$ share edges, then it would mean either $\mathcal{C}$ or $\mathcal{C}'$ is not a cactus, which is a contradiction.

The next two theorems capture the additive behaviour of optimal cactus subgraphs when a connected graph is decomposed at a cut vertex. When such a decomposition separates the graph into two connected parts, any maximal cactus structure respecting this separation consists of optimal cactus subgraphs on each part, and its size is the sum of their sizes.

Theorem 6. *Let $G = (V, E)$ be a connected graph, and let $X \subseteq V$ be a set such that $|X| > 5$ and $G[X]$ is connected. Then*

$$I(X) = \max_{x \in X,\, A \subseteq X} J_X[x, A].$$

Proof. We first show that

$$I(X) \leq \max_{x \in X,\, A \subseteq X} J_X[x, A].$$

Let $\mathcal{C}$ be a spanning cactus subgraph of $G[X]$ with the maximum possible number of edges. By definition,

$$|E(\mathcal{C})| = I(X).$$

By Lemma 3, we can assume that the cactus $\mathcal{C}$ has at least one cut vertex; fix such a vertex and denote it by x. Let A be one of the connected components of $\mathcal{C} - x$, and set

$$B := X \setminus (A \cup \{x\}).$$

Definitely both A and B are non-empty and $\deg_{G[X]}(x) \geq 2$. Consider the induced subgraphs $\mathcal{C}[A \cup \{x\}]$ and $\mathcal{C}[B \cup \{x\}]$. These are cactus subgraphs of $G[A \cup \{x\}]$ and $G[B \cup \{x\}]$, respectively, and hence both $G[A \cup \{x\}]$ and $G[B \cup \{x\}]$ are connected. Moreover, $\mathcal{C}$ is exactly the union of $\mathcal{C}[A \cup \{x\}]$ and $\mathcal{C}[B \cup \{x\}]$. It follows that the number of edges of $\mathcal{C}$ is bounded above by $J_X[x, A]$ for this choice of x and A, and therefore

$$I(X) \leq \max_{x \in X,\, A \subseteq X} J_X[x, A].$$

We now prove the reverse inequality. Let $x' \in X$ and let $A' \subseteq X \setminus \{x'\}$ be such that $A' \neq \emptyset$ and the induced subgraphs $G[A' \cup \{x'\}]$ and $G[B' \cup \{x'\}]$ are connected, where

$$B' := X \setminus (A' \cup \{x'\}).$$

Let $B' \neq \emptyset$. Let $\mathcal{C}'$ be a subgraph of $G[X]$ with the maximum number of edges among all spanning subgraphs of $G[X]$ that have x' as a cut vertex and for which $\mathcal{C}'[A' \cup \{x'\}]$ and $\mathcal{C}'[B' \cup \{x'\}]$ are cacti. By construction,

$$|E(\mathcal{C}')| = J_X[x', A'].$$

Since $\mathcal{C}'$ is a cactus subgraph of $G[X]$ by Lemma 4, its number of edges is at most $I(X)$. And x' and A' were chosen arbitrarily, we have

$$\max_{x \in X,\, A \subseteq X} J_X[x, A] \leq I(X).$$

Combining the two inequalities completes the proof.

Theorem 7. *Let $G = (V, E)$ be a connected graph, and let $X \subseteq V$ be such that $|X| > 5$ and $G[X]$ is connected. Let $x \in X$ satisfy $\deg_{G[X]}(x) \geq 2$, and let $A \subseteq X \setminus \{x\}$ be a nonempty set. Set*

$$B := X \setminus (A \cup \{x\}),$$

and suppose that B is nonempty and that both $G[A \cup \{x\}]$ and $G[B \cup \{x\}]$ are connected. Then

$$J_X[x, A] = I(A \cup \{x\}) + I(B \cup \{x\}).$$

Proof. Let $\mathcal{C}$ be a maximum-sized connected subgraph of $G[X]$ such that x is a cut vertex of $\mathcal{C}$, and such that $\mathcal{C}[A \cup \{x\}]$ and $\mathcal{C}[B \cup \{x\}]$ are cacti. By definition,

$$|E(\mathcal{C})| = J_X[x, A].$$

Since both $\mathcal{C}[A \cup \{x\}]$ and $\mathcal{C}[B \cup \{x\}]$ are cacti, we have

$$|E(\mathcal{C}[A \cup \{x\}])| \leq I(A \cup \{x\}) \quad \text{and} \quad |E(\mathcal{C}[B \cup \{x\}])| \leq I(B \cup \{x\}).$$

Therefore,

$$J_X[x, A] \leq I(A \cup \{x\}) + I(B \cup \{x\}). \tag{1}$$

Conversely, let $\mathcal{C}'$ be a maximum-sized spanning cactus subgraph of $G[A \cup \{x\}]$, and let $\mathcal{C}''$ be a maximum-sized spanning cactus subgraph of $G[B \cup \{x\}]$. Define

$$\mathcal{C}^\star := \mathcal{C}' \cup \mathcal{C}''.$$

By Lemma 4, $\mathcal{C}^\star$ is a cactus subgraph of $G[X]$ and spans $G[X]$. Moreover, deleting x from $\mathcal{C}^\star$ disconnects the graph, since there is no edge with one endpoint in $\mathcal{C}'$ and the other in $\mathcal{C}''$. Hence, x is a cut vertex of $\mathcal{C}^\star$.

Thus $|E(\mathcal{C}^\star)| \leq J_X[x, A]$, and hence

$$I(A \cup \{x\}) + I(B \cup \{x\}) \leq J_X[x, A]. \tag{2}$$

Combining the inequalities 1 and 2, we get the proof.

Combining Theorems 6 and 7, we obtain a procedure for computing $I(V)$, and hence the number of edges in a maximum-sized spanning cactus subgraph of G. Intuitively, the algorithm performs dynamic programming over all vertex subsets $X \subseteq V$ that induce connected subgraphs, and for each such set it considers all possible ways of choosing a cut vertex and splitting the remaining vertices into two connected parts. This leads to a branching behaviour in which each vertex may belong to one of three roles (outside the current set, in the first part, or in the second part), yielding a total running time of order 3^n.

Thus we have established the following theorem.

Theorem 3. *There exists an algorithm that solves* EDGE DELETION TO CACTUS *in time $\mathcal{O}^\star(3^n)$.*

5   Conclusion

We studied two closely related graph modification problems whose objective is to transform a connected graph into a cactus.

For the EDGE DELETION TO CACTUS problem, which was known to be NP-hard, we presented an exact algorithm running in time $\mathcal{O}^{\star}(3^n)$. We also resolved the previously open problem SPANNING TREE TO CACTUS by designing a polynomial-time algorithm based on an auxiliary EPT graph construction and the computation of a maximum independent set. As a consequence, this result yields an alternative exact algorithm for EDGE DELETION TO CACTUS with running time $\mathcal{O}^{\star}(n^{n-2})$. Although this algorithm is less efficient than our $\mathcal{O}^{\star}(3^n)$ algorithm, it still improves substantially over brute-force enumeration of all subsets of edges. An obvious open question would be to obtain an algorithm for EDGE DELETION TO CACTUS which runs in time $\mathcal{O}^{\star}(c^n)$, for some $c < 3$.

Another important direction for future research is the parameterised version of EDGE DELETION TO CACTUS, where the parameter k is the number of deleted edges. We observe that if an input graph G is a YES-instance of the parameterised version of EDGE DELETION TO CACTUS, then the treewidth of G is at most $2k+2$ [10]. Since cactus graphs are exactly those graphs that exclude the diamond as a minor, and since minor-closed graph classes are expressible in monadic second-order logic, the problem admits a fixed-parameter tractable algorithm by Courcelle's Theorem [4]. However, it remains an open challenge to design an explicit FPT algorithm with a practical and small dependence on k.

Further directions include weighted variants of the problem, where edges have costs and the goal is to minimise the total deletion cost, as well as extensions to related graph modification problems targeting cactus-like graph classes.

We hope that the structural insights and algorithmic techniques developed in this work will stimulate further progress on cactus modification problems and related graph editing questions.

References

1. Aoike, Y., Gima, T., Hanaka, T., Kiyomi, M., Kobayashi, Y., Kobayashi, Y., Kurita, K., Otachi, Y.: An improved deterministic parameterized algorithm for cactus vertex deletion. Theory Comput. Syst. **66**(2), 502–515 (2022)
2. Bonnet, É., Brettell, N., Kwon, O., Marx, D.: Parameterized vertex deletion problems for hereditary graph classes with a block property. In: Heggernes, P. (ed.) WG 2016. LNCS, vol. 9941, pp. 233–244. Springer, Heidelberg (2016). https://doi.org/10.1007/978-3-662-53536-3_20
3. Brimkov, B., Hicks, I.V.: Memory efficient algorithms for cactus graphs and block graphs. Discret. Appl. Math. **216**, 393–407 (2017). https://www.sciencedirect.com/science/article/pii/S0166218X15005193, graph-theoretic and Polyhedral Combinatorics Issues and Approaches in Imaging Sciences, https://doi.org/10.1016/j.dam.2015.10.032
4. Courcelle, B., Engelfriet, J.: Graph structure and monadic second-order logic. a language-theoretic approach. Encycl. Math. Appl. **138** (2012)

5. El-Mallah, E., Colbourn, C.: The complexity of some edge deletion problems. IEEE Tran. Cir. Syst. **35**(3), 354–362 (1988). https://doi.org/10.1109/31.1748, funding Information: Manuscript received May 27, 1987; revised October 19, 1987. C. J. Colbourn's research was supported by NSERC Canada under Grant A0579. This paper was recommended by K. Thulasiraman and N. Deo, Guest Editors for this issue
6. Golumbic, M.C., Jamison, R.E.: Edge and vertex intersection of paths in a tree. Discret. Math. **55**(2), 151–159 (1985)
7. Golumbic, M.C., Jamison, R.E.: The edge intersection graphs of paths in a tree. J. Combinat. Theory, Series B **38**(1), 8–22 (1985)
8. Harary, F., Uhlenbeck, G.E.: On the number of husimi trees. Proc. Natl. Acad. Sci. **39**(4), 315–322 (1953)
9. Hopcroft, J., Tarjan, R.: Algorithm 447: efficient algorithms for graph manipulation. Commun. ACM **16**(6), 372–378 (1973)
10. Koch, I., Pardal, N., dos Santos, V.F.: Edge deletion to tree-like graph classes. Discret. Appl. Math. **348**, 122–131 (2024)
11. Misra, P., Raman, V., Ramanujan, M.S., Saurabh, S.: Parameterized Algorithms for EVEN CYCLE TRANSVERSAL. In: Golumbic, M.C., Stern, M., Levy, A., Morgenstern, G. (eds.) WG 2012. LNCS, vol. 7551, pp. 172–183. Springer, Heidelberg (2012). https://doi.org/10.1007/978-3-642-34611-8_19
12. Tarjan, R.E.: Decomposition by clique separators. Discret. Math. **55**(2), 221–232 (1985)
13. Tsur, D.: Faster deterministic algorithm for cactus vertex deletion. Inf. Process. Lett. **179**, 106317 (2023)
14. Winter, P.: An algorithm for the enumeration of spanning trees. BIT Numer. Math. **26**(1), 44–62 (1986)

Minimizing the Weighted Makespan
with Restarts on a Single Machine

Aflatoun Amouzandeh[1]([✉]) [iD], Klaus Jansen[2]([✉]) [iD], Lis Pirotton[2]([✉]) [iD], Rob
van Stee[1]([✉]) [iD], and Corinna Wambsganz[2]([✉]) [iD]

[1] Department of Mathematics, University of Siegen, Siegen, Germany
{aflatoun.amouzandeh,rob.vanstee}@uni-siegen.de
[2] Department of Computer Science, Kiel University, Kiel, Germany
{kj,lpi,cwa}@informatik.uni-kiel.de

Abstract. We consider the problem of minimizing the weighted makespan on a single machine with restarts. Restarts are similar to preemptions but weaker: a job can be interrupted, but then it has to be run
again from the start instead of resuming at the point of interruption
later. The objective is to minimize the weighted makespan, defined as
the maximum weighted completion time of jobs.

We establish a lower bound of 1.46557 on the competitive ratio achievable by deterministic online algorithms. For the case where all jobs have
identical processing times, we design and analyze a deterministic online
algorithm that improves the competitive ratio to better than 1.3098.
Finally, we prove a lower bound of 1.2344 for this case.

1 Introduction

Makespan minimization is a fundamental and extensively studied problem in
scheduling theory, with a significant body of research devoted to it. A sequence
of jobs must be scheduled non-preemptively on m identical machines. Each job
j is characterized by a processing time p_j and a non-negative arrival time r_j,
where $1 \leq j \leq n$. In this paper, we assume that jobs arrive over time, i.e.,
each job becomes available for processing only at its release time. The goal is to
minimize the *makespan*, defined as the maximum completion time of any job in
the schedule. The problem is obviously NP-hard as it generalizes the makespan
problem where all jobs arrive at time 0. Regarding the online setting, the best
competitive ratio currently known, for general m, is equal to 1.5 and attained
by the LPT (Longest Processing Time First) algorithm [5].

A more general objective is the *weighted makespan,* denoted as $WC_{\max} =
\max_j\{w_j C_j\}$, which measures the maximum *weighted completion time* of all jobs.
Each job j is assigned a positive weight w_j, reflecting its relative importance. The
classical makespan minimization problem is a special case in which all weights
are equal. This objective function was first considered only fairly recently and has
received some attention [4,13,17,20]. Even on a single machine, it is nontrivial
to optimize the weighted makespan non-preemptively if jobs arrive over time. In

F. Foucaud and A. Parreau (Eds.): IWOCA 2026, LNCS 16587, pp. 45–59, 2026.
https://doi.org/10.1007/978-3-032-27732-9_4

the standard three-field notation introduced by Graham et al. [10], this problem is denoted as $1 \mid r_j \mid WC_{\max}$. If all jobs arrive at time 0 or preemptions are allowed, then the algorithm Largest Weight (LW) which always runs the job that has the largest weight is 1-competitive [4,13].

In the *offline* setting, all job parameters (processing times, release times, and weights) are known in advance, allowing the algorithm to make better decisions. In contrast, the *online setting* assumes that jobs arrive over time and decisions must be made without knowledge of the future input. Once a job arrives, the online algorithm learns its parameters and must decide whether to process it immediately or delay its execution. The effectiveness of an online algorithm is measured by its *competitive ratio* [3,8,11]. For a given instance I, the objective values of the online and optimal offline algorithms are denoted as $\mathrm{ALG}(I)$ and $\mathrm{OPT}(I)$, respectively. The competitive ratio $\mathcal{R}$ of an online algorithm is defined as:

$$\mathcal{R} = \sup_I \frac{\mathrm{ALG}(I)}{\mathrm{OPT}(I)}.$$

This ensures that, in the worst case, the online algorithm's cost does not exceed $\mathcal{R}$ times the optimal cost. An online algorithm may use as much computation as needed to make decisions, but it lacks knowledge of the input.

Feng and Yuan [17] were the first to consider the weighted makespan in the context of single-machine scheduling. Li [13] explored the online version of this problem and established a lower bound of 2 on the competitive ratio for deterministic algorithms. Moreover, for the problem on a single machine, he provided an online algorithm with a competitive ratio of 3. For the case of identical machines where all jobs have the same processing time, Li proposed an online algorithm with the best-possible competitive ratio of $\frac{\sqrt{5}+1}{2}$.

Chai et al. [4] further improved the results on a single machine, by developing two deterministic online algorithms with a competitive ratio of 2. Lu et al. [16] considered the offline single-machine problem with job rejection but without release dates. They showed that this problem is NP-hard and proposed a fully polynomial-time approximation scheme (FPTAS). More recently, Sun [20] proposed a $(2 - \frac{1}{m})$-approximation algorithm for the weighted makespan scheduling problem on m identical machines, again in the setting without release dates $(P||WC_{\max})$. The algorithm runs List Scheduling on the jobs after sorting them by decreasing weight. Sun also introduced a randomized efficient polynomial-time approximation scheme (EPTAS) for this problem, as well as a randomized FPTAS for the case when m is fixed.

One important extension to the classical scheduling framework is the use of *restarts*, which forms a central focus of this paper. In this model, a running job can be interrupted when a more urgent job arrives (e.g., a smaller or heavier job in the case of weighted jobs). However, any work done on the interrupted job is lost, and it must be restarted from the beginning. This model, also known as *preemption with restarts*, differs from the more commonly studied preemption with resume model, where interrupted jobs can be resumed from the point of interruption. The concept of restarts was first introduced by Shmoys et al. [18]

in the context of makespan minimization and was later shown to improve the performance of online algorithms for other objectives on a single machine [1,2, 12,19], including for equal-length jobs [6]. However, for the important objective function of total weighted flow time, where the flow time of a job is the time between its arrival and its completion, it is known that restarts do not help [7]: any online algorithm that can restart jobs has a competitive ratio of $\Omega(n)$.

Variants of the restart model, including *limited restarts*, which means that each job may be restarted at most once, and k-limited restart where each job can be restarted at most k times and k can be either a positive integer or infinity, have also been studied, particularly in the context of parallel batch machines [9,15,21].

Liang et al. [14] recently explored the weighted makespan problem in the restart model for online single-machine scheduling. They proved that when only a single restart is allowed across all jobs, no online algorithm can achieve a competitive ratio lower than 2, matching the lower bound for the problem without restarts [13]. However, in the case where all jobs have unit processing times, they presented the best possible online algorithm with a competitive ratio of approximately 1.46557. The best known upper bound on the competitive ratio for deterministic online algorithms in the general setting with restarts is 2 resulting from [4].

In this paper, we consider the problem of online scheduling with restarts for minimizing the weighted makespan $WC_{\max}$. Specifically, we study the problem $1|r_j, \text{online}, \text{restart}|WC_{\max}$ and establish a new lower bound of 1.46557 on the competitive ratio for deterministic online algorithms in the general setting with restarts. Remarkably, this is the same value that Liang et al. [14] achieved for jobs of unit size and a single allowed restart. However, the problems are different.

Our main result is a deterministic online algorithm that achieves a competitive ratio better than 1.3098 for the case where all jobs have identical processing times. The algorithm balances a greedy approach, favoring heavier jobs, with the consideration that not too much time should be wasted on jobs that are eventually interrupted anyway. If a job arrives that is heavier than the currently running job, the new job is started instead unless the current job has already been running for sufficiently long. What sufficiently long is depends on the current time: we do not interrupt the running job in the case that the new job, if it is started after the current job completes, still completes within a factor of 1.3098 of its smallest possible completion time (which is achieved if the new job starts immediately when it arrives).

We also prove a lower bound of 1.2344 for this case. In determining both lower bounds, we first identified some difficult inputs and then used a computer to optimize the parameters. For the optimized values we then considered in which cases the goal competitive ratio was achieved. This resulted in a system of equalities that we solved to get exact lower bounds.

2   Lower Bound

We now consider the online problem $1|r_j, \text{online}, \text{restart}|WC_{\max}$. The lower bound presented here shows that the problem maintains its difficulty, even when unlimited restarts are allowed.

Theorem 1. *Any algorithm for the problem $1|r_j, \text{online}, \text{restart}|WC_{\max}$ is at least $R_1 \approx 1.46557$ competitive which is the real root of the equation $x^3 - x^2 - 1 = 0$.*

Proof. For contradiction, suppose that there exists an online algorithm ALG with a competitive ratio of less than R_1.

We consider the following job instance I. At time $r_1 = 0$, job 1 with $w_1 = 1$ and $p_1 = 1$ arrives. Then job 2 with $w_2 = \frac{1}{R_1-1} \approx 2.1478$ and $p_2 = 0$ arrives at time $r_2 = \frac{1}{R_1(R_1-1)} - 1 \approx 0.46557$.

Case 1. If ALG runs job 1 before job 2, no more jobs arrive. Then we have $\text{ALG}(I) \geq (p_1 + p_2)w_2 = w_2$ and $\text{OPT}(I) \leq \max(r_2 w_2, 1 + r_2) = 1 + r_2$ and we get

$$\frac{\text{ALG}(I)}{\text{OPT}(I)} \geq \frac{w_2}{1 + r_2} = \frac{\frac{1}{R_1-1}}{\frac{1}{R_1(R_1-1)}} = R_1.$$

Case 2. If ALG runs job 2 first, then job 1 is restarted no earlier than at time r_2. Now, two more jobs arrive at time $r_3 = r_4 = 1$: job 3 with $w_3 = w_2 \approx 2.1478$ and $p_3 = 0$ and job 4 with $w_4 = 1$ and $p_4 = w_2 - 1 \approx 1.1478$. Without loss of generality, any algorithm processes job 3 before job 4, since job 3 and job 4 are released at the same time and $p_3 = 0$ holds.

OPT schedules first job 1 and then the jobs 2,3 and 4 in this order, thus $\text{OPT}(I) = \max(1, w_2, w_3, p_4 + 1) = w_2 = \frac{1}{R_1-1}$.

Case 2.1. If ALG runs job 3 before job 1, then job 1 and job 4 are interchangeable without affecting the weighted makespan since they have equal weight. We have

$$\frac{\text{ALG}(I)}{\text{OPT}(I)} \geq \frac{\max(r_2 w_2, w_3, p_4 + 1, p_4 + 2)}{w_2} = \frac{w_2 + 1}{w_2} = R_1.$$

Case 2.2. If ALG runs and finishes job 1 directly after job 2, then job 3 starts and finishes not before $r_2 + 1$. In this case we have $\text{ALG}(I) \geq (r_2 + 1)w_3$. Thus,

$$\frac{\text{ALG}(I)}{\text{OPT}(I)} \geq \frac{(r_2 + 1)w_3}{w_2} = r_2 + 1 = \frac{1}{R_1(R_1 - 1)} = R_1$$

using $R_1^3 - R_1^2 - 1 = 0$. $\qquad\square$

3   Unit-Sized Jobs

In this section, we consider the problem of online scheduling with restarts for minimizing the weighted makespan, assuming all jobs have identical processing times. In this case, the problem is equivalent to considering unit-sized jobs, by scaling the release times accordingly. We first present our deterministic online algorithm Limited Largest Weight (LLW) which has a competitive ratio of $R \approx 1.3098$, the real root of $3x^3 - 2x^2 - x - 2 = 0$. We also provide a lower bound for this problem.

We first define two versions of the Largest Weight (LW) strategy.

- LW *with interruptions:* If the machine is idle, the next arriving job is started when it arrives. A running job is interrupted if a new job with higher weight arrives. If a job is interrupted or finishes and there are jobs waiting, the heaviest waiting job starts processing immediately.
- LW *without interruptions:* The same rules apply, except that once a job has started it is never interrupted.

The idea of our algorithm is as follows. In principle, we prefer to run the heaviest job. This would mean running LW with interruptions. However, if a significant part of the currently running job c has already been completed, it can be better to let it finish first, rather than interrupt it for an incoming job j. This depends on what the completion time of j will be when it completes immediately following c versus what its completion time will be if it starts immediately, and also on the relative weights of c and j. If c started at time t and j arrives at time t', then if c is allowed to complete, job j will not complete before time $t + 2$. We use the following very simple rule: we will interrupt c if j is heavier than c and $t + 2$ is more than R times the completion time of j if j starts immediately, which is $t' + 1$ (assuming no further interruptions occur).

We will see that this condition will never be satisfied after time $(2 - R)/(R - 1)$; from then on we simply run LW without interruptions. A complete description of our algorithm follows.

Our algorithm starts the first arriving job when it arrives. Once the algorithm starts a job at time $\frac{2-R}{R-1}$ or later, it runs LW without interruptions until all jobs have finished. Whenever it starts a job at time $t < \frac{2-R}{R-1} \approx 2.2279$ (Fig. 1), it runs LW with interruptions until time

$$t' = \frac{t+2}{R} - 1. \tag{1}$$

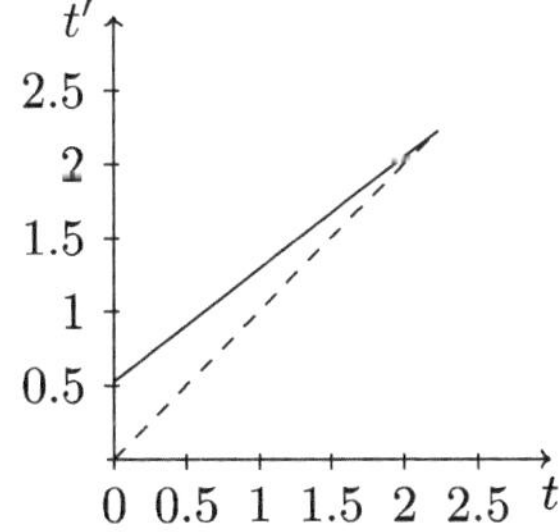

Fig. 1. Plot of t' as a function of t.

Note that for all $t < \frac{2-R}{R-1}$ it holds that $t < t'$ since

$$t < \frac{2-R}{R-1} \Leftrightarrow tR - t < 2 - R \Leftrightarrow t < \frac{t+2}{R} - 1.$$

We also have that t' never exceeds the boundary $(2-R)/(R-1)$. During the interval (t, t'), if an interruption occurs at time t_i, then we update $t := t_i$ and set t' according to (1). If time t' is ever reached without an interruption, the running job is allowed to continue until completion without any interruptions. After this, the algorithm resumes running LW either with or without interruptions, depending on the current time as described above.

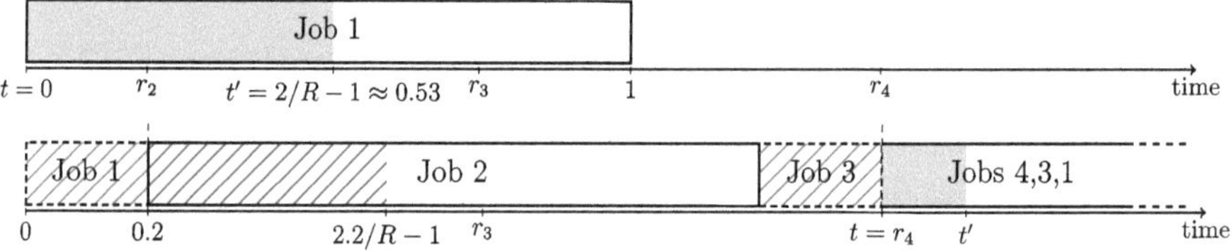

Fig. 2. This example shows the behavior of our algorithm.

For any job that starts at some time $t < \frac{2-R}{R-1}$, the interval (t, t') is called an LW-*phase*; however, if an interruption occurs in such an LW-phase, the LW-phase ends at that time, and the next LW-phase starts immediately (because this only happens before time $\frac{2-R}{R-1}$). In Fig. 2, we sketched in an example the behavior of the algorithm for the following job sequence: $((0, 1), (0.2, 1.1), (0.7, 1.6), (1.4, 2.3))$. Here each pair specifies first a release date and then the weight of the job released at this time. The top figure shows the situation at time $t = 0$ and the first LW-phase with values t and t'. In this LW-phase job 1 gets interrupted because job 2 is heavier. The bottom figure shows the schedule of the algorithm at time $t = r_4$. Job 2 is not interrupted during its LW-phase, so it continues without interruption. At its completion time the algorithm starts the waiting job 3. Job 3 gets interrupted in its LW-phase and after that the remaining jobs are scheduled with LW in the order $4, 3, 1$.

Analysis

We now prove that LLW is R-competitive. Recall that R is the positive real root of $3x^3 - 2x^2 - x - 2 = 0$.

We call a job *critical* for an input I if its weighted completion time is equal to $\text{LLW}(I)$. That is, no other job in the schedule of LLW for input I has a higher weighted completion time. An input I may contain multiple critical jobs.

We assume the existence of a *minimal counterexample I*. A minimal counterexample has ratio strictly larger than R and consists of the smallest number of jobs. We show by contradiction that such a counterexample does not exist. Without loss of generality, we can assume that no two jobs arrive at the same time, as otherwise, we could postpone the lighter job(s) by an arbitrarily small amount.

Consider an arbitrary critical job in I and let S be the maximal set of jobs that run consecutively without interruptions and that ends with this critical job. We scale the input such that this critical job has weight 1. Thus, $\text{LLW}(I)$ is simply the completion time of the critical job.

We now derive a sequence of properties that a counterexample must have, which will eventually lead to the conclusion that no counterexample exists. We begin by showing that the critical job is the lightest job in S (Property 1), and that OPT starts running its first job in S relatively early (Property 2). LLW starts its first job in S (which we will call job 1, see text below Property 1) when it arrives (Property 3), and the arrival of job 1 actually caused the interruption of another job $h \in S$ (Property 4). We then show that LLW does not complete any job before job 1 (Property 9). This requires some preparation: we first show that LLW does not complete two or more jobs before job 1 (Property 5) and then give a structural property for the case where LLW completes one job before job 1 (Property 7).

Property 1. The critical job is the lightest job in S.

Proof. Suppose not. Let a job be *heavy* if its weight is at least that of the critical job. We number the jobs in S according to the order in which LLW completes them and starting from 0 for the last non-heavy job in S (any jobs in S before this job have negative indices). Then after job 0, all jobs in S are heavy up to and including the critical job. Let H be the set of these heavy jobs in S, and let $k = |H|$. Then

$$\text{LLW}(I) = s_0 + k + 1,$$

where s_0 denotes the time at which LLW starts job 0 (for the last time). Since LLW runs job 0 before any job in H, all jobs in H arrive after time s_0, and they are all at least as heavy as the critical job. We have

$$\text{OPT}(I) \geq s_0 + k.$$

The competitive ratio is at most R as long as $s_0 + k \geq 1/(R-1)$, so the only problematic case is $s_0 < \frac{2-R}{R-1}$. By the rules of the algorithm, this means no jobs from H arrive before time $(s_0 + 2)/R - 1$, so we also have the lower bound

$$\text{OPT}(I) \geq \frac{s_0 + 2}{R} + k - 1.$$

Therefore

$$\frac{\text{LLW}(I)}{\text{OPT}(I)} \leq \frac{s_0 + k + 1}{(s_0 + 2)/R + k - 1}.$$

Setting $s_0 = 0$ and $k = 1$ we get $2/(2/R) = R$. This shows that the competitive ratio is at most R for all values of k and s_0.   □

Given Property 1, from this point on we simply again number the jobs in S according to their order in S, with the first job in S having index 1 and starting (for the last time) at time s_1 (Fig. 3). We define $k = |S|$, so we have

$$\text{LLW}(I) = s_1 + k.$$

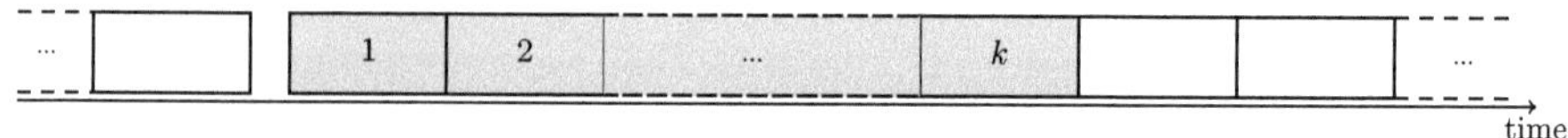

Fig. 3. This figure illustrates the set S with jobs 1 up to k. Job k is the critical job, and the jobs after k are not in S and are lighter.

Note that there may be additional jobs that are run consecutively without interruption immediately after S, but not immediately before S. Let r_S be the earliest arrival time of a job in S, and let t^* be the time OPT starts running a job in S for the first time. Then $t^* \geq r_S$. If OPT runs the jobs in S in (essentially) the same order as LLW and does not start them too much earlier than LLW, then the competitive ratio is maintained. We formalize this intuition in the next definition.

Definition 1 (Good set). *Suppose $t^* \geq (s_1 + k)/R - k$ and OPT completes a job which is at least as heavy as the critical job at time $t^* + k$ or later. Then S is* good.

Property 2. S is not good, and $t^* < (s_1 + k)/R - k$.

Proof. Suppose S is good. Then $\text{OPT}(I) \geq t^* + k \geq (s_1 + k)/R$, so I is not a counterexample. Second, suppose OPT starts running the first job in S at time $(s_1 + k)/R - k$ or later. Then again $\text{OPT}(I) \geq (s_1 + k)/R$. □

Property 3. No job completes at time s_1. Job 1 in S arrives at time s_1. If a job in S arrives before s_1, it is not heavier than job 1.

Proof. The first statement follows from the definition of S. It means that job 1 did not wait for another job to complete before starting. Since LLW also does not idle the machine if there is a job available, it starts job 1 immediately when it arrives, at time s_1.

Suppose some job $j \in S$ arrives at time $r_j < s_1$ and is heavier than job 1. When j arrives, LLW may continue the job it is running, but LLW will not start any new job that is lighter than j until it has completed j.

By the definition of S, LLW does not complete any job in S before time s_1. So it completes j at some time $c_j > s_1$. In the interval $[r_j, c_j)$, LLW only starts jobs of weight at least w_j. Since $s_1 \in (r_j, c_j)$, this contradicts the fact that LLW starts job 1 at time s_1. This proves the last statement. □

Property 4. A job $h \in S$ was interrupted and then job 1 starts at time

$$k(R-1) < s_1 < \frac{2-R}{R-1}, \text{ and } 2 \le k \le 7.$$

Proof. If OPT starts its first job from S at time $\frac{s_1}{R} - k(1 - \frac{1}{R})$ or later, then S is good by Definition 1, as $|S| = k$. Hence, OPT starts at least one job in S at some time $s' < (s_1 + k)/R - k < s_1$, implying that at least one job in S arrives at or before time s'. From $(s_1 + k)/R - k > s' \ge 0$ we get $s_1 > k(R-1)$.

Since LLW does not complete any job in S before time s_1 but also does not idle if at least one job is available, LLW does not idle in the interval $(s', s_1]$. By Property 3, LLW does not complete a job at time s_1. Since LLW was not idle at time s_1, LLW interrupted some job $h \ne 1$ at time s_1.

If $h \notin S$, then h is lighter than the critical job, because otherwise h would be started not later than s_k by the rules of the algorithm, implying that $h \in S$ after all: a contradiction. Since h is lighter than the critical job which is the lightest job in S, all jobs in S are heavier than h. Then all jobs in S arrived at or after time s_1 because otherwise LLW would not still have been running h at time s_1, contradicting $s' < s_1$.

Thus, $h \in S$ and with this, we have $k \ge 2$. Moreover, since LLW interrupted a job h at time s_1, we get $s_1 < \frac{2-R}{R-1}$ by the rules of the algorithm. From $(R-1)k < s_1 < \frac{2-R}{R-1}$ we get $k \le 7$. $\qquad\square$

Property 5. LLW does not complete two jobs before time s_1.

Proof. We prove this by contradiction. Suppose LLW completes two jobs before time s_1, called q_1 and q_2 in order. Then $s_1 \in [2, \frac{2-R}{R-1})$. By Properties 2 and 4, OPT starts running its first job j in S (say $j \ne 1$) before time

$$\frac{s_1 + k}{R} - k < \frac{\frac{2-R}{R-1} + 2}{R} - 2 < 1.23.$$

Note that for the starting time of job q_2 we have $s_{q_2} \ge 1$ and with this, we get that the LW-phase of job q_2 ends not before

$$\frac{s_{q_2} + 2}{R} - 1 \ge \frac{1+2}{R} - 1 = \frac{3}{R} - 1 > 1.29.$$

Suppose j is heavier than q_2. Then either j is started before q_2 or q_2 is interrupted, as j arrives before or during its LW-phase. In both scenarios LLW completes j before q_2, which contradicts our assumption. Therefore, j is not heavier than q_2.

Now, if j is not heavier than q_1, OPT needs to complete $k+2$ jobs that are at least as heavy as the critical job, so $\text{OPT}(I) \ge k+2$ whereas $\text{LLW}(I) < \frac{2-R}{R-1} + k \approx 2.2279 + k$. Therefore, the competitive ratio is less than R for $k \ge 2$. Hence, j is heavier than q_1.

Job j must arrive no earlier than time $2/R - 1$ as j is heavier than q_1 but does not cause q_1 to be interrupted. This implies $\text{OPT}(I) \ge 2/R - 1 + k + 1 \approx$

$1.5270 + k$ as OPT must also complete q_2 which is heavier than the critical job and heavier than q_1 and can therefore also not arrive before time $2/R - 1$. Since $\text{LLW}(I) < \frac{2-R}{R-1} + k \approx 2.2279 + k$, the competitive ratio is less than R for $k \geq 2$.

Consequently, in both cases the competitive ratio is less than R for $k \geq 2$, which contradicts the assumption that I is a counterexample. $\qquad\square$

Property 6. After the first job arrives, LLW does not idle until the last job completes.

Proof. Let t be the time at which LLW completes the critical job. If there is idle time after time t, all jobs following that idle time (these jobs all arrive after time t) can be removed. If there is idle time before time t but after the first job arrives, all jobs preceding that idle time can be removed. In both cases, the schedule of LLW for the critical job is unaffected, and the optimal weighted makespan can only decrease by the removal of jobs. Thus idle time does not happen in a minimal counterexample. $\qquad\square$

Property 7. If LLW completes a job q before time s_1, then q is lighter than the critical job, and OPT starts the first job in S at time $(s_q + 2)/R - 1$ or later.

Proof. We prove this by contradiction. Suppose LLW completes a job q before time s_1 which is at least as heavy as the critical job. Let s_q be the starting time of job q. Since OPT must complete q as well, OPT must start its first job in $S \cup \{q\}$ before time $(s_1 + k)/R - k - 1 \approx 0.764s_1 - 0.237k - 1$, so this value must be larger than 0. Since $s_1 < \frac{2-R}{R-1} < 2.228$, this implies $k \leq 2.97$, so $k = 2$ by Property 4.

Suppose there are at least two interruptions after q completes: the penultimate (maybe the only) one at some time t and one at time $s_1 > t$. No job completes after q and before time s_1 by Property 5, so each successive interruption is of a heavier job than the previous one by the rules of LLW. Then the jobs in S are exactly job 1 and the job that arrived at time t, since any previous jobs (if they even exist) are lighter. This implies that S is good, as $t^* \geq r_S = t$ and $s_1 < (t + 2)/R - 1$. Since $t > s_q + 1$, this means S is good.

Because LLW does not complete another job at time s_q by Property 5, LLW starts job q when it arrives. Then, as we have seen, it interrupts the next job it starts at time s_1, and then it completes two jobs. On the other hand, if q is at least as heavy as the critical job, OPT must complete three jobs that are at least as heavy as the critical job starting from time s_q. After completing q, LLW does not idle by Property 6. Therefore, we have $s_1 < (s_q + 3)/R - 1$ and this means that I is not a counterexample in this case: the competitive ratio is at most

$$\frac{(s_q + 3)/R + 1}{s_q + 3} \leq \frac{3/R + 1}{3} < R.$$

The first statement of the lemma follows, and it implies that all jobs in S are heavier than q. Any such job would cause q to be interrupted if it arrived before time $(s_q + 2)/R - 1$, which proves the second statement. $\qquad\square$

We define $n_1(t) := (t + 2)/R - 1$ and $n_{i+1}(t) := (n_i(t) + 2)/R - 1$ for $i \geq 1$.

Property 8. Suppose some job j starts at time $t \in [r_S, s_1]$. Then $j \in S$ and each interruption in $[t, s_1)$ (if any) is for a job in S, so there are at most $k - 1$ interruptions. Only jobs in S are running in $[t, s_1]$ (and beyond), and up to and including at time s_1, all these jobs get interrupted. The ith interruption after time t takes place before time $n_i(t)$.

Proof. At time t, at least one job in S is available to be run, and LLW starts job j. This means j is at least as heavy as the lightest job in S, which is the critical job. Hence j completes before the critical job (if these two jobs have the same weight, the last one of the two to complete is the critical job). So $j \in S$, and if $t < s_1$, j is interrupted before or at time s_1 since that is when the first job in S that completes without interruption starts. Jobs are only interrupted for heavier jobs, so each job that arrives in (t, s_1) and causes an interruption is also in S and is interrupted itself by the same reasoning as above. The bound on the number of interruptions follows from $|S| = k$ and the bounds on the times of the interruptions follow from (1). $\qquad\square$

Property 9. LLW does not complete any job before time s_1.

Proof. By definition, LLW does not complete a job at time s_1. Let q be the last job that LLW completes before time s_1. By Property 7, q is lighter than the critical job. After completing q, LLW does not idle by Property 6. Any job started after q but before time s_1 is interrupted by Property 5. If there is exactly one interruption after time $s_q + 1$ (namely at time s_1), then $\text{LLW}(I) \leq n_1(s_q + 1) + k$ and $\text{OPT}(I) \geq (s_q + 2)/R + k - 1$ by Property 7; the ratio is

$$\frac{\text{LLW}(I)}{\text{OPT}(I)} \leq \frac{(s_q + 3)/R + k - 1}{(s_q + 2)/R + k - 1} \leq \frac{(0 + 3)/R + 2 - 1}{(0 + 2)/R + 2 - 1} < 1.303,$$

for $k \geq 2$ and $s_q \geq 0$.

If there are additional interruptions, all but the first are of jobs that arrived after time $s_q + 1$, and they all happen before or at time s_1. We use Property 8.

If $k = 2$, we find that I is not a counterexample: we have $\text{OPT}(I) \geq r_S + 2$ and $\text{LLW}(I) \leq n_1(r_S) + 2$ as at most one interruption takes place after time t. The ratio $(n_1(r_S) + 2)/(r_S + 2)$ is at most $(2/R + 1)/2 < 1.27$. For the remaining bounds we always either use the bound $\text{OPT}(I) \geq r_S + k =: L_1$ or $\text{OPT}(I) \geq (s_q + 2)/R + k - 1 =: L_2$.

If $k \geq 3$ and there are exactly two restarts, then $\text{LLW}(I) \leq n_2(s_q + 1) + k$, and $\text{LLW}(I)/L_2 \leq 4.513/3.526 \approx 1.280$ for $k \geq 3$ and $s_q \geq 0$.

Else if $k = 3$ we have $\text{LLW}(I) \leq n_2(r_S) + 3$, and $(n_2(r_S) + 3)/(r_S + 3) \leq R$.

If $k \geq 4$ and there are exactly three restarts, then $\text{LLW}(I) < n_3(s_q + 1) + k$, and the ratio $\text{LLW}(I)/L_2$ is less than $5.682/4.526 \approx 1.256$ for $k \geq 4$ and $s_q \geq 0$.

Finally, if $k = 4$ we have $\text{LLW}(I) \leq n_3(r_S) + 4$, and the ratio is less than R.

In all remaining cases, we use $\text{LLW}(I) \leq \frac{2-R}{R-1} + k$ and $\text{OPT}(I) \geq \frac{s_q+2}{R} + k - 1$. The ratio is less than $7.22815/5.52698 \approx 1.30776$ for $k \geq 5$ and $s_q \geq 0$. $\qquad\square$

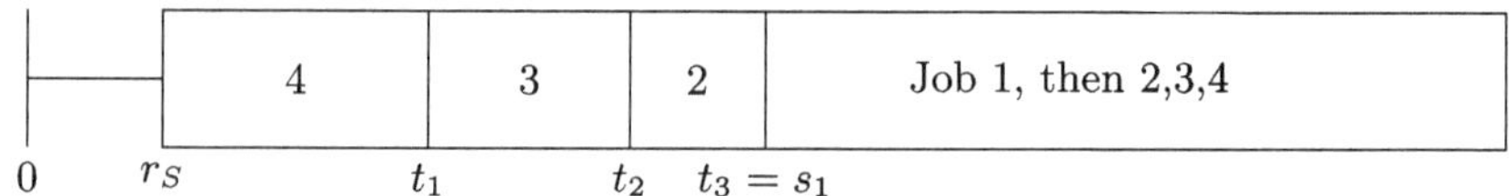

Fig. 4. The final shape of the hypothetical counterexample. The first job in S that arrives is job 4, and it arrives at time r_S. Then there are interruptions for increasingly heavier jobs, until time s_1 when job 1 starts and is not interrupted. Jobs are then completed in order of decreasing weight. We have $t_i < n_i(r_S)$ for $i = 1, 2, 3$.

By Property 9 and Property 4, in I it happens one or more times that a job is interrupted before any job completes. All interrupted jobs started when they arrived. The final interruption before the critical job completes is at time s_1. Let m be the total number of interruptions. Then $m \leq k - 1$ by Property 8. Recall that $2 \leq k \leq 7$ by Property 4. We now show for all values m by contradiction that I is not a counterexample (Fig. 4).

If $m = 1$, then, since $r_S < (s_1 + k)/R - k$, we have $s_1 < n_1\left(\frac{s_1+k}{R} - k\right) = n_1\left(\frac{s_1}{R} - k(1-\frac{1}{R})\right) =: N_1(s_1, k)$. The function $n_1(t)$ is increasing in t and $\frac{s_1}{R} - k(1-\frac{1}{R})$ is decreasing in k, so $N_1(s_1, k)$ is decreasing in k. The bound $s_1 < N_1(s_1, 2)$ implies $s_1 \leq 0.4$, contradicting $s_1 \geq 2(R - 1)$ which holds by Property 4.

If $m = 2$, $\mathrm{OPT}(I) \geq \max(0, ((s_1 + 1)R - 1)R - 2) + k$ and $k \geq 3$. The ratio is at most R since $s_1 > 3(R - 1)$ by Property 4. It is exactly R for $s_1 = 3(R - 1)$ and $k = 3$, since R is the positive real root of $3x^3 - 2x^2 - x - 2 = 0$.

It is easily checked that the ratio is at most R for larger values of m as well. In conclusion, we have proved the following theorem.

Theorem 2. *The LLW algorithm is R-competitive.*

In the next theorem, we complement Theorem 2 with a lower bound, leaving a small gap for further improvement.

Theorem 3. *Any algorithm for the problem $1|r_j, online, p_j = 1, restart|WC_{\max}$ is at least $R_2 \approx 1.2344$ competitive which is the real root of the equation $4x^3 - x^2 - 6 = 0$.*

Proof. For a contradiction, suppose that there exists an online algorithm ALG with a competitive ratio of less than R_2. We consider the following job instance. At time $r_1 = 0$, job 1 with $w_1 = 1$ arrives. Then job 2 arrives with $w_2 = \frac{4}{3}$ at time $r_2 = \frac{3}{R_2} - 2 \approx 0.430$.

Case 1. If ALG completes job 1 before job 2, a final job 3 with $w_3 = \frac{3+r_2}{2+r_2} \approx 1.411$ arrives at time $r_3 = 1 + r_2 = \frac{3}{R_2} - 1 \approx 1.430$. The following schedule is feasible: job 2 is completed at time $1 + r_2$, job 3 is completed at time $2 + r_2$ and job 1 is completed at time $3 + r_2$. Thus, $\mathrm{OPT}(I) \leq 3 + r_2 = 3/R_2 + 1$.

If ALG completes job 2 before job 3, the job order is $1, 2, 3$. Thus, we have

$$\frac{\text{ALG}(I)}{\text{OPT}(I)} \geq \frac{\max(1, 2w_2, 3w_3)}{3/R_2 + 1} = \frac{\max(8/3, (9+3r_2)/(2+r_2))}{3/R_2 + 1}$$

$$= \frac{\frac{3(3+(3/R_2-2))}{3/R_2}}{3/R_2 + 1}$$

$$= \frac{\frac{3(3/R_2+1)}{3/R_2}}{3/R_2 + 1}$$

$$= R_2.$$

If ALG completes job 3 before job 2, then ALG would schedule job 1, then job 3 and finally job 2. Since $w_3(r_3 + 1) \leq w_2(r_3 + 2)$, we obtain

$$\frac{\text{ALG}(I)}{\text{OPT}(I)} \geq \frac{\max(1, w_3(r_3 + 1), w_2(r_3 + 2))}{3/R_2 + 1} = \frac{w_2(r_3 + 2)}{3/R_2 + 1}$$

$$= \frac{w_2((3/R_2 - 1) + 2)}{3/R_2 + 1} = \frac{4}{3} \geq R_2.$$

Case 2. If ALG completes job 2 before job 1, then a (different) third job 3 with $w_3 = 2$ arrives at time $r_3 = \frac{6}{R_2^2} - 3 \approx 0.938$.

If now ALG completes job 2 before job 3, no more jobs arrive and OPT could first schedule job 3, then job 2 and finally job 1. Therefore, $\text{OPT}(I) \leq \max(w_3(r_3 + 1), w_2(r_3 + 2), r_3 + 3) = r_3 + 3$. However, after completing job 2, in the best case, ALG schedules job 3 after job 2 and before job 1, since job 3 has higher weight. Thus, $\text{ALG}(I) \geq \max(w_2(r_2 + 1), w_3(r_2 + 2), r_2 + 3) = w_3(r_2 + 2)$ and we have

$$\frac{\text{ALG}(I)}{\text{OPT}(I)} \geq \frac{w_3(r_2 + 2)}{r_3 + 3} = \frac{2((3/R_2 - 2) + 2)}{(6/R_2^2 - 3) + 3} = \frac{6/R_2}{6/R_2^2} = R_2.$$

Now, if ALG completes job 3 before job 2, a final fourth job 4 with $w_4 = 1$ arrives at time $r_4 = 3$. A feasible schedule s would be the job order $1, 3, 2, 4$ with $WC_s = 4$. However, if ALG completes job 3 before job 2, at best, ALG first schedules job 3 at time r_3, then job 2 and finally the two jobs with weight 1 in arbitrary order, since job 2 has higher weight than jobs 1 and 4 which have the same weight. Therefore, we have

$$\frac{\text{ALG}(I)}{\text{OPT}(I)} \geq \frac{\max((r_3 + 1)w_3, (r_3 + 2)w_2, r_3 + 3, r_3 + 4)}{4}$$

$$= \frac{r_3 + 4}{4} = \frac{6/R_2^2 + 1}{4} = R_2,$$

using $4R_2^3 - R_2^2 - 6 = 0$. $\qquad\square$

4   Conclusion

We studied the problem of online scheduling with restarts for minimizing the weighted makespan $WC_{\max}$. We presented two new lower bounds on the competitive ratio for deterministic algorithms with restarts, one in the general setting and another in the case where all jobs have identical processing times. We designed a deterministic online algorithm for unit processing times that improves the competitive ratio to better than 1.3098. This leaves a small gap to our lower bound for further improvement. Closing this gap appears to be nontrivial and is an interesting topic for future research. This result may also lead the way to an improved result for general processing times.

Acknowledgments. Aflatoun Amouzandeh was financially supported by the Deutsche Forschungsgemeinschaft (DFG, German Research Foundation) - Project Number STE 1727/4-1. Klaus Jansen, Rob van Stee and Corinna Wambsganz were funded by the Deutsche Forschungsgemeinschaft (DFG, German Research Foundation) - Project Number 453769249. Klaus Jansen and Lis Pirotton were funded by the Deutsche Forschungsgemeinschaft (DFG, German Research Foundation) - Project Number 528381760.

Disclosure of Interests. The authors have no competing interests to declare that are relevant to the content of this article.

References

1. van den Akker, M., Hoogeveen, H., Vakhania, N.: Restarts can help in the on-line minimization of the maximum delivery time on a single machine. In: Paterson, M.S. (ed.) ESA 2000. LNCS, vol. 1879, pp. 427–436. Springer, Heidelberg (2000). https://doi.org/10.1007/3-540-45253-2_39
2. Amouzandeh, A., van Stee, R.: Improved online scheduling with restarts on a single machine. In: Bieńkowski, M., Englert, M. (eds.) Approximation and Online Algorithms. WAOA 2024. LNCS, vol. 15269. Springer, Cham (2025). https://doi.org/10.1007/978-3-031-81396-2_2
3. Borodin, A., El-Yaniv, R.: Online computation and competitive analysis, Cambridge University Press (1998)
4. Chai, X., Lu, L., Li, W., Zhang, L.: Best-possible online algorithms for single machine scheduling to minimize the maximum weighted completion time. Asia Pac. J. Oper. Res. **35**(6), 1850048:1–1850048:11 (2018). https://doi.org/10.1142/S0217595918500483
5. Chen, B., Vestjens, A.: Scheduling on identical machines: How good is LPT in an on-line setting? Oper. Res. Lett. **21**(4), 165–169 (1997). https://doi.org/10.1016/S0167-6377(97)00040-0
6. Chrobak, M., Jawor, W., Sgall, J., Tichý, T.: Online scheduling of equal-length jobs: randomization and restarts help. SIAM J. Comput. **36**(6), 1709–1728 (2007). https://doi.org/10.1137/S0097539704446608
7. Epstein, L., van Stee, R.: Lower bounds for on-line single-machine scheduling. Theor. Comput. Sci. **299**(1–3), 439–450 (2003). https://doi.org/10.1016/S0304-3975(02)00488-7

8. Fiat, A., Woeginger, G.J. (eds.): Online Algorithms: The State of the Art. LNCS, vol. 1442. Springer, Heidelberg (1998). https://doi.org/10.1007/BFb0029561
9. Fu, R., Cheng, T.C.E., Ng, C.T., Yuan, J.: Online scheduling on two parallel-batching machines with limited restarts to minimize the makespan. Inf. Process. Lett. **110**(11), 444–450 (2010). https://doi.org/10.1016/j.ipl.2010.04.008
10. Graham, R.L., Lawler, E.L., Lenstra, J.K., Kan, A.H.G.R.: Optimization and approximation in deterministic sequencing and scheduling: a survey. Ann. Discrete Math. **5**, 287–326 (1979)
11. Komm, D.: An Introduction to Online Computation - Determinism, Randomization, Advice. Texts in Theoretical Computer Science. An EATCS Series. Springer (2016). https://doi.org/10.1007/978-3-319-42749-2
12. Laalaoui, Y., M'Hallah, R.: Enhancing the best-first-search F with incremental search and restarts for large-scale single machine scheduling with release dates and deadlines. Ann. Oper. Res. **351**(1), 303–332 (2025). https://doi.org/10.1007/S10479-024-06386-7
13. Li, W.: A best possible online algorithm for the parallel-machine scheduling to minimize the maximum weighted completion time. Asia Pac. J. Oper. Res. **32**(4), 1550030:1–1550030:10 (2015). https://doi.org/10.1142/S021759591550030X
14. Liang, X., Lu, L., Sun, X., Yu, X., Zuo, L.: Online scheduling on a single machine with one restart for all jobs to minimize the weighted makespan. AIMS Math. **9**(1), 2518–2529 (2024)
15. Liu, H.L., Lu, X.W.: Online scheduling on a parallel batch machine with delivery times and limited restarts. J. Operat. Res. Soc. China, 1–19 (2022)
16. Lu, L., Zhang, L., Ou, J.: Single machine scheduling with rejection to minimize the weighted makespan. In: Wu, W., Du, H. (eds.) AAIM 2021. LNCS, vol. 13153, pp. 96–110. Springer, Cham (2021). https://doi.org/10.1007/978-3-030-93176-6_9
17. Qi, F., Jinjiang, Y.: Np-hardness of a multicriteria scheduling on two families of jobs. Operat. Res. Trans., 121–126 (2007)
18. Shmoys, D.B., Wein, J., Williamson, D.P.: Scheduling parallel machines on-line. In: McGeoch, L.A., Sleator, D.D. (eds.) DIMACS Series in Discrete Mathematics and Theoretical Computer Science, vol. 7, pp. 163–166. DIMACS/AMS (1991). https://doi.org/10.1090/DIMACS/007/13
19. van Stee, R., Poutré, J.A.L.: Minimizing the total completion time on-line on a single machine, using restarts. J. Algorithms **57**(2), 95–129 (2005). https://doi.org/10.1016/J.JALGOR.2004.10.001
20. Sun, R.: Randomized approximation schemes for minimizing the weighted makespan on identical parallel machines. J. Comb. Optim. **47**(3), 34 (2024). https://doi.org/10.1007/s10878-024-01118-w
21. Tian, J., Fu, R., Yuan, J.: Online over time scheduling on parallel-batch machines: a survey. J. Operat. Res. Soc. China **2**(4), 445–454 (2014)

Parameterized Algorithms for k-Inversion

Dhanyamol Antony[1], L. Sunil Chandran[2], Dalu Jacob[3], and R. B. Sandeep[4(✉)]

[1] School of Data Science, Indian Institute of Science Education and Research Thiruvananthapuram, Thiruvananthapuram, India
`dhanyamolantony@iisertvm.ac.in`
[2] Department of Computer Science and Automation, Indian Institute of Science, Bengaluru, Bengaluru, India
`sunil@iisc.ac.in`
[3] Department of Mathematics, Indian Institute of Technology Delhi, New Delhi, India
`dalujacob@iitd.ac.in`
[4] Department of Computer Science and Engineering, Indian Institute of Technology Dharwad, Dharwad, India
`sandeeprb@iitdh.ac.in`

Abstract. Inversion of a directed graph D with respect to a vertex subset Y is the directed graph obtained from D by reversing the direction of every arc whose endpoints both lie in Y. More generally, the inversion of D with respect to a tuple $(Y_1, Y_2, \ldots, Y_\ell)$ of vertex subsets is defined as the directed graph obtained by successively applying inversions with respect to $Y_1, Y_2, \ldots, Y_\ell$. Such a tuple is called a *decycling family* of D if the resulting graph is acyclic. In the k-INVERSION problem, the input consists of a directed graph D and an integer k, and the task is to decide whether D admits a decycling family of size at most k. Alon et al. (SIAM J. Discrete Math., 2024) proved that the problem is NP-complete for every fixed value of k, thereby ruling out XP algorithms, and presented a fixed-parameter tractable (FPT) algorithm parameterized by k for tournament inputs. In this paper, we generalize their algorithm to a broader variant of the problem on tournaments and subsequently use this result to obtain an FPT algorithm for k-INVERSION when the underlying undirected graph of the input is a block graph. Furthermore, we obtain an algorithm for k-INVERSION on general directed graphs with running time $2^{O(\mathrm{tw}(k+\mathrm{tw}))} \cdot n^{O(1)}$, where tw denotes the treewidth of the underlying graph.

Keywords: Parameterized algorithms · Inversion · Block graphs

1   Introduction

Inversion of a directed graph D with respect to a vertex subset Y, denoted by $D \oplus Y$, is the directed graph obtained from D by reversing the direction of every

Supported by ANRF research grants ANRF/ECRG/2024/001049/ENS, CRG/2022/006770, and MTR/2022/000692.

arc whose endpoints both lie in Y. More generally, the inversion of D with respect to a tuple $\mathcal{Y} = (Y_1, Y_2, \ldots, Y_\ell)$ of vertex subsets, denoted by $D \oplus \mathcal{Y}$, is obtained by successively applying inversions with respect to $Y_1, Y_2, \ldots, Y_\ell$. Such a tuple is called a *decycling family* of D if the resulting graph is acyclic. The minimum cardinality of a decycling family of D is known as the *inversion number* of D and is denoted by $\mathrm{inv}(D)$.

Note that the order in which the inversions are applied has no effect on the final graph. Indeed, an arc uv is present in $D \oplus \mathcal{Y}$ if and only if either uv is an arc in D and the number of sets in $\mathcal{Y}$ that contain $\{u, v\}$ is even, or vu is an arc in D and the number of sets in $\mathcal{Y}$ that contain $\{u, v\}$ is odd.

Belkhechine et al. [7] initiated the study of inversion in 2010 and investigated the inversion number in connection with the Boolean dimension of graphs. Bang-Jensen, da Silva, and Havet [4] studied the inversion number in relation to the cycle transversal number, the cycle arc-transversal number, and the cycle packing number. They conjectured that for any two oriented graphs L and R, $\mathrm{inv}(L \to R) = \mathrm{inv}(L) + \mathrm{inv}(R)$, where $L \to R$ denotes the *dijoin* of L and R, that is, the oriented graph obtained from L and R by adding all possible arcs from L to R. This conjecture, known as the *dijoin conjecture*, led to a series of subsequent works [1,2,5,6,14]. Recently, Alon et al. [1] and Aubian et al. [2] disproved the conjecture. The former also showed that $\mathrm{inv}(D) \leq (1 + o(1))n$ for every directed graph D on n vertices.

Havet, Hörsch, and Rambaud [12] studied the diameter of a graph whose vertices correspond to oriented graphs, with an arc from one vertex to another if the second oriented graph can be obtained from the first by a single inversion. They established connections between this diameter and several graph parameters, including the star chromatic number, acyclic chromatic number, oriented chromatic number, and treewidth. The connection with treewidth was further investigated by Wang et al. [14].

Yuster [15] obtained bounds on the inversion number of tournaments. There are also recent works that consider inversion operations with respect to different target properties; see, for example, Duron et al. [11] and Belkhechine and Ben Salha [8].

In the k-INVERSION problem, the input consists of a directed graph D and an integer k, and the objective is to decide whether $\mathrm{inv}(D) \leq k$.

Known Algorithmic Results. Algorithmic results related to inversion are relatively sparse in comparison with the extensive non-algorithmic literature. Bang-Jensen, da Silva, and Havet [4] proved that k-INVERSION is NP-complete for $k = 1$, ruling out XP algorithms. Alon et al. [1] extended this result and established NP-completeness for every fixed value of k. They also showed that the problem is fixed-parameter tractable when parameterized by k on tournament inputs, using an iterative-compression-based algorithm. Very recently, Bang-Jensen et al. [3] conducted an algorithmic study of a variant of k-INVERSION in which each set used for inversion is required to have bounded size. To the best of our knowledge, no other algorithmic results are known for the problem. We address this gap by presenting three fixed-parameter tractable (FPT) algorithms.

Our Results. Our first contribution is an FPT algorithm for a generalization of k-INVERSION ON TOURNAMENTS.

A decycling family $\mathcal{Y} = (Y_1, Y_2, \ldots, Y_k)$ of a directed graph D induces a *characteristic vector* $\mathbf{u}_i \in \{0,1\}^k$ for each vertex u_i of D. The jth coordinate of $\mathbf{u}_i$ is 1 if $u_i \in Y_j$ and 0 otherwise. The *weight* of $\mathbf{u}_i$ with respect to $\mathcal{Y}$, denoted by $\mathrm{wt}_{\mathcal{Y}}(\mathbf{u}_i)$, is the number of 1s in $\mathbf{u}_i$. We omit the subscript when the decycling family is clear from the context.

We define the WEIGHT-RESTRICTED k-INVERSION ON TOURNAMENTS problem as follows. The input consists of a tournament T, an integer k, and, for each vertex u_i of T, a set A_i of allowed weights. The objective is to decide whether there exists a decycling family $\mathcal{Y}$ of size k such that $\mathrm{wt}(\mathbf{u}_i) \in A_i$ for every vertex u_i of T.

Theorem 1. WEIGHT-RESTRICTED k-INVERSION ON TOURNAMENTS *admits an FPT algorithm when parameterized by* k.

The algorithm adapts the iterative-compression-based algorithm of Alon et al. [1] for k-INVERSION ON TOURNAMENTS. Next, we consider directed graphs whose underlying undirected graph is a block graph. Using Theorem 1 as a subroutine, we obtain an FPT algorithm for k-INVERSION on this class by dynamic programming over the block tree of the underlying graph.

Theorem 2. k-INVERSION, *parameterized by* k, *admits an FPT algorithm when the underlying undirected graph of the input digraph is a block graph.*

The k-INVERSION problem can be expressed by an MSO_2 formula whose length depends only on k. By Courcelle's theorem [9], this implies that k-INVERSION, parameterized by $k+\mathrm{tw}$, admits an FPT algorithm. Here tw denotes the treewidth of the underlying undirected graph. Our final result provides an explicit and more efficient algorithm for this parameterization, which is a dynamic programming over a tree decomposition.

Theorem 3. k-INVERSION *admits an algorithm with running time* $2^{O(\mathrm{tw}(k+\mathrm{tw}))} \cdot n^{O(1)}$, *where* tw *is the treewidth of the underlying undirected graph of the input digraph.*

Moreover, with standard techniques, our algorithms can be made constructive and output a decycling family for yes-instances. As noted in [4], it suffices to prove our results for oriented graphs, that is, directed graphs obtained by orienting the edges of simple undirected graphs. Therefore, throughout the remainder of the paper, we restrict our attention to oriented input graphs.

We follow standard graph-theoretic and set-theoretic notations. For background on the parameterized algorithmic techniques and related concepts used in this paper, we refer the reader to the book by Cygan et al. [10]. The number of vertices in the input graph is always denoted by n. By $[t]$, we denote the set $\{1, 2, \ldots, t\}$. For technical reasons, we allow a decycling family of a subgraph of a directed graph to include vertices outside the subgraph; however, such vertices do not induce any arc changes within the subgraph. Notation and terminology not introduced in this section are defined just before their first use. Due to space constraints, many of the proofs are not included in this paper. In particular, the proof of Theorem 1 and some part of the proof of Theorem 3 are not included.

2   Block Graphs

In this section, we consider oriented input graphs for k-INVERSION such that the underlying undirected graph is a block graph. We obtain an FPT algorithm using the algorithm we developed for WEIGHT-RESTRICTED k-INVERSION ON TOURNAMENTS (WKIT) and by performing dynamic programming over the block tree graph of the underlying graph. We assume that the algorithm provided by Theorem 1 is WKIT$(T, k, \mathcal{A})$, which returns `True` for a yes-instance, and `False` for a no-instance. Recall that a graph is a *block graph* if each of its 2-connected components is a clique. Each such clique is known as a *block*.

Let D be an oriented graph and let G be its underlying undirected graph. If G is disconnected, then we can apply our algorithm separately on each connected component and combine the results straightforwardly. Therefore, we assume that G is connected.

Let $B_1, B_2, \ldots, B_p$ denote the blocks of G, and let $c_1, c_2, \ldots, c_q$ denote the cut vertices of G. The block tree of G, denoted by $BT(G)$, is defined as follows. We introduce a set of p vertices $B = \{b_1, b_2, \ldots, b_p\}$ corresponding to the blocks $B_1, B_2, \ldots, B_p$, and a set of q vertices $C = \{c_1, c_2, \ldots, c_q\}$ corresponding to the cut vertices of G with the same labels. The block tree $BT(G)$ has vertex set $U = B \cup C$. There is an edge $c_i b_j$ in $BT(G)$ if and only if $c_i \in B_j$. Note that $BT(G)$ is a tree and that B and C are independent sets.

We root the tree $BT(G)$ at b_p. The *parent* and *children* of a vertex in $BT(G)$ are defined in the usual way. Let the vertex set U of $BT(G)$ be $\{u_1, u_2, \ldots, u_{p+q}\}$ such that the parent of a vertex is numbered higher than the vertex. We assume such an ordering for B and C as well. That is, a vertex in B comes before its ancestors in B. Similarly for C. Note that the choice of the root respects the above ordering.

By G_{u_i}, we denote the subgraph of G induced by all vertices in the blocks corresponding to the block vertices in the subtree rooted at u_i. By D_{u_i}, we denote the graph G_{u_i} with edge orientations inherited from D.

For a cut vertex $c_i \in C$, we let $W(c_i)$ denote the set of all integers $w \in \{0, 1, \ldots, k\}$ such that there exists a decycling family $\mathcal{Y}$ of size k of D_{c_i} with $\mathrm{wt}(\mathbf{c}_i) = w$. Slightly differently, for a block vertex $b_i \in B \setminus \{b_p\}$ with $\mathrm{parent}(b_i) = c_j \in C$, we let $W(b_i)$ denote the set of all integers $w \in \{0, 1, \ldots, k\}$ such that there exists a decycling family $\mathcal{Y}$ of size k of D_{b_i} with $\mathrm{wt}(\mathbf{c}_j) = w$. As a boundary case, for the root block vertex b_p, we let $W(b_p) = \{0, 1, \ldots, k\}$ if (D, k) is a yes-instance, and $W(b_p) = \emptyset$ otherwise.

We now show how to compute $W(u_i)$ in a bottom-up manner.

Lemma 1. *For a cut vertex $c_i \in C$, we have*

$$W(c_i) = \bigcap_{b_j \in \text{children}(c_i)} W(b_j).$$

Proof. Let $W' = \bigcap_{b_j \in \text{children}(c_i)} W(b_j)$. Let $w \in W(c_i)$. Then there exists a decycling family $\mathcal{Y}$ of size k of D_{c_i} such that $\text{wt}(\mathbf{c}_i) = w$. Recall that, for each $b_j \in \text{children}(c_i)$, the graph D_{b_j} is an induced subgraph of D_{c_i} and contains c_i. Therefore, $\mathcal{Y}$ induces a vector of weight w on $\text{parent}(b_j) = c_i$, implying that $w \in W'$.

For the converse direction, let $w \in W'$. Then, for every $b_j \in \text{children}(c_i)$, there exists a decycling family $\mathcal{Y}_j = (Y_{j,1}, Y_{j,2}, \ldots, Y_{j,k})$ of D_{b_j} that induces a vector of weight w on $\text{parent}(b_j) = c_i$. Without loss of generality, assume that $c_i \in Y_{j,\ell}$ for all $\ell \in \{1, 2, \ldots, w\}$ and that $c_i \notin Y_{j,\ell}$ for all $\ell \in \{w+1, w+2, \ldots, k\}$. We construct a decycling family $\mathcal{Y} = (Y_1, Y_2, \ldots, Y_k)$ for D_{c_i} by defining $Y_\ell = \bigcup_{b_j \in \text{children}(c_i)} Y_{j,\ell}$ for each $\ell \in [k]$. Note that for each $\ell \in \{1, \ldots, w\}$, the sets $\{Y_{j,\ell}\}_{b_j \in \text{children}(c_i)}$ pairwise intersect exactly at c_i, and for each $\ell \in \{w+1, w+2, \ldots, k\}$, the sets in $\{Y_{j,\ell}\}_{b_j \in \text{children}(c_i)}$ are pairwise disjoint. Hence, the effect of applying $\mathcal{Y}$ on D_{b_j} is identical to that of applying $\mathcal{Y}_j$. It follows that $\mathcal{Y}$ is a decycling family of D_{c_i} inducing weight w on c_i. $\square$

Lemma 2. *Let $b_i \in B \setminus \{b_p\}$. Then $W(b_i)$ is the set W' of all $w \in \{0, 1, \ldots, k\}$ such that $\text{WKIT}(B_i, k, \mathcal{A})$ is a yes-instance, where $\mathcal{A}$ consists of the sets $A(v)$ for $v \in B_i$, defined as follows: $A(v) = \{w\}$ if $v = \text{parent}(b_i)$, $A(v) = W(c_j)$ if $v = c_j \in \text{children}(b_i)$, and $A(v) = \{0, 1, \ldots, k\}$ otherwise.*

Proof. Let $c_\ell = \text{parent}(b_i)$. Let $w \in W(b_i)$. This implies that there exists a decycling family $\mathcal{Y} = (Y_1, Y_2, \ldots, Y_k)$ of D_{b_i} such that it induces weight w on c_ℓ. Assume that it induces weight w_j on $c_j \in \text{children}(b_i)$. Clearly, $w_j \in W(c_j)$. Then $\text{WKIT}(B_i, k, \mathcal{A})$ is a yes-instance, where $\mathcal{A}$ is defined as in the lemma statement. Indeed, $\mathcal{Y}$ is a decycling family of B_i satisfying the weight constraints. Therefore, $w \in W'$.

For the other direction, assume that $w \in W'$. This implies that $\text{WKIT}(B_i, k, \mathcal{A})$ is a yes-instance, where $\mathcal{A}$ is as defined in the lemma statement. Then there exists a decycling family $\mathcal{X} = (X_1, X_2, \ldots, X_k)$ of B_i satisfying the weight constraints, that is, $\mathcal{X}$ induces a weight in $A(v)$ for each vertex $v \in B_i$. We obtain a decycling family $\mathcal{Y} = (Y_1, Y_2, \ldots, Y_k)$ of D_{b_i} which induces weight w on c_ℓ as follows.

Let w_j be the weight induced by $\mathcal{X}$ on $c_j \in \text{children}(b_i)$. Since $A(c_j) = W(c_j)$, it follows that $w_j \in W(c_j)$. Let $\mathcal{Y}_j = (Y_{j,1}, Y_{j,2}, \ldots, Y_{j,k})$ be a decycling family of D_{c_j} inducing weight w_j on c_j, for each $c_j \in \text{children}(b_i)$. Assume without loss of generality that $c_j \in Y_{j,a}$ if and only if $c_j \in X_a$ (renumber the sets in $\mathcal{Y}_j$ accordingly). Now define

$$Y_a = X_a \cup \bigcup_{c_j \in \text{children}(b_i)} Y_{j,a}.$$

Algorithm 1: An FPT algorithm for k-INVERSION for oriented graphs with underlying block graphs

---

Input: An oriented graph D where the underlying undirected graph G is a connected block graph, and an integer k

Output: True if (D, k) is a yes-instance, False otherwise

(1) Compute $BT(G)$ with $U = V(BT(G)) = \{u_1, u_2, \ldots, u_{p+q}\}$

(2) Let $B = \{b_1, b_2, \ldots, b_p\}$ be the block vertices in U corresponding to the blocks $B_1, B_2, \ldots, B_p$

(3) Let $C = \{c_1, c_2, \ldots, c_q\}$ be the vertices in U corresponding to the cut vertices in G with the same labels

(4) **for** $i \leftarrow 1$ **to** $p + q$ **do**

(5)   **if** $u_i = c_j$ *for some* $j \in [q]$ **then**

(6) $W(c_j) \leftarrow \displaystyle\bigcap_{b_\ell \in \mathrm{children}(c_j)} W(b_\ell)$

(7)   **else**

(8)   Let $u_i = b_j$ for some $j \in [p]$

(9) $W(b_j) \leftarrow \emptyset$

(10)   **foreach** $w \in \{0, 1, \ldots, k\}$ **do**

(11)   **foreach** $v \in B_j$ **do**

(12)   **if** $v = \mathrm{parent}(b_j)$ **then**

(13) $A(v) \leftarrow \{w\}$

(14)   **else**

(15)   **if** $v \in \mathrm{children}(b_j)$ **then**

(16) $A(v) \leftarrow W(v)$

(17)   **else**

(18) $A(v) \leftarrow \{0, 1, \ldots, k\}$

(19)   Let $\mathcal{A}$ be the tuple $(A(v))_{v \in B_j}$

(20)   **if** $\mathrm{WKIT}(B_j, k, \mathcal{A})$ **then**

(21) $W(b_j) \leftarrow W(b_j) \cup \{w\}$

(22)   **if** $i = p + q$ **then**

(23)   **if** $W(b_j) \neq \emptyset$ **then**

(24)   **return** *True*

(25)   **return** *False*

Clearly, $\mathcal{Y}$ induces weight w on c_ℓ, since none of the sets in $\mathcal{Y}_j$ (for $c_j \in \mathrm{children}(b_i)$) contains c_ℓ. It remains to prove that $\mathcal{Y}$ is a decycling family of D_{b_i}. Note that no two vertices in B_i come together in a set in $\mathcal{Y}_j$. Therefore, the sets of $\mathcal{X}$ added to the sets in $\mathcal{Y}$ are solely responsible for changing the adjacency of edges inside B_i. Then the sets of $\mathcal{X}$ added to the sets in $\mathcal{Y}$ make sure that no cycle remains in $D_{b_i} \oplus \mathcal{Y}$. Note also that, for $c_j, c_{j'} \in \mathrm{children}(b_i)$ with $j \neq j'$, the graphs D_{c_j} and $D_{c_{j'}}$ do not share any common vertex. Therefore, any set in $\mathcal{Y}_j$ is pairwise disjoint with any set in $\mathcal{Y}_{j'}$. Moreover, the sets in $\mathcal{Y}_j$ added to the sets in $\mathcal{Y}$ kill the cycles in G_{c_j}. Therefore, $\mathcal{Y}$ is a decycling family of G_{b_i}. $\square$

Lemma 3. *Algorithm 1 returns* **True** *if the instance* (D, k) *is a yes-instance, and returns* **False** *otherwise.*

Proof. Lemma 1 essentially states that the algorithm computes $W(c_i)$, for $c_i \in C$, correctly and Lemma 2 implies that the algorithm computes $W(b_i)$, for $b_i \in B \setminus \{b_p\}$, correctly.

Assume that the algorithm returns **True**. This implies that the algorithm computes $W(b_p) \neq \emptyset$. Note that, in the algorithm, we compute $W(b_p)$ in exactly the same way as for the other block vertices. The only difference in execution arises from the fact that b_p has no parent. Therefore, $A(v)$ is set to $\{0, 1, \ldots, k\}$ for every vertex other than the cut vertices in B_p. This implies that $\mathrm{WKIT}(B_p, k, \mathcal{A})$ returns **True** for at least one choice of w, and hence for all choices of w. Indeed, the value of w has no effect on the iteration for b_p. Let $\mathcal{X}$ be any decycling family for which $\mathrm{WKIT}(B_p, k, \mathcal{A})$ returns **True**. By combining $\mathcal{X}$ with the decycling families of D_{c_j} having matching weights on c_j (as induced by $\mathcal{X}$), as done in the proof of Lemma 2, we obtain a decycling family for $D = D_{b_p}$. Therefore, (D, k) is a yes-instance.

Now assume that the algorithm returns **False**. This implies that, for no choice of w, $\mathrm{WKIT}(B_p, k, \mathcal{A})$ returns **True**. Consequently, there is no decycling family for B_p such that the weight induced on each c_j belongs to $W(c_j)$, for all $c_j \in \mathrm{children}(b_p)$. It follows that there is no decycling family for D. $\qquad\square$

Now, Theorem 2 follows from Lemma 3, Theorem 1, and the fact that the running time is dominated by polynomial many executions of the WKIT algorithm.

3   An FPT Algorithm for k-Inversion

In this section, we obtain an FPT algorithm for k-INVERSION parameterized by $k + \mathrm{tw}$, where tw denotes the treewidth of the underlying undirected graph of the input graph. The algorithm is based on dynamic programming over a nice tree decomposition. We remark that the problem can be expressed by an MSO_2 formula whose length depends only on k. By Courcelle's theorem [9], this implies that k-INVERSION, parameterized by $k + \mathrm{tw}$, admits an FPT algorithm. Our algorithm is explicit and runs in time $2^{O(\mathrm{tw}(k+\mathrm{tw}))}$.

The key idea is simple. To decide whether there exists a solution for the subgraph induced by V_t (the union of vertices appearing in the bags of the subtree rooted at t), it suffices to know whether solutions exist for the child nodes (or children, in the case of join nodes) that impose certain reachability constraints on pairs of vertices in the bag X_t.

To capture this information, we maintain a table $c(t, P, \mathcal{S})$, where t is a node of the nice tree decomposition, P is a subset of ordered pairs of vertices from X_t, and $\mathcal{S}$ is a k-tuple of subsets of X_t. Intuitively, the table entry $c(t, P, \mathcal{S})$ is set to **True** if and only if there exists a solution $\hat{\mathcal{S}}$, which is a k-tuple of subsets of V_t, such that the componentwise intersection of $\hat{\mathcal{S}}$ with X_t equals $\mathcal{S}$, and for every pair (u, v) of vertices in X_t, there exists a directed path from u to v in $D[V_t] \oplus \hat{S}$ if and only if $(u, v) \in P$.

Let $(T, \mathcal{X})$ be a nice tree decomposition of the underlying undirected graph of D. We denote the bags in $\mathcal{X}$ by $X_1, X_2, \ldots$. For a node $X_t \in V(T)$, let V_t denote the union of the vertex sets in the bags corresponding to the nodes in the subtree of T rooted at X_t.

Throughout this section, $\mathcal{S}$ and $\hat{\mathcal{S}}$ denote k-tuples. The ith set in $\mathcal{S}$ and $\hat{\mathcal{S}}$ is denoted by S_i and $\hat{S}_i$, respectively. We write $\mathcal{S} \subseteq \hat{\mathcal{S}}$ (equivalently, $\hat{\mathcal{S}} \supseteq \mathcal{S}$) if $S_i \subseteq \hat{S}_i$ holds for all $1 \leq i \leq k$. Likewise, for a vertex set X, we define $\hat{\mathcal{S}} \cap X :=$ $(\hat{S}_1 \cap X, \hat{S}_2 \cap X, \ldots, \hat{S}_k \cap X)$. Similarly, $\hat{\mathcal{S}} \cup X := (\hat{S}_1 \cup X, \hat{S}_2 \cup X, \ldots, \hat{S}_k \cup X)$, and $\hat{\mathcal{S}} \setminus X := (\hat{S}_1 \setminus X, \hat{S}_2 \setminus X, \ldots, \hat{S}_k \setminus X)$. By $\hat{\mathcal{S}} \cup \mathcal{S}$ we denote the k-tuple, $(\hat{S}_1 \cup S_1, \hat{S}_2 \cup S_2, \ldots, \hat{S}_k \cup S_k)$. That is, the operation is applied componentwise to each set in the tuple. The set of all k-tuples formed by the subsets of a set Y is denoted by $\mathcal{P}(Y)^k$.

Let $\text{even}(Z, \mathcal{S})$ be $\texttt{True}$ if Z is a subset of even number of sets in $\mathcal{S}$, and $\texttt{False}$ otherwise. On a similar note, let $\text{odd}(Z, \mathcal{S})$ be the negation of $\text{even}(Z, \mathcal{S})$. Let P_t be the set of all ordered pairs of distinct vertices in X_t. For every subset $P \subseteq P_t$, and for every k-tuple $\mathcal{S}$ of subsets of X_t, we define $c(t, P, \mathcal{S})$ as follows:

$$
c(t, P, \mathcal{S}) = \begin{cases} \texttt{True,} & \text{if there exists a } k\text{-tuple } \hat{\mathcal{S}} \in \mathcal{P}(V(D))^k \text{ such that } \hat{\mathcal{S}} \supseteq \mathcal{S}, \\ & \hat{\mathcal{S}} \cap X_t = \mathcal{S}, \text{ and } D[V_t] \oplus \hat{\mathcal{S}} \text{ is a DAG in which } (u, v) \in P \iff \\ & \text{there is a directed path from } u \text{ to } v \text{ in } D[V_t] \oplus \hat{\mathcal{S}}. \\ \texttt{False,} & \text{otherwise.} \end{cases}
$$

If there exists a k-tuple $\hat{\mathcal{S}}$ satisfying the conditions for $c(t, P, \mathcal{S})$ to be $\texttt{True}$, we say that $\hat{\mathcal{S}}$ *witnesses* $c(t, P, \mathcal{S}) = \texttt{True}$.

Before getting into the technical details, we recall the following observation.

Observation 4 (folklore). *The graph $D \oplus \mathcal{S}$ has an edge (u, v) if and only if either of the following conditions is true.*

1. *(u, v) is an arc in D and $\text{even}(\{u, v\}, \mathcal{S})$, or*
2. *(v, u) is an arc in D and $\text{odd}(\{u, v\}, \mathcal{S})$.*

We now derive recurrence formulations for $c(t, P, \mathcal{S})$ corresponding to each type of node in the tree decomposition T. Let X_t be a leaf node in T. Recall that leaf nodes are empty bags. Therefore, $c(t, \emptyset, \mathcal{S}) = \texttt{True}$, for the k-tuple $\mathcal{S}$ of empty sets.

Let t be an introduce node and let t' be its child. Let $X_t \setminus X_{t'} = \{w\}$. Assume that there exists a k-tuple $\hat{\mathcal{S}}'$ of subsets of $V_{t'}$ witnessing that $c(t', P'', \mathcal{S}') = \texttt{True}$.

Since the vertex w is introduced at node t, the presence of w may create additional reachability relations in the graph $D[V_t] \oplus \hat{\mathcal{S}}'$. Therefore, in order to determine the value of $c(t, P, \mathcal{S})$, we consider, for each subset $P'' \subseteq P$ that contains no ordered pair involving w, the effect of introducing w. For this purpose, we construct an auxiliary directed graph $A_{t, P'', \mathcal{S}}$ on the vertex set X_t that captures the reachability relation induced by $D[V_t] \oplus \hat{\mathcal{S}}$, where $\hat{\mathcal{S}}$ is a k-tuple of subsets of V_t such that $\hat{\mathcal{S}} \cap X_t = \mathcal{S}$ componentwise, the reachability relation on

$X_{t'}$ in $D[V_{t'}] \oplus \hat{S}$ is exactly P'', and the reachability relation on X_t in $D[V_t] \oplus \hat{S}$ is exactly P. Let $P' \subseteq P_{t'}$, and let $S \in (\mathcal{P}(X_t))^k$.

We define the auxiliary directed graph $A_{t,P',S}$ as follows. The vertex set of $A_{t,P',S}$ is X_t and the arc set is:

$$
\begin{aligned}
E(A_{t,P',S}) = P' \; \cup \; &\{(w,v) \mid v \in X_{t'}, \; ((w,v) \in E(D[X_t]) \text{ and even}(\{w,v\},S) \\
&\text{or } ((v,w) \in E(D[X_t]) \text{ and odd}(\{w,v\},S))\} \\
\cup \; &\{(u,w) \mid u \in X_{t'}, \; ((u,w) \in E(D[X_t]) \text{ and even}(\{u,w\},S) \\
&\text{or } ((w,u) \in E(D[X_t]) \text{ and odd}(\{u,w\},S))\}.
\end{aligned}
$$

Let $\hat{S} \in \mathcal{P}(V_t)^k$ be such that $\hat{S} \cap X_t = S$. Let $\hat{S}' = \hat{S} \setminus \{w\}$ (equivalently, $\hat{S} \cap V_{t'}$, since $V_t = V_{t'} \cup \{w\}$). Let $\hat{D} = D[V_t] \oplus \hat{S}$ and $\hat{D}' = D[V_{t'}] \oplus \hat{S}'$. Observe that $\hat{D}'$ is obtained from $\hat{D}$ by deleting vertex w. Let $P' \subseteq P_{t'}$ be the set of pairs (a,b) such that there is a directed path from a to b in $\hat{D}'$.

Observation 5. *Let $u,v \in X_{t'}$. The arc (u,w) is present in $\hat{D}$ if and only if it is an arc of $A_{t,P',S}$. Similarly, the arc (w,v) is present in $\hat{D}$ if and only if it is an arc of $A_{t,P',S}$.*

Proof. Let $A = A_{t,P',S}$. For the forward direction, assume that (u,w) is an arc of $\hat{D}$. Then by Observation 4, either of the following cases holds.

1. (u,w) is an arc of $D[X_t]$ and $\{u,w\}$ is contained in an even number of sets in $\hat{S}$, or
2. (w,u) is an arc of $D[X_t]$ and $\{u,w\}$ is contained in an odd number of sets in $\hat{S}$.

Since $u,w \in X_t$ and $S = \hat{S} \cap X_t$, the parity of the number of sets containing $\{u,w\}$ is the same in S and in $\hat{S}$. Therefore, in either case, $(u,w) \in E(A)$.

For the backward direction, assume that $(u,w) \in E(A)$. Since no pair in P' contains w, the construction of A implies that one of the two cases above holds. Then Observation 4 implies that (u,w) is an arc of $\hat{D}$. The statement for arcs of the form (w,v) is proved analogously. $\square$

Observation 6. *For every pair $(u,v) \in P_{t'}$, there is a directed path from u to v in $\hat{D}'$ if and only if there is a directed path from u to v in $A_{t,P',S} - w$.*

Proof. Let $A = A_{t,P',S}$. For the forward direction, assume that there is a directed path from u to v in $\hat{D}'$. Then by the definition of P', we have $(u,v) \in P'$. Hence $(u,v) \in E(A)$, and since $w \notin \{u,v\}$, the arc (u,v) is present in $A - w$, yielding a directed path from u to v in $A - w$.

For the backward direction, assume that there is a directed path $ux_1x_2\ldots x_q v$ from u to v in $A - w$. Every arc on this path has both endpoints in $X_{t'}$. By construction, the only arcs of A with both endpoints in $X_{t'}$ are those in P'. Therefore,

$$\{(u,x_1),(x_1,x_2),\ldots,(x_{q-1},x_q),(x_q,v)\} \subseteq P'.$$

By the definition of P', for each arc (y,z) in this set there is a directed path from y to z in $\hat{D}'$. Concatenating these directed paths yields a directed path from u to v in $\hat{D}'$. $\square$

Algorithm 2: Compute $c(t, P, \mathcal{S})$ for an introduce node X_t

---

Input: $t, P, \mathcal{S}$

Output: $c(t, P, \mathcal{S})$

(1) $P^\star \leftarrow \{\, (u, v) \in P \mid u \neq w \text{ and } v \neq w \,\}$

(2) $\overline{P} \leftarrow P_t \setminus P$

(3) $\mathcal{S}' \leftarrow \mathcal{S} \setminus \{w\}$

(4) **for** *each* $P'' \subseteq P^\star$ **do**

(5)   **if** *not* $c(t', P'', \mathcal{S}')$ **then**

(6)   skip (this choice of P'' and continue with the next iteration)

(7)   Create the auxiliary graph $A = A_{t, P'', \mathcal{S}}$

(8)   **if** *A is not a DAG* **then**

(9)   skip (this choice of P'' and continue with the next iteration)

(10)   **for** *each pair* $(u, v) \in P$ **do**

(11)   **if** *there is no directed path from u to v in A* **then**

(12)   skip (this choice of P'' and continue with the next iteration of the outer loop)

(13)   **for** *each pair* $(u, v) \in \overline{P}$ **do**

(14)   **if** *there is a directed path from u to v in A* **then**

(15)   skip (this choice of P'' and continue with the next iteration of the outer loop)

(16)   **return** *True*

(17) **return** *False*

Lemma 4. *For every pair $(u, v) \in P_t$, there is a directed path from u to v in $\hat{D}$ if and only if there is a directed path from u to v in $A_{t, P', \mathcal{S}}$.*

Proof. Let $A = A_{t, P', \mathcal{S}}$.

Forward Direction. Assume that there is a directed path from u to v in $\hat{D}$. Since all directed paths are simple, the vertex w appears on the path at most once.

Case 1: $u, v \in X_{t'}$. If $(u, v) \in P'$, then by construction (u, v) is an arc of A, and we are done. Now assume that $(u, v) \notin P'$. Then there is no directed path from u to v in $\hat{D}'$. Hence the (simple) directed path from u to v in $\hat{D}$ must pass through w exactly once. Let $u', v' \in X_{t'}$ be such that (u', w) and (w, v') are arcs of $\hat{D}$, and there is a directed path from u to u' in $\hat{D}'$ and a directed path from v' to v in $\hat{D}'$ (where $u' = u$ or $v' = v$ is allowed). By definition of P', we have $u' = u$ or $(u, u') \in P'$, and $v' = v$ or $(v', v) \in P'$. By Observation 5, the arcs (u', w) and (w, v') belong to A. Hence A contains a directed path from u to v.

Case 2: $u = w$. Since the path is simple, it starts with an arc (w, v') for some $v' \in X_{t'}$. Either $v' = v$ or there is a directed path from v' to v in $\hat{D}'$. By Observation 5, (w, v') is an arc of A, and if $v' \neq v$, then $(v', v) \in P'$ and hence there is a directed path from v' to v in $A - w$. Thus there is a directed path from u to v in A.

Case 3: $v = w$. This case is symmetric to Case 2.

Backward Direction. Assume that there is a directed path from u to v in A.

Case 1: $u, v \in X_{t'}$. If $(u, v) \in P'$, then by definition of P' there is a directed path from u to v in $\hat{D}'$, and hence also in $\hat{D}$. Now assume that $(u, v) \notin P'$. Then by Observation 6, there is no directed path from u to v in $A - w$. Hence the directed path from u to v in A must pass through w exactly once. Let $u', v' \in X_{t'}$ be such that (u', w) and (w, v') are arcs of A, and there is a directed path from u to u' in $A - w$ and from v' to v in $A - w$ (allowing $u' = u$ or $v' = v$). By Observation 6, the paths in $A - w$ correspond to directed paths in $\hat{D}'$, and by Observation 5, the arcs (u', w) and (w, v') are arcs of $\hat{D}$. Concatenating these yields a directed path from u to v in $\hat{D}$.

Case 2: $u = w$. Then the directed path in A starts with an arc (w, v') for some $v' \in X_{t'}$. By Observation 5, this arc belongs to $\hat{D}$. If $v' \neq v$, then by Observation 6 there is a directed path from v' to v in $\hat{D}'$. Hence there is a directed path from u to v in $\hat{D}$.

Case 3: $v = w$. This case is symmetric to Case 2. □

Algorithm 2 computes (and returns) the value of $c(t, P, \mathcal{S})$ from the values computed for $X_{t'}$.

Lemma 5. *If the correct value of $c(t', Q, \mathcal{Z})$ is available for every $Q \subseteq P_{t'}$ and every $\mathcal{Z} \in \mathcal{P}(X_{t'})^k$, then Algorithm 2 correctly computes $c(t, P, \mathcal{S})$ for an introduce node X_t in time $2^{|X_t|^2} \cdot |X_t|^{O(1)}$.*

Proof. We prove that $c(t, P, \mathcal{S})$ is **True** if and only if the algorithm returns **True**.

Forward direction. Assume that $c(t, P, \mathcal{S})$ is **True**. Let $\hat{\mathcal{S}}$ be a k-tuple that witnesses this, and let $\hat{\mathcal{S}}' = \hat{\mathcal{S}} \setminus \{w\}$. Set $\hat{D} = D[V_t] \oplus \hat{\mathcal{S}}$ and $\hat{D}' = D[V_{t'}] \oplus \hat{\mathcal{S}}'$. Let P'' be the set of pairs $(u, v) \in P_{t'}$ such that there is a directed path from u to v in $\hat{D}'$. Then $P'' \subseteq P^*$ because $\hat{D}'$ is an induced subgraph of $\hat{D}$, and P^* is exactly P restricted to pairs avoiding w.

Since $\hat{\mathcal{S}}'$ witnesses $c(t', P'', \mathcal{S}') = $ **True**, the first skip is not taken. Moreover, since $\hat{D}$ is a DAG, Lemma 4 implies that $A_{t, P'', \mathcal{S}}$ is a DAG, so the DAG-check skip is not taken.

Finally, by Lemma 4, for every $(u, v) \in P$ there is a directed path from u to v in $\hat{D}$ if and only if there is a directed path from u to v in A, and similarly for $(u, v) \in \overline{P}$. Hence both path-consistency loops pass, and the algorithm returns **True**.

Backward Direction. Assume that the algorithm returns **True**. We prove that $c(t, P, \mathcal{S})$ is **True**.

Since the algorithm returns **True**, there exists a set $P'' \subseteq P'$ such that none of the **skip** statements is executed. Since the skip in line 6 is not taken, we have $c(t', P'', \mathcal{S}') = $ **True**. Let $\hat{\mathcal{S}}' \in \mathcal{P}(V_{t'})^k$ be a k-tuple witnessing this fact, and let $\hat{D}' = D[V_{t'}] \oplus \hat{\mathcal{S}}'$. Then $\hat{D}'$ is a DAG. Define $\hat{\mathcal{S}} = \hat{\mathcal{S}}' \cup \mathcal{S}$. Since $\hat{\mathcal{S}}' \cap X_{t'} = \mathcal{S}'$, it follows that $\hat{\mathcal{S}} \cap X_t = \mathcal{S}$, recalling that $X_t = X_{t'} \cup \{w\}$ and $\mathcal{S}' = \mathcal{S} \setminus \{w\}$. Let $\hat{D} = D[V_t] \oplus \hat{\mathcal{S}}$ and let $A = A_{t, P'', \mathcal{S}}$. Since the skip in line 9 is not taken, the auxiliary graph A is a DAG. Moreover, $\hat{D}' = \hat{D} - w$ is a DAG. Hence, if

$\hat{D}$ contains a directed cycle, then every such cycle must contain the vertex w. Assume for contradiction that $\hat{D}$ contains a directed cycle C. Let C be a shortest directed cycle in $\hat{D}$. Then $w \in V(C)$, and there exist vertices $u, v \in X_{t'}$ such that (v, w) and (w, u) are arcs of C. Let Q be the directed subpath of C from u to v that avoids w. By minimality of C, Q is a simple directed path entirely contained in $\hat{D}'$. Thus, $(u, v) \in P''$ by definition of P''. By Observation 5, the arcs (v, w) and (w, u) belong to A. Since $(u, v) \in P'' \subseteq E(A)$, the graph A contains the directed triangle $u \to v \to w \to u$, contradicting the fact that A is a DAG. Hence, $\hat{D}$ is a DAG. Now, since the skip in line 12 is not taken, for every pair $(u, v) \in P$ there is a directed path from u to v in A. By Lemma 4, this implies that there is a directed path from u to v in $\hat{D}$. Similarly, since the skip in line 15 is not taken, for every pair $(u, v) \in \overline{P}$ there is no directed path from u to v in A. Again by Lemma 4, there is no directed path from u to v in $\hat{D}$. Therefore, $\hat{S}$ witnesses that $c(t, P, \mathcal{S}) = \texttt{True}$. The complexity follows from the fact that P^* has at most $|X_t|^2$ pairs and everything inside the main loop can be done in $|X_t|^{O(1)}$-time. This completes the proof. $\square$

We have similar lemmas for forget nodes and join nodes.

Lemma 6. *If the correct value of $c(t', Q, \mathcal{Z})$ is available for every $Q \subseteq P_{t'}$ and every $\mathcal{Z} \in \mathcal{P}(X_{t'})^k$, then there is an algorithm which correctly computes $c(t, P, \mathcal{S})$ for an forget node X_t in time $O(2^{2|X_t|+k})$.*

Lemma 7. *If correct values of $c(t_1, Y_1, \mathcal{Z})$ and $c(t_2, Y_2, \mathcal{Z})$ are available for all $Y_1, Y_2 \subseteq P_t$ and all $\mathcal{Z} \in \mathcal{P}(X_t)^k$, then there is an algorithm which correctly computes $c(t, P, \mathcal{S})$ for a join node X_t in time $2^{|X_t|^2} \cdot |X_t|^{O(1)}$.*

Now we are ready to prove Theorem 3.

Proof (Proof of Theorem 3). There exists an algorithm running in time $2^{O(\mathrm{tw})} \cdot n^{O(1)}$ that computes a tree decomposition of width at most $2 \cdot \mathrm{tw} + 1$ [13]. Given such a tree decomposition, it is well known that it can be transformed in polynomial time into a nice tree decomposition of the same width and with $O(\mathrm{tw} \cdot n)$ nodes. As discussed earlier, for a leaf node there is only a single table entry to compute, which can be done in constant time. For every other node, we apply an algorithm depending on the type of the node, in a bottom-up manner. For each node t, we compute $2^{O(\mathrm{tw}^2)} \cdot 2^{O(k \cdot \mathrm{tw})}$ table entries: there are $2^{O(\mathrm{tw}^2)}$ possibilities for P and $2^{O(k \cdot \mathrm{tw})}$ possibilities for $\mathcal{S}$. Filling a single entry takes $2^{|X_t|^2} \cdot |X_t|^{O(1)}$ time for an introduce node, $O(2^{2|X_t|+k})$ time for a forget node, and $2^{2|X_t|} \cdot |X_t|^{O(1)}$ time for a join node. Since $|X_t| \leq \mathrm{tw} + 1$, all these bounds are subsumed by $2^{O(\mathrm{tw}^2 + k \cdot \mathrm{tw})}$. As the nice tree decomposition has $O(\mathrm{tw} \cdot n)$ nodes, the overall running time is $2^{O(\mathrm{tw}(k + \mathrm{tw}))} \cdot n^{O(1)}$. By Lemmas 5, 6, and 7, the input graph D is k-invertible if and only if $c(r, \emptyset, \emptyset) = \texttt{True}$, where X_r is the root bag of the nice tree decomposition. $\square$

Concluding Remarks. We conclude by posing the following question. In Sect. 2, we obtained a fixed-parameter tractable algorithm for k-INVERSION when the underlying undirected graph of the input digraph is a block graph. It is unclear how to adapt this approach even to the case where the underlying graph can be covered by two cliques with a large intersection. Does k-INVERSION, parameterized by k, admit an FPT algorithm on graphs with this structure?

References

1. Alon, N., Powierski, E., Savery, M., Scott, A.D., Wilmer, E.: Invertibility of digraphs and tournaments. SIAM J. Discret. Math. **38**(1), 327–347 (2024). https://doi.org/10.1137/23M1547135
2. Aubian, G., et al.: Problems, proofs, and disproofs on the inversion number. Electron. J. Comb. **32**(1) (2025). https://doi.org/10.37236/12983
3. Bang-Jensen, J., Havet, F., Hörsch, F., Rambaud, C., Reinald, A., Silva, C.: Making an oriented graph acyclic using inversions of bounded or prescribed size. arXiv preprint arXiv:2511.22562 (2025). https://doi.org/10.48550/arXiv.2511.22562
4. Bang-Jensen, J., da Silva, J.C.F., Havet, F.: On the inversion number of oriented graphs. Discret. Math. Theor. Comput. Sci. **23**(2) (2021). https://doi.org/10.46298/DMTCS.7474
5. Behague, N., Gaudart-Wifling, P.: A case of the dijoin conjecture on inverting oriented graphs. arXiv preprint arXiv:2509.10232 (2025). https://doi.org/10.48550/arXiv.2509.10232
6. Behague, N., Johnston, T., Morrison, N., Ogden, S.: A note on inverting the dijoin of oriented graphs. Electron. J. Comb. **32**(1) (2025). https://doi.org/10.37236/13018
7. Belkhechine, H., Bouaziz, M., Boudabbous, I., Pouzet, M.: Inversion dans les tournois. Comptes Rendus. Mathématique **348**(13–14), 703–707 (2010)
8. Belkhechine, H., Salha, C.B.: Making a tournament indecomposable by one subtournament-reversal operation. Graphs Comb. **37**(3), 823–838 (2021). https://doi.org/10.1007/S00373-021-02282-0
9. Courcelle, B.: The monadic second-order logic of graphs. I. recognizable sets of finite graphs. Inf. Comput. **85**(1), 12–75 (1990). https://doi.org/10.1016/0890-5401(90)90043-H
10. Cygan, M., et al.: Parameterized Algorithms. Springer (2015). https://doi.org/10.1007/978-3-319-21275-3
11. Duron, J., Havet, F., Hörsch, F., Rambaud, C.: On the minimum number of inversions to make a digraph k-(Arc-) strong. J. Graph Theory **111**(2), 31–62 (2026). https://doi.org/10.1002/jgt.23290
12. Havet, F., Hörsch, F., Rambaud, C.: Diameter of the inversion graph. arXiv preprint arXiv:2405.04119 (2024). https://doi.org/10.48550/arXiv.2405.04119
13. Korhonen, T.: A single-exponential time 2-approximation algorithm for treewidth. In: 62nd IEEE Annual Symposium on Foundations of Computer Science, FOCS 2021, Denver, CO, USA, 7–10 February 2022, pp. 184–192. IEEE (2021). https://doi.org/10.1109/FOCS52979.2021.00026
14. Wang, H., Yang, Y., Lu, M.: The inversion number of dijoins and blow-up digraphs. arXiv preprint arXiv:2404.14937 (2024). https://doi.org/10.48550/arXiv.2404.14937
15. Yuster, R.: On tournament inversion. J. Graph Theory **110**(1), 82–91 (2025). https://doi.org/10.1002/JGT.23251

On the Rank and the General Position Number in Cycle Convexity

Júlio Aráujo[1] , Samuel N. Aráujo[1,2,3], Pedro P. Medeiros[2],
Nicolas Nisse[2(✉)] , and Caroline Silva[2,4]

[1] ParGO, Universidade Federal do Ceará, Fortaleza, Brazil
[2] Université Côte d'Azur, CNRS, Inria, I3S, Nice, France
`nicolas.nisse@inria.fr`
[3] Instituto Federal de Educação Ciência e Tecnologia do Ceará, Fortaleza, Brazil
[4] Universidade Estadual de Campinas, São Paulo, Brazil

Abstract. Graph convexities have been widely investigated through combinatorial parameters that capture different aspects of convexity-related structures. In the cycle convexity, the interval $I(S)$ of a subset S of vertices of a graph $G = (V, E)$ is the set of all vertices in S union the vertices with two neighbors in a same connected component of $G[S]$. The hull of S is the closure $H(S) = I^*(S)$. The general position number of G is the maximum size of a set S such that $v \notin I(S \setminus \{v\})$ for every $v \in S \subseteq V(G)$, and the rank of G is the maximum size of a set S such that $v \notin H(S \setminus \{v\})$ for every $v \in S \subseteq V(G)$.

Here, we first show that the general position number of G equals the maximum order of an induced forest in G. Then, we focus on the computational complexity of computing the rank of G. We show that it is NP-hard, even if the input graph is known to be bipartite, and W[1]-hard parameterized by the size of the solution. Then, we prove that it is polynomial-time computable in various graph classes such as forests, cycles, complete graphs, complete bipartite graphs, cographs, Cartesian grids, chordal graphs, starlike graphs, and $(q, q-4)$-graphs. Surprisingly, the cases of the classes of Cartesian grids and of not 2-connected chordal graphs are quite involved. We conclude by presenting fixed-parameter tractable algorithms for computing the rank of a given graph when parameterized by the neighborhood diversity, by vertex cover, or by the treewidth of the input graph.

Keywords: graph convexity · cycle convexity · rank · general position number

Most of the proofs of this paper are omitted due to space constraints, but we refer to [5] for a full version of the paper.

Research supported by the CAPES-Cofecub project Ma 1004/23, by the Inria Associated Team CANOE, by the French government, through the EUR DS4H Investments in the Future project managed by the National Research Agency (ANR) with the reference number ANR-17-EURE-0004. Júlio Araújo was also supported by CNPq (Brazil) projects 313153/2021-3, 404613/2023-3 and 308939/2025-5. Caroline Silva was supported by FAPESP Proc. 2023/16755-2.

F. Foucaud and A. Parreau (Eds.): IWOCA 2026, LNCS 16587, pp. 73–87, 2026.
https://doi.org/10.1007/978-3-032-27732-9_6

1   Introduction

A *convexity space*, or simply *convexity* [12], is an ordered pair $(V, \mathcal{C})$ where V is a set and $\mathcal{C}$ is a subset family of V, called *convex sets*, such that $V, \emptyset \in \mathcal{C}$ and that $\mathcal{C}$ is closed under intersection. Let $(V, \mathcal{C})$ be a convex space, the *convex hull* of a set $S \subseteq V$, denoted by $\mathrm{H}(S)$, is the inclusion-wise minimal convex set $C \in \mathcal{C}$ such that $C \supseteq S$. If $\mathrm{H}(S) = V$, we say that S is a *hull set* of $(V, \mathcal{C})$.

When V is the vertex set of a given graph G, we say that $(V, \mathcal{C})$ is a *graph convexity*. Most graph convexities are defined by an *interval function*, i.e., a function $\mathrm{I} : 2^V \to 2^V$, which satisfies: a) $S \subseteq \mathrm{I}(S)$ (extensive law), b) $S \subseteq S'$ implies $\mathrm{I}(S) \subseteq \mathrm{I}(S')$ (monotone law) and c) $\mathrm{I}(\emptyset) = \emptyset$ (normalization law) [8]. Therefore, in a convexity $(V, \mathcal{C})$ associated with the interval function I, a set S is convex if and only if $\mathrm{I}(S) = S$. Usually, in a graph convexity $(V, \mathcal{C})$ defined by an interval function I, the set $\mathrm{I}(S)$ contains all vertices in S, as well as all vertices lying on paths between vertices in S that satisfy some specific property. The most studied graph convexities are the *geodesic convexity* [23,32], the P_3 *convexity* [13, 19], P_3^* *convexity* [9], and the *monophonic convexity* [17,25].

Here, we are interested in *cycle convexity*, first introduced in [4]. The primary inspiration for the development of cycle convexity arose from its relevance in studying the tunnel number of knots and links in the field of Knot Theory [7]. In cycle convexity, the interval function is defined as

$$\mathrm{I}(S) = S \cup \{u \mid G[S \cup \{u\}] \text{ contains a cycle that contains } u\}$$
$$= S \cup \{u \mid u \text{ has two neighbours in a same component of } G[S]\}.$$

In [4], the *interval number* of a graph in the cycle convexity is investigated. The interval number of a graph G, denoted by $\mathrm{in}(G)$, is the size of a smallest set S of G, such that $\mathrm{I}(S) = V(G)$. They showed that the problem to determine whether $\mathrm{in}(G) \leq k$ is NP-complete, even if G is a split graph or a bounded-degree planar graph, and W[2]-hard even if G is a bipartite graph when the problem is parameterized by k. On the other hand, they showed that the problem is polynomial-time solvable for outerplanar graphs, cobipartite graphs and interval graphs. In [3], Araujo *et al.* examined the *hull number* in the cycle convexity. The hull number of a graph G, denoted by $\mathrm{hn}(G)$, is the size of a smallest hull set S of G. They showed that $\mathrm{hn}(G) \leq n/2$ when G is a 4-regular planar graph on n vertices. Furthermore, they proved that the problem to determine whether $\mathrm{hn}(G) \leq k$ for a given graph G and a positive integer k is NP-complete, even when G is a planar graph, but the problem is solvable in polynomial time when G is a chordal graph, P_4-sparse, or a grid.

In this paper, we are interested in the rank and general position numbers with respect to cycle convexity.

General Position. The general position number [30,31] of a graph G in any convexity with interval function I is the maximum cardinality of a subset $S \subseteq V(G)$ such that for every $s \in S$, $s \notin \mathrm{I}(S \setminus \{s\})$. The general position number problem has its roots in the classical No-Three-in-Line problem [18], which asks for the maximum number of points that can be placed on an $n \times n$ grid such that no three lie on a common straight line. Despite decades of investigation, the exact value of this number remains unknown for $n > 46$ [21]. Inspired by this combinatorial question, Manuel and Klavžar [30] introduced the general position number as a graph parameter in geodesic convexity. More recently, Araújo et al. [6] investigated this parameter for monophonic convexity. We first show that, in the case of the cycle convexity, the general position number of a graph G, denoted by $\mathrm{gp}(G)$, coincides with the *forest number* $\mathrm{MIF}(G)$ of G [20], the maximum number of vertices that induces a forest in G, as shown below.

Proposition 1. *For every graph G, $\mathrm{gp}(G) = \mathrm{MIF}(G)$.*

Proof. Let $S \subseteq V(G)$ such that S is in general position, i.e., for every $s \in S$, $s \notin \mathrm{I}(S \setminus \{s\})$. This implies that s is not in any cycle in $G[S \cup \{s\}]$. Therefore, S induces a forest in G. Conversely, if $G[S]$ is an induced forest, every subset of S induces a forest. Hence for every $s \in S$, $s \notin S \setminus \{s\} = \mathrm{I}(S \setminus \{s\})$.   ◇

An induced forest in a graph corresponds to the complement of a *feedback vertex set*, whose removal results in an acyclic graph. Therefore, all known results about $\mathrm{MIF}(G)$ and minimum feedback vertex set transfer to the general position in the cycle convexity.

The Rank. A closely related problem to the general position number is the rank of a graph, which is the main topic of our paper. Given a graph $G = (V, E)$, we say that $S \subseteq V$ is *convexly dependent* in G if there exists $v \in S$ such that $v \in \mathrm{H}(S \setminus \{v\})$. Otherwise, we say that S is *convexly independent* in G. The *rank* of a graph G [26,33], denoted here by $\mathrm{rk}(G)$ in the cycle convexity, is the largest cardinality of a convexly independent set in G. The notion of convex rank was first introduced by Jamison [26] in one of the first works on graph convexity. With respect to the monophonic and P_3 convexities, it was established in [33] that computing the rank is NP-hard. For the geodesic convexity, it was shown in [27] that the problem admits a polynomial-time algorithm when restricted to distance-hereditary graphs, but it remains NP-hard even when restricted to bipartite graphs, highlighting its computational intractability in general.

Our Contribution. In this paper, we mainly focus on the computational complexity of the RANK Problem in cycle convexity, i.e., given a graph G and an integer $k \in \mathbb{N}$, deciding if $\mathrm{rk}(G) \leq k$. In Sect. 2, we start with preliminary results showing that the rank can be trivially computed in forests, cycles, complete graphs, and complete bipartite graphs. Then, we characterize its behavior under union and join operations (Lemma 3). As straightforward consequences, we show that the rank of cographs can be computed in linear time (Corollary 1), but in the class of graphs obtained by joining two identical arbitrary graphs that the RANK Problem is NP-complete and W[1]-hard (Corollary 1). We conclude this section by showing that the RANK Problem is NP-complete even in the class of bipartite graphs (Theorem 2).

The Sect. 3 is devoted to polynomial-time algorithms to solve the RANK Problem in several graph classes. We characterize the rank of any $n \times m$ Cartesian grid, depending on the parity of n and m (Theorem 3). We design a polynomial-time dynamic programming algorithm that computes the rank of chordal graphs

(Theorem 4) generalizing some results in [2]. Then, we show that the rank can be computed in polynomial time in starlike graphs (Theorem 5) and in $(q, q-4)$-graphs (Theorem 7).

The Sect. 4 focuses on the parameterized complexity of the RANK Problem. We show that the rank is FPT when parameterized by the neighborhood diversity, by the vertex cover or by the treewidth of the input graph plus the size of the solution (Theorem 8 and Theorem 9).

2   Preliminaries and Hardness Results

In this section, we start with some basic results on the rank of a graph in cycle convexity. We then provide exact values of this parameter in basic graph classes. Finally, we discuss its complexity in general graphs. Let us start with some basic observations that will be useful throughout the paper.

Lemma 1. *Let S be a convexly independent set in any graph G. For any $S' \subseteq S$, S' is convexly independent.*

Lemma 2. *Let G be a graph. If $S \subseteq V$ is a convexly independent set in G, then $G[S]$ induces a forest (i.e., $G[S]$ is acyclic).*

Let $\alpha(G)$ be the largest size of a *stable set* (set of pairwise non-adjacent vertices) of G. The next straightforward theorem gives the general position and rank of several graph classes.

Theorem 1. *Let $n, m \in \mathbb{N}^*$. Let F_n (resp., C_n, resp., K_n) denote any forest (resp., cycle, resp., complete graph) with n vertices, and let $K_{n,m}$ be the complete bipartite graph:*

(a) $\mathrm{gp}(F_n) = \mathrm{rk}(F_n) = n$;
(b) $\mathrm{gp}(C_n) = \mathrm{rk}(C_n) = n - 1$, for any $n \geq 3$;
(c) $\mathrm{gp}(K_n) = \mathrm{rk}(K_n) = 2$, for any $n \geq 3$;
(d) $\mathrm{gp}(K_{n,m}) = \mathrm{rk}(K_{n,m}) + 1 = \max\{n, m\} + 1$, for any $n, m > 2$.

Given two disjoint graphs G_1 and G_2, the union of G_1 and G_2 is the graph $G_1 \cup G_2 = (V(G_1) \cup V(G_2), E(G_1) \cup E(G_2))$. The join of G_1 and G_2, denoted by $G_1 \oplus G_2$, is the graph obtained from $G_1 \cup G_2$ by adding every possible edge having one endpoint in $V(G_1)$ and the other in $V(G_2)$.

Lemma 3. *Let $G_1 = (V_1, E_1)$ and $G_2 = (V_2, E_2)$ be two disjoint graphs. Then $\mathrm{rk}(G_1 \cup G_2) = \mathrm{rk}(G_1) + \mathrm{rk}(G_2)$ and*

(a) $\mathrm{rk}(G_1 \oplus G_2) = 2$ if $G_1 \oplus G_2$ is a complete graph, else
(b) $\mathrm{rk}(G_1 \oplus G_2) = 3$ if G_1 is the disjoint union of 2 vertices and G_2 is the disjoint union of one vertex and one complete graph, else
(c) $\mathrm{rk}(G_1 \oplus G_2) = \max\{\alpha(G_2); t + 1\}$, if G_1 is a one single vertex and G_2 has $t \geq 1$ connected components. Else,
(d) $\mathrm{rk}(G_1 \oplus G_2) = \max\{\alpha(G_1), \alpha(G_2)\}$.

Recall that a cograph is any graph without P_4 (path on 4 vertices) as an induced subgraph. The class of cographs can be defined recursively: one vertex is a cograph, the disjoint union of two cographs is a cograph and the join of two cographs is a cograph [15]. Moreover, such a decomposition can be computed in linear time [14]. Note that computing $\alpha(G)$ if G is a cograph can be done in linear time. Moreover, note that Lemma 3(c) implies that $\mathrm{rk}(G) = \alpha(G)$ for any graph G obtained from a connected graph (not a complete graph) by adding a universal vertex. Therefore, Lemma 3 implies the following.

Corollary 1. *Let G be a graph and $k \in \mathbb{N}$. Deciding if $\mathrm{rk}(G) \leq k$ is*

- *linear time solvable in the class of cographs;*
- NP-*complete and* W[1]-*hard when parameterized by k, even if G is obtained from any connected graph by adding a universal vertex.*

To conclude this section, let us show that deciding whether $\mathrm{rk}(G) \geq k$ is NP-complete, even if we know that the input graph G is bipartite. Our reduction is from the well-known SET PACKING problem, which NP-complete [28].

The SET PACKING Problem has as input a triple $(U, \mathcal{S}, k)$ being U an arbitrary ground set, $\mathcal{S}$ a family of subsets of U, and k a positive integer, and aims to decide whether there are at least k pairwise disjoint sets of $\mathcal{S}$. This is a well-known NP-complete problem [28].

Theorem 2. *Deciding if* $\mathrm{rk}(G) \leq k$ *is* NP-*complete, even if the input graph G belongs to the class of bipartite graphs.*

Sketch of Proof. Let $\mathcal{I} = (U, \mathcal{S}, k)$ be an instance of the SET PACKING Problem (which is NP-complete [28]) with $U = \{u_1, \cdots, u_n\}$ and $\mathcal{S} = \{S_1, \cdots, S_m\} \subseteq 2^U$. We build an equivalent instance $(G(\mathcal{I}), k')$ as follows.

For every $1 \leq i \leq m$, let $I_i = \{j \mid 1 \leq j \leq m, i \neq j, S_i \cap S_j \neq \emptyset\}$ be the set of indices of subsets in $\mathcal{S}$ that intersect with S_i and let $I = \bigcup_{1 \leq i \leq m} \{\{i, j\} \mid j \in I_i\}$ be the family composed of pairs of intersecting sets of $\mathcal{S}$.

To build $G(\mathcal{I})$ (see Fig. 1), start with three pairwise disjoint stable sets: R having cardinality $|R| = 2|I| + n + m$, $U' = \{u'_1, \cdots, u'_n\}$ and $S' = \{s'_1, \cdots, s'_m\}$. Then, add a join between R and U', i.e., add an edge between every $r \in R$ and every $u'_i \in U'$. Then, for every $1 \leq i \leq n$ and every $1 \leq j \leq m$, add the edge $\{u'_i, s'_j\}$ if and only if $u_i \in S_j$.

Then, for every $1 \leq i < j \leq m$, such that $S_i \cap S_j \neq \emptyset$, add two disjoint paths $P^{ij} - (a^{ij}, b^{ij}, c^{ij})$ and $P^{ji} - (a^{ji}, b^{ji}, c^{ji})$ and the edges $\{\{s'_i, a^{ij}\}, \{s'_i, c^{ij}\}, \{s'_i, a^{ji}\}, \{s'_j, a^{ji}\}, \{s'_j, c^{ji}\}, \{s'_j, a^{ij}\}\}$. For every $1 \leq i \leq m$, let $A^i = \{a^{ij} \mid j \in I_i\}$ and define B^i and C^i analogously. Note that $\{A^i, B^i, C^i\}$ is a partition of $\bigcup \{V(P^{ij}) \mid j \in I_i\}$.

Note that the resulting graph $G(\mathcal{I})$ is bipartite and has $\mathcal{O}(\mathrm{poly}(n + m))$ vertices. Finally, set $k' = |R| + 4|I| + k$. We will show that $\mathcal{I} = (U, \mathcal{S}, k)$ is a yes instance of the SET PACKING problem if and only if $\mathrm{rk}(G(\mathcal{I})) \geq k'$.

The intuition one may have, provided by the value of k', is that a convexly independent set on at least k' vertices, must necessarily be composed of all

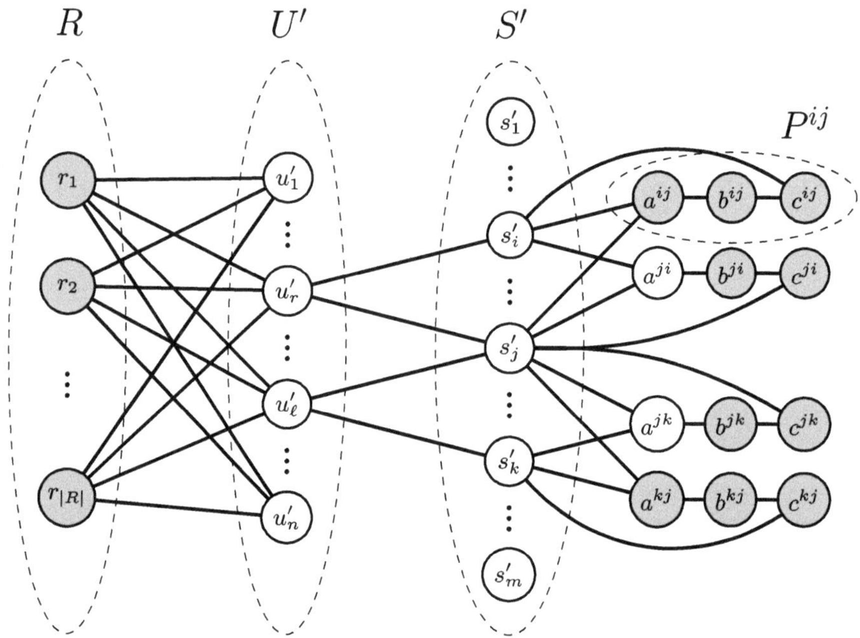

Fig. 1. Representation of the reduction for the RANK problem in bipartite graphs. The corresponding instance of the set packing problem is such that $u_r \in S_i \cap S_j$ and $u_\ell \in S_j \cap S_k$. If S_i and S_k are part of a solution to SET PACKING, we depict in gray the corresponding selected vertices used to form a convexly independent set.

vertices of R, none of U', union four or five vertices among $\{s'_i, s'_j\} \cup V(P^{ij}) \cup V(P^{ji})$, for every $\{i, j\} \in I$. The possible choice of such fifth vertex is what indicates the corresponding choice of a set S_i to be part of a solution to the SET PACKING problem. $\qquad\square$

3   Exact Values and Efficient Algorithms in Graph Classes

This section is devoted to proving that the rank can be computed in polynomial time in several graph classes such as Cartesian grids, chordal graphs, starlike graphs, and $(q, q - 4)$-graphs.

3.1   Cartesian Grids

The *cartesian product* of G and H, written $G\square H$ is the graph with vertex set $V(G) \times V(H)$ such that (u, v) is adjacent to (u', v') if and only if (1) $u = u'$ and $\{v, v'\} \in E(H)$, or (2) $v = v'$ and $\{u, u'\} \in E(G)$. The n-by-m grid $G_{n \times m} = P_n \square P_m$ is defined as the Cartesian product of the paths P_n and P_m on n and m vertices, respectively. Precisely, $G_{m \times n}$ is the grid with m rows (with n vertices) and n columns (with m vertices). If G is a grid, we use $L_i(G)$ (respectively,

$C_i(G)$) to denote the set of vertices of row i (respectively, column i). Moreover, we use $L_i^j(G)$ (respectively, $C_i^j(G)$) with $j \geq i$ to denote the set of vertices of G from row i to row j (respectively, from column i to column j). If G is clear in the context, we may use directly L_i, C_i, L_i^j, C_i^j.

Lemma 4. *For any* $n \in \mathbb{N}^*$, $\mathrm{rk}(G_{1 \times n}) = n$. *If* $n \geq 2$, *then* $\mathrm{rk}(G_{2 \times n}) = n + 1$.

Proof. The first statement is trivial (see Theorem 1(a)). Let $n \geq 2$.

First, to prove that $\mathrm{rk}(G_{2 \times n}) \geq n + 1$, observe that the set S consisting of one row with n vertices and one vertex in the other row is convexly independent.

The proof that $\mathrm{rk}(G_{2 \times n}) \leq n + 1$ is by induction on n. If $n = 2$, then the result holds since, clearly (Theorem 1(b)), $\mathrm{rk}(G_{2 \times 2}) = 3$. Let $n > 2$ and assume by induction that the result holds for any grid $G_{2 \times n'}$ with $2 \leq n' < n$. Let S be convexly independent in $G_{2 \times n}$.

First, assume that there exists $1 \leq i \leq n$ such that no vertex of S belongs to C_i. Let G_L (resp., G_R) be the subgrid consisting of the $i - 1$ first columns (resp, of the last $n - i$ columns). Possibly G_L (if $i = 1$) or G_R (if $i = n$) may be empty. By Lemma 1, $S_L = S \cap G_L$ and $S_R = S \cap G_R$ are convexly independent and so, by induction, $|S_L| \leq i$ and $|S_R| \leq n - i + 1$. Hence, $|S| = |S_L| + |S_R| \leq n + 1$.

Second, assume that every column (with 2 vertices) intersects S. If $|S| \geq n+2$, then at least two columns are fully contained in S. Let v be a vertex of a column fully included in S. Then, $S \setminus \{v\}$ contains at least one vertex per column, and at least one column has both its vertices in $S \setminus \{v\}$. Hence, observe that $\mathrm{H}(S \setminus \{v\}) = V(G_{2 \times n})$ and so $v \in \mathrm{H}(S \setminus \{v\})$, contradicting that S is convexly independent. $\square$

Let $G = G_{m \times n}$ be a grid with $n \geq m \geq 2$, and let S be a convexly indepen-dent set of G. Define $X = \{i \mid C_i \cap S = \emptyset\}$ to be the set of column indices i such that column C_i contains no vertex from S. We say that S is a *nice set* of G if X satisfies the following properties.

(i) $i \notin X$, for $i \in \{1, n\}$,
(ii) for every $2 \leq i < n$, if $i \in X$ then $\{i - 1, i + 1\} \cap X = \emptyset$ (no two consecutive integers belong to X), and
(iii) $\mathrm{H}(S) = V(G) \setminus (\bigcup_{i \in X} C_i)$.

Observe that (i) and (ii) imply that $|X| \leq \lceil \frac{n}{2} \rceil - 1$; that is, there are at most $\lceil \frac{n}{2} \rceil - 1$ columns of G containing no vertices of a nice set S. Let $X = \{x_1, \ldots, x_k\}$. We may assume without loss of generality that $x_1 < \ldots < x_k$. Note that, for a nice set S, $\mathrm{H}(S)$ induces $|X| + 1$ connected components, $G_1, \ldots, G_{k+1}$, such that

$$V(G_1) = C_1^{x_1 - 1}, V(G_2) = C_{x_1 + 1}^{x_2 - 1}, \ldots, V(G_{k+1}) = C_{x_k + 1}^n.$$

We also say that $G_1, \ldots, G_{k+1}$ are the *S-components* of G.

The proof of Theorem 3 relies on the notion of nice sets and on Lemma 5.

Lemma 5. *Let* $G = G_{m \times n}$ *be a grid with* $n \geq m \geq 3$. *Let* S *be a convexly independent set of* G *such that* $|S \cap (L_1 \cup L_2)| = n + 1$. *Then either there is* $3 \leq k \leq m$ *such that* $|L_k \cap S| < \lceil \frac{n}{2} \rceil$ *or*

(a) $|L_k \cap S| = \lceil \frac{n}{2} \rceil$ *for every* $3 \le k \le m$,
(b) S *is a nice set of* G, *and*
(c) G *has exactly* $\lceil \frac{n}{2} \rceil$ *S-components.*

Theorem 3. *Let* $G_{m \times n}$ *be a grid with* $n \ge m \ge 2$. *Then,*

$$
\mathrm{rk}(G_{m \times n}) = \begin{cases} \frac{mn}{2} + 1, & \text{for } m \text{ and } n \text{ both even,} \\ \frac{m(n+1)}{2} & \text{for } m \text{ even, and } n \text{ odd} \\ \frac{(m+1)n}{2} & \text{for } m \text{ odd.} \end{cases}
$$

Proof. **Lower Bounds.** If n (number of columns) and m (number of rows) are even, let $S = \{v\} \cup \bigcup_{1 \le i \le \frac{n}{2}} C_{2i-1}$ where v is any vertex of C_n, then S is convexly independent of size $\frac{nm}{2} + 1$. If m is even and n is odd, let $S = \bigcup_{1 \le i \le \frac{n+1}{2}} C_{2i-1}$, then S is convexly independent of size $\frac{(n+1)m}{2}$. If m is odd, let $S = \bigcup_{1 \le i \le \frac{m+1}{2}} L_{2i-1}$, then S is convexly independent of size $\frac{(m+1)n}{2}$. See Fig. 2.

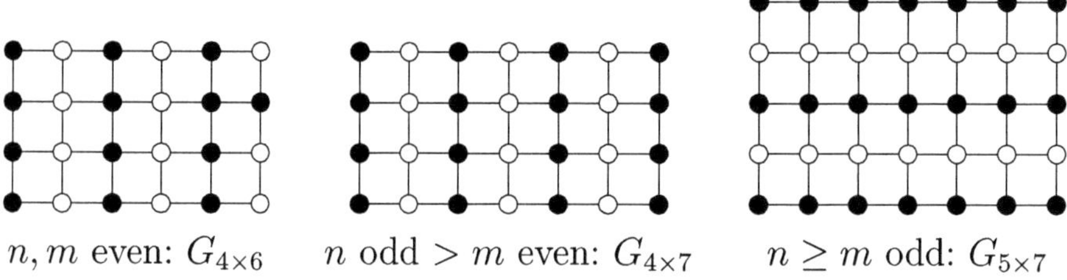

n, m even: $G_{4 \times 6}$ n odd $> m$ even: $G_{4 \times 7}$ $n \ge m$ odd: $G_{5 \times 7}$

Fig. 2. Grids $G_{4 \times 6}, G_{4 \times 7}$ and $G_{5 \times 7}$ and their respective convexly independent set colored black. On the left, with m and n even, $S = \{v\} \cup \bigcup_{1 \le i \le \frac{n}{2}} C_{2i-1}$. On center, for m even and n odd, $S = \bigcup_{1 \le i \le \frac{n+1}{2}} C_{2i-1}$. On the right, with m odd, $S = \bigcup_{1 \le i \le \frac{m+1}{2}} L_{2i-1}$.

Upper Bounds. We prove the result by induction on m. In $m = 2$, then the result follows by Lemma 4. We may thus assume that $n \ge m \ge 3$. We shall analyze the following cases depending on the parity of m and n.

Case 1. m is even and n is odd. Note that G can be partitioned into $k = \frac{m}{2}$ subgrids $G_1, \ldots, G_k$ of size $2 \times n$. By induction hypothesis and Lemma 1, $|S \cap G_i| \le n + 1$ for every $1 \le i \le k$. Therefore, $|S| \le \frac{m(n+1)}{2}$.

Case 2. m and n are both odd. Suppose first that $|S \cap (L_1 \cup L_2)| \le n$. Let G' be the subgrid induced by L_3^m. By induction hypothesis, $|S \cap V(G')| \le \frac{(m-1)n}{2}$. So,

$$
|S| \le \frac{(m-1)n}{2} + n = \frac{(m+1)n}{2}.
$$

Hence, we may assume that $|S \cap (L_1 \cup L_2)| = n + 1$. By Lemma 5 applied on the subgrid induced by L_1^3 and $S \cap L_1^3$, it follows that $|S \cap L_3| \le \lceil \frac{n}{2} \rceil$. So $G - L_3$

can be partitioned into $k = \frac{m-1}{2}$ subgrids $G_1, \ldots, G_k$ of size $2 \times n$. By induction hypothesis and Lemma 1, $|S \cap G_i| \leq n+1$ for every $1 \leq i \leq k$. Therefore,

$$|S| \leq \frac{(m-1)(n+1)}{2} + |S \cap L_3| \leq \frac{(m-1)(n+1)}{2} + \frac{n+1}{2}$$
$$= \frac{m(n+1)}{2} \leq \frac{(m+1)n}{2}, \text{ since } n \geq m.$$

Case 3. m *and* n *are both even.* We first prove the following claim.

Claim. There exists an even k with $2 \leq k \leq m$ such that $|S \cap L_1^k| \leq \frac{kn}{2}$.

Proof of Claim. Towards a contradiction, suppose the opposite. This implies that $|S \cap (L_1 \cup L_2)| = n+1$ (because $|S \cap (L_1 \cup L_2)| > n$). By Lemma 1 and Lemma 4 and, by symmetry, $|S \cap (L_{m-1} \cup L_m)| = n+1$. Note that all arguments are symmetric by relabeling the rows of G. Suppose first that there is even $k \in \{4, \ldots, m-2\}$ such that $|S \cap (L_{k-1} \cup L_k)| \neq n$. Choose k to be the smallest even integer satisfying such condition. If $|S \cap (L_{k-1} \cup L_k)| < n$, then

$$|S \cap L_1^k| = |S \cap L_1^2| + |S \cap L_3^{k-2}| + |S \cap L_{k-1}^k| \leq n+1+\frac{(k-4)}{2}n+n-1 \leq \frac{kn}{2},$$

a contradiction to our assumption. Otherwise, $|S \cap (L_{k-1} \cup L_k)| = n+1$. In this case, let G' be the subgrid induced by L_1^k. Note that

$$|S \cap V(G')| = |S \cap L_1^2| + |S \cap L_3^{k-2}| + |L_{k-1}^k| = n+1+\frac{(k-4)n}{2}+n+1 > \frac{kn}{2}+1,$$

a contradiction with our inductive hypothesis (since k is even).

So we may assume that for every even $k \in \{4, \ldots, m-2\}$, $|S \cap (L_{k-1} \cup L_k)| = n$. Now, we prove that there is no i such that $|L_i \cap S| > \frac{n}{2}$ for $3 \leq i \leq m-2$. Towards a contradiction, suppose that such an i exists. By symmetry of relabeling the rows of G, we may assume that i is odd, and subject to that, i is minimum. If $|S \cap L_j| = \frac{n}{2}$ for every $j < i$, then there is a contradiction to Lemma 5 (applied to L_1^i). Otherwise (because i is minimum subject to being odd), there is even $4 \leq j < i$ such that $|L_j \cap S| > \frac{n}{2}$. Let $\ell = i-j+1$. Note that ℓ is even and L_j^i induces a subgrid H of size $\ell \times n$. Moreover, $|H \cap S| > 2(\frac{n}{2}+1)+\frac{(\ell-2)n}{2} > \frac{\ell n}{2}+1$, a contradiction with the induction hypothesis.

Hence, we may assume that $|S \cap L_i| = \frac{n}{2}$ for every $3 \leq i \leq m-2$. By Lemma 5 applied to $G[L_1^{m-2}]$, G has exactly $\frac{n}{2}$ S-components. Since $|S \cap (L_1 \cup L_2)| = n+1$ and $|S \cap (L_{m-1} \cup L_m)| = n+1$, it follows that either (a) $|L_{m-1} \cap S| \geq \frac{n}{2}+1$ or (b) $|L_{m-1} \cap S| = \frac{n}{2}$ and $|L_m \cap S| = \frac{n}{2}+1$. If (a), then $S \cap L_1^{m-1}$ is a nice set of the subgrid induced by L_1^{m-1} which gives us a contradiction to Lemma 5. If (b), then G and S give a contradiction to Lemma 5. $\diamond$

By the previous claim, we may assume that there is even k such that $|S \cap L_1^k| \leq \frac{kn}{2}$. Let G' be the subgrid induced by L_{k+1}^m. By induction hypothesis, $|S \cap V(G')| \leq \frac{(m-k)n}{2} + 1$. Hence,

$$|S| \leq \frac{(m-k)n}{2} + 1 + \frac{kn}{2} \leq \frac{mn}{2} + 1.$$

$\square$

3.2 Chordal Graphs

In this section, we show that the rank of chordal graphs can be computed in polynomial time. Note that, the cycle convexity coincides with Δ-convexity in chordal graphs (in the Δ-convexity, the interval of a set $S \subseteq V$ consists of the union of S with every vertex that is adjacent to two adjacent vertices in S [1]). The rank of Δ-convexity has been shown to be polynomial solvable in 2-connected chordal graphs and block graphs (where each 2-connected component induces a clique) in [2]. Here, we solve the problem in any chordal graph.

Lemma 6. *[2] Let G be a 2-connected chordal graph. Then,* $\mathrm{rk}(G) = \max\{2, \alpha(G)\}$.

The key point for 2-connected chordal graphs that will be widely used in this subsection is the following.

Lemma 7. *For any convexly independent set S of a 2-connected chordal graph G, the following holds:*

- $\mathrm{H}(S) = V(G)$ *if and only if S induces a K_2, and*
- $\mathrm{H}(S) \neq V(G)$ *if and only if S is a stable set. In particular, in this case, $\mathrm{H}(S) = S$.*

Let G be any connected chordal graph and let B be any block of it. Let $v_1, \ldots, v_d$ be the cut-vertices of G that belong to B. For every $1 \leq i \leq d$, let C_i' be the union of the vertex sets of the connected components of $G - v_i$ that do not intersect B and let G_i be the subgraph of G induced by $C_i' \cup \{v_i\}$. With a slight abuse of notation, in the sequel sometimes we use only G_i to actually refer to $V(G_i)$. Let Q be any convexly independent set of G. For every $1 \leq i \leq d$, let $Q_i = Q \cap V(G_i)$ and let $Q_B = Q \cap (B \setminus \{v_1, \cdots, v_d\})$. Note that $\{Q_i \mid 1 \leq i \leq d\} \cup \{Q_B\}$ is a partition of Q. To give some intuition of our algorithm, let us list the possibly different cases for Q_B, ensuring that Q is convexly independent.

1. Either there exist $1 \leq i < j \leq d$, $v_i, v_j \in E(G)$ and $v_i \in \mathrm{H}_{G_i}(Q_i)$ (where $\mathrm{H}_{G_i}(Q_i)$ denote the convex hull of Q_i in G_i) and $v_j \in \mathrm{H}_{G_j}(Q_j)$, then $Q \cap (B \setminus \{v_i, v_j\}) = \emptyset$. Moreover, for every $h \in \{1, \cdots, d\} \setminus \{i, j\}$ and for every $w \in Q_h \setminus \{v_h\}$, we have $w \notin \mathrm{H}_{G_h}(Q_h \cup \{v_h\} \setminus \{w\})$, or
2. there exists $1 \leq i \leq d$, such that $v_i \in \mathrm{H}_{G_i}(Q_i)$ and v_i is adjacent to some vertex of Q_B. In this case, $|Q \cap B \setminus \{v_i\}| = 1$. Moreover, for every $h \in \{1, \cdots, d\} \setminus \{i\}$ and for every $w \in Q_h \setminus \{v_h\}$, we have $w \notin \mathrm{H}_{G_h}(Q_h \cup \{v_h\} \setminus \{w\})$, or

3. there exist $u, v \in Q_B$ with $u, v \in E(G)$; in this case, $Q_B = \{u, v\}$ and, for every $h \in \{1, \cdots, d\}$ and for every $w \in Q_h \setminus \{v_h\}$, we have $w \notin H_{G_h}(Q_h \cup \{v_h\} \setminus \{w\})$, or

4. let $I \subseteq \{1, \cdots, d\}$ such that $i \in I$ if and only if $v_i \in H_{G_i}(Q_i)$, then $Q_B \cup \{v_i \mid i \in I\}$ is a stable set of B.

Observe that the first three cases correspond to the cases when $B \subseteq H(Q)$ (by Lemma 7). In the discussion above, it naturally appears a notion that we formalize here. Given a graph $G = (V, E)$, a set $Q \subset V$ and a vertex $v \notin Q$, we say that Q is v-*protected* if Q is convexly independent and, for every $w \in Q$, it holds that $w \notin H(Q \cup \{v\} \setminus \{w\})$.

To compute the rank of chordal graphs, we design a dynamic programming algorithm that will need to compute more information at every step. For this, we introduce the following notation that is guided by the above cases. For every graph G and $v \in V(G)$, we define $r_v(G, v)$, $r_{\bar{v}h}(G, v)$, $r_{\bar{v}\bar{h}}(G, v)$, $r_{\bar{v}hp}(G, v)$, and $r_{\bar{v}\bar{h}p}(G, v)$ as the maximum size of a convexly independent set Q of G satisfying the constraints represented by its corresponding indexes: v, h, p denote respectively whether v belongs to Q, whether v belongs to the convex hull of Q and whether Q is v-protected, or not, if there is a "bar" above the letter. Note that, by definition,

$$\mathrm{rk}(G) = \max\{r_v(G, v), r_{\bar{v}h}(G, v), r_{\bar{v}\bar{h}}(G, v)\}.$$

Before going further, the following claim states that we can focus on bridgeless graphs. Given a graph $G = (V, E)$, an edge $e \in E$ is a *bridge* if $G' = G - e = (V, E \setminus \{e\})$ has more connected components than G.

Claim. Let $G = (V, E)$ be any graph with a bridge $e = \{u, v\} \in E$ and let G_1, G_2 be the two components of $G - e$, then $\mathrm{rk}(G) = \mathrm{rk}(G_1) + \mathrm{rk}(G_2)$.

We design a dynamic programming algorithm that, given a bridgeless chordal graph G and a vertex $v \in V(G)$, computes in polynomial time each value $\beta \in \{r_v(G, v), r_{\bar{v}h}(G, v), r_{\bar{v}\bar{h}}(G, v), r_{\bar{v}hp}(G, v), r_{\bar{v}hp}(G, v)\}$; in particular, our algorithm keeps a convexly independent set of size β that satisfies the corresponding constraints. This algorithm relies on the fact that, in chordal graphs, its block decomposition and the size of a maximum (weighted) independent set problem can be computed in polynomial time [22].

Theorem 4. *The rank can be computed in polynomial time in the class of chordal graphs.*

3.3   Starlike Graphs

A connected graph G is *starlike* if G is obtained from the disjoint union of $t + 1$ complete graphs $K_0, \cdots, K_t$ by making the join between K_i and some $K_0^i \subseteq K_0$ for every $1 \leq i \leq t$. If $|K_l| \leq k$ for $l \in \{1, \ldots, t\}$, then G is k-starlike. The 1-starlike graphs are exactly the split graphs. Recall that a vertex v is *simplicial* if $N(v)$ induces a complete subgraph. We first prove that, if v is a degree-one vertex, then a set S is convexly independent in any graph if and only if $S - v$ is convexly independent. Then, by a simple case analysis, we obtain the following theorem.

Theorem 5. *Let $G = (K_0, K_1, \cdots, K_t)$ be a starlike graph with minimum degree at least 2. Then,*

1. $\mathrm{rk}(G) = t + 2$ *if* $|N(K_i) \cap K_0| = 1$, *for all* $1 \le i \le t$ *and* $|V(K_0)| \ge 2$, *else*
2. $\mathrm{rk}(G) = t+1$, *if* $\exists v \in K_0$ *such that* $\forall u \in K_0, u \ne v, N(v) \cap N(u) \cap V(G) \backslash K_0 = \emptyset$ *(v is a cut-vertex or v is a simplicial vertex in K_0), else*
3. $\mathrm{rk}(G) = t.$

3.4 The $(q, q - 4)$-Graphs

A graph G is $(q, q - 4)$ for some integer $q \ge 4$ if every subset with at most q vertices induces at most $q - 4$ P_4's [11]. Observe that the cographs are exactly the $(4, 0)$-graphs and the P_4-sparse [24] are the $(5, 1)$-graphs. It is well known that $(q, q - 4)$-graphs admit a tree structure that can be computed in linear time[1].

Theorem 6. *[10] Let $q \ge 4$ be fixed. If G is a $(q, q - 4)$-graph, then one of the following holds:*

- *G is a single vertex;*
- *$G = G_1 \cup G_2$ is the disjoint union of two $(q, q - 4)$-graphs G_1 and G_2;*
- *$G = G_1 \oplus G_2$ is the join of two $(q, q - 4)$-graphs G_1 and G_2;*
- *G is a (R, K, S)-spider where $G[R]$ is a $(q, q - 4)$-graph if $R \ne \emptyset$;*
- *G is a q-pseudo-spider (R, K, S, E) where $G[R]$ is a $(q, q-4)$-graph if $R \ne \emptyset$.*

We first prove that the rank of spiders can be computed in polynomial time.

Lemma 8. *Let G be a (R, K, S)-spider. Therefore,*

$$\mathrm{rk}(G) = \begin{cases} \alpha(G[R]) + |S|, & \text{if } G \text{ is a thick spider;} \\ \max\{\alpha(G[R]), 2\} + |S|, & \text{if } G \text{ is a thin spider} \end{cases}$$

Using Theorem 6, Lemma 8, Lemma 3 and exhaustive search (FPT in q) in the case of q-pseudo-spiders, we get a recursive polynomial-time algorithm.

Theorem 7. *Let G a $(q, q - 4)$-graph with fixed $q \ge 4$, $\mathrm{rk}(G)$ can be computed in polynomial time.*

4 FPT Algorithms

This section is devoted to FPT algorithms for computing the rank, parameterized by neighbourhood diversity and treewidth.

---

[1] Due to lack of space, we omit the formal definition of spiders and q-pseudo-spiders.

4.1 Neighborhood Diversity

Two vertices u, v of a graph G have the same *type* if $N(u)\setminus\{v\} = N(v)\setminus\{u\}$. Two vertices with the same type are *true twins* (resp. *false twins*) if they are adjacent (resp. non-adjacent). The *neighborhood diversity* of a graph G is the minimum integer k such that $V(G)$ can be partitioned into k sets (classes) such that all the vertices in each set have the same type [29]. We first show that, for deciding the rank, any graph can be reduced to an equivalent instance where each class of false twins has size at most 4. Then, since in each class of true twins, at most 2 vertices can be considered (because any set containing a triangle cannot be convexly independent), we get:

Theorem 8. *Computing* $\mathrm{rk}(G)$ *can be done in* $\mathcal{O}(5^k k^3 + n^3)$*-time if the input graph G has n vertices and neighborhood diversity k.*

4.2 Treewidth

We prove that deciding if $\mathrm{rk}(G) \leq k$ may be formulated using an MSOL formula and so, by Courcelle's theorem [16]:

Theorem 9. *Given a graph G and a positive integer k, deciding whether* $\mathrm{rk}(G) \geq k$ *is* FPT *parameterized by* $\mathrm{tw}(G) + k$.

5 Further Work

In the view of Theorem 2, since such a reduction is not a parameterized one, even though the SET PACKING problem is W[1]-hard when parameterized by the value of the solution, a natural question is:

Problem 1. Given a bipartite graph G and a positive integer k, deciding whether $\mathrm{rk}(G) \geq k$ is FPT when parameterized by k?

Let $\mathrm{vc}(G)$ be the minimum cardinality of a vertex cover of G. If a graph G has a vertex cover S of size k, $V(G)$ can be easily partitioned into at most $2^k + k$ equivalence classes of types, thus implying that, for any graph G, $\mathrm{nd}(G) \leq 2^{\mathrm{vc}(G)} + \mathrm{vc}(G)$. Consequently, from Theorem 8 we obtain the problem of computing the rank parameterized by the vertex cover is FPT, but such an algorithm will be double exponential in $\mathrm{vc}(G)$. Hence, we pose the following natural questions.

Problem 2. Does the problem of computing the rank admit a polynomial kernel when parameterized by the vertex cover of the input graph?

Problem 3. Given a graph G and a positive integer k, is deciding whether $\mathrm{rk}(G) \geq |V(G)| - k$ an FPT problem, when the parameter is k?

References

1. Anand, B.S., Anil, A., Changat, M., Dourado, M.C., Ramla, S.S.: Computing the hull number in Δ-convexity. Theor. Comput. Sci. **844**, 217–226 (2020). https://doi.org/10.1016/j.tcs.2020.08.024. https://www.sciencedirect.com/science/article/pii/S0304397520304849
2. Anand, B.S., Anil, A., Changat, M., Nair, R.S., Narasimha-Shenoi, P.G.: Helly number, radon number and rank in Δ-convexity on graphs. In: Gaur, D., Mathew, R. (eds.) Algorithms and Discrete Applied Mathematics, pp. 292–306. Springer Nature Switzerland, Cham (2025)
3. Araujo, J., Campos, V., Girão, D., Nogueira, J., Salgueiro, A., Silva, A.: On the hull number on cycle convexity of graphs. Inf. Process. Lett. **183**, 106420 (2024)
4. Araujo, J., Ducoffe, G., Nisse, N., Suchan, K.: On interval number in cycle convexity. Discrete Math. Theor. Comput. Sci. **20**(1), 1–28 (2018)
5. Aráujo, J., Aráujo, S.N., Medeiros, P.P., Nisse, N., Silva, C.: On the rank and the general position number in cycle convexity. Technical report, Inria & Université Cote d'Azur, CNRS, I3S, Sophia Antipolis, France (2025). https://inria.hal.science/hal-05378038
6. Araujo, J., Dourado, M.C., Protti, F., Sampaio, R.: The iteration time and the general position number in graph convexities. Appl. Math. Comput. **487**, 129084 (2025)
7. Aráujo, J., Campos, V., Girão, D., Nogueira, J., Salgueiro, A., Silva, A.: Cycle convexity and the tunnel number of links (2020). https://arxiv.org/abs/2012.05656
8. Aráujo, J., Dourado, M.C., Protti, F., Sampaio, R.M.: Introduction to Graph Convexity - An Algorithmic Approach. Springer, Cham (2025)
9. Aráujo, R., Sampaio, R., dos Santos, V., Szwarcfiter, J.: The convexity of induced paths of order three and applications: complexity aspects. Discrete Appl. Math. **237**, 33–42 (2018)
10. Babel, L., Olariu, S.: On the isomorphism of graphs with few P4s. In: Nagl, M. (ed.) Graph-Theoretic Concepts in Computer Science, pp. 24–36. Springer, Heidelberg (1995)
11. Babel, L., Olariu, S.: On the structure of graphs with few P4s. Discret. Appl. Math. **84**(1), 1–13 (1998)
12. Bryant, V., Webster, R.: Convexity spaces. I. the basic properties. J. Math. Anal. Appl. **37**(1), 206–213 (1972)
13. Centeno, C.C., Dourado, M.C., Szwarcfiter, J.L.: On the convexity of paths of length two in undirected graphs. Electron. Notes Discrete Mathe. **32**, 11–18 (2009)
14. Corneil, D.G., Perl, Y., Stewart, L.K.: A linear recognition algorithm for cographs. SIAM J. Comput. **14**(4), 926–934 (1985)
15. Corneil, D., Lerchs, H., Burlingham, L.: Complement reducible graphs. Discret. Appl. Math. **3**(3), 163–174 (1981)
16. Courcelle, B., Mosbah, M.: Monadic second-order evaluations on tree-decomposable graphs. Theor. Comput. Sci. **109**(1), 49–82 (1993). https://doi.org/10.1016/0304-3975(93)90064-Z. https://www.sciencedirect.com/science/article/pii/030439759390064Z
17. Duchet, P.: Convex sets in graphs, II. Minimal path convexity. J. Comb. Theory Ser. B **44**(3), 307–316 (1988)
18. Dudeney, H.: Amusements in Mathematics, vol. 473. Courier Corporation (1958)
19. Erdős, P.: Some remarks on simple tournaments. Algebra Universalis **2**, 238–245 (1972)

20. Erdös, P., Saks, M., Sós, V.T.: Maximum induced trees in graphs. J. Comb. Theory Ser. B **41**(1), 61–79 (1986)
21. Flammenkamp, A.: Progress in the no-three-in-line problem, II. J. Comb. Theory Ser. A **81**(1), 108–113 (1998)
22. Frank, A.: Some polynomial algorithms for certain graphs and hypergraphs. In: Proceedings of the Fifth British Combinatorial Conference (1975)
23. Harary, F., Nieminen, J.: Convexity in graphs. J. Differ. Geom. **16**(2), 185–190 (1981)
24. Jamison, B., Olariu, S.: Linear time optimization algorithms for P4-sparse graphs. Discret. Appl. Math. **61**(2), 155–175 (1995)
25. Jamison, R.: A perspective on abstract convexity: classifying alignments by varieties, convexity and related combinational geometry. In: Proceedings University of Oklahoma Conference Marcel Dekker, Inc., New York (1982)
26. Jamison, R.: Partition numbers for trees and ordered sets. Pac. J. Math. **96** (1981)
27. Kanté, M., Sampaio, R., dos Santos, V., Szwarcfiter, J.: On the geodetic rank of a graph. J. Comb. **8**, 323–340 (2017)
28. Karp, R.: Reducibility among combinatorial problems. In: Proceedings of a Symposium on the Complexity of Computer Computations, vol. 40, pp. 85–103 (1972)
29. Lampis, M.: Algorithmic meta-theorems for restrictions of treewidth. Algorithmica **64**(1), 19–37 (2012). https://doi.org/10.1007/S00453-011-9554-X
30. Manuel, P., Klavzar, S.: A general position problem in graph theory. Bull. Aust. Math. Soc. **98**(2), 177–187 (2018)
31. Manuel, P., Klavžar, S.: The graph theory general position problem on some interconnection networks. Fund. Inform. **163**(4), 339–350 (2018)
32. Pelayo, I.: Geodesic Convexity in Graphs. Springer (2013)
33. Ramos, I., dos Santos, V., Szwarcfiter, J.: Complexity aspects of the computation of the rank of a graph. Discrete Math. Theor. Comput. Sci. **16** (2014)

Hardness Results on Bondage
and Reinforcement Problems in Chordal
Graphs

Deepak M. Bakal[1,2(✉)] and Y. M. Borse[2]

[1] Department of Applied Sciences, COEP Technological University, Pune, India
`iamdeepakbakal@gmail.com`
[2] Department of Mathematics, Savitribai Phule Pune University, Pune, India

Abstract. Let $G = (V, E)$ be a graph with domination number $\gamma(G)$. The BONDAGE problem asks whether removing at most k edges increases $\gamma(G)$, while the REINFORCEMENT problem asks whether adding at most k non-edges decreases $\gamma(G)$. Although both problems are known to be NP-hard on general and bipartite graphs, their complexity on chordal graphs has remained open. We prove that both problems are NP-hard and coNP-hard for $k = 1$ even on split graphs, a subclass of chordal graphs. This confirms a conjecture of Hu and Xu by showing that, unless NP = coNP, both problem do not belong to NP. Further, we show that, unless P = NP, both problems do not admit a polynomial-time approximation algorithm with performance ratio smaller than 2, even for split graphs.

Keywords: Chordal graphs · Domination · Bondage · Reinforcement

1 Introduction

We consider finite, undirected, and simple graphs. For a graph G, let $V(G)$ and $E(G)$ denote its vertex set and edge set, respectively. For graph-theoretic terminology and notation, we refer the reader to [15], and for complexity-theoretic terminology and notation, to [6,12].

A subset $D \subseteq V(G)$ is a *dominating set* of G if every vertex of G either belongs to D or is adjacent to a vertex in D. The *domination number* of G, denoted by $\gamma(G)$, is the minimum cardinality of a dominating set of G. A dominating set D is called a γ-set of G if $|D| = \gamma(G)$.

Domination is a classical graph parameter with applications in communication networks, social networks, facility location, resource placement, and network monitoring. In many applications, networks are subject to small changes and thus understanding the behaviour of graph parameters under local perturbations of the underlying graph is important. In particular, domination under edge deletion [3,7,10], edge addition [8,19], edge contraction [11,18], and edge subdivision [2,9,14] has been studied extensively in the literature.

F. Foucaud and A. Parreau (Eds.): IWOCA 2026, LNCS 16587, pp. 88–102, 2026.
https://doi.org/10.1007/978-3-032-27732-9_7

In 1990, Fink et al. [10], and Kok and Mynhardt [19] introduced the notions of *bondage* and *reinforcement*, respectively, to study the effect of edge deletion and edge addition on domination number. The *bondage number* of a graph G, denoted by $b(G)$, is the minimum number of edges whose removal increases the domination number of G. The *reinforcement number* of G, denoted by $r(G)$, is the minimum number of non-edges whose addition decreases the domination number of G. These two parameters are well studied from combinatorial aspects, see [21–26].

The decision version of the bondage and reinforcement problems can be stated as follows. The BONDAGE problem asks whether, given a graph G and an integer k, there exists a set of at most k edges whose removal increases the domination number of G.

BONDAGE PROBLEM

Instance: A graph G and a positive integer k.

Question: Whether $b(G) \leq k$?

The REINFORCEMENT problem asks whether, given a graph G and an integer k, there exists a set of at most k non-edges whose addition decreases the domination number of G.

REINFORCEMENT PROBLEM

Instance: A graph G and a positive integer k.

Question: Whether $r(G) \leq k$?

The BONDAGE and REINFORCEMENT problems capture the robustness and vulnerability of a graph by studying the stability of the domination number under edge deletion and non-edge addition, respectively.

Although these concepts were introduced in 1990, algorithmic and complexity-theoretic aspects were studied only later. Hu and Xu [17] initiated the algorithmic study of BONDAGE and REINFORCEMENT problems. They showed that both problems are NP-hard for general graphs. Hu and Sohn [16] subsequently proved NP-hardness of both problems for bipartite graphs. Recently, Bouquet [4] studied the complexity of BONDAGE problem for the planar graphs.

In this paper, we study BONDAGE and REINFORCEMENT problems on chordal graphs, in particular, on split graphs. A graph is *chordal* if it has no induced cycle of length at least four. Chordal graphs form a fundamental graph class in both structural and algorithmic graph theory. They are also known as *triangulated graphs.* Chordal graphs and bipartite graphs are important subclasses of perfect graphs and their intersection is precisely the class of forests. In algorithmic graph theory, it is customary to study complexity of graph theoretic problems on both these classes. A *split graph* is a graph whose vertex set can be partitioned into a clique and an independent set. It is easy to see that every split graph is chordal.

We first prove that the BONDAGE problem is NP-hard even when restricted to split graphs and $k = 1$. We further show that the problem is also coNP-hard under these restrictions. Next, we establish analogous results for the REINFORCE-MENT problem by proving that it is NP-hard and coNP-hard on split graphs for $k = 1$. In our proofs, we reduce from the classical 3-SAT problem. An instance of 3-SAT consists of a set U of variables and a collection $\mathscr{C} = \{C_1, \ldots, C_m\}$ of clauses containing exactly three literals over U.

As a consequence of our hardness results, we affirmatively resolve the following conjecture of Hu and Xu [17] under the standard complexity assumption $\text{NP} \neq \text{CoNP}$. We restate the conjecture in their words:

We cannot determine whether the bondage and reinforcement problems belong to NP, *since for any subset* $B \subseteq E(G)$, *it is unclear whether one can verify in polynomial time that* $\gamma(G - B) > \gamma(G)$ *(or* $\gamma(G + B) < \gamma(G)$*). Since computing the domination number is* NP-*complete, we conjecture that these problems are not in* NP.

Also, our results establish that no polynomial-time approximation algorithm for both BONDAGE and REINFORCEMENT problem achieves a performance ratio strictly smaller than 2, unless $\text{P} = \text{NP}$. This provides a partially answer to the question "*find approximation polynomial algorithms with performance ratio as small as possible*" raised by Hu and Xu [17].

Further, we consider the parameterized version of both problems and establish that these problems do not admit an XP algorithm parameterized by the solution size k, unless $\text{P} = \text{NP}$. Moreover, we show that for fixed k, both problems are fixed-parameter tractable on graphs of bounded clique-width via a brute-force approach.

The remainder of the paper is organized as follows. Basic definitions and preliminary results are introduced in Sect. 2. The hardness results for the BONDAGE problem are established in Sect. 3. Section 4 presents the hardness results for REINFORCEMENT problem. Complexity-theoretic consequences, including non-membership in NP and approximation lower bound are discussed in Sect. 5. Section 6 studies the parameterized complexity of these problems parameterized by solution size and clique-width. We conclude with directions for further research.

2   Preliminaries

For a vertex $v \in V(G)$, the open neighborhood of v is the set $N_G(v) = \{u \in V(G) \mid uv \in E(G)\}$. A vertex $u \in V(G)$ is said to be *dominated* by a vertex v if $u = v$ or $u \in N_G(v)$. For any graph G and any edge e, let $G - e$ denotes the graph obtained from G by deleting the edge e and $G + e$ denote the graph obtained from G by adding the non-edge e. It is easy to note that $\gamma(G - e) \geq \gamma(G)$ and $\gamma(G + e) \leq \gamma(G)$, since every dominating set of $G - e$ is also a dominating set of G, and every dominating set of G is also a dominating set of $G + e$.

The bondage number of G is the minimum cardinality of a set $F \subseteq E(G)$ such that $\gamma(G - F) > \gamma(G)$. Let $\overline{E(G)}$ denote the set of non-edges of G. Then the reinforcement number of G is the minimum cardinality of a set $F \subseteq \overline{E(G)}$ such that $\gamma(G + F) < \gamma(G)$.

A split graph is a graph whose vertex set can be partitioned into a clique and an independent set. In particular, every split graph is chordal, since it has no induced cycle of length at least four. We begin with a folklore observation about dominating sets in split graphs.

Lemma 1. *Let $G = (K \sqcup I, E)$ be a connected split graph with clique K and independent set I. For every dominating set $D \subseteq V(G)$, there exists a dominating set $D' \subseteq K$ with $|D'| \leq |D|$.*

Proof. Let D be a dominating set of G. If $D \cap I = \emptyset$, then we are done. Otherwise, let $v \in D \cap I$. Since G is connected and I is an independent set, the vertex v has at least one neighbor $u \in K$. Define $D' = (D \setminus \{v\}) \cup \{u\}$. It is easy to see that D' is a dominating set of G since every vertex previously dominated by v is also dominated by u. $\square$

Consequently, without loss of generality, every dominating set of a split graph is contained in its clique. Throughout the paper, we assume that any γ-set of a split graph is a subset of its clique. For a positive integer k, we use the standard notation $[k]$ to denote the set $\{1, 2, 3, \ldots, k\}$.

3   Hardness of Bondage Problem

In this section, we establish the computational hardness of determining the bondage number on chordal graphs. We prove that the problem remains NP-hard even for split graphs and $k = 1$.

Theorem 1. *The BONDAGE problem is NP-hard even when restricted to split graphs and $k = 1$.*

Proof. Let $U = \{u_1, u_2, \ldots, u_n\}$ and $\mathscr{C} = \{C_1, C_2, \ldots, C_m\}$ be an arbitrary instance of 3-SAT. We construct a split graph G from the instance $(U, \mathscr{C})$ of 3-SAT such that $\mathscr{C}$ is satisfiable if and only if $b(G) = 1$ as follows.

- For each variable $u_i \subset U$, introduce two vertices u_i and $\overline{u_i}$, corresponding to the positive and negative literals, respectively and a third vertex v_i adjacent to both u_i and $\overline{u_i}$.
- For each clause $C_j \in \mathscr{C}$, introduce a vertex c_j. For every literal $u_i \in C_j$, add the edge $c_j u_i$, and similarly, for every $\overline{u_i} \in C_j$, include the edge $c_j \overline{u_i}$.
- We introduce the vertices p, q, r. The vertex p is made adjacent to all clause vertices. We introduce the vertices a, b and add edges pa, qa, qb, rb.
- Finally, we form a clique K of the vertices in the set $\{u_i, \overline{u_i}, p, q, r \mid i \in [n]\}$.

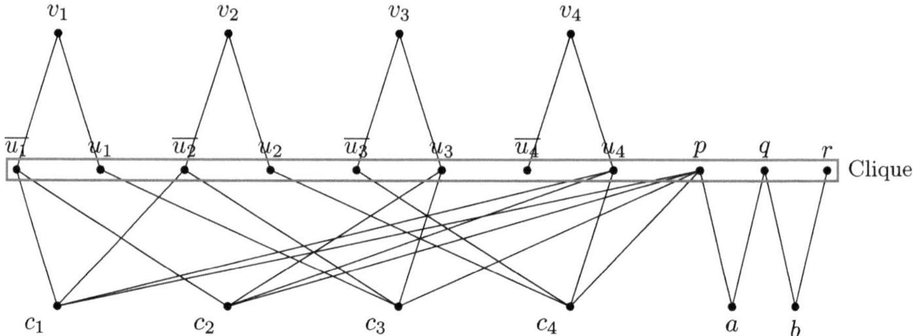

Fig. 1. An illustration of the construction of graph G in the proof of Theorem 1

It is easy to see that G is a split graph with clique $K = \{u_i, \overline{u_i}, p, q, r \mid i \in [n]\}$ and independent set $I = \{v_i, c_j, a, b \mid i \in [n], j \in [m]\}$.

The graph depicted in Fig. 1 illustrates the above construction for the instance $(U, \mathscr{C})$ with $U = \{u_1, u_2, u_3, u_4\}$ and $\mathscr{C} = \{C_1, C_2, C_3, C_4\}$, where $C_1 = \{\overline{u_1}, \overline{u_2}, u_4\}, C_2 = \{\overline{u_1}, u_3, u_4\}, C_3 = \{u_1, \overline{u_2}, u_3\}, C_4 = \{u_2, \overline{u_3}, u_4\}$.

Claim 1.1. *For the graph G, $n + 1 \leq \gamma(G) \leq n + 2$.*

Proof. It is easy to see that, in order to dominate the vertex v_i, we must include either the literal vertex u_i or $\overline{u_i}$ in a dominating set D of G, for each $i \in [n]$. Similarly, in order to dominate the vertex b, we must include at least one of the vertices q or r in D. Hence, $\gamma(G) \geq n + 1$. Again $D = \{u_i, q, p \mid i \in [n]\}$ is a dominating set of G. Hence, $\gamma(G) \leq |D| \leq n + 2$.

Claim 1.2. *For the graph G, $\gamma(G) = n + 1$ if and only if $\mathscr{C}$ is satisfiable.*

Proof. Suppose D is a γ-set of G with $\gamma(G) = n + 1$. By Claim 1.1, D includes either u_i or $\overline{u_i}$ for each $i \in [n]$. Thus, $|D \cap \{u_i, \overline{u_i} \mid i \in [n]\}| \geq n$. None of the vertices u_i or $\overline{u_i}$, for $i \in [n]$, dominates the vertices a and b. The only vertex dominating both a and b is the vertex q. Thus, D includes the vertex q. Consequently, no vertex c_j belongs to D, and the vertex $p \notin D$. Let us define a mapping

$$t : U \to \{T, F\} \quad \text{by} \quad t(u_i) = \begin{cases} T & \text{if } u_i \in D, \\ F & \text{otherwise.} \end{cases}$$

We prove that t is a satisfying truth assignment for $\mathscr{C}$. Let C_j be any clause in $\mathscr{C}$. Since $p \notin D$ and the corresponding clause vertex $c_j \notin D$, the vertex c_j must be dominated by some literal vertex. Hence, there exists $k \in [n]$ such that either $u_k \in D$ or $\overline{u_k} \in D$ and this literal dominates c_j. By the definition of the assignment t, the corresponding literal is set to T, and therefore C_j is satisfied. As the clause C_j is chosen arbitrarily, the assignment t satisfies all clauses in $\mathscr{C}$, and thus $\mathscr{C}$ is satisfiable.

Conversely, suppose that $\mathscr{C}$ is satisfiable. Let $t : U \to \{T, F\}$ be a satisfying truth assignment for $\mathscr{C}$. We construct a dominating set $D \subseteq V(G)$ with $|D| = n + 1$ as follows. For each $i \in [n]$, if $t(u_i) = T$, then include u_i in D; otherwise, include $\overline{u_i}$ in D. Since t satisfies $\mathscr{C}$, every clause $C_j \in \mathscr{C}$ contains a literal that evaluates to T, and hence the corresponding clause vertex c_j is adjacent to a vertex in D. Therefore, each clause vertex c_j, $j \in [m]$, is dominated by D. Finally, add the vertex q to D in order to dominate the vertices a and b. Thus, D is a dominating set of G with $|D| = n + 1$. By Claim 1.1, $\gamma(G) \geq n + 1$, and therefore $\gamma(G) = n + 1$.

Claim 1.3. *For any $e \in E(G)$, $\gamma(G - e) \leq n + 2$.*

Proof. We consider the deletion of an arbitrary edge $e \in E(G)$.

(i) If $e \in \{rb, pa\}$, then $D = \{u_i \mid i \in [n]\} \cup \{p, q\}$ is a dominating set of $G - e$.
(ii) If $e \in \{qa, qb\}$, then $D = \{u_i \mid i \in [n]\} \cup \{p, r\}$ is a dominating set of $G - e$.
(iii) If e is incident to a clause vertex c_j for some $j \in [m]$, we distinguish two cases. If the clause C_j contains a positive literal u_k for some $k \in [n]$, then $D = \{u_i \mid i \in [n]\} \cup \{p, q\}$ is a dominating set of $G - e$. Otherwise, C_j contains only negated literals, and $D = \{\overline{u_i} \mid i \in [n]\} \cup \{p, q\}$ is a dominating set of $G - e$.
(iv) If the edge $v_i \overline{u_i}$ (resp. $v_i u_i$) is deleted for some $i \in [n]$, then $D = \{u_i \mid i \in [n]\} \cup \{p, q\}$ (resp. $D = \{\overline{u_i} \mid i \in [n]\} \cup \{p, q\}$) is a dominating set of $G - e$.
(v) If an edge in the clique K is deleted, then $D = \{u_i \mid i \in [n]\} \cup \{p, q\}$ is a dominating set of $G - e$.

In all cases, we obtain a dominating set of $G - e$ of size at most $n + 2$; hence $\gamma(G - e) \leq n + 2$.

Claim 1.4. $\gamma(G) = n + 1$ *if and only if* $b(G) = 1$.

Proof. Suppose first that $\gamma(G) = n + 1$. Consider the graph $G - qb$. In $G - qb$, the vertex b can only be dominated by r, and hence $r \in D$ for any dominating set D of $G - qb$. Moreover, to dominate the vertex a, at least one of the vertices p or q must belong to D. Finally, at least n literal vertices are required to dominate the vertices v_i, for all $i \in [n]$. Therefore, $\gamma(G - qb) \geq n + 2$. On the other hand, $D = \{u_i \mid i \in [n]\} \cup \{p, r\}$ is a dominating set of $G - qb$, and hence $\gamma(G - qb) = n + 2$. Thus, $\gamma(G - qb) > \gamma(G)$, which implies $b(G) = 1$.

Conversely, suppose that $b(G) = 1$. Let $e \in E(G)$ be an arbitrary edge such that $\gamma(G - e) > \gamma(G)$. By Claim 1.3, we have $n + 1 \leq \gamma(G) < \gamma(G - e) \leq n + 2$. Since $b(G) = 1$, it follows that $\gamma(G - e) = n + 2$, and hence $\gamma(G) = n + 1$.

By Claims 1.2 and 1.4, it follows that $b(G) = 1$ if and only if the collection $\mathscr{C}$ is satisfiable. The construction of the bondage instance is straightforward from the given 3-SAT instance, and the size of the resulting graph is bounded above by a polynomial function of the size of the 3-SAT instance. Therefore, the reduction can be carried out in polynomial time, and the NP-hardness of the bondage problem in split graphs follows. □

The BONDAGE problem asks whether $b(G) \leq k$, while its complementary decision problem asks whether $b(G) > k$. We show that, even for split graphs and $k = 1$, deciding whether $b(G) > 1$ is NP-hard.

Theorem 2. *The BONDAGE problem is coNP-hard, even when restricted to split graphs and $k = 1$.*

Proof. We give a polynomial time reduction from 3-SAT. Although the construction is similar to that used in the proof of Theorem 1, we present the construction explicitly. Let $U = \{u_1, \ldots, u_n\}$ and $\mathscr{C} = \{C_1, \ldots, C_m\}$ be an instance of 3-SAT. We construct a split graph G such that $\mathscr{C}$ is satisfiable if and only if $b(G) > 1$.

- For each variable $u_i \in U$, add literal vertices u_i, $\overline{u_i}$, and a vertex v_i adjacent to both literals. For each clause $C_j \in \mathscr{C}$, add a vertex c_j and join it to all vertices corresponding to literals appearing in C_j.
- Introduce three additional vertices p, q and a. The vertex p is adjacent to all clause vertices c_j, $j \in [m]$, and the edges pa, qa are added.
- Finally, make the set $\{u_i, \overline{u_i}, p, q \mid i \in [n]\}$ a clique in G.

It is easy to see that G is a split graph with clique $K = \{u_i, \overline{u_i}, p, q \mid i \in [n]\}$ and independent set $I = \{v_i, c_j, a \mid i \in [n], j \in [m]\}$. See Fig. 2 for an illustration.

Claim 2.1. *For the graph G, $\gamma(G) = n + 1$.*

Proof. Let D be a dominating set of G. For each $i \in [n]$, the vertex v_i is adjacent only to the literal vertices u_i and $\overline{u_i}$; hence, at least one of these vertices must belong to D. Moreover, the vertex a is adjacent only to p and q, and therefore D must contain at least one of these vertices. It follows that $|D| \geq n + 1$, and thus $\gamma(G) \geq n + 1$. On the other hand, the set $D = \{u_i \mid i \in [n]\} \cup \{p\}$ is a dominating set of G. Hence, $\gamma(G) \leq |D| = n + 1$. Therefore, $\gamma(G) = n + 1$.

Claim 2.2. *For the graph G, $b(G) > 1$ if and only if the collection $\mathscr{C}$ is satisfiable.*

Proof. Suppose $\mathscr{C}$ is satisfiable. We consider the deletion of an arbitrary edge $e \in E(G)$.

(i) If e is incident to a vertex v_s for some $s \in [n]$, then $D = \{u_i \mid i \in [n]\} \cup \{p\}$ is a dominating set of $G - e$.
(ii) If $e = qa$, then $D = \{u_i \mid i \in [n]\} \cup \{p\}$ is a dominating set of $G - e$.
(iii) If $e = pa$, then the vertices corresponding to literals evaluating to true under the assignment, together with q, form a dominating set of $G - e$.

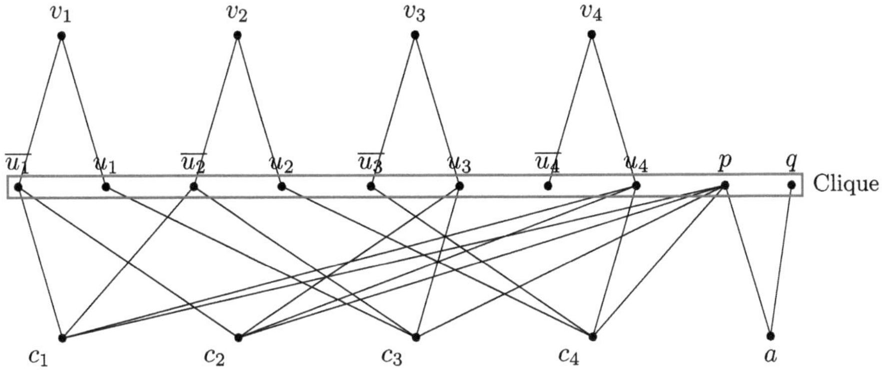

Fig. 2. An illustration of the construction of graph G in the proof of Theorem 2.

(iv) If e is incident to a clause vertex c_j for some $j \in [m]$, we distinguish the following cases: If $e = c_j u_k$ or $e = c_j \overline{u_k}$ for some $k \in [n]$, then $D = \{u_i \mid i \in [n]\} \cup \{p\}$; If $e = c_j p$, then $D = \{u_i \mid i \in [n]\} \cup \{p\}$ if C_j contains a positive literal, and $D = \{\overline{u_i} \mid i \in [n]\} \cup \{p\}$ otherwise.

(v) If e is an edge of the clique K, then $D = \{u_i \mid i \in [n]\} \cup \{p\}$ is a dominating set of $G - e$.

Thus, in all cases, $\gamma(G - e) \le |D| = n + 1$. As $\gamma(G) \le \gamma(G - e)$, it follows from Claim 2.1 that $\gamma(G - e) = n + 1$. Hence, $b(G) > 1$.

On the other hand, suppose that $\mathscr{C}$ is not satisfiable, and let $G' = G - pa$. In G', the vertex a can be dominated only by q; hence, every dominating set D of G' must contain q. Moreover, for each $i \in [n]$, the vertex v_i can be dominated only by u_i or $\overline{u_i}$, and therefore D must contain at least one literal vertex from $\{u_i, \overline{u_i}\}$. Thus, any dominating set of G' must contain at least $n + 1$ vertices, namely q together with one literal vertex for each variable. However, no such set of size $n+1$ can dominate all clause vertices; otherwise, the chosen literals would induce a satisfying truth assignment for $\mathscr{C}$, which is a contradiction. Hence, $\gamma(G') > n + 1$, and therefore $b(G) = 1$.

The graph G can be obtained from an instance of 3-SAT in polynomial time. Hence, by Claim 2.2, the BONDAGE problem is coNP-hard even for split graphs and $k = 1$. $\qquad\square$

4 Hardness of REINFORCEMENT Problem

In this section, we study the complexity of the REINFORCEMENT problem on chordal graphs. We begin by showing that this problem is NP-hard, even when restricted to split graphs and $k = 1$.

Theorem 3. *The REINFORCEMENT problem is NP-hard even when restricted to split graphs and $k = 1$.*

Proof. Let $U = \{u_1, \ldots, u_n\}$ and $\mathscr{C} = \{C_1, \ldots, C_m\}$ be an instance of 3-SAT. We construct a split graph G such that $\mathscr{C}$ is satisfiable if and only if $r(G) = 1$.

- For each $i \in [n]$, add vertices u_i and $\overline{u_i}$, and vertices v_i, w_i, x_i adjacent to both literal vertices.
- For each $j \in [m]$, add a vertex c_j and join it to all vertices corresponding to literals appearing in C_j.
- Add vertices p and a, and make p adjacent to all clause vertices c_j. Add the edge pa.
- Add a vertex q and three vertices b_1, b_2, b_3, and add the edges qb_1, qb_2, and qb_3.
- Finally make the set $K = \{u_i, \overline{u_i}, p, q \mid i \in [n]\}$ a clique in G.

The resulting graph G is a split graph with clique $K = \{u_i, \overline{u_i}, p, q \mid i \in [n]\}$ and independent set $I = \{x_i, w_i, v_i, c_j, a, b_1, b_2, b_3 \mid i \in [n], j \in [m]\}$. Let H_i denote the subgraph induced by the vertices $\{u_i, \overline{u_i}, v_i, w_i, x_i\}$. See Fig. 3 for an illustration.

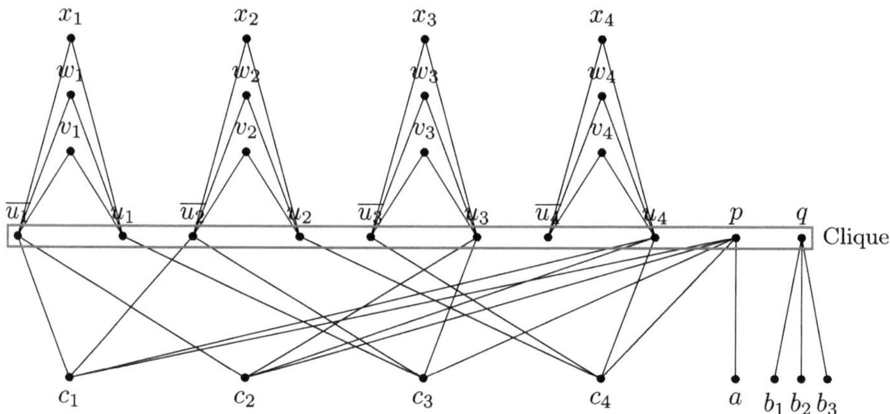

Fig. 3. An illustration of the construction of graph G in the proof of Theorem 3.

Claim 3.1. *For the graph G, $\gamma(G) = n + 2$.*

Proof. It is easy to see that $D = \{u_i \mid i \in [n]\} \cup \{p, q\}$ is a γ-set of G.

Claim 3.2. *For the graph G, $r(G) = 1$ if and only if $\mathscr{C}$ is satisfiable.*

Proof. Assume that $\mathscr{C}$ is satisfiable. Add the non-edge qa. Let D be a set consisting of exactly one literal vertex from each pair $\{u_i, \overline{u_i}\}$, which evaluates to true in a satisfying assignment of $\mathscr{C}$, together with vertex q. Since the assignment satisfies every clause, each clause vertex c_j is adjacent to at least one literal in D. Thus D is a dominating set of the graph $G + qa$ with $|D| = n + 1$. By Claim 3.1, $\gamma(G) = n + 2$. Thus, $\gamma(G + qa) \leq n + 1 < \gamma(G)$ leading to $r(G) = 1$.

On the other hand, assume that $r(G) = 1$. Then there exists a non-edge $e \notin E(G)$ such that $\gamma(G + e) < n + 2$. Let D be a minimum dominating set of $G + e$. Observe that the vertex q is the unique neighbor of vertices b_1, b_2, b_3. Since any vertex other than q can dominate at most two vertices in $\{b_1, b_2, b_3\}$ in $G + e$, at least one of these vertices would remain undominated unless $q \in D$. Hence, $q \in D$. Similarly, for each $i \in [n]$, the vertices v_i, w_i, x_i are adjacent only to u_i and $\overline{u_i}$. Thus, D must contain at least one of u_i or $\overline{u_i}$. Since the addition of a single non-edge can make at most one of v_i, w_i, x_i adjacent to a vertex other than $\{u_i, \overline{u_i}\}$, at least one of these three vertices would remain undominated unless D contains u_i or $\overline{u_i}$. Since $|D| < n + 2$ and D contains at least one vertex from each pair $\{u_i, \overline{u_i}\}$ for $i \in [n]$ as well as vertex q, it follows that $|D| = n + 1$. In particular, $p, a, c_j \notin D$ and one endpoint of non-edge e is the vertex a. Since $p \notin D$, each clause vertex c_j must be dominated by a literal vertex in D. Let us

define a mapping $t : U \to \{T, F\}$ by $t(u_i) = \begin{cases} T, & \text{if } u_i \in D, \\ F, & \text{otherwise.} \end{cases}$

Clearly, t is a satisfying truth assignment for the collection $\mathscr{C}$. Consequently, the collection $\mathscr{C}$ is satisfiable, and the claim follows.

Since the graph G can be obtained from an instance of 3-SAT in polynomial time, by Claim 3.2, the REINFORCEMENT problem is NP-hard even for split graphs and $k = 1$. $\square$

Next, we prove the CoNP-harndess of REINFORCEMENT problem in chordal graphs.

Theorem 4. *The* REINFORCEMENT *problem is coNP-hard even when restricted to split graphs and $k = 1$.*

Proof. Let $U = \{u_1, \ldots, u_n\}$ and $\mathscr{C} = \{C_1, \ldots, C_m\}$ be an instance of 3-SAT. We construct a split graph G such that $\mathscr{C}$ is satisfiable if and only if $r(G) > 1$. We modify the graph constructed in Fig. 3 as follows. The resulting graph $G = (V, E)$ is defined by the following steps.

- For each $i \in [n]$, add vertices u_i and $\overline{u_i}$, and vertices v_i, w_i, x_i adjacent to both u_i and $\overline{u_i}$.
- For each $j \in [m]$, add a vertex c_j and make it adjacent to all vertices corresponding to literals appearing in clause C_j.
- Add vertices p, q and vertices b_1, b_2, b_3 and make each of b_1, b_2, b_3 adjacent to both p and q. Add a vertex a and add the edge aq.
- Make vertex p adjacent to every clause vertex c_j, for all $j \in [m]$.
- Finally, the set $K = \{u_i, \overline{u_i} \mid i \in [n]\} \cup \{p, q\}$ induces a clique in G.

The resulting graph G is a split graph with clique $K = \{u_i, \overline{u_i} \mid i \in [n]\} \cup \{p, q\}$, and independent set $I = \{v_i, w_i, x_i, c_j, a, b_1, b_2, b_3 \mid i \in [n], j \in [m]\}$. Let H_i denote the subgraph induced by $\{u_i, \overline{u_i}, v_i, w_i, x_i\}$. See Fig. 4 for an illustration.

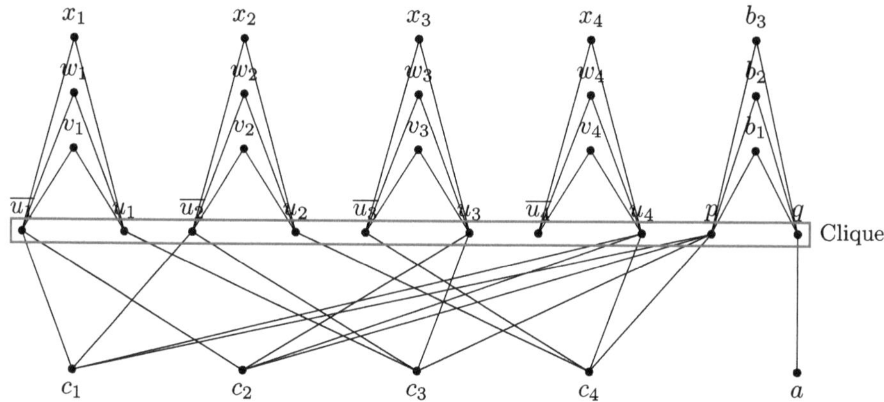

Fig. 4. An illustration of the construction of graph G in the proof of Theorem 4.

Claim 4.1. *For the graph G, $n + 1 \leq \gamma(G) \leq n + 2$.*

Proof. For each $i \in [n]$, the vertices v_i, w_i, x_i are adjacent only to u_i and $\overline{u_i}$. Hence, any dominating set of G must contain at least one vertex from each pair $\{u_i, \overline{u_i}\}$. Moreover, the vertex a is adjacent only to the vertex q. So, q must belong to any dominating set of G. Thus, $\gamma(G) \geq n + 1$. On the other hand, selecting exactly one vertex from each pair $\{u_i, \overline{u_i}\}$, together with both vertices p and q, forms a dominating set of size $n + 2$. Therefore, $\gamma(G) \leq n + 2$.

Claim 4.2. *For the graph G, $\gamma(G) = n + 1$ if and only if the collection $\mathscr{C}$ is satisfiable.*

Proof. Assume that $\gamma(G) = n + 1$, and let D be a minimum dominating set of G. By the previous claim, every dominating set of G has size at least $n + 1$, and hence $|D| = n + 1$. For each $i \in [n]$, the vertices v_i, w_i, x_i are adjacent only to u_i and $\overline{u_i}$, implying that D contains at least one vertex from each pair $\{u_i, \overline{u_i}\}$. Moreover, vertex a is adjacent only to q, and therefore $q \in D$. Since $|D| = n + 1$, it follows that $p \notin D$, and consequently every clause vertex c_j must be dominated by a literal vertex in D. Define a truth assignment $t : U \to \{T, F\}$ by setting $t(u_i) = T$ if $u_i \in D$, and $t(u_i) = F$ otherwise. Then each clause vertex c_j is adjacent to at least one literal vertex in D, implying that each clause C_j is satisfied under t. Hence, $\mathscr{C}$ is satisfiable.

Conversely, assume that $\mathscr{C}$ is satisfiable, and let $t : U \to \{T, F\}$ be a satisfying truth assignment. Let $D = \{q\} \cup \{u_i \mid t(u_i) = T\} \cup \{\overline{u_i} \mid t(u_i) = F\}$. Clearly, $|D| = n + 1$. For each $i \in [n]$, the vertices v_i, w_i, x_i are dominated by the corresponding literal vertex in D, the vertices b_1, b_2, b_3 and a are dominated by q, and each clause vertex c_j is dominated by a literal vertex since t satisfies $\mathscr{C}$. Thus, D is a dominating set of G, implying that $\gamma(G) \leq n + 1$. Together with the bound $\gamma(G) \geq n + 1$, we conclude that $\gamma(G) = n + 1$.

Claim 4.3. *The collection $\mathscr{C}$ is satisfiable if and only if $r(G) > 1$.*

Proof. Assume first that $\mathscr{C}$ is satisfiable. By Claim 4.2, we have $\gamma(G) = n + 1$. Let $e \notin E(G)$ be any non-edge and consider the graph $G + e$. For each $i \in [n]$, the vertices v_i, w_i, x_i are adjacent only to u_i and $\overline{u_i}$. Since the addition of a single non-edge can make at most two of v_i, w_i, x_i dominated by a vertex outside $\{u_i, \overline{u_i}\}$, any dominating set of $G + e$ must contain at least one vertex from each pair $\{u_i, \overline{u_i}\}$. Moreover, the vertices b_1, b_2, b_3 are adjacent only to p and q, and hence any dominating set of $G + e$ must contain at least one of p or q. Consequently, $\gamma(G + e) \geq n + 1$. Since $\gamma(G) = n + 1$, it follows that $\gamma(G + e) = \gamma(G)$ for every $e \notin E(G)$, and therefore $r(G) > 1$.

Conversely, assume that $\mathscr{C}$ is not satisfiable. Again by Claim 4.2, we have $\gamma(G) = n + 2$. In this case, adding the non-edge pq, we get $\gamma(G + pq) = n + 1$ leading to $r(G) = 1$. Therefore, $\mathscr{C}$ is satisfiable if and only if $r(G) > 1$.

Since the construction of graph G can done in polynomial-time, the CoNP-hardness of the REINFORCEMENT problem restricted to split graphs and $k = 1$ follows from Claim 4.3.   □

5   Consequences of Hardness Results

In this section, we discuss two consequences of the hardness results from the perspectives of classical and approximation, and thereby settling a conjecture and partially answering a question due to Hu and Xu [17]. First, we obtain the following consequence proving the non-membership of BONDAGE andREINFORCEMENT in NP.

Proposition 1. *Unless $NP = coNP$, the BONDAGE and REINFORCEMENT problems do not belong to NP even when restricted to split graphs and $k = 1$.*

Proof. By Theorems 1 and 2, BONDAGE problem is NP-hard and coNP-hard on split graphs for $k = 1$. By Theorems 3 and 4, REINFORCEMENT problem is NP-hard and coNP-hard on split graphs for $k = 1$. If either problem belonged to NP, then its complement would also belong to NP, implying NP = coNP (see, for instance, [1]). Therefore, unless NP = coNP, neither problem admits a polynomial-time verifiable certificate.   □

Hu and Xu [17] conjectured that BONDAGE and REINFORCEMENT problems do not belong to NP. The above proposition settles this conjecture affirmatively.

We now turn our attention to the approximability of these two problems. Recall that a polynomial-time approximation algorithm with performance ratio ρ returns, for every instance G, a feasible solution of value $\mathcal{A}(G)$ satisfying $\mathrm{OPT}(G) \leq \mathcal{A}(G) \leq \rho \cdot \mathrm{OPT}(G)$ (see, for instance, [20]). The next proposition shows that the optimization versions of both problems do not admit a $(2 - \epsilon)$-approximation algorithm for any $\epsilon > 0$.

Proposition 2. *Unless $P = NP$, MINIMUM BONDAGE and MINIMUM REINFORCEMENT problems do not admit a polynomial-time approximation algorithm with performance ratio smaller than 2, even on split graphs.*

Proof. Suppose that there exists a polynomial-time approximation algorithm $\mathcal{A}$ for MINIMUM BONDAGE with ratio $\rho < 2$. Then, for every split graph G, $\mathcal{A}(G)$ satisfies $b(G) \le \mathcal{A}(G) \le \rho \cdot b(G)$. If $b(G) = 1$, then $\mathcal{A}(G) = 1$, while if $b(G) \ge 2$, then $\mathcal{A}(G) \ge 2$. Hence, $\mathcal{A}$ decides whether $b(G) = 1$ in polynomial time, a contradiction unless $P = NP$. The same gap argument applies to the MINIMUM REINFORCEMENT problem. $\qquad\square$

6   Parameterized Complexity of BONDAGE and REINFORCEMENT Problems

In this section, we consider the parameterized versions of the BONDAGE and REINFORCEMENT problems, where the parameter is the solution size k. Recall that a parameterized problem is said to be fixed-parameter tractable (FPT) if it can be solved in time $f(k) \cdot n^{O(1)}$, and belongs to the class XP if it can be solved in time $n^{f(k)}$, for some computable function f (see, for instance, [6]).

Proposition 3. *Unless $P = NP$, the BONDAGE and the REINFORCEMENT problems do not admit fixed-parameter tractable algorithms, or even XP algorithms, parameterized by the solution size, even when restricted to split graphs.*

Proof. Since problems are NP-hard on split graphs already for the fixed constant value $k = 1$, if either problem admitted an XP (resp. FPT) algorithm running in time $n^{f(k)}$ (resp. $f(k) \cdot n^{O(1)}$), then for the fixed value $k = 1$ this would give a polynomial-time algorithm. $\qquad\square$

We now turn to the positive side. For the case $k = 1$, we show that both the BONDAGE and REINFORCEMENT problems are fixed-parameter tractable parameterized by clique-width via a brute-force approach. It is well-known fact that the DOMINATING SET problem is fixed-parameter tractable parameterized by clique-width [5] (see also [6]). A crucial property for edge-modification problems is that adding or deleting a single edge changes the clique-width by at most two. Recall that, for any graph G and any edge e, the clique-width $\mathrm{cw}(G)$ satisfies the inequality $\mathrm{cw}(G) - 2 \le \mathrm{cw}(G \pm e) \le \mathrm{cw}(G) + 2$; (see Theorem 9 [13]).

Proposition 4. *Deciding whether $b(G) = 1$ (resp. $r(G) = 1$) is fixed-parameter tractable when parameterized by the clique-width of G.*

Proof. First compute $\gamma(G)$, and then for each edge $e \in E(G)$ compute $\gamma(G - e)$ (resp. for each non-edge $e \notin E(G)$ compute $\gamma(G+e)$). Since $\mathrm{cw}(G-e) \le \mathrm{cw}(G)+2$ (resp. $\mathrm{cw}(G+e) \le \mathrm{cw}(G)+2$), each such computation can be performed in time $O\big(f(\mathrm{cw}(G) + 2) \cdot n^{O(1)}\big)$, (see [6]). As there are $O(n^2)$ edges (resp. non-edges), this gives an fixed-parameter tractable algorithm with respect to $\mathrm{cw}(G)$. $\qquad\square$

The above proposition extends for any fixed integer $k \geq 1$ (not part of the input) as follows.

Proposition 5. *For each fixed integer k, deciding whether $b(G) \leq k$ (resp. $r(G) \leq k$) is fixed-parameter tractable parameterized by the clique-width of G.*

Proof. Let k be a fixed positive integer. By brute-force approach, we enumerate by all subsets $F \subseteq E(G)$ (resp. $F \subseteq \overline{E(G)}$) with $|F| = k$ and check whether $\gamma(G - F) > \gamma(G)$ (resp. $\gamma(G + F) < \gamma(G)$). Since $\mathrm{cw}(G - F) \leq \mathrm{cw}(G) + 2k$ (resp. $\mathrm{cw}(G + F) \leq \mathrm{cw}(G) + 2k$), each such check can be performed in time $O\big(f(\mathrm{cw}(G) + 2k) \cdot n^{O(1)}\big)$. The resulting running time is $O\big(|E(G)|^{2k} \cdot f(\mathrm{cw}(G) + 2k) \cdot n^{O(1)}\big) = O\big(f(\mathrm{cw}(G)) \cdot n^{O(1)}\big)$ for the constant value of k and thus, the proposition follows. $\qquad\square$

Remark 1. In the above result, if k is part of the input (rather than a fixed constant), the brute-force enumeration of k-element subsets is no longer polynomial in n. Nevertheless, for the combined parameter $\kappa = k + \mathrm{cw}(G)$, the same enumeration gives an XP algorithm. Whether the BONDAGE or REINFORCEMENT problems admit a fixed-parameter tractable algorithm for this combined parameter remains an interesting open problem.

Remark 2. Another natural direction is to consider the problems parameterized by the domination number $\gamma(G)$. We believe that BONDAGE and REINFORCEMENT are W[1]-hard parameterized by $\gamma(G)$.

Acknowledgements. The authors gratefully acknowledge the anonymous reviewers for their careful reading and constructive feedback. This work was done while the first author was visiting the Institute of Mathematical Sciences (IMSc), Chennai (January–July 2025). The first author thanks Professor Sushmita Gupta for the opportunity and her mentorship.

References

1. Arora, S., Barak, B.: Computational Complexity: A Modern Approach. Cambridge University Press (2009)
2. Bakal, D.M., Borse, Y.M., Waphare, B.N.: Algorithmic complexity of three domination subdivision number problems in graphs. Commun. Comb. Optim. (2025)
3. Bauer, D., Harary, F., Nieminon, J., Suffel, C.L.: Domination alteration sets in graphs. Discrete Math. **47**(2–3), 153–161 (1983)
4. Bouquet, V.: The complexity of the bondage problem in planar graphs. arXiv preprint arXiv:2107.11216 (2021)
5. Courcelle, B., Makowsky, J.A., Rotics, U.: Linear time solvable optimization problems on graphs of bounded clique-width. Theory Comput. Syst. **33**(2), 125–150 (2000)
6. Cygan, M., et al.: Parameterized Algorithms. Springer, Cham (2015)
7. Desormeaux, W.J., Haynes, T.W., Henning, M.A.: Total domination critical and stable graphs upon edge removal. Discrete Appl. Math. **158**(15), 1587–1592 (2010)

8. Desormeaux, W.J., Haynes, T.W., Henning, M.A.: Total domination stable graphs upon edge addition. Discrete Math. **310**(24), 3446–3454 (2010)
9. Dettlaff, M., Raczek, J., Topp, J.: Domination subdivision and domination multisubdivision numbers of graphs. Discuss. Math. Graph Theory **39**(4), 829–839 (2019)
10. Fink, J.F., Jacobson, M.S., Kinch, L.F., Roberts, J.: The bondage number of a graph. Discrete Math. **86**(1–3), 47–57 (1990)
11. Galby, E., Lima, P.T., Ries, B.: Reducing the domination number of graphs via edge contractions and vertex deletions. Discrete Math. **344**(1), 112169 (2021)
12. Garey, M.R., Johnson, D.S.: Computers and Intractability: A Guide to the Theory of NP-Completeness. W. H. Freeman and Co., San Francisco (1979)
13. Gurski, F.: The behavior of clique-width under graph operations and graph transformations. Theory Comput. Syst. **60**(2), 346–376 (2017)
14. Haynes, T.W., Hedetniemi, S.M., Hedetniemi, S.T.: Domination and independence subdivision numbers of graphs. Discuss. Math. Graph Theory **20**(2), 271–280 (2000)
15. Haynes, T.W., Hedetniemi, S.T., Henning, M.A.: Domination in Graphs: Core Concepts. Springer Monographs in Mathematics, Springer, Cham (2023)
16. Hu, F.T., Sohn, M.Y.: The algorithmic complexity of bondage and reinforcement problems in bipartite graphs. Theoret. Comput. Sci. **535**, 46–53 (2014)
17. Hu, F.T., Xu, J.M.: On the complexity of the bondage and reinforcement problems. J. Complexity **28**(2), 192–201 (2012)
18. Huang, J., Xu, J.M.: Domination and total domination contraction numbers of graphs. Ars Combin. **94**, 431–443 (2010)
19. Kok, J., Mynhardt, C.M.: Reinforcement in graphs. Congr. Numer. **79**, 225–231 (1990)
20. Williamson, D.P., Shmoys, D.B.: The Design of Approximation Algorithms. Cambridge University Press (2011)
21. Blair, J.R.S., Goddard, W., Hedetniemi, S.T., Horton, S., Jones, P., Kubicki, G.: On domination and reinforcement numbers in trees. Discrete Math. **308**(7), 1165–1175 (2008)
22. Carlson, K., Develin, M.: On the bondage number of planar and directed graphs. Discrete Math. **306**(8–9), 820–826 (2006)
23. Kang, L., Sohn, M.Y., Kim, H.K.: Bondage number of the discrete torus $C_n \times C_4$. Discrete Math. **303**(1–3), 80–86 (2005)
24. Zhao, W., Wang, F., Gao, X., Li, H.: Bondage number of the strong product of two trees. Discrete Appl. Math. **230**, 133–145 (2017)
25. Domke, G.S., Laskar, R.C.: The bondage and reinforcement numbers of γ_f for some graphs. Discrete Math. **167**, 249–259 (1997)
26. Bouquet, V.: The bondage number of chordal graphs. arXiv preprint arXiv:2203.09256 (2022)

Degree Realization with Maximum Matching

Amotz Bar-Noy[1], Igor Kalinichev[2(✉)], David Peleg[2], and Dror Rawitz[3]

[1] City University of New York (CUNY), New York, USA
amotz@sci.brooklyn.cuny.edu
[2] Weizmann Institute of Science, Rehovot, Israel
{igor.kalinichev,david.peleg}@weizmann.ac.il
[3] Bar Ilan University, Ramat-Gan, Israel
dror.rawitz@biu.ac.il

Abstract. The MAXIMUM MATCHING DEGREE REALIZATION problem requires, for a sequence d of n positive integers, to compute a realizing graph (whose degree sequence corresponds to d) with the maximum possible matching. We present a polynomial-time algorithm for this problem.

Keywords: Degree sequences · Degree realization · Graph algorithms · Maximum matching

1 Introduction

The DEGREE REALIZATION (DR) problem requires to decide, for a given non-increasing sequence $d = (d_1, \ldots, d_n)$ of positive integers, if there exists some n-vertex (simple undirected) graph $G = (V, E)$ such that $\deg_G(i) = d_i$ for every $i \in \{1, \ldots, n\}$, and to construct such a graph if it exists. Such a sequence d is called *graphic*.

Two central questions on realizations concern characterizations (i.e., necessary and sufficient conditions) for a sequence to be graphic, and efficient algorithms for finding realizing graphs. A necessary and sufficient condition for a given sequence of integers to be graphic (also implying an $O(n)$ decision algorithm) was presented by Erdős and Gallai [7]. Havel [16] and Hakimi [15] described an $O(\sum_i d_i)$-time algorithm that given a sequence d of integers proves that the given sequence is not graphic or computes an m-edge graph realizing it, where $m = \frac{1}{2}\sum_{i=1}^{n} d_i$.

Here, we study the variant of the problem where, given a sequence with possibly many realizing graphs, the goal is to find a realizing graph that also optimizes some quality measure. Hereafter, we refer to such problems as *optimized realization* problems. Concretely, we consider the classical MAXIMUM MATCHING (MM) problem. For a graph $G = (V, E)$, a matching is an edge set $M \subseteq E$, such

---

Supported by US-Israel BSF grant 2022205.

F. Foucaud and A. Parreau (Eds.): IWOCA 2026, LNCS 16587, pp. 103–117, 2026.
https://doi.org/10.1007/978-3-032-27732-9_8

that e and e' do not share a vertex, for every two distinct $e, e' \in M$. The size of the maximum matching in G is denoted by

$$\mathrm{MM}(G) = \max\{|M| \mid M \text{ is a matching in } G\} \,.$$

Given a graphic sequence d, we look for a realizing graph G with the maximum possible matching. For example, for $d = (2, 2, 2, 2, 2, 2)$, a realization G_1 consisting of a 6-vertex cycle is better that a realization G_2 consisting of two triangles, since $\mathrm{MM}(G_1) = 3$ whereas $\mathrm{MM}(G_2) = 2$. The resulting optimized realization problem is formally defined as follows. Letting

$$\mathrm{MM}(d) = \max\{\mathrm{MM}(G) \mid G \text{ is a realization of } d\} \,,$$

the MM DEGREE REALIZATION (MM-DR) problem requires finding a realizing graph G attaining $\mathrm{MM}(d)$.

Similarly, one may define optimized realization problems for other graph optimization problems, such as MAXIMUM CLIQUE, MAXIMUM INDEPENDENT SET, MINIMUM VERTEX COVER, and MINIMUM DOMINATING SET. Their optimal values on a given graph G are defined as $\mathrm{MC}(G)$, $\mathrm{MIS}(G)$, $\mathrm{MVC}(G)$, and $\mathrm{MDS}(G)$, respectively. The corresponding optimized realization problems on degree sequences are named MAXIMUM CLIQUE DEGREE REALIZATION (MC-DR), MIS DEGREE REALIZATION (MIS-DR), MVC DEGREE REALIZATION (MVC-DR), and MDS DEGREE REALIZATION (MDS-DR), and their optimal values on a given d are denoted by $\mathrm{MC}(d)$, $\mathrm{MIS}(d)$, $\mathrm{MVC}(d)$, and $\mathrm{MDS}(d)$, respectively.

Interestingly, while the graph versions of most of these optimization problems are NP-hard, the optimized realization versions of MC, MIS, MVC, and MDS have been shown to be polynomial-time solvable. Rao [23] gave a characterization of sequences realizable by a graph containing a clique K_ℓ of size ℓ. This result was based on a phenomenon known as the *prefix lemma*, which for MC says that if d has a realization containing K_ℓ, then it has a realization such that the ℓ vertices of maximum degree induce a clique K_ℓ. This result implies a polynomial-time algorithm for MC-DR (see, e.g., [13]). Consequently, it follows readily that polynomial-time algorithms also exist for MIS-DR and MVC-DR. Note that a prefix lemma also applies to MVC. This prefix property is satisfied by MDS [13], and it was recently used to obtain a characterization and a realization algorithm for MDS-DR [3].

As for MM-DR, the prefix property is satisfied also by MM [14]. However, so far it was not known how to exploit this property in order to derive a polynomial-time algorithm for MM-DR. This problem was handled for some restricted graph families in [5], but the general problem was left open.

Recently, Erdös et al. [9] studied MM-DR. Given a sequence d and an integer ν, they presented an Erdös-Gallai type characterization, based on a system of $O(n)$ inequalities, which is satisfied if and only if d has a realization with a matching of size ν. Applying these characterizations to a given degree sequence, the maximum size of a matching can be computed efficiently, although a specific realization is not provided.

1.1   Our Results

Here, we present a polynomial-time realization algorithm for MM-DR. Our algorithm makes use of a stronger version of an existing prefix lemma for general graphs [14], termed the *inverted prefix lemma,* established in the current paper. This lemma is analogous to the inverted prefix lemma established in [5] for *bipartite* realizations of two sequences (or a bi-sequence). We also utilize an algorithm due to Fulkerson, Hoffman and McAndrew [10] for a special case of the f-factor problem (discussed later), which is based on the following three phases:

(1) Realization of the bi-sequence (d, d) by a *bipartite* graph $\hat{G}$,
(2) Conversion of $\hat{G}$ to a half-integral general (non-bipartite) realization G^ω for d, where in a half-integral realization an edge with weight half contributes half to the degree of its end-points.
(3) Rounding G^ω to an integral general realization G.

1.2   Related Work

Several papers studied network realization with a given subgraph. Kundu [19] showed that the sequences d and the component-wise difference $d - d'$, such that $d'_i \in \{k, k+1\}$ and $d' \le d$ (component-wise), are graphic only if there exists a graph G realizing d that has a subgraph G' realizing d'. This result can be used to decide whether a given sequence d has a perfect matching by assigning $d'_i = 1$, for every i. Kleitman and Wang [18] gave an algorithm for computing a realization of d that contains a subgraph which realizes d'. Their algorithm can be used to compute a realization of d that contains a perfect matching, if such a realization exists. Extensions of the above result were presented in [18,21].

Rao and Rao [24] and Kundu [19] gave a characterization of sequences that can be realized by a Hamiltonian graph. Chungphaisan [6] gave an algorithm, that given a sequence d, constructs a realization with a Hamiltonian cycle (or path), if such a realization exists. Rao [23] characterized sequences that can be realized by a graph containing K_ℓ (a clique of size ℓ). This result was based on the Prefix Lemma for MC. An alternative proof was given in [17], and a constructive proof was presented in [26]. It follows that MC-DR can be solved in polynomial time by checking if d has a realization containing K_ℓ, for $\ell \in \{2, \ldots, n\}$.

Gould, Jackson and Lehel [14] extended Rao's result by showing that if d has a realization containing H as a subgraph (but not necessarily an induced subgraph), then there exists a realization of d containing H where the vertices of H have the $|V(H)|$ largest degrees. Yin [25] used this to extend Rao's result [23] by giving a characterization of graphic sequences that contains a split graph $S_{r,s}$ composed of a clique is of size r and an independent set of size s. (Observe that $S_{r,1}$ is a clique of size $r + 1$.)

Gentner, Henning and Rautenbach [13] studied MIS-DR and MDS-DR. They proved the prefix lemma for MDS-DR and gave a realization algorithm for MDS-DR on sequences with $d_1 = O(1)$, leaving the general case open. They also provided characterizations for MIS-DR and MDS-DR in forests.

Gentner, Henning and Rautenbach [12] gave characterizations to realizations that minimize the maximum size of an independent set and that maximize the minimum size of a dominating set in forests. Recently, Bar-Not et al. [3] gave a characterization and a realization algorithm for MDS-DR in general graphs.

Bock and Rautenbach [5] studied MM-DR in trees and bipartite graphs, where the partition is given. Their result on bipartite graphs is also based on a stronger version of the prefix lemma, which is essentially an *inverted prefix lemma for bipartite graphs.*

Fulkerson, Hoffman and McAndrew [10] provided a characterization for the realizability of a sequence d with respect to multi-graphs that satisfy the *odd cycle condition* (discussed later). Building on the result of [10], Kundu [20] presented conditions for the factorization of graphs that satisfy the odd cycle condition. Anstee relied on the technique of [10] to obtain an algorithmic proof of the f-factor theorem [1] and a simplified characterization for the existence of a (g, f)-factor for special cases [2]. The result of Yin [25], mentioned above, was also based on [10]. The results for MDS-DR [3] are also based on this technique.

2   Inverted Prefix Lemma

In this section, we establish the inverted prefix lemma. Recall that Gould, Jacobson and Lehel [14] gave a general prefix lemma that applies to any subgraph. In the following lemma we provide the special case of a matching.

Lemma 1 (Arbitrary Prefix Lemma for [14]). *If a sequence d has a realization with a matching of size ν, then d has a realization with a matching of the same size such that 2ν vertices participating in the matching have the highest 2ν degrees in d, i.e., $d_1, d_2, \ldots, d_{2\nu}$.*

For our purposes, however, we need a stronger type of prefix lemma, similar to the one used in [5] for solving MM-DR over bipartite graphs. Towards proving this stronger claim, we introduce some notation. Consider a realization $G = (V, E)$ of a degree sequence d, with four vertices $x, y, u, v \in V$ such that $(x, u), (y, v) \in E$ and $(x, y), (u, v) \notin E$. Then the *FLIP operation* transforms G into a graph $G' = (V, E')$ by replacing the former two edges with the latter two, i.e., setting

$$E' = E \cup \{(x, y), (u, v)\} \setminus \{(x, u), (y, v)\} \, .$$

This operation preserves the degrees of individual vertices, but could lead to non-isomorphic realization. It was widely studied and used in different contexts, see [4,8,11]. Finally, define $[i, j] = \{i, \ldots, j\}$, for $i \leq j$.

Lemma 2. *Let $G = (V, E)$ be a realization of d with $V = \{v_1, \ldots, v_n\}$, such that $\deg_G(v_i) = d_i$ for every $i \in [1, n]$. If there is a matching $M \subseteq E$ covering the first 2ν vertices $v_1, \ldots, v_{2\nu}$, then there is a realization $G' = (V, E')$ of d with the "inverted" matching*

$$M^{inv} = \{(v_i, v_{2\nu-i+1}) \mid i \in [1, \nu]\} \, ,$$

such that $M^{inv} \subseteq E'$.

Proof. Call the pair (G, M) *valid* if $G = (V, E)$ is defined on $V = \{v_1, \ldots, v_n\}$ with $\deg(v_i) = d_i$, $M \subseteq E$, and M is on $\{v_1, v_2, \ldots, v_{2\nu}\}$. Consider a valid pair (G, M) and let

$$I(M) = \{i \mid (v_i, v_{2\nu-i+1}) \notin M, \ i \in [1, \nu]\}$$

and $f(M) = \min I(M)$. If $I(M) = \emptyset$, then $M = M^{inv}$, and we are done.

We describe procedure **Invert** that given a valid pair (G, M) violating the claim produces a new valid pair (G', M'), such that either (G', M') satisfies the claim or $f(M') > f(M)$. Repeatedly applying **Invert** gradually modifies (G, M) while preserving validity and strictly increasing $f(M)$. Since f is bounded above by ν, the process terminates in at most ν steps and the obtained pair (G', M') satisfies the claim.

Given a pair (G, M) not satisfying the claim, let $i = f(M)$ and $j = 2\nu - i + 1$. Since $(v_i, v_j) \notin M$, there are edges $(v_i, v_x), (v_y, v_j) \in M$, and the involved vertices satisfy $\deg_G(v_i) \geq \deg_G(v_y)$ and $\deg_G(v_j) \leq \deg_G(v_x)$ by the definition of f. Let

$$M' = M \cup \{(v_i, v_j), (v_x, v_y)\} \setminus \{(v_i, v_x), (v_y, v_j)\} \ .$$

If there is a realization G' such that $M' \subseteq G'$, then the pair (G', M') is valid and either satisfies the claim or increases f. Proceed according to one of the following cases to obtain G'. See Fig. 1.

Case 1: $(v_i, v_j), (v_x, v_y) \in E(G)$.
  Then simply take $G' = G$.
Case 2: $(v_i, v_j), (v_x, v_y) \notin E(G)$.
  Then perform a FLIP operation on G replacing $(v_i, v_x), (v_y, v_j)$ with $(v_i, v_j), (v_x, v_y)$. G' obtained in this way satisfies $M' \subseteq E(G')$.
Case 3: $(v_i, v_j) \in E(G)$ and $(v_x, v_y) \notin E(G)$.
  Then there is a vertex $v_z \in V(G) \setminus \{v_i, v_j, v_x, v_y\}$, such that $(v_x, v_z) \in E(G)$ and $(v_j, v_z) \notin E(G)$. Indeed, $\deg_G(v_x) \geq \deg_G(v_j)$ and v_y is connected to v_j, but not to v_x, so there exists v_z compensating for this. Obtain G' by doing a FLIP replacing $(v_x, v_z), (v_j, v_y)$ with $(v_x, v_y), (v_j, v_z)$.
Case 4: $(v_i, v_j) \notin E(G)$ and $(v_x, v_y) \in E(G)$.
  Then there is a vertex $v_z \in V(G) \setminus \{v_i, v_j, v_x, v_y\}$, such that $(v_i, v_z) \in E(G)$ and $(v_y, v_z) \notin E(G)$. Similarly to the previous case, this is because $\deg_G(v_i) \geq \deg_G(v_y)$. Obtain G' by doing a FLIP replacing $(v_i, v_z), (v_j, v_y)$ with $(v_i, v_j), (v_y, v_z)$.

The lemma follows. $\qquad\square$

Combining Lemma 1 and Lemma 2 gives a more specific Prefix Lemma for MAXIMUM MATCHING.

Lemma 3 (Inverted Prefix Lemma for MM). *If a sequence d has a realization $G = (V, E)$ where $V = \{v_1, \ldots, v_n\}$ such that $\deg_G(v_i) = d_i$ for every $i \in [1, n]$, with a matching M of size ν, then d has a realization $G' = (V, E')$ with the matching $M' = \{(v_i, v_{2\nu-i+1}) \mid i \in [1, \nu]\}$.*

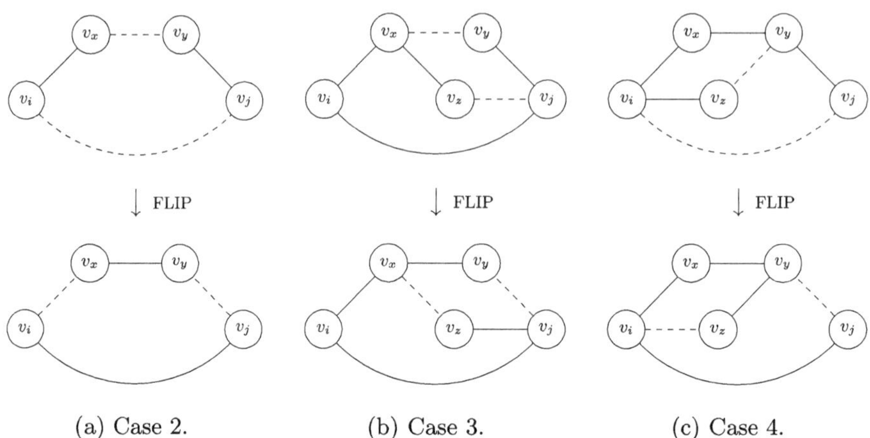

(a) Case 2.   (b) Case 3.   (c) Case 4.

Fig. 1. Illustration of the cases in the proof of Lemma 2. All the illustrating figures maintain the convention that solid lines represent edges existing in the graph and dashed lines represent edges not in the graph.

3   Algorithm for Realization with Maximum Matching

Fulkerson, Hoffman and McAndrew [10] presented an algorithm for a special case of the f-factor problem. More specifically, they considered the following degree realization problem. Let d be a sequence, and let $c \in (\mathbb{Z}_+)^{n \times n}$ be a symmetric integral matrix. Intuitively, one can think of c as representing a multigraph K_c over $V = \{1, \ldots, n\}$ with $c_{i,j}$ parallel edges between i and j for every $1 \le i, j \le n$. Fulkerson et al. [10] studied the problem of finding a multigraph $H = (V, E)$, such that $\deg_H(i) = d_i$ and the number of parallel edges between i and j is bounded above by $c_{i,j}$. Intuitively, the sought realizing multigraph H needs to be a sub-multigraph, or factor, of K_c. We refer to such a realization as a *realization of d with respect to c*.

Let $G_c = (V, E_c)$ be a graph derived from c, where

$$E_c = \{(i, j) \mid i, j \in V \text{ and } c_{i,j} > 0\} \ .$$

That is, G_c is obtained from K_c by replacing all multiedges by edges.

The matrix c is said to satisfy the *odd-cycle* condition if and only if the distance between any two disjoint odd cycles in the derived graph G_c is at most one. In other words, any two disjoint odd cycles in G_c must be connected by an edge.

Fulkerson et al. [10] presented a characterization for the realizability of a sequence d with respect to c that satisfies the odd-cycle condition. Their proof is constructive in the sense that it implies a realization algorithm for instances, where c satisfies the odd-cycle condition. The details of this algorithm for the specific context of realization with maximum matching are given in Sect. 4.

Theorem 1 ([10]). *Let d be a sequence of size n, and let $c \in (\mathbb{Z}_+)^{n \times n}$ be a symmetric integral matrix. There exists a polynomial-time algorithm that computes a realization of d with respect to c, assuming that c satisfies the odd-cycle condition.*

We show that we can use Theorem 1 to obtain a realization algorithm for MM-DR with a matching of size $\nu = \mathrm{MM}(d)$.

Given a sequence d, our first step is to compute $\mathrm{MM}(d)$ using the Erdős-Gallai type characterization of Erdős et al. [9].

Next, Based on the Inverted Prefix Lemma (Lemma 3), we construct an instance d' and c' as follows. First,

$$d'_i = \begin{cases} d_i - 1 & i \le 2\nu, \\ d_i & i > 2\nu. \end{cases}$$

Also, let

$$c'_{i,j} = \begin{cases} 0 & i = j, \\ 0 & i + j = 2\nu + 1 \\ 1 & \text{otherwise.} \end{cases}$$

Observation 4. *There is a realization of d with a matching of size ν if and only if there is a realization of d' with respect to c'.*

Observation 5 (Odd-cycle condition). *c' satisfies the odd-cycle condition. That is, for any pair of disjoint odd cycles C_1, C_2 in $G_{c'}$, there exist $i \in C_1$ and $j \in C_2$, such that $(i, j) \in E_{c'}$.*

Proof. Let $i \in C_1$, $j' \in C_2$ be arbitrary vertices. If $i + j' = 2\nu + 1$, then let $j \ne j'$ be any other vertex in C_2. Otherwise, let $j = j'$. $\qquad\square$

This implies the following.

Theorem 2. *There exists a polynomial-time algorithm for constructing a realization of a given graphic sequence d that also has a matching of maximum size (among all possible realizations of d).*

4   Detailed Realization Algorithm

In this section we present the full details of the realization algorithm that was mentioned in the previous section. The algorithm proceeds in several steps, presented in the coming subsections. First, the algorithm computes $\nu = \mathrm{MM}(d)$ using the Erdős-Gallai type characterization of Erdős et al. [9]. Next, the MM-DR problem over general graph is reduced to the same problem over bipartite graphs, by applying the realization algorithm of Fulkerson et al. [10] while ensuring preservation of a matching. Then, MM-DR over bipartite graphs is reduced to a maximum flow problem. To be more specific, given the size ν of the maximum matching and a degree sequence pair (d, d), we construct a bipartite flow graph $G_{d,\nu}$, such that if the maximum flow attained in $G_{d,\nu}$ equals $\sum_{i=1}^{n} d_i - 2\nu$, then it corresponds to a realization $\hat{G}$ of (d, d) with a matching of size 2ν. (This is where we make use of the Inverted Prefix Lemma 3). The next step is to convert $\hat{G}$ to a half integral general (non-bipartite) realization G^{ω} for d. Finally, G^{ω} is rounded to an integral general realization G.

4.1 Computation of $\mathbf{MM}(d)$

The characterization of Erdös et al. [9] contains $O(n)$ inequalities. Given a candidate value, each inequality can be checked in $O(n)$ time. To compute $\mathrm{MM}(d)$, we can conduct a binary search for the maximum value ν which satisfies the $O(n)$ inequalities in the above mentioned characterization. Hence, a naive implementation would result in a running time of $O(n^2 \log n)$.

4.2 Reduction to a Bipartite Sequence Pair

Given a sequence d, a bipartite graph $\hat{G} = (V, W, \hat{E})$, where $V = \{v_1, \ldots, v_n\}$ and $W = \{w_1, \ldots, w_n\}$, is a ν-*matched* realization for the bi-sequence, (d, d) if it satisfies the properties:

(M1) $\hat{G}$ realizes the sequence pair (d, d).
(M2) $(v_i, w_i) \notin \hat{E}$ for every $i \in [1, n]$.
(M3) $\hat{M} = \{(v_i, w_{2\nu - i + 1}) \mid i \in [1, 2\nu]\} \subseteq \hat{E}$.

First, we describe an algorithm that given a ν-matched realization $\hat{G}$ of (d, d), produces a graph G realizing d with a matching of size ν.

1. Compute a half-integral solution.
   (a) For all $i, j \in [1, n]$, let

$$y_{ij} = \begin{cases} 1, & \{v_i, w_j\} \in \hat{E}, \\ 0, & \text{otherwise}, \end{cases} \qquad \text{and} \qquad w(i, j) = \frac{1}{2}(y_{ij} + y_{ji}).$$

   (b) Define a weighted graph $G^{\omega} = (V^{\omega}, E^{\omega}, \omega)$ with vertex set $V^{\omega} = [1, n]$ and an edge $e = (i, j)$ of weight $\omega(e) = \omega(i, j)$ for every $i, j \in V^{\omega}$. Clearly, ω is half-integral.
   (c) Let $M^{\omega} = \{(i, 2\nu - i + 1) \mid i \in [1, \nu]\}$ be the inverted matching in G^{ω}. According to (M3)

$$\omega(e) = 1, \text{ for every } e \in M^{\omega}. \tag{1}$$

   (d) Define the *weighted degree* of a vertex $i \in V^{\omega}$ to be

$$d^{\omega}(i) = \sum_{j \in V^{\omega}} \omega(i, j) \ .$$

   Note that G^{ω} realizes d in the *weighted* sense, namely,

$$d^{\omega}(i) = \sum_{j \in V^{\omega}} \omega(i, j) = \frac{1}{2}\left(\sum_{j=1}^{n} y_{ij} + \sum_{j=1}^{n} y_{ji}\right) = d_i, \quad \text{for any } i \in V^{\omega}.$$

2. Preparing for discarding non-integral weights while keeping the degrees unchanged.
   Construct a graph $G^{1/2} = (V^{1/2}, E^{1/2})$ obtained by removing from G^ω the edges of integral weight and keeping only those of weight $1/2$. Formally, $V^{1/2} = V^\omega$ and $E^{1/2} = \{e \in E^\omega \mid \omega(e) = 1/2\}$.
   In later stages of the construction, whenever G^ω is modified (by changing the weight of some edge e from $1/2$ to 0 or 1), $G^{1/2}$ is modified accordingly (by removing the edge e).

3. Partition into cycles.
   (a) Partition the edge set of $G^{1/2}$ into disjoint (not necessarily simple) cycles each covering an entire connected component of the graph. Since $G^{1/2}$ is an even graph by Observation 6, this can be done in polynomial number of steps.
   (b) Let $\mathscr{C}$ be a set of cycles in the partition.
   (c) Let $\mathscr{C}^{even} \leftarrow \{C \in \mathscr{C} \mid C \text{ is of even length}\}$, $\mathscr{C}^{odd} \leftarrow \mathscr{C} \setminus \mathscr{C}^{even}$.

4. Eliminate even cycles.
   For every cycle $C \in \mathscr{C}^{even}$, do the following.
   (a) Traverse C starting from an arbitrary vertex $x \in C$ and continuing along the cycle until returning to x. Denote the resulting sequence of edges by $E(C) = (e_1, e_2, \ldots, e_\ell)$.
   (b) Increase (resp., decrease) edge weights on even (resp., odd) positions in the sequence $E(C)$ by $1/2$. That is, for every $i \in [1, \ell]$, $\omega(e_i)$ is set to 1 if i is even and 0 otherwise. (This does not change the weighted degrees in G^ω and does not affect other cycles in $\mathscr{C}$ or edges in M^ω.)

5. Eliminate odd cycles.
   Arrange the cycles in $\mathscr{C}^{odd}$ in pairs (by Observation 7 their number is even). For every pair of cycles (C, C') choose vertices $x \in C$, $y \in C'$ according to Observation 8. Then do the following.
   (a) Starting from $x \in C$, traverse C via edges $E(C) = (e_1, e_2, \ldots, e_\ell)$. Starting from vertex $y \in C'$, traverse C' via edges $E(C') = (e'_1, \ldots, e'_k)$.
   (b) Let $\xi = \omega(x, y)$ (by Observation 8, this weight must be either 0 or 1).
   (c) Set $\omega(x, y) \leftarrow 1 - \xi$.
   (d) For every $i \in [1, \ell]$ and $j \in [1, k]$, modify the edge weights in the cycles C and C' as follows:

$$\omega(e_i) \leftarrow \begin{cases} 1 - \xi, & i \text{ is even,} \\ \xi, & i \text{ is odd,} \end{cases} \qquad \omega(e'_j) \leftarrow \begin{cases} 1 - \xi, & j \text{ is even,} \\ \xi, & j \text{ is odd.} \end{cases}$$

   By Observation 9, at this stage each edge in G^ω has weight 1 or 0, G^ω realizes d, and Eq. (1) holds.

6. Generate the output G.
   Transform G^ω into a simple graph G, with an edge (i, j) whenever $\omega(i, j) = 1$ in G^ω. Note that $M = M^\omega$ forms a matching in G by Eq. (1).

   The algorithm is justified by the following Observations.

Observation 6. *The graph $G^{1/2}$ is even (i.e., all its vertex degrees are even).*

A connected even graph has an Euler cycle, implying the following.

Observation 7. *The number of cycles in $\mathscr{C}^{odd}$ is even.*

Proof. Observe that $\sum_{e \in E^\omega} \omega(e) = \frac{1}{2} \sum_{i=1}^n d_i = m$, where m is the number of edges, which is an integer. Therefore, the number of edges with weight $1/2$ must be even. Since the cycles in $\mathscr{C}$ cover all of $E^{1/2}$ and are disjoint, the observation follows. $\square$

The following observation is implied by Observation 5.

Observation 8. *For any pair of disjoint odd cycles $C, C' \in \mathscr{C}^{odd}$, there exist $x \in C$ and $y \in C'$, such that $(x, y) \notin M^\omega$ and $\omega(x, y) \neq 1/2$.*

Proof. Observation 5 implies that for any pair of disjoint odd cycles $C, C' \in \mathscr{C}^{odd}$, there exist $x \in C$ and $y \in C'$, such that $(x, y) \notin M^\omega$.

Assume, towards contradiction, that $\omega(x, y) = 1/2$. Then the edge (x, y) belongs to $G^{1/2}$, implying that x and y belong to the same connected component in $G^{1/2}$. But this contradicts the fact that $x \in C$ and $y \in C'$ where C and C' cover different connected components of $G^{1/2}$. $\square$

Observation 9. *At the end of Step 5 , each edge in G^ω has weight 1 or 0, G^ω realizes d and Eq. (1) holds.*

Proof. After Step 5 , G^ω contains no more edges of weight $1/2$. Moreover, neither of the modifications performed in Step 5 change the weighted degrees in G^ω, affect other cycles in $\mathscr{C}$ or edges in M^ω. The observation follows. $\square$

The reduction correctness follows from the next lemma.

Lemma 10. *There exists a ν-matched realization of the bi-sequence (d, d) if and only if $\mathrm{MM}(d) \geq \nu$.*

Proof. If there is a ν-matched realization of (d, d), then the aforementioned algorithm produces a realization of d with a matching of size ν.

For the other direction, given a realization $G = (V, E)$ of d with vertices $V = [1, n]$ and a matching $M \subseteq E$ of size ν, there is a realization $G' = (V, E')$ with a matching $M' = \{(v_i, v_{2\nu - i + 1}) \mid i \in [1, \nu]\}$ by the Inverted Prefix Lemma (Lemma 3). Consider the following graph $\hat{G} = (\hat{V}, \hat{W}, \hat{E})$ with $\hat{V} = \{v_1, \ldots, v_n\}$, $\hat{W} = \{w_1, \ldots, w_n\}$ and $(v_i, w_j) \in \hat{E}$ if and only if $(i, j) \in E'$. Clearly, $\hat{G}$ realizes (d, d) and $(v_i, w_i) \notin \hat{E}$ for every $i \in [1, n]$, so it satisfies (M1) and (M2). Moreover, the set $\hat{M} = \{(v_i, w_j) \mid (i, j) \in M'\}$ is a matching of size 2ν in $\hat{G}$ satisfying (M3). Indeed, since M' is a matching, the edges in $\hat{M}$ form a matching too and each edge $(i, j) \in M'$ leads to two edges $(v_i, w_j), (v_j, w_i) \in \hat{M}$. The lemma follows. $\square$

4.3   Reduction to Flow

According to the previous section (in particular, Lemma 10) given a degree sequence d all we need is to find a bipartite ν-matched realization of (d, d) for maximum possible ν. Next we describe how to construct a flow graph $G_{d,\nu}$ for any ν, such that ν-matched realization of (d, d) exists if and only if the maximum flow in $G_{d,\nu}$ is $\sum_{i=1}^{n} d_i - 2\nu$. Moreover, given an integer flow of this value we describe how to construct a desired realization.

First we describe the construction of $G_{d,\nu}$. The node set X contains nodes x_i (for $i \in [1, n]$), corresponding to the vertices of V, and the source s is connected to them. Similarly, the node set Y contains nodes y_j (for $j \in [1, n]$), corresponding to the vertices in W, and these are connected to the sink t. The nodes are organized into the following sets:

- Candidates for the matching:

$$X_M = \{x_i \mid i \in [1, 2\nu]\} \qquad \text{and} \qquad Y_M = \{y_j \mid j \in [1, 2\nu]\} \,,$$

- Rest of the nodes:

$$X_R = \{x_i \mid i \in [2\nu + 1, n]\} \qquad \text{and} \qquad Y_R = \{y_j \mid j \in [2\nu + 1, n]\} \,,$$

- Source and sink: $\{s,t\}$.

The edges are capacitated and directed from left to right, i.e., an edge (α, β, γ) leads from the node α to the node β and can carry up to γ units of flow. The source s has edges leading to every node x_i, with capacity $d_i - 1$ for $x_i \in X_M$ and d_i for $x_i \in X_R$. Similarly, there are edges leading from every node y_j to the sink t, with capacity $d_j - 1$ for $y_j \in Y_M$ and d_j for $y_j \in Y_R$. This enforces the degree constraint for the vertices v_i and w_i in G. Finally, there's a unit capacity edge $(x_i, y_j, 1)$ for every $i, j \in [1, n]$, such that $i \neq j$ and $j \neq 2\nu - i + 1$. Overall, the edge set $\tilde{E}$ of the flow graph is structured as a complete (X, Y) biclique minus an inverted matching, with a source s connected to X and a sink t connected to Y. Formally, $\tilde{E}$ (along with the edge weights) is defined as follows:

$$
\begin{aligned}
\tilde{E} \;=\; & \{(s, x_i, d_i - 1) \mid i \in [1, 2\nu]\} \\
& \cup \{(s, x_i, d_i) \mid i \in [2\nu + 1, n]\} \\
& \cup \{(x_i, y_j, 1) \mid i, j \in [1, n], \ i \neq j, j \neq 2\nu - i + 1\} \\
& \cup \{(y_j, t, d_j - 1) \mid j \in [1, 2\nu]\} \\
& \cup \{(y_j, t, d_j) \mid j \in [2\nu + 1, n]\} \;.
\end{aligned}
$$

The construction is justified by the following lemma.

Lemma 11. *There exists a ν-matched bipartite graph $\hat{G} = (V, W, \hat{E})$ that realizes the (d, d) if and only if the value of the maximum $s - t$ flow in $G_{d,\nu}$ is equal to $\sum_{i=1}^{n} d_i - 2\nu$.*

Proof. Suppose the maximum value of an $s - t$ flow in $G_{d,\nu}$ is $\sum_{i=1}^{n} d_i - 2\nu$. Denote a value of the flow from a vertex x to a vertex y by $\mathsf{flow}(x, y)$. Define the bipartite graph $\hat{G} = (V, W, \hat{E})$ as follows:

1. For each node $x_i \in X$, create a corresponding node $v_i \in V$.
2. For each node $y_j \in Y$, create a corresponding node $w_j \in W$.
3. For every $1 \leq i, j \leq n$, if $\mathsf{flow}(x_i, y_j) = 1$, then create an edge (v_i, w_j) and add it to $\hat{E}$.
4. For every $1 \leq i \leq 2\nu$ create an edge $(v_i, w_{2\nu-i+1})$ and add it to $\hat{E}$.

It remains to verify that $\hat{G}$ correctly realizes (d, d) and is ν-matched. The degree of v_i in $\hat{G}$ equals to the number of edges (x_i, y_j) that carry flow if $i \in [2\nu + 1, n]$ and one more than the corresponding flow if $i \in [1, 2\nu]$. Since x_i conserves flow, and since all the edges (s, x_i) are saturated, $\deg(v_i) = d_i$. Similarly, $\deg(w_j) = d_j$ for every $j \in [1, n]$, so $\hat{G}$ satisfies (M1). Since $G_{d,\nu}$ does not have edges (x_i, y_i) for any $i \in [1, n]$, it follows that $\hat{G}$ does not have edges (v_i, w_i), satisfying (M2). Finally, the matching $\hat{M} = \{(v_i, w_{2\nu-i+1}) \mid 1 \leq i \leq 2\nu\}$ was added in step 4, so $\hat{G}$ satisfies (M3).

Conversely, suppose $\hat{G} = (V, W, \hat{E})$, where $V = \{v_1, \ldots, v_n\}$ and $W = \{w_1, \ldots, w_n\}$, is a ν-matched bipartite graph realizing (d, d) with the matching $\hat{M} = \{(v_i, w_{2\nu-i+1}) \mid 1 \leq i \leq 2\nu\}$. We need to define a flow function on $G_{d,\nu}$ and prove that its value equals $\sum_{i=1}^{n} d_i - 2\nu$. Since the sum of capacities of outgoing edges from s is exactly this sum, maximum flow cannot exceed it, thus it is enough to construct a flow with the aforementioned value.

1. **Saturate Edges from Source and to Sink:**

$$\mathsf{flow}(s, x_i) \leftarrow \begin{cases} d_i - 1 & i \in [1, 2\nu], \\ d_i & i \in [2\nu + 1, n]. \end{cases}$$

$$\mathsf{flow}(y_j, t) \leftarrow \begin{cases} d_j - 1 & j \in [1, 2\nu], \\ d_j & j \in [2\nu + 1, n]. \end{cases}$$

2. **Flow Through Edges:**
   For every $i \in [1, n]$ and $j \in [1, n]$ such that $(v_i, w_j) \in \hat{E} \setminus \hat{M}$, set $\mathsf{flow}(x_i, y_j) \leftarrow 1$,
3. **Completing the flow function:**
   Set the flows of all other edges to 0.

We need to argue that the defined flow is legal. Note that by construction, the total flow out of the source s is $\sum_{i=1}^{n} d_i - 2\nu$ and the total flow into the sink t is $\sum_{j=1}^{n} d_j - 2\nu$. It remains to verify that for every vertex except s and t, the incoming and outgoing flows are equal.

- **Node x_i:** By construction, the incoming flow at x_i is $\mathsf{flow}(s, x_i) = d_i - 1$ if $x_i \in X_M$ and $\mathsf{flow}(s, x_i) = d_i$ if $x_i \in X_R$. As for the outgoing flow, recall that there are exactly $d_i - 1$ indices j for which $(v_i, w_j) \in \hat{E} \setminus \hat{M}$ if $i \in [1, 2\nu]$ and d_i such indices if $i \in [2\nu + 1, n]$. Hence, x_i conserves flow.

- **Node** y_j: By construction, the outgoing flow at y_j is $\mathsf{flow}(y_j, t) = d_j - 1$ if $x_j \in X_M$ and $\mathsf{flow}(y_j, t) = d_j$ if $y_j \in X_R$. As for the incoming flow, recall that there are exactly $d_j - 1$ indices i for which $(v_i, w_j) \in \hat{E} \setminus \hat{M}$ if $j \in [1, 2\nu]$ and d_j such indices if $j \in [2\nu + 1, n]$. Hence, y_j conserves flow.

The claim follows. $\qquad\qquad\qquad\qquad\qquad\qquad\qquad\qquad\qquad\qquad\qquad\qquad\qquad\qquad$ $\square$

We conclude the section with the following result.

Theorem 3. *There exists an $O(n^3)$ time algorithm for constructing a realization of a given graphic sequence d that also has a matching of maximum size (among all possible realizations of d).*

Proof. As was shown in the beginning of this section a maximum size of a matching ν can be found in a running time of $O(n^2 \log n)$. As for the realization algorithm, given ν, to find a ν-matched realization of d we first build the flow graph $G_{d,\nu}$ and find a maximum s-t flow in it. Since $G_{d,\nu}$ has $O(n)$ vertices and $O(n^2)$ edges, a maximum flow can be computed in $O(n^3)$ time using the maximum flow algorithm by Orlin [22]. Given a maximum flow, adjacency matrix for ν-matched bipartite realization $\hat{G}$ for the sequence pair (d, d) can be computed in $O(n^2)$ time. Next we go through all the steps of transformation of $\hat{G}$ into ν-matched realization G for d and show that each step takes $O(n^3)$ time.

Step 1: The adjacency matrix of the weighted graph G^ω can be computed from the adjacency matrix of $\hat{G}$ in $O(n^2)$.

Step 2: It takes $O(n^2)$ to compute the adjacency matrix of $G^{1/2}$ using the adjacency matrix of G^ω.

Step 3: Since an Eulerian cycle can be found by linear time in the number of edges and $G^{1/2}$ is an even graph, a partition of $G^{1/2}$ into disjoint cycles $\mathscr{C}$ takes $O(n^2)$ time.

Step 4: Separating even cycles in $\mathscr{C}$ and applying the corresponding modification to them takes $O(n^2)$ time.

Step 5: For every pair of odd cycles (C, C') it takes constant time to find vertices $x \in C$ and $y \in C'$ satisfying Observation 8 following the proof of Observation 5. Given x and y it takes linear time $O(|C| + |C'|)$ to apply corresponding modifications. Therefore, the complexity for all pairs of cycles is $O(n^2)$.

Step 6: At this step G^ω does not have edges of weight $1/2$, so it can be transformed into G in $O(n^2)$ time.

Disclosure of Interests. The authors have no competing interests to declare that are relevant to the content of this article.

References

1. Anstee, R.P.: An algorithmic proof of Tutte's f-factor theorem. J. Alg. **6**(1), 112–131 (1985)
2. Anstee, R.P.: Simplified existence theorems for (g, f)-factors. Discr. Appl. Math. **27**(1–2), 29–38 (1990)
3. Bar-Noy, A., Kalinichev, I., Peleg, D., Rawitz, D.: Degree realization with minimum dominating set. In: 27th IPCO (2026, to appear)
4. Barrus, M.D.: On realization graphs of degree sequences. Discr. Math. **339**, 2146–2152 (2016)
5. Bock, F., Rautenbach, D.: On matching numbers of tree and bipartite degree sequences. Discr. Math. **342**(6), 1687–1695 (2019)
6. Chungphaisan, V.: Construction of Hamiltonian graphs and bigraphs with prescribed degrees. J. Comb. Theory B **24**(2), 154–163 (1978)
7. Erdős, P., Gallai, T.: Graphs with prescribed degrees of vertices [Hungarian]. Mat. Lapok (N.S.) **11**, 264–274 (1960)
8. Erdos, P.L., Kiraly, Z., Miklos, I.: On the swap-distances of different realizations of a graphical degree sequence. Comb. Probab. Comput. **22**, 366–383 (2013)
9. Erdös, P.L., Kharel, S.R., Mezei, T.R., Toroczkai, Z.: New results on graph matching from degree-preserving growth. Mathematics **12**(22) (2024)
10. Fulkerson, D.R., Hoffman, A.J., McAndrew, M.H.: Some properties of graphs with multiple edges. Canad. J. Math. **17**, 166–177 (1965)
11. Fulkerson, D.: Zero-one matrices with zero trace. Pacific J. Math. **12**, 831–836 (1960)
12. Gentner, M., Henning, M.A., Rautenbach, D.: Largest domination number and smallest independence number of forests with given degree sequence. Discr. Appl. Math. **206**, 181–187 (2016)
13. Gentner, M., Henning, M.A., Rautenbach, D.: Smallest domination number and largest independence number of graphs and forests with given degree sequence. J. Graph Theory **88**(1), 131–145 (2018)
14. Gould, R.J., Jackson, M.S., Lehel, J.: Potentially g-graphical degree sequences. In: Combinatorics, Graph Theory, and Algorithms, vol. 1, pp. 451–460 (1999)
15. Hakimi, S.L.: On realizability of a set of integers as degrees of the vertices of a linear graph - I. SIAM J. Appl. Math. **10**(3), 496–506 (1962)
16. Havel, V.: A remark on the existence of finite graphs. Casopis Pest. Mat. **80**, 477–480 (1955)
17. Kézdy, A., Lehel, J.: Degree sequences of graphs with prescribed clique size. In: Graph Theory, Combinatorics, Algorithms, and Applications, pp. 535–544 (1998)
18. Kleitman, D.J., Wang, D.L.: Algorithms for constructing graphs and digraphs with given valences and factors. Discr. Math. **6**, 79–88 (1973)
19. Kundu, S.: The k-factor conjecture is true. Discr. Math. **6**(4), 367–376 (1973)
20. Kundu, S.: A factorization theorem for a certain class of graphs. Discr. Math. **8**(1), 41–47 (1974)

21. Kundu, S.: Generalizations of the k-factor theorem. Discr. Math. **9**(2), 173–179 (1974)
22. Orlin, J.B.: Max flows in $o(nm)$ time, or better. In: 45th STOC, pp. 765–774 (2013)
23. Rao, A.R.: The clique number of a graph with a given degree sequence. ISI Lect. Notes **4**, 251–267 (1979)
24. Rao, A.R., Rao, S.B.: On factorable degree sequences. J. Comb. Theoey B **13**, 185–191 (1972)
25. Yin, J.H.: A Rao-type characterization for a sequence to have a realization containing a split graph. Discr. Math. **311**(21), 2485–2489 (2011)
26. Yin, J.H.: A short constructive proof of A. R. Rao's characterization of potentially k_{r+1}-graphic sequences. Discr. Appl. Math. **160**(3), 352–354 (2012)

Vertex-Critical Graphs in Subfamilies of $(P_4 + \ell P_1)$-Free Graphs

Iain Beaton[1] and Ben Cameron[2(✉)]

[1] Acadia University, Wolfville, NS, Canada
[2] University of Prince Edward Island, Charlottetown, PE, Canada
brcameron@upei.ca

Abstract. A graph G is k-vertex-critical if $\chi(G) = k$ but $\chi(G - v) < k$ for all $v \in V(G)$. In this paper we make progress on the open problem of the finiteness of k-vertex-critical $(P_4 + \ell P_1)$-free graphs by showing that there are only finitely many k-vertex-critical graphs in the following subfamilies of $(P_4 + \ell P_1)$-free graphs for all $k \geq 1$ and $\ell \geq 0$:

- $(P_4 + \ell P_1, 2P_2)$-free graphs,
- $(P_4 + \ell P_1, \text{chair})$-free graphs,
- $(P_4 + \ell P_1, P_5, \text{bull})$-free graphs, and
- $(P_4 + \ell P_1, P_5, \text{cricket})$-free graphs.

In fact, all but the first of these are special cases of our general result that there are only finitely many k-vertex-critical $(P_4 + \ell P_1, B_4(m), B_3(m)^+)$-free graphs for all $k \geq 1$ and $\ell, m \geq 0$. Here $B_n(m)$ is the graph obtained from a path of order n by identifying one of its leaves with the centre vertex of $K_{1,m}$ and $B_n(m)^+$ is the graph obtained by identifying an edge of K_3 with the edge of $B_n(m)$ with endpoints of degrees 2 and m, respectively. Our results imply the existence of simple polynomial-time certifying algorithms to decide the k-colourability of all graphs in these subfamilies for every fixed k.

We also show that $\chi(G) \leq \ell + 2$ for all $(P_4 + \ell P_1, K_3)$-free graphs and all $\ell \geq 0$, improving the previously known upper bound of $2\ell + 2$ that followed from Randerath and Schiermeyer's 2004 result on (P_t, K_3)-free graphs. More generally, we provide a χ-bound in $O(\ell^{\omega - 1})$ for $(P_4 + \ell P_1)$-free graphs which improves the bound of $(2\ell + 2)^{\omega - 1}$ which followed from Gravier, Hoàng and Maffray in 2003 for P_t-free graphs.

Keywords: Graph coloring · k-critical graphs · Polynomial-time algorithms

1 Introduction

The problem of determining the k-colourability of a graph for fixed k, which we will denote as k-Colouring, has seen a very active area of research since it was shown to be NP-complete for all $k \geq 3$ as one of Karp's original 21 problems [30]. While there are many directions one can take in this field, we are following a long line of research on developing polynomial-time algorithms to

© The Author(s), under exclusive license to Springer Nature Switzerland AG 2026
F. Foucaud and A. Parreau (Eds.): IWOCA 2026, LNCS 16587, pp. 118–130, 2026.
https://doi.org/10.1007/978-3-032-27732-9_9

solve k-COLOURING when the input graphs are restricted to be from a certain family of graphs. The restricted families of graphs we are interested in are those defined by forbidden induced subgraphs. A graph G is H-free if no induced subgraph of G is isomorphic to H. For graphs $H_1, H_2, \ldots, H_m$, a graph is $(H_1, H_2, \ldots H_m)$-free if it is H_i-free for each $i \in \{1, \ldots, m\}$. k-COLOURING H-free remains NP-complete when H contains an induced claw [23,31] and when H contains an induced cycle [28]. Thus, assuming P$\neq$NP, if a polynomial-time k-COLOURING algorithm exists for H-free graphs, then H must be a linear forest, that is, a forest where each component is a path. When forbidding connected graphs, it is known that k-COLOURING P_t-free graphs is NP-complete for all $k \geq 5$ and $t \geq 6$ and for $k = 4$ and $t \geq 7$ [24]. On the other hand, 4-COLOURING P_6-free graphs is polynomial-time solvable [15,16], and k-COLOURING P_5-free graphs is polynomial-time solvable *for all k* [21]. When forbidding a disconnected graph, it is known that k-COLOURING H-free graphs remains NP-complete for all $k \geq 5$ when H is $P_5 + P_2$ [13] or $2P_4$ [20], but is polynomial-time solvable for all $k, \ell \geq 0$ when H is an induced subgraph of $P_5 + \ell P_1$ [17] or of ℓP_3 [13].

In H-free graphs with known polynomial-time k-COLOURING algorithms, there is very active research on determining which of these have the stronger property that they contain only finitely many $(k + 1)$-vertex-critical graphs. This implies that k-COLOURING is polynomial-time solvable as one can search the input graph for any of the finitely many $(k + 1)$-vertex-critical graphs as an induced subgraph to determine its k-colourability, where, if found, a $(k + 1)$-vertex-critical induced subgraph can be returned as a certificate for a negative answer (see [5], for example). This is especially helpful when paired with the polynomial-time k-COLOURING algorithms for $(P_5 + \ell P_1)$-free graphs that also return a k-colouring of the graph if one exists. Thus, in any subfamily of $(P_5 + \ell P_1)$-free graphs where there are only finitely many $(k + 1)$-vertex-critical graphs, there exists a polynomial-time *certifying* k-COLOURING algorithm for the subfamily. See [33] for a survey on certifying algorithms and strong arguments for their importance. Thus, there has been lots of recent attention paid to proving various families of graphs contain only finitely many k-vertex-critical graphs for all k. For a positive integer k and graphs $H_1, H_2, \ldots, H_m$, let $\operatorname{crit}_k(H_1, H_2, \ldots, H_m)$ denote the set of all k-vertex-critical $(H_1, H_2, \ldots, H_m)$-free graphs, so $|\operatorname{crit}_k(H_1, H_2, \ldots, H_m)|$ will denote the number of k-vertex-critical $(H_1, H_2, \ldots, H_m)$-free graphs.

In the least restrictive setting with only a single forbidden induced subgraph, the most comprehensive result is that $|\operatorname{crit}_4(H)| < \infty$ if and only if H is an induced subgraph of P_6, $2P_3$, or $P_4 + \ell P_1$ for some $\ell \geq 0$ [12]. Further, it was shown in [1] that $|\operatorname{crit}_k(P_3 + \ell P_1)| < \infty$ for all $k \geq 1$ and $\ell \geq 0$. It follows from these results (and some we will cite later) that the only graphs H for which the finiteness of $|\operatorname{crit}_k(H)|$ is unknown for all k is when $H = P_4 + \ell P_1$ for all $\ell \geq 1$. Thus, determining exactly which graphs H admit $|\operatorname{crit}_k(H)| < \infty$ for all k is equivalent to answering the open question from [7]: For which values of $k \geq 5$ and $\ell \geq 1$ is $|\operatorname{crit}_k(P_4 + \ell P_1)| < \infty$? While there are no known infinite families of k-vertex-critical $(P_4 + \ell P_1)$-free graphs for any k or ℓ, progress on resolving this open problem is limited to subfamilies of the smallest open case (i.e., when $\ell = 1$); it is known that $|\mathrm{crit}_k(P_4 + P_1, H)| < \infty$ when H is gem [1], paw+P_1, or any graph of order 4 [2].

The limited progress on vertex-critical $(P_4 + \ell P_1)$-free graphs is especially surprising in contrast to the abundance of work on vertex-critical P_5-free graphs. Unlike for $(P_4 + \ell P_1)$-free graphs, there are known infinite families of k-vertex-critical graphs; while $|\mathrm{crit}_4(P_5)| = 12$ [32], $|\mathrm{crit}_k(2P_2, K_3 + P_1)| = \infty$ for all $k \geq 5$ [22] (note that $(2P_2, K_3 + P_1)$-free graphs is a proper subfamily of P_5-free graphs). Therefore, attention shifted to considering the finiteness of k-vertex-critical graphs in further-restricted subfamilies of P_5 free graphs. In [10], it was shown that $|\mathrm{crit}_k(P_5, H)| < \infty$ when H is a graph of order 4 for all k if and only if H is not $2P_2$ or $K_3 + P_1$. In the same paper, an open problem was posed to determine for which graphs H of order 5 is $|\mathrm{crit}_k(P_5, H)| < \infty$ for all $k \geq 1$. Toward an answer to this question, it has been shown that $|\mathrm{crit}_k(P_5, H)| < \infty$ for all k when H is banner [4], $K_{2,3}$ and $K_{1,4}$ [29], $P_2 + 3P_1$ [7], $P_3 + 2P_1$ [1], $\overline{P_5}$ [18], $\overline{P_3 + P_2}$ and gem [6] (see also [8]), dart [41], $K_{1,3} + P_1$ and $\overline{K_3 + 2P_1}$ [42], and W_4 [40]. Further, $|\mathrm{crit}_k(P_5, C_5)| = \infty$ for all $k \geq 6$ [9]. Thus, the open problem on $|\mathrm{crit}_k(P_5, H)|$ when H is of order 5 only remains open when H is one of the 9 graphs in Fig. 1.

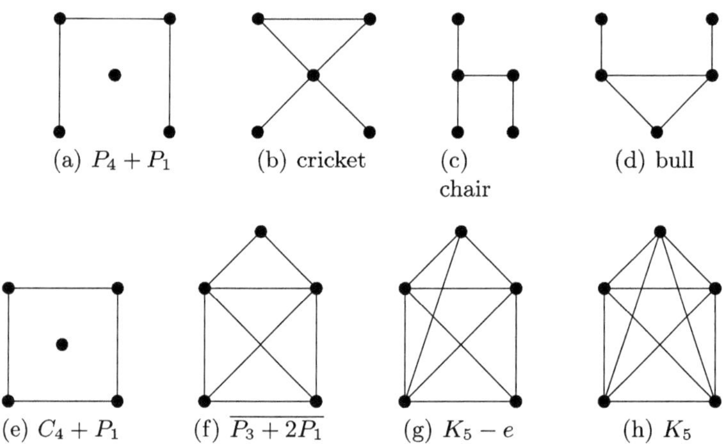

Fig. 1. Graphs H of order 5 where the finiteness of k-vertex-critical (P_5, H)-free graphs is unknown.

Note that the finiteness of $|\mathrm{crit}_k(P_5, P_4 + P_1)|$ is not known for any $k \geq 5$, thus linking interest in the two open problems discussed above. In this paper, we provide the first results on k-vertex-critical $(P_4 + \ell P_1, H)$-free graphs for $\ell > 1$ and, in fact, all of our results hold for all $\ell \geq 0$. To state our results, we will first require the definitions of two graph families. Let $B_n(m)$ be the graph obtained from a path of order n by identifying one of its leaves with the centre vertex of $K_{1,m}$ (see Figs. 2(a) and 2(b) for examples with $n = 3$ and 4). Let $B_n(m)^+$ be the graph obtained by identifying an edge of K_3 with the edge of $B_n(m)$ with endpoints of degrees 2 and m, respectively (see Fig. 2(c)).

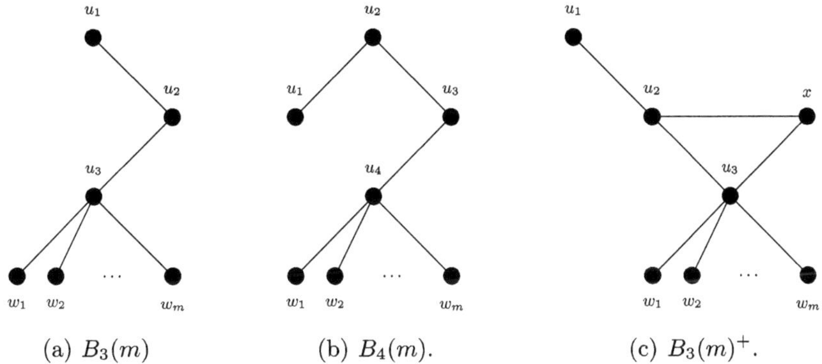

(a) $B_3(m)$ (b) $B_4(m)$.   (c) $B_3(m)^+$.

Fig. 2. The forbidden induced graphs in Theorems 1 and Corollary 1.

We are now ready to state the two main results of our paper, and discuss some corollaries.

Theorem 1. *For all $k \geq 1$ and $\ell, m \geq 0$, $|\mathrm{crit}_k(P_4 + \ell P_1, B_4(m), B_3(m)^+)| < \infty$.*

Theorem 2. *For all $k \geq 1$ and $\ell \geq 0$, $|\mathrm{crit}_k(P_4 + \ell P_1, 2P_2)| < \infty$.*

Despite its generality, we give a very short and easy to follow proof of Theorem 1. We achieve this by a new technical lemma that imposes helpful structure on the non-neighbours of large independent sets of $(P_4 + \ell P_1)$-free graphs that we use in conjunction with the aforementioned result that $|\mathrm{crit}_k(P_3 + \ell P_1)| < \infty$ for all $k \geq 1$ and $\ell \geq 0$. The generality of Theorem 1 leads to many specific corollaries of interest, some of which we now list. First, since $B_3(m)$ is an induced subgraph of both $B_3(m)^+$ and $B_4(m)$, we get.

Corollary 1. *For all $k \geq 1$ and $\ell, m \geq 0$, $|\mathrm{crit}_k(P_4 + \ell P_1, B_3(m))| < \infty$.*

Since $B_3(2)$ is chair, we get the even more specific corollary that follows.

Corollary 2. *For all $k \geq 1$ and $\ell, m \geq 0$, $|\mathrm{crit}_k(P_4 + \ell P_1, \mathrm{chair})| < \infty$.*

Since P_5 is $B_4(1)$, bull is $B_3(1)^+$ and cricket is the graph obtained from $B_3(2)^+$ by deleting the leaf adjacent to the vertex of degree 3, we obtain the following immediate corollaries.

Corollary 3. *For all $k \geq 1$, $|\mathrm{crit}_k(P_4 + \ell P_1, P_5, \mathrm{bull})| < \infty$.*

Corollary 4. *For all $k \geq 1$, $|\mathrm{crit}_k(P_4 + \ell P_1, P_5, \mathrm{cricket})| < \infty$.*

Theorems 1 and 2 provide the best evidence to date that $|\mathrm{crit}_k(P_4 + \ell P_1, P_5)|$ might be finite for all k. Further, Corollaries 2, 3, and 4 are likely to be helpful for resolving three of the 9 remaining open cases of the open problem stated from [10] stated above, as a now only graphs containing an induced $P_4 + \ell P_1$ for any fixed ℓ depending only on the chromatic number need to be considered. We note that $|\mathrm{crit}_k(P_5, \mathrm{bull})|$ and $|\mathrm{crit}_k(P_5, \mathrm{chair})|$ are of particular interest as it is known that each is finite for $k = 5$ as shown in [25], and [26], respectively (and in fact it was further shown that $|\mathrm{crit}_5(P_6, \mathrm{bull})| < \infty$ [27]).

1.1   Outline

The rest of the paper is organized as follows. We first present some preliminary results and terminology in Sect. 2 that will be used throughout our proofs of our main results. In Sect. 3, we prove Theorem 1, and in Sect. 4, we prove Theorem 2. In Sect. 5, where we show that for all $\ell \geq 0$ and $k \geq 3$, every $(P_4 + \ell P_1, K_k)$-free graph G has

$$\chi(G) \leq \ell^{k-2} + 2\ell^{k-3} + 3\ell^{k-4} \cdots + (k-2)\ell + (k-1).$$

We conclude in Sect. 6 with a discussion on future directions.

2   Preliminaries

For standard graph theory terminology and notation, we follow [39]. We denote to adjacent vertices u and v by $u \sim v$. Two vertices u and v are *twins* if $N(u) = N(v)$ or $N[u] = N[v]$. We say u and v are *false twins* if $N(u) = N(v)$ and *true twins* if $N[u] = N[v]$. Two disjoint sets of vertices X and Y are *complete* to each other if every vertex in X is adjacent to every vertex in Y. We say X and Y are *anti-complete* if there are no edges between the two sets. A vertex is *mixed* on a set of vertices X if it is not complete nor anti-complete to X, that is, it is adjacent to at least one vertex in X and non-adjacent to at least one other.

We now state key lemma and theorem that will be applied in our main theorems.

Lemma 1 ([10])**.** *Let G be a k-vertex-critical graph. There does not exist two vertex subsets $X, Y \subseteq V(G)$ satisfying all of the following conditions.*

- *X and Y are anticomplete to each other.*
- *$\chi(G[X]) \leq \chi(G[Y])$.*
- *Y is complete to $N(X)$.*

Theorem 3 ([1])**.** *There are only finitely many k-vertex-critical $(P_3 + \ell P_1)$-free graphs for all $k \geq 1$ and $\ell \geq 0$.*

3 $(P_4 + \ell P_1,\ B_4(m),\ B_3(m)^+)$-Free

Before we can prove Theorem 1, we require the following technical lemma on the structure of P_4-free graphs that we will apply to the non-neighbours of a large independent set. Note that every component in a P_4-free graph can be constructed from a single vertex and iteratively choosing a vertex v and adding a copy of v which is either a false twin or true twin to v (see [3], Theorem 11.3.3).

Lemma 2. *Let G be a connected P_4-free graph. If G contains an induced P_3 then there exists disjoint sets X and Y which are anticomplete to each other such that Y is complete to $N(X) \neq \emptyset$ and X is complete to $N(Y)$ (and therefore, $N(X) = N(Y) \neq \emptyset$).*

Proof. We will proceed by induction on the order of G, a connected P_4-free graph with an induced P_3. If G has three vertices, then $G \cong P_3$, and $X = \{x\}$ and $Y = \{y\}$ satisfies the conclusion when x and y are the leaves of the induced P_3. Suppose that our claim is true for all connected P_4-free graphs of order k with an induced P_3.

Suppose G has order $k + 1$. As G is P_4-free, there exist two twin vertices u and v. If u and v are false twins, then $X = \{u\}$ and $Y = \{v\}$ satisfies the conclusion. So suppose u and v are true twins. By induction, there exist disjoint and anticomplete sets X and Y in $G - v$ such that $N_{G-v}(X) \neq \emptyset$ and X is complete to $N_{G-v}(Y)$. Let $W = N_{G-v}(X) = N_{G-v}(Y)$. If $u \in X$ (resp. $u \in Y$) then $N(v) \subseteq W \cup X$ (resp. $N(v) \subseteq W \cup Y$). Thus $X \cup \{v\}$ and Y (resp. X and $Y \cup \{v\}$) would have the desired properties in G.

So suppose $u \notin X$ and $u \notin Y$. If $u \in W$ then v is complete to X and Y and thus X and Y have the desired property in G. If $u \notin W$ then v is anticomplete to X and Y and thus X and Y have the desired property in G. This completes the proof. $\qquad\blacksquare$

With this lemma in hand, we can now proceed with the proof of our first main theorem.

Proof (Proof of Theorem 1).
If $\ell = 0$ or $m = 0$, then we are only dealing with P_4-free graphs. It follows from the Strong Perfect Graph Theorem [14] that P_4-free graphs are perfect. Moreover, there is exactly one perfect k-vertex-critical graph, K_k, so the result is immediate. So, let k, ℓ and m be given positive integers and let G be a k-vertex-critical $(P_4 + \ell P_1,\ B_4(m),\ B_3(m)^+)$-free graph. We will show the equivalent statement that G necessarily has finite order. Let $c = \ell + m + 1$. If $\alpha(G) < c$, then Ramsey's Theorem [35] implies that there exists an n such that all graphs with at least n vertices contain an independent set of order c or a clique of order $k + 1$. Therefore there are only finitely many k-colorable graphs with $\alpha(G) < c$. So we may suppose $\alpha(G) \geq c$.

By Theorem 3 there are only finitely many k-vertex-critical $(P_3 + cP_1)$-free graphs. Thus, if G does not have finite order, there must exist an independent set S with $|S| = c$ such that $G - N[S]$ contains an induced P_3. As G is $(P_4 + \ell P_1)$-free, then $G - N[S]$ is P_4-free. Let C be a component of $G - N[S]$ that contains an induced P_3.

By Lemma 2, there must be anticomplete sets $X, Y \subseteq G - N[S]$ such that Y is complete to $N_C(X) \neq \emptyset$ and X is complete to $N_C(Y)$. Without loss of generality suppose $\chi(X) \leq \chi(Y)$. By Lemma 1, we must have $x \in X$ and $y \in Y$ such that there are vertices x', z such that $x' \in N(x) \setminus N(y)$ and $z \in N_C(x) \cap N_C(y)$. By our assumption, we know that $x' \in N(S)$. Without loss of generality, let s_1 be a neighbour of x' in S. See Fig. 3 for an illustration of G. Note the dashed lines are potential adjacencies of x'.

Let S' be the set $S \setminus N(x')$. If $|S'| \geq \ell$, then $\{z, x, x', s_1\} \cup S'$ induces a graph containing an induced $P_4 + \ell P_1$ if $z \nsim x'$ and $\{y, z, x', s_1\} \cup S'$ induces a graph containing an induced $P_4 + \ell P_1$ if $z \sim x'$. Therefore, $|S'| \leq \ell - 1$, and $|N(x') \cap S| \geq c - \ell + 1 = m$. Now, the union of $\{y, z, x, x'\}$ and any subset of S containing m neighbours of x' induces $B_4(m)$ if $z \nsim x'$ and $B_3(m)^+$ if $z \sim x'$.

Since we reach a contradiction in both cases, we obtain the desired result.

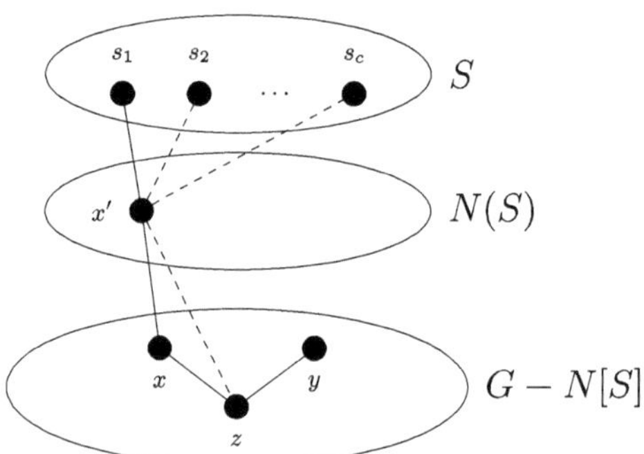

Fig. 3. Illustration of G in proof of Theorem 1.

4 $(P_4 + \ell P_1, 2P_2)$-Free

In this section, part of our proof (Claim 1) is an adaptation of part of the proof in [2] that $|\mathrm{crit}_k(P_4 + P_1, P_5, P_3 + cP_2)| < \infty$. The majority of our proof uses new techniques that allow us to get our result for $(P_4 + \ell P_1)$-free graphs for all $\ell > 1$ (as opposed to $\ell = 1$ in [2]).

Proof (Proof of Theorem 2). If $\ell = 0$, then the graph is P_4-free and the result is immediate (see comment at the beginning of the Proof of Theorem 1). Therefore, let $k \geq 1$ and $\ell \geq 1$ be given positive integers and let G be a k-vertex-critical $(P_4 + \ell P_1, 2P_2)$-free graph. We will prove that G is $(P_3 + cP_1)$-free for

$$c = \binom{k\ell}{\lfloor \frac{k\ell}{2} \rfloor}.$$

Suppose, by way of contradiction, that G contains an induced $P_3 + cP_1$ and let $P = \{p_1, p_2, p_3\}$ and $S = \{s_1, s_2, \ldots, s_c\}$ such that $P \cup S$ induces a $P_3 + cP_1$ where the P_3 is in order of the indices. Let M be the set of all vertices in $V(G)$ that are mixed on S. Partition M into sets such that all vertices with the exact same neighbours and non-neighbours in S belong to the same set of the partition. Let U be a subset of M defined by taking exactly one vertex from each of the sets in the partition. The following claim is a consequence of Sperner's Theorem [37] and a series of claims in [2] which only depend on G being k-vertex-critical and containing an induced $P_3 + cP_1$. To keep this paper self-contained, we will include its proof here.

Claim 1 ([2], Claim 3.4). $|U| \geq k\ell$.

Proof (Proof of Claim 1). We will show that $A = \{N(s) \cap U : s \in S\}$ is an antichain in the partial order $(\mathcal{P}(U), \subseteq)$. Then

$$|A| = c = \binom{k\ell}{\lfloor \frac{k\ell}{2} \rfloor},$$

and follows from Sperner's Theorem [37] that $|U| \geq k\ell$.

Suppose A is not an antichain. Then there exists $s_1, s_2 \in S$ such that $N(s_1) \cap U \subseteq N(s_2) \cap U$. As S is an independent set, then, by Lemma 1, there must be a vertex $u_1 \sim s_1$ and $u_1 \nsim s_2$. So u_1 is mixed on S and hence $u_1 \in M$. By the construction of U, there is a vertex in $u_1' \in U$ with the same neighbors and non-neighbors as u_1 in S. Thus $u_1' \sim s_1$ and $u_1' \nsim s_2$. Hence $u_1' \in N(s_1) \cap U$ but $u_1' \notin N(s_1) \cap U$ so $N(s_1) \cap U \nsubseteq N(s_2) \cap U$ which is a contradiction.

Claim 2. $\alpha(G[U]) \leq \ell$.

Proof (Proof of Claim 2). Suppose by way of contradiction that U contains an independent set of order at least $\ell + 1$ and let $U' = \{u_1, u_2, \ldots u_{\ell+1}\}$ be an independent subset of U. Moreover, without loss of generality, let

$$|N(u_1) \cap S| \leq |N(u_2) \cap S| \leq \cdots \leq |N(u_{\ell+1}) \cap S|.$$

We claim that $N(u_i) \cap S \subset N(u_j) \cap S$ for each $1 \leq i < j \leq \ell + 1$. To show a contradiction, suppose not. By our definition of U and U', clearly $N(u_i) \cap S \neq N(u_j) \cap S$ and $N(u_j) \cap S \not\subset N(u_i) \cap S$. So there exists $s_i, s_j \in S$ such that $s_i \in (N(u_i) \cap S) \setminus (N(u_j) \cap S)$ and $s_j \in (N(u_j) \cap S) \setminus (N(u_i) \cap S)$. Furthermore U' and S are each independent sets so $u_i \nsim u_j$ and $s_i \nsim s_j$. However, $\{u_i, s_i, u_j, s_j\}$ induce a $2P_2$ which is a contradiction. Therefore $N(u_i) \cap S \subset N(u_j) \cap S$ for each $1 \leq i < j \leq \ell + 1$ and hence

$$|N(u_1) \cap S| < |N(u_2) \cap S| < \cdots < |N(u_{\ell+1}) \cap S|.$$

By the definition of U, each u_i is mixed on S. Therefore $|N(u_{\ell+2}) \cap S| \leq |S| - 1$ and hence $1 \leq |N(u_1) \cap S| < |N(u_2) \cap S| \leq |S| - \ell$. Let $T \subset S$ be a set of ℓ vertices not adjacent to u_2. Note T is an independent set. Moreover, no vertex in T is adjacent to u_1 or any other vertex in S. Now let $s_1 \in N(u_1) \cap S$ and $s_2 \in (N(u_2) \cap S) \setminus (N(u_1) \cap S)$. As $N(u_1) \cap S \subset N(u_2) \cap S$ then $s_1 \in N(u_2) \cap S$ and hence $\{u_1, s_1, u_2, s_2\}$ induces a P_4. Moreover $\{u_1, s_1, u_2, s_2\} \cup T$ induces a $P_4 + \ell P_1$ which is a contradiction.

Claim 3. $\chi(G[U]) \geq k$.

Proof (Proof of Claim 3). We have $\frac{|U|}{\alpha(G[U])} \geq \frac{k\ell}{\ell} = k$ by Claims 1 and 3. Further, $\frac{|U|}{\alpha(G[U])} \leq \chi(G[U])$ by an elementary bound on the chromatic number. Therefore, we have $\chi(G[U]) \geq k$.

We now complete the proof of the theorem. Claim 3 contradicts G being k-vertex-critical since we will have $\chi(G - v) \geq k$ for all $v \in V(G) \setminus U$. So, it must be that G is $(P_3 + cP_1)$-free and therefore that there are only finitely many $(P_4 + \ell P_1, 2P_2)$-free graphs for all $k \geq 1$ by Theorem 3.

5 $(P_4 + \ell P_1, K_k)$-Free

In this section, we explore the possibility of removing the restriction of $B_4(m)$ from Theorem 1. In order for $|\mathrm{crit}_k(P_4 + \ell P_1, B_3(m)^+)| < \infty$ there must necessarily be only finitely many k-vertex-critical $(P_4 + \ell P_1, K_3)$-free graphs. We will resolve this for $k \geq \ell + 3$ through a colourability result on $(P_4 + \ell P_1, K_k)$-free graphs.

Note that it is known that every (P_t, K_3)-free graphs is $(t-2)$-colourable [36], so it follows that every $(P_4 + \ell P_1, K_3)$-free graph is $(2\ell + 2)$-colourable. We improve this result with a corollary of the following general result.

Theorem 4. *Let H be a finite graph, $\ell \geq 0$ and $k \geq 1$. If every (H, K_3)-free graph is k-colourable, then every $(H + \ell P_1, K_3)$-free graph is $(k + \ell)$-colourable*

Proof. Suppose for some finite graph H that every (H, K_3)-free graph is k-colourable. Let G be a $(H + \ell P_1, K_3)$-free graph. Note if $\ell = 0$ then G is (H, K_3)-free and hence k-colourable. So suppose $\ell \geq 1$ Let S be an independent set of size at most $\ell \geq 1$ such that either $|S| = \ell$ or, if no such independent set exists, $|S|$ is maximal. It suffices to show that G is $(|S| + k)$-colourable. Label the vertices such that $S = \{v_1, v_2, \ldots, v_{|S|}\}$. For each $i \in \{1, 2, \ldots |S|\}$ let S_i be the vertices adjacent to v_i in S but not adjacent to v_j for any $j < i$. Note S together with each S_i partitions $N[S]$. Moreover as G is K_3-free then each S_i is an independent set. Thus we can assign S and each S_i their own colour to colour $N[S]$ with $|S|+1$ colours. Now consider $G - N[S]$. Note that if $|S|$ is maximal then $G - N[S]$ is empty so G is $(|S| + 1)$-colourable and hence $(\ell + k)$-colourable as $|S| < \ell$ and $1 \leq k$. Thus we may assume $|S| = \ell$ in which case $G - N[S]$ is H-free. Therefore $G - N[S]$ is (H, K_3)-free and hence k-colourable We can reuse the the colour assigned to S and colour G with $|S| + k$ colours.

Thus, immediate corollaries of this are that every $(H + \ell P_1, K_3)$-free graph is $(\ell + 4)$-colourable when H is $2P_3$ or $P_4 + P_2$, as the 4-colourability of (H, K_3)-free graphs for each H was shown in [34] and [11], respectively. Further, from the result that every (P_t, K_3)-free graph is $(t - 2)$-colourable [36] we have the following corollary.

Corollary 5. *For all $\ell \geq 0$ and $t \geq 4$, every $(P_t + \ell P_1, K_3)$-free graph is $(t + \ell - 2)$-colourable.*

It was shown in [19] that $\chi(G) \leq (t - 2)^{\omega(G)-1}$ for P_t-free graphs. This χ-bound implies that every $(P_4 + \ell P_1, K_k)$-free graph is $(2\ell + 2)^{k-2}$−colourable. We improve this in the next result by generalizing Corollary 5 for $t = 4$.

Theorem 5. *For all $\ell \geq 0$ and $k \geq 3$, every $(P_4 + \ell P_1, K_k)$-free graph G has*

$$\chi(G) \leq \ell^{k-2} + 2\ell^{k-3} + 3\ell^{k-4} \cdots + (k-2)\ell + (k-1).$$

Proof. We proceed by induction on k. The case of $k = 3$ follows from Corollary 5. So suppose for all $\ell \geq 0$ and some $k \geq 3$ that every graph is $f(k)$-colourable where $f(k) = \ell^{k-2} + 2\ell^{k-3} + \cdots + (k-1)$. Let G be a $(P_4 + \ell P_1, K_{k+1})$-free graph. Note if $\ell = 0$ then G is P_4-free and hence perfect. Therefore $\chi(G) = \omega(G) \leq k$ as G is K_{k+1}-free. Let S be an independent set of size at most $\ell \geq 1$ such that either $|S| = \ell$ or S is maximal. It suffices to show that G is $(|S| \cdot f(k) + k)$-colourable. Label the vertices such that $S = \{v_1, v_2, \ldots, v_{|S|}\}$. Moreover, for each $i \in \{1, 2, \ldots |S|\}$, let S_i be the vertices adjacent to v_i in S but not adjacent to v_j for any $j < i$. Note S together with each S_i partition $N[S]$. Since G is K_{k+1}-free, each S_i must be K_k-free and hence $(P_4 + \ell P_1, K_k)$-free graph. Thus, by induction, each S_i is $f(k)$-colourable. So we can colour $N[S]$ with $|S| \cdot f(k) + 1$ colours by also assigning S its own colour. Now consider $G - N[S]$. Note that if S is maximal then $G - N[S]$ is empty so G is $(|S| \cdot f(k) + 1)$-colourable. Thus we may assume $|S| = \ell$ in which case $G - N[S]$ is P_4-free. Therefore $G - N[S]$ is perfect and $\chi(G - N[S]) = \omega(G - N[S]) \leq k$ as G is K_{k+1}-free. We can reuse the the colour assigned to S and colour G with $|S| \cdot f(k) + k$ colours.

6   Conclusion

In this work we gave the first results on the number of vertex-critical graphs in subfamilies of $(P_4 + \ell P_1)$-free graphs with no restrictions on ℓ.

A natural open question is if it is possible to remove the restriction of either $B_4(m)$ or $B_3(m)^+$ from Theorem 1. Removing the former would result in the highly desirable corollary that $|\mathrm{crit}_k(P_4 + \ell P_1, P_5)| < \infty$ for all $k \geq 1$ and $\ell \geq 0$. Removing the latter would have, as its most specific new corollary, that $|\mathrm{crit}_k(P_4 + \ell P_1, K_3)| < \infty$ which was discussed in Sect. 5. We think this is a particularly interesting open problem, especially given the very recent results that all $(P_5 + P_1, K_3)$-free graphs are 3-colourable and $|\mathrm{crit}_4(P_{11}, K_3)| = \infty$ [43].

Problem 1. Determine if $|\mathrm{crit}_k(P_4 + \ell P_1, K_3)|$ is finite for all $\ell \geq 2$.

Beyond the number of vertex-critical graphs in subfamilies, the colourability in general is often of interest. Our results in Sect. 5 imply that every $(P_4 + \ell P_1, K_k)$-free graph is $\mathcal{O}(\ell^{k-2})$-colourable. However, a conjecture in [38] posits that (P_t, K_k)-free graphs are polynomial-bounded by k. Which leads to the following analogous problem.

Problem 2. For each $k \geq 3$, determine the precise value M_k such that $\chi(G) \leq M_k$ for all $(P_4 + \ell P_1, K_k)$-free graphs G.

Another direction for future research would be to use our results to prove the finiteness of k-vertex-critical (P_5, H)-free graphs when H is any of chair, bull, or cricket. The idea would be that from our results, it is known that there are only finitely many that are $(P_4 + \ell P_1)$-free, so it remains only to prove that there are only finitely many that contain an induced $P_4 + \ell P_1$. This could be a promising technique as ℓ can be made as large as possible as long as it only depends on k, and thereby giving lots of control to the interested researcher.

Acknowledgments. The research of Ben Cameron was supported by the Natural Sciences and Engineering Research Council of Canada (NSERC), grants RGPIN-2022-03697 and DGECR-2022-00446. The research of Iain Beaton was also supported by NSERC grants RGPIN-2025-06012 and DGECR-2025-00001.

Disclosure of Interests. The author(s) have no competing interests to declare that are relevant to the content of this article.

References

1. Abuadas, T., Cameron, B., Hoàng, C.T., Sawada, J.: Vertex-critical $(P_3 + \ell P_1)$-free and vertex-critical (gem, co-gem)-free graphs. Discrete Appl. Math. **344**, 179–187 (2024). https://doi.org/10.1016/j.dam.2023.11.042
2. Beaton, I., Cameron, B.: Vertex-critical graphs in co-gem-free graphs. Theoret. Comput. Sci. **1042**, 115234 (2025). https://doi.org/10.1016/j.tcs.2025.115234
3. Brandstädt, A., Le, V.B., Spinrad, J.P.: Graph classes: a survey. SIAM (1999)
4. Brause, C., Geißer, M., Schiermeyer, I.: Homogeneous sets, clique-separators, critical graphs, and optimal χ-binding functions. Discrete Appl. Math. **320**, 211–222 (2022). https://doi.org/10.1016/j.dam.2022.05.014
5. Cai, Q., Huang, S., Li, T., Shi, Y.: Vertex-critical $(P_5,$ banner)-free graphs. In: Chen, Y., Deng, X., Lu, M. (eds.) FAW 2019. LNCS, vol. 11458, pp. 111–120. Springer, Cham (2019). https://doi.org/10.1007/978-3-030-18126-0_10
6. Cai, Q., Goedgebeur, J., Huang, S.: Some results on k-critical P_5-free graphs. Discrete Appl. Math. **334**, 91–100 (2023). https://doi.org/10.1016/j.dam.2023.03.008
7. Cameron, B., Hoàng, C.T., Sawada, J.: Dichotomizing k-vertex-critical H-free graphs for H of order four. Discrete Appl. Math. **312**, 106–115 (2022). https://doi.org/10.1016/j.dam.2021.11.001
8. Cameron, B., Hoàng, C.T.: A refinement on the structure of vertex-critical $(P_5,$ gem)-free graphs. Theoret. Comput. Sci. **961**, 113936 (2023). https://doi.org/10.1016/j.tcs.2023.113936
9. Cameron, B., Hoàng, C.T.: Infinite families of k-vertex-critical (P_5, C_5)-free graphs. Graphs Combin. **40**, 30 (2024)
10. Cameron, K., Goedgebeur, J., Huang, S., Shi, Y.: k-Critical graphs in P_5-free graphs. Theoret. Comput. Sci. **864**, 80–91 (2021). https://doi.org/10.1016/j.tcs.2021.02.029
11. Chen, R., Wu, D., Zhang, X.: Structure, perfect divisibility and coloring of $(P_2 \cup P_4, C_3)$-free graphs. preprint, arXiv:2509.14135 [math.CO] (2025)

12. Chudnovsky, M., Goedgebeur, J., Schaudt, O., Zhong, M.: Obstructions for three-coloring and list three-coloring H-free graphs. SIAM J. Discrete Math. **34**(1), 431–469 (2020)

13. Chudnovsky, M., Hajebi, S., Spirkl, S.: List-k-Coloring H-Free Graphs for All $k>4$. Combinatorica **44**(5), 1063–1068 (2024). https://doi.org/10.1007/s00493-024-00106-2

14. Chudnovsky, M., Robertson, N., Seymour, P., Thomas, R.: The strong perfect graph theorem. Ann. Math. **164**(1), 51–229 (2006). https://doi.org/10.4007/annals.2006.164.51

15. Chudnovsky, M., Spirkl, S., Zhong, M.: Four-coloring P_6-free graphs. I. Extending an excellent precoloring. SIAM J. Comput. **53**(1), 111–145 (2024). https://doi.org/10.1137/18M1234837

16. Chudnovsky, M., Spirkl, S., Zhong, M.: Four-coloring P_6-free graphs. II. Finding an excellent precoloring. SIAM J. Comput. **53**(1), 146–187 (2024). https://doi.org/10.1137/18M1234849

17. Couturier, J.F., Golovach, P.A., Kratsch, D., Paulusma, D.: List coloring in the absence of a linear forest. Algorithmica **71**(1), 21–35 (2015). https://doi.org/10.1007/s00453-013-9777-0

18. Dhaliwal, H.S., Hamel, A.M., Hoàng, C.T., Maffray, F., McConnell, T.J.D., Panait, S.A.: On color-critical $(P_5,$co-$P_5)$-free graphs. Discrete Appl. Math. **216**, 142–148 (2017). https://doi.org/10.1016/j.dam.2016.05.018

19. Gravier, S., Hoàng, C.T., Maffray, F.: Coloring the hypergraph of maximal cliques of a graph with no long path. Discrete Math. **272**(2), 285–290 (2003). https://doi.org/10.1016/S0012-365X(03)00197-3. https://www.sciencedirect.com/science/article/pii/S0012365X03001973

20. Hajebi, S., Li, Y., Spirkl, S.: Complexity dichotomy for list-5-coloring with a forbidden induced subgraph. SIAM J. Discrete Math. **36**(3), 2004–2027 (2022). https://doi.org/10.1137/21M1443352

21. Hoàng, C.T., Kamiński, M., Lozin, V., Sawada, J., Shu, X.: Deciding k-colorability of P_5-free graphs in polynomial time. Algorithmica **57**, 74–81 (2010). https://doi.org/10.1007/s00453-008-9197-8

22. Hoàng, C.T., Moore, B., Recoskie, D., Sawada, J., Vatshelle, M.: Constructions of k-critical P_5-free graphs. Discrete Appl. Math. **182**, 91–98 (2015). https://doi.org/10.1016/j.dam.2014.06.007

23. Holyer, I.: The NP-completeness of edge-coloring. SIAM J. Comput. **10**(4), 718–720 (1981). https://doi.org/10.1137/0210055

24. Huang, S.: Improved complexity results on k-coloring P_t-free graphs. European J. Combin. **51**, 336–346 (2016). https://doi.org/10.1016/j.ejc.2015.06.005

25. Huang, S., Li, J., Xia, W.: Critical $(P_5,$ bull)-free graphs. Discrete Appl. Math. **334**, 15–25 (2023). https://doi.org/10.1016/j.dam.2023.02.019

26. Huang, S., Li, Z.: Vertex-critical $(P_5, chair)$-free graphs. Discrete Appl. Math. **341**, 9–15 (2023). https://doi.org/10.1016/j.dam.2023.07.014

27. Jua, Y., Jooken, J., Goedgebeur, J., Huang, S.: There are finitely many 5-vertex-critical $(P_6,$ bull)-free graphs [math.CO] (2025)

28. Kamiński, M., Lozin, V.: Coloring edges and vertices of graphs without short or long cycles. Contrib. Discrete Math. **2**(1), 61–66 (2007). https://doi.org/10.11575/cdm.v2i1.61890

29. Kamiński, M., Pstrucha, A.: Certifying coloring algorithms for graphs without long induced paths. Discrete Appl. Math. **261**, 258–267 (2019). https://doi.org/10.1016/j.dam.2018.09.031

30. Karp, R.M.: Reducibility among combinatorial problems. In: Complexity of computer computations (Proc. Sympos., IBM Thomas J. Watson Res. Center, Yorktown Heights, N.Y., 1972), pp. 85–103 (1972)
31. Leven, D., Gail, Z.: NP completeness of finding the chromatic index of regular graphs. J. Algorithms **4**, 35–44 (1983)
32. Maffray, F., Morel, G.: On 3-colorable P_5-free graphs. SIAM J. Discrete Math. **26**(4), 1682–1708 (2012). https://doi.org/10.1137/110829222
33. McConnell, R., Mehlhorn, K., Näher, S., Schweitzer, P.: Certifying algorithms. Comput. Sci. Rev. **5**(2), 119–161 (2011). https://doi.org/10.1016/j.cosrev.2010.09.009
34. Pyatkin, A.V.: Triangle-free $2P_3$-free graphs are 4-colorable. Discret. Math. **313**(5), 715–720 (2013). https://doi.org/10.1016/j.disc.2012.10.019
35. Ramsey, F.P.: On a problem of formal logic. Proc. Lond. Math. Soc. (s2–30), 264–286 (1928)
36. Randerath, B., Schiermeyer, I.: Vertex colouring and forbidden subgraphs–a survey. Graphs Combin. **20**(1), 1–40 (2004). https://doi.org/10.1007/s00373-003-0540-1
37. Sperner, E.: Ein Satz über Untermengen einer endlichen Menge. Math. Z. **27**(1), 544–548 (1928). https://doi.org/10.1007/BF01171114
38. Trotignon, N., Pham, L.A.: χ-bounds, operations, and chords. J. Graph Theory **88**(2), 312–336 (2018)
39. West, D.B.: Introduction to Graph Theory. Prentice Hall Inc, Upper Saddle River (1996)
40. Xia, W., Jooken, J., Goedgebeur, J., Beaton, I., Cameron, B., Huang, S.: Vertex-critical (P_5, W_4)-free graphs. In: Fomin, F.V., Xiao, M. (eds.) Computing and Combinatorics, pp. 139–152. Springer, Singapore (2026)
41. Xia, W., Jooken, J., Goedgebeur, J., Huang, S.: Critical $(P_5, dart)$-free graphs. In: Combinatorial Optimization and Applications: 16th International Conference, COCOA 2023, Hawaii, HI, USA, 15–17 December 2023, Proceedings, Part II, pp. 390–402. Springer, Heidelberg (2023). https://doi.org/10.1007/978-3-031-49614-1_29
42. Xia, W., Jooken, J., Goedgebeur, J., Huang, S.: Some results on critical (P_5, H)-free graphs. Theoret. Comput. Sci. **1051**, 115411 (2025). https://doi.org/10.1016/j.tcs.2025.115411
43. Zhou, Y., Jooken, J., Shan, B., Goedgebeur, J., Huang, S.: Three-coloring triangle-free graphs without long forbidden paths [math.CO] (2025)

Breadth-First Search Trees with Many
or Few Leaves

Jesse Beisegel[1], Ekkehard Köhler[1], Robert Scheffler[1(✉)], and Martin Strehler[2]

[1] Institute of Mathematics, Brandenburg University of Technology,
Cottbus, Germany
{jesse.beisegel,ekkehard.koehler,robert.scheffler}@b-tu.de
[2] Department of Mathematics, Westsächsische Hochschule Zwickau,
Zwickau, Germany
martin.strehler@whz.de

Abstract. The Maximum (Minimum) Leaf Spanning Tree problem asks for a spanning tree with the largest (smallest) number of leaves. As spanning trees are often computed using graph search algorithms, it is natural to restrict this problem to the set of search trees of some particular graph search, e.g., find the Breadth-First Search (BFS) tree with the largest number of leaves. We study this problem for Generic Search (GS), BFS and Lexicographic Breadth-First Search (LBFS) using search trees that connect each vertex to its first neighbor in the search order (*first-in trees*) just like the classic BFS tree. In particular, we analyze the complexity of these problems, both in the classical and in the parameterized sense.

Among other results, we show that the minimum and maximum leaf problems are in FPT for the first-in trees of GS, BFS and LBFS when parameterized by the number of leaves in the tree. However, when these problems are parameterized by the number of internal vertices of the tree, they are W[1]-hard for the first-in trees of GS, BFS and LBFS.

Keywords: graph search · spanning tree · parameterized complexity

1   Introduction

The study of the number of leaves in a spanning tree is a well-established area within graph theory. On the one hand, the problem of minimizing the number of leaves is related to the HAMILTONIAN PATH problem: a spanning tree of a graph G with exactly one leaf corresponds to a Hamiltonian path in G.[1] More broadly, this problem is equivalently known as the MAXIMUM INTERNAL SPANNING TREE problem (MIST). For this problem, there are both Fixed-Parameter Tractable (FPT) results when parameterized by the number of leaves [24,38,41] as well as polynomial-time algorithms for specific graph classes. Linear-time algorithms

---

[1] Note that we only consider rooted spanning trees here and we do not count the root as a leaf even if it has degree 1, following the convention in [7].

© The Author(s), under exclusive license to Springer Nature Switzerland AG 2026
F. Foucaud and A. Parreau (Eds.): IWOCA 2026, LNCS 16587, pp. 131–145, 2026.
https://doi.org/10.1007/978-3-032-27732-9_10

exist for, e.g., interval graphs, block graphs, cactus graphs, or chain graphs [37, 50].

On the other hand, maximizing the number of leaves is known as the MAXIMUM LEAF SPANNING TREE problem (MLST). This particular problem is NP-hard not only for general graphs (problem ND2 in [26]), but also for specific graph classes such as planar cubic graphs [43]. When viewed from the perspective of parameterized complexity, there is a wide range of FPT algorithms for this problem when parameterized by the number of leaves [9,10,22,23,34,42,53]. The parametric dual of the problem, i.e., finding a spanning tree with minimal number of internal vertices, is equivalent to the W[2]-complete problem of finding a minimum connected dominating set [25].

Another line of research has focused on studying the properties of search orderings and their associated search trees (see [18] for a survey on graph searches). Graph searches[2] like *Breadth-First Search* (BFS), *Depth-First Search* (DFS), or their variants, are crucial components in a multitude of graph algorithms. For example, *Lexicographic BFS* (LBFS) or *Maximum Cardinality Search* (MCS) can be used to recognize chordal graphs in linear time [46,51]. Furthermore, the search ordering derived from LBFS or MCS on a chordal graph can also be used to compute an optimal coloring, a maximum independent set or a maximum clique in linear time. Here, the end vertex of the search plays a pivotal role, as it is proven to be simplicial and a valid starting vertex for a perfect elimination ordering (see [27] for details). As a consequence, the end vertex problem has become a key area of study within this context, see, e.g., [3,15,17,36,44,45].

Another structure related to graph searches are their search trees. The complexity of deciding whether a spanning tree can be constructed by a certain graph search was first studied in the 1980s for BFS and DFS [29,30,35,40]. More recently, this problem has also been considered for searches such as LBFS and MCS [4,5]. As observed in [48], this research unveiled a strong relation to the end vertex problem. This relation also lead to the study of leaf recognition, i.e., the question whether a vertex can be a leaf in a search tree [49].

Recently, Bergougnoux et al. [7] have combined the study of spanning trees with few or many leaves with the concept of search trees. They examine the algorithmic problems of finding DFS trees with either few or many leaves, highlighting potential applications in network design. They showed that the problems are hard when parameterized by the numbers of leaves but in FPT when parameterized by the number of internal vertices. It therefore seems logical to explore similar questions for other searches and their corresponding trees.

Note that BFS trees and DFS trees differ significantly in their structure. While DFS trees connect each vertex v to the last neighbor visited before v, BFS trees connect each vertex to its first visited neighbor. Following [4], we call the first concept the *last-in tree* (or $\mathcal{L}$-tree) of the search ordering, whereas the second is called the *first-in tree* (or $\mathcal{F}$-tree). DFS trees tend to have few leaves as, for example, every Hamiltonian path is a DFS tree. In contrast, BFS trees

---

[2] Note that in this paper, the term *graph searches* always refers to *connected searches*, i.e., each prefix of a search ordering induces a connected graph.

tend to have many leaves. For example, if we start the BFS in a universal vertex, then all vertices except the root are leaves. It therefore seems to be a reasonable hypothesis that the parameterized complexity of finding BFS trees with many or few leaves differs significantly from that of DFS trees for the natural parameters *number of leaves* and *number of internal vertices*.

Our results. We study the problem of finding $\mathcal{F}$-trees with few or many leaves. For a given graph G and a search paradigm $\mathcal{A}$ such as BFS, we call an ordering constructed by $\mathcal{A}$ when applied to G an $\mathcal{A}$-*ordering* of G. We consider both the minimizing and maximizing versions of the following two problems, which differ in their use of parameter k. The first is the MIN-LEAF (MAX-LEAF) $\mathcal{F}$-TREE of search $\mathcal{A}$, where given a graph G and a parameter k we ask for an $\mathcal{A}$-ordering whose $\mathcal{F}$-tree has at most (least) k leaves.

The second is the MIN-INTERNAL (MAX-INTERNAL) $\mathcal{F}$-TREE of search $\mathcal{A}$, where given a graph G and a parameter k we ask for an $\mathcal{A}$-ordering of G whose $\mathcal{F}$-tree has at most (least) k internal vertices. Note that even though the parameter k does not appear in the problem name, it is essential in differentiating these two problems, as – without parameterization – minimizing the number of leaves is the same as maximizing the number of the internal vertices and vice versa.

In particular, we consider GS, BFS, LBFS and related graph searches that share the so-called clique-starter property. Our results presented in this paper are summarized in Table 1. We are not aware of any previous work on these problems for these searches. Omitted proofs can be found in the full version [6].

Table 1. Results given in this paper. The first part of an entry concerns hardness results and the second entry concerns algorithms. For all considered cases, W[1]- and W[2]-hardness also implies NP-hardness of the non-parameterized problem.

results	GS	BFS	LBFS	other clique starters
MIN-LEAF	NP-compl. / FPT	NP-compl. / FPT	NP-compl. / FPT	NP-hard / ?
MAX-LEAF	NP-compl. / FPT	NP-compl. / FPT	NP-compl. / FPT	NP-hard / ?
MAX-INTERNAL	W[1]-compl. / XP	W[1]-hard / XP	W[1]-hard / ?	W[1]-hard / ?
MIN-INTERNAL	W[2]-compl. / XP	W[2]-hard / XP	para-NP-hard	W[2]-hard / ?

2   Preliminaries

All graphs in this paper are undirected, simple, connected, and non-empty. Given a graph G, we denote the set of *vertices* by $V(G)$ and the set of *edges* by $E(G)$. As is customary, we use $n = |V|$ and $m = |E|$ to represent the number of vertices and edges, respectively. We use $N_G(v)$ to denote the *(open) neighborhood* of a vertex v in the graph G, while $N_G[v] = N_G(v) \cup \{v\}$ is the *closed neighborhood* For further graph theoretic concepts, we refer to [52].

A *graph search* $\mathcal{A}$ is a procedure to traverse all vertices of a connected graph. The associated *search ordering* $\sigma = (v_1, \ldots, v_n)$ represents the order in which the vertices are visited. The *bandwidth of a vertex ordering* σ is the maximum distance of adjacent vertices in the ordering, i.e., $\mathrm{bw}(\sigma) = \max_{v_i v_j \in E(G)} |i - j|$. The *bandwidth of a graph* $\mathrm{bw}(G)$ is the minimum bandwidth over all orderings.

In this paper, we consider the following searches $\mathcal{A}$. *Generic Search* (GS) imposes no other conditions on the search than connectivity, meaning that after the first node is visited each newly visited node must be a neighbor of an already visited node. BFS refers to the standard *Breadth-First Search* implemented via a queue data structure [33]. *Lexicographic Breadth-First Search* (LBFS) is similar to BFS, but uses partition refinement as an improved tie-breaker (see [28,46] for details). An $\mathcal{A}_\rho^+$-search is a search following the search paradigm $\mathcal{A}$ using a given linear ordering ρ of the vertices as a tie-breaker. Whenever $\mathcal{A}$ faces a tie, i.e., several vertices are valid choices for the next vertex, it uses the left-most of these vertices in ρ to proceed. Such +-searches usually appear in multi-sweep algorithms where ρ was obtained through a preceding search.

We say that a graph search $\mathcal{A}$ is a *clique starter* if for every graph G, every clique C of G and every ordering σ of the vertices in C, there is an $\mathcal{A}$-ordering of G that starts with σ. Most of the graph searches considered in the literature, in particular GS, BFS, and LBFS, are clique starters.

A spanning tree of G rooted at a vertex $r \in V(G)$ is denoted by (T, r) or simply by T whenever the root is clear from context. A vertex $v \in V(T)$ is a *leaf* if it has no descendants. Otherwise, it is called an *internal vertex*. Note that we do not consider the root to be a leaf, even if it has degree one. A spanning tree T of G is an $\mathcal{F}$-*tree* associated with search ordering $\sigma = (r = v_1, v_2, \ldots, v_n)$ if it is rooted in r and for any vertex v_i, $1 < i \leq n$, among all vertices in $N(v_i)$, the parent of v_i in the tree T is left-most in σ.

For a detailed presentation of the concepts of parameterized complexity and the definition of the considered width-parameters, we refer the reader to [19,21]; for the definition of graph classes, we refer to [11].

3   Number of Leaves as Parameter

Here, we will present parameterized algorithms for both MIN-LEAF $\mathcal{F}$-TREE and MAX-LEAF $\mathcal{F}$-TREE for several search paradigms. Note that deciding whether a particular vertex can be a leaf of an $\mathcal{F}$-tree is NP-complete for almost all considered searches [49]. This also holds for sets of k fixed vertices with $k \geq 2$ by just appending leaves to the graph. Therefore, this problem seems to require a more sophisticated approach than simply checking all possible sets of k vertices.

To see the difference between the general spanning tree problems and the $\mathcal{F}$-tree problems, we compare them on the examples given in Fig. 1. By definition, any $\mathcal{F}$-tree is also a spanning tree. Therefore, when considering the minimization problems, any minimum leaf number for spanning trees is a lower bound for the number of leaves of an $\mathcal{F}$-tree. However, the solutions can be arbitrarily far apart from each other: In the left graph of Fig. 1, we give a family of graphs where

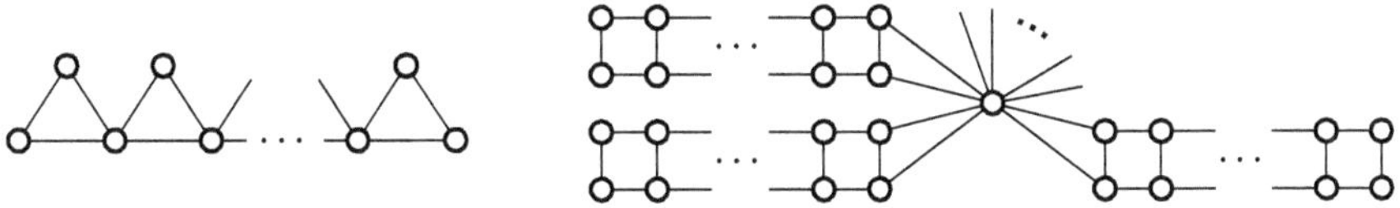

Fig. 1. In the left graph the minimum number of leaves in a spanning tree is 1, whereas any BFS-tree has at least $n/2$ leaves. The right graph is a star of k ladders with $2k$ vertices each. This graph has a spanning tree with k^2 leaves while every BFS-tree has at most $3k$ leaves.

the minimum number of leaves in a general spanning tree is 1 and the minimum number of leaves in a BFS-tree is $\Omega(n)$. For the maximization problems, any solution of the $\mathcal{F}$-tree problem is clearly a lower bound for the spanning tree problem. However, the right graph of Fig. 1 shows an example where the maximum number of leaves in a BFS-tree is $\mathcal{O}(\sqrt{n})$, while there are spanning trees with $\Omega(n)$ leaves. We leave open whether this gap can be increased, i.e., whether there is a family of graphs in which the maximum number of leaves in any $\mathcal{F}$-tree is constant and there are spanning trees with $\Omega(n)$ leaves.

To start with, we focus on *layered* searches [16], i.e., graph searches that traverse the distance layers of the start vertex in increasing order. Well-known examples of such searches are BFS and LBFS.

Observation 3.1. *For graphs with at least two vertices, the number of leaves of the $\mathcal{F}$-tree of some layered search ordering is at least the maximum number of vertices in a layer.*

It is a well-known fact that for layered searches the edges of a graph only connect vertices in the same layer or in consecutive layers. Thus, the observation above implies the following.

Corollary 3.2. *Let G be a graph with at least two vertices. If the $\mathcal{F}$-tree of a layered graph search ordering σ of G has at most k leaves, then the bandwidth of σ is at most $2k - 1$.*

We can improve this bound when we consider BFS orderings.

Theorem 3.3. *If the $\mathcal{F}$-tree T of a BFS ordering σ of graph G has at most k leaves, then the bandwidth of σ is at most k.*

Proof. Let v be a vertex in G and let w be the rightmost neighbor of v in σ. Let $\sigma' = (v_1, \ldots, v_\ell = w)$ be the subordering of σ between v and w, i.e., v_1 is the direct successor of v in σ. The parents of all vertices in σ' lie to the left of v_1 in σ. Thus, none of the vertices in σ' can have a descendant in T that is also in σ'. Therefore, there are at least ℓ leaves in T. Thus, $\ell \leq k$. $\qquad\square$

As we will see later (Theorems 4.1 and 4.3), both the MIN-LEAF $\mathcal{F}$-TREE and the MAX-LEAF $\mathcal{F}$-TREE problem is NP-hard for BFS and for LBFS. However, we can show that these problems are in FPT.

Theorem 3.4. *Min-Leaf $\mathcal{F}$-Tree and Max-Leaf $\mathcal{F}$-Tree of BFS and LBFS can be solved in* FPT *time.*

Proof. We first consider Min-Leaf $\mathcal{F}$-Tree. For every possible start vertex r, we do the following. We first compute the layers of the BFS. If there is a layer with more than k vertices, then we can reject that start vertex, due to Observation 3.1. Otherwise, we use dynamic programming to solve the problem. Let $V_{\leq i}$ be the union of all layers with index $\leq i$. For every ordering σ of the i-th layer, we have the value $M[i,\sigma]$ that contains ∞ if there is no (L)BFS ordering starting with r that contains σ as subordering. Otherwise, it contains the minimal number of leaves in $V_{\leq i}$ of the $\mathcal{F}$-tree of an (L)BFS ordering of $V_{\leq i+1}$ starting in r that has σ as subordering.

Layer 0 has exactly one ordering (r) and we have $M[0,(r)] = 0$. So assume that for some value i, we have computed the values $M[i-1,\sigma]$ for all orderings σ of layer $(i-1)$. Now consider an entry $M[i,\tau]$. If layer i is the last layer, then all vertices of the layer are leaves in the $\mathcal{F}$-tree of every BFS ordering starting in r. If it is not the last layer, then the ordering τ directly implies which of the vertices of layer i has a child in layer $i+1$. If some vertex does not have a child, then it has to be a leaf in the $\mathcal{F}$-tree of every BFS ordering starting in r that contains τ as subordering. Therefore, we compute the number $\ell(i,\tau)$ of leaves in layer i for the ordering τ. To compute $M[i,\tau]$, we now check for every ordering σ of layer $i-1$ whether τ can be the ordering of layer i when σ is the ordering of layer $i-1$. Note that this can be done in polynomial time. If this is the case, then the value $M[i-1,\sigma] + \ell(i,\tau)$ gives the number of leaves in $V_{\leq i}$ if σ and τ are the orderings of layer $i-1$ and layer i, respectively. We minimize this value over all suitable orderings σ. The minimum value is chosen for $M[i,\tau]$.

If we have computed all entries of M, we check for the last layer i whether there is an entry $M[i,\sigma] \leq k$. If so, then we return "Yes". If no start vertex works, then we return "No". Maximization works analogously with the only difference being that $M[i,\sigma]$ contains the maximum value of leaves or 0 if there is no possible (L)BFS ordering. Note that we can stop the maximization procedure directly with a positive answer as soon as we find a layer with $\geq k$ elements or an entry $M[i,\sigma] \geq k$. The minimization procedure for a fixed start vertex can be stopped if every entry $M[i,\sigma]$ of some layer i is larger than k.

Finally, let us consider the running time of our algorithm. We have to check at most n start vertices. For every of the at most n layers, there are at most $k!$ entries of M. To compute such an entry, we have to check at most $k!$ other entries and have to do a computation polynomial in k. So the final running time is in $\mathcal{O}(k!^2 \cdot k^{\mathcal{O}(1)} \cdot n^2)$. $\qquad\square$

Next we show that Max-Leaf $\mathcal{F}$-Tree of GS is equivalent to the problem Max-Leaf Spanning Tree.

Theorem 3.5. *Let G be a connected graph and let S be a set of vertices of G. The following statements are equivalent.*

1. There is a GS $\mathcal{F}$-tree where all elements of S are leaves.

2. *There is a spanning tree of G where all elements of S are leaves.*
3. *The set $V(G) - S$ forms a connected dominating set of G.*

Note that the equivalence of the MAX-LEAF SPANNING TREE problem and the connected dominating set problem has been already observed earlier (see, e.g., [25]). Theorem 3.5 implies the following.

Corollary 3.6. *Let G be a connected graph with n vertices. The following statements are equivalent.*

1. *There is a GS $\mathcal{F}$-tree of G with $\geq k$ leaves.*
2. *There is a spanning tree of G with $\geq k$ leaves.*
3. *There is a connected dominating set of G with $\leq n - k$ vertices.*

As already mentioned in the introduction, there is a wide range of FPT results for MAX-LEAF SPANNING TREE when parameterized by the number of leaves [9,10,22,23,34,42]. The currently best known algorithm runs in $\mathcal{O}^*(3.188^k)$ time [53]. By Corollary 3.6, this implies the following.

Corollary 3.7. MAX-LEAF $\mathcal{F}$-TREE *of GS is in* FPT.

To complement this result, we will prove next that MIN-LEAF $\mathcal{F}$-TREE of GS is also in FPT. We start with the observation that graphs having GS $\mathcal{F}$-trees with a small number of leaves have also small pathwidth. The proof follows the same lines as the proofs of Theorem 2.2.1 in [1] and Lemma 1 in [8].

Lemma 3.8. *If a graph G has an GS $\mathcal{F}$-tree with at most k leaves, then the pathwidth of G is at most k. This bound is tight.*

Note that – in contrast to layered searches – we cannot bound the bandwidth of such graphs. Consider the path with n vertices and an additional universal vertex. This graph has a GS $\mathcal{F}$-tree with two leaves, but its bandwidth is $\geq \lceil \frac{n}{2} \rceil$.

Due to Lemma 3.8, it is sufficient to show that MIN-LEAF $\mathcal{F}$-TREE of GS can be solved in FPT time when parameterized by pathwidth to also show that MIN-LEAF $\mathcal{F}$-TREE of GS can be solved in FPT time when parameterized by the number of leaves. Even more general, we will solve MIN-LEAF $\mathcal{F}$-TREE of GS parameterized by the treewidth. To this end, we relate the $\mathcal{F}$-trees of GS to two concepts that are well studied in the literature.

The first concept are *dominating sequences* which are sequences $(v_1, \ldots, v_k)$ of vertices where every vertex dominates some vertex that was not dominated by any vertex to the left of it. More formal, for all $i \in [k]$ it holds that $N\langle v_i \rangle_1 \setminus \bigcup_{j=1}^{i-1} N\langle v_j \rangle_2 \neq \emptyset$, where $N\langle \cdot \rangle_1$ and $N\langle \cdot \rangle_2$ are replaced by $N(\cdot)$ or $N[\cdot]$, depending on the type of dominating sequence (see [12–14]). Here, we are interested in so-called *Z-sequences*, where $N\langle \cdot \rangle_1 = N(\cdot)$ and $N\langle \cdot \rangle_2 = N[\cdot]$. These sequences were introduced in [12]. We adapt them in such a way that the sequence forms a prefix of a GS ordering.

Definition 3.9. *Let G be a graph. A sequence $(v_1, \ldots, v_k)$ of vertices is a generic Z-sequence of G if it is the prefix of a GS ordering of G and if for all $i \in [k]$ it holds that $N(v_i) \setminus \bigcup_{j=1}^{i-1} N[v_j]$ is not empty.*

A concept strongly related to dominating sequences are *zero forcing sets*. Such a set is a subset of vertices which are initially colored blue while all other vertices are colored white. Further, there is a set of certain color change rules that allow to change the color of white vertices to blue. There are many different color change rule schemes in the literature [2,31,39]. One of these rules, called *Z-rule* in [47], allows a blue vertex v to color a white neighbor w blue if and only if w is the only white neighbor of v. We adapt this rule as follows.

Z*-rule If v is a blue vertex and v has exactly one white neighbor w and w is either the unique white vertex in G or w has at least one white neighbor x, then change the color of w to blue. We write $v \xrightarrow{Z*} w$.

Given a graph G, we define a set $S \subseteq V(G)$ to be an Z^*-*forcing set* of G if the following procedure is able to color all vertices of G blue:

1. Color the vertices of S blue and all vertices of $V(G) \setminus S$ white.
2. Iteratively apply the Z^*-rule to the vertices of G.

The concepts generic Z-sequences, Z^*-forcing sets and minimum leaf $\mathcal{F}$-trees of GS are strongly related.

Theorem 3.10. *The following statements are equivalent for an n-vertex graph G with $n \geq 2$:*

1. *There is a GS $\mathcal{F}$-tree of G with $\leq k$ leaves.*
2. *There is a Z^*-forcing set of size $\leq k$.*
3. *There is a generic Z-sequence of length $\geq n - k$.*

The problem ZERO FORCING SET which uses the Z-rule mentioned above can be solved in FPT time when parameterized by the treewidth of G [8,47]. We adapt the algorithm given in [47] to also find a minimal Z^*-forcing set in FPT time when parameterized by the treewidth of G.

Theorem 3.11. *Given an n-vertex graph G of treewidth d, we can compute a smallest Z^*-forcing set of G in $2^{\mathcal{O}(d^2)} \cdot n$ time.*

As pointed out above, this implies an FPT algorithm for MIN-LEAF $\mathcal{F}$-TREE.

Theorem 3.12. *MIN-LEAF $\mathcal{F}$-TREE of GS can be solved in $2^{\mathcal{O}(d^2)} \cdot n$ time where d is the treewidth of the graph; it can be solved in $2^{\mathcal{O}(k^2)} \cdot n$ time where k is the number of leaves of the tree.*

4   Number of Internal Vertices as Parameter

4.1   Hardness for Search Trees with Few Internal Vertices

As we have seen in Corollary 3.6, MIN-INTERNAL $\mathcal{F}$-TREE is equivalent to CONNECTED DOMINATING SET. This problem is known to be NP-complete and W[2]-complete when parameterized by the size of the dominating set [19]. This implies that MIN-INTERNAL $\mathcal{F}$-TREE of GS is W[2]-complete and NP-complete and MAX-LEAF $\mathcal{F}$-TREE is NP-complete. As shown next, we can generalize these hardness results to all connected graph searches that are clique starters.

Theorem 4.1. *Let $\mathcal{A}$ be a connected graph search that is a clique starter. Then*

1. *MAX-LEAF $\mathcal{F}$-TREE of $\mathcal{A}$ is NP-hard on split graphs and*
2. *MIN-INTERNAL $\mathcal{F}$-TREE of $\mathcal{A}$ is W[2]-hard and NP-hard on split graphs.*

Proof. We reduce from SET COVER. An instance of this problem consists of a finite set $U = \{u_1, \ldots, u_n\}$ and a family $\mathcal{W} = \{W_1, \ldots, W_p\}$ of subsets of U as well as an integer $k \leq p$. The question is whether there are $\ell \leq k$ sets $W_{i_1}, \ldots, W_{i_\ell}$ such that $\bigcup_{j=1}^{\ell} W_{i_j} = U$. This problem is both NP-complete [32] and W[2]-complete [20] when parameterized by k. W.l.o.g. we may assume that $\bigcup_{j=1}^{p} W_j = U$, i.e., every element of U is in at least one set W_j. Furthermore, we assume that no element of U is contained in every set $W_i \in \mathcal{W}$ since such an element has no influence on the size of the minimum set cover.

We build the following split graph G. For every W_i, we construct a vertex c_i and for every u_i we construct a vertex a_i. We call the set of c-vertices C and the set of a-vertices A. We add edges to G such that the set C induces a clique while the set A induces an independent set. Furthermore, vertex c_i is made adjacent to vertex a_j if and only if the set W_i contains the element u_j. We claim that there is a set cover of size less than or equal to k if and only if there is an $\mathcal{F}$-tree of search $\mathcal{A}$ on G with no more than k internal vertices.

First assume there is a set cover $W_{i_1}, \ldots W_{i_\ell}$ with $\ell \leq k$. Since $\mathcal{A}$ is a clique starter, there is an $\mathcal{A}$-ordering σ with the prefix $(c_{i_1}, \ldots, c_{i_\ell})$. Since we consider a set cover, all vertices that are not part of this prefix have a neighbor in the prefix. This implies that all these vertices are leaves in the $\mathcal{F}$-tree of σ. Hence, this $\mathcal{F}$-tree has at most k internal vertices.

Now assume that there is an $\mathcal{A}$-ordering σ whose $\mathcal{F}$-tree T has at most k internal vertices. Let S be the set of internal vertices of T that are part of C. We claim that the family $\mathcal{W}' = \{W_i \mid c_i \in S\}$ forms a set cover of U. Let u_j be an arbitrary element of U. First assume that the vertex a_j is not the root of T. Then the parent of a_j is in C. Let us call the parent c_i. As c_i is an internal vertex of T, it is contained in S. Thus, $W_i \in \mathcal{W}'$ and u_j is covered by $\mathcal{W}'$. Now assume that a_j is the root of T. As $\mathcal{A}$ is a connected search, the second vertex of σ is a neighbor of a_j and, thus, part of C. Due to our assumption, u_j is not contained in every set W_i and, thus, there is at least one vertex $c_\ell \in C$ that is not adjacent to a_j. Let c_s be the second vertex of σ. Vertex c_s is different from c_ℓ and has c_ℓ as its child in the $\mathcal{F}$-tree of σ. Hence, c_s is contained in S and u_j is covered by W_s. $\qquad\square$

Note that we leave open whether there are clique starters other than CS for which MIN-INTERNAL $\mathcal{F}$-TREE is not just W[2]-hard but also W[2]-complete. The main difference between many of these searches and GS is the fact that the previously visited vertices have a significant influence on the ordering of the following vertices. In particular, the ordering of leaves of the $\mathcal{F}$-tree might influence the ordering of some internal vertices (see end of Sect. 4.3). Nevertheless, the problem is NP-complete for a search as long as its orderings can be recognized in polynomial time. This holds, e.g., for BFS. Adapting proofs given in [49], we can strengthen this result for LBFS.

Theorem 4.2. *For every fixed value $k \geq 3$, MIN-INTERNAL $\mathcal{F}$-TREE of LBFS is* NP-*complete on weakly chordal graphs.*

Note that the bound for k is tight. For all clique starters, it is easy to see that they have an $\mathcal{F}$-tree with at most $k \leq 2$ internal vertices if and only if the graph has a connected dominating set of size at most k.

4.2 Hardness for Search Trees with Many Internal Vertices

To prove that MAX-INTERNAL $\mathcal{F}$-TREE is W[1]-hard for all clique starters, we consider special vertex orderings of graphs that are related to the generic Z-sequences used in Sect. 3. A sequence $\sigma = (v_1, \ldots, v_k)$ of vertices of a graph G is called *total dominating sequence*[3] of G if for every $i \in [k]$ it holds that $N(v_i) \setminus \bigcup_{j=1}^{i-1} N(v_j)$ is not empty. This notion was introduced by Brešar et al. [14] where it is shown that the corresponding optimization problem that looks for a total dominating sequence of size at least k is NP-complete. Scheffler [47] considered a bipartite version called ONE-SIDED GRUNDY TOTAL DOMINATION. Here, a bipartite graph G is given with $V(G) = X \dot\cup Y$ and one asks whether there is a total dominating sequence of size k that only contains vertices of X. It is shown in [47] that ONE-SIDED GRUNDY TOTAL DOMINATION is both W[1]-complete and NP-complete. This allows us to prove the following result.

Theorem 4.3. *Let $\mathcal{A}$ be a connected graph search that is a clique starter. Then*

1. *MAX-INTERNAL $\mathcal{F}$-TREE of $\mathcal{A}$ is* W[1]-*hard and* NP-*hard on split graphs and*
2. *MIN-LEAF $\mathcal{F}$-TREE of $\mathcal{A}$ is* NP-*hard on split graphs.*

Proof. We reduce from ONE-SIDED GRUNDY TOTAL DOMINATION. Let G be a bipartite graph with $V(G) = X \dot\cup Y$ and $k \in \mathbb{N}$. W.l.o.g. we may assume that X does not contain isolated vertices as such vertices can never be part of a total dominating sequence. We construct the graph G' from G as follows: We add a vertex r to X and make X to a clique. Note that G' is a split graph. We claim that G has a total dominating sequence containing k vertices of X if and only if there is an $\mathcal{A}$-ordering of G' whose $\mathcal{F}$-tree has $\geq k + 1$ internal vertices.

First assume that $(v_1, \ldots, v_k)$ is a total dominating sequence of G containing k vertices of X. Consider an $\mathcal{A}$-ordering σ starting with the prefix $(r, v_1, \ldots, v_k)$. Since $\mathcal{A}$ is a clique starter, there exists such an $\mathcal{A}$-ordering. Let T be the $\mathcal{F}$-tree of σ. Obviously, r is an internal vertex of T. As $(v_1, \ldots, v_k)$ is a total dominating sequence, for each vertex v_i with $i \in [k]$, there is a vertex in the set $N(v_i) \setminus (N(r) \cup \bigcup_{j=1}^{i-1} N(v_i))$. This vertex is a child of v_i in T and, thus, v_i is an internal vertex of T.

Now assume that there is an $\mathcal{A}$-ordering σ of G' whose $\mathcal{F}$-tree has at least $k + 1$ internal vertices. If the start vertex is an element of X, then all vertices of X are its children. Thus, no vertex of Y can be an internal vertex of T. If

---

[3] Note that a total dominating sequence does not necessarily dominate the graph. Nevertheless, it can always be extended to such a sequence.

the start vertex is an element of $Y \cup \{r\}$, then the second vertex must be an element of $X \setminus \{r\}$ since $\mathcal{A}$ is a connected graph search. Again, no other vertex of Y can be an internal vertex of T as all vertices of X are children of either the start vertex or the second vertex. Summarizing, there are at least k internal vertices in $X \setminus \{r\}$. Let $(v_1, \ldots, v_k)$ be the ordering of the first k internal vertices of $X \setminus \{r\}$ in σ. We claim that $(v_1, \ldots, v_k)$ is a total dominating sequence of G. First consider v_1. Since v_1 is not isolated in G, it has at least one neighbor in Y and, thus, $N_G(v_1) \setminus \bigcup_{j=1}^{0} N_G(v_j)$ is not empty. So consider a vertex v_i with $i \geq 2$. This vertex has a child x in T. Vertex x is not adjacent to any vertex v_j with $j < i$. Thus, x cannot be an element of $X \cup \{r\}$ since these vertices are adjacent to v_1. So $x \in Y$ and, thus, $x \in N_G(v_i) \setminus \bigcup_{j=1}^{i-1} N_G(v_j)$. $\qquad\square$

Similar as for the minimization problem, we can show that MAX-INTERNAL $\mathcal{F}$-TREE is W[1]-complete at least for GS.

Theorem 4.4. *MAX-INTERNAL $\mathcal{F}$-TREE of GS is* W[1]*-complete.*

Again, we have to leave open whether MAX-INTERNAL $\mathcal{F}$-TREE of other clique starters is in W[1]. Nevertheless, NP-completeness holds as long as the orderings of the search can be recognized in polynomial time.

4.3 Algorithms

The results of the last section imply that we do not have to look for FPT algorithms for MIN-INTERNAL $\mathcal{F}$-TREE or MAX-INTERNAL $\mathcal{F}$-TREE for any graph search that is a clique starter. Here, we will consider the existence of XP algorithms for these problems. As seen above (see Theorem 3.5), MIN-INTERNAL $\mathcal{F}$-TREE of GS is equivalent to CONNECTED DOMINATING SET which can be solved straightforwardly in $n^{\mathcal{O}(k)}$ time. The proof of Theorem 4.4 presents a reduction of MAX-INTERNAL $\mathcal{F}$-TREE of GS to WEIGHTED CIRCUIT SATISFIABILITY that increases the parameter only linearly. Since WEIGHTED CIRCUIT SATISFIABILITY can be solved in $n^{\mathcal{O}(k)}$ time, this implies the following.

Theorem 4.5. *MIN-INTERNAL $\mathcal{F}$-TREE and MAX-INTERNAL $\mathcal{F}$-TREE of GS can be solved in $n^{\mathcal{O}(k)}$ time.*

Next we will give a similar result for BFS. To this end, we show that a BFS $\mathcal{F}$-tree with a certain set of internal vertices S can be computed using only the knowledge about the ordering of the vertices in S.

Lemma 4.6. *Let T_σ be the $\mathcal{F}$-tree of some BFS ordering σ and let S be the set of the first k internal vertices of T_σ. Let ρ' be the ordering of S in σ and let ρ be a vertex ordering of G starting with ρ'. Let τ be the BFS_ρ^+ ordering of G and let T_τ be its $\mathcal{F}$-tree. Then, every $v \in S$ has the same children in T_σ and T_τ.*

Proof. First observe that the start vertex of τ is the first vertex of ρ and the first vertex of ρ is the first vertex of ρ'. As the start vertex is by definition an internal vertex of the $\mathcal{F}$-tree, it holds that the start vertex of σ is an element of S. As σ contains ρ' as subordering, it must hold that σ also starts with the first vertex of ρ'. Thus, σ and τ start with the same vertex and, consequently, the layers of σ and τ are identical.

We show by induction that for every layer the following statements are true:

(T1) If v and w are in layer i, $v \in S$ and $v \prec_\sigma w$, then $v \prec_\tau w$.
(T2) Every vertex of S in layer i has the same set of children in T_σ as in T_τ.

Note that these statements are trivially true for the zeroth layer as σ and τ have the same start vertex r and all elements of the first layer are the children of r. So we may assume that the two statements hold for some layer i. First, we prove statement (T1) for layer $i + 1$. Let v and w be in layer $i + 1$ such that $v \in S$ and $v \prec_\sigma w$. Since layer $i + 1$ contains at least one of the first k internal vertices of T, the parents of v and w are also in S. Since (T2) holds for layer i, v has the same parent p_v in T_τ and T_σ and w has the same parent p_w in T_τ and T_σ. If p_v and p_w are not identical, then p_v is before p_w in σ. As (T1) holds for layer i, p_v is before p_w also in τ and, thus, it follows that $v \prec_\tau w$. So we may assume that p_v and p_w are the same vertex. Since $v \prec_\sigma w$ and $v \in S$, it holds that $v \prec_\rho w$ (either w is to the right of v in ρ' or w is not in S and, thus, to the right of v in σ). Thus, the tie-breaking of BFS_ρ^+ puts v before w in τ.

It remains to show that (T2) holds for layer $i + 1$. Let x be an element of layer $i + 2$ and assume, for contradiction, that the parent of x in T_σ is p_σ, the parent of x in T_τ is $p_\tau \neq p_\sigma$ and at least one of p_σ and p_τ is in S. Observe, that p_σ has to be an element of S, because either p_σ is the only element of $\{p_\sigma, p_\tau\}$ contained in S, or, if p_τ is in S, then p_σ has to be in S as well, because it is an internal vertex of T_σ, preceding $p_\tau \in S$ in σ. Furthermore, $p_\sigma \prec_\sigma p_\tau$ but $p_\tau \prec_\tau p_\sigma$. This contradicts (T1) for layer $i + 1$.   □

Using this result, we can give XP algorithms for MIN-INTERNAL $\mathcal{F}$-TREE and MAX-INTERNAL $\mathcal{F}$-TREE of BFS.

Theorem 4.7. *MIN-INTERNAL $\mathcal{F}$-TREE and MAX-INTERNAL $\mathcal{F}$-TREE of BFS can be solved in $\mathcal{O}(n^{k+2})$ time.*

Due to Theorem 4.2, we cannot give such an algorithm for MIN-INTERNAL $\mathcal{F}$-TREE of LBFS, unless P = NP. This difference in the complexity is in line with the complexity of the $\mathcal{F}$-tree recognition problem which is polynomial-time solvable for GS and BFS [30,48] but NP-hard for LBFS [4].

An adaptation of Theorem 4.7 for MAX-INTERNAL $\mathcal{F}$-TREE of LBFS seems also to be difficult. In contrast to BFS, the ordering of the leaves of the tree has a significant influence on the ordering of the following layers (see Fig. 2).

5   Conclusion

We have studied the Maximum (Minimum) Leaf Spanning Tree problem for various search algorithms. The achieved results can be contrasted with the results of Bergougnoux et al. [7], who studied the same problems for DFS trees. The

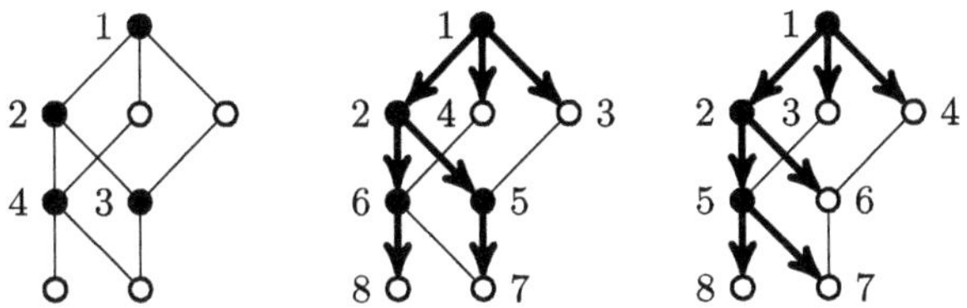

Fig. 2. Example that shows that the idea of Lemma 4.6 is not valid for LBFS. The left figure shows the input graph. The filled vertices are the elements of S and the given numbers describe the fixed ordering of S. The graph in the middle shows an LBFS ordering whose $\mathcal{F}$-tree fulfills the condition on its internal vertices. However, the right graph shows an LBFS ordering that could be computed by LBFS$^+$ but has less internal vertices. In particular, vertices 3 and 4 of the left graph have different children in the trees of the middle and the right graph.

complexity is effectively inverted, i.e., the problems for DFS are W[1]-hard when parameterized by the number of leaves, but in FPT for GS and BFS-type searches and vice versa. This coincides with the intuitive assumption that a DFS will commonly lead to few leaves while a BFS yields many leaves.

Table 1 shows some remaining questions: We are rather sure that we can adapt the para-NP-hardness proof of MIN-INTERNAL $\mathcal{F}$-TREE to other clique starters, such as Maximum Cardinality Search (MCS) or Maximal Neighborhood Search (MNS). We also expect that MAX-INTERNAL $\mathcal{F}$-TREE is para-NP-hard for LBFS for larger values of k (the case of $k = 3$ can be solved by enumeration).

All of the results shown here are for first-in trees only. It is possible to state the same problems for last-in trees as well. This problem is addressed for a variety of searches in a separate article which is in preparation.

References

1. Aazami, A.: Hardness Results and Approximation Algorithms for Some Problems on Graphs. Ph.D. Thesis, University of Waterloo (2008)
2. Barioli, F., et al.: Parameters related to tree-width, zero forcing, and maximum nullity of a graph. J. Graph Theory **72**(2), 146–177 (2013)
3. Beisegel, J., et al.: On the end-vertex problem of graph searches. Discrete Math. Theor. Comput. Sci. **21** (2019)
4. Beisegel, J., et al.: The recognition problem of graph search trees. SIAM J. Discrete Math. **35**(2), 1418–1446 (2021)
5. Beisegel, J., Köhler, E., Ratajczak, F., Scheffler, R., Strehler, M.: Graph search trees and the intermezzo problem. In: MFCS 2024. LIPIcs, vol. 306, pp. 22:1–22:18 (2024)
6. Beisegel, J., Köhler, E., Scheffler, R., Strehler, M.: Breadth-first search trees with many or few leaves (2026). arXiv:2604.00691
7. Bergougnoux, B., Blaser, N., Fellows, M.R., Golovach, P.A., Rosamond, F.A., Sam, E.: On the parameterized complexity of lineal topologies (depth-first spanning trees) with many or few leaves. J. Comput. Syst. Sci. **154**, 103680 (2025)

8. Bhyravarapu, S., Kanesh, L., Kundu, M., Lokshtanov, D., Saurabh, S.: Parameterized algorithms for power edge set and zero forcing set. In: IWOCA 2025, pp. 349–361 (2025)
9. Bodlaender, H.L.: On linear time minor tests with depth-first search. J. Algorithms **14**(1), 1–23 (1993)
10. Bonsma, P., Zickfeld, F.: Spanning trees with many leaves in graphs without diamonds and blossoms. In: Laber, E.S., Bornstein, C., Nogueira, L.T., Faria, L. (eds.) LATIN 2008. LNCS, vol. 4957, pp. 531–543. Springer, Heidelberg (2008). https://doi.org/10.1007/978-3-540-78773-0_46
11. Brandstädt, A., Le, V.B., Spinrad, J.P.: Graph Classes: A Survey. SIAM, Discrete Mathematics and Applications (1999)
12. Brešar, B., et al.: Grundy dominating sequences and zero forcing sets. Discrete Optim. **26**, 66–77 (2017)
13. Brešar, B., Gologranc, T., Milanič, M., Rall, D.F., Rizzi, R.: Dominating sequences in graphs. Discrete Math. **336**, 22–36 (2014)
14. Brešar, B., Henning, M.A., Rall, D.F.: Total dominating sequences in graphs. Discrete Math. **339**(6), 1665–1676 (2016)
15. Charbit, P., Habib, M., Mamcarz, A.: Influence of the tie-break rule on the end-vertex problem. Discrete Math. Theor. Comput. Sci. **16**(2), 57–72 (2014)
16. Corneil, D.G., Dusart, J., Habib, M., Mamcarz, A., de Montgolfier, F.: A tie-break model for graph search. Discrete Appl. Math. **199**, 89–100 (2016)
17. Corneil, D.G., Köhler, E., Lanlignel, J.M.: On end-vertices of Lexicographic Breadth First Searches. Discrete Appl. Math. **158**(5), 434–443 (2010)
18. Corneil, D.G., Krueger, R.: A unified view of graph searching. SIAM J. Discrete Math. **22**(4), 1259–1276 (2008)
19. Cygan, M., et al.: Parameterized Algorithms. Springer (2015)
20. Downey, R.G., Fellows, M.R.: Fixed-parameter tractability and completeness I: basic results. SIAM J. Comput. **24**(4), 873–921 (1995)
21. Downey, R.G., Fellows, M.R.: Fundamentals of Parameterized Complexity. Texts in Computer Science, Springer (2013)
22. Fellows, M.R., Langston, M.A.: On well-partial-order theory and its application to combinatorial problems of VLSI design. SIAM J. Discrete Math. **5**(1), 117–126 (1992)
23. Fellows, M.R., McCartin, C., Rosamond, F.A., Stege, U.: Coordinatized kernels and catalytic reductions: an improved FPT algorithm for max leaf spanning tree and other problems. In: FST TCS 2000. LNCS, vol. 1974, pp. 240–251 (2000)
24. Fomin, F.V., Gaspers, S., Saurabh, S., Thomassé, S.: A linear vertex kernel for maximum internal spanning tree. J. Comput. Syst. Sci. **79**(1), 1–6 (2013)
25. Fujie, T.: An exact algorithm for the maximum leaf spanning tree problem. Comput. Oper. Res. **30**(13), 1931–1944 (2003)
26. Garey, M.R., Johnson, D.S.: Computers and Intractability: A Guide to the Theory of NP-completeness. Freeman, W. H (1979)
27. Golumbic, M.C.: Algorithmic Graph Theory and Perfect Graphs, Annals of Discrete Mathematics, vol. 57. Elsevier, 2nd edn. (2004)
28. Habib, M., McConnell, R.M., Paul, C., Viennot, L.: Lex-BFS and partition refinement, with applications to transitive orientation, interval graph recognition and consecutive ones testing. Theor. Comput. Sci. **234**(1–2), 59–84 (2000)
29. Hagerup, T.: Biconnected graph assembly and recognition of DFS trees. Tech. Rep. A 85/03, Universität des Saarlandes (1985)
30. Hagerup, T., Nowak, M.: Recognition of spanning trees defined by graph searches. Tech. Rep. A 85/08, Universität des Saarlandes (1985)

31. Hogben, L., Lin, J.C.H., Shader, B.L.: Inverse Problems and Zero Forcing For Graphs, Mathematical Surveys and Monographs, vol. 270. AMS (2022)
32. Karp, R.M.: Reducibility among combinatorial problems. In: Complexity of Computer Computations, pp. 85–103. Plenum Press (1972)
33. Kleinberg, J., Tardos, E.: Algorithm Design. Addison Wesley (2006)
34. Kneis, J., Langer, A., Rossmanith, P.: A new algorithm for finding trees with many leaves. Algorithmica $\mathbf{61}$(4), 882–897 (2011)
35. Korach, E., Ostfeld, Z.: DFS tree construction: algorithms and characterizations. In: van Leeuwen, J. (ed.) WG 1988. LNCS, vol. 344, pp. 87–106. Springer, Heidelberg (1989). https://doi.org/10.1007/3-540-50728-0_37
36. Kratsch, D., Liedloff, M., Meister, D.: End-vertices of graph search algorithms. In: Paschos, V.T., Widmayer, P. (eds.) CIAC 2015. LNCS, vol. 9079, pp. 300–312. Springer, Cham (2015). https://doi.org/10.1007/978-3-319-18173-8_22
37. Li, P., Shang, J., Shi, Y.: A simple linear time algorithm to solve the MIST problem on interval graphs. Theor. Comput. Sci. $\mathbf{930}$, 77–85 (2022)
38. Li, W., Cao, Y., Chen, J., Wang, J.: Deeper local search for parameterized and approximation algorithms for maximum internal spanning tree. Inf. Comput. $\mathbf{252}$, 187–200 (2017)
39. Lin, J.C.H.: Zero forcing number, Grundy domination number, and their variants. Linear Algebra Appl. $\mathbf{563}$, 240–254 (2019)
40. Manber, U.: Recognizing breadth-first search trees in linear time. Inf. Process. Lett. $\mathbf{34}$(4), 167–171 (1990)
41. Prieto, E., Sloper, C.: Either/Or: Using Vertex Cover structure in designing FPT-algorithms – The case of k- Internal Spanning Tree. In: WADS 2003. LNCS, vol. 2748, pp. 474–483 (2003)
42. Raible, D., Fernau, H.: An amortized search tree analysis for k-leaf spanning tree. In: SOFSEM 2010. LNCS, vol. 5901, pp. 672–684 (2010)
43. Reich, A.: Complexity of the maximum leaf spanning tree problem on planar and regular graphs. Theor. Comput. Sci. $\mathbf{626}$, 134–143 (2016)
44. Rong, G., Cao, Y., Wang, J., Wang, Z.: Graph searches and their end vertices. Algorithmica $\mathbf{84}$(9), 2642–2666 (2022)
45. Rong, G., Yuan, B., Yang, Y., Zhang, Z.: A linear-time algorithm for the MCS end-vertex problem on chordal graphs: a bonus-driven search strategy. In: SOSA 2026. pp. 298–311 (2026)
46. Rose, D.J., Tarjan, R.E., Lueker, G.S.: Algorithmic aspects of vertex elimination on graphs. SIAM J. Comput. $\mathbf{5}$(2), 266–283 (1976)
47. Scheffler, R.: On the parameterized complexity of Grundy domination and zero forcing problems (2025). arXiv:2508.18104
48. Scheffler, R.: The partial search order problem. Electron. J. Comb. $\mathbf{32}$(4) (2025)
49. Scheffler, R.: On the leaves of graph search trees. Ars Math. Contemp. $\mathbf{26}$ (2026)
50. Sharma, G., Pandey, A., Wigal, M.C.: Algorithms for maximum internal spanning tree problem for some graph classes. J. Comb. Optim. $\mathbf{44}$(5), 3419–3445 (2022)
51. Tarjan, R.E., Yannakakis, M.: Simple linear-time algorithms to test chordality of graphs, test acyclicity of hypergraphs, and selectively reduce acyclic hypergraphs. SIAM J. Comput. $\mathbf{13}$(3), 566–579 (1984)
52. West, D.B.: Introduction to Graph Theory. Prentice Hall (2001)
53. Zehavi, M.: The k-leaf spanning tree problem admits a Klam value of 39. Eur. J. Comb. $\mathbf{68}$, 175–203 (2018)

The Parameterized Complexity of Scheduling with Precedence Delays: Shuffle Product and Directed Bandwidth

Hans L. Bodlaender[1] and Maher Mallem[2]($\boxtimes$)

[1] Department of Information and Computing Sciences, Utrecht University, Utrecht, The Netherlands
h.l.bodlaender@uu.nl

[2] Inria, CNRS, ENS de Lyon, Université Claude Bernard Lyon 1, LIP, UMR 5668, 69342 Lyon Cedex 07, France
maher.mallem@ens-lyon.fr

Abstract. In this paper, we study the parameterized complexity of several variants of scheduling with precedence constraints between jobs. Namely, we consider the single machine setting with delay values on top of the precedence constraints. Such scheduling problems are related to several decades-old problems with open parameterized complexity status, notably SHUFFLE PRODUCT and DIRECTED BANDWIDTH. We obtain XNLP-completeness results for both problems, and derive implications to scheduling with minimum (resp. maximum) delays parameterized by the width of the directed acyclic graph giving the precedence constraints, and/or by the maximum delay value in the input. Regarding DIRECTED BANDWIDTH, we also settle the case of trees by showing XNLP-completeness parameterized by the target value. Beyond these results, we believe that SHUFFLE PRODUCT is an unusual and promising addition to the list of XNLP-complete problems.

Keywords: Parameterized Complexity · XNLP · Scheduling · Shuffle Product · Directed Bandwidth

1 Introduction

In this paper, we study the parameterized complexity of several scheduling problems where there are precedence constraints with delays between jobs. Such delays are handy in that they can model a wide range of real-life problems. For instance, minimum delays mimic how instructions are handled in pipelined processors [9], and maximum delays can model products of short shelf life. Even theoretically speaking, several decades-old problems with open parameterized complexity status like UNARY BIN PACKING [19] and DIRECTED BANDWIDTH [2] are equivalent to scheduling settings with precedence delays - more on this in the later sections.

© The Author(s), under exclusive license to Springer Nature Switzerland AG 2026
F. Foucaud and A. Parreau (Eds.): IWOCA 2026, LNCS 16587, pp. 146–160, 2026.
https://doi.org/10.1007/978-3-032-27732-9_11

Several of the problems studied in this paper have the same complexity: they are known to be in the class XP—that is, for each fixed value of the parameter, there is a polynomial time algorithm, but the exponent of the polynomial grows when the parameter grows, and the problems are not known to be fixed-parameter tractable. For many problems, the precise complexity is unknown. In this paper, we show that several parameterized scheduling problems with precedence constraints and delays are complete for a complexity class now known as XNLP.

The class XNLP was originally introduced by Elberfeld et al. [12], and renamed to XNLP by Bodlaender et al. [5]. In [5], several problems were shown to be complete for this class, including SCHEDULING OF JOBS WITH PRECEDENCE CONSTRAINTS with both the number of machines and the width as parameters—the first example where a scheduling problem was shown to be XNLP-complete. Other recent work that shows that scheduling problems are complete for the class XNLP or for the related class XALP [4] can be found in [6,22]. Showing that a problem is XNLP-complete is interesting for several reasons: it pinpoints the exact complexity class of the problem and, assuming a conjecture by Pilipczuk and Wrochna [25], we have that there is no algorithm in XP for the problem that uses only $f(k)n^{O(1)}$ space—indeed, about all known XP algorithms for these problems employ dynamic programming and have space usage of $\Omega\left(n^{f(k)}\right)$ for f a divergent function. Also, XNLP-hardness implies hardness for all classes $W[i]$ with i positive integer (e.g., see Lemma 2 in [5]).

Among the several parameterized problems whose complexity we resolve by showing XNLP-completeness, we have the SHUFFLE PRODUCT problem. Given a target word t and source words $s_1, \ldots, s_k$, one asks whether t can be obtained by interleaving the letters of $s_1, \ldots, s_k$, while keeping the order of the letters from the original words. We strengthen the W[2]-hardness proof for SHUFFLE PRODUCT from [1] to XNLP-completeness (Sect. 3.1). We believe that this result is of independent interest, as SHUFFLE PRODUCT can be used as the starting point for several XNLP-hardness proofs. From this result, in Sect. 3.2, we notably derive XNLP-completeness of single machine scheduling with precedence constraints, constant minimum delays and processing time at most two parameterized by the width of the directed acyclic graph.

In Sect. 4, we show XNLP-completeness of a scheduling problem with precedence constraints and maximum delays, namely the single machine scheduling of unit-time jobs under precedence constraints with equal maximum delays. This problem is equivalent to the directed variant of the well studied BANDWIDTH problem. Formulating it as a graph problem, we are given in the DIRECTED BANDWIDTH problem a directed acyclic graph, and ask for a topological ordering f with for each arc vw, $f(v) < f(w)$ and $f(w) - f(v) \leq k$ for some target value k. We give two XNLP-completeness results for DIRECTED BANDWIDTH in Sect. 4, namely for general directed acyclic graphs with both the target bandwidth and the width as parameters, and for directed trees with the target bandwidth as parameter.

Table 1. Summary of our main results. Connections between SHUFFLE PRODUCT and scheduling with minimum precedence delays are discussed in Sect. 3.2. DIRECTED BANDWIDTH is equivalent to single machine scheduling of unit-time jobs with equal-length maximum precedence delays - see the beginning of Sect. 4.

Problem	Section	Parameter	Result
(BINARY) SHUFFLE PRODUCT	3.1	number of source words	XNLP-complete
DIRECTED BANDWIDTH	4.1	target value plus width	XNLP-complete for directed acyclic graphs
	4.2	target value	XNLP-complete for directed trees

Our main results are summarized in Table 1. In many cases, we improve upon or complement existing results. We delay the discussion of related literature to the separate sections. Figure 1 summarizes the XNLP-hardness reductions proposed in this paper.

2   Preliminaries

Given two integers a, b, $[a, b]$ denotes set $\{a, a + 1, \ldots, b\}$.

Graphs. We assume the reader to be familiar with standard notions from graph theory. Throughout the paper, given a graph G, n denotes the number of vertices of G.

We consider two types of directed trees: downwards directed trees, where each arc is directed from a vertex to one of its children, and upward directed trees, where each arc is directed from a vertex to its parent. Note that a hardness proof for one of these two trees directly transforms to a hardness proof for the other type, by reversing the direction of all arcs.

Scheduling. The goal of scheduling is to complete a set of jobs over time under a set of constraints. Unless otherwise stated, n is the number of jobs, $\mathcal{J} = [1, n]$ is the set of jobs and $G_{prec} = (\mathcal{J}, A)$ is a directed acyclic graph representing precedence constraints between jobs. Given two jobs $i, j \in \mathcal{J}$, if ij is an arc in G_{prec} then job i must be completed before job j is started. For the sake of simplicity, we restrict ourselves to the parallel identical machine setting, where m machines are available to execute jobs. Each machine can execute at most one job at a time. In our base setting, every job takes unit time to be processed. The formal definition is the following:

SCHEDULING OF JOBS WITH PRECEDENCE CONSTRAINTS
Given: positive integers n, m, D and a directed acyclic graph G_{prec}.
Question: is there a schedule completing all jobs by time D and meeting all precedence and machine constraints? In other words, is there a function $f : \mathcal{J} \to [0, D - 1]$ assigning start times to jobs such that:
(*i*) for all $t \in [0, D - 1]$, $|\{i \in \mathcal{J}, f(i) = t\}| \leq m$, and
(*ii*) for every arc ij in G_{prec}, $f(i) + 1 \leq f(j)$?

Natural extensions include giving two properties to each job j: a positive integer p_j representing its processing time, and a positive integer $size_j$ representing the number of required machines to process job j. In the base setting, both p_j and $size_j$ are equal to one. If values $size_j$ and/or p_j can be non-unit, then condition (i) (resp. (ii)) becomes: $\left(\sum_{j \in \mathcal{J}, f(j) \leq t \leq f(j)+p_j-1} size_j \right) \leq m$ (resp. $f(i) + p_i \leq f(j)$).

In order to denote these problem variants in a concise way, we use the standard three-field notation $\alpha|\beta|\gamma$ introduced in [16]. Fields α, β and γ respectively describe the machine environment, the job properties and the objective function. In this paper, the latter will always be denoted by '$C_{max} \leq D$', meaning that all jobs must be completed by time D. In field α, the parallel identical machine setting is denoted by 'P'. If m is a constant then we denote 'Pm'; if $m = 1$ then we are in the single machine setting, which is denoted by '1'. In field β, '$prec$' denotes the presence of directed acyclic graph G_{prec} in the input. If there is no mention of $size_j$ (resp. p_j), then by default they are equal to 1 (resp. they can be any positive integer). For instance, SCHEDULING OF JOBS WITH PRECEDENCE CONSTRAINTS is denoted by $P|prec, p_j = 1|C_{max} \leq D$.

Precedence delays specify additional waiting time constraints on top of existing precedence relations. Given an arc ij in G_{prec}, we say that there is a minimum (resp. maximum, exact) precedence delay of value $\ell_{i,j}$ when job j can only start at least (resp. at most, exactly) $\ell_{i,j}$ time units after job i is completed. In particular, note that minimum precedence delays of value zero have a matching requirement to regular precedence constraints. When scheduling problems allow for precedence delays, we extend the three-field notation the same way as in [23]. Namely, in field β, the presence of minimum (resp. maximum, exact) precedence delays is denoted by '$prec(\ell^{min})$' (resp. '$prec(\ell^{max})$', '$prec(\ell^{ex})$'). In this case, all arcs in G_{prec} are equipped with a delay of the respective nature.

In this paper, we focus on two parameters for scheduling problems: the *width* of the instance, which is defined as the width of G_{prec} (see the beginning of Sect. 4), and, in the presence of precedence delays, the *maximum delay value* in the input.

XNLP. We assume the reader to be familiar with the most standard notions of parameterized complexity; otherwise, e.g., see [11]. In this paper, we focus on showing problems to be complete for the class XNLP.

XNLP is the class of parameterized problems, that can be solved with a non-deterministic Turing Machine in $f(k)n^{O(1)}$ time and $O(f(k)\log n)$ space, for a computable function f, with k the parameter and n the input size. A *parameterized logspace* reduction from parameterized problem A to parameterized problem B is an algorithm, that, when given an input (x, k) gives an output (x', k'), with the following properties: (1) $A(x, k) \Leftrightarrow B(x', k')$; (2) $k' \leq g(k)$ for a computable function g; (3) the algorithm uses $f(k)+O(\log n)$ space for a computable function f; (4) the algorithm uses $n^{O(1)}$ time.

The class XNLP was first introduced by Elberfeld et al. [12], and was renamed to XNLP in [5]. Also, Bodlaender et al. [5] introduced a problem that they named

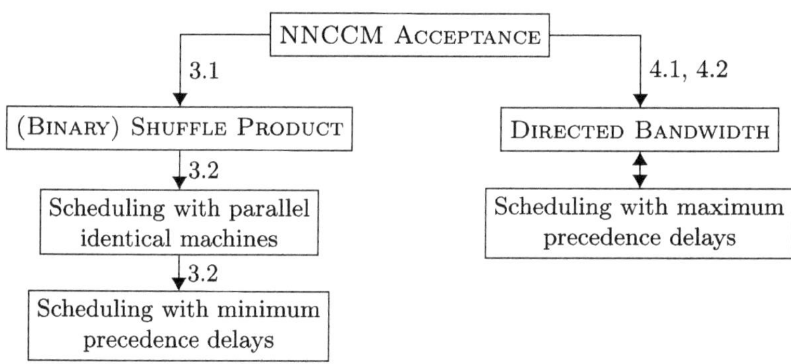

Fig. 1. Reductions between XNLP-complete problems in this paper. See the corresponding sections for the exact definitions of the problem variants.

NNCCM ACCEPTANCE, and showed it to be XNLP-complete. This problem, defined below, is used as starting point for our reductions.

In an instance of NNCCM ACCEPTANCE we are given: an integer k, that tells the number of counters we have; an integer n (the maximum value of each counter); and a sequence of r *checks*. Each check is a 4-tuple $(c_1, n_1, c_2, n_2) \in [1, k] \times [0, n] \times [1, k] \times [0, n]$. These describe a (somewhat artificial but useful for hardness proof) machine model, that works as follows.

- There are k counters, each with an initial value of 0.
- Each of r rounds consists in the following two steps. In round $i \in [1, r]$:
  1. In this non-deterministic step, some, one or none of the counters are increased, but never to a value larger than n.
  2. The ith check is done: if the c_1st counter has value n_1 and the c_2th counter has value n_2, then the machine halts and rejects.
- If the machine did not reject in any of the r rounds, it accepts.

NNCCM ACCEPTANCE
Given: description of an NNCCM: integers k, n, r and a sequence of r 4-tuples in $([1, k] \times [0, n] \times [1, k] \times [0, n])^r$.
Parameter: number of counters k.
Question: is there an accepting run of the NNCCM?

Theorem 1 (Bodlaender, Groenland, Nederlof, Swennenhuis [5]).
NNCCM ACCEPTANCE *is complete for* XNLP.

Finally, note that the relations between the parameters considered in this paper are discussed in the full version [7].

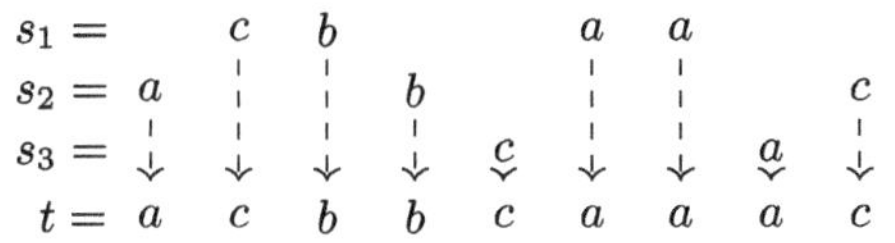

$$
\begin{array}{ccccccccc}
s_1 = & & c & b & & & a & a & \\
s_2 = & a & & & b & & & & c \\
s_3 = & & & & & c & & a & \\
t = & a & c & b & b & c & a & a & a & c
\end{array}
$$

Fig. 2. Example of a word $t = acbbcaaac$ in shuffle product $s_1 \sqcup\!\sqcup s_2 \sqcup\!\sqcup s_3$ with source words $s_1 = cbaa$, $s_2 = abc$ and $s_3 = ca$.

3 Shuffle Product and its Consequences

Given a word s, its length is denoted by $|s|$, and its pth letter in s is denoted by $s\langle p \rangle$ $(1 \le p \le |s|)$.

Definition 1 (Shuffle product). *A word t is in the* shuffle product *of words $s_1, \ldots, s_k$ if and only if t can be obtained by interleaving the letters of $s_1, \ldots, s_k$, while keeping the order of the letters from the original words.*

Formally, we denote $s_1 \sqcup\!\sqcup s_2 \sqcup\!\sqcup \cdots \sqcup\!\sqcup s_k$ the set of words t for which there is be a bijective mapping $f : (\cup_{1 \le i \le k}(\{i\} \times [1, |s_i|])) \to [1, |t|]$ such that, for all $i \in [1, k]$, sequence $(f(i, p))_{1 \le p \le |s_i|}$ is increasing and, for all $p \in [1, |s_i|]$, $s_i \langle p \rangle = t \langle f(i, p) \rangle$.

 (Binary) Shuffle Product
Given: source words $s_1, \ldots, s_k$ and target word t over a (binary) alphabet Σ.
Parameter: number of source words k.
Question: is t in $s_1 \sqcup\!\sqcup s_2 \sqcup\!\sqcup \cdots \sqcup\!\sqcup s_k$?

Figure 2 gives an example. Shuffle Product is NP-hard for unbounded k even over alphabets of size 3 (see Theorem 3.1 by Warmuth and Haussler [29]), while it is polynomial-time solvable for constant k using dynamic programming [24]. Rizzi and Vialette [26] asked about the parameterized complexity of Shuffle Product. This was partially answered by van Bevern et al., who showed the W[2]-hardness of Binary Shuffle Product [1]. In the next subsection, we give a definite answer to this open question by showing XNLP-completeness of (Binary) Shuffle Product.

3.1 XNLP-Completeness of (Binary) Shuffle Product

In this subsection, we show our main result regarding Shuffle Product:

Theorem 2. (Binary) Shuffle Product *is XNLP-complete.*

 XNLP-membership is straightforward: we propose a non-deterministic algorithm where we memorize the advancement on the target word and each of the source words. For each letter in t in order, we guess which of the source words it corresponds to. Note that we only keep track of position indices in the words, and

$$s_1 = a\ b \qquad a\ b\ a\ bb \qquad a\ b\ a\ b\ a\ bb\ a\ b \qquad bb\ a\ b$$
$$s_2 = a\ b\ a\ b\ a\ bb\ a\ b \quad (\ldots) \quad a\ b\ a\ b\ a\ b\ a\ b \quad (\ldots) \quad b$$
$$s_3 = a\ b\ a\ b\ a\ b\ a\ b \qquad a \qquad b\ a\ b\ a\ bb \qquad b\ a\ bb$$
$$t = a^3\ b^5\ a^3\ b^5\ a^3\ b^4\ a^3\ b^5 \qquad a^3\ b^5\ a^3\ b^5\ a^3\ b^4\ a^3\ b^5 \qquad b^4\ a^3\ b^5\ a^3\ b^5$$

Fig. 3. Illustration of the reduction with $k = 3$ source words, $n = 2$ and $r = 2$ checks $(1, 0, 2, 0)$ and $(1, 1, 3, 0)$. In the proposed block mapping, gray areas represent counter checks, where counters have values $(1, 0, 0)$ and $(1, 0, 1)$ respectively.

we do not need to write down any letter in the working space. So, the alphabet size does not appear in the working space, and the latter is in $\mathcal{O}(k \log(|t|))$.

To show XNLP-hardness, we adapt the W[2]-hardness proof by van Bevern et al. [1]. We reduce from NNCCM ACCEPTANCE. We are given an instance $\mathcal{I}$ with integers k, n, r and a sequence of r 4-tuples in $[1, k] \times [0, n] \times [1, k] \times [0, n]$. Recall that k is the number of counters and our parameter, n is the maximum value of the counter, r is the number of checks and each 4-tuple represents a check, done in the same order as in the sequence. For $j \in [1, r]$, if the jth check is (c_1, n_1, c_2, n_2) then, for all couples $(i, p) \in [1, k] \times [0, n]$, we define:

$$\ell_{i,j,p} = \begin{cases} 2 & \text{if } (i, n - p) = (c_1, n_1) \text{ or } (i, n - p) = (c_2, n_2), \\ 1 & \text{otherwise.} \end{cases} \tag{1}$$

Given a sequence of words $(w_i)_{1 \le i \le q}$, we denote $(\prod_{i=1}^{q} w_i)$ the concatenated word $w_1 \cdots w_q$. If $w_1 = w_2 = \cdots = w_q$, then this word is denoted by $(w_1)^q$. Let $N = 2kn + 1$. We propose the following instance $\mathcal{I}'$ of BINARY SHUFFLE PRODUCT with $k + 3$ words over $\Sigma = \{a, b\}$:

$$s_i = \prod_{j=1}^{r}\left(\left(\prod_{p=0}^{n} ab^{\ell_{i,j,p}}\right)^{N}\right), \qquad t = \left((a^k b^{k+2})^n a^k b^{k+1}\right)^{Nr} (a^k b^{k+2})^n,$$
$$s_{k+1} = a^{|t|_a - \sum_{i=1}^{k} |s_i|_a}, \text{ and} \qquad s_{k+2} = b^{|t|_b - \sum_{i=1}^{k} |s_i|_b}.$$

Similarly to van Bevern et al. [1], we define positions and long and short blocks the following way.

Definition 2. *A* block *(resp. a* c-block*) in a word s is defined as a maximal consecutive subword using only one letter (resp. only letter $c \in \Sigma$). A block has position i in s if it is the ith successive block in s.*

In t, we call b-blocks of length $k + 1$ short and b-blocks of length $k + 2$ long. For $i \in [1, k]$, in s_i, we call b-blocks of length 1 short and b-blocks of length 2 long.

Remark 1. Note that t has $2(n+1)Nr + 2n$ blocks and, for $i \in [1, k]$, s_i has exactly $2n$ less blocks—i.e., $2(n+1)Nr$ blocks. Essentially, we dedicate $2(n+1)N$ blocks to each of the r checks successively.

Intuition. Figure 3 illustrates the reduction. We track the offset in block positions between t and every s_i. By Remark 1 and the alternation of a-blocks and b-clocks in all the words, for each s_i these offsets are even integers in $\{0,\dots,2n\}$ and form a non-decreasing sequence. After a division by two, these values are meant to represent the sequence of values for counter i.

Checks are performed in sequence, along $2(n+1)N$ consecutive blocks each. A check (c_1,n_1,c_2,n_2) is represented by a short b-block in which no 'block skip' happens—i.e., every s_i contributes at least a letter to this block. Since short b-blocks in t are of length $k+1$, source words s_{c_1} and s_{c_2} cannot contribute both a long b-block of length two to this short b-block. This reflects that we cannot have both $c_1 = n_1$ and $c_2 = n_2$ during the corresponding check of the counter machine. Note that the existence of such a short b-block for each check is ensured by setting N to a sufficiently large value—i.e., more than the total number of possible block skips which, by Remark 1, is $2kn$.

The proof of equivalence between instances $\mathcal{I}$ and $\mathcal{I}'$ is available in the full version [7].

3.2   Consequences

In this subsection, from the XNLP-completeness of BINARY SHUFFLE PRODUCT, we derive several implications to scheduling with precedence constraints.

First, we consider SCHEDULING OF JOBS WITH PRECEDENCE CONSTRAINTS, denoted by $P|prec, p_j = 1|C_{max}$ in the three-field notation. This problem is NP-complete [28], and it was recently shown to be XNLP-complete parameterized by the number of machines m plus width [8]. Although this implies XNLP-hardness parameterized by m, it is open whether the problem is NP-hard for constant $m \geq 3$ [28].

In contrast, when p_j or $size_j$ are allowed to be non-unit, the problem is known to be NP-complete for a constant number of machines [10,28]. In fact, van Bevern et al. showed that problems $P3|prec, p_j = 1, size_j \in \{1,2\}|C_{max} \leq D$ and $P2|prec, p_j \in \{1,2\}|C_{max} \leq D$ are W[2]-hard parameterized by width (see Sect. 2.3 in [1]). They first proved the W[2]-hardness of BINARY SHUFFLE PRODUCT, then proposed reductions to both scheduling problems. It is not hard to see that these reductions only require logarithmic working space and thus can be reused as is to infer XNLP-hardness. Furthermore, XNLP-membership can be derived from straightforward adaptations of the algorithm proposed in the full version of [8, Section 4.3].

Corollary 1. *Scheduling problems* $P3|prec, p_j = 1, size_j \in \{1,2\}|C_{max} \leq D$ *and* $P2|prec, p_j \in \{1,2\}|C_{max} \leq D$ *are XNLP-complete parameterized by width.*

Now, we consider scheduling with minimum precedence delays. On a single machine, the problem is already NP-complete with unit-time jobs and equal-length precedence delays—which, in our extended three-field notation, can be denoted by $1|prec(\ell^{min}), \ell_{i,j} = \ell, p_j = 1|C_{max} \leq D$ [21] —, albeit it can be solved in polynomial time if G_{prec} is an upward (or downward) directed forest [9].

Regarding (maximum) delay value ℓ as parameter, the situation is similar to the parallel identical machine parameterized by the number of machines. In [23], it was shown that $1|prec(\ell^{min}), \ell_{i,j} = \ell, p_j = 1|C_{max} \leq D$ is XNLP-complete parameterized by ℓ plus width. Although this implies XNLP-hardness parameterized by ℓ, it is open whether the problem is NP-hard for constant $\ell \geq 3$. However, when processing times can be equal to one or two, the problem is known to be NP-complete for constant $\ell \geq 2$ (see Sect. 8.2.4 in [13]); of note that the reduction required the width to be unbounded.

From Corollary 1, we can deduce that the restriction to constant $\ell \geq 3$ and processing times in $\{1, 2\}$ is XNLP-complete parameterized by width. Indeed, starting from problem $P3|prec, p_j = 1, size_j \in \{1, 2\}|C_{max} \leq D$, the reduction from SCHEDULING OF JOBS WITH PRECEDENCE CONSTRAINTS proposed in [23, Section 5] can be easily adapted to the case of jobs with non-unit $size_j$: we reduce in the same manner, i.e., every time unit in the parallel machine problem corresponds to a time interval of length m in our single machine instance with minimum precedence delays. Then, jobs of size $size_j$ in the original instance are turned into jobs of processing time $p'_j = size_j$. Plus the delay value is $\ell = m = 3$, and jobs of size two are turned into jobs of processing time two. This shows XNLP-hardness. Finally, note that XNLP-membership was already established in [23, Section 4].

Corollary 2. *Scheduling problem* $1|prec(\ell^{min}), \ell_{i,j} = 3, p_j \in \{1, 2\}|C_{max} \leq D$ *is XNLP-complete parameterized by width.*

This improves upon the results in [13,23] about single machine scheduling with equal-length minimum precedence delays.

4   Directed Bandwidth

This section shows that DIRECTED BANDWIDTH is XNLP-complete. More precisely, we give two results: we show that DIRECTED BANDWIDTH parameterized by the target bandwidth and the width is complete for XNLP, and that DIRECTED BANDWIDTH parameterized by the target bandwidth is complete for XNLP when the input is a downward directed tree or an upward directed tree.

A directed graph $G = (V, A)$ is *acyclic* (i.e., a *DAG*), if it has no directed cycle. A *topological ordering* of a directed acyclic graph $G = (V, A)$ with $|V| = n$ is a bijective function $f : V \to [0, n - 1]$, such that for each arc $vw \in A$: $f(v) < f(w)$. The *bandwidth* of a topological ordering f is $\max_{vw \in A} f(w) - f(v)$. The *directed bandwidth* of a directed acyclic graph G is the minimum bandwidth over all topological borderings of G. The *width* of a directed acyclic graph $G = (V, A)$ is the maximum size of a set $S \subseteq V$, such that there is no pair of vertices $v, w \in S$, $v \neq w$ with a directed path from v to w in G, i.e., it is the maximum size of an antichain in the partial order that is formed by the closure of G.

In the DIRECTED BANDWIDTH problem, given a directed acyclic graph $G = (V, A)$, and an integer k, we ask whether G has directed bandwidth at most k.

The DIRECTED BANDWIDTH problem is the natural directed variant of the well-studied BANDWIDTH problem, where we ask for an undirected graph $G = (V, E)$ and integer k, if there is a bijective mapping $f : V \rightarrow [0, n-1]$, with for each $\{v, w\} \in E$: $|f(v) - f(w)| \leq k$. Already in 1978, Garey et al. [14] showed that DIRECTED BANDWIDTH is NP-complete, for directed trees with no vertices of indegree more than two, and they showed that the problem is polynomial for $k = 2$. In this paper, we focus on its parameterized complexity. BANDWIDTH was shown first to be in XP by Saxe [27], with a faster variant given by Gurari and Sudborough [17]; these algorithms can be easily modified to give algorithms for DIRECTED BANDWIDTH with the same running time. In [2], it was shown that BANDWIDTH and DIRECTED BANDWIDTH are both W[t]-hard for all t, even if the input graph (resp. the undirected version of the input graph) is a caterpillar—i.e., a tree with one path that contains all vertices of degree at least three. Later, in [5], it was shown that BANDWIDTH is complete for XNLP, again over caterpillars.

As observed by Mallem (Section 3.1 in [22]), DIRECTED BANDWIDTH is equivalent to scheduling jobs with unit times on one processor with equal-length maximum precedence delays - denoted by $1|prec(\ell^{max}), \ell_{i,j} = \ell, p_j = 1|C_{max} \leq D$ in our extended three-field notation for scheduling problems. Mallem poses as an open problem whether DIRECTED BANDWIDTH is XNLP-complete parameterized by the target bandwidth; this paper solves this open problem, even for directed trees.

Below, in Sect. 4.1, we first give the complexity result for directed acyclic graphs with target bandwidth and width of the DAG as parameter. Then, in Sect. 4.2, we show how we can modify this proof to show XNLP-hardness for DIRECTED BANDWIDTH for trees. Our proofs are inspired by the XNLP-completeness proof for the undirected case from [5].

4.1 Directed Bandwidth for Directed Acyclic Graphs

In this subsection, we show the following result.

Theorem 3. DIRECTED BANDWIDTH *is* XNLP-*complete for directed acyclic graphs, with as parameter target value plus width.*

Membership can be shown the same way as membership for the undirected variant of the problem (e.g., see [5]): with target value k, repeatedly guess the next vertex in the ordering, and keep the last k vertices in the ordering in memory.

Intuition. To show XNLP-hardness, we use a transformation from NNCCM ACCEPTANCE. Given an instance of NNCCM ACCEPTANCE with k counters, we build a directed acyclic graph G with width $k+O(1)$, such that G has directed bandwidth at most $k + 6$, if and only if the NNCCM has an accepting run. We have a properly chosen value L. G has $(k + 6)L + 1$ vertices with exactly one vertex $v_{f,0}$ of indegree 0, and exactly one vertex $v_{f,L}$ of outdegree 0. $v_{f,0}$ must be the first, and $v_{f,L}$ the last vertex in the ordering. Then, we have a *floor path* with length exactly L from $v_{f,0}$ to $v_{f,L}$, which means that every $(k+6)$th vertex in the ordering must be a vertex in the floor path.

There are $k+1$ other paths from $v_{f,0}$ to $v_{f,L}$. k of these paths represent a counter; each counter gadget path has length $n+L$. Between each pair of successive floor path vertices, we must have at least one vertex of each counter gadget path. The increase of a counter is modelled by having more than one vertex of the counter gadget path between two successive vertices—thus a value of a counter is the difference between the number of steps of the counter gadget path and the floor path. Then, the floor path, and the counter gadget path have '*between*' vertices: additional vertices which are between two successive vertices of these paths. These are positioned in such a way, that these *between* vertices fit, precisely if and only if the NNCCM accepts—at a position that resembles a check, we have 'many' *between* vertices of the floor path, and each counter whose value matches the check has one *between* vertex—and we have space to only fit one such. The last path is the *filler path*: this path has precisely the length to ensure that the total number of vertices equals $(k+6)L+1$. Figure 4 illustrates part of the construction.

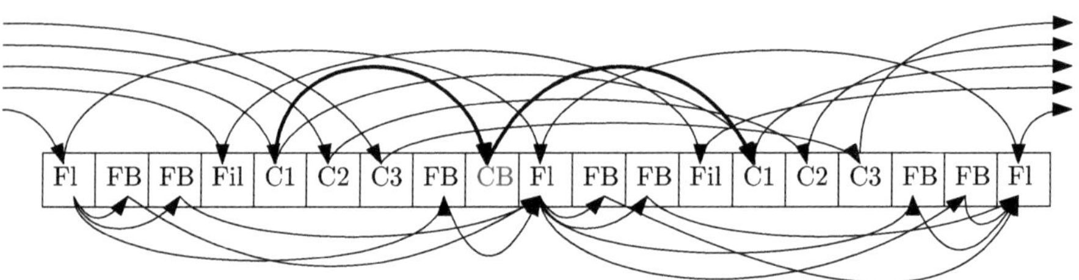

Fig. 4. Part of the construction of G for DIRECTED BANDWIDTH for *DAGs*. Fl = floor. Fil = filler path. Ci = Counter path i. FB = *between* vertex of floor. CB = *between* vertex of counter gadget. In the example, counter 1 is part of a check and has the value given in the check; thus counter path 1 has a *between* vertex. If another counter would also give the check value, there would be a second CB vertex which does not fit in between the two same consecutive floor vertices—halting the machine corresponds to not being able to realize bandwidth B.

The details of the construction of G, and the formal proof of equivalence between G and the instance of NNCCM ACCEPTANCE are available in the full version [7].

4.2   XNLP-Hardness for Trees

In this subsection, we show the following result.

Theorem 4. DIRECTED BANDWIDTH *is* XNLP-*complete, even for inputs that are either a downward directed tree or an upward directed tree, with as parameter target value.*

Membership follows the same way as the XNLP-membership for DAGs. We now show XNLP-hardness for downward directed trees, and the result for upward directed trees follows directly (just reverse the direction of all arcs).

Intuition. The graph created in the proof of Theorem 3 can be turned into an downward directed tree by removing the following arcs: for each *between* vertex, its outgoing arc, and for each counter gadget path and the filler path, the arc to $v_{f,L}$. The outgoing arcs of *between* vertices are actually not needed to make the reduction work, but keep the width of directed acyclic graph small—without these, the width would no longer be bounded by a function of k. As in Theorem 4, the width of the tree is not a parameter, we can thus afford to remove these arcs. The other type of arcs however disables the proof that a topological ordering of bandwidth at most B implies an accepting run of the NNCCM—in particular, we cannot guarantee that the last vertex of the floor path $v_{f,L}$ is the last vertex of the ordering, and we cannot guarantee that the last vertex of each counter gadget path and the filler path is in the last batch, i.e., between $v_{f,L-1}$ and $v_{f,L}$ in the ordering. So, what we do below is to add new gadgets, called the *tails*, to ensure these: for each of the floor, filler path, and counter gadgets, we add 'tails'—a new subtree whose root has as parent the last vertex of the floor path, filler path, or a counter gadget path.

The construction of the proposed instance of DIRECTED BANDWIDTH and the full proof of Theorem 4 are available in the full version [7].

5   Discussion

We end the paper with a few open problems.

First, regarding scheduling with minimum delays, current reduction techniques from SHUFFLE PRODUCT to scheduling problems require jobs of either non-unit processing time or non-unit size. As such, it is still open whether $1|prec(\ell^{min}), \ell_{i,j} = \ell, p_j = 1|C_{max} \leq D$ is NP-complete for constant $\ell \geq 3$. In fact, following the relation with the parallel identical machine setting discussed in Sect. 3.2, this mirrors the longstanding open question about whether SCHEDULING OF JOBS WITH PRECEDENCE CONSTRAINTS is NP-complete for a constant number of machines $m \geq 3$ [28].

Second, regarding scheduling with maximum delays, settling the case of directed trees parameterized by width would complement both results about DIRECTED BANDWIDTH in this paper. In this context, the parameter bounds both the number of leaves and the number of branching internal nodes in the tree. We believe that a fixed-parameter algorithm could stem from enumerating all sequences of these nodes which are compatible with the bandwidth constraints then, for each one of these sequences, solve an integer linear program featuring variables for the positions of these nodes in the topological order. This would imply fixed-parameter tractability, as long as the number of variables in each integer linear program is bounded by a function of the width [20].

Additionally, in 1978 Garey et al. [14] showed that DIRECTED BANDWIDTH is NP-complete for directed trees with each vertex indegree at most two. In this paper, we only showed XNLP-completeness for general directed trees; we conjecture that this result could be strengthened to directed trees with indegree two, but expect that a proof of such a result would be very technical. In [2,3], it was shown that BANDWIDTH is XNLP-complete for caterpillars with hair length at most three. We do not expect this result to hold for the directed variant—i.e., if we have a downward or upward directed tree with all vertices of degree more than two on one path—but conjecture this case to be polynomial time solvable.

Finally, regarding single machine scheduling with exact delays, results seem to differ significantly between maximum delay value taken alone and combined with the width as parameter (see Tables 1 and 2 in [23]). In particular, on a single machine with chains of unit-time jobs and equal-length delays, the problem is W[1]-hard with the former parameter and fixed-parameter tractable with the latter one. Crucially, note that this problem parameterized by maximum delay value is equivalent to UNARY BIN PACKING (see [7]). As such, improving upon the W[1]-hardness of scheduling with exact delays would also help settling the open question in [19]. So far, to the best of our knowledge, only XP-membership is known (e.g., a polynomial-time algorithm for a constant number of bins is mentioned in [15] and detailed in [18, Section 5.2]). In contrast, slight generalizations, such as having either $\ell_{i,j} \in \{0, \ell\}$ or $p_j \in \{1, 2\}$, are known to be NP-complete even when the maximum delay value is equal to one [23].

Disclosure of Interests. The authors have no competing interests to declare that are relevant to the content of this article.

References

1. van Bevern, R., Bredereck, R., Bulteau, L., Komusiewicz, C., Talmon, N., Woeginger, G.J.: Precedence-constrained scheduling problems parameterized by partial order width. In: Kochetov, Y., Khachay, M., Beresnev, V., Nurminski, E., Pardalos, P. (eds.) DOOR 2016. LNCS, vol. 9869, pp. 105–120. Springer, Cham (2016). https://doi.org/10.1007/978-3-319-44914-2_9
2. Bodlaender, H.L.: Parameterized complexity of BANDWIDTH of caterpillars and WEIGHTED PATH EMULATION. In: Kowalik, L., Pilipczuk, M., Rzazewski, P. (eds.) WG 2021. LNCS, vol. 12911, pp. 15–27. Springer, Cham (2021). https://doi.org/10.1007/978-3-030-86838-3_2
3. Bodlaender, H.L., Groenland, C., Jacob, H., Jaffke, L., Lima, P.T.: XNLP-completeness for parameterized problems on graphs with a linear structure. In: Dell, H., Nederlof, J. (eds.) 17th International Symposium on Parameterized and Exact Computation, IPEC 2022. LIPIcs, vol. 249, pp. 8:1–8:18. Schloss Dagstuhl - Leibniz-Zentrum für Informatik (2022). https://doi.org/10.4230/LIPIcs.IPEC.2022.8
4. Bodlaender, H.L., Groenland, C., Jacob, H., Pilipczuk, M., Pilipczuk, M.: On the complexity of problems on tree-structured graphs. In: Dell, H., Nederlof, J. (eds.) 17th International Symposium on Parameterized and Exact Computation, IPEC 2022. LIPIcs, vol. 249, pp. 6:1–6:17. Schloss Dagstuhl - Leibniz-Zentrum für Informatik (2022). https://doi.org/10.4230/LIPIcs.IPEC.2022.6

5. Bodlaender, H.L., Groenland, C., Nederlof, J., Swennenhuis, C.M.F.: Parameterized problems complete for nondeterministic FPT time and logarithmic space. Inf. Comput. **300**, 105195 (2024). https://doi.org/10.1016/J.IC.2024.105195

6. Bodlaender, H.L., Hermelin, D., van Leeuwen, E.J.: Concurrency constrained scheduling with tree-like constraints. In: Fernau, H., Kindermann, P. (eds.) Proceedings 51th International Workshop on Graph-Theoretic Concepts in Computer Science, WG 2025. Lecture Notes in Computer Science, pp. 105–120. Springer (2025). https://doi.org/10.1007/978-3-032-11835-6_8

7. Bodlaender, H.L., Mallem, M.: The parameterized complexity of scheduling with precedence delays: shuffle product and directed bandwidth (2026). https://arxiv.org/abs/2605.03727

8. Bodlaender, H.L., van der Wegen, M.: Parameterized complexity of scheduling chains of jobs with delays. In: Cao, Y., Pilipczuk, M. (eds.) 15th International Symposium on Parameterized and Exact Computation, IPEC 2020, pp. 4:1–4:15. LIPIcs, Schloss Dagstuhl - Leibniz-Zentrum für Informatik (2020). https://doi.org/10.4230/LIPICS.IPEC.2020.4, full version: https://arxiv.org/abs/2007.09023

9. Bruno, Jones, So, K.: Deterministic scheduling with pipelined processors. IEEE Trans. Comput. **C-29**(4), 308–316 (1980). https://doi.org/10.1109/TC.1980.1675569

10. Błażwicz, J., Liu, Z.: Scheduling multiprocessor tasks with chain constraints. Eur. J. Oper. Res. **94**(2), 231–241 (1996). https://doi.org/10.1016/0377-2217(96)00126-9

11. Cygan, M., et al.: Parameterized Algorithms. Springer (2015). https://doi.org/10.1007/978-3-319-21275-3

12. Elberfeld, M., Stockhusen, C., Tantau, T.: On the space and circuit complexity of parameterized problems: classes and completeness. Algorithmica **71**(3), 661–701 (2015). https://doi.org/10.1007/s00453-014-9944-y

13. Engels, D.W.: Scheduling for Hardware/Software Partitioning in Embedded System Design. Ph.D. thesis, Massachusetts Institute of Technology (2000). https://dspace.mit.edu/bitstream/handle/1721.1/86443/46804459-MIT.pdf

14. Garey, M.R., Graham, R.L., Johnson, D.S., Knuth, D.E.: Complexity results for bandwidth minimization. SIAM J. Appl. Math. **34**(3), 477–495 (1978). https://doi.org/10.1137/0134037

15. Garey, M.R., Johnson, D.S.: Computers and Intractability, vol. 174. Freeman, San Francisco (1979)

16. Graham, R., Lawler, E., Lenstra, J., Kan, A.: Optimization and approximation in deterministic sequencing and scheduling: a survey. In: Hammer, P., Johnson, E., Korte, B. (eds.) Discrete Optimization II, Annals of Discrete Mathematics, vol. 5, pp. 287–326. Elsevier (1979). https://doi.org/10.1016/S0167-5060(08)70356-X

17. Gurari, E.M., Sudborough, I.H.: Improved dynamic programming algorithms for bandwidth minimization and the MinCut linear arrangement problem. J. Algorithms **5**, 531–546 (1984). https://doi.org/10.1016/0196-6774(84)90006-3

18. Gurski, F., Rehs, C., Rethmann, J.: Knapsack problems: a parameterized point of view. Theor. Comput. Sci. **775**, 93–108 (2019). https://doi.org/10.1016/J.TCS.2018.12.019

19. Jansen, K., Kratsch, S., Marx, D., Schlotter, I.: Bin packing with fixed number of bins revisited. J. Comput. Syst. Sci. **79**(1), 39–49 (2013). https://doi.org/10.1016/j.jcss.2012.04.004

20. Lenstra, H.W., Jr.: Integer programming with a fixed number of variables. Math. Oper. Res. **8**(4), 538–548 (1983). https://doi.org/10.1287/moor.8.4.538

21. Leung, J.T., Vornberger, O., Witthoff, J.: On some variants of the bandwidth minimization problem. SIAM J. Comput. **13**(3), 650–667 (1984). https://doi.org/10.1137/0213040
22. Mallem, M.: Parameterized Complexity and New Efficient Enumerative Schemes for RCPSP. Ph.D. thesis, Sorbonne Université (2024). https://theses.hal.science/tel-04910718v1/file/146301_MALLEM_2024_archivage.pdf
23. Mallem, M., Hanen, C., Munier-Kordon, A.: Single machine scheduling with precedence constraints and bounded maximum delay value. J. Comb. Optim. **51**(3), 33 (2026). https://doi.org/10.1007/s10878-026-01411-w
24. Mansfield, A.: On the computational complexity of a merge recognition problem. Discret. Appl. Math. **5**(1), 119–122 (1983). https://doi.org/10.1016/0166-218X(83)90021-5
25. Pilipczuk, M., Wrochna, M.: On space efficiency of algorithms working on structural decompositions of graphs. ACM Trans. Comput. Theory **9**(4), 18:1–18:36 (2018). https://doi.org/10.1145/3154856
26. Rizzi, R., Vialette, S.: On recognizing words that are squares for the shuffle product. In: Bulatov, A.A., Shur, A.M. (eds.) CSR 2013. LNCS, vol. 7913, pp. 235–245. Springer, Heidelberg (2013). https://doi.org/10.1007/978-3-642-38536-0_21
27. Saxe, J.B.: Dynamic-programming algorithms for recognizing small-bandwidth graphs in polynomial time. SIAM J. Algebraic Discrete Methods **1**(4), 363–369 (1980). https://doi.org/10.1137/0601042
28. Ullman, J.: NP-complete scheduling problems. J. Comput. Syst. Sci. **10**(3), 384–393 (1975). https://doi.org/10.1016/S0022-0000(75)80008-0
29. Warmuth, M.K., Haussler, D.: On the complexity of iterated shuffle. J. Comput. Syst. Sci. **28**(3), 345–358 (1984). https://doi.org/10.1016/0022-0000(84)90018-7

Parameterized Algorithms for Computing MAD Trees

Tom-Lukas Breitkopf[1], Vincent Froese[1], Anton Herrmann[1(✉)],
André Nichterlein[1], and Camille Richer[2,3]

[1] Algorithmics and Computational Complexity, Technische Universität Berlin,
Berlin, Germany
{t.breitkopf,vincent.froese,a.herrmann,andre.nichterlein}@tu-berlin.de
[2] Université Paris-Dauphine, PSL Research University, CNRS, UMR 7243,
LAMSADE, Paris, France
camille.richer@dauphine.eu
[3] Orange Research, Châtillon, France

Abstract. We consider the well-studied problem of finding a spanning tree with minimum average distance between vertex pairs (called a MAD tree). This is a classic network design problem which is known to be NP-hard. While approximation algorithms and polynomial-time algorithms for some graph classes are known, the parameterized complexity of the problem has not been investigated so far.

We start a parameterized complexity analysis with the goal of determining the border of algorithmic tractability for the MAD tree problem. To this end, we provide a linear-time algorithm for graphs of constant modular width and a polynomial-time algorithm for graphs of bounded treewidth; the degree of the polynomial depends on the treewidth. That is, the problem is in FPT with respect to modular width and in XP with respect to treewidth. Moreover, we show it is in FPT when parameterized by vertex integrity or by an above-guarantee parameter. We complement these algorithms with NP-hardness on split graphs.

Keywords: Optimum Distance Spanning Trees · Wiener Index · Network Design · Routing Cost · Width Parameters

1 Introduction

Computing a spanning tree is a classic algorithmic graph problem in computer science. There are numerous objectives to optimize, e.g. the weight, the diameter, the radius or the number of leaves of the spanning tree. In this work, we consider the problem of finding a spanning tree of an undirected unweighted graph with *minimum average distance* (called a *MAD tree*), that is, a spanning tree that minimizes the average distance of all vertex pairs. Note that this is equivalent to minimizing the total sum of pairwise vertex distances also known as the Wiener index; an important concept in chemical graph theory [31].

F. Foucaud and A. Parreau (Eds.): IWOCA 2026, LNCS 16587, pp. 161–174, 2026.
https://doi.org/10.1007/978-3-032-27732-9_12

The problem of computing a MAD tree (also known as a *minimum routing cost spanning tree*) was introduced by Hu [20] (there called *optimum distance spanning tree*) and has various applications, e.g. in network design for transportation and communication networks, facility location, and also in genome alignment in biology [15]. It is known to be NP-hard [23]. On the algorithmic side, a PTAS was shown by Wu et al. [32]. Regarding exact algorithms, a cubic algorithm is known for series-parallel graphs [13]. The problem is also known to be linear-time solvable on complete r-partite graphs [4], distance hereditary graphs [7], and interval graphs [8] and solvable in quadratic time on permutation graphs [21], trapezoid graphs [28] and circular-arc graphs [22]. Moreover, several mixed integer programming formulations were developed [15] (for an overview we refer to Zetina et al. [34]). Beyond that, many heuristic approaches have been considered (see the survey by Masone et al. [27]).

We extend these algorithmic results by adopting a *parameterized complexity* view, that is, we develop algorithms where the exponential part of the running time can be confined to certain parameters of the input graph. This leads to efficient algorithms for inputs where the parameter is small (constant). Concretely, we obtain the following results: The problem is *fixed-parameter tractable* (in FPT) for the vertex integrity and the modular width of the input graph and it is in XP for the treewidth. Moreover, we give a stronger hardness result showing NP-hardness on split graphs.

Further Related Work. The structure of MAD trees is well studied in the graph theory literature [2, 9, 14, 26]. They appear in many different contexts, e.g. in computational geometry [1], distributed computing [19], computational biology [15], and operations research [34]. There are also many other variants of special spanning trees of interest, for example, shortest-path trees [17], k-leaf spanning trees [33], or minimum-diameter spanning tree [18].

2   Preliminaries

For $n \in \mathbb{N}$, we define $[n] := \{1, 2, \ldots, n\}$ with $[0] := \emptyset$.

Graphs. All graphs in this work are simple and unweighted. For a graph G, we denote the set of vertices by $V(G)$ and the set of edges by $E(G)$. For an edge $\{u, v\} \in E(G)$, we write uv as a shorthand. We set $n_G := |V(G)|$ and $m_G := |E(G)|$. We denote the degree of a vertex $v \in V(G)$ by $\deg_G(v) := |N_G(v)|$, where $N_G(v) := \{u \in V(G) \mid uv \in E(G)\}$. The set $N_G(v)$ is called the open neighborhood of v and $N_G[v] := N_G(v) \cup \{v\}$ is called the closed neighborhood. If the considered graph is clear from the context, then we drop the subscripts. We write $H \subseteq G$ to denote that H is a subgraph of G. For a subset $X \subseteq V(G)$, we write $G - X$ for the subgraph of G obtained by deleting the vertices from X and all edges which have nonempty intersection with X. For an edge $uv \in E(G)$, we write $G - uv$ for the subgraph of G obtained by deleting uv from G.

A tree T is an acyclic and connected graph. A subgraph $H \subseteq G$ is a spanning tree of G if H is a tree and $V(H) = V(G)$. Throughout the rest of this work, we assume G to be connected since we are interested in spanning trees. We define the distance $\mathrm{dist}_G(u, v)$ between the two vertices u and v in G as the length of the shortest u-v-path in G. For a vertex v, we write $\mathrm{dist}_G(v) := \sum_{u \in V(G)} \mathrm{dist}_G(u, v)$. A vertex v of G is said to be *median* if it minimizes $\mathrm{dist}_G(v)$. For a vertex set $A \subseteq V$, we define $\mathrm{dist}_G(A, A) := \frac{1}{2} \sum_{u,v \in A} \mathrm{dist}_G(u, v)$ and if $B \subseteq V$ is another vertex set with $A \cap B = \emptyset$, then we define $\mathrm{dist}_G(A, B) := \sum_{u \in A, v \in B} \mathrm{dist}_G(u, v)$. For an edge uv of a tree T, we define T^u_{uv} to be the subtree of T that contains u when removing the edge uv from T. Finally, $P^T_{u,v}$ denotes the unique path between vertices u and v in the tree T and if T is clear from the context we drop the superscript.

Wiener Index. The Wiener index of a graph G is the sum of all pairwise distances, that is, $\mathrm{W}(G) := \sum_{u,v \in V} \mathrm{dist}_G(u, v)$. Several formulas for the Wiener index of a tree T are known [29]. In particular, we heavily use the following formula: For an edge uv, let $w_T(uv) := |V(T^u_{uv})| \cdot |V(T^v_{uv})|$. Then $\mathrm{W}(T) = \sum_{uv \in E(T)} w_T(uv)$. A spanning tree with minimum Wiener index is called a *minimum average distance* (MAD) tree. Dahlhaus et al. [8] proved the following useful lemma about median vertices in MAD trees.

Lemma 1 (Dahlhaus et al. [8]). *If c is a median vertex of a MAD tree T of a graph G, then every path in T starting at c is an induced path in G. Moreover, the tree T and the median c can be chosen such that there is no vertex $c' \neq c$ with $N_G[c] \subset N_G[c']$.*

We study the following decision problem:

MAD SPANNING TREE (MADST)
Input: An undirected connected graph $G = (V, E)$ and an integer b.
Question: Is there a spanning tree T of G with $\mathrm{W}(T) \leq b$?

Parameterized Complexity. We assume the reader to be familiar with basic notions from complexity theory like P and NP. Parameterized complexity is a multivariate approach to measure the time complexity of computational problems [6,11]. An instance (x, k) of a parameterized problem consists of a classical instance x together with a number k called *parameter*. A parameterized problem is called *fixed-parameter tractable* (that is, it is contained in the class FPT) if there is an algorithm deciding an instance (x, k) in time $f(k) \cdot |x|^{O(1)}$, where f is an arbitrary function solely depending on k. The class XP contains all parameterized problems which are polynomial-time solvable for constant parameter values, that is, in time $|x|^{f(k)}$. It is known that FPT $\subset$ XP.

3   Width Parameters

In this section, we analyze the complexity of MADST with respect to treewidth and modular width. Our first result is a dynamic program running in polynomial time on graphs of bounded treewidth. Interestingly, our second algorithm for MADST parameterized by modular width is a branching algorithm.

Treewidth. Our result generalizes for example the polynomial-time algorithm on series-parallel graphs [13], which have treewidth two. We use a bottom-up dynamic programming approach on a nice tree-decomposition, which can be computed in $2^{O(k^3)}n$ time [6]. To this end, whenever an edge uv of the graph is seen for the last time, we compute its contribution $w_T(uv)$ to each potential solution T. Recall that $w_T(uv) := |V(T_{uv}^u)| \cdot |V(T_{uv}^v)|$ where $|V(T_{uv}^u)|$ denotes the number of vertices in the tree T that are closer to u than to v. As for a tree T the Wiener index is $W(T) = \sum_{uv \in E(T)} w_T(uv)$, we can simply add all the contributions to a potential solution T and store the cost of all partial solutions in a table. Here, consistency in keeping the number of vertices closer to one vertex than a neighboring vertex is the technically involved part responsible for the majority of information we store in the table. We achieve the following:[1]

Theorem 1 ($\star$). *MADST is solvable in $2^{O(2^k)}n^{O(k)}$ time on treewidth-k graphs.*

Modular Width. Although less prominent than treewidth, modular width is a by now well-established width parameter. An advantage over treewidth is that modular width can be computed in $O(n + m)$ time [30]. As usual, modular width comes with a decomposition encoding the input graph and algorithms typically process this decomposition in a bottom-up manner while maintaining partial solutions for the problem at hand [5,16,24]. Our approach breaks with this standard approach: we only use a small part of the structure provided by modular width in our branching algorithm as we can show that MAD trees have a very special structure in graphs of low modular width.

Before describing the high-level idea of our algorithm, we need to introduce some basics on modular width and modular decompositions first. We follow the notation of Kratsch and Nelles [24]. A *module* of a graph G is a set of vertices $M \subseteq V(G)$, such that for any vertex $x \in V(G) \setminus M$, either $M \subseteq N(x)$ or $M \cap N(x) = \emptyset$, meaning all vertices of M share the same neighborhood in $V(G) \setminus M$. A partition $P = \{M_1, M_2, \ldots, M_\ell\}$ of the vertices $V(G)$ into $\ell \geq 2$ modules of G is called a *modular partition*. If there are $v \in M_i$ and $u \in M_j$ with $uv \in E(G)$, then all vertices in M_i and M_j are adjacent and we say modules M_i and M_j are *adjacent*. The *quotient graph* $G_{/P}$ of G has vertices $\{q_1, q_2, \ldots, q_\ell\}$ and two vertices q_i and q_j are adjacent if and only if the modules M_i and M_j are adjacent.

The quotient graph $G_{/P}$ stores whether or not two vertices $u, v \in V(G)$ from *different* modules are adjacent. What is missing is the information on edges within modules; this is typically covered recursively: for a module M_i one considers a modular partition for $G[M_i]$ until the modules are single vertices. This leads to a modular decomposition and the modular width is the largest number of nodes in any employed quotient graph in the decomposition; we refer to Gajarský et al. [16] for a formal definition.

---

[1] The $\star$ indicates that the proof is deferred to the full version [3].

Our approach to find a MAD tree only needs a modular partition P of the input graph G with the parameter k being the number of modules in P. Thus, k is upper bounded by the modular width but can be much smaller as we only need a single modular partition P of G with a minimum number of modules and its corresponding quotient graph $G_{/P}$. For an intuition why considering $G_{/P}$ suffices, recall that the n-vertex trees with minimum and maximum Wiener index are the star $K_{1,n-1}$ and the path P_n [10]. Intuitively, we want our MAD tree to be as close as possible to being a star. This is where the modularity is beneficial: a vertex v is either adjacent to all or none of the vertices in another module. Thus, for two adjacent modules M_1 and M_2, a MAD tree can be very "star-like". More precisely, if the modular partition consists of only two modules M_1 and M_2, then it is not too difficult to verify that an optimal solution is either a star or what we call a "double-star"—a tree with exactly two inner vertices (see Fig. 1). The internal edges within $G[M_1]$ and $G[M_2]$ only influence which vertices can be inner vertices of the double-star (namely the vertices with highest degree in each module). This means a MAD tree can be found without access to optimal sub-solutions of M_1 and M_2. It is therefore sufficient to only consider a single quotient graph $G_{/P}$ of G.

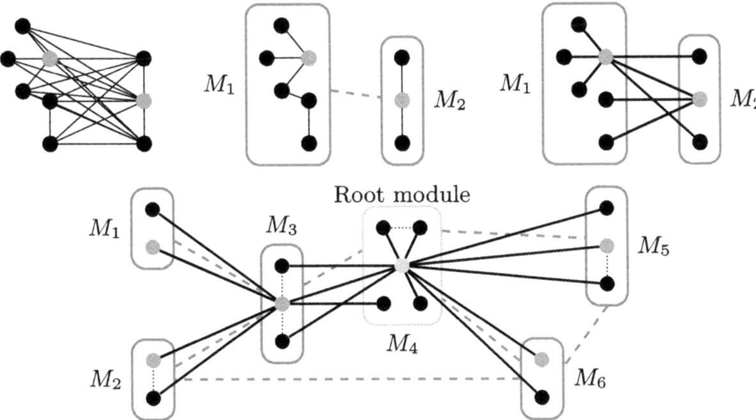

Fig. 1. The three top figures represent the case with two modules: on the left is the input graph, in the middle is the quotient graph (in grey dashed edges) and the internal edges of each module, and on the right is a double-star with centers highlighted in blue. The bottom figure represents a poly-star in solid edges. The root and the root module are highlighted in orange. One maximum-degree vertex per module is highlighted in blue. Internal edges unused in the poly-star are dotted. (Color figure online)

We generalize this approach to more than two modules by showing that we can always find a MAD tree that is a *poly-star*, that is, a tree in which there is at most one non-leaf in any module. We restrict the structure of the optimal solution even further by showing several properties: Crucially, there is

---

Algorithm 1. Solving MADST in FPT time with respect to modular width

---

Input: A connected graph G with modular partition $P = \{M_1, \ldots, M_k\}$ and quotient graph $G_{/P}$ with vertices $\{q_1, \ldots, q_k\}$; budget $b \in \mathbb{N}$.

Output: TRUE if there exists a MAD tree T of G with $W(T) \leq b$ and FALSE otherwise.

1: For a vertex $v \in V(G)$, define $\mathrm{mdl}(v) := i \in [k]$, where $v \in M_i$.
2: opt $:= \infty$
3: **for each** $i \in [k]$ **do**
4:     Choose a vertex $r \in M_i$ with maximum degree in G as root
5:     **for each** spanning tree T' of $G_{/P}$ with $N_{T'}(q_i) = N_{G_{/P}}(q_i)$ **do**
6:         Initialize $T := (V(G), \emptyset)$
7:         Choose an arbitrary vertex $d_j \in M_j$ for every $j \in [k] \setminus \{i\}$ and set $d_i := r$
8:         Let $D := \{d_\ell \mid \ell \in [k]\}$
9:         For each edge $q_\ell q_{\ell'}$ in $E(T')$, add $d_\ell d_{\ell'}$ to $E(T)$
10:        **for each** module M_j with $j \neq i$ **do**
11:            Let $c_j \in D$ be such that $q_{\mathrm{mdl}(c_j)}$ is the unique vertex in $P^{T'}_{q_i, q_j} \cap N_{T'}(q_j)$
12:            For every $u \in M_j \setminus \{d_j\}$, add the edge $u c_j$ to $E(T)$
13:        **end for**
14:        For every $u \in N_G(r) \cap M_i$, add the edge ur to $E(T)$
15:        Let $R := T - (M_i \setminus N_T[r])$ and choose some $c_i \in \underset{w \in N_T(r) \cap D}{\arg\min} \ \mathrm{dist}_R(w)$
16:        For every $u \in M_i \setminus N_G[r]$, add the edge $u c_i$ to $E(T)$
17:        opt $:= \min(\mathrm{opt}, W(T))$
18:    **end for**
19: **end for**
20: **return** opt $\leq b$

---

one dedicated module, called the *root module*. For one of the maximum-degree vertices in the root module (called the *root*) the entire neighborhood (in the input graph) is contained in the poly-star. Beside these edges, the poly-star does not contain any edges between vertices that are in the same module in the input graph. Furthermore, we show that all vertices of a non-root module are adjacent to a single vertex in the adjacent module that is closest to the root module in the quotient graph. All vertices in the root module that are not adjacent to the root are adjacent to a single vertex in an adjacent module. An example of such a poly-star can be seen in Fig. 1.

Our algorithm is given a modular partition (which is computable in linear-time [30]) and finds such a poly-star as follows: First, enumerate all possible spanning trees of the quotient graph and each option for the root module. For each option, create a poly-star satisfying the additional criteria. Finally, choose the best poly-star among all considered options. Our algorithm is shown in Algorithm 1 and is the main result of this section:

Theorem 2. *MADST can be solved in $O(2^{k^2} k^2 (m + n))$ time where k is the minimum number of modules in any modular partition of the input graph G.*

For the correctness of Algorithm 1, we first prove a general lemma which does not rely on a modular partition and which describes the change in the

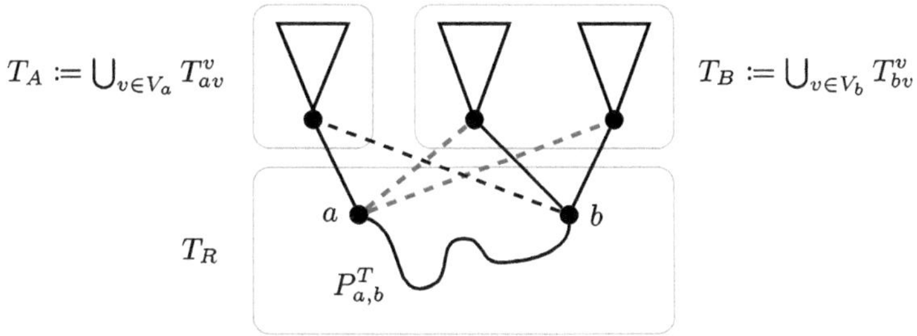

Fig. 2. A sketch of Lemma 2. Given two options a and b to connect a family of subtrees of a spanning tree T, it is always the best to choose the same option based on the distance to the rest T_R of the tree. The figure shows the case $\mathrm{dist}_{T_R}(a) \leq \mathrm{dist}_{T_R}(b)$, this means it would be better to replace the two edges from b to T_B by the two dashed green edges.

Wiener index of a tree when repositioning a family of subtrees. Lemma 2 is visualized in Fig. 2. Using this and, crucially, Lemma 1 we then step by step show the structural observations we informally stated above relying on a modular partition.

Lemma 2 ($\star$). *Let G be a graph with two distinct vertices a, b and two nonempty vertex sets V_a, V_b such that $\{a, b\} \subseteq N_G(v)$ for every $v \in V_a \cup V_b$. Let T be a spanning tree of G such that $V_a \subseteq N_T(a)$, $V_b \subseteq N_T(b)$ and $V(P_{a,b}^T) \cap (V_a \cup V_b) = \emptyset$. Let T_1 and T_2 be obtained from T by setting $V(T_1) = V(T_2) := V(T)$ and*

$$E(T_1) := (E(T) \setminus \{bv \mid v \in V_b\}) \cup \{av \mid v \in V_b\},$$
$$E(T_2) := (E(T) \setminus \{av \mid v \in V_a\}) \cup \{bv \mid v \in V_a\}.$$

Then, T_1 and T_2 are spanning trees of G and $\min(\mathrm{W}(T_1), \mathrm{W}(T_2)) < \mathrm{W}(T)$.

We call a vertex $r \in V(G)$ the *root* of MAD tree T of G if every T-path starting in r is induced in G. Such a vertex always exists according to Lemma 1. We call $M_i \ni r$ the *root module* of G. We call a spanning tree T of G a *poly-star* if for every module $M_j \in P$ there is at most one vertex d_j with $\deg_T(d_j) \geq 2$. In Lemmas 3 to 5 we now establish that a poly-star MAD tree always exists. These lemmas are depicted in Figs. 3 and 4.

Lemma 3. *Let T be a MAD tree of G with root $r \in M_i$. There is no edge $uv \in E(T)$ such that $u, v \in M_j$ for any $M_j \in P \setminus \{M_i\}$.*

Proof. Assume towards a contradiction that T contains an edge uv inside a module M_j with $j \neq i$. W.l.o.g. assume $\mathrm{dist}_T(r, u) < \mathrm{dist}_T(r, v)$ and consider the path $P_{r,v}^T$. This path contains vertices x and y such that $x \notin M_j, y \in M_j$ and $xy \in E(T)$. By modularity, it holds that also $xv \in E(G)$. Hence, $P_{r,v}^T$ is not induced in G which contradicts the assumption that r is a root of T. $\qquad\square$

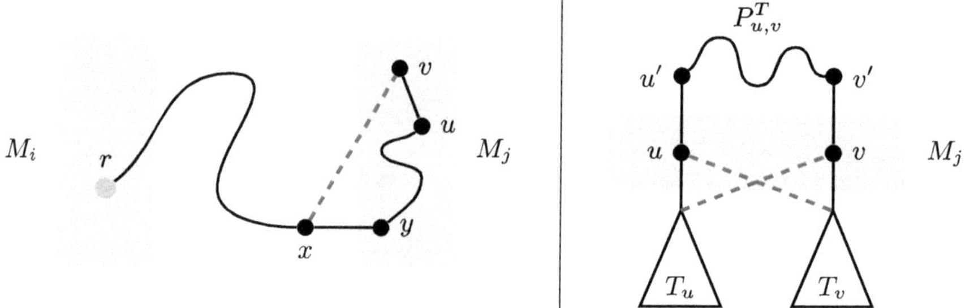

Fig. 3. On the left is a sketch of Lemma 3. If we have a MAD tree T with root r (highlighted in orange), then T contains no internal edge of any other module than the root module M_i. As sketched in the figure, if there would be such an edge uv in T, then the path from r to v would not be induced in G due to the dashed red edge which contradicts the assumption that r is a root. On the right is a sketch of Lemma 4. If we have a MAD tree T with two non-leaves u and v in a module M_j, which is not the root module, then repositioning one of the subtrees T_u or T_v by one of the dashed green edges yields a better spanning tree. Note that u and v have the same neighborhood outside their module and by Lemma 3 they have no neighbors inside their module.

Lemma 4. *Let T be a MAD tree of G with root $r \in M_i$. Then in any module $M_j \in P \setminus \{M_i\}$ there is at most one vertex $d \in M_j$ with $\deg_T(d) \geq 2$.*

Proof. Assume towards a contradiction that there are two vertices $u, v \in M_j$ with $\deg_T(u) \geq 2$ and $\deg_T(v) \geq 2$. Since T is connected, there is the path $P^T_{u,v}$ in T. Let $u', v' \in V(P^T_{u,v})$ such that $u \in N_T(u')$ and $v \in N_T(v')$ and let $V_u := N_T(u) \setminus \{u'\}$ and $V_v := N_T(v) \setminus \{v'\}$. By Lemma 3, we know that $V_u \cap M_j = V_v \cap M_j = \emptyset$ and therefore, by modularity, $V_u, V_v \subseteq N_G(v)$ and $V_u, V_v \subseteq N_G(u)$. Using Lemma 2, we immediately get a contradiction, since there exists a spanning tree T' with $W(T') < W(T)$. $\qquad\square$

Lemma 5. *There exists a MAD tree T^* of G that is a poly-star, meaning in every module of G there is at most one vertex with degree larger than one in T^*.*

Proof. By Lemma 4, we know that there is a MAD tree T of G with root $r \in M_i$ such that the claim holds for any module $M_j \in P \setminus \{M_i\}$.

We now consider the root module M_i and show the existence of a MAD tree T^* with the desired properties. Let $v \in M_i \setminus \{r\}$ be a vertex with $\deg_T(v) \geq 2$. We consider the two cases $V(P^T_{r,v}) \cap M_i \neq V(P^T_{r,v})$ and $V(P^T_{r,v}) \cap M_i = V(P^T_{r,v})$.

For the first case, we show that no such v can exist. Assume the contrary. Then there must be some $w \in V(P^T_{r,v}) \setminus M_i$ and $u \in V(P^T_{r,v}) \cap M_i$ with $wu \in E(P^T_{r,v})$. Consider some $v' \in N_T(v) \setminus V(P^T_{r,v})$, which must exist since by assumption $\deg_T(v) \geq 2$. If $v' \in M_i$, then the path $P^T_{r,v'}$ is not induced in G because of the edge wv', which exists by modularity. If $v' \notin M_i$, then the path $P^T_{r,v'}$ is not induced in G because of the edge rv', which exists by modularity. Either case produces a contradiction to the fact that r is root of T.

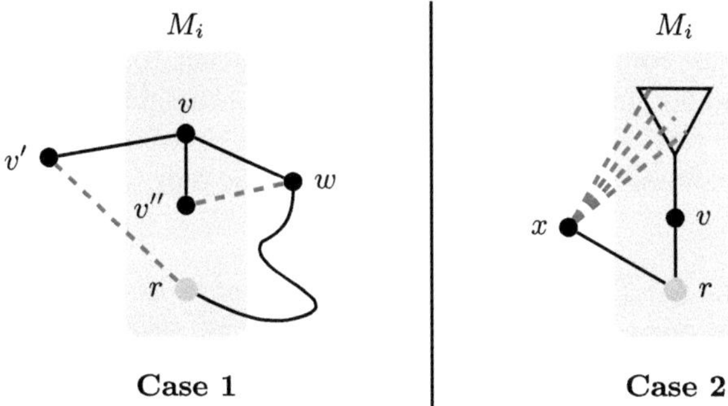

Fig. 4. A sketch of Lemma 5. Suppose we have a MAD tree T with root r (highlighted in orange) and another vertex v from the root module M_i which also has degree at least two in T. Then there are two cases. **Case 1:** If the path from r to v in T leaves the root module M_i, then the paths from r to v' and to v'' are not induced as indicated by the dashed red edges. **Case 2:** Otherwise, we can assume that v is a neighbor of r in T. In this case we obtain another MAD tree by attaching all vertices from T_Y in a star-like manner to a neighbor x of r outside of M_i. This is indicated by the dashed green edges.

For the second case, we show how T^* can be obtained from T. We assume that $v \in N_T(r)$ as otherwise we can consider the unique vertex from $N_T(r) \cap P_{r,v}^T$ instead. Let $Y := N_T(v) \setminus \{r\}$ and $T_Y := \bigcup_{u \in Y} T_{vu}^u$. If T_Y contains a vertex w which is not contained in the root module M_i, then $P_{r,w}^T$ is not induced in G. This implies $T_Y \subseteq M_i$. Since $k \geq 2$, there exists $x \in N_T(r) \setminus M_i$, because T is connected and $P_{r,x}^T$ is induced in G. In particular, if the module of x contains an internal vertex of T, then let x be this internal vertex (it still holds that $x \in N_T(r) \setminus M_i$, as otherwise $P_{r,x}^T$ would be non-induced in G). Using modularity, we obtain T^* from T by repositioning all vertices from T_Y as leaves to x:

$$E(T^*) := \Big(E(T) \setminus \big(E(T_Y) \cup \{uv \mid u \in Y\}\big)\Big) \cup \{ux \mid u \in V(T_Y)\}.$$

Let $A := V(T_Y) \cup \{x, v\}$ and $B := V(G) \setminus A$. Observing that a star attains minimum Wiener index; that in T the distance to x is no larger than the distance to v for any vertex in B; and that distances within B remain unaffected by the changes, we get:

$$\operatorname{dist}_{T^*}(A, A) \leq \operatorname{dist}_T(A, A),$$
$$\operatorname{dist}_{T^*}(A, B) \leq \operatorname{dist}_T(A, B),$$
$$\operatorname{dist}_{T^*}(B, B) = \operatorname{dist}_T(B, B),$$

and conclude $W(T^*) \leq W(T)$. Hence T^* is a MAD tree. By iteratively applying this argument to all neighbors of r from M_i in T with degree at least two, we obtain the desired MAD tree. $\square$

We now further restrict the structure of a poly-star MAD tree, by specifying adjacency of the leaves. We show that all vertices of a non-root module are adjacent to a single vertex in the adjacent module closest to the root module. All vertices in the root module that are not adjacent to the root are connected to a single neighbor in an adjacent module – the neighbor that minimizes the distance to "the rest" of the graph. Note that the proof of Lemma 5 implies that we can always assume that in the root module the only vertex of degree at least two is the root.

Lemma 6 ($\star$). *Let T be a poly-star MAD tree of G with root $r \in M_i$. Let $d_j \in M_j$ be such that $\deg_T(d_j) \geq \deg_T(v)$ for all $v \in M_j$ and let $D := \{d_1, \dots, d_i = r, \dots, d_k\}$. Then the following holds:*

i) $T[D]$ is a tree.
ii) For every non-root module $M_j \in P \setminus \{M_i\}$, it holds for all leaves $v \in M_j \setminus \{d_j\}$
   that $N_T(v) = \{c_j\}$, where c_j is the unique vertex in $P_{r,d_j}^{T[D]} \cap N_T(d_j)$.
iii) Let $R = T - (M_i \setminus N_T[r])$. For all leaves $v \in M_i \setminus N_T[r]$ from the root module
   which are not adjacent to the root, it holds that $N_T(v) = \{c_i\}$, where $c_i \in$
   $\underset{w \in N_T(r) \cap D}{\arg\min} \ \mathrm{dist}_R(w).$

To determine the exact layout of the poly-star MAD tree it remains to consider which vertices to connect to the root and how to choose the root vertex within the root module. For the first part, note that for a MAD tree T of G with root $r \in M_i$ it holds that $N_T(r) = N_G(r)$, as any $v \in N_G(r) \setminus N_T(r)$ would create a non-induced T-path starting in r. For the second part, we argue that the root r has maximum degree in G among the vertices of M_i.

Lemma 7 ($\star$). *There exists a poly-star MAD tree T^* of G with root $r^* \in M_{i^*}$, such that $\deg_G(r^*) \geq \deg_G(v)$ for all $v \in M_{i^*}$.*

Combining the structural observations made so far finally proves Theorem 2($\star$).

4   Parameters Larger Than Treewidth

In Sect. 3, the question whether MADST is fixed-parameter tractable with respect to treewidth is left open. Exploring the border of tractability for MADST, we looked for larger parameters where we can show fixed-parameter tractability. We found two such parameterizations: vertex integrity and an above guarantee parameter; which we briefly discuss now.

Vertex Integrity. A graph has small vertex integrity if it can be broken into small components by removing a small number of vertices. Formally, the vertex integrity k of a graph G is $k := \min_{S \subseteq V(G)} \{|S| + \max_{C \in cc(G-S)} |C|\}$ where $cc(G - S)$ denotes the set of all connected components in $G - S$. We remark that computing a witness S for k is fixed-parameter tractable with respect to k [12].

Algorithms exploiting vertex integrity are often based on the observation that the number of different (i.e. non-isomorphic) components in $G - S$ is bounded in a function of k. This is often combined with a problem-specific observation stating that a "solution" can be assumed to be the same in two isomorphic components in $G - S$. While such an observation seems plausible for MADST as well, proving it turns out to be non-trivial. To this end, assume that the edges of an optimal MAD tree T within $G[S]$ are given (this can be brute-forced in $f(k)$ time). Further assume for simplicity that T induces a tree on $G[S]$ and that two connected components C_1 and C_2 in $G - S$ are isomorphic and have exactly the same neighbors in S. A natural approach is to show that if the solution T is different within C_1 and C_2, then we simply change the solution within one component so that they are identical. The crux here is that this changes the distances between the vertices from C_1 and C_2, and one needs to argue that the sum of these distances will not increase; this seems non-trivial.

We encountered similar situations when designing our algorithm for modular width. The difference is that for two adjacent modules *each* vertex in one module is adjacent to *all* vertices in the other module. We do not have such strong structural properties here. Thus, we employ an integer quadratic program to compute the best way to extend the tree on S to the connected components in $G - S$, obtaining the following.

Theorem 3 ($\star$). *MADST is in FPT with respect to the vertex integrity.*

Above Guarantee Parameterization. For a given MADST-instance (G, b), we define the parameter $k := b - W(G)$. We next observe that MADST is in FPT with respect to this "above guarantee" parameter.

Proposition 1. *MADST is solvable in $(k + 2)^k n^{O(1)}$ time.*

Proof. Let (G, b) be a MADST instance and consider the following branching algorithm. If G is a tree, then return "yes" if $b \geq W(G)$ and otherwise return "no". If G is not a tree, then find a shortest cycle C in G. If C has length at least $k + 3$, then return "no". If C has length at most $k + 2$, then branch for every edge uv from C by calling the algorithm on $(G - uv, b)$.

The correctness of the algorithm comes from the fact that a shortest cycle of length at least $k + 3$ implies that (G, b) is a no-instance since deleting any edge uv from the cycle increases the distance between u and v by at least $k + 1$.

To see that the algorithm runs in FPT time with respect to k, note that we only branch into at most $k + 2$ cases and that we branch at most k times since $W(G - uv) > W(G)$ (and hence k decreases in each branching). In particular, the running time is $(k + 2)^k n^{O(1)}$. $\qquad\square$

If b equals the actual Wiener index of a MAD tree for G, then k, as a graph parameter, is larger than the feedback edge number ℓ (number of edges to remove in order to obtain a tree): Any MAD tree has exactly ℓ edges less than G and for any edge $uv \in E(G) \setminus E(T)$ the distance between u and v is at least one larger in T than in G. However, k is still incomparable to other large parameters such as the vertex cover number or the maximum leaf number.

One obstacle on improving on Proposition 1 to fixed-parameter tractability with respect to the feedback edge number or maximum leaf number is the following: Given a graph with at most k vertices of degree at least three, at most k paths between each pair of them and without degree-one vertices, how to find an optimal MAD tree in $f(k) \cdot n^{O(1)}$ time? Similarly to our approach behind Theorem 3, we can formulate the problem as an integer quadratic program with simple box constraints where the number of variables and constraints is bounded in the maximum leaf number. However, the coefficients in the objective are unbounded. Nevertheless, the problem would be in FPT if INTEGER QUADRATIC PROGRAMMING is in FPT with respect to the number of variables and constraints; this, however, is an open question posed by Lokshtanov [25].

5 Discussion and Conclusion

We initiated a parameterized analysis of MAD SPANNING TREE, providing several parameterized algorithms. For future work, we discuss two lines of possible research in more detail.

Lower Bounds. To the best of our knowledge, there is only one known NP-hardness result for MADST dating back to 1978 [23]. We are able to adapt the construction of Johnson et al. [23] to obtain NP-hardness on split graphs, which are graphs that can be partitioned into a clique and an independent set.

Theorem 4 ($\star$). *MADST is NP-hard on split graphs.*

Note that split graphs cannot contain induced P_5's. Thus, the NP-hardness contrasts the polynomial-time algorithm for P_4-free graphs (that is, cographs) implied by Theorem 2 (cographs have modular width 2).

In general, the structure of MAD trees makes it difficult not only to design algorithms but also to design reductions. Notably, our NP-hardness result (as well as the existing one) actually shows that it is NP-hard to find a MAD tree which is a shortest-path tree. In general, however, a MAD tree is not always a shortest-path tree [9]. Maybe one needs to use such special MAD trees for hardness constructions. Further lower bounds in terms of hardness results could shed some more light on what makes computing MAD trees challenging.

Pushing the Border of Tractability. Given the long list of algorithmic and structural insights [4, 7, 8, 13, 21, 22, 28, 32], it seems easier to further extend the list of known tractable special cases than providing lower bounds. From the viewpoint of parameterized algorithmics, the following questions arise:

- Is the problem in FPT for treewidth? As a first step, one might show FPT for the tree-depth or the maximum leaf number.
- Is it in XP for the cliquewidth? (An open question by Dahlhaus et al. [7])
- What about other parameterizations such as "distance to polynomial-time solvability" (for example, distance to cluster or distance to outerplanar)?

References

1. Abu-Affash, A.K., Carmi, P., Luwisch, O., Mitchell, J.S.B.: Geometric spanning trees minimizing the wiener index. Comput. Geom. Topol. **3**(1), 2:1–2:15 (2024). https://doi.org/10.57717/CGT.V3I1.52
2. Barefoot, C.A., Entringer, R.C., Székely, L.A.: Extremal values for ratios of distances in trees. Discret. Appl. Math. **80**(1), 37–56 (1997). https://doi.org/10.1016/S0166-218X(97)00068-1
3. Breitkopf, T.L., Froese, V., Herrmann, A., Nichterlein, A., Richer, C.: Parameterized algorithms for computing MAD trees (2026). https://arxiv.org/abs/2603.29381
4. Burns, K., Entringer, R.C.: A graph-theoretic view of the united states postal service. In: Alavi, Y., Schwenk, A.J. (eds.) Graph Theory, Combinatorics, and Algorithms: Proceedings of the Seventh International Conference on the Theory and Applications of Graphs, pp. 323–334. Wiley (1995)
5. Coudert, D., Ducoffe, G., Popa, A.: Fully polynomial FPT algorithms for some classes of bounded clique-width graphs. ACM Trans. Algorithms **15**(3), 33:1–33:57 (2019). https://doi.org/10.1145/3310228
6. Cygan, M., et al.: Parameterized Algorithms. Springer, Cham (2015). https://doi.org/10.1007/978-3-319-21275-3
7. Dahlhaus, E., Dankelmann, P., Goddard, W., Swart, H.C.: MAD trees and distance-hereditary graphs. Discret. Appl. Math. **131**(1), 151–167 (2003). https://doi.org/10.1016/S0166-218X(02)00422-5
8. Dahlhaus, E., Dankelmann, P., Ravi, R.: A linear-time algorithm to compute a MAD tree of an interval graph. Inf. Process. Lett. **89**(5), 255–259 (2004). https://doi.org/10.1016/J.IPL.2003.11.009
9. Dankelmann, P.: A note on MAD spanning trees. J. Comb. Math. Comb. Comput. **32**, 93–96 (2000)
10. Dobrynin, A.A., Entringer, R., Gutman, I.: Wiener index of trees: theory and applications. Acta Applicandae Mathematica **66**(3), 211–249 (2001). https://doi.org/10.1023/A:1010767517079
11. Downey, R.G., Fellows, M.R.: Fundamentals of Parameterized Complexity. Springer, New York (2013). https://doi.org/10.1007/978-1-4471-5559-1
12. Drange, P.G., Dregi, M.S., van 't Hof, P.: On the computational complexity of vertex integrity and component order connectivity. Algorithmica **76**(4), 1181–1202 (2016). https://doi.org/10.1007/S00453-016-0127-X
13. El-Mallah, E.S., Colbourn, C.J.: Optimum communication spanning trees in series-parallel networks. SIAM J. Comput. **14**(4), 915–925 (1985). https://doi.org/10.1137/0214064
14. Entringer, R.: Distance in graphs: Trees. J. Comb. Math. Comb. Comput. **24**, 65–84 (1997)
15. Fischetti, M., Lancia, G., Serafini, P.: Exact algorithms for minimum routing cost trees. Networks **39**(3), 161–173 (2002). https://doi.org/10.1002/NET.10022
16. Gajarský, J., Lampis, M., Ordyniak, S.: Parameterized algorithms for modular-width. In: Gutin, G., Szeider, S. (eds.) IPEC 2013. LNCS, vol. 8246, pp. 163–176. Springer, Cham (2013). https://doi.org/10.1007/978-3-319-03898-8_15
17. Hansen, P., Zheng, M.: Shortest shortest path trees of a network. Discret. Appl. Math. **65**(1), 275–284 (1996). https://doi.org/10.1016/0166-218X(95)00038-S
18. Hassin, R., Tamir, A.: On the minimum diameter spanning tree problem. Inf. Process. Lett. **53**(2), 109–111 (1995). https://doi.org/10.1016/0020-0190(94)00183-Y

19. Hochuli, A., Holzer, S., Wattenhofer, R.: Distributed approximation of minimum routing cost trees. In: Halldórsson, M.M. (ed.) SIROCCO 2014. LNCS, vol. 8576, pp. 121–136. Springer, Cham (2014). https://doi.org/10.1007/978-3-319-09620-9_11
20. Hu, T.C.: Optimum communication spanning trees. SIAM J. Comput. **3**(3), 188–195 (1974). https://doi.org/10.1137/0203015
21. Jana, B., Mondal, S.: Computation of a minimum average distance tree on permutation graphs. Ann. Pure Appl. Math. **2**(1), 74–85 (2012)
22. Jana, B., Mondal, S., Pal, M.: Computation of a minimum average distance tree on circular-arc graphs. Int. J. Electron. Commun. Comput. Eng. **6**(3), 384–390 (2015)
23. Johnson, D.S., Lenstra, J.K., Kan, A.H.G.R.: The complexity of the network design problem. Networks **8**(4), 279–285 (1978). https://doi.org/10.1002/net.3230080402
24. Kratsch, S., Nelles, F.: Efficient and adaptive parameterized algorithms on modular decompositions. In: Azar, Y., Bast, H., Herman, G. (eds.) 26th Annual European Symposium on Algorithms, ESA 2018, Helsinki, Finland, 20–22 August 2018. LIPIcs, vol. 112, pp. 55:1–55:15. Schloss Dagstuhl - Leibniz-Zentrum für Informatik (2018). https://doi.org/10.4230/LIPICS.ESA.2018.55
25. Lokshtanov, D.: Parameterized integer quadratic programming: variables and coefficients. CoRR abs/1511.00310 (2015). http://arxiv.org/abs/1511.00310
26. Lyaudet, L., Yonta, P.M., Tchuenté, M., Ndoundam, R.: Distance preserving subtrees in minimum average distance spanning trees. Discret. Math. Algorithms Appl. **5**(3) (2013). https://doi.org/10.1142/S1793830913500109
27. Masone, A., Nenni, M.E., Sforza, A., Sterle, C.: The minimum routing cost tree problem – state of the art and a core-node based heuristic algorithm. Soft Comput. **23**(9), 2947–2957 (2019). https://doi.org/10.1007/S00500-018-3557-3
28. Mondal, S.: An efficient algorithm for computation of a minimum average distance tree on trapezoid graphs. J. Sci. Res. Rep. **2**(2), 598–611 (2013)
29. Schmuck, N.S.: The wiener index of a graph (2010)
30. Tedder, M., Corneil, D., Habib, M., Paul, C.: Simpler linear-time modular decomposition via recursive factorizing permutations. In: Aceto, L., Damgård, I., Goldberg, L.A., Halldórsson, M.M., Ingólfsdóttir, A., Walukiewicz, I. (eds.) ICALP 2008. LNCS, vol. 5125, pp. 634–645. Springer, Heidelberg (2008). https://doi.org/10.1007/978-3-540-70575-8_52
31. Wiener, H.: Structural determination of paraffin boiling points. J. Am. Chem. Soc. **69**(1), 17–20 (1947)
32. Wu, B.Y., Lancia, G., Bafna, V., Chao, K.M., Ravi, R., Tang, C.Y.: A polynomial-time approximation scheme for minimum routing cost spanning trees. SIAM J. Comput. **29**(3), 761–778 (2000). https://doi.org/10.1137/S009753979732253X
33. Zehavi, M.: The k-leaf spanning tree problem admits a Klam value of 39. Eur. J. Comb. **68**, 175–203 (2018). https://doi.org/10.1016/J.EJC.2017.07.018
34. Zetina, C.A., Contreras, I.A., Fernández, E., Luna-Mota, C.: Solving the optimum communication spanning tree problem. Eur. J. Oper. Res. **273**(1), 108–117 (2019). https://doi.org/10.1016/J.EJOR.2018.07.055

Dominating Set with Quotas: Balancing Coverage and Constraints

Sobyasachi Chatterjee[1], Sushmita Gupta[1], Saket Saurabh[1,2],
Sanjay Seetharaman[1(✉)], and Anannya Upasana[1]

[1] The Institute of Mathematical Sciences, HBNI, Chennai, India
{sobyasachic,sushmitagupta,saket,sanjays,anannyaupas}@imsc.res.in
[2] University of Bergen, Bergen, Norway

Abstract. We study a natural generalization of the classical DOMINATING SET problem, called DOMINATING SET WITH QUOTAS (DSQ). In this problem, we are given a graph G, an integer k, and for each vertex $v \in V(G)$, a lower quota lo_v and an upper quota up_v. The goal is to determine whether there exists a set $S \subseteq V(G)$ of size at most k such that for every vertex $v \in V(G)$, the number of vertices in its closed neighborhood that belong to S, i.e., $|N[v] \cap S|$, lies within the range $[\mathrm{lo}_v, \mathrm{up}_v]$. This richer model captures a variety of practical settings where both under- and over-coverage must be avoided—such as in fault-tolerant infrastructure, load-balanced facility placement, or constrained communication networks.

While DS is already known to be computationally hard, we show that the added expressiveness of per-vertex quotas in DSQ introduces additional algorithmic challenges. In particular, we prove that DSQ becomes W[1]-hard even on structurally sparse graphs—such as those with degeneracy 2, or excluding $K_{3,3}$ as a subgraph—despite these classes admitting FPT algorithms for DS. On the positive side, we show that DSQ is fixed-parameter tractable when parameterized by solution size and treewidth, and more generally, on nowhere dense graph classes. Furthermore, we design a subexponential-time algorithm for DSQ on apex-minor-free graphs using the bidimensionality framework. These results collectively offer a refined view of the algorithmic landscape of DSQ, revealing a sharp contrast with the classical DS problem and identifying the key structural properties that govern tractability.

Keywords: dominating set · quotas · sparse graph classes

1 Introduction

DOMINATING SET (DS, in short) is a prototypical covering problem studied on graphs. In this problem, we are given a graph G and an integer k, and the goal is to find a subset of vertices S of size at most k such that for each vertex $v \in V(G)$, either $v \in S$ or a neighbor of v is in S. This is one of the classic NP-complete problems, listed in Garey and Johnson [22], and it remains intractable even on

© The Author(s), under exclusive license to Springer Nature Switzerland AG 2026
F. Foucaud and A. Parreau (Eds.): IWOCA 2026, LNCS 16587, pp. 175–189, 2026.
https://doi.org/10.1007/978-3-032-27732-9_13

planar graphs of maximum degree 3. It could be viewed as a graph version of the SET COVER problem. Since the problem is intractable, it has been studied extensively using algorithmic approaches meant for coping with NP-hardness, including approximation [30], exact [29], and parameterized algorithms [6].

Together with VERTEX COVER, the DOMINATING SET problem can be considered the Drosophila of parameterized complexity. These problems provide a fertile ground for testing new ideas and tools. This has led to the study of several variants and generalizations of DS. This includes adding constraints on solutions (such as independence or connectivity) or constraints on how vertices outside the solution interact with the solution vertices. These led to problems such as CONNECTED DOMINATING SET, INDEPENDENT DOMINATING SET, $[1, j]$-TOTAL DOMINATING SET, and (σ, ρ)-DOMINATING SET [25,26,34,37], to name a few.

In this paper, we initiate a systematic study of another variant of DOMINATING SET, called DOMINATING SET WITH QUOTAS: a generalization of DS in which each vertex has a lower and an upper-quota on domination, motivated by applications in resource/fair allocation [4]. This richer model captures situations where having too little or too much coverage is undesirable. Such cases often appear in real-world systems that need to manage redundancy, balance load, or limit communication costs. Formally, the problem is stated as follows.

DOMINATING SET WITH QUOTAS (DSQ, in short)
Input: An undirected graph G, an integer k, and domination quotas $f_{lq}, f_{uq} : V(G) \to \mathbb{N} \cup \{0\}$.
Parameter: k
Question: A vertex v is said to be dominated *properly* by a set $S \subseteq V(G)$ if $f_{lq}(v) \leq |N[v] \cap S| \leq f_{uq}(v)$. Is there a subset of vertices S such that $|S| \leq k$ and S properly dominates $V(G)$?

In this paper, we study DSQ in the framework of parameterized complexity. It is well known that DS is W[2]-complete and, moreover, hard to approximate even within FPT time [9,31]. Since DSQ is a generalization of DS, these hardness results naturally extend to DSQ as well. Therefore, to obtain positive algorithmic results for DSQ, we must restrict our attention to graph classes where DS is known to be fixed-parameter tractable (FPT)—that is, where it admits an algorithm with running time of the form $f(k) \cdot n^{\mathcal{O}(1)}$, with n denoting the number of vertices in the input graph.

Although DS is W[2]-complete on general graphs, it is known to be fixed-parameter tractable on several sparse graph classes, including planar graphs, graphs of bounded genus, graphs excluding a fixed graph H as a (topological) minor, graphs of bounded expansion, nowhere dense graphs, and biclique-free graphs. Notably, biclique-free graphs subsume all of these classes and form the largest known family of graphs on which DS admits an FPT algorithm.

Research on DS in sparse graph classes has been one of the most fruitful directions in parameterized complexity, yielding several powerful tools and techniques [1,8,10,12,13,17–21,33]. This line of work is not only theoretically rich but also practically motivated. For many real-world applications, it is reasonable to assume that the solution size parameter k is small and that the input graphs are structurally sparse—such as having bounded degree, low degeneracy, or constrained topological features. These sparse classes capture a broad range of practical settings. For instance, planar graphs naturally model geographic networks or physical layouts, making them highly relevant in applied contexts.

Motivated by this, our main objective in this paper is to investigate the parameterized complexity of DSQ by drawing upon the extensive work on DS and its behavior on structurally sparse graph classes. Our goal is to chart the boundary between tractable and intractable cases for DSQ across this rich landscape (Table 1).

Table 1. An overview of the FPT status of DS and DSQ in various graph classes.

Graph class	DS	DSQ
Bounded degeneracy	FPT(k, d) [1]	W-hard on 2-degenerate graphs (Theorem 3)
Bounded treewidth	FPT(tw) [3]	W-hard par. tw (Theorem 1)
		FPT(k, tw) (Theorem 2)
Apex-minor-free	$2^{\mathcal{O}(\sqrt{k})}$ [16]	$2^{\mathcal{O}(\sqrt{k}\log k)}$ (Theorem 5)
Nowhere dense	FPT(k) [7]	FPT(k) (Theorem 4)
Biclique-free	FPT(k, t) [38]	W-hard on $K_{3,3}$-free graphs (from Theorem 3)

1.1   Our Results

We start by showing that DSQ parameterized by solution size is W[1]-hard even on graphs of degeneracy 2, Theorem 3. A graph G has degeneracy d if every subgraph of G has a vertex of degree at most d. This is in sharp contrast to DS which has been shown to be FPT parameterized by $k+d$ on graphs of degeneracy d by Alon and Gutner [1]. Moreover, DS is FPT parameterized by $k+t$ on $K_{t,t}$-free graphs (i.e., graphs that exclude the biclique $K_{t,t}$ as a subgraph) [38]. Since graphs with degeneracy d are $K_{d+1,d+1}$-free, it follows that DSQ is W[1]-hard even on $K_{3,3}$-free graphs. Our next result shows that DSQ parameterized by treewidth (tw) is W[1]-hard. This is unlike DS which is known to admit an algorithm with running time $3^{\mathrm{tw}}n^{\mathcal{O}(1)}$ on graphs of treewidth tw [32,36], that is tight under SETH [28]. We complement this intractability result by showing that DSQ can be solved in time $2^{\mathrm{tw}}(k+1)^{2\mathrm{tw}} \cdot n^{\mathcal{O}(1)}$ on graphs of treewidth tw using standard dynamic programming on tree decompositions. Using a "subset convolution"-like method, we improve this to a $((2k+2)^{\mathrm{tw}} + 2^{\mathcal{O}(k)}) \cdot n^{\mathcal{O}(1)}$ time algorithm. For this, we follow the underlying proof of the subset convolution method given in [35] rather than using the tool in a black-box fashion.

Using the *bidimensionality* approach of Fomin et al. [16], which works based on the structure of graphs of large treewidth, in conjunction with our algorithm on graphs of bounded treewidth, we show that DSQ can be solved in time

$2^{\mathcal{O}(\sqrt{k}\log k)} \cdot n^{\mathcal{O}(1)}$ on apex-minor-free graphs. The largest graph class for which we show that DSQ is FPT is the class of nowhere dense graphs, which includes several graph classes such as (topological) minor-free graphs and graphs of bounded expansion. We prove this result by expressing DSQ using First-Order (FO) logic and then invoking the result of Grohe et al. [24] that FO model-checking is FPT in the size of the formula on nowhere dense graphs.

In Section 6, we study SET COVER WITH QUOTAS, a generalization of DSQ, and show that it is FPT parameterized by $k + d$ when the sets are of size at most d. While the same is easy to see for SET COVER (for e.g., by dynamic programming), the presence of elements with lower-quota zero makes the result interesting. Since DSQ on graphs of maximum degree d is a special case of SCQ with sets of bounded size, we also obtain an algorithm for DSQ on graphs of bounded degree.

Related Work. The THRESHOLD DOMINATING SET problem is a special case of DSQ, where we are given a graph G, positive integers k and r, and the objective is to find a set $S \subseteq V(G)$ such that $|S| \leq k$ and for every vertex $v \in V(G)$, it holds that $|N[v] \cap S| \geq r$. Golovach and Villanger [23] showed that this problem is fixed-parameter tractable on graphs of bounded degeneracy. A related problem, studied by Meybodi et al. [34], is the $[1, j]$-TOTAL DOMINATING SET: given a graph G and an integer k, the goal is to find a set $S \subseteq V(G)$ of size at most k such that for each vertex $v \in V(G)$, it holds that $1 \leq |N(v) \cap S| \leq j$, for some fixed $j \in \mathbb{N}$. Another closely related problem is the $[\sigma, \rho]$-DOMINATING SET, also known as GENERALIZED DOMINATION, introduced by Telle [37]. In this model, the quota conditions for each vertex depend on whether the vertex is included in the solution. This contrasts with DSQ, where each vertex has a fixed quota that is independent of whether it belongs to the solution, and different vertices may have distinct lower and upper bounds. Gupta et al. [27] studied the complexity of d-HITTING SET WITH QUOTAS, a problem equivalent to SET COVER WITH QUOTAS where each element appears in at most d sets. We note that neither DSQ (where the degree is unbounded) nor SET COVER WITH QUOTAS subsumes the other. The literature on variants of DOMINATING SET is vast; to the best of our knowledge, the problems mentioned above are the most closely related to ours. Additional related work is discussed in the full version of the paper.[1]

Preliminaries. We say that a vertex v has quota $\langle i, j \rangle$ if $f_{lq}(v) = i$ and $f_{uq}(v) = j$. For integers i and j, we denote the sets $\{i, \ldots, j\}$ and $\{1, \ldots, j\}$ by $[i, j]$ and $[j]$, resp. We use n and m to denote the number of vertices and edges in the input graph, respectively. We use $N_G^d(v)$ to denote the set of vertices at distance at most d from v in G. The rest of the preliminaries can be found in the full version.

---

[1] Missing details and proofs of statements marked † are proved in the full version of the paper [5].

2 DSQ on Graphs of Bounded Treewidth

In this section, we study DSQ on graphs of bounded treewidth. Firstly, we show that DSQ is hard when parameterized by tw. Following that, we show that DSQ is FPT when parameterized by $\text{tw} + k$.

2.1 Hardness of DSQ Parameterized by Treewidth

We establish hardness by a reduction from EXACT-EQUITABLE COLORING (E-EC): given a graph H and an integer r, does H admit a proper vertex coloring with exactly r colors such that the sizes of any two color classes differ by at most 1? The problem is W[1]-hard parameterized by $\text{tw}(H)+r$ [14]. Fellows et al. [14] define EQUITABLE COLORING (EC) as the problem in which an equitable coloring of H using *at most* r colors is found instead of exactly r colors. However, they present a reduction that establishes the hardness of E-EC, which also establishes the hardness of EC. Next, we observe a property of the color class sizes in an equitable coloring, which forms a key component in our reduction.

Proposition 1 (†). *Given an equitable coloring in a graph with n vertices using r colors, each color class is of size either $\lfloor n/r \rfloor$ or $\lceil n/r \rceil$.*

By Proposition 1, we know the smallest and largest sizes of the color classes. The basic idea in the reduction is to express this fact (along with expressing proper vertex coloring) using domination lower and upper-quotas. Given an instance $\mathcal{I} = (H, r)$ of E-EC, we construct an instance $\mathcal{J} = (G, k, \text{f}_{\text{lq}}, \text{f}_{\text{uq}})$ of DSQ in polynomial time as given in Algorithm 1. See Figure 1 for an illustration.

Any DSQ of size at most k contains $\{M\} \cup \{\alpha_v\}_{v \in V(H)} \cup \{c^i\}_{i \in [r]}$ since each of it is adjacent to $k+1$ pendants with quota $\langle 1, 1 \rangle$. Moreover, no solution contains those pendants. From now on, we call those vertices *newly added pendants*.

We show the correctness of the reduction in the full version. To establish W[1]-hardness parameterized by treewidth, we also show that $\text{tw}(G)$ is bounded by a function of $\text{tw}(H)$ and r, by exhibiting a tree decomposition of G of such width.

Theorem 1. DSQ *parameterized by treewidth is* W[1]-*hard.*

2.2 Algorithms

Using standard dynamic programming over tree decompositions, similar to that for DS (see [6]), we first establish that DSQ is FPT parameterized by $\text{tw} + k$.

Theorem 2 (†). *Given a tree decomposition of G of width* tw, DSQ *can be solved in time* $2^{\text{tw}} (k+1)^{2\text{tw}} \cdot n^{\mathcal{O}(1)}$.

Using a "subset convolution" like method in the join node, we can get an improvement in our algorithm. Let $\hat{u} = \max_{v \in V(G)} \text{f}_{\text{uq}}(v)$, the maximum upper-quota of a vertex. Since we are interested in finding a solution of size at most k, we

---

Algorithm 1: Reduction from E-EC to DSQ

---

1 For each vertex $v \in V(H)$, create r vertices $x_v^1, \ldots, x_v^r$ and set $\mathsf{f}_{\mathsf{lq}}(x_v^i) = 1$ and $\mathsf{f}_{\mathsf{uq}}(x_v^i) = 2$ for each $i \in [r]$.

2 For each edge $uv \in E(H)$ and each $i \in [r]$, create a vertex y_{uv}^i with $\mathsf{f}_{\mathsf{lq}}(y_{uv}^i) = 1$, $\mathsf{f}_{\mathsf{uq}}(y_{uv}^i) = 2$, and make it adjacent to x_u^i and x_v^i.

3 Create a vertex M adjacent to all vertices created in the previous step: $\{y_{uv}^i : i \in [r], uv \in E(H)\}$, with $\mathsf{f}_{\mathsf{lq}}(M) = \mathsf{f}_{\mathsf{uq}}(M) = 1$.

4 Create $2n+r+2$ pendants $\{p_j^M\}_{j=1}^{2n+r+2}$ adjacent to M with $\mathsf{f}_{\mathsf{lq}}(p_j^M) = \mathsf{f}_{\mathsf{uq}}(p_j^M) = 1$ for each $j \in [2n+r+2]$.

5 Let l and h denote the smallest and the largest sizes of the color classes: that is, $\ell = \lfloor n/r \rfloor$ and $h = \lceil n/r \rceil$. For each $i \in [r]$, create a vertex c^i (to "check" the color class size) that is adjacent to all vertices $\{x_v^i\}_{v \in V(H)}$, with $\mathsf{f}_{\mathsf{lq}}(c^i) = \ell+1$ and $\mathsf{f}_{\mathsf{uq}}(c^i) = h+1$.

6 For each vertex c^i, create $2n+r+2$ pendants $\{p_j^i\}_{j=1}^{2n+r+2}$ adjacent to c^i, with $\mathsf{f}_{\mathsf{lq}}(p_j^i) = \mathsf{f}_{\mathsf{uq}}(p_j^i) = 1$.

7 For each vertex $v \in V(H)$, create a vertex α_v (to "force" a vertex among its neighbors to be picked) adjacent to all vertices $\{x_v^i\}_{i \in [r]}$, with $\mathsf{f}_{\mathsf{lq}}(\alpha_v) = \mathsf{f}_{\mathsf{uq}}(\alpha_v) = 2$.

8 For each α_v, create $2n+r+2$ pendants $\{p_j^v\}_{j=1}^{2n+r+2}$ adjacent to α_v, with $\mathsf{f}_{\mathsf{lq}}(p_j^v) = \mathsf{f}_{\mathsf{uq}}(p_j^v) = 1$.

9 Set $k = 2n+r+1$.

---

can assume that $\hat{u} \leq k$ without loss of generality (each vertex can be dominated at most k times). We would like to note that if we change the set of states in a node t in the DP table by considering the functions $f : X_t \to [\hat{u}]$ instead of $f : X_t \to [k]$, we get an algorithm for DSQ that runs in time $2^{\mathrm{tw}}(\hat{u}+1)^{2\mathrm{tw}} \cdot n^{\mathcal{O}(1)}$. If $\hat{u}$ is a fixed constant, then we can use the techniques of [35] to obtain an algorithm for computing all the entries in the join node that runs in total time $2^{\mathrm{tw}}(\hat{u}+1)^{\mathrm{tw}} \cdot n^{\mathcal{O}(1)} = (2\hat{u}+2)^{\mathrm{tw}} \cdot n^{\mathcal{O}(1)}$. If $\hat{u}$ is not a fixed constant, then the techniques of [35] run in time $((2\hat{u}+2)^{\mathrm{tw}} + 2^{\mathcal{O}(\hat{u})}) \cdot n^{\mathcal{O}(1)}$ (see Remark 4.20 in [15]). We have discussed this in the full version.

3   DSQ on Graphs of Bounded Degeneracy

The degeneracy of a graph G is the smallest number d such that any subgraph of G has a vertex of degree at most d. Note that the degeneracy is bounded by the maximum degree. In this section, we show the W[1]-hardness of DSQ even in constant degeneracy graphs by a parameterized reduction from the INDEPENDENT SET problem (IS): given a graph H and an integer r, does there exist a set $S \subseteq V(H)$ such that S is an r-sized independent set in H (i.e., $H[S]$ is edgeless)?

The IS problem parameterized by r is W[1]-hard [6]. The basic idea of the following reduction is to express the fact that any independent set contains at most one endpoint of an edge as a domination upper-quota. Given an instance $\mathcal{I} = (H,r)$ of IS, we construct an instance $\mathcal{J} = (G, k, \mathsf{f}_{\mathsf{lq}}, \mathsf{f}_{\mathsf{uq}})$ of DSQ in polynomial time as given in Algorithm 2. See Figure 2 for an illustration. We would like

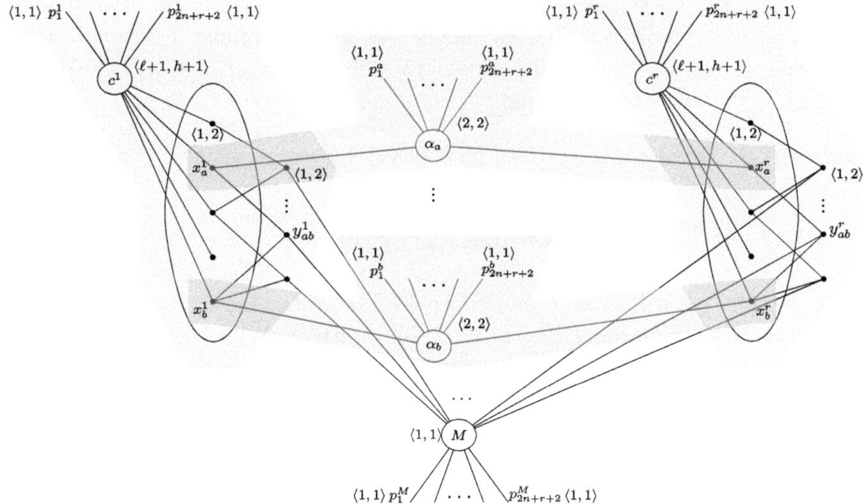

Fig. 1. The instance $\mathcal{J}$ constructed in the reduction from E-EC. One can view the construction as r verticals: one for each color and n horizontals: one for each vertex.

Algorithm 2: Reduction from IS to DSQ

---

1 For each vertex $v \in V(H)$, create a vertex x_v in G and set $f_{lq}(x_v) = 0$ and $f_{uq}(x_v) = 1$.
2 For each edge $uv \in E(H)$, create a vertex y_{uv} with $f_{lq}(y_{uv}) = 0$, $f_{uq}(y_{uv}) = 1$, and make it adjacent to x_u and x_v.
3 Create a vertex α (to "force" r vertices in H to be picked) that is adjacent to all of $\{x_v\}_{v \in V(H)}$, with $f_{lq}(\alpha) = f_{uq}(\alpha) = r$.
4 Create a pendant vertex p^α adjacent to α, with $f_{lq}(p^\alpha) = f_{uq}(p^\alpha) = 0$.
5 Set $k = r$.

---

to mention that the basic construction in the reduction is similar to that in a reduction presented in [34], which establishes the hardness of a different related problem called $[1, j]$-DOMINATING SET on graphs of degeneracy $j + 1$.

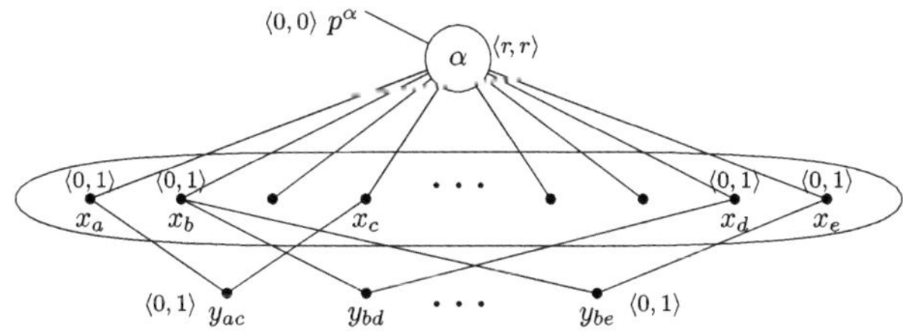

Fig. 2. The instance $\mathcal{J}$ constructed in the reduction from IS.

We show the correctness of the reduction in the full version. The following ordering of vertices (in which each vertex has degree at most 2 into the vertices succeeding it) shows that G has degeneracy 2: p^α, $\{y_{uv}\}_{uv \in E(H)}$, $\{x_v\}_{v \in V(H)}$, α. Overall, we have the following.

Theorem 3. DSQ *parameterized by k is* W[1]-*hard on graphs of degeneracy 2.*

4   DSQ on Nowhere Dense Graphs

In this section, we will show that DSQ is $\mathsf{FPT}(k)$ if the input graph belongs to a graph class that is *nowhere dense*. We briefly introduce the required definitions for our result. For a more extensive presentation, we refer to the paper [24], which contains the algorithmic meta-theorem of our interest. A *model* of a graph H in G is a function ψ that maps vertices in H to vertex disjoint connected subgraphs in G (called *branch sets*) such that $\psi(u)$ and $\psi(v)$ are adjacent if $uv \in E(H)$. It is known that H is a minor of G if and only if there is a model of H in G. The graph H is said to be an *r-shallow minor* of G if there is a model ψ of H in G such that for each $v \in H$, the radius of $\psi(v)$ (that is, the minimum maximum distance from a vertex, given by $\min_{x \in \psi(v)} \max_{y \in \psi(v)} \mathrm{dist}(x, y)$) is at most r. A class $\mathcal{C}$ of graphs is said to be *nowhere dense* if there is a function $f \colon \mathbb{N} \to \mathbb{N}$ such that for all r, $K_{f(r)}$ is not an r-shallow minor of G for all $G \in \mathcal{C}$. We focus on the case where f is computable: such classes are called *effectively* nowhere dense. Note that this is a reasonable restriction since the functions corresponding to all natural nowhere dense graph classes are computable [24]. Next, we argue that DSQ is $\mathsf{FPT}(k)$ if the input graph is nowhere dense. On a high level, we construct a new graph that belongs to a nowhere dense graph class and "contains" G along with the quota information of each vertex, and use first-order model-checking on this graph to determine the existence of a solution.

We follow the notation and terminology of Grohe et al. [24]. For a detailed overview, see the full version of the paper.

Proposition 2 (Theorem 8.1 in [24]). *Let $\mathcal{C}$ be an effectively nowhere dense class of graphs. There is a computable function f and an algorithm that, given $\epsilon > 0$, a formula $\varphi(x) \in \mathrm{FO}(\sigma)$ for some colored-graph vocabulary σ and a σ-colored graph $G \in \mathcal{C}$, computes the set of all $v \in V(G)$ such that $G \models \varphi(v)$ in time $f(|\varphi|, \epsilon) \cdot n^{1+\epsilon}$.*

Suppose that the input graph G belongs to a nowhere dense graph class $\mathcal{C}$ such that each graph in $\mathcal{C}$ excludes $K_{f(r)}$ as an r-shallow minor for each $r \in \mathbb{N}$. In particular, it excludes $K_{f(0)}$ as a subgraph (which is equivalently a 0-shallow minor). The basic idea is to construct another nowhere dense graph G' in which the quota information of each vertex is embedded in the graph structure itself; and then work on G'. One can consider adding $2(k+1)$ different unary relations to the vocabulary for expressing the lower and upper-quotas of the vertices in the graph and then writing an FO formula based on that to solve DSQ^2. But this results in different vocabularies for different values of k. We present an algorithm using only the edge relation in the vocabulary that decides whether G has a DSQ of size at most k.

For each vertex v, we attach a *quota gadget* to it: $(v) - (K_{f(0)}) - $ (path on $f_{lq}(v)+1$ vertices) $- (K_{f(0)}) - $ (path on $f_{uq}(v)+1$ vertices) $- (K_{f(0)})$, where the hyphens denote an edge incident to an arbitrary vertex in the complete graph). We call a vertex a *gadget vertex* if it is part of a quota gadget, and call it an *original vertex* otherwise. Let G' be the graph obtained from a copy of G by attaching the quota gadgets to each vertex. Since $K_{f(0)}$ is not a subgraph of any graph in $\mathcal{C}$, we have that $G' \notin \mathcal{C}$. We claim that G' belongs to a different nowhere dense graph class.

Lemma 1 (†). *The graph G' excludes $K_{g(r)}$ as an r-shallow minor for each $r \in \mathbb{N}$, where $g(r) = \max(f(r), 3f(0) + 2))$. Hence, it belongs to the nowhere dense graph class characterized by the function g.*

Next, we use FO model-checking on G' to solve DSQ in G. Here, the universe A is the vertices of the constructed graph G' and σ contains the binary relation E that captures the edges in G'. Since we have shown that G' belongs to a nowhere dense graph class, we are left with showing the first order expressibility of the problem. By construction and properties of G and G', we have the following.

Lemma 2. *A vertex v is present in G with quota $\langle \ell, u \rangle$ if and only if in G', v is attached to the following: a clique on $f(0)$ vertices, a path on $\ell + 1$ vertices, a clique on $f(0)$ vertices, a path on $u + 1$ vertices, a clique on $f(0)$ vertices; in the order as listed with each part connected by an edge.*

We show that there is a formula of length $\mathcal{O}(2^k)$ on G' that solves DSQ in G. Consequently, we have that DSQ is FPT parameterized by k if $G \in \mathcal{C}$ for a nowhere dense class $\mathcal{C}$. Without loss of generality, we assume that $k > 0$ and at least one vertex has non-zero lower-quota. If $k = 0$, then the input is a yes-instance if and only if all vertices have lower-quota 0. Moreover, if all vertices have lower-quota 0, then the input is a yes-instance. The detailed description of each formula is in the full version.

- **Dom**(u, v) checks the domination of a vertex by another.
- **Distinct**$(x_1, \ldots, x_s)$ checks whether the input arguments are distinct.
- **Clique**$(q_1, \ldots, q_r)$ checks if $\{q_1, \ldots, q_r\}$ form a clique.
- **Path**$(q_1, \ldots, q_r)$ checks if $\langle q_1, \ldots, q_r \rangle$ form a path.
- **Check**$_{\ell,u}(w)$ checks if w is an original vertex with quota $\langle \ell, u \rangle$.

---

² As pointed out by an anonymous reviewer, it appears that this approach of adding new unary relations can be used to obtain the more general result that DSQ is FPT(k) in monadically stable graph classes, where FO model checking is FPT in the formula size due to the result by Dreier et al., [11]. Since we are unclear about how adding unary relations affects the overall running time of the algorithm, we have refrained from doing that.

- **SatLQ**$_\ell(v, x_1, \ldots, x_s)$ checks if v is dominated by at least ℓ vertices in $x_1,...,x_s$.
- **SatUQ**$_u(v, x_1, \ldots, x_s)$ checks if v is dominated by at most u vertices in $x_1,...,x_s$.
- Formula φ_s that combines everything and checks for a solution of size s:

$$\varphi_s := \exists x_1, \ldots, x_s \, \forall v \left[\mathbf{Distinct}(x_1, \ldots, x_s) \wedge \bigwedge_{i \in [s]} \left(\bigvee_{\substack{\ell \in [0,k] \\ u \in [\ell,k]}} \mathbf{Check}_{\ell,u}(x_i) \right) \right]$$

$$\wedge \left[\bigwedge_{\substack{\ell \in [0,k] \\ u \in [0,k]}} \left(\neg \mathbf{Check}_{\ell,u}(v) \vee \left(\mathbf{SatLQ}_\ell(v, x_1, \ldots, x_s) \wedge \mathbf{SatUQ}_u(v, x_1, \ldots, x_s) \right) \right) \right]$$

Observe that the last part of the above formula checks if v is an original vertex with quota $\langle \ell, u \rangle$ and if so, whether v is properly dominated by $x_1, \ldots, x_k$.
- Formula φ that checks if there is a solution of size at most k: $\varphi := \bigvee_{s \in [k]} \varphi_s$.

By construction, we have that $G' \models \varphi$ if and only if there is a DSQ of size k in $\mathcal{I}$. Since the size of ψ is bounded by a function of k (†), we have the following.

Theorem 4. DSQ *parameterized by k is* FPT *on nowhere dense graphs.*

5   DSQ on Apex-Minor-Free Graphs

A graph is said to be *planar* if it can be drawn on a plane with no two edges crossing each other. By Wagner's theorem, a graph is planar if and only if it does not contain K_5 or $K_{3,3}$ as a minor. A graph is an *apex* graph if deleting a vertex results in a planar graph. A graph class $\mathcal{C}$ is said to be *apex-minor-free* if all graphs in $\mathcal{C}$ avoid a fixed apex graph H as a minor. Observe that planar graphs are also apex-minor-free. We show that DSQ is FPT(k) in apex-minor-free graphs using the *bidimensionality* approach [16].

On a high level, the bidimensionality approach exploits the structure of graphs with large treewidth. Towards that, we use the algorithm in Sect. 2.2 for DSQ on graphs of bounded treewidth.

In this section, we design a sub-exponential time FPT algorithm for DSQ on apex-minor-free graphs. For that, we observe that, in polynomial time, we can discard some lower-quota 0 vertices depending on their neighborhoods.

Reduction Rule 1 *If there is a vertex v with lower-quota 0 such that all vertices at a distance at most 2 from v have lower-quota 0, then we delete v.*

We establish the correctness of Reduction Rule 1 in the full version. A vertex satisfying the above condition can be found in polynomial time. Thus, exhaustive application of the above reduction rule can also be done in polynomial time. From now on, we will assume so: for any vertex $v \in V(G)$, there is a vertex $w \in N_G^2[v]$ such that $f_{lq}(w) \geq 1$. Any solution in $\mathcal{I}$ contains a vertex w_d that dominates w and w_d is at distance at most 3 from v. Consequently, any solution in $\mathcal{I}$ is also a

3-dominating set (3-DS, in short) in G: a set S such that any vertex $v \in V(G)$ is at distance at most 3 from S.

For an integer q, the $(q \times q)$-grid is a graph with vertices $\{(i,j)\}_{i,j\in[q]}$ and edges $\{((i_1,j_1),(i_2,j_2)): |i_1 - i_2| + |j_1 - j_2| = 1\}$. The graph Γ_q is obtained from a $(q \times q)$-grid by triangulating all inner faces so that all internal vertices have degree 6 (say by adding the edges $\{((x+1,y),(x,y+1)): x,y \in [q-1]\}$) and then connecting one corner vertex of degree 2 with all vertices of the external face (say by adding the edges $\{((q,q),(x,y)): x \in \{1,q\}$ or $y \in \{1,q\}\}$). For obtaining an algorithm for DSQ, we use the following result due to Fomin et al.

Proposition 3 (Theorem 1 in [16]). *Let H be an apex graph. There is a constant c_H such that every connected graph G excluding H as a minor and of treewidth at least $c_H \cdot q$, contains Γ_q as a contraction.*

Let D_3^G denote a minimum 3-DS in G. Observe that for any graph C that is a contraction of G, we have $|D_3^G| \geq |D_3^C|$. This is because contracting any edge cannot increase the size of a minimum 3-DS. Now we are ready to present the algorithm for the case where G is H-minor-free for a fixed apex graph H.

It is known that there is a $2^{\mathcal{O}(s)}n^{\mathcal{O}(1)}$ algorithm that, given the graph G and an integer s, either returns a tree decomposition of G of width at most $4s + 4$ or concludes that $\mathrm{tw}(G) > s$ [6]. We set $s = c_H \cdot 15\sqrt{k}$ and run the algorithm: we consider two cases depending on its output. If it concludes that $\mathrm{tw}(G) > s$, then by Proposition 3, G contains $\Gamma_{15\sqrt{k}}$ as a contraction. Observe that any internal vertex can 3-dominate at most 49 vertices (in the 7x7 grid that it is the center of). Thus, any DSQ in G, which is also a 3-DS, is of size at least $|D_3^{\Gamma_{15\sqrt{k}}}| \geq (225k - (4 \cdot 3 \cdot (15\sqrt{k} - 1)))/49 > k$ (the last inequality holds for all $k \geq 1$; in the first inequality, we use an upperbound on the number of vertices that the corner vertex adjacent to all external vertices can 3-dominate in $\Gamma_{15\sqrt{k}}$) and we return that there is no DSQ of size k. In the other case, the algorithm returns a tree decomposition $\mathcal{T}$ of G. The width of $\mathcal{T}$ is at most $4 \cdot c_H \cdot 15\sqrt{k} + 4 = \mathcal{O}(\sqrt{k})$ and we run the DP algorithm (in Sect. 2.2) on $\mathcal{T}$ and obtain an answer in time $2^{\mathcal{O}(\sqrt{k}\log k)} \cdot n^{\mathcal{O}(1)}$. Thus, we have the following.

Theorem 5. DSQ *can be solved in time* $2^{\mathcal{O}(\sqrt{k}\log k)} \cdot n^{\mathcal{O}(1)}$ *on apex-minor-free graphs.*

6   SET COVER WITH QUOTAS When All Sets are *Small*

The SET COVER problem (SC, in short) is a generalization of DS in which we are given a universe $\mathcal{U}$, a family $\mathcal{F} \subseteq 2^{\mathcal{U}}$, and an integer k, and the goal is to find a subfamily $S \subseteq \mathcal{F}$ of size at most k such that $\bigcup_{R \in S} R = \mathcal{U}$. While it is W[2]-hard parameterized by k, it is FPT w.r.t. k and d when the sets are of size at most d [6,9]. We define SET COVER WITH QUOTAS (SCQ, in short) as follows. Given a universe $\mathcal{U}$, a family $\mathcal{F} \subseteq 2^{\mathcal{U}}$, coverage quotas $\mathsf{f}_{\mathrm{lq}}, \mathsf{f}_{\mathrm{uq}} \colon \mathcal{U} \to \mathbb{N} \cup \{0\}$, and an integer k, determine whether there is a subfamily $S \subseteq \mathcal{F}$ of size at most k such that for each $u \in \mathcal{U}$ we have $\mathsf{f}_{\mathrm{lq}}(u) \leq |\{R: R \in S \text{ and } u \in R\}| \leq \mathsf{f}_{\mathrm{uq}}(u)$. Two natural restrictions that we study are the following:

(1) SCQ when each element occurs in at most f sets (bounded frequency);
(2) SCQ when each set contains at most d elements (bounded size).

Observation 1 (†). SCQ *is* NP-*hard even if* $f=2$; *and is* FPT *par. by* (k, f).

In this section, we will consider instances of SCQ where each set in the set family has size at most d. We first present a randomized algorithm based on the *color-coding* technique [2,6]. Existing results on derandomizations of the technique lead to deterministic algorithms for the problem we are interested in.

We call an element $u \in \mathcal{U}$ *interesting* if $f_{lq}(u) \geq 1$, and *uninteresting* otherwise. Suppose that the input admits a solution $S = \{R_i\}_{i=1}^{k}$. Observe that the number of elements covered by any k sized solution (even when counting multiplicity) is at most kd. This gives an upper-bound on the number of interesting elements (the number of uninteresting elements is unbounded though). The idea is to randomly color the universe with kd colors so that the elements covered by S are colorful (i.e., no two covered elements have the same color) and then determine the existence of such a solution based on the coloring.

Let $C = (C_1, \ldots, C_{kd})$ be a random kd-partition of $\mathcal{U}$; such a partition/coloring can be obtained by independently performing the following for each element $u \in \mathcal{U}$: choose $i \in [kd]$ uniformly at random and place u in C_i. From now on, we will condition on the event that the elements covered by S are colorful according to C: this happens with probability at least $\binom{kd}{\ell} \ell! / (kd)^{\ell} \geq e^{-\ell} \geq e^{-kd}$, where $\ell = |\bigcup_{R \in S} R| \leq kd$. Then, any part C_i contains at most one interesting element: if there are two interesting elements in C_i, then since both are covered by S, the coloring C does not satisfy the event condition). If a part C_i contains an interesting element u, then delete from $\mathcal{U}$ all elements $e \neq u \in C_i$ and the sets that contain e. This deletion is safe since the unique element in C_i covered by S is u (i.e., $C_i \cap \bigcup_{R \in S} R = \{u\}$). Thus, we have the following.

Observation 2. *Any part* C_i *is either empty, or is of the form* $\{e_i\}$ *for some interesting element* e_i, *or is a set of uninteresting elements.*

Using dynamic programming, we obtain a deterministic algorithm that checks in time $2^{\mathcal{O}(k \log d)}(|\mathcal{U}| + |\mathcal{F}|)^{\mathcal{O}(1)}$ whether $\mathcal{F}$ contains an SCQ of size at most k (where the elements covered are colorful according to C) and, if this is the case, returns one such solution. Using a standard pseudorandom object called *perfect hash family*, we extend our algorithm on the colored instance and obtain a deterministic algorithm for SCQ.

Theorem 6 (†). SCQ *with sets of size at most* d *can be solved in time* $2^{\mathcal{O}(kd)} \cdot (|\mathcal{U}| + |\mathcal{F}|)^{\mathcal{O}(1)}$.

7   Conclusion and Open Questions

Our results leave several intriguing questions open for future investigation. Firstly, our positive result for nowhere dense graphs relies on a reduction to FO model-checking. A natural direction for future work is to develop more direct and combinatorial FPT algorithms for DSQ on these graph classes, avoiding the reliance on logical machinery. It would also be interesting to explore whether similar logic-based techniques can be extended to characterize even broader classes of graphs that admit efficient algorithms for DSQ.

Another promising direction concerns the approximability of DSQ, which remains largely unexplored. Can one design FPT-approximation algorithms for DSQ, especially in settings where exact solutions are unlikely due to strong hardness results?

Finally, the question of preprocessing is open. In particular, can we obtain efficient kernelization algorithms for DSQ, at least on restricted graph classes? Given the success of kernelization techniques for DS on sparse graphs, extending these ideas to DSQ could yield significant algorithmic insights.

References

1. Alon, N., Gutner, S.: Linear time algorithms for finding a dominating set of fixed size in degenerated graphs. Algorithmica **54**(4), 544–556 (2009). https://doi.org/10.1007/s00453-008-9204-0
2. Alon, N., Yuster, R., Zwick, U.: Color-coding. J. ACM (JACM) **42**(4), 844–856 (1995). https://doi.org/10.1145/210332.210337
3. Arnborg, S., Proskurowski, A.: Linear time algorithms for NP-hard problems restricted to partial k-trees. Discret. Appl. Math. **23**(1), 11–24 (1989). https://doi.org/10.1016/0166-218X(89)90031-0
4. Banerjee, S., Eichhorn, M., Kempe, D.: Allocating with priorities and quotas: algorithms, complexity, and dynamics. In: Proceedings of the 24th ACM Conference on Economics and Computation (EC), pp. 209–240 (2023). https://doi.org/10.1145/3580507.3597733
5. Chatterjee, S., Gupta, S., Saurabh, S., Seetharaman, S., Upasana, A.: Dominating set with quotas: balancing coverage and constraints (2026). https://arxiv.org/abs/2604.04912
6. Cygan, M.: Parameterized Algorithms. Springer. Cham (2015). https://doi.org/10.1007/978-3-319-21275-3
7. Dawar, A., Kreutzer, S.: Domination problems in nowhere-dense classes. In: Proceedings of the IARCS Annual Conference on Foundations of Software Technology and Theoretical Computer Science (FSTTCS). LIPIcs, vol. 4, pp. 157–168. Schloss Dagstuhl - Leibniz-Zentrum für Informatik (2009). https://doi.org/10.4230/LIPICS.FSTTCS.2009.2315
8. Demaine, E.D., Hajiaghayi, M.: The bidimensionality theory and its algorithmic applications. Comput. J. **51**(3), 292–302 (2008) https://doi.org/10.1093/COMJNL/BXM033
9. Downey, R.G., Fellows, M.R.: Fixed-parameter tractability and completeness I: basic results. SIAM J. Comput. **24**(4), 873–921 (1995). https://doi.org/10.1137/S0097539792228228
10. Drange, P.G., et al.: Kernelization and sparseness: the case of dominating set. In: Ollinger, N., Vollmer, H. (eds.) 33rd Symposium on Theoretical Aspects of Computer Science, STACS 2016. LIPIcs, vol. 47, pp. 31:1–31:14. Schloss Dagstuhl - Leibniz-Zentrum für Informatik (2016). https://doi.org/10.4230/LIPICS.STACS.2016.31

11. Dreier, J., Eleftheriadis, I., Mählmann, N., McCarty, R., Pilipczuk, M., Torunczyk, S.: First-order model checking on monadically stable graph classes. In: 65th IEEE Annual Symposium on Foundations of Computer Science, FOCS 2024, pp. 21–30. IEEE (2024). https://doi.org/10.1109/FOCS61266.2024.00012

12. Einarson, C., Reidl, F.: A general kernelization technique for domination and independence problems in sparse classes. In: Cao, Y., Pilipczuk, M. (eds.) 15th International Symposium on Parameterized and Exact Computation, IPEC 2020, 14–18 December 2020. LIPIcs, vol. 180, pp. 11:1–11:15. Schloss Dagstuhl - Leibniz-Zentrum für Informatik (2020). https://doi.org/10.4230/LIPICS.IPEC.2020.11

13. Fabianski, G., Pilipczuk, M., Siebertz, S., Torunczyk, S.: Progressive algorithms for domination and independence. In: Niedermeier, R., Paul, C. (eds.) 36th International Symposium on Theoretical Aspects of Computer Science, STACS 2019. LIPIcs, vol. 126, pp. 27:1–27:16. Schloss Dagstuhl - Leibniz-Zentrum für Informatik (2019). https://doi.org/10.4230/LIPICS.STACS.2019.27

14. Fellows, M.R., et al.: On the complexity of some colorful problems parameterized by treewidth. Inf. Comput. **209**(2), 143–153 (2011). https://doi.org/10.1016/J.IC.2010.11.026

15. Focke, J., et al.: Tight complexity bounds for counting generalized dominating sets in bounded-treewidth graphs part I: algorithmic results. CoRR abs/2211.04278 (2022). https://doi.org/10.48550/ARXIV.2211.04278

16. Fomin, F.V., Golovach, P., Thilikos, D.M.: Contraction bidimensionality: the accurate picture. In: Fiat, A., Sanders, P. (eds.) ESA 2009. LNCS, vol. 5757, pp. 706–717. Springer, Heidelberg (2009). https://doi.org/10.1007/978-3-642-04128-0_63

17. Fomin, F.V., Lokshtanov, D., Saurabh, S., Thilikos, D.M.: Linear kernels for (connected) dominating set on H-minor-free graphs. In: Rabani, Y. (ed.) Proceedings of the Twenty-Third Annual ACM-SIAM Symposium on Discrete Algorithms, SODA 2012, pp. 82–93. , SIAM (2012). https://doi.org/10.1137/1.9781611973099.7

18. Fomin, F.V., Lokshtanov, D., Saurabh, S., Thilikos, D.M.: Linear kernels for (connected) dominating set on graphs with excluded topological subgraphs. In: Portier, N., Wilke, T. (eds.) 30th International Symposium on Theoretical Aspects of Computer Science, STACS 2013. LIPIcs, vol. 20, pp. 92–103. Schloss Dagstuhl - Leibniz-Zentrum für Informatik (2013). https://doi.org/10.4230/LIPICS.STACS.2013.92

19. Fomin, F.V., Lokshtanov, D., Saurabh, S., Thilikos, D.M.: Bidimensionality and kernels. SIAM J. Comput. **49**(6), 1397–1422 (2020). https://doi.org/10.1137/16M1080264

20. Fomin, F.V., Thilikos, D.M.: Dominating sets in planar graphs: branch-width and exponential speed-up. SIAM J. Comput. **36**(2), 281–309 (2006). https://doi.org/10.1137/S0097539702419649

21. Gajarský, J., et al.: Kernelization using structural parameters on sparse graph classes. J. Comput. Syst. Sci. **84**, 219–242 (2017). https://doi.org/10.1016/J.JCSS.2016.09.002

22. Garey, M.R., Johnson, D.S.: Computers and Intractability: A Guide to the Theory of NP-Completeness. W. H. Freeman, New York (1979)

23. Golovach, P.A., Villanger, Y.: Parameterized complexity for domination problems on degenerate graphs. In: Broersma, H., Erlebach, T., Friedetzky, T., Paulusma, D. (eds.) WG 2008. LNCS, vol. 5344, pp. 195–205. Springer, Heidelberg (2008). https://doi.org/10.1007/978-3-540-92248-3_18

24. Grohe, M., Kreutzer, S., Siebertz, S.: Deciding first-order properties of nowhere dense graphs. J. ACM (JACM) **64**(3), 17:1–17:32 (2017). https://doi.org/10.1145/3051095

25. Guha, S., Khuller, S.: Approximation algorithms for connected dominating sets. In: Diaz, J., Serna, M. (eds.) ESA 1996. LNCS, vol. 1136, pp. 179–193. Springer, Heidelberg (1996). https://doi.org/10.1007/3-540-61680-2_55

26. Guha, S., Khuller, S.: Improved methods for approximating node weighted Steiner trees and connected dominating sets. In: Arvind, V., Ramanujam, S. (eds.) FSTTCS 1998. LNCS, vol. 1530, pp. 54–65. Springer, Heidelberg (1998). https://doi.org/10.1007/978-3-540-49382-2_6

27. Gupta, S., Jain, P., Petety, A., Singh, S.: Parameterized complexity of d-Hitting set with quotas. In: Bureš, T. (ed.) SOFSEM 2021. LNCS, vol. 12607, pp. 293–307. Springer, Cham (2021). https://doi.org/10.1007/978-3-030-67731-2_21

28. Impagliazzo, R., Paturi, R.: On the complexity of k-sat. J. Comput. Syst. Sci. **62**(2), 367–375 (2001). https://doi.org/10.1006/JCSS.2000.1727

29. Iwata, Y.: A faster algorithm for dominating set analyzed by the potential method. In: Marx, D., Rossmanith, P. (eds.) IPEC 2011. LNCS, vol. 7112, pp. 41–54. Springer, Heidelberg (2012). https://doi.org/10.1007/978-3-642-28050-4_4

30. Johnson, D.S.: Approximation algorithms for combinatorial problems, **9**, 256–278 (1974). https://doi.org/10.1016/S0022-0000(74)80044-9

31. Karthik, C.S., Laekhanukit, B., Manurangsi, P.: On the parameterized complexity of approximating dominating set. In: Proceedings of the 50th Annual ACM Symposium on Theory of Computing (STOC), pp. 1283–1296. ACM (2018). https://doi.org/10.1145/3188745.3188896

32. Lokshtanov, D., Marx, D., Saurabh, S.: Known algorithms on graphs of bounded treewidth are probably optimal. ACM Trans. Algorithms **14**(2), 13:1–13:30 (2018). https://doi.org/10.1145/3170442

33. Lokshtanov, D., Mnich, M., Saurabh, S.: A linear kernel for planar connected dominating set. Theor. Comput. Sci. **412**(23), 2536–2543 (2011). https://doi.org/10.1016/J.TCS.2010.10.045

34. Meybodi, M.A., Fomin, F.V., Mouawad, A.E., Panolan, F.: On the parameterized complexity of [1, j]-domination problems. Theor. Comput. Sci. (TCS) **804**, 207–218 (2020). https://doi.org/10.1016/J.TCS.2019.11.032

35. Rooij, J.M.M.: A generic convolution algorithm for join operations on tree decompositions. In: Santhanam, R., Musatov, D. (eds.) CSR 2021. LNCS, vol. 12730, pp. 435–459. Springer, Cham (2021). https://doi.org/10.1007/978-3-030-79416-3_27

36. van Rooij, J.M.M., Bodlaender, H.L., Rossmanith, P.: Dynamic programming on tree decompositions using generalised fast subset convolution. In: Fiat, A., Sanders, P. (eds.) ESA 2009. LNCS, vol. 5757, pp. 566–577. Springer, Heidelberg (2009). doi: https://doi.org/10.1007/978-3-642-04128-0_51

37. Telle, J.A.: Complexity of domination-type problems in graphs. Nordic J. Comput. (NJC) **1**(1), 157–171 (1994)

38. Telle, J.A., Villanger, Y.: FPT algorithms for domination in biclique-free graphs. In: Epstein, L., Ferragina, P. (eds.) ESA 2012. LNCS, vol. 7501, pp. 802–812. Springer, Heidelberg (2012). https://doi.org/10.1007/978-3-642-33090-2_69

Reachability in Graphs with Polynomially Many Surface Non-separating Cycles is in UL

Neelabjo Shubhashis Choudhury[1]([✉]), Chetan Gupta[2]([✉]),
and Raghunath Tewari[1]([✉])

[1] Indian Institute of Technology Kanpur, Kanpur, India
`neelabjo@cse.iitk.ac.in`, `rtewari@cse.iitk.ac.in`
[2] Indian Institute of Technology Roorkee, Roorkee, India
`chetan.gupta@cs.iitr.ac.in`

Abstract. In this work, we consider the class of surface-embedded, directed graphs that have at most polynomially many surface non-separating cycles. We show that for this class of graphs, there is a logspace computable, polynomially bounded, skew-symmetric edge weight function, with respect to which any cycle in the graph gets a non-zero weight.

As a consequence of our result, we show that reachability in graphs with polynomially many surface non-separating cycles is in UL. Moreover, we also show that computing perfect matching in bipartite graphs that have polynomially many surface non-separating cycles is in SPL.

Keywords: Surface Non-separating Cycles · Unambiguous Logspace · Genus · Reachability · Bipartite Perfect Matching

1 Introduction

In the reachability problem, given a graph $G = (V, E)$, and two vertices s and t in G, we need to find whether there exists a path from s to t. Deciding reachability between a pair of vertices is an important problem in complexity theory. Reachability in directed graphs precisely captures the complexity class nondeterministic logspace (NL). Also, the seminal result of Reingold showed that reachability in an undirected graph characterises the class deterministic logspace (L) [5]. Whether these two classes are equal is a big open question in complexity theory. Unambiguous computation is a restricted version of nondeterministic computation. We know that in a nondeterministic computation, given any input, the answer is *yes* if there exists at least one accepting computation path and the answer is no if all computation paths reject. However, in an unambiguous computation, given any input, there is at most one accepting computation path. The input is accepted if exactly one computation path that terminates in yes, and no if all the paths terminate in no. The class UL was first introduced formally by Buntrock et al. in

F. Foucaud and A. Parreau (Eds.): IWOCA 2026, LNCS 16587, pp. 190–203, 2026.
https://doi.org/10.1007/978-3-032-27732-9_14

[2] and then studied in more detail by Àlvarez and Jenner in [4]. Since the class UL lies between NL and L, resolving $NL \stackrel{?}{=} UL$ will be a big intermediate step toward the NL vs. L question. In 2000, Reinhardt and Allender took the first step in this direction and proved that these classes are equal in a non-uniform computation model [7]. In a later work, it was shown that if deterministic linear space functions cannot be computed by circuits of size $2^{\epsilon n}$, then $NL = UL$ [3]. These results suggest that the classes might be equal unconditionally. Since directed reachability is a complete problem for NL, giving UL algorithm for reachability will prove that $NL = UL$. Although solving general directed reachability has been elusive so far, a lot of progress has been made in this direction by solving reachability in restricted classes of graphs, for example, using isolating weight assignment to establish an upper bound of UL for planar graphs, bounded genus graphs, logarithmic genus graphs, $K_{3,3}$ and K_5 free graphs, single crossing minor free graphs [1,13,16,17,19,25] and proving an L upper bound for bounded tree-width graphs [20]. For the general case, Kalampally and Tewari showed that the problem can be solved by an unambiguous algorithm that takes $O(\log^2 n)$ space and polynomial time [8]. The result was subsequently improved by Prakriya and Melkebeek by reducing the space bound to $O(\log^{1.5} n)$ [9].

Perfect Matching is one of the problems that is closely related to the reachability problem. In the perfect matching problem, we are give an undirected graph, and the goal is to find a disjoint subset of edges of the graph that covers all the vertices of the graph. First polynomial time algorithm to solve this problem dates back to 1965 by Edmonds [23]. However, the parallel complexity of this problem is not clear, i.e., whether there exists an efficient parallel algorithm (an NC algorithm) for the problem. Lovász took the first step towards this question and proved that the problem belongs to the class randomized (NC). Later, Mulmuley et al. also gave a randomized (NC) algorithm to solve the problem using the isolation lemma. In particular, they proved that if we can efficiently construct a polynomial-size weight assignment for the edges of a graph such that there exists a unique minimum weight perfect matching in the graph, then we can solve the problem efficiently by an NC algorithm. Allender et al. improved this result in [3] and proved that if we can construct such a weight function in logarithmic space, then the problem will belong to the class SPL, which is a subclass of NC. In the last few years, people have constructed such a weight assignment for various restricted classes of graphs [16–18,22]. For a general graph, recently Fenner et al. gave a construction of a weight function of size $O(\log^2 n)$-bits, proving that the problem can be solved deterministically by a quasi-NC algorithm. As we see that if we can efficiently construct a weight assignment that isolates the path and perfect matching efficiently, then we can obtain the desired bound for both problems. In most previous results, people have constructed a single weight assignment for a class of graphs that isolates a path in the graphs of the class and a perfect matching in the corresponding undirected bipartite graphs. These weight functions satisfy two properties:(i) it should be *skew symmetric*, and it should give nonzero weight to all the cycles in the graphs. We will also construct such a weight function for a restricted graph class here.

In this work, we focus on directed graphs that can be embedded on surfaces. A cycle in a directed graph refers to a *directed cycle* and one in an undirected graph refers to an *undirected cycle*. We do not consider 2-cycles in this work. A surface has genus g if it can be obtained by attaching g handles to a sphere. The *genus of a graph* is the minimum genus surface on which the graph can be embedded without any edge crossings (For example, planar graphs can be embedded on a surface of genus 0, i.e. a sphere, K_5 can be embedded on a surface of genus 1). It is easy to see that any graph can be embedded on a surface of polynomially ordered (in the number of vertices) genus. Therefore, solving the problem for such kind of graphs has been another direction for making progress towards this question. If we consider the underlying undirected version of directed graphs embedded on surfaces, then the cycles in such a graph can be partitioned into two classes - *surface separating cycles* (SSC) and *surface non-separating cycles* (SNSC). It has been proved that for graphs that contain an arbitrary number of surface separating cycles and up to $O(\log n)$ many topologically different surface non-separating cycles, both reachability and perfect matching can be isolated efficiently. The next step is to handle an arbitrary number of surface non-separating cycles. In this paper, we construct path-isolating as well as bipartite perfect matching-isolating weight assignments for a graph that has polynomial number of surface non-separating cycles.

1.1   Our Results

In this paper, we study reachability and perfect matching in graphs embedded on surfaces that have a polynomial number of surface non-separating cycles and prove the following results.

Theorem 1. *In directed graphs embedded on surfaces that contain at most a polynomial number of surface non-separating cycles, the reachability problem can be solved in unambiguous logspace* (UL).

We complement this result by establishing an analogous complexity bound for the bipartite perfect matching problem.

Theorem 2. *In undirected bipartite graphs embedded on surfaces that contain at most a polynomial number of surface non-separating cycles, the bipartite perfect matching problem lies in* SPL.

2   Preliminaries

Given a directed graph $G = (V, E)$, a *weight assignment* (alternatively function) $w : E \to \mathbb{Z}$ is a function that assigns integer weights to all the edges of G. Consequently, the weight of a path in G is defined as the sum of the weights of edges that form the path. Graph G is said to be *min-unique* with respect to a weight function w, if between every pair of vertices, there exists at most one minimum weight path. The weight function w is called a min-isolating (or path-isolating) weight function. A *weight function w* is called *skew-symmetric* if for all edges $e = (u, v) \in E$, $w(e) = -w(e^r)$, where $e^r = (v, u)$. Edge e^r may not necessarily be present in the given graph. In other words, if edge e in a directed graph G has weight t, we create another directed graph G' where edge e has the reverse direction, specifically e^r, and is assigned a weight of $-t$. A weight function is called *polynomially bounded* (or of polynomial size) if it assigns $O(\log n)$-bit weights to all the edges of a graph. One of the ways to obtain isolation of paths and perfect matchings in a directed graph is by designing a skew-symmetric weight function that assigns non-zero weights to all the cycles in the graph.

We know that every graph can be embedded on a *surface* such that no two edges of the graph intersect each other. The *genus of a graph* is defined as the minimum genus surface on which it can be embedded without intersecting its edges. A surface having genus g has g holes in it. Cycles in graphs embedded on a surface of genus > 0 can be divided into two classes: (i) *surface separating cycles (SSCs)* and, (ii) *surface non-separating cycles (SNSCs)*.

Definition 1. *Let G be a directed graph embedded on a surface Σ. A directed cycle C is surface-separating if its underlying undirected cycle separates Σ, i.e., $\Sigma \setminus C$ is disconnected. Otherwise, C is surface non-separating. Alternatively, a cycle in the graph is called surface-separating if removing the cycle from the surface splits the surface into two disconnected regions and it is called non-separating if cutting along the cycle does not disconnect the graph and reduces the genus by 1.*

As the name suggests, a cycle in a surface embedded graph is called a surface non-separating cycle if cutting along the cycle does not divide the surface into two parts and vice versa (see Fig. 1). SNSCs are important because they are related to the genus of graphs. Genus of graph G is at most the number of disjoint SNSCs in G, such that the surface does not divide even after cutting it along all cycles together. In other words, if a graph is embedded on a minimum possible genus surface, then the genus of the graph (and surface) will be equal to the number of SNSCs present in the graph such that cutting the surface along those cycles does not separate the surface. The importance of the class of graphs with poly-many SNSCs is explained in detail in Sect. 4. Our objective therefore is to construct a weight function for the above class of graphs.

Definition 2. *Let* PolySNSC *be the class of surface-embedded directed graphs such that they contain at most polynomially many surface non-separating cycles.*

For a nondeterministic Turing machine M, let $acc_M(x)$ and $rej_M(x)$ denote the number of accepting computations and the number of rejecting computations, respectively, on an input x. Denote $gap_M(x) = acc_M(x) - rej_M(x)$.

Definition 3. *[16] A language L is in* SPL *if there exists a logspace bounded nondeterministic machine M so that for all inputs $x, gap_M(x) \in \{0, 1\}$ and $x \in L$ if and only if $gap_M(x) = 1$.*

Complexity class SPL was initially studied by Allender et al. in [3]. With respect to matching, we define the following problem - BPMDecision: Given a bipartite graph G, checking if G has a perfect matching [16]. Although the weight function constructed by Gupta et al. [1] isolates paths in $O(\log n)$ genus graphs, it is worth noticing that it is a skew-symmetric weight function that gives non-zero weights to all surface separating cycles (but not necessarily to any surface non-separating cycles) in graphs of arbitrary genus. We reuse their weight function as part of our final weight function and encapsulate their result in the following lemma.

Lemma 1. *[1] Let G be a graph given along with its embedding on a surface. There exists a logspace computable polynomial-size skew-symmetric weight function w_{pl} that gives nonzero weight to every surface-separating cycle in G.*

We now state the hashing result by Fredman, Komlós and Szemerédi that we use later. We refer to it as FKS Hashing Lemma in this text.

Lemma 2. *[6] Let $S = \{x_1, x_2, \ldots, x_k\}$ be a set of n-bit integers. Then there exists an $O(\log n + \log k)$ bit prime number p so that for all $x_i \neq x_j \in S, x_i$ mod $p \neq x_j$ mod p.*

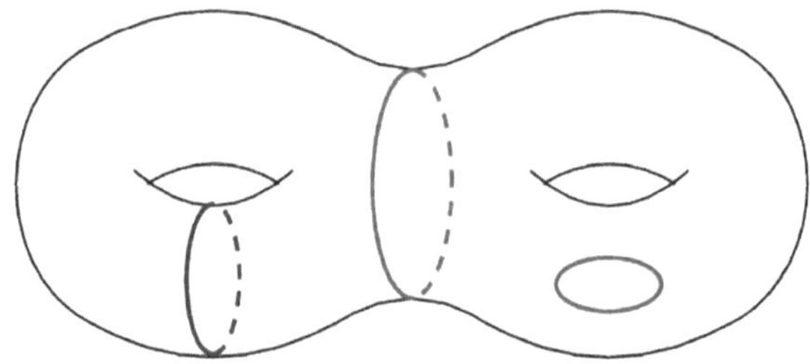

Fig. 1. Surface separating cycles (marked in red) and surface non-separating cycle (denoted in blue). Figure Source (Color figure online)

We will use the following Theorem from Reinhardt and Allender [7] to prove Theorem 1.

Theorem 3. *[7] There is a nondeterministic logspace Turing machine M that takes $\langle G, s, t, w \rangle$ as input where G is a directed graph on n vertices, s and t are vertices in G and w is an $O(\log n)$-bit edge weight function and outputs the following along a unique computation path while all other computation paths halt and reject:*

- *Not Min-unique if G is not min-unique with respect to w,*
- *Yes, if G is min-unique with respect to w and there is a path from s to t in G, and*
- *No, if G is min-unique with respect to w and there is no path from s to t in G.*

3   Overview of the Paper

In the rest of the paper, we show that directed reachability and bipartite perfect matching in PolySNSC graphs are in UL and SPL, respectively. In Sect. 4, we show why the class PolySNSC is important to study as it also contains graphs having polylogarithmic and even linear genus.

In Sect. 5, we construct path-isolating and perfect matching-isolating weight functions for graphs in PolySNSC class. Specifically, in Sect. 5.1, we show that for the class of graphs PolySNSC, there exists an $O(\log n)$-bit weight function which assigns non-zero weights to all the cycles. This is done by combining two weight functions: one assigning non-zero weights to all SSCs (using the weight function of Gupta et al. [1]) and the other assigning non-zero weights to all SNSCs in the graph. Assigning exponential weights ensure that every SNSC gets a non-zero weight and applying FKS Hashing Lemma help bring edge weights down from exponential to polynomial bound. We also prove the same result for the class of undirected g-genus graphs corresponding to PolySNSC.

Finally in Sect. 5.2, we show the existence of weight functions computable in logarithmic space for graphs in PolySNSC and undirected bipartite graphs containing poly-many SNSCs, such that the minimum weight perfect matching with respect to those weight functions are individually unique.

4   Motivation

The class of graphs that contain polynomially many SNSCs is of particular interest due to its intricate relationship with graph embeddings and surface topology. Surface non-separating cycles are closely tied to the genus of a graph (or equivalently, the embedding surface). As the genus increases, more complex graphs can be embedded, and in fact, any graph can be embedded on a surface of polynomial genus. Interestingly, we observe that graphs with poly-many SNSCs can still exhibit very high genus, say of $O(n)$. Conversely, there exist graphs of low genus that nevertheless contain exponentially many SNSCs. This demonstrates that the class PolySNSC is neither a strict superset nor a strict subset of the classes for which polynomial-size path-isolating weight functions are currently known. Consequently, identifying such a weight function for this class becomes a compelling and nontrivial direction. Below we give an example of a graph that has a very high genus but only polynomially many SNSCs.

4.1   Constructing a Graph with High Genus Having Poly-Many SNSCs

Let us assume we have i copies of the graph K_5, namely graphs $G_1, G_2, \ldots G_i$, where each copy is embedded separately on a genus 1 surface (i.e., a torus). Let us assume that G_i contains vertices $\{v, v_1^i, v_2^i, v_3^i, v_4^i\}$. 1-*clique sum* is a special version of k-clique sum that is obtained by gluing two graphs together at a single vertex. This is done by selecting one vertex each from two separate graphs and identifying them as one in the resultant graph, keeping all edges incident to both vertices. Now we repeatedly take 1-clique sum of all the G_is on v. Let G be the final graph thus constructed. We have given an informal proof that the genus of G is i and it contains only polynomially many SNSCs. We will prove it for the case when $i = 2$, and the same argument can be extended for arbitrary i. We have G_1 and G_2 each embedded on a torus. We will prove that the genus of G is 2. For the sake of contradiction, assume that G can be embedded on a surface S of genus 1.

Definition 4. *For a graph G embedded in an orientable or non-orientable closed surface Σ, the embedding is called a 2-cell embedding or a cellular embedding where every face is homeomorphic to an open disk, i.e., each face is connected, and the boundary of each face is a closed walk in the graph.*

The subgraph of G induced by the vertices of G_1 is a non-planar graph, say H. The embedding of H on the surface S will be a *2-cell embedding*. No matter how we embed this subgraph on S, each face in the embedding will be homeomorphic to an open disc. Notice that the remaining subgraph of G, i.e., $(G - H)$, formed by removing only the edges, is also non-planar. Therefore, it cannot be embedded inside any face of H, hence contradicting the assumption. This proves that G cannot be embedded on the surface of genus 1. Similar argument can be extended for arbitrary i.

We now show that the number of SNSCs in this graph is linear in i. Let us consider a cycle starting from and ending at some $v_p^i \in G_i$ and covering v. No vertex v_p^i can form a cycle containing another vertex $v_q^j \in G_j$, since there is only a single way to traverse between them, through v. v has to be visited in any path going from v_p^i to v_q^j. While completing the cycle back to v_p^i, v has to be visited again, as there does not exist any other way to connect them. But a unique vertex cannot be visited twice in a cycle. Hence, each G_i can have at most 5 SNSCs. If we consider G to be of linear genus, i.e., if $i = n$, then the total number of surface non-separating cycles is at most $5i = O(n)$. Thus G is a graph with high genus (of order linear to n), yet having poly-many SNSCs.

5 Isolating Paths and Perfect Matchings in Graphs with Polynomially Many SNSCs

Let us consider a directed acyclic graph (DAG) $G' = (V_{G'}, E_{G'})$ and its corresponding underlying undirected graph, H'. Suppose the number of directed SNSCs in G' is t. We know that reachability in G' is NL-complete because reachability in a DAG belongs to the class NL and is also NL-hard. On the other hand, H' can contain exponentially many undirected cycles. For example, there can be a K_5 without any directed cycles but having exponentially many undirected cycles. Therefore, putting a polynomial bound on the number of directed cycles does not imply a polynomial bound on the number of undirected cycles.

We briefly explain why constructing a skew-symmetric weight function that assigns non-zero weights to all the cycles in a graph is enough to isolate paths in that graph.

Proposition 1. *For a class of directed graphs, if there exists a skew-symmetric weight function that gives non-zero weight to all the cycles in graphs of that class, then the weight function is path-isolating for that class of directed graphs.*

Proof. Let w_s be a skew-symmetric weight function that gives nonzero weights to all the cycles in a directed graph G. Let us assume there are two different paths P_1 and P_2, from u to v that have the same minimum weight. Now there are two cases, either they intersect or they do not intersect. Let us first assume that they do not intersect. Let us construct a graph G', that is obtained from G by only reversing the direction of the edges along path P_2. Let us call that path P_2^r in G'. Let C be simple cycle formed by paths P_1 and P_2^r, in G. We know that $w_s(C) = w_s(P_1) + w_s(P_2^r)$. Since w_s is skew-symmetric, we have $w_s(P_2) = w_s(P_2^r)$. This implies that $w_s(C) = w_s(P_1) - w_s(P_2)$. Since P_1 and P_2 are equal weight we can conclude that $w_s(C) = 0$, which contradicts the assumption. Similarly, when P_1 and P_2 intersect, we can repeat the argument by considering the two sub-paths of P_1 and P_2 that start from u and end at their first intersection point.

5.1   Deciding Reachability in **PolySNSC**

Theorem 4. *Let $G = (V, E)$ be a g-genus directed graph containing polynomially many simple surface non-separating cycles. Then there exists an $O(\log n)$-bit skew-symmetric weight function $w : E \rightarrow \mathbb{Z}$ computable in logarithmic space that assigns non-zero weights to all the surface non-separating cycles in G.*

Proof. Let $C = \{C_1, C_2, \cdots, C_t\}$ be the set of t many surface non-separating cycles, $t = O(n^k)$, for some constant k. Let us assume that each cycle C_i consists of two paths, namely P_i from u_i to v_i and P_i' from v_i to u_i, respectively. Let us index all the edges in the graph as $e_1, e_2, \cdots, e_m$. We define a skew-symmetric weight function w_{exp} which assigns 2^i to edge e_i, i.e., $w_{exp}(e_i) = 2^i$. Weights of paths P_i and P_i' are defined as,

$$w_{exp}(P_i) = \sum_{\forall e_i \in P_i} 2^i, \qquad w_{exp}(P_i') = \sum_{\forall e_j \in P_i'} 2^j$$

Let $P_i'^r$ be the path obtained by reversing the edges along path P_i'. Let $\mathcal{P} = \{P_1, P_1'^r, P_2, P_2'^r \ldots P_t'^r\}$. We can see that every path $\mathcal{P}$ gets a different weight with respect to w_{exp}, but the weight of a path is exponential. Since the cardinality of $\mathcal{P}$ is $O(n^k)$, from Lemma 2, we know that there exists an $O(\log n)$-bit prime p, such that with respect to the weight function $w_{fks}(e_i) = w_{exp}(e_i) \bmod p$, each path in $\mathcal{P}$ gets a different weight. w_{fks} here is of polynomial size.

$$w_{fks}(C_i) = w_{fks}(P_i) - w_{fks}(P_i'r) \tag{1}$$

Notice that, since each path in $\mathcal{P}$ gets a different weight, we can conclude that each C_i get a nonzero weight with respect to w_{fks}. Also, from Lemma 1, we know that each surface separating cycle in G gets a nonzero weight with respect to a skew-symmetric weight function w_{pl}. Therefore, taking a linear combination of these two weight functions, each cycle in the graph will now have zero weight with respect to the new weight function. In particular, we define the final weigh function as follows,

$$w_{final} := w_{pl} \cdot n^{l} + w_{fks}, \text{for some constant } l > 0. \tag{2}$$

Since both w_{pl} and w_{fks} are skew-symmetric, w_{final} is also skew-symmetric. This settles the proof.

Corollary 1. *Deciding reachability in* PolySNSC *is in* UL.

Given directed graph $G \in$ PolySNSC on n vertices and 2 vertices s and t, we iterate over all primes and for each of these primes, the weight function w is computed according to Theorem 5. Thereby using Theorem 3, we verify if G is min-unique with respect to w. If yes, we check if t is reachable from s. If G is not min-unique with respect to w, then we move to the next prime. The existence of such a prime with respect to which G is min-unique is guaranteed by Lemma 2. Hence, along a unique computation path, we get a *Yes* or a *No* answer depending on whether s is reachable from t or not respectively, while all other computation paths halt and reject. This completes the proof of Theorem 1.

Now, consider a graph H having $O(n^k)$ many undirected cycles. We show that this implies that the number of directed cycles is bounded polynomially in n for any directed version of H, say G'.

Theorem 5. *For an undirected g-genus graph H having polynomially many SNSCs, there exists an $O(\log n)$-bit weight function $w' : E \to \mathbb{Z}$ computable in logarithmic space that assigns non-zero weights to all the surface non-separating cycles in H. Moreover, for every graph $H \in$ PolySNSC, the minimum weight path in H with respect to w' is unique.*

Proof. Let t be the number of undirected simple cycles in H (each undirected cycle is a closed simple sequence of distinct vertices and edges). We create a directed graph G' by replacing every undirected edge in H with two directed edges in opposite directions. It is clear that each undirected simple cycle $C \in H$ produces exactly two distinct simple directed cycles in G' - one following the edges of C in one direction and the other following edges in the opposite direction. Let us denote one of those directions as *clockwise* and the other as *counterclockwise*. Then each undirected cycle contributes one clockwise directed simple cycle and one counterclockwise directed simple cycle. While considering clockwise orientation for every undirected cycle, let the number of directed simple cycles oriented clockwise thus formed in G' is t. Similarly, the number of counterclockwise oriented directed simple cycles in G' also equals t, and the total number of directed cycles in G' is $2t$ (polynomial in n, since $t = O(n^k)$).

Given the g-genus graph G' having $2t$ many cycles, let $\mathcal{P}$ be the set of paths that form these cycles. We now apply Theorem 4 on the directed graph G' to get a $O(\log n)$-bit weight function w' that assigns non-zero weights to all the SNSCs in G'. In the first stage, each edge is assigned w'_{exp} with respect to which, each path in $\mathcal{P}$ get a different weight. In the second stage, the FKS Hashing scheme is used to compute an $O(\log n + \log t)$ bit weight function w'_{fks} with respect to which only one among the many paths of the first stage gets the minimum weight value. Applying FKS Hashing Lemma is possible because the number of directed cycles is still polynomial in t, thereby also polynomial in n. Thus, with respect to an $O(\log n)$ bit weight function w'_{final} created accordingly, G' is min-unique.

Combining the above argument with Proposition 1, if there exists a skew-symmetric weight function w'_{final} that is path-isolating for G', then for the underlying undirected graph H of G', there exists some $O(\log n)$-bit weight function w', with respect to which, the minimum weight path in H is unique. This concludes the proof of Theorem 5.

5.2   Isolating Bipartite Perfect Matchings in **PolySNSC**

Definition 5. *[14] Given an undirected graph H, let $\mathcal{O}_H$ be the set of all directed graphs H', such that the underlying undirected graph of H' is H. For a class of undirected graphs $\mathcal{H}$, define $\mathcal{O}_\mathcal{H} = \bigcup_{H \in \mathcal{H}} \mathcal{O}_H$.*

Theorem 6. *[14] Let $\mathcal{G}$ be a class of directed graphs and let w be an edge weight function defined for every $G \in \mathcal{G}$ such that w is skew-symmetric and for any cycle $C \in G$, $w(C) \neq 0$. Let $\mathcal{H}$ be a class of undirected bipartite graphs, such that $\mathcal{O}_\mathcal{H} \subseteq \mathcal{G}$. Then we can construct a weight function w' in log-space, such that for every graph $H \in \mathcal{H}$, the minimum weight perfect matching in H with respect to w' is unique.*

Theorem 7. *For the class of undirected bipartite graph $\mathcal{H}$ with polynomially many SNSCs, there exists an $O(\log n)$-bit weight function $w'_{pm} : E \to \mathbb{Z}$ computable in logarithmic space that assigns non-zero weights to all the surface non-separating cycles in H such that with respect to w'_{pm}, for every graph $H' \in \mathcal{H}$, the minimum weight perfect matching is unique.*

Let $\hat{H}$ be the directed version of the underlying undirected bipartite graph $H \in \mathcal{H}$ with (L, R) being the bipartition. For two perfect matchings M_1 and M_2 in $\hat{H}$, we define $E_{M_1 \square M_2} = (M_1 \cup M_2) \setminus (M_1 \cap M_2)$. The weight function w_{pm} is the same as the weight function w_{final} derived in the previous subsection for isolating minimum weight unique paths.

Lemma 3. *For a directed bipartite graph $\hat{H} \in$ PolySNSC, there exists at most one minimum weight perfect matching in $\hat{H}$ with respect to the weight function w_{pm}.*

Proof. Let us assume that M_1 and M_2 are two minimum weight perfect matching in $\hat{H}$ with respect to the weight function w_{pm}. The set of edges in $E_{M_1 \square M_2}$ form vertex disjoint cycles, say $C_1, C_2, \ldots C_k$, as they contain edges which are either in M_1 or in M_2 but not both. Note that any cycle C_i formed thus contain even number of edges alternating between M_1 and M_2 and all edges in this cycle are directed from L to R.

Let E_{i1}, E_{i2} be the set of edges in M_1 and M_2 respectively for a cycle C_i. We see that $w_{pm}(E_{j1}) = w_{pm}(E_{j2})$ for some cycle C_j. Otherwise, without loss of generality, if $w_{pm}(E_{j1}) > w_{pm}(E_{j2})$, then a new perfect matching $(M_1 \setminus E_{k1}) \cup E_{k2}$ can be formed having strictly lesser weight. This is a contradiction to the assumption that M_1 is a minimum weight perfect matching.

Now consider another graph $\hat{H}'$ which is same as $\hat{H}$ but the direction of edges in $M_2 \in \hat{H}$ are reversed. Let $M_1', M_2' \in \hat{H}'$ be the corresponding matchings to $M_1, M_2 \in \hat{H}$ such that the underlying undirected edges of M_1 and M_2 are the same as that of M_1' and M_2'. Let $C_1', C_2', \ldots C_k'$ be the set of vertex disjoint cycles formed from the edges in $E_{M_1 \square M_2}$ and E_{i1}', E_{i2}' be the edges in M_1', M_2' respectively in a cycle C_i'.

We have seen that $w_{pm}(E_{i1}) = w_{pm}(E_{i2})$, for all $i \in [k]$. The set of edges E_{i1} equals E_{i1}' and the set of edges E_{i2} equals the reversed set of edges E_{i2}'. Since w_{pm} is skew-symmetric, combining the previous assertion that $w_{pm}(E_{j1}) = w_{pm}(E_{j2})$, we get,

$$w_{pm}(E_{i1}') = -w_{pm}(E_{i2}'),$$
$$w_{pm}(E_{i1}') + w_{pm}(E_{i2}') = 0,$$
$$w_{pm}(C_i') = 0.$$

It is known that w_{pm}, like w_{final}, assigns non-zero weights to all the SNSCs in $\hat{H}$. Therefore, $w_{pm}(C_i') \neq 0$ and this is a contradiction. Thus, there exists at most one minimum weight perfect matching in $\hat{H}$ with respect to w_{pm}. Note that, since the edges in E_{i2}' are reversed they have a direction from R to L whereas edges in E_{i1}' are from L to R. Therefore, the cycles $C_1', C_2', \ldots C_k'$ that were formed are directed cycles in $\hat{H}'$.

We also know from the proof of Theorem 5 that if the number of SNSCs in the class of undirected bipartite graphs is polynomial in n, then the number of directed cycles in the corresponding directed version of the graphs, including cycles oriented both clockwise and anti-clockwise, is also bounded polynomially in n.

For the class of directed graphs $\hat{H} \in \mathsf{PolySNSC}$, there exists a skew-symmetric weight function w_{pm}, that gives non-zero weights to all the cycles in any graph $H \in \hat{H}$. Then using Theorem 6, we can say that for the class of undirected bipartite graph $\mathcal{H}$, another weight function, say w_{pm}', that can be constructed in logspace, such that for every $H' \in \mathcal{H}$, if there exists a matching, the minimum weight perfect matching in H' with respect to w_{pm}' is unique. Thus, we see that for an undirected g-genus bipartite graph $H' \in \mathcal{H}$, with respect to the weight function w_{pm}', if there is a perfect matching in H', the minimum weight perfect matching in H' is unique. This finishes the proof of Theorem 7.

Lemma 4. *[16] For any weighted graph G assume that the minimum weight perfect matching in G is unique and also for any subset of edges $E' \subseteq E$, the* *minimum weight perfect matching in $(G \setminus E')$ is also unique. Then deciding if G has a perfect matching is in* SPL.

Corollary 2. BPMDecision *in* PolySNSC *is in* SPL.

We have the weight function w'_{pm} according to which if there exists a matching in $H' \in \mathcal{H}$, then the minimum weight perfect matching in H' is unique. Then by simply applying Lemma 4 for the weighted graph H', the problem of deciding if H' has a perfect matching lies in SPL. This completes the proof of Theorem 2.

Acknowledgments. This work was supported by the PM ECRG grant, India, awarded to Chetan Gupta (grant no: ANRF/ECRG/2024/004480/ENS).

References

1. Gupta, C., Sharma, V.R., Tewari, R.: Reachability in o (log n) genus graphs is in unambiguous logspace. In: 36th International Symposium on Theoretical Aspects of Computer Science (STACS 2019), pp. 34:1–34:13 (2019). https://doi.org/10. 4230/LIPIcs.STACS.2019.34
2. Buntrock, G., Jenner, B., Lange, K.-J., Rossmanith, P.: Unambiguity and fewness for logarithmic space. In: Budach, L. (ed.) FCT 1991. LNCS, vol. 529, pp. 168–179. Springer, Heidelberg (1991). https://doi.org/10.1007/3-540-54458-5_61
3. Allender, E., Reinhardt, K., Zhou, S.: Isolation, matching, and counting: uniform and nonuniform upper bounds. J. Comput. Syst. Sci. **59**, 164–181 (1999)
4. Àlvarez, C., Jenner, B.: A very hard log-space counting class. Theoret. Comput. Sci. **107**, 3–30 (1993)
5. Reingold, O.: Undirected connectivity in log-space. J. ACM **55**(4), 1–24 (2008). https://doi.org/10.1145/1391289.1391291
6. Fredman, M.L., Komlós, J., Szemerédi, E.: Storing a sparse table with 0(1) worst case access time. J. ACM **31**(3), 538–544 (1984). https://doi.org/10.1145/828.1884
7. Reinhardt, K., Allender, E.: Making nondeterminism unambiguous. SIAM J. Comput. **29**(4), 1118–1131 (2000). https://doi.org/10.1137/S0097539798339041
8. Kallampally, V.A.T., Tewari, R.: Trading determinism for time in space bounded computations. In: 41st International Symposium on Mathematical Foundations of Computer Science, MFCS 2016, August 22–26, 2016 - Kraków, Poland, pp. 10:1–10:13 (2016). https://doi.org/10.4230/LIPIcs.MFCS.2016.10
9. van Melkebeek, D., Prakriya, G.: Derandomizing isolation in space-bounded settings. In: 32nd Computational Complexity Conference, CCC 2017, July 6–9, 2017, Riga, Latvia, pp. 5:1–5:32 (2017). https://doi.org/10.4230/LIPIcs.CCC.2017.5
10. Pavan, A., Tewari, R., Vinodchandran, N.V.: On the power of unambiguity in logspace. Comput. Complex. **21**(4), 643–670 (2012) https://doi.org/10.1007/s00037-012-0047-3
11. Allender, E., Barrington, D.A., Chakraborty, T., Datta, S., Roy, S.: Planar and grid graph reachability problems. Theory Comput. Syst. **45**(4), 675–723 (2009). https://doi.org/10.1007/s00224-009-9172-z
12. Thierauf, T., Wagner, F.: Reachability in $K_{3,3}$-free graphs and K_5-free graphs is in unambiguous log-space. In: 17th International Conference on Foundations of Computation Theory (FCT), Lecture Notes in Computer Science 5699, pp. 323–334. Springer, Berlin, Heidelberg (2009). https://doi.org/10.1007/978-3-642-03409-1_29

13. Bourke, C., Tewari, R., Vinodchandran, N.V.: Directed planar reachability is in unambiguous log-space. ACM Trans. Comput. Theory **1**(1), 1–17 (2009). https://doi.org/10.1145/1490270.1490274
14. Tewari, R., Vinodchandran, N.V.: Green's theorem and isolation in planar graphs. Inf. Comput. **215**, 1–7 (2012). https://doi.org/10.1016/j.ic.2012.03.002
15. Kyncl, J., Vyskocil, T.: Logspace reduction of directed reachability for bounded genus graphs to the planar case. ACM Trans. Comput. Theory **1**(3), 1–11 (2010). https://doi.org/10.1145/1714450.1714451
16. Datta, S., Kulkarni, R., Tewari, R., Vinodchandran, N.V.: Space complexity of perfect matching in bounded genus bipartite graphs. J. Comput. Syst. Sci. **78**(3), 765–779 (2012). https://doi.org/10.1016/j.jcss.2011.11.002
17. Datta, S., Gupta, C., Jain, R., Mukherjee, A., Sharma, V.R., Tewari, R.: Reachability and matching in single crossing minor free graphs. In: 41st IARCS Annual Conference on Foundations of Software Technology and Theoretical Computer Science (FSTTCS 2021). Leibniz International Proceedings in Informatics (LIPIcs), Volume 213, pp. 16:1–16:16, Schloss Dagstuhl – Leibniz-Zentrum für Informatik (2021). https://doi.org/10.4230/LIPIcs.FSTTCS.2021.16
18. Datta, S., Kulkarni, R., Roy, S.: Deterministically isolating a perfect matching in bipartite planar graphs. Theory Comput. Syst. **47**, 737–757 (2010). https://doi.org/10.1007/s00224-009-9204-8
19. Datta, S., Kumar, P., Mukherjee, A., Tawari, A., Vortmeier, N., Zeume, T.: Dynamic complexity of reachability: how many changes can we handle? In: 47th International Colloquium on Automata, Languages, and Programming, ICALP 2020, July 8–11, 2020, Saarbrücken, Germany (Virtual Conference), pp. 122:1–122:19 (2020)
20. Das, B., Datta, S., Nimbhorkar, P.: Log-space algorithms for paths and matchings in k-trees. In: 27th International Symposium on Theoretical Aspects of Computer Science. Leibniz International Proceedings in Informatics (LIPIcs), vol. 5, pp. 215–226, Schloss Dagstuhl – Leibniz-Zentrum für Informatik (2010). https://doi.org/10.4230/LIPIcs.STACS.2010.2456
21. Mahajan, M., Varadarajan, K.R.: A new NC-algorithm for finding a perfect matching in bipartite planar and small genus graphs (extended abstract). In: Proceedings of the Thirty-Second Annual ACM Symposium on Theory of Computing (STOC '00). Association for Computing Machinery, New York, NY, USA, pp. 351–357 (2000). https://doi.org/10.1145/335305.335346
22. Gupta, C., Sharma, V.R., Tewari, R.: Efficient isolation of perfect matching in o(log n) genus bipartite graphs. In: 45th International Symposium on Mathematical Foundations of Computer Science (MFCS 2020). Leibniz International Proceedings in Informatics (LIPIcs), vol. 170, pp. 43:1–43:13, Schloss Dagstuhl – Leibniz-Zentrum für Informatik (2020). https://doi.org/10.4230/LIPIcs.MFCS.2020.43
23. Edmonds, J.: Paths, trees and flowers. Can. J. Math. 449–467 (1965)

24. Lovász, L.: On determinants, matchings, and random algorithms. In: FCT, pp. 565–574 (1979)
25. Arora, R., Gupta, A., Gurjar, R., Tewari, R.: Derandomizing isolation lemma for $k_{3,3}$-free and k_5-free bipartite graphs. In: 33rd Symposium on Theoretical Aspects of Computer Science, STACS 2016, February 17–20, 2016, Orléans, France, pp. 10:1–10:15 (2016)

Enumerating Spanners in Directed Temporal Graphs

Lapo Cioni[1]([✉]) [iD], Andrea Marino[1] [iD], Jason Schoeters[1] [iD], and Takeaki Uno[2] [iD]

[1] Dipartimento di Statistica, Informatica, Applicazioni, Università degli Studi di
Firenze, Firenze, Italy
`{lapo.cioni,andrea.marino,jason.schoeters}@unifi.it`
[2] National Institute of Informatics, Tokyo, Japan
`uno@nii.jp`

Abstract. A spanner of a temporal graph is a subset of arcs that allows a given set of source vertices to reach every vertex. A spanner is minimal if no arc can be removed without destroying this reachability. Our focus is on enumerating all minimal spanners of a given directed temporal graph. We study several variants of this problem, depending on the choice of sources: ONE2ALL, where a single vertex must reach all others; MANY2ALL, where a constant number of vertices must reach all others; and ALL2ALL, where every vertex must reach every other vertex. We focus on directed temporal graphs and analyze how the complexity of these enumeration problems depends on both the number of sources and the temporality of the input graph, that is, the number of labels per arc. For the case of two sources and temporality one, we present a reverse search algorithm that efficiently enumerates all minimal spanners. In contrast, we show that increasing the number of sources and the temporality makes the enumeration problem harder than enumerating hypergraph transversals. Moreover, we prove that even for temporality one, the ALL2ALL variant admits no output-sensitive unless P=NP.

Keywords: temporal graphs · temporal spanners · temporally connected subgraph enumeration

1 Introduction

In graph theory, *spanners* of a given (directed) graph G are subgraphs that maintain connectivity and approximately preserve the distances between vertices of G while using fewer edges. A spanner is not necessarily a tree. The main goal of a spanner is to balance sparsity and connectivity/distance preservation: it significantly reduces the number of edges while ensuring that connectivity is maintained and distances do not increase arbitrarily. Spanners play an important role in network design, distributed algorithms, and approximation algorithms, where efficient graph representations are essential. For a recent survey on graph spanners, see [1].

F. Foucaud and A. Parreau (Eds.): IWOCA 2026, LNCS 16587, pp. 204–219, 2026.
https://doi.org/10.1007/978-3-032-27732-9_15

Temporal Graphs and Temporal Paths. Temporal graphs model networks in which edges are active only at specific time steps, and reachability is defined via *temporal paths*, namely paths whose edge labels form an increasing sequence. Formally, a temporal graph $\mathcal{G}$ consists of an underlying graph $G = (V, E)$ together with an edge-labeling function λ that assigns to each edge a non-empty set of discrete time labels from $[\tau]$, where τ is referred to as the *lifetime* of the graph (see Sect. 2 for full definitions). In the so-called *strict* model, temporal paths are required to have strictly increasing edge labels. In contrast, the *non-strict* model allows edge labels along a temporal path to be merely non-decreasing, meaning that an arbitrary number of edges with time label i may be traversed at time step i. A fundamental parameter of temporal graphs we will massively use in this paper is *temporality* ℓ, defined as $\ell = \max_{e \in E} |\lambda(e)|$, which captures the maximum number of times any edge is available and has been widely studied in the literature [11,21].

Temporal Spanners. The notion of spanning structures in temporal graphs was first formalized by Kempe, Kleinberg, and Kumar [16], and simply aims to select a subset of the edges of the underlying graph to maintain *temporal connectivity*, i.e., connectivity via temporal paths. These objects are now commonly referred to as *temporal spanners*. A temporal spanner is called *minimal* if the removal of any single edge destroys temporal connectivity; throughout this work, we restrict attention to minimal spanners unless explicitly stated otherwise. A lot of attention has been devoted to temporal spanners in the literature. In this paper, we use the definition of temporal spanners in [19], reported below. The only difference wrt [19] is the fact that we are dealing with directed temporal graphs rather than undirected ones. The differences wrt to the Definition in [19] are underlined in Definition 1.

Definition 1 (Directed version of Definition 1 in [19]). *Given a <u>directed</u> temporal graph $\mathcal{G} = (G, \lambda)$, $E' \subseteq E(G)$ is an X spanner of $\mathcal{G}$ if it is a minimal set[1] such that in $\mathcal{G}$ restricted to edges E':*

- **if $X =$ ONE2ALL:** *given $s \in V(G)$, s reaches v for all $v \in V$;*
- **if $X =$ MANY2ALL:** *given $s_1, \ldots, s_k$, s_i reaches v for all $v \in V$, and all $i \in [k]$;*
- **if $X =$ ALL2ALL:** *u reaches v and v reaches u for all $u, v \in V$.*

Although the difference between Definition 1 and the Definition in [19] may appear minor, we show that this difference is crucial as neither the positive nor the negative results from [19] can extend to our setting.[2]

---

[1] No proper subset satisfies the same property.

[2] For the sake of completeness, we note that [19] also considers the ONE2ALL2ONE variant. For simplicity and space constraints, we do not address this variant in this paper.

Problem Definition. In this paper, we study the problem of enumerating, i.e., listing exactly once, all the temporal spanners in Definition 1 for a given directed temporal graph in input.

Problem 1 (ENUMERATION). Given a directed temporal graph $\mathcal{G}$ and connectivity X, enumerate all X spanners of $\mathcal{G}$.

As closely related to enumeration, we also study the following problem.

Problem 2 (EXTENSION PROBLEM). Given directed temporal graph $\mathcal{G}$, connectivity X, and edge set $E'' \subseteq E(G)$, decide whether there is $E' \supseteq E''$ that is an X spanner of $\mathcal{G}$.[3]

Indeed, being able to solve the EXTENSION PROBLEM efficiently   implies being able to solve the corresponding enumeration problem [13,20] efficiently. We study the problems ENUMERATION and EXTENSION PROBLEM, focusing on the number of considered sources k and the temporality ℓ of the input temporal directed graph, and show that these two parameters play a central role in determining the computational complexity of the problems.

These problems were investigated in [19] for *undirected temporal graphs*; here we restate them for the *directed* setting. Although an undirected temporal graph can be viewed as a directed temporal graph by replacing each edge with two oppositely oriented arcs carrying the same time label, neither the positive nor the negative results for the undirected case extend to the directed one. The failure of positive results to extend is immediate, since undirected temporal graphs form a special case of directed ones. More surprisingly, negative results also do not carry over. To illustrate this, consider Fig. 1, which concerns MANY2ALL spanners with two sources, s_1 and s_2. The construction on the left has only one spanner, while the one on the right has two. It can be easily generalized to obtain an exponential gap between the two cases, by introducing multiple "twin" copies of the graphs with identical adjacency and time labels, and two new dummy sources, namely s^* connected to all the copies of s_1 and t^* connected to all copies of s_2, both at times 0 and 5. Intuitively, each such copy introduces an independent binary choice in the directed setting, while no such freedom exists in the undirected one.

We study the complexity of enumeration using the classical hierarchy of efficiency guarantees. We consider algorithms that run in *output-polynomial time*, meaning that their total running time is bounded by a polynomial in the combined size of the input and the produced solutions. We also examine *incremental polynomial-time* algorithms, which ensure that the first i solutions can be generated within time polynomial in the input size and in i, as well as algorithms with *polynomial delay*, where the time elapsed between any two consecutive outputs is polynomially bounded in the input size. Beyond these standard notions, a central open question in enumeration complexity concerns the problem known as HYPERGRAPH DUALIZATION (Dual), which asks for the enumeration of all minimal hitting sets of a given hypergraph [15]. Despite extensive study, it remains

---

[3] For $X = $ ONE2ALL, MANY2ALL, we assume that s or all s_i are contained in $G[E'']$..

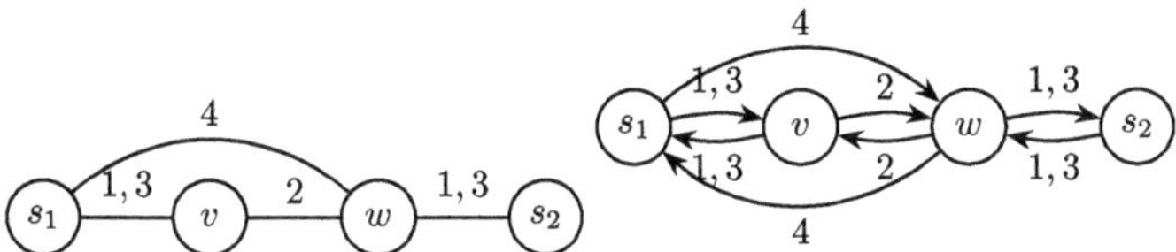

Fig. 1. Undirected spanners vs directed ones. The undirected temporal graph on the left has exactly one spanner. Indeed, the edges $\{s_1, v\}$ and $\{w, v\}$ are both necessary: the former is required to guarantee reachability from s_1 to v, and the latter to guarantee reachability from s_2 to v. Since edges can be traversed in both directions, these two edges, together with $\{s_2, w\}$, also ensure all required reachability. By minimality, no additional edge can be included, and in particular edge $\{s_1, w\}$ is never selected. Now consider the directed temporal graph on the right, obtained by replacing each undirected edge with two oppositely oriented arcs. While the same reachability constraints still force the inclusion of the arcs (s_1, v) and (w, v), their reverse arcs are no longer required. Thus, s_2 may reach s_1 either via (w, s_1) or via (v, s_1), whereas s_1 still has a unique way to reach s_2. This asymmetry yields two distinct spanners in the directed graph.

unknown whether Dual can be solved in output-polynomial time in full generality. The importance of this problem stems from its deep connections to a wide variety of enumeration tasks, to which it is known to be polynomially or quasi-polynomially interreducible [15]. Accordingly, when we establish a reduction from Dual to another enumeration problem, we describe the latter as Dual-hard, indicating that an output-polynomial algorithm for it would imply the same for Dual. For comprehensive overviews of these concepts and relationships, we refer the reader to [14, 22]

Our Contribution. In this paper, we establish the results summarized in Table 1. All the results, except the one marked with *, apply both to strict and non-strict model. In the single-source setting (see the ONE2ALL row in Table 1), we observe that EXTENSION PROBLEM can be solved in polynomial time even when the partial solution is not connected. This stands in sharp contrast to the undirected case studied in [19], where the same extension problem is shown to be NP-complete. The hardness result for undirected graphs crucially relies on the ambiguity in choosing the direction in which an edge should be traversed. In directed graphs, however, arc orientations are fixed, which eliminates this source of complexity and leads to a polynomial-time algorithm, as established in Lemma 1. Moreover, the ability to efficiently decide whether a given *partial spanner*—that is, a set of arcs—can be extended to a full spanner immediately yields a polynomial-delay enumeration algorithm via binary partition, as formalized in Corollary 1 (see also [13] for background on this technique). In contrast, the polynomial-delay result for ENUMERATION in the undirected setting of [19] requires solving a more restrictive version of the extension problem, where the partial solution is assumed to be connected.

Table 1. We use m for the number of edges, and τ for the lifetime (i.e., the maximum assigned time to the edges) of the input temporal graph. All the negative results hold even when lifetime τ is constant. k refers to the number of sources in MANY2ALL. Note that, for MANY2ALL ENUMERATION, if $k = n$ (resp. $k = 1$), then the problem becomes ONE2ALL ENUMERATION). We use the $\Leftarrow$ sign to mean that the corresponding result is implied by the result on the right-hand side in the table. All the results, except the one starred, i.e., marked with *, applies to both strict and non-strict models.

X	EXTENSION	ENUMERATION
ONE2ALL	Poly, even if partial sol. is not connected (Proposition 1)	Poly-delay (Corollary 1)
MANY2ALL	NP-c, even if $k = 2$, $\ell = 1$, and partial sol. is connected (Proposition 2)	Poly-delay if $k = 2$, and $\ell = 1$ (Theorem 2)*
		Dual-hard, even if $k = 3$, and $\ell = 2$ (Theorem 1)
ALL2ALL	NP-c $\Leftarrow$ (Theorem 4)	No output poly algorithm unless $\mathsf{P} = \mathsf{NP}$, even if $\ell = 1$ (Theorem 3)

As the number of sources increases (see the MANY2ALL row in Table 1), the computational difficulty of the problems rises significantly. In particular, EXTENSION PROBLEM becomes NP-complete even in a highly restricted setting: when each arc carries a single time label, only two sources are present (i.e., $k = 2$ and $\ell = 1$), and the partial solution E'' is connected. This hardness result indicates that, for $k = 2$ and $\ell = 1$, an output-sensitive algorithm for ENUMERATION is unlikely to be based on backtracking or binary partition techniques, as these approaches fundamentally rely on efficiently deciding whether a partial solution can be extended without encountering dead ends. Nevertheless, despite this barrier, we show that polynomial-delay enumeration remains achievable in this setting by employing a different paradigm, namely the reverse search technique. Reverse search is a general enumeration framework that avoids explicitly storing all previously generated solutions. It relies on an implicit directed graph over the solution space, in which each solution—except for a designated root—is assigned a unique parent. Starting from the root, the algorithm performs a depth-first traversal of this graph, systematically outputting each solution when it is reached via its parent relation. Because only the current solution must be maintained, reverse search typically requires only polynomial space while ensuring complete and duplicate-free enumeration (see also [13] for background on this technique). In our case, the design of the algorithm is non-trivial as it requires careful case analysis based on temporal reachability from the sources to define parent-child relationship among solutions and avoid duplication. We further show that increasing both the number of sources k and the temporality ℓ leads to a sharp rise in complexity: already for $k = 3$ and $\ell = 2$, ENUMERATION becomes Dual-hard. Specifically, we establish a reduction from Dual in which the spanners enumerated by ENUMERATION are in bijection with the solutions of Dual. Consequently, the existence of an output-sensitive enumeration algorithm for ENUMERATION would directly imply an output-sensitive enumeration algorithm for Dual.

Finally, in the ALL2ALL setting (see the last row of Table 1), we prove that ENUMERATION cannot admit an output-polynomial algorithm unless $P = NP$, even when $\ell = 1$. Our reduction, inspired by [19], transforms a SAT instance into an instance of ENUMERATION with the following property: among all spanners, only a polynomial number are *spurious*, meaning that they do not correspond to any valid truth assignment, whereas every remaining spanner encodes a satisfying assignment. Hence, an output-sensitive enumeration algorithm could be used to solve SAT, by waiting polynomial time to see whether there is a non-spurious spanner, and hence get a valid assignment. For ALL2ALL, this lower bound also implies hardness for EXTENSION PROBLEM, since a polynomial-time algorithm for EXTENSION PROBLEM would yield an output-polynomial enumeration algorithm for ENUMERATION. The structure of the paper follows the rows of Table 1.

Related Works. Enumeration problems involving connectivity constraints have been extensively studied in static graphs, both for maintaining and destroying connectivity. Classic examples include the enumeration of minimal Steiner trees and related variants, for which polynomial-delay or incremental polynomial-time algorithms are known under suitable restrictions [8,18]. Several connectivity-related enumeration problems are intractable, e.g., the enumeration of minimal edge sets ensuring reachability in directed graphs, unless $P = NP$ [17]. In contrast to this extensive literature on static graphs, the enumeration of temporal spanners has been addressed in only one prior work, namely [19], which focuses exclusively on undirected temporal graphs. As discussed earlier, the techniques and results developed there do not carry over to the directed setting. In this paper, we therefore study temporal spanner enumeration in directed temporal graphs, with particular emphasis on the role played by the number of sources and the temporality. In this setting, we establish an output-sensitive enumeration algorithm for the case $k = 2$ and $\ell = 1$. Existing research on temporal spanners has explored their complexity and approximability [2], sparsity, and lower bounds [5,12], robustness, stretch, and structural properties [6,7,10]. Other works focus on temporal spanning trees under additional optimality criteria, such as weight minimization or distance optimization [9,23].

2   Preliminaries

A temporal graph $\mathcal{G}$ is a pair (G, λ), where G is a (static) graph, and $\lambda : E(G) \to \mathcal{P}(\mathbb{N})$ the function which assign times to the edges of G, with $\mathcal{P}$ denoting the power set. $G = (V(G), E(G))$ is a directed, simple graph. We refer to the sets of vertices $V(G)$ and edges $E(G)$ as V and E respectively when there is no ambiguity on graph G. Each edge $e \in E(\mathcal{G})$ has an associated set of times (or labels or timestamps) $\lambda(e)$. We assume $\lambda(e)$ to be finite for each $e \in E$, and call *lifetime* the maximum timestamp appearing in the graph, $\tau = \max\{\bigcup_{e \in E} \lambda(e)\}$. We call *temporality* greatest amount of timestamps on the edges of the graph, $\ell = \max_{e \in E} |\lambda(e)|$. Also, we usually drop the set notation for the times on an edge, especially in figures and when the edge only has one timestamp.

Connectivity in temporal graphs takes time into account. This means that a path in G might not result in a temporal path in $\mathcal{G}$ (but the reverse is always true). A *strict* (or *non-strict*) *temporal path* in $\mathcal{G}$ is a sequence of pairs $((e_1, t_1), \ldots, (e_k, t_k))$ such that $(e_1, \ldots, e_k)$ is a path in G, $t_i \in \lambda(e_i)$ for each $1 \leq i \leq k$, and $t_i < t_{i+1}$ (or $t_i \leq t_{i+1}$) for each $1 \leq i \leq k-1$. That is, the times at which one passes through an edge must be strictly increasing for strict temporal paths, and weakly increasing for non-strict ones. When a temporal path exists from a vertex s to a vertex v, we say that s can reach v, and v is reachable from s. Our interested is in reachability from a set of k *sources* $\{s_1, \ldots, s_k\}$. Given $S \subseteq E(G)$, we denote with $R_i(S)$ the set of vertices that can be reached from s_i in $\mathcal{G}$ using only edges in S.

For some tractability result we use a search algorithm which is commonly used in temporal graphs. This algorithm prioritizes earliest arrival time, and one of the first papers to introduce, analyze, and use it is [23]. We refer to it as **TFS** (time-first search). Using **TFS** we can check in $O(m\tau)$ time if a source can reach all the vertices of a temporal graph, also obtaining the earliest time at which each vertex can be reached. Therefore checking whether a set is a minimal spanner can be answered in polynomial time, by first checking reachability for every source, and then deleting one edge at a time and checking reachability again. If first check succeeds and the latter fail for each edge, then it is a minimal spanner.

For other tractability results we use reductions from the SATISFIABILITY PROBLEM (or SAT) and from HYPERGRAPH DUALIZATION (or Dual). The former is a decision problem with input a CNF logic formula ϕ of n variables x_i and m clauses C_j, asking if there exists some assignment σ of the variables such that ϕ is assigned true. The latter is an enumeration problem with input an hypergraph $G = (U, H)$ on n vertices and m hyperedges, and it is required to enumerate all minimal transversals (or hitting sets) of the hypergraph, i.e., all minimal subsets $U' \subseteq U$ such that, the intersection $U' \cap h$ is nonempty for all $h \in H$. For space constraints, most of the claims have sketched proofs. Full proofs will appear in the full version of the paper.

3   ONE2ALL Spanners

For space constraints, we only give a brief overview of the case ONE2ALL. We observe that any ONE2ALL spanner is a directed rooted tree (also known as arborescence or branching). To solve ONE2ALL EXTENSION PROBLEM, we consider the edges in E'' and delete all other edges incident to the same end vertices. The answer to EXTENSION PROBLEM is yes iff the resulting graph grants reachability to all vertices and no two edges of E'' are incident to the same end vertex.

Proposition 1. ONE2ALL EXTENSION PROBLEM *is polynomial.*

Using a binary search approach, as the extension problem is polynomial because of Proposition 1 it is now easy to enumerate all ONE2ALL spanners. This is a standard approach (see [13]).

Corollary 1. ONE2ALL ENUMERATION *can be solved in polynomial-delay time.*

4   Negative Results for **MANY2ALL** Spanners

In this section, we show that MANY2ALL EXTENSION PROBLEM is NP-complete even if $k = 2$, $\ell = 1$, and the partial solution E'' is connected (Proposition 2). We then prove MANY2ALL ENUMERATION is Dual-hard for k at least of 3 and ℓ at least 2 (Theorem 1).

We show a reduction from SAT to MANY2ALL EXTENSION PROBLEM. Let ϕ be a SAT formula with variables $x_1, \ldots, x_n$ and clauses $C_1, \ldots, C_m$. Construct the temporal graph $\mathcal{G}$ with two sources s_1, s_2, two vertices v_{x_i}, $v_{x_i'}$ for each variable x_i, and one vertex v_{C_i} for each clause. The sources are connected, (s_2, s_1) at time 1 and (s_1, s_2) at time 9. Then, for each variable x_i, we have edges (s_1, v_{x_i}), (s_2, v_{x_i}), $(s_1, v_{x_i'})$, $(v_{x_i}, v_{x_i'})$ and $(v_{x_i'}, v_{x_i})$ at times 2, 4, 7, 5, and 8, respectively. Also, for each clause C_j such that x_i appears positive in C_j, connect v_{x_i} to v_{C_j} at time 3. For each clause $C_{j'}$ such that v_{x_i} appears negated in C_k, connect $v_{x_i'}$ to $v_{C_{j'}}$ at time 6. Finally, connect s_1 to each v_{C_j} at time 0.

Proposition 2. MANY2ALL EXTENSION PROBLEM *is* NP-*complete even if we have 2 sources, temporality 1, and the partial solution is connected.*

To show that MANY2ALL is Dual-hard for three sources and temporality 2, we first describe a reduction from an instance $(\{u_i\}_{i=1,\ldots,n}, \{h_j\}_{j=1,\ldots,m})$ of Hypergraph Dualization to an instance of MANY2ALL.

Let V_j be the set of vertices incident to h_j, for each hyperedge h_j, $V_j = \{u_{j,1}, \ldots, u_{j,|V_j|}\}$. The graph is essentially divided in layers. The first layer consists of the sources s_1, s_2, s_3. The second layer consists of vertices u_1', $\ldots$, u_n', $u_1, \ldots, u_n$, two for each vertex of the hypergraph. Then is the hyperedges layer. Specifically, for each hyperedge h_j we have a path containing one vertex $u_{j,i}$ for each element of V_j, plus one final vertex labeled V_j. We now describe the edges between the vertices. The sources s_1 and s_2 are connected by two edges, both with timestamp 4. s_3 is connected to s_2 at time 3 and s_2 is connected to s_3 at time $n+5$. The vertex layer is as follows: s_1 and s_2 are both connected to each u_i' by edges with times 1 and 2, respectively. u_i' is then connected to u_i by an edge with time 3. Also, s_1 is connected to each v_{u_i} by an edge with time 1. Finally, s_3 is connected to each u_i', u_i with an edge with time $n + 4$. For each pair u_i, h_j such that $u_i \in V_j$, u_i is connected at times $\{2, n + 3\}$ to one of the vertices in the path that reach h_j, such that no two u_i, u_k are connected to the same vertex. Finally, the times in the path $u_{i,1}, \ldots, u_{i,|V_i|}$ are increasing, from 3 to $|V_{h_j}| + 2$, and $u_{i,|V_i|}$ is connected to V_i at time $n + 2$, for each hyperedge h_i. s_3 is connected at time 1 to each vertex $u_{i,1}$ at the beginning of each path, while s_2 is connected at time 1 to each vertex V_i.

Theorem 1. MANY2ALL *is* Dual-*hard for at least 3 sources and temporality at least 2.*

Remark 1. For non-strict spanners, the reduction can be easily modified in one with constant lifetime, by having the same temporal label (for example 3) on each edge in the paths relative to each h_j, and changing the times from u_i to $u_{j,k}$ and from s_3 to u'_i and u_i to some suitable number (e.g., $2, 4$ and 5, respectively).

5   Positive Results for MANY2ALL ENUMERATION

In this section we show that MANY2ALL spanner enumeration is output-polynomial for temporality 1 and 2 sources, in the strict case.

Recall that MANY2ALL EXTENSION PROBLEM is NP-complete even for two sources and temporality 1, as shown in Proposition 2. This implies that, even for this restricted case, a binary partition algorithm will not give us polynomial delay. Instead, we use the technique known as reverse search [3,4].

For simplicity, we present the algorithm in the case where the temporal graph has at most one edge at any given time, i.e., $|E(G_t)| \leq 1$ for each $t \in [\tau]$. In this way we have a unique way to order the edges by their timestamp. To apply the algorithm in the general case it is sufficient to choose an arbitrary ordering between the edges having the same timestamp.

The reverse search approach works by establishing a partial order relation between solutions. In our case, the solutions are spanners and the relation gives a tree-like structure in the solution space. First, we define decorated partial spanners, which are used as intermediate steps for generating spanners and for defining the partial order. Then we define the partial order and show that the structure of the solution space is a tree. Finally, we show how to enumerate all the spanners.

Recall that $R_i(S)$ denotes set of vertices that can be reached from source s_i using only edges in S. A *partial spanner* S is a set of edges such that, for every $e \in S$, there exists a source s_i such that $R_i(S) \neq R_i(S \setminus \{e\})$. That is, a partial spanner is a set of edges that are all necessary for some reachability. This implies that a partial spanner is a set of connected edges, but there are no guarantees neither on the reachability of the whole graph. Indeed, the partial spanner only possesses the property of minimality.

Partial decorated spanners are partial spanners where each edge e has two associated sets: a set of *colors* C_e and a set of *dependencies* D_e. The colors of an edge represent which sources use that edge for reachability. To explain the meaning of dependencies, we refer to the example in Fig. 2. The graph represented has five edges with timestamps 1 to 5, which we call $e_1, \ldots, e_5$. e_1, e_2 and e_3 are colored red (the color for s_1), while e_4 and e_5 are colored blue (the color for s_2). No edge has both colors. Note that $D_1 = \{e_2\}$. Indeed, e_1 is not needed to reach vertex v, which can also be reached from s_1 using e_3 and e_5, but e_1 is needed to reach w by using it and e_2. In other words, e depends on D_e when the added reachability that it provides can be obtained only by also using some following edge in D_e.

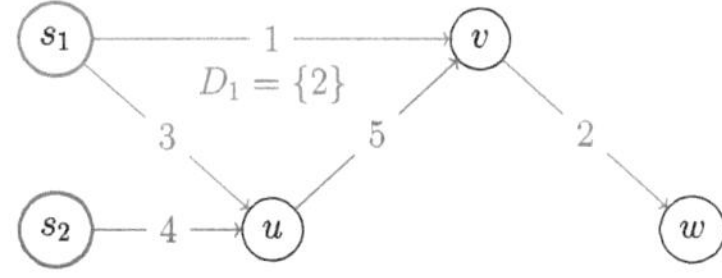

Fig. 2. Decorated partial spanner. e_1 and e_2 are used by s_1, while e_5 is used only by s_2, even if s_1 could reach v with it. e_1 is dependent on e_2.

The dependencies are used when we delete an edge e from a decorated partial spanner $\mathcal{S}$. In fact, there are some steps to follow to ensure that we obtain a decorated partial spanner. We call these steps *cleaning*, and they work as follows: after removing $\{e, C_e, D_e\}$, find all the dependency sets $D_{(u,v)}$ that contain e, and remove it from them. If $D_{(u,v)}$ becomes empty, then there must be exactly one other edge $(w,v) \in S$ such that $|C_{(w,v)}| = 1$. Update $C_{(w,v)}$ to $\{1,2\}$. Then delete $\{(u,v), C_{(u,v)}, D_{(u,v)}\}$ from the partial decorated spanner and iterate this procedure for (u,v).

In a general case of deletion it might also be needed to delete edges that have timestamp greater than $\lambda(e)$, but in our procedures this never happens. For example, referring to Fig. 2, removing e_2 would also cause the cleaning to remove e_1. Indeed, $\{e_1, e_3, e_4, e_5\}$ is not a partial spanner because e_1 can be removed without changing reachability for both sources.

We describe a procedure that builds a decorated partial spanner from a (eventually empty) decorated partial spanner $\mathcal{S}$ and an edge $e = (x, y) \notin S$ such that $\lambda(e) > \lambda(e')$ for each $e' \in S$. The resulting set is called the *augmentation* of $\mathcal{S}$ with e, denoted $A_e(\mathcal{S})$, and is defined by these rules:

- if $x \notin R_1(S) \cup R_2(S)$, then $A_e(\mathcal{S}) = \mathcal{S}$;
- if $x \in R_1(S) \setminus R_2(S)$ (analogous for the opposite case), then:
  - if $y \in R_1(S)$, then $A_e(\mathcal{S}) = \mathcal{S}$;
  - else, $A_e(\mathcal{S}) = \mathcal{S} \cup \{(e, \{1\}, \emptyset)\}$;
- if $x \in R_1(S) \cap R_2(S)$, then:
  - if $y \notin R_1(S) \cup R_2(S)$, then $A_e(\mathcal{S}) = \mathcal{S} \cup \{(e, \{1,2\}, \emptyset)\}$;
  - if $y \in R_1(S) \cap R_2(S)$, then $A_e(\mathcal{S}) = \mathcal{S}$;
  - if $y \in R_1(S) \setminus R_2(S)$ (analogous for the opposite case), then let f be the edge incoming to y such that $C_f = \{1\}$, and:
    * if there are outgoing edges $(y, w_1), \ldots (y, w_k) \in S$ with $C_{(y,w_i)} = \{1\}$ for each i, then let $A_e(\mathcal{S}) = \mathcal{S} \cup \{(e, \{2\}, \emptyset)\}$ but with $D_f = \{(y, w_1), \ldots, (y, w_k)\}$;
    * else, delete f from $\mathcal{S}$ and apply the cleaning described above, obtaining $\mathcal{T}$. Let $A_e(\mathcal{S}) = \mathcal{T} \cup \{(e, \{1,2\}, \emptyset)\}$.

Essentially, the augmentation greedily adds an edge whenever it improves the reachability from some source, and never reduces the reachability. It is not an extension algorithm, since it can delete edges with case ⋆. The augmentation gives a decorated partial spanner, and can be applied in increasing time order to all the edges with greater timestamp than those already considered. We call this procedure *completion* $\mathcal{C}$ of $\mathcal{S}$, and it produces a spanner. This can be checked using a case-by-case approach.

The completion induces a parent-child relationship from which we build the tree of spanners of $\mathcal{G}$. Let S_j be the set of edges of S with time at most j.

Definition 2. *Let S be a spanner, and $\mathcal{S}$ its associated decorated spanner. Let $i_S = \max\{j \leq \tau \mid \mathcal{C}(S_j) \neq \mathcal{S}\}$. We define the* parent *of S as $P(S) = \mathcal{C}(S_{i_S})$.*

Finding the parent of a spanner is easy, but we need to do the opposite, i.e., find all children of a spanner. To do this we distinguish between two kind of children. Let $S = \{e_1, \ldots, e_{|S|}\}$ be a spanner, with $\lambda(e_1) < \cdots < \lambda(e_{|S|})$. We call $L(S)$ the ordered sequence $\lambda(e_1) \ldots \lambda(e_{|S|})$ of times of the edges of S.

Definition 3. *Let S and T be spanners such that $S = P(T)$. If $L(T)$ is lexicographically smaller than $L(S)$, we call T* special. *Otherwise, it is* regular.

Figure 3 shows an example of a regular and a special children. P is the parent of R because $\mathcal{C}(\{1,2,3\}) = P$, while $\mathcal{C}(\{1,2,3,7\} = R)$, and R is a regular children because $1235\cdots$ is lexicographically smaller than $1237\cdots$. P is also the parent of S because $\mathcal{C}(\{1,2,3,4\}) = P$ and $\mathcal{C}(\{1,2,3,4,6\}) = S$, but S is a special children because $1234\cdots$ is lexicographically smaller than $1235\cdots$.

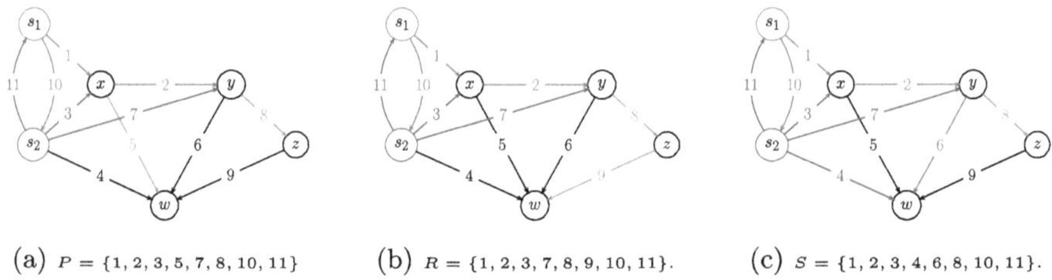

(a) $P = \{1, 2, 3, 5, 7, 8, 10, 11\}$ (b) $R = \{1, 2, 3, 7, 8, 9, 10, 11\}$.   (c) $S = \{1, 2, 3, 4, 6, 8, 10, 11\}$.

Fig. 3. Three decorated spanners on the same temporal graph. P is the parent of R and S. R is a regular children, S is a special children.

Note that when completing $\{1,2,3,4\}$ and obtaining P, edge 4 is deleted by rule $\star$. Since S contains 4, it ends up being a special children. We can see that all the special children can be obtained by ignoring the deleting instruction of rule $\star$, but first we show how to obtain all the regular children of a node. This is an immediate consequence of the definition of regular children.

Remark 2. Let $P = \{p_1, \ldots, p_k\}$ be a spanner whose edges are indexed in increasing time order. Let R be a regular children of P. Then there exist $i < k$ and $e \in E(G)$ such that $\lambda(e) > \lambda(p_i)$, with $R = \mathcal{C}(\{p_1, \ldots, p_{i-1}, e\})$.

Thus we can obtain all the regular children of a spanner by checking whether $\mathcal{C}(\{p_1, \ldots, p_{i-1}, e\})$ is a spanner for each combination of $i < k$ and $e \in E(G)$, with $\lambda(e) > \lambda(p_{i-1})$. Since $i < k \leq \binom{n}{2}$ and $|E(G)| \leq \binom{n}{2}$, then all the regular children of S can be found in polynomial time.

We now give an algorithm that finds special children. We define the special augmentation $A_e^*(\mathcal{S})$ as the augmentation of $\mathcal{S}$ with edge e, but where rule $\star$ is not considered. That is, if the case required by rule $\star$ holds, then $A_e^*(\mathcal{S}) = \mathcal{S}$. We define the *e-special completion* of $\mathcal{S}$ as $\mathcal{C}_e^*(\mathcal{S}) = A_E(A_e^*(\mathcal{S}))$, where E is the set of edges with timestamp greater than $\lambda(e)$. Note that $A_e^*(\mathcal{S})$ is a decorated partial spanner, and so is $\mathcal{C}_e^*(\mathcal{S})$. Still, $\mathcal{C}_e^*(\mathcal{S})$ might not be a (full) spanner, but can be easily checked.

Proposition 3. *Let $P = \{p_1, \ldots, p_k\}$ be a spanner whose edges are indexed in increasing time order. Let S be a special children of P. Then there exist an edge $e \in R \setminus P$ such that, during the generation of $P = P(S)$, e was deleted by rule $\star$.*

Proof. Let $S = \{s_1, \ldots, s_{k'}\}$, with the edges indexed in increasing order. Since S is a children of P, then there exist a $j \leq k'$ such that $\mathcal{C}(\{s_1, \ldots, s_j\}) = P$ and $\mathcal{C}(\{s_1, \ldots, s_{j+1}\}) = S$. By definition of special children there exists an $i \leq k'$ such that $S = \{p_1, \ldots, p_{i-1}, s_i, s_{i+1}, \ldots s_{k'}\}$, with $\lambda(s_i) < \lambda(p_i)$, $s_i \notin P$. It holds $i \leq j$, because otherwise we would have $S = \mathcal{C}(\{s_1, \ldots, s_{j+1}\}) = \mathcal{C}(\{p_1, \ldots, p_{j+1}\}) \neq S$, which is a contradiction. We obtain $P = \mathcal{C}(\{s_1, \ldots, s_j\}) = \mathcal{C}(\{p_1, \ldots, p_{i-1}, s_i, \ldots, s_j\})$, with $s_i \notin P$, hence s_i is deleted during the completion process, and this can only be done by rule $\star$. $\square$

Now we know how to search for special children. Indeed, every time the completion does an augmentation $A_e(\mathcal{T})$ where rule $\star$ is applied, there could exist some spanner that instead keeps the deleted edge and does not add the newly added edge e. We need to check all these instances.

Let $\mathcal{S}$ be a decorated partial spanner and S its set of edges. Consider the steps at which deletions happen in $\mathcal{C}(S_{i_S})$, and store $\star(S) = \{(\mathcal{T}, e) \mid A_e(\mathcal{T})$ is computed during $\mathcal{C}(S_{i_S})$ and rule $\star$ is applied$\}$. Given a pair $(\mathcal{T}, e)$, we can compute $\mathcal{C}_e^*(\mathcal{T})$. That is, we can continue the completion but ignoring e. If the resulting set is a spanner, then it is a special children of S by definition. Also, by Proposition 3, all the special children of S can be found in this way.

Now Algorithm 1 finds all the children of a spanner S.

Theorem 2. MANY2ALL ENUMERATION *can be solved in output-polynomial time for 2 sources and temporality 1, in strict graphs.*

6   ALL2ALL SPANNERS

This section is concerned with ALL2ALL spanners, which is where all vertices play the role of source, i.e., all vertices need to reach each other.

We present the sketch of a reduction inspired by the ones used in [19] to prove that their problems of enumerating ALL2ALL spanners in undirected graphs cannot be done in output-polynomial time unless $\mathsf{P} = \mathsf{NP}$.

$$\begin{array}{|l|}
\hline
\textbf{Algorithm 1: } \text{enumerateTwo2AllSpanners} \\
\text{Input: a spanner } S = \{s_1, \ldots, s_k\} \qquad \text{Output: the children of } S \\
\hline
\end{array}$$

Algorithm 1: enumerateTwo2AllSpanners
Input: a spanner $S = \{s_1, \ldots, s_k\}$ Output: the children of S

```
1  *(S) ← {(T,e) | A_e(T) is computed during C(S_{i_S}) and rule * is applied};
2  foreach (T,e) ∈ *(S) do
3  │   compute T = C*_e(T);
4  │   if T is a spanner then
5  │   │   output T as a special children of S;
6  │   foreach j = 1,...,k do
7  │   │   foreach e ∈ E(G) such that λ(e) > λ(s_j) do
8  │   │   │   if C({s_1,...,s_{j-1},e}) is a spanner then
9  │   │   │   │   output it as a regular children of S;
```

Table 2. Temporal edges of $\mathcal{G}$, except those at times 5, 6, 7 and 8. In the table, $1 \leq i \leq n$, $1 \leq j, \ell, k \leq m$, and $\ell \neq k$.

$t=1$	$t=2$	$t=3$	$t=4$	$t=9$	$t=10$	$t=11$	$t=12$	$t=13$	$t=14$	$t=15$
(w,z)	(C_j,w)	(z,C_j)	(C_j,S^*)	(D_j,S^*)	(T^*,u)	(u,S^*)	(D_j,u)	(u,v_i)	(z,g)	(z,w)
(S^*,z)	(D_j,g)	(T^*,w)	(C_j,z)	(u,C_j)	(C_ℓ,D_k)	(D_j,C'_j)		(u,T^*)	(w,C_j)	
(v_i,z)	(D_ℓ,C'_k)	(T^*,g)	(u,z)	(T^*,C'_j)				(C_j,D_j)		
(C'_j,z)			(D_j,z)	(D_j,w)				(g,D_j)		
(g,z)			(T^*,z)							

We introduce the auxiliary problem ANOTHER ALL2ALL SPANNER, which is the following decision problem: given a temporal graph $\mathcal{G}$ and a set of k ALL2ALL spanners S, with k polynomial in the size of $\mathcal{G}$, does there exist another ALL2ALL spanner of $\mathcal{G}$ which is not in S? We first prove this problem to be NP-complete, through a technical reduction. Afterwards, we use this result to show that if an output-polynomial time algorithm exists for ALL2ALL ENUMERATION, it would imply that a polynomial time algorithm exists for ANOTHER ALL2ALL SPANNER. Indeed, since k is polynomial, an output sensitive algorithm for ENUMERATION running in poly time $O(n^c)$ (for some constant c) per solution would terminate in $O(k \cdot n^c)$ iff there are no more ALL2ALL spanners, thus solving ANOTHER ALL2ALL SPANNER in polynomial time.

Theorem 3. *ALL2ALL ENUMERATION cannot be done in output-polynomial time unless* $\mathsf{P} = \mathsf{NP}$*, even if the temporal graph has temporality 1.*

As a consequence, the extension problem for ALL2ALL is NP-complete.

Theorem 4. *ALL2ALL EXTENSION PROBLEM is* NP*-complete.*

Construction of the Temporal Graph. To prove the NP-completeness for ANOTHER ALL2ALL SPANNER, we reduce from SAT. We present in the following how to construct a temporal graph $\mathcal{G}$ for the ANOTHER ALL2ALL SPANNER instance from a given SAT instance ϕ. Figure 4 is provided to follow the construction.

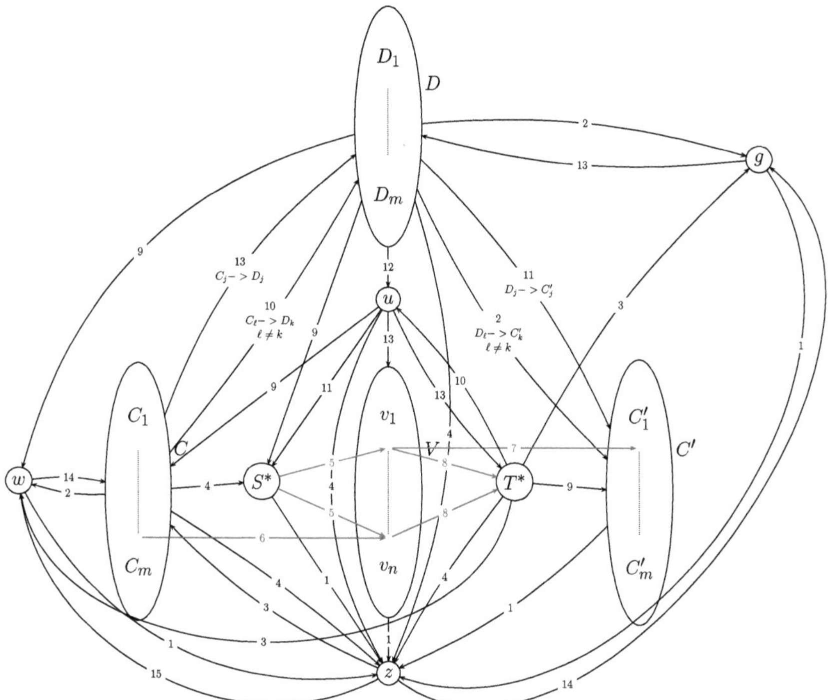

Fig. 4. Graph for ANOTHER ALL2ALL SPANNER. The red edges are the only optional ones. When there are edges between all the vertices in a group (such as C, V, C' or D) and other vertices, only one edge is represented.

Let ϕ be a CNF formula with n variables $x_1,\ldots,x_n$ and m clauses $C_1,\ldots,C_m$. Let $\mathcal{G}$ be the temporal graph defined as follows. $V(G)$ contains vertices $v_1,\ldots,v_n$ (one for each variable of ϕ), $C_1,\ldots,C_m$, $C'_1,\ldots,C'_m$, $D_1,\ldots,D_m$ (three for each clause of ϕ), and S^*,T^*,g,u,w,z (auxiliary vertices). The vertices are connected with edges having one timestamp each. We have some edges that are present independently from the formula; these are detailed in Table 2. Also, we have edges which depend on ϕ, in the same manner as [19]: for each pair (i,j) such that literal $x_i \in C_j$, connect C_j to v_i at time 6. On the other hand, if $\bar{x}_i \in C_j$, then connect v_i to C'_j at time 7. Finally, connect, for every $1 \leq i \leq n$, S^* to v_i at time 5 and v_i to T^* at time 8. This concludes the construction of $\mathcal{G}$. Let $\mathcal{G}''$ be the temporal graph which contains the edges of $\mathcal{G}$ with timestamps 5, 6, 7, 8, and $\mathcal{G}'$ the temporal graph containing the other edges, i.e., those in Table 2.

We give in input to ANOTHER ALL2ALL SPANNER the n spanners $E(G') \cup \{(S^*,v_i),(v_i,T^*)\}$, $1 \leq i \leq n$, and ask if there exists another spanner. The answer is yes iff the original formula was satisfiable.

Acknowledgments. L.C. and A.M. were partially funded by MUR of Italy, under PRIN Project n. 2022ME9Z78 - NextGRAAL. L.C. was partially funded by Finanziamento Dipartimenti di Eccellenza 2023-2027, CUP: B13C22004500001. A.M. was partially funded by Italian PNRR CN4 Centro Nazionale per la Mobilità Sostenibile, NextGeneration EU - CUP, B13C22001000001. A.M. and J.S. were partially funded by PRIN PNRR Project n. P2022NZPJA - DLT-FRUIT.

References

1. Ahmed, R., et al.: Graph spanners: a tutorial review. Comput. Sci. Rev. **37**, 100253 (2020)
2. Akrida, E.C., Gąsieniec, L., Mertzios, G.B., Spirakis, P.G.: The complexity of optimal design of temporally connected graphs. Theory Comput. Syst. **61**(3) (2017)
3. Avis, D., Fukuda, K.: A pivoting algorithm for convex hulls and vertex enumeration of arrangements and polyhedra. In: Proceedings of the Seventh Annual Symposium on Computational Geometry, pp. 98–104 (1991)
4. Avis, D., Fukuda, K.: Reverse search for enumeration. Discret. Appl. Math. **65**(1), 21–46 (1996)
5. Axiotis, K., Fotakis, D.: On the size and the approximability of minimum temporally connected subgraphs. In: 43rd International Colloquium on Automata, Languages, and Programming (ICALP 2016) (2016)
6. Bilò, D., D'Angelo, G., Gualà, L., Leucci, S., Rossi, M.: Sparse temporal spanners with low stretch. In: 30th Annual European Symposium on Algorithms (ESA 2022) (2022)
7. Bilò, D., D'Angelo, G., Gualà, L., Leucci, S., Rossi, M.: Blackout-tolerant temporal spanners. J. Comput. Syst. Sci. **141**, 103495 (2024)
8. Boros, E., Borys, K., Elbassioni, K., Gurvich, V., Makino, K., Rudolf, G.: Generating minimal k-vertex connected spanning subgraphs. In: Computing and Combinatorics, pp. 222–231 (2007)
9. Bubboloni, D., Catalano, C., Marino, A., Silva, A.: On computing optimal temporal branchings and spanning subgraphs. J. Comput. Syst. Sci. **148**, 103596 (2025)
10. Casteigts, A., Corsini, T.: In search of the lost tree: hardness and relaxation of spanning trees in temporal graphs. In: International Colloquium on Structural Information and Communication Complexity, pp. 138–155 (2024)
11. Casteigts, A., Corsini, T., Sarkar, W.: Simple, strict, proper, happy: a study of reachability in temporal graphs. Theoret. Comput. Sci. **991**, 114434 (2024)
12. Casteigts, A., Peters, J.G., Schoeters, J.: Temporal cliques admit sparse spanners. J. Comput. Syst. Sci. **121**, 1–17 (2021)
13. Conte, A., Kurita, K., Marino, A., Punzi, G., Uno, T., Wasa, K.: Designing output sensitive algorithms for subgraph enumeration. In: From Strings to Graphs, and Back Again. OASIcs, vol. 132, pp. 19:1–19:40 (2025)
14. Eiter, T., Makino, K., Gottlob, G.: Computational aspects of monotone dualization: a brief survey. Discrete Appl. Math. **156**(11), 2035–2049 (2008), in Memory of Leonid Khachiyan (1952 - 2005)
15. Elbassioni, K., Rauf, I., Ray, S.: A global parallel algorithm for enumerating minimal transversals of geometric hypergraphs. Theoret. Comput. Sci. **767**, 26–33 (2019)

16. Kempe, D., Kleinberg, J., Kumar, A.: Connectivity and inference problems for temporal networks. In: Proceedings of the Thirty-Second annual ACM Symposium on Theory of Computing, pp. 504–513 (2000)
17. Khachiyan, L., Boros, E., Elbassioni, K., Gurvich, V., Makino, K.: Enumerating disjunctions and conjunctions of paths and cuts in reliability theory. Discret. Appl. Math. **155**(2), 137–149 (2007)
18. Kobayashi, Y., Kurita, K., Wasa, K.: Linear-delay enumeration for minimal steiner problems. In: PODS '22: International Conference on Management of Data, Philadelphia, PA, USA, June 12–17, 2022, pp. 301–313. ACM (2022)
19. Kurita, K., Marino, A., Schoeters, J., Uno, T.: Spanner enumeration for temporal graphs. In: 4th Symposium on Algorithmic Foundations of Dynamic Networks, SAND 2025, Liverpool, UK, June 9–11, 2025. LIPIcs, vol. 330, pp. 9:1–9:21 (2025)
20. Mary, A., Strozecki, Y.: Efficient enumeration of solutions produced by closure operations. Discret. Math. Theor. Comput. Sci. **21**(3) (2019)
21. Michail, O.: An introduction to temporal graphs: an algorithmic perspective. Internet Math. **12**(4), 239–280 (2016)
22. Strozecki, Y.: Enumeration complexity. Bull. EATCS **129** (2019)
23. Xuan, B.B., Ferreira, A., Jarry, A.: Computing shortest, fastest, and foremost journeys in dynamic networks. Int. J. Found. Comput. Sci. **14**(02), 267–285 (2003)

Beer Path Problems in Temporal Graphs

Andrea D'Ascenzo[1], Giuseppe F. Italiano[2], Sotiris Kanellopoulos[3,4],
Anna Mpanti[2], Aris Pagourtzis[3,4], and Christos Pergaminelis[3,4]($\boxtimes$)

[1] Gran Sasso Science Institute, L'Aquila, Italy
`andrea.dascenzo@gssi.it`
[2] LUISS University, Rome, Italy
`{gitaliano,ampanti}@luiss.it`
[3] National Technical University of Athens, Athens, Greece
`pagour@cs.ntua.gr`
[4] Archimedes, Athena Research Center, Marousi, Greece
`s.kanellopoulos@athenarc.gr, pagour@cs.ntua.gr,`
`chr.pergaminelis@athenarc.gr,s.kanellopoulos@athenarc.gr`

Abstract. Computing paths in graph structures is a fundamental operation in a wide range of applications, from transportation networks to data analysis. The *beer path* problem, which captures the option of visiting points of interest, such as gas stations or convenience stores, prior to reaching the final destination, has been recently introduced and extensively studied in static graphs. However, existing approaches do not account for temporal information (for example, transit service schedules), which is often crucial in real-world scenarios. In this work, we introduce the notion of beer paths in temporal graphs, where edges are time-dependent and *beer vertices* are active only at specific times. We formally define the problems of computing earliest-arrival, latest-departure, fastest, and shortest temporal beer paths and propose efficient algorithms for all of them under both edge stream and adjacency list representations. We show that the time complexity of each of our algorithms is aligned with that of the corresponding standard temporal pathfinding algorithm, thus preserving efficiency. We also present preprocessing techniques that enable efficient query answering under dynamic conditions, for example new openings or closings of shops. We achieve this either through appropriate precomputation of selected paths or by transforming a temporal graph into an equivalent static graph.

1 Introduction

Answering path queries is a fundamental operation on networks, with numerous optimization and information retrieval tasks relying on an efficient response to such queries. Bacic et al. [2] recently introduced the concepts of *beer vertices* and *beer paths* to model the option of taking a detour towards a set of *points of interest* when travelling from a source point to a destination point. In fact, choosing detours is an essential part of daily life and modern computing systems.

The full version of this paper is available at https://arxiv.org/abs/2507.08685 [8].

F. Foucaud and A. Parreau (Eds.): IWOCA 2026, LNCS 16587, pp. 220–235, 2026.
https://doi.org/10.1007/978-3-032-27732-9_16

When travelling from point A to point B, we often need to adjust our route, perhaps to refuel at a gas station or pick up a beer. This can be modelled as follows: Given a weighted graph $G = (V, E)$ and a subset $B \subseteq V$ of beer vertices, a beer path between two vertices u and v is a path that starts from u, ends at v, and visits at least one beer vertex in B. The *beer distance* between two vertices is the shortest length of any beer path connecting them.

Since their introduction, beer paths have been extensively studied in different graph classes and frameworks [2,4,6,7,10,11]. All existing techniques, however, are applicable to *static* graphs. Because of the inherent nature of beer path applications, temporal information is, in fact, a fundamental feature to consider. For example, bus routes may operate only at specific times, while beer shops may be subject to opening and closing hours. Temporal graphs [1,13,15,18–20] extend the concept of static graphs to include such temporal information, with each edge being associated with a starting time and a travelling time. In this work we introduce and study beer path problems in temporal graphs.

1.1 Related Work

Temporal graphs have been the focus of research in several fields due to their ability to model a wide range of real-world scenarios. Comprehensive overviews on temporal graphs are provided in [12,27]. The concepts of earliest-arrival, latest-departure, fastest, and shortest temporal paths were introduced in [28] and further studied in [29]. Beyond regular pathfinding, temporal walks have been studied in [3,9,21,23,30].

The problem of computing *shortest beer paths* in static graphs was recently introduced by Bacic et al. [2]. Since the problem is polynomially solvable, researchers have focused on specific graph classes to design data structures that allow answering beer path queries faster than the baseline method. Bacic et al. developed an index for *outerplanar graphs* [2]; Das et al. [7] studied the beer path problem on *interval graphs*; Hanaka et al. [11] achieved optimal query time on *series-parallel graphs* and linear preprocessing time on graphs having *bounded-size triconnected components*; Gudmundsson and Sha [10] focused on bounded treewidth graphs. Additionally, Bilò et al. [4] address the problem of constructing graph spanners w.r.t. the *group Steiner metric*, which generalizes the beer distance metric. Note that the shortest beer path problem is a special case of the *Generalized Shortest Paths* [25] problem and the NP-hard *Generalized Travelling Salesperson Problem* [16,24,26]. To the best of our knowledge, this is the first study of beer path problems in temporal graphs.

1.2 Our Contribution

In this paper, we introduce the concept of beer paths in (directed) temporal graphs and several related optimization problems, namely *earliest-arrival, latest-departure, fastest,* and *shortest temporal beer paths* (EABP, LDBP, FBP, SBP). In order to further incorporate temporal information that accurately reflects real-life applications, our framework also considers timestamped beer vertices, i.e., the case where each beer vertex is augmented with the time instants in which it is active as a point of interest. Note that our algorithms can also work for active time intervals instead of instants with minor modifications (by using binary searches inside time intervals instead of lists of time instants). Most of our algorithms receive a sorted *edge stream* as input, as is standard for temporal graphs [17,22,28]. However, we also consider algorithms with random access to the edges of the graph (through e.g. an adjacency list with temporal logs [5]) when they can offer a better time complexity.

In Sect. 3, we propose efficient algorithms for the aforementioned objectives, with each one differing both in methods and in complexity (see Table 1 for a synopsis). As intermediate steps for computing EABPs and LDBPs, we define *multiple-source* earliest-arrival paths and *multiple-target* latest-departure paths and propose algorithms for computing them, which can be of independent interest. For FBP and SBP we rely on the computation of certain non-dominated paths via various domination criteria. For SBP in particular, we define a new path domination criterion and propose an algorithm for computing respective non-dominated paths. The time complexity of our algorithms generally aligns with that of respective standard temporal pathfinding algorithms [28]. All our algorithms except the one for FBP preserve the original graph structure, which is beneficial because adding new edges to a temporal graph may be a costly operation.

Table 1. Overview of the algorithms presented in Sect. 3. Some complexity terms that are negligible under reasonable assumptions are omitted from this table for simplicity. See Sect. 2, as well as the full version of this paper [8], for a detailed explanation of all variable symbols used here and throughout the paper.

Objective	Source/Target	Input Structure	Time Complexity
EABP	One-to-all	Edge Stream	$\mathcal{O}(n + M)$
	One-to-all	Adjacency List	$\mathcal{O}(m \log \pi + n \log n)$
LDBP	All-to-one	Edge Stream (reverse)	$\mathcal{O}(n + M)$
	All-to-one	Adjacency List	$\mathcal{O}(m \log \pi + n \log n)$
FBP	One-to-all	Edge Stream	$\mathcal{O}(n + (M + kc) \log d_{in})$
SBP	One-to-one	Edge Stream (regular and reverse)	$\mathcal{O}(n + M \log d_{max} + k d_{in} \log d_{out})$

In Sect. 4, we propose preprocessing methods for answering beer path queries faster. For EABP and LDBP, this is achieved through the precomputation of selected non-dominated paths in the graph. For FBP and SBP, we utilize the transformation proposed by [28] to convert a temporal graph to a static graph. We remark that the query time achieved for SBP is not faster than the respective algorithm of Sect. 3, however, we obtain a one-to-all algorithm (instead of one-to-one). See Table 2 for a synopsis of these results.

Table 2. Overview of our preprocessing methods in Sect. 4.

Objective	Method	Structure Time	Space	Source/Targ.	Query Time
EABP	Path precomputation	$\mathcal{O}(n^2 + nM \log c)$	$\mathcal{O}(nkc)$	One-to-one	$\mathcal{O}(k \log c)$
LDBP	Path precomputation	$\mathcal{O}(n^2 + nM \log c)$	$\mathcal{O}(nkc)$	One-to-one	$\mathcal{O}(k \log c)$
FBP	Conversion to static	$\mathcal{O}(M \log d_{max})$	$\mathcal{O}(M)$	One-to-all	$\mathcal{O}(M)$
SBP	Conversion to static	$\mathcal{O}(M \log d_{max})$	$\mathcal{O}(M)$	One-to-all	$\mathcal{O}(M \log M)$

2   Preliminaries and Problem Definitions

2.1   Preliminaries

Let $G = (V, E)$ be a directed temporal graph, with $|V| = n$ and $|E| = M$. Each temporal edge $e \in E$ is defined as a quadruple (u, v, t, λ), with $u, v \in V$, where t denotes the *starting time*, λ denotes the *traversal time* needed to travel from u to v starting at time t, and $t + \lambda$ represents the *ending time*. Most of our algorithms receive as input an *edge stream*, containing all edges sorted by ascending t-value. This is standard for temporal graphs [12,17,22,28].

The set of temporal edges from u to v $(u, v \in V)$ is denoted by $E(u, v)$. Let $\pi = \max\limits_{u,v \in V} \{|E(u, v)|\}$ (maximum number of parallel edges) and $m = |\{u, v \in V : |E(u, v)| > 0\}|$ (number of connected pairs of vertices). We denote with $N_{out}(u) = \{v : (u, v, t, \lambda) \in E\}$ the set of out-neighbours of $u \in V$, and with d_{out} the maximum out-degree in G. We similarly define $N_{in}(u)$ and d_{in}. Finally, d_{max} denotes the maximum between d_{out} and d_{in}.

A *temporal path* P is defined as a sequence of edges $e_i = (v_i, v_{i+1}, t_i, \lambda_i) \in E$, with $(t_i + \lambda_i) \leq t_{i+1}$, for $1 \leq i \leq \ell$, where ℓ is the number of edges in the path. Throughout the paper we may informally denote a path by the sequence of vertices it traverses (instead of its edges), with the notation $P = \langle v_1, v_2, \ldots, v_{\ell+1} \rangle$. The quantity $(t_\ell + \lambda_\ell)$ is known as the *ending time* of P, denoted by end(P), while t_1 is the *starting time* of P, start(P). The *duration* of P is defined as dura$(P) = $ end$(P) - $ start(P). Finally, the sum over the traversal times of the edges of P is known as the *distance* of the path, denoted as dist(P).

Definition 1 (Dominated edge). *We say that a temporal edge $e = (u, v, t, \lambda)$ is* dominated *if there exists another edge $e' = (u, v, t', \lambda')$ such that $t' \geq t$ and $t' + \lambda' \leq t + \lambda$, with at least one of the two inequalities being strict.*

Definition 2 (Dominated paths). *Let P be a temporal u–v path. We say that P is* dominated *if there exists another temporal u–v path P' such that start$(P') \geq $ start(P) and end$(P') \leq $ end(P), with at least one of the inequalities being strict. We say that P is* distance-wise dominated *if there exists a temporal u–v path P' such that dist$(P') \leq $ dist(P) and end$(P') \leq $ end(P), with at least one of the inequalities being strict.*

In some of our algorithms we will use subroutines that compute all non-dominated paths (excluding ties) from one node to all others within a given time interval. This is useful for finding fastest or shortest paths (depending on which domination criterion is used), as well as for precomputing certain temporal paths to answer queries faster. These subroutines (Algorithms 1 and 2) stem from the fastest and shortest path algorithms of [28]. We refer the reader to the full version of this paper [8] for a detailed description of these algorithms.

Algorithm 1 `Non-dom_paths`$(G, x, [t_\alpha, t_\omega])$

Input: A temporal graph $G = (V, E)$ in edge stream representation, source vertex x, time interval $[t_\alpha, t_\omega]$.
Output: For each $v \in V$, the list L_v of non-dominated (s, a) pairs (from x to v) within $[t_\alpha, t_\omega]$.

Algorithm 2 `Dist_non-dom_paths`$(G, x, [t_\alpha, t_\omega])$

Input: A temporal graph $G = (V, E)$ in edge stream representation, source vertex x, time interval $[t_\alpha, t_\omega]$.
Output: For each $v \in V$, the list S_v of distance-wise non-dominated (d, a) pairs (from x to v) within $[t_\alpha, t_\omega]$.

Algorithms 1 and 2 run in time $\mathcal{O}(n + M \log c)$ and $\mathcal{O}(n + M \log d_{in})$, respectively, where $c = \min\{d_{in}, d_{out}\}$. See [8] for more details.

2.2 The Temporal Beer Path Problem

We assume a (directed) temporal graph with a set $B = \{b_1, \ldots, b_k\} \subseteq V$ of k *beer vertices*. For each beer vertex $b \in B$, we are given a set T_b of timestamps denoting the time instants in which b is *active*. We assume that beer vertices can still be traversed in time units in which they are inactive, functioning as regular vertices. We define $T = \max_{b \in B}\{|T_b|\}$.

Given a temporal graph G, vertices $x, y \in V$ and a time interval $[t_\alpha, t_\omega]$, we denote the set of temporal beer paths between two vertices x and y starting at or after t_α and ending no later than t_ω as $P_B(x, y, [t_\alpha, t_\omega]) = \{P = \langle x = v_1, v_2, \ldots, v_\ell = y \rangle : P$ is a temporal path from x to y such that $start(P) \geq t_\alpha, \ end(P) \leq t_\omega, \ \exists \ (i \in [\ell], t \in T_{v_i}) : (v_i \in B, \ t \in [t_{i-1} + \lambda_{i-1}, t_i])\}$. Note that a beer path must traverse at least one beer vertex at a time in which it is active (e.g. the x–b_5–y path in Fig. 1 is not a valid temporal beer path).

Definition 3 (Objectives). *A temporal x–y path $P \in P_B(x, y, [t_\alpha, t_\omega])$ is:*

- *An* earliest-arrival beer path *(**EABP**) if $end(P) = \min\{end(P') : P' \in P_B(x, y, [t_\alpha, t_\omega])\}$.*

- A latest-departure beer path *(**LDBP**) if* $start(P) = \max\{start(P') : P' \in P_B(x, y, [t_\alpha, t_\omega])\}$.
- A fastest beer path *(**FBP**) if* $dura(P) = \min\{dura(P') : P' \in P_B(x, y, [t_\alpha, t_\omega])\}$.
- A shortest beer path *(**SBP**) if* $dist(P) = \min\{dist(P') : P' \in P_B(x, y, [t_\alpha, t_\omega])\}$.

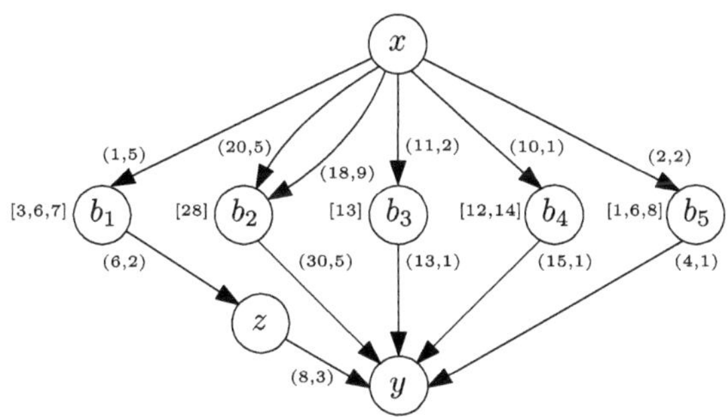

Fig. 1. Example in which the x–y EABP/LDBP/FBP/SBP use $b_1, \ldots, b_4$ respectively. Numbers in parentheses denote (t, λ) of edges and numbers in brackets denote beer active times. The x–y path that uses b_5 is not a valid temporal beer path. The edge with $(t, \lambda) = (18, 9)$ is dominated and is irrelevant for all objectives.

For examples for each objective, see Fig. 1. In all our algorithms, we focus on the computation of the objective function (e.g., the earliest arrival time) instead of reporting the corresponding optimal path. This path can be easily recovered by storing parent information. The proofs of all statements in this paper can be found in the respective full version [8].

3 Algorithms for Temporal Beer Path Problems

3.1 Algorithms for Earliest-Arrival Beer Paths

In this subsection we propose an EABP algorithm (from source x to all nodes), based on the earliest-arrival time algorithm of [28] with a modification for computing *multiple-source* earliest-arrival paths (MSEAP - Definition 4). We also present an alternative algorithm for graphs containing no dominated edges (a reasonable assumption, since dominated edges are irrelevant for all commonly considered objectives). The latter is similar in spirit to Dijkstra's algorithm and uses binary search for parallel edges, achieving a running time independent of M. Although faster than the former, it requires random access to the edges of the graph through an adjacency list [5], instead of scanning an edge stream in a specific order. We remark that our algorithms do not modify the graph (in contrast to, e.g., using dummy vertices), which is important because inserting new edges in temporal graphs may be a costly operation.

An Earliest-Arrival Beer Path Algorithm in Edge Stream Representation. We define Multiple-Source Earliest-Arrival Paths (MSEAPs), which will prove useful as an intermediate step for computing EABPs. Intuitively, an MSEAP is a path that can reach a target node in the earliest possible time unit, starting from any node of the graph, with restrictions for the starting time of each node. If some $init[v]$ is ∞, we cannot start a path from v.

Definition 4 (MSEAP). *Given a temporal graph $G = (V, E)$ and initial times $init[v] \in \mathbb{R}^+ \cup \{\infty\}$, $v \in V$, we say that a temporal path x–y starting from x no earlier than $init[x]$ is a* multiple-source earliest-arrival path *to y if its arrival time at y is minimum among all temporal paths v–y, $v \in V$, starting from v no earlier than $init[v]$. We call this arrival time the* multiple-source earliest-arrival path time *of y.*

We first show that a generalization of the earliest-arrival path algorithm of [28] can compute MSEAPs.

Algorithm 3 $\mathtt{MSEAP}(G, [t_\alpha, t_\omega], \mathtt{init})$

Input: A temporal graph G in its edge stream representation, a time window $[t_\alpha, t_\omega]$ and initial times $init[v]$, such that $t_\alpha \leq init[v] \leq t_\omega$ or $init[v] = \infty$, for all $v \in V$.
Output: The MSEAP times within $[t_\alpha, t_\omega]$ to every vertex, assuming we can start at time $init[v]$ from each $v \in V$.
1: $\tau[v] \leftarrow init[v]$ **for all** $v \in V$
2: **for** $e = (u, v, t, \lambda)$ in the edge stream **do**
3:     **if** $\tau[u] \leq t$ **and** $t + \lambda \leq t_\omega$ **then**
4:         $\tau[v] \leftarrow \min\{\tau[v], t + \lambda\}$
5: **return** $\tau[v]$ for each $v \in V$

Algorithm 3 scans each temporal edge once and thus runs in time $\mathcal{O}(n + M)$. Note that with initialization $init[x] = t_\alpha$ for some $x \in V$ and $init[v] = \infty$ for $v \in V \setminus \{x\}$, Algorithm 3 coincides with the earliest-arrival time algorithm of [28] with source node x and computes single-source earliest arrival paths.

Lemma 1. *For $G = (V, E)$ and $init[v] \in \mathbb{R}^+ \cup \{\infty\}$, $v \in V$, it holds that for every $y \in V$ with a finite MSEAP time there exists some MSEAP $\langle v_0, v_1, \ldots, v_p = y \rangle$ to y for which every prefix-subpath $\langle v_0, v_1, \ldots, v_i \rangle$ is an MSEAP to v_i.*

Lemma 2. *Algorithm 3 computes the MSEAP times to each vertex, within time window $[t_\alpha, t_\omega]$. If time ∞ is returned for some vertex, then that vertex is unreachable from all $v \in V$ (starting no earlier than $init[v]$) within $[t_\alpha, t_\omega]$.*

We now present our EABP algorithm, which calls Algorithm 3 twice, computing earliest-arrival path times in the first phase and EABP times in the second phase.

Algorithm 4 EABP$(G, x, [t_\alpha, t_\omega])$

Input: A temporal graph G, source vertex x, time window $[t_\alpha, t_\omega]$, a set of beer vertices B and a list T_i of time units for each of them.
Output: The EABP time from x to every vertex $v \in V$ within $[t_\alpha, t_\omega]$.
1: $\text{init}[x] \leftarrow t_\alpha$, $\text{init}[v] \leftarrow \infty$ **for all** $v \in V \setminus \{x\}$
2: $\tau[v] \leftarrow \text{MSEAP}(G, [t_\alpha, t_\omega], \text{init})$, for $v \in V$
3: $\text{init}'[v] \leftarrow \infty$ **for all** $v \in V$
4: **for all** $b \in B$ **do**
5:      $beer\_time \leftarrow \min\{time \in T_b \mid time \geq \tau[b]\}$ $\triangleright \infty$ if it does not exist
6:      **if** $beer\_time \leq t_\omega$ **then**
7:         $\text{init}'[b] \leftarrow beer\_time$
8: $\tau'[v] \leftarrow \text{MSEAP}(G, [t_\alpha, t_\omega], \text{init}')$, for $v \in V$
9: **return** $\tau'[v]$ for each $v \in V$

Theorem 1. *Algorithm 4 computes the EABP times from x to all $v \in V$ in time $\mathcal{O}(n + M + k \log T)$.*

A Faster Earliest-Arrival Beer Path Algorithm with Adjacency List Representation. We present an alternative algorithm to be used as the MSEAP subroutine of Algorithm 4, in the place of Algorithm 3. Although faster for graphs with no dominated edges, this algorithm requires an adjacency list instead of an edge stream, which may be costly to produce (see e.g., [5]). The algorithm is similar in spirit to Dijkstra's algorithm, using a min-priority queue for the vertices and their arrival times, while utilizing binary search on parallel edges to achieve a speedup. Note that the input graph of Algorithm 4 can be either in edge stream or in adjacency list representation, depending on which MSEAP subroutine is used.

Algorithm 5 MSEAP_alternative$(G, [t_\alpha, t_\omega], \text{init})$

Input: A temporal graph G in adjacency list representation, a time window $[t_\alpha, t_\omega]$ and initial times $\text{init}[v]$, such that $t_\alpha \leq \text{init}[v] \leq t_\omega$ or $\text{init}[v] = \infty$, for all $v \in V$.
Output: The MSEAP times within $[t_\alpha, t_\omega]$ to every vertex, assuming we can start at time $\text{init}[v]$ from each $v \in V$.
1: $\tau[v] \leftarrow \text{init}[v]$ **for all** $v \in V$
2: $PQ \leftarrow \emptyset$ $\triangleright$ a min-priority queue
3: $PQ.enqueue(\tau[v], v)$ **for all** $v \in V$ $\triangleright$ Add v to priority queue with $\tau[v]$ priority
4: **while** $PQ \neq \emptyset$ **do**
5:      $(\tau[u], u) \leftarrow PQ.extractMin()$
6:      **for** $v \in N_{out}(u)$ **do**
7:         Let $(u, v, t, \lambda) \in E(u, v)$ be a temporal edge with minimum t s.t. $t \geq \tau[u]$
8:         **if** $t + \lambda \leq t_\omega$ **and** $t + \lambda < \tau[v]$ **then**
9:            $\tau[v] \leftarrow t + \lambda$
10:        $PQ.decreaseKey(v, \tau[v])$
11: **return** $\tau[v]$ for each $v \in V$

Theorem 2. *For graphs with no dominated edges, Algorithm 4 computes the EABP times from x to all $v \in V$ in time $\mathcal{O}(m \log \pi + n \log n + k \log T)$, if Algorithm 5 is used as its MSEAP subroutine.*

The $n \log n$ term in the complexity stems from the usage of a Fibonacci heap as min-priority queue. The $m \log \pi$ term stems from binary search on parallel edges, which is possible by the assumption that G contains no dominated edges. Note that M is bounded by $m\pi$ and may be arbitrarily larger than m, which renders $\mathcal{O}(m \log \pi)$ a significant improvement over $\mathcal{O}(M)$.

3.2　Algorithms for Latest-Departure Beer Paths

We propose two LDBP algorithms (from all nodes to some node y), based on our earliest-arrival algorithms with modifications for *multiple-target* latest departure paths (MTLDP). The main idea is to reverse the direction of time, starting from y in t_ω and traversing temporal edges in reverse, while storing the latest time in which we can depart from each node to reach y. In these algorithms, node times are initialized to $-\infty$ and the edge stream is scanned in reverse. Due to their similarity to the algorithms of Sect. 3.1, we omit these algorithms and we refer the reader to the full version of this paper [8] for their analysis.

3.3　An Algorithm for Fastest Beer Paths

A significant difference for FBP compared to EABP/LDBP is that no property similar to Lemma 1 holds, meaning that prefix-subpaths of an FBP are not necessarily fastest paths. Hence, computing a fastest path from the source to a beer vertex is not useful as an intermediate step for computing FBPs. The following lemma is the key to obtaining an FBP algorithm with time complexity comparable to that of our EABP/LDBP algorithms in the previous subsections (excluding some log factors), despite the aforementioned complication.

Lemma 3. *If there is a temporal x–y beer path within time interval $[t_\alpha, t_\omega]$, then there is an x–y FBP $\langle v_0 = x, v_1, \ldots, v_j = b, \ldots, v_p = y \rangle$ within $[t_\alpha, t_\omega]$, traversing $b \in B$ at a time $t \in T_b$, such that its subpaths $\langle v_0 = x, v_1, \ldots, v_j = b \rangle$ and $\langle v_j = b, v_{j+1}, \ldots, v_p = y \rangle$ are non-dominated paths.*

Based on Lemma 3, we propose a two-phase FBP algorithm (from source x to all nodes), utilizing Algorithm 1 to compute non-dominated paths. In the first phase, all non-dominated paths (excluding ties) from x to each $b \in B$ are computed. Similar to Sect. 3.1, we find the minimum eligible active time for the respective beer vertex for each of these paths. For the second phase, a dummy vertex is used as source, connected with each $b \in B$ by (multiple) dummy edges with starting and traversal times dictated by the values previously computed for each path. Algorithm 1 is then used again for the new graph, computing eligible non-dominated paths from the dummy vertex to all vertices. The pseudocode of the described algorithm can be found in the full version of this paper [8].

Recall that Algorithm 1 runs in $\mathcal{O}(n + M \log c)$ time [28], where $c = \min\{d_{in}, d_{out}\}$. Hence, our FBP algorithm runs in time $\mathcal{O}(n + (M + kc) \log d_{in} + kc \log T)$, since $\mathcal{O}(kc)$ temporal edges are introduced for the second phase and d_{in} becomes (at most) double its initial value. The term $kc \log T$ stems from binary searches over beer times. The correctness of our algorithm follows directly from Lemma 3.

3.4   An Algorithm for Shortest Beer Paths

SBPs are significantly more complicated, in part because, somewhat unexpectedly, Lemma 3 does not hold verbatim for *distance-wise* non-dominated paths. To overcome this, we define the following alternative domination criterion.

Definition 5 (Inverse-distance-wise dominated path). *Let P be a temporal u–v path. We say that P is* inverse-distance-wise dominated *if there exists another temporal u–v path P' such that $dist(P') \leq dist(P)$ and $start(P') \geq start(P)$, with at least one of the inequalities being strict.*

This domination criterion allows us to prove the following lemma, which is analogous to Lemma 3 and serves as the backbone of our SBP algorithm.

Lemma 4. *If there is a temporal x–y beer path within time interval $[t_\alpha, t_\omega]$, then there is a x–y SBP $\langle v_0 = x, v_1, \ldots, v_j = b, \ldots, v_p = y \rangle$ within $[t_\alpha, t_\omega]$, traversing $b \in B$ at a time $t \in T_b$, such that $\langle v_0 = x, v_1, \ldots, v_j = b \rangle$ is a distance-wise non-dominated path (within $[t_\alpha, t_\omega]$) and $\langle v_j = b, v_{j+1}, \ldots, v_p = y \rangle$ is an inverse-distance-wise non-dominated path (within $[t_\alpha, t_\omega]$).*

We propose a modification of Algorithm 2 to compute inverse-distance-wise non-dominated paths from all vertices to some vertex y, by scanning the edge stream in reverse. Due to space constraints, the full description of this algorithm is omitted; see the full version of this paper [8] for details. Algorithm 6 runs in time $\mathcal{O}(n + M \log d_{out})$.

Algorithm 6 `Inv_dist_non-dom_paths`$(G, y, [t_\alpha, t_\omega])$

Input: A temporal graph $G = (V, E)$ in reverse edge stream representation, target vertex y, time interval $[t_\alpha, t_\omega]$.
Output: For each $v \in V$, the list I_v of inverse-distance-wise non-dominated (d, s) pairs (from v to y) within $[t_\alpha, t_\omega]$.

Theorem 3. *Algorithm 6 computes all inverse-distance-wise non-dominated (d, s) pairs from each $v \in V$ to y within $[t_\alpha, t_\omega]$, in time $\mathcal{O}(n + M \log d_{out})$.*

We propose an SBP algorithm, utilizing Algorithms 2 and 6 to compute (inverse) distance-wise non-dominated paths. The pseudocode of this algorithm can be found in the full version of this paper [8]. Unlike most of our algorithms, this SBP algorithm only computes an SBP between two given vertices and it requires both the regular and the reverse edge streams as input; this is necessary because it uses two different domination criteria (see Lemma 4), with each one imposing different restrictions via its respective algorithm (i.e., Algorithm 2 is one-to-all and Algorithm 6 is all-to-one). Our SBP algorithm calls Alg. 2 and 6 once each. The size of each L_v is bounded by d_{in} and the size of each I_v by d_{out}. Hence, it runs in time $\mathcal{O}(n + M \log(d_{max}) + kd_{in}(\log T + \log d_{out}))$. Its correctness follows from Lemma 4 and Theorem 3.

4   Temporal Beer Path Queries with Preprocessing

In this section we present preprocessing techniques that help improve the asymptotic complexity of beer path queries. This is useful for applications where multiple queries need to be answered, each with different source and target vertices, different time intervals and even different active beer shops. A natural application would be to quickly recompute a beer path assuming some beer shops just closed (or some new ones opened), without computing everything from scratch.

We assume that each query contains as input a source node x, a target node y, a time interval $[t_\alpha, t_\omega]$ and a Boolean array $A = [a_1, \ldots, a_k]$, with $a_i = \texttt{true}$ iff beer vertex b_i is *active*. For simplicity, we assume that each beer vertex $b \in B$ is either active or inactive for the entire duration of a query; note that our methods can be easily extended to the case where active beer times are given as input in each query, by incorporating binary searches in lists T_b, as in Sect. 3. We also assume that the set B containing the k beer vertices is known during preprocessing; these vertices may be active or inactive for different queries, but no vertex $v \notin B$ may become a beer vertex.

4.1   EABP and LDBP with Precomputation of Non-Dominated Paths

We rely on the observation that Lemma 3 also holds for EABPs and LDBPs, with the same proof (although not for SBPs). This allows us to run Algorithm 1 to precompute non-dominated paths to and from beer vertices. To compute all non-dominated paths to each beer vertex (starting from any vertex) and all non-dominated paths from each beer vertex (to any vertex), it suffices to run Algorithm 1 $\mathcal{O}(n)$ times. Recall that the amount of non-dominated paths (excl. ties) between any two vertices is bounded by $c = \min\{d_{in}, d_{out}\}$. We infer that this preprocessing requires $\mathcal{O}(n^2 + nM \log c)$ time and $\mathcal{O}(nkc)$ space.

With this preprocessing, any EABP query can be answered as follows. We need a binary search for each $b_i \in B$ with $a_i = \texttt{true}$ to find the non-dominated x–b_i path that starts as early as possible (but no earlier than t_α). Then, for each $b_i \in B$ with $a_i = \texttt{true}$ we need a binary search to find the non-dominated b_i–y

path that starts as early as possible (but no earlier than the arrival time of the respective x–b_i path). Concatenating these paths with the previous ones yields k candidate EABPs, out of which we pick the best one. This process runs in $\mathcal{O}(k \log c)$ time. Identical arguments hold for LDBP queries.

4.2   Preprocessing Through Graph Transformation

In this subsection we adopt the graph transformation approach proposed in [28] as preprocessing. This technique converts a temporal graph $G = (V, E)$ into a static, weighted, directed acyclic graph $\tilde{G} = (\tilde{V}, \tilde{E})$ that explicitly encodes temporal constraints. The key idea is to model the timing of events in the temporal graph by creating time-stamped copies of each node and connecting them with appropriately weighted edges. The transformed graph $\tilde{G}$ contains $\mathcal{O}(M)$ vertices and edges and can be produced from G in $\mathcal{O}(M \log d_{max})$ time. For each $v \in V$, the transformation creates two sets of time-stamped nodes: $\tilde{V}_{in}(v)$, which includes a node (v, t) for every unique time t at which an edge arrives at v, and $\tilde{V}_{out}(v)$, which includes a node (v, t) for every time t at which an edge departs from v. These sets are sorted in descending time order. Figure 2(a),(b) show a temporal graph G and its transformed graph $\tilde{G}$, respectively. We refer the reader to the full version of this paper [8] for more details regarding this transformation.

Fastest Beer Paths on the Transformed Graph. Let $S = \{(x, t) : (x, t) \in \tilde{V}_{out}(x), t_\alpha \leq t \leq t_\omega\}$ be the set of time-stamped copies of x, sorted in descending order of t. In order to compute fastest temporal paths from x to every $v \in V$, it suffices to start a BFS from each $(x, t) \in S$ (in descending order of t), skipping nodes already explored in previous BFSs to avoid redundancy, since reaching (v, t) from (x, t_1) results in smaller duration compared to reaching it from (x, t_2) with $t_2 < t_1$. The duration of each path is computed as the difference between the timestamp of the target and the timestamp of the source [28]. We will modify this technique to compute FBPs on $\tilde{G}$.

For each vertex $(v, t) \in \tilde{G}$, we keep track of a 3-valued flag, indicating whether (v, t) has been visited by a beer path (value 2) or a regular path (value 1) or not at all (value 0), with higher values superseding smaller ones. We run BFSs from each $(x, t) \in S$ in descending order of t (as described above), setting the flag of (x, t) to 1. We only visit a vertex if its flag is smaller than the parent's flag; in this case, the child obtains the parent's flag. If the child is a copy of some $b_i \in B$ with $a_i = \texttt{true}$, we always update its flag to 2. When the flag of a vertex becomes 2, we calculate the duration of the respective beer path as the difference between the timestamp of that vertex and the timestamp of the source.

The above answers FBP queries on $\tilde{G}$ in $\mathcal{O}(M)$ time, since each $v \in \tilde{G}$ is visited at most twice. The correctness of the algorithm follows from the fact that $\tilde{G}$ fully encodes all possible temporal walks in G, combined with arguments similar to the ones for regular fastest paths.

We remark that EABP and LDBP queries can also be answered in $\tilde{G}$ in $\mathcal{O}(M)$ time with similar techniques. However, this is not particularly useful, since the algorithms proposed in Sects. 3.1 and 3.2 require no preprocessing and are just as fast (or even faster) under reasonable assumptions.

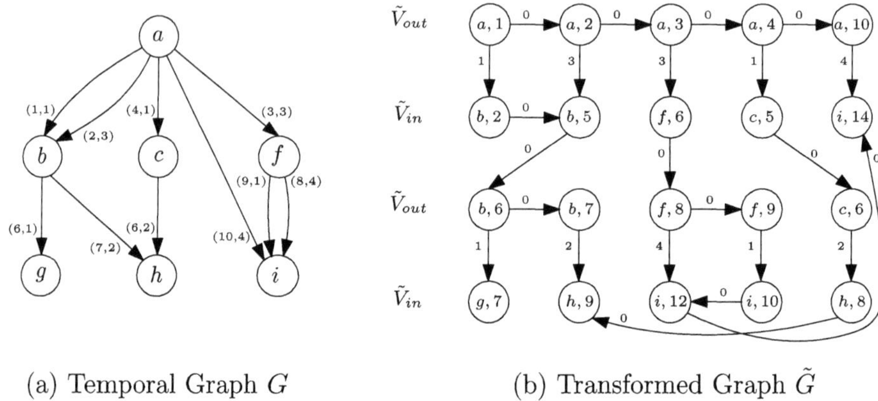

(a) Temporal Graph G (b) Transformed Graph $\tilde{G}$

Fig. 2. Graph transformation, from G in (a) to $\tilde{G}$ in (b).

Shortest Beer Paths on the Transformed Graph. To compute shortest temporal paths from $x \in V$, it suffices to run Dijkstra's algorithm in $\tilde{G}$ from a dummy vertex x', connected via directed edges of weight 0 to every $(x, t) \in \tilde{V}_{out}(x)$ s.t. $t_\alpha \leq t \leq t_\omega$, thus representing all valid departure times from x [28].

We modify Dijkstra's algorithm by storing a 3-valued flag for each $(v, t) \in \tilde{G}$ and updating it in the same manner described for FBP. This algorithm visits each $(v, t) \in \tilde{G}$ at most twice, thus calculating SBP distances in $\mathcal{O}(M \log M)$ time. Although this SBP algorithm is not faster than that of Sect. 3.4 under reasonable assumptions, it computes SBPs from a source to *all* other nodes.

5   Open Questions and Future Work

Our work opens several directions for further research, both on the algorithmic side and on the design of data structures for temporal beer paths. We briefly discuss some of them below.

FBP and SBP Under Adjacency List Representation. In this work we showed that it is possible to achieve a speedup for EABP and LDBP if an adjacency list is given as input (compared to an edge stream), assuming non-existence of dominated edges. A natural open question is whether the same is possible for FBP and SBP.

One-to-All SBP. In this work we have shown that SBP admits a *one-to-one* algorithm running in time (roughly) $\mathcal{O}(M \log d_{max})$, with an edge stream as input. We have also presented a *one-to-all* SBP algorithm that utilizes the graph transformation of [28], running in time $\mathcal{O}(M \log M)$, which is slower than the aforementioned one under reasonable assumptions. We leave as an open question whether a one-to-all SBP algorithm can be achieved with running time comparable to $\mathcal{O}(M \log d_{max})$.

Multiple Types of Points of Interest. In many applications there may be several classes of different points of interest (e.g., gas stations, charging points and rest areas), possibly with different constraints or perhaps even priorities. Extending our framework to handle such cases appears to be non-trivial, particularly for FBP and SBP.

Interval Temporal Graphs. An appealing generalization is that of interval temporal graphs, where edge availabilities (or beer vertex active times) are described by continuous time intervals instead of discrete instants. Adapting our algorithms to interval representations similar to [14] seems possible with minor modifications (such as binary searches inside each interval).

Data Structures for Special-Case Sublinear Beer Queries. In the static setting, several works propose data structures that achieve sublinear time queries for shortest beer path queries on restricted graph classes, such as outerplanar graphs [2]. It is natural to ask for analogous temporal graph classes (e.g. sparse temporal graphs) for which such a data structure would be possible.

Acknowledgements. Andrea D'Ascenzo and Giuseppe Italiano have been funded by the Italian Ministry of University and Reseach under PRIN Project n. 2022TS4Y3N - EXPAND: scalable algorithms for EXPloratory Analyses of heterogeneous and dynamic Networked Data. Sotiris Kanellopoulos, Aris Pagourtzis and Christos Pergaminelis have been supported by project MIS 5154714 of the National Recovery and Resilience Plan Greece 2.0 funded by the European Union under the NextGenerationEU Program. Anna Mpanti has been funded by the European Union - Next Generation EU, Mission 4 Component 2 CUP: I83C22000990001.

References

1. Akrida, E.C., Czyzowicz, J., Gasieniec, L., Kuszner, L., Spirakis, P.G.: Temporal flows in temporal networks. J. Comput. Syst. Sci. **103**, 46–60 (2019). https://doi.org/10.1016/J.JCSS.2019.02.003
2. Bacic, J., Mehrabi, S., Smid, M.: Shortest beer path queries in outerplanar graphs. Algorithmica **85**(6), 1679–1705 (2022). https://doi.org/10.1007/s00453-022-01045-4

3. Bentert, M., Himmel, A., Nichterlein, A., Niedermeier, R.: Efficient computation of optimal temporal walks under waiting-time constraints. Appl. Netw. Sci. **5**(1), 73 (2020)
4. Bilò, D., Gualà, L., Leucci, S., Straziota, A.: Graph spanners for group steiner distances. In: ESA 2024. LIPIcs, vol. 308, pp. 25:1–25:17. Schloss Dagstuhl - Leibniz-Zentrum für Informatik (2024). https://doi.org/10.4230/LIPICS.ESA.2024.25
5. Caro, D., Rodríguez, M.A., Brisaboa, N.R.: Data structures for temporal graphs based on compact sequence representations. Inf. Syst. **51**, 1–26 (2015). https://doi.org/10.1016/J.IS.2015.02.002
6. Coudert, D., D'Ascenzo, A., D'Emidio, M.: Indexing graphs for shortest beer path queries. In: ATMOS 2024. OASIcs, vol. 123, pp. 2:1–2:18. Schloss Dagstuhl - Leibniz-Zentrum für Informatik (2024). https://doi.org/10.4230/OASICS.ATMOS.2024.2
7. Das, R., He, M., Kondratovsky, E., Munro, J.I., Naredla, A.M., Wu, K.: Shortest beer path queries in interval graphs. In: ISAAC, 2022. LIPIcs, vol. 248, pp. 59:1–59:17. Schloss Dagstuhl - Leibniz-Zentrum für Informatik (2022). https://doi.org/10.4230/LIPICS.ISAAC.2022.59
8. D'Ascenzo, A., Italiano, G.F., Kanellopoulos, S., Mpanti, A., Pagourtzis, A., Pergaminelis, C.: Beer path problems in temporal graphs. arXiv:2507.08685 (2025)
9. Erlebach, T., Hoffmann, M., Kammer, F.: On temporal graph exploration. J. Comput. Syst. Sci. **119**, 1–18 (2021). https://doi.org/10.1016/J.JCSS.2021.01.005
10. Gudmundsson, J., Sha, Y.: Shortest beer path queries in digraphs with bounded treewidth. In: ISAAC, 2023. LIPIcs, vol. 283, pp. 35:1–35:17. Schloss Dagstuhl - Leibniz-Zentrum für Informatik (2023). https://doi.org/10.4230/LIPICS.ISAAC.2023.35
11. Hanaka, T., Ono, H., Sadakane, K., Sugiyama, K.: Shortest beer path queries based on graph decomposition. In: ISAAC, 2023. LIPIcs, vol. 283, pp. 37:1–37:20. Schloss Dagstuhl - Leibniz-Zentrum für Informatik (2023). https://doi.org/10.4230/LIPICS.ISAAC.2023.37
12. Holme, P.: Modern temporal network theory: a colloquium. Eur. Phys. J. B **88**, 1–30 (2015)
13. Hosseinzadeh, M.M., Cannataro, M., Guzzi, P.H., Dondi, R.: Temporal networks in biology and medicine: a survey on models, algorithms, and tools. Netw. Model. Anal. Health Inf. Bioinf. **12**(1), 10 (2022)
14. Jain, A., Sahni, S.: Polynomial time algorithm for shortest paths in interval temporal graphs. Algorithms **17**(10), 468 (2024). https://doi.org/10.3390/A17100468
15. Kostakos, V.: Temporal graphs. XXPhys. A **388**(6), 1007–1023 (2009)
16. Laporte, G., Nobert, Y.: Generalized travelling salesman problem through n sets of nodes: an integer programming approach. INFOR Inf. Syst. Oper. Rese. **21**(1), 61–75 (1983)
17. Lee, J.B., Nguyen, G., Rossi, R.A., Ahmed, N.K., Koh, E., Kim, S.: Dynamic node embeddings from edge streams. IEEE Trans. Emerg. Top. Comput. Intell. **5**(6), 931–946 (2021). https://doi.org/10.1109/TETCI.2020.3011432
18. Mertzios, G.B., Michail, O., Spirakis, P.G.: Temporal network optimization subject to connectivity constraints. Algorithmica **81**(4), 1416–1449 (2019). https://doi.org/10.1007/S00453-018-0478-6
19. Michail, O.: An introduction to temporal graphs: an algorithmic perspective. Internet Math. **12**(4), 239–280 (2016). https://doi.org/10.1080/15427951.2016.1177801
20. Mertzios, G.B., Michail, O., Spirakis, P.G.: Temporal network optimization subject to connectivity constraints. Algorithmica **81**(4), 1416–1449 (2019). https://doi.org/10.1007/S00453-018-0478-6

21. Naima, M.: Temporal betweenness centrality on shortest walks variants. Appl. Netw. Sci. **10**(1), 11 (2025). https://doi.org/10.1007/S41109-024-00685-5
22. Oettershagen, L., Kriege, N.M., Mutzel, P.: A higher-order temporal h-index for evolving networks. In: Proceedings of the 29th ACM SIGKDD, KDD 2023, pp. 1770–1782. ACM (2023). https://doi.org/10.1145/3580305.3599242
23. Oettershagen, L., Mutzel, P., Kriege, N.M.: Temporal walk centrality: ranking nodes in evolving networks. In: WWW '22, pp. 1640–1650. ACM (2022). https://doi.org/10.1145/3485447.3512210
24. Pop, P.C., Cosma, O., Sabo, C., Sitar, C.P.: A comprehensive survey on the generalized traveling salesman problem. Eur. J. Oper. Res. **314**(3), 819–835 (2024). https://doi.org/10.1016/J.EJOR.2023.07.022
25. Rice, M.N., Tsotras, V.J.: Engineering generalized shortest path queries. In: ICDE, pp. 949–960 (2013). https://doi.org/10.1109/ICDE.2013.6544888
26. Srivastava, S., Kumar, S., Garg, R., Sen, P.: Generalized traveling salesman problem through n sets of nodes. CORS J. **7**(2), 97 (1969)
27. Wang, Y., Yuan, Y., Ma, Y., Wang, G.: Time-dependent graphs: Definitions, applications, and algorithms. Data Sci. Eng. **4**(4), 352–366 (2019). https://doi.org/10.1007/S41019-019-00105-0
28. Wu, H., Cheng, J., Huang, S., Ke, Y., Lu, Y., Xu, Y.: Path problems in temporal graphs. Proc. VLDB Endow. **7**(9), 721–732 (2014). https://doi.org/10.14778/2732939.2732945
29. Wu, H., Cheng, J., Huang, S., Ke, Y., Lu, Y., Xu, Y.: Path problems in temporal graphs. Proc. VLDB Endow. **7**(9), 721–732 (2014). https://doi.org/10.14778/2732939.2732945
30. Zhang, T., et al.: Efficient exact and approximate betweenness centrality computation for temporal graphs. In: Proceedings of the ACM Web Conference 2024, pp. 2395–2406 (2024)

On $(1, \leq l)$-Locating-Dominating Codes in Infinite Triangular Grid

Soura Sena Das[1]([✉]) [iD], Tuomo Lehtilä[2] [iD], and Sagnik Sen[3]

[1] Indian Statistical Institute, Kolkata, India
`sourasenad@gmail.com`
[2] University of Turku, Turku, Finland
`tualeh@utu.fi`
[3] Indian Institute of Technology Dharwad, Dharwad, India
`sen@iitdh.ac.in`

Abstract. Let G be a simple (possibly infinite) graph. Our focus is on finding a vertex subset $C \subseteq V(G)$, called a code, that has certain special properties, and minimize the size (or density) of C. Given a graph G, a fixed code C, and an arbitrary vertex subset F of G, the (I, r)-set of F is defined as $I_r(F) = B_r(F) \cap C$, where $B_r(F)$ is the set of vertices at distance at most r from some vertex of F. The code C is an $(r, \leq l)$-LDA code if $I_r(F_1) \neq I_r(F_2)$ whenever $F_1 \neq F_2$, $F_1 \cap C = F_2 \cap C$, and $|F_1|, |F_2| \leq l$ for all $F_1, F_2 \subseteq V(G)$. Similarly, C is an $(r, \leq l)$-LDB code if $I_r(F_1) \neq I_r(F_2)$ whenever $F_1 \neq F_2$, and $|F_1|, |F_2| \leq l$ for all $F_1, F_2 \subseteq V(G) \setminus C$. In particular, $(1, \leq 1)$-LDA and $(1, \leq 1)$-LDB codes are the same, and are popularly known as locating-dominating codes. In this article, we find the exact minimum densities of $(1, \leq l)$-LDA and $(1, \leq l)$-LDB codes in the infinite triangular grid for all $l \geq 2$.

Keywords: Locating-dominating sets · Infinite triangular grid · Density · $(r \cdot \leq l)$-LDA code · $(r \cdot \leq l)$-LDB code

1 Introduction

In this paper, we consider infinite undirected graphs. A vertex set S is *dominating* if every vertex is in S or adjacent to a vertex in S. Dominating set S is *locating-dominating* in graph G if every vertex in $V(G) \setminus S$ has a unique neighborhood in S (that is, S is a dominating set with an additional separation property). Locating-dominating codes (or sets) were introduced by Rall and Slater in 1980 s [16,21,22] and since then they and their variants have been studied in a wide number of articles, see an online bibliography containing over 500 articles on the topic in [8].

A major branch of research on the topic has been studying the optimal (smallest) density they can obtain in infinite graphs such as triangular, king, square or hexagonal grids (see for example [2,6,9]). One may find over 40 such articles considering grids in the online bibliography [8]. While the optimal density is known for many combinations of grids and variants of location-domination,

F. Foucaud and A. Parreau (Eds.): IWOCA 2026, LNCS 16587, pp. 236–250, 2026.
https://doi.org/10.1007/978-3-032-27732-9_17

some have remained open or proven hard to solve, such as, identifying codes in the hexagonal grid [2,4,18]. In this paper, we give the exact optimal cardinalities for $(1, \leq l)$-locating-dominating codes of types A and B ($(1, \leq l)$-LDA and -LDB for short) for each value of l in the infinite triangular grid. Previously only the value with $l = 1$ was known [5]. In the cases of other grids, the optimal densities for $(1, \leq l)$-LDA and -LDB codes are known for king grids [13–15].

The early motivation for studying location-domination stemmed from locating in structures described by graphs. In [21] for *safeguard analysis of critical facilities* and in [10], the authors introduced concept of identifying codes (which can be understood as a special type of locating-dominating code) for the purpose of locating malfunctioning processors in multiprocessor systems. One interesting structure for the study from this perspective, were the infinite grids (here, the infinite grid could be considered to approximate finite architectures based on the grid structures). More recently, these concepts have been connected to neighborhood complexity [1] which is useful for kernelization algorithms.

A perceived weakness of both locating-dominating and identifying codes was their inability to simultaneously locate more than a single object, for example, in the case of multiple simultaneously malfunctioning processors or in sensor networks. For this purpose, $(1, \leq l)$-identifying codes were formally introduced in [11] and briefly discussed in [10] to locate at most l objects simultaneously. Later, in [7], $(1, \leq l)$-locating-dominating codes were introduced. However, as there were two different natural definitions for these concepts, two types, type A and type B were considered. When $l = 1$, the definitions give the usual locating-dominating codes. When $l \geq \Delta(G) + 1$, where $\Delta(G)$ is the maximum degree of graph G, a $(1, \leq l)$-LDB code is equivalent with a complement of an open-irredundant set and also with a multiple intruder locating-dominating code (MILD code) [12].

Preliminaries: We follow West [26] for standard graph theoretic notation and terminology. The graphs considered in this article are undirected and do not have any loops or parallel edges, but they can be infinite.

Given a graph G and a vertex v, the ball with center v and radius r is given by

$$B_r(v) = \{u \mid d(u, v) \leq r\},$$

where $d(u, v)$ denotes the distance between u and v in G. For a vertex subset $F \subseteq V(G)$, the ball around F with radius r is given by

$$B_r(F) = \bigcup_{v \in F} B_r(v).$$

A non-empty vertex subset $C \subseteq V(G)$ is called a *code* and its vertices are called *codewords* while the vertices of $V(G) \setminus C$ are called *non-codewords*. For each vertex $u \in V(G)$, the *(I,r)-set* of u is defined as

$$I_r(u) = B_r(u) \cap C.$$

Furthermore, the *(I, r)-set* of a vertex set F is defined as

$$I_r(F) = B_r(F) \cap C.$$

A generalization of locating-dominating codes was introduced by Honkala et al. [7]. In particular, C is called an $(r, \leq l)$-*locating-dominating code of type A*, or simply $(r, \leq l)$-*LDA code* if $I_r(F_1) \neq I_r(F_2)$ whenever $F_1 \neq F_2$, $F_1 \cap C = F_2 \cap C$, and $|F_1|, |F_2| \leq l$ for all $F_1, F_2 \subseteq V(G)$. Similarly, C is called an $(r, \leq l)$-*locating-dominating code of type B*, or simply $(r, \leq l)$-*LDB code* if $I_r(F_1) \neq I_r(F_2)$ whenever $F_1 \neq F_2$, and $|F_1|, |F_2| \leq l$ for all $F_1, F_2 \subseteq V(G) \setminus C$. Since the objective of our article is to study the $(1, \leq l)$-LDA and the $(1, \leq l)$-LDB codes of infinite triangular grids, for simplicity, we will refer the $(I, \leq 1)$-sets as I-sets, and will use the notation $I(\cdot)$ instead of $I_1(\cdot)$.

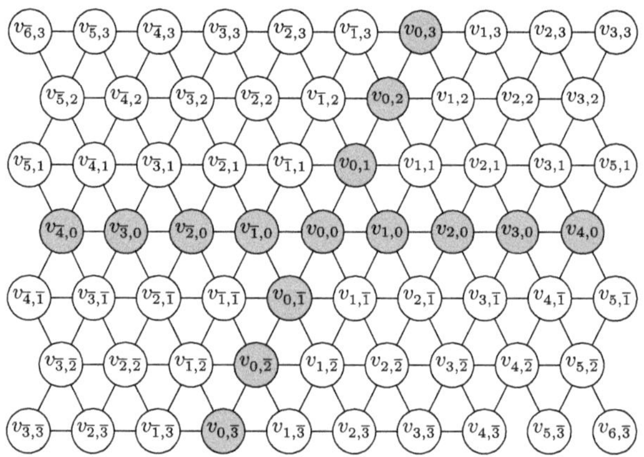

Fig. 1. The vertices of the infinite triangular grid is denoted by $v_{i,j}$ where $i, j \in \mathbb{Z}$. Moreover, we use the notation $\bar{i}$ instead of $-i$ in the subscript in the index of v. The infinite triangular grid is embedded in $\mathbb{R}^2$ with the coordinates of the vertex $v_{i,j}$ being $\left(i + \frac{1}{2}\,j, \frac{\sqrt{3}}{2}\,j\right)$. The above figure depicts this particular embedding, which we will follow in this article as a convention, unless otherwise stated.

When $l = 1$, both $(1, \leq l)$-LDA and -LDB codes give an equivalent definition with the usual locating-dominating codes: a code C is a $(1, \leq 1)$-LDA or -LDB code if and only if C is dominating and $I(x) \neq I(y)$ for each $x \neq y$ where $x, y \in V(G) \setminus C$. Note that here the domination condition is due to allowing an empty set F in the definitions of LDA and LDB codes. Moreover, we know that any $(1, \leq l)$-LDA code is also a $(1, \leq l)$-LDB code. Similarly, we have that any $(1, \leq l + 1)$-LDA $((1, \leq l + 1)$-LDB) code is also a $(1, \leq l)$-LDA $((1, \leq l)$-LDB) code [7].

In this article, our focus is on $(1, \leq l)$-locating-dominating codes of types A and B on the infinite triangular grid, denoted by $\mathbf{G_6}$.

Definition 1. *Let* $\mathbf{G_6}$ *be the infinite triangular grid with vertex set* $V(\mathbf{G_6})$ *and edge set* $E(\mathbf{G_6})$. *We define* $V(\mathbf{G_6})$ *as a collection of the following points of the* *plane* $\mathbb{R}^2$

$$V(\mathbf{G_6}) = \left\{ v_{i,j} = \left(i + \frac{1}{2} \, j, \, \frac{\sqrt{3}}{2} \, j \right) \mid i, j \in \mathbb{Z} \right\},$$

and two vertices $v_{i,j}, v_{i',j'}$ *are adjacent iff their Euclidean distance is exactly one. In other words,* $v_{i,j}$ *is adjacent to* $v_{i',j'}$ *in* $\mathbf{G_6}$ *if one of the following holds:*

(i) $i = i'$, $|j - j'| = 1$,
(ii) $j = j'$, $|i - i'| = 1$,
(iii) $(i - i') = (j' - j) = 1$,
(iv) $(i' - i) = (j - j') = 1$.

For notational convenience, the negative subscripts of the type $-i$ *of the vertices of* $\mathbf{G_6}$ *are replaced by* $\bar{i}$. *For instance, we write* $v_{1,\bar{1}}$ *instead of* $v_{1,-1}$. *See Fig. 1 for a reference illustration of a portion of the infinite triangular grid.*

Our interest is on the relative size of vertex sets in the infinite graphs. To measure this, we introduce the following notion of density.

Definition 2. *The density of a code* C *of* $\mathbf{G_6}$ *is given by*

$$D(C) = \limsup_{n \to \infty} \frac{|C \cap B_n(v_{0,0})|}{|B_n(v_{0,0})|}.$$

Remark 1. We note that in [20], it has been shown for connected graphs with the slow-growth property [20] (such as the infinite triangular grid), that the density of a set can be defined using any center point (in our case, we choose the center point as $v_{0,0}$). While Definition 2 is quite technical, often the simplest method for finding the density of a given periodic vertex set in an infinite grid is to find a periodic tiling covering the entire grid in which each tile has a constant size and contains a constant number of vertices in C. For example, if each tile has n vertices out of which c belong to code C, then we have $D(C) = \frac{c}{n}$, [20, Lemma 3.4].

An $(r, \leq l)$-LDA code (resp., $(r, \leq l)$-LDB code) is *minimum* if there does not exist any $(r, \leq l)$-LDA code (resp., $(r, \leq l)$-LDB code) with a strictly smaller density. Moreover, the density of a minimum $(r, \leq l)$-LDA code (resp., $(r, \leq l)$-LDB code) for the infinite triangular grid $\mathbf{G_6}$ is denoted by $D_{r,l}^A$ (resp., $D_{r,l}^B$). Our objective is to compute the $D_{1,l}^A$ and $D_{1,l}^B$ for all $l \geq 1$.

Our Contributions: We summarize our findings in the following two theorems.

Theorem 1. *The density of a minimum* $(1, \leq l)$-*LDA code for the infinite triangular grid* $\mathbf{G_6}$ *for all* $l \geq 1$ *is given by:*

(a) $D_{1,1}^A = \frac{13}{57}$ *[5],*
(b) $D_{1,2}^A = \frac{1}{3}$,
(c) $D_{1,l}^A = 1$, *for all* $l \geq 3$.

Theorem 2. *The density of a minimum* $(1, \leq l)$*-LDB code for the infinite triangular grid* $\mathbf{G_6}$ *for all* $l \geq 1$ *is given by:*

(a) $D_{1,1}^B = \frac{13}{57}$ *[5],*
(b) $D_{1,2}^B = \frac{1}{3}$,
(c) $D_{1,3}^B = \frac{4}{7}$,
(d) $D_{1,l}^B = \frac{2}{3}$, *for all* $l \geq 4$.

Organization: The rest of the article presents the proofs of Theorems 1 and 2 in Sect. 2. We present upper bound constructions for Theorems 1(b) and 2(b) in Sect. 2.1, then in Sect. 2.2 we prove Theorem 1(c). We continue in Sect. 2.3 by giving a tight construction for Theorem 2(c) and in Sect. 2.4 for Theorem 2(d). Section 2.5 discusses our general proof technique for the remaining lower bounds. Then in the remaining sections we prove the lower bounds. Most of the Sects. 2.7 and 2.8 are omitted due to space constraints. Finally, we conclude in Sect. 3.

2   Proofs of Theorems 1 and 2

In this section, we present all the proofs. The proofs use similar techniques. The upper bounds are obtained through providing explicit examples. The lower bounds uses discharging method or its variants. The proof of Theorem 1 is short, and partially gets implied by Theorem 2. Notice that, Theorem 1(a) and Theorem 2(a) were already proved by Honkala [5]. Thus, our focus is to prove the rest. We present each of the proofs in a separate subsection for the convenience of the readers.

2.1   Proof of Theorems 1(b) and 2(b) (Upper Bound)

A $(1, \leq 2)$-LDA code of $\mathbf{G_6}$ with density $\frac{1}{3}$ is given in Fig. 2. Since any $(1, \leq 2)$-LDA code is also a $(1, \leq 2)$-LDB code, this construction provides an upper bound also for $D_{1,l}^B$. One may use Remark 1 to verify that the density of the code is indeed $\frac{1}{3}$ by observing the tiling provided in Fig. 2. Specifically, each tile has 6 vertices, among which 2 are codewords.

We omit the proof that the construction in Fig. 2 is a $(1, \leq 2)$-LDA code due to space constraints. However, this is easy (although tedious) to see by considering $I(X)$ for different vertex sets X of size at most 2 and first checking that each case with $|I(X)| = 3$ gives a unique set and continuing to $|I(X)| = 4$, then $|I(X)| = 5$ and ending the considerations with $|I(X)| = 6$. $\qquad\square$

2.2   Proof of Theorem 1(c)

It is enough to show that every vertex of the infinite triangular grid $\mathbf{G_6}$ must be part of any $(1, \leq 3)$-LDA code as any $(1, \leq l + 1)$-LDA code is also a $(1, \leq l)$-LDA code. Moreover, note that if C is a $(1, \leq 3)$-LDA code, then any superset

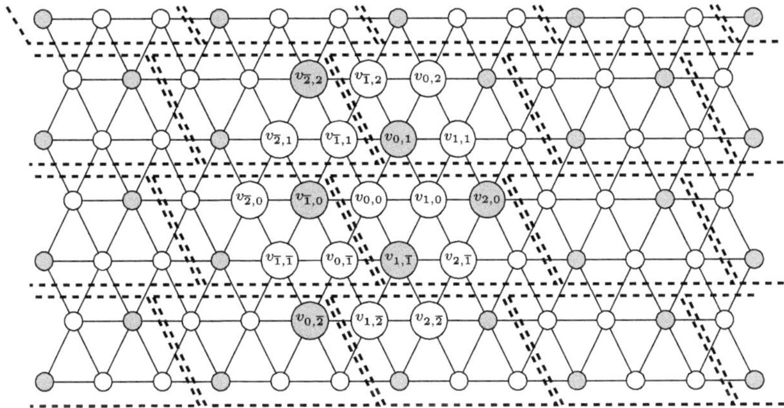

Fig. 2. The gray vertices form a $(1, \leq 2)$-LDA code with density $\frac{1}{3}$

C' of C is also a $(1, \leq 3)$-LDA code. Thus, without loss of generality, if we can show that $C \setminus \{v_{0,0}\}$ is not a $(1, \leq 3)$-LDA code, then we can conclude the proof. Therefore, consider the code $C \setminus \{v_{0,0}\}$ (see Fig. 3). Notice that $I(\{v_{\bar{1},1}, v_{1,\bar{1}}\}) = I(\{v_{\bar{1},1}, v_{1,\bar{1}}, v_{0,0}\})$ in this case. Since $C \cap \{v_{\bar{1},1}, v_{1,\bar{1}}\} = C \cap \{v_{\bar{1},1}, v_{1,\bar{1}}, v_{0,0}\}$, code $C \setminus \{v_{0,0}\}$ is not $(1, \leq 3)$-LDA. This concludes the proof. $\square$

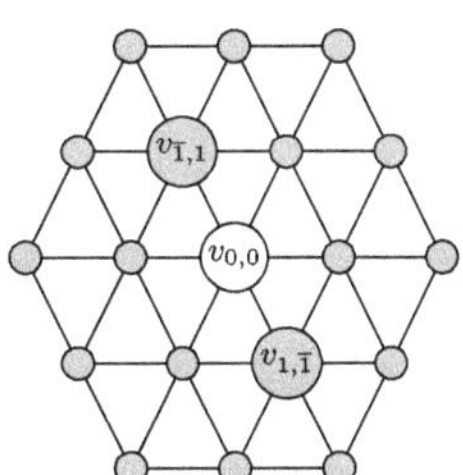

Fig. 3. A reference figure for the proof of Theorem 1(c) where gray vertices are codewords

2.3   Proof of Theorem 2(c) (Upper Bound)

A $(1, \leq 3)$-LDB code of $\mathbf{G_6}$ with density $\frac{4}{7}$ is given in Fig. 4. One may use Remark 1 to verify that the density of the code is indeed $\frac{4}{7}$ by observing the tiling provided in Fig. 4. Specifically, each hexagonal tile has 7 vertices, among which 4 are codewords.

Again, we omit the proof that the given construction is a $(1, \leq 3)$-LDB code. This establishes the upper bound $D_{1,3}^B \leq \frac{4}{7}$. $\square$

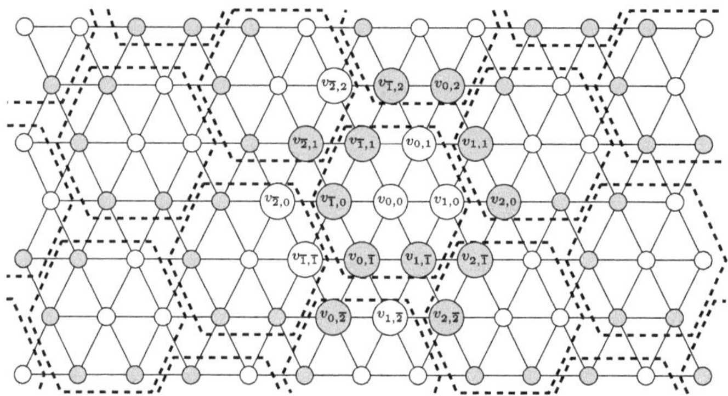

Fig. 4. The gray vertices form a $(1, \leq 3)$-LDB code with density $\frac{4}{7}$

2.4   Proof of Theorem 2(d) (Upper Bound)

A $(1, \leq 4)$-LDB code of $\mathbf{G_6}$ with density $\frac{2}{3}$ is given in Fig. 5. One may use Remark 1 to verify that the density of the code is indeed $\frac{2}{3}$ by observing the tiling provided in Fig. 5. Specifically, each tile has 6 vertices, among which 4 are codewords.

The given construction is a $(1, \leq 4)$-LDB code since each non-codeword is adjacent to a codeword which is adjacent to no other non-codewords. For example, if we have $v_{0,0} \in I(X)$, then we immediately know that $v_{0,1} \in X$. This establishes the upper bound $D_{1,4}^{B} \leq \frac{2}{3}$.

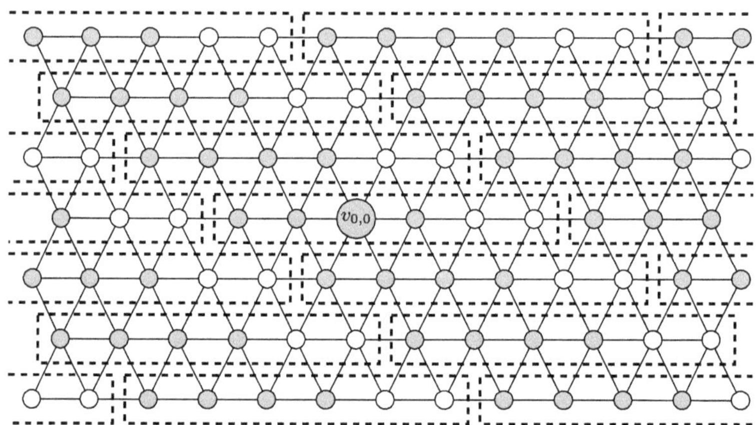

Fig. 5. The gray vertices form a $(1, \leq 4)$-LDB code with density $\frac{2}{3}$

2.5   A General Discharging Method for Proving the Lower Bounds

In this section, we establish the lower bound proof technique which we use to show the lower bounds of Theorems 2(b), (c), and (d). To avoid duplicating certain specifics for those proofs, we are presenting a general overview of them here. Our proofs are based on the discharging method. An interested reader may see [3] to read more on the discharging method.

Let C be a $(1, \leq l)$-LDB code of $\mathbf{G_6}$. We wish to establish a lower bound $D_{1,l}^B \geq \alpha$, for some $\alpha \in (0, 1)$. The proof strategy is as follows. Initially, we assign a charge function to the vertices of $\mathbf{G_6}$ as follows:

$$ch(x) = \begin{cases} 1, & \text{if } x \in C \\ 0, & \text{if } x \in V(\mathbf{G_6}) \setminus C. \end{cases}$$

In the next step, we define certain *discharging rules* and relocate the charges. In this way, the total charge of the graph is conserved. We use the following discharging rules in each of our proofs. Thus, let us present the discharging rules here as a common step of the proofs.

(R1) Each codeword $c \in C$ donates a charge of $\frac{1}{7-|I(c)|}$ to its non-codeword neighbors. The updated charge of a vertex x after applying (R1) is denoted by $ch_1(x)$.

(R2) Suppose a non-codeword v has charge $ch_1(v) = \frac{\alpha}{1-\alpha} + \beta$, for some $\beta > 0$, after (R1). If v has $t > 0$ non-codeword neighbors $u_1, u_2, \cdots, u_t$ with $ch_1(u_i) < \frac{\alpha}{1-\alpha}$, for all $i \in \{1, 2, \ldots, t\}$, then v donates a charge of $\frac{\beta}{t}$ to each of them. The updated charge of the vertex x, after applying (R2), is denoted by $ch_2(x)$.

Note that we apply rule (R1) simultaneously to all the vertices. After that, we apply rule (R2) to every vertex simultaneously.

Suppose that after applying (R1) and (R2) we have

$$ch_2(x) \geq \begin{cases} 0, & \text{if } x \in C \\ \frac{\alpha}{1-\alpha}, & \text{if } x \in V(\mathbf{G_6}) \setminus C. \end{cases}$$

By Lemma 1, this will prove that the density of C is at least α, and hence prove the lower bound $D_{1,l}^B \geq \alpha$.

Let c be a codeword, that is, $c \in C$. Notice that, if $|I(c)| = 7$, then c does not have a non-codeword neighbor to donate any charge in (R1), and thus the donation amount $\frac{1}{7-|I(c)|}$ according to (R1), whenever applicable, is defined. If c has $k \geq 1$ non-codewords in its neighborhood, then c donates a total charge of

$$k \times \frac{1}{7 - |I(c)|} = k \times \frac{1}{7 - (7 - k)} = 1$$

due to (R1). That means, $ch_1(c) \geq 1 - 1 = 0$. Moreover, application of (R2) does not change the charge of any codewords. This implies that after applying (R1) and (R2), the updated charge of each codeword c is $ch_1(c) = ch_2(c) \geq 0$.

Furthermore, let u be a non-codeword. Notice that, if $ch_1(u) \geq \frac{\alpha}{1-\alpha}$ then, even after applying (R2), we will still have $ch_2(u) \geq \frac{\alpha}{1-\alpha}$. That means, to prove $D_{1,l}^B \geq \alpha$ we need to show that $ch_2(u) \geq \frac{\alpha}{1-\alpha}$, for all $u \in V(\mathbf{G_6}) \setminus C$. In other words, the conditions of the following lemma hold.

Lemma 1. *If $ch_2(u) \geq \frac{\alpha}{1-\alpha}$, for all $u \in V(\mathbf{G_6}) \setminus C$, then $D_{1,l}^B \geq \alpha$.*

Proof. Observe that we move charge to at most distance 2 from a codeword. Hence, when $ch_2(u) \geq \frac{\alpha}{1-\alpha}$, for all $u \in V(\mathbf{G_6}) \setminus C$, we obtain $|C \cap B_n(v_{0,0})| \geq \frac{\alpha}{1-\alpha}(|B_{n-2}(v_{0,0})| - |C \cap B_{n-2}(v_{0,0})|) \geq \frac{\alpha}{1-\alpha}(|B_{n-2}(v_{0,0})| - |C \cap B_n(v_{0,0})|)$. This implies $|C \cap B_n(v_{0,0})| \geq |B_{n-2}(v_{0,0})|\alpha$. Therefore, we have

$$\limsup_{n \to \infty} \frac{|C \cap B_n(v_{0,0})|}{|B_n(v_{0,0})|} \geq \limsup_{n \to \infty} \frac{|B_{n-2}(v_{0,0})|\alpha}{|B_n(v_{0,0})|} = \alpha \limsup_{n \to \infty} \frac{|B_{n-2}(v_{0,0})|}{|B_n(v_{0,0})|} = \alpha.$$

$\square$

Remark 2. Notice that, the above process and Lemma 1 can be generalized further by changing the discharging rules. Moreover, it is possible to replace $\mathbf{G_6}$ by a square, hexagonal, or a King's grid keeping the method the same (as long as value 7 in rule (R1) is chosen well). Furthermore, the technique can be readily extended to prove lower bounds of the minimum density of $(r, \leq l)$-LDA and $(r, \leq l)$-LDB codes for general values of r and l. Similar techniques have been used in [21] for $r = l = 1$ case. Another similar concept is share, introduced in [23].

2.6   Proof of Theorems 1(b) and 2(b) (Lower Bound)

In this section, we show that $D_{1,2}^A, D_{1,2}^B \geq \frac{1}{3}$. Recall that, any $(1, \leq 2)$-LDA code is, in particular, a $(1, \leq 2)$-LDB code. Thus, it is enough to prove that any $(1, \leq 2)$-LDB code has density at least $\frac{1}{3}$. It is enough to show that $ch_2(v_{0,0}) \geq \frac{1}{2} = (1/3)/(2/3)$ assuming $v_{0,0}$ is a non-codeword due to Lemma 1 along with the discussion in Sect. 2.5.

Hence let us calculate $ch_2(u)$ for each non-codeword u based on how many adjacent codewords they have, and their positions. For this part, without loss of generality, we will assume $v_{0,0}$ to be a non-codeword, and consider all the cases under this assumption. That is, under different scenarios, we want to show that $ch_2(v_{0,0}) \geq \frac{1}{2}$, which will conclude the proof.

Case 1: $|\mathbf{I}(\mathbf{v_{0,0}})| \geq 3$. Note that, every codeword donates a charge of at least $\frac{1}{6}$ to $v_{0,0}$. Thus, $ch_1(v_{0,0}) \geq 3 \times \frac{1}{6} = \frac{1}{2}$ as $|I(v_{0,0})| \geq 3$. Therefore, $ch_2(v_{0,0}) \geq \frac{1}{2}$.

Case 2: $|\mathbf{I}(\mathbf{v_{0,0}})| = 2$. Suppose that $|N(v_{0,0}) \cap C| = 2$. Observe that, the two codeword neighbors of $v_{0,0}$ can be at distance 1, 2 or 3 in $\mathbf{G_6}[N(v_{0,0})]$, that is, the 6-cycle induced by the neighbors of $v_{0,0}$.

Subcase 2.1: The Two Codeword Neighbors of $\mathbf{v_{0,0}}$ are at Distance 1 in $\mathbf{G_6}[\mathbf{N(v_{0,0})}]$. Without loss of generality, suppose that the two codeword neighbors of $v_{0,0}$ are $v_{0,1}$ and $v_{1,0}$ (depicted in Fig. 6(a) for reference). That is, capturing the scenario when the two codeword neighbors are at distance 1 in $\mathbf{G_6}[N(v_{0,0})]$. In this scenario, $I(\{v_{0,0}, v_{1,1}\}) = I(v_{1,1})$, and thus, $v_{1,1}$ must be a codeword. That implies $|I(v_{0,1})|, |I(v_{1,0})| \geq 3$. Therefore, both $v_{0,1}$ and $v_{1,0}$ donates $\frac{1}{4}$ each to $v_{0,0}$. Thus, $ch_1(v_{0,0}) \geq \frac{1}{2}$, which implies $ch_2(v_{0,0}) \geq \frac{1}{2}$.

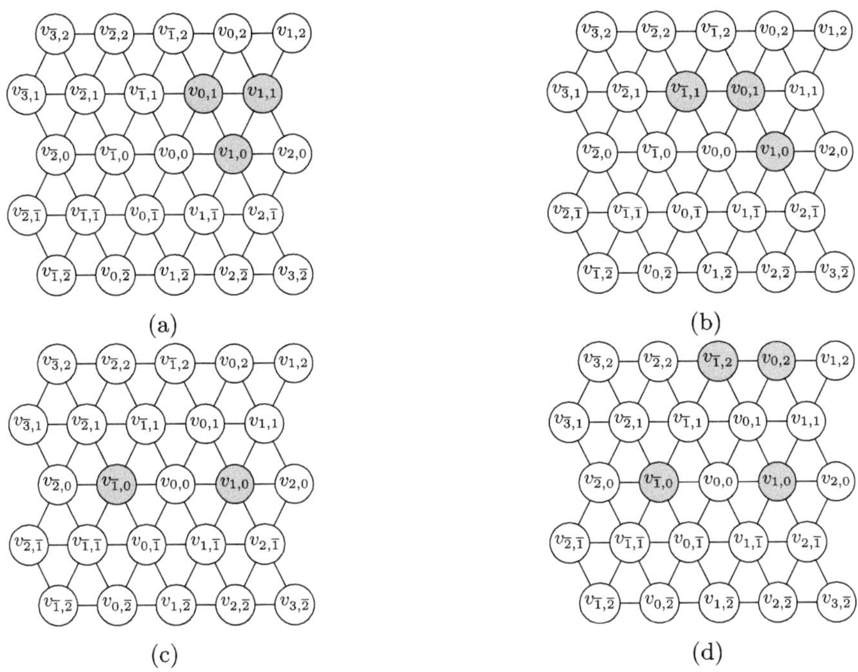

Fig. 6. Reference figures for Case 2 in the proof of Theorem 1(b) and Theorem 2(b), where gray vertices are codewords

Subcase 2.2: The Two Codeword Neighbors of $\mathbf{v_{0,0}}$ are at Distance 2 in $\mathbf{G_6[N(v_{0,0})]}$.

Without loss of generality, suppose that the two codeword neighbors of $v_{0,0}$ are $v_{\bar{1},1}$ and $v_{1,0}$ (depicted in Fig. 6(b) for reference). That is, capturing the scenario when the two codeword neighbors are at distance 2 in $\mathbf{G_6}[N(v_{0,0})]$, without loss of generality. In this scenario, $I(\{v_{0,0}, v_{0,1}\}) = I(v_{0,1})$, and thus, $v_{0,1}$ must be a codeword, contradicting the fact that $v_{0,0}$ has exactly two codeword neighbors. Thus, if $v_{0,0}$ has exactly two codeword neighbors, it is not possible for them to be at distance 2 in $\mathbf{G_6}[N(v_{0,0})]$. In general, we can conclude that a non-codeword u cannot have exactly two codeword neighbors at distance 2 in $\mathbf{G_6}[N(u)]$.

Subcase 2.3: The Two Codeword Neighbors of $\mathbf{v_{0,0}}$ are at Distance 3 in $\mathbf{G_6[N(v_{0,0})]}$.

Finally, without loss of generality, suppose that the two codeword neighbors of $v_{0,0}$ are $v_{1,0}$ and $v_{\bar{1},0}$ (depicted in Fig. 6(c) for reference). That is, capturing the scenario when the two codeword neighbors are at distance 3 in

$\mathbf{G_6}[N(v_{0,0})]$, without loss of generality. We will analyze this scenario further, starting with a claim.

Claim 1. If $v_{1,1}$ (resp., $v_{\bar{2},1}, v_{\bar{1},\bar{1}}, v_{2,\bar{1}}$) is a non-codeword, then $v_{0,1}$ (resp., $v_{\bar{1},1}$, $v_{0,\bar{1}}$, $v_{1,\bar{1}}$) donates a charge of $\frac{1}{15}$ to $v_{0,0}$ according to (R2).

Proof of Claim: Observe that, it is enough to show that if $v_{1,1}$ is a non-codeword, then $v_{0,1}$ donates a charge of $\frac{1}{15}$ to $v_{0,0}$ according to (R2) due to symmetry. Notice that, if $v_{1,1}$ is a non-codeword, then $v_{0,2}$ is a codeword, as otherwise $I(\{v_{0,0}, v_{\bar{1},1}\}) = I(\{v_{0,1}, v_{\bar{1},1}\})$. Now, notice that, $v_{0,1}$ has two codeword neighbors at distance exactly 2 in $\mathbf{G_6}[N(v_{0,1})]$, unless $v_{\bar{1},2}$ is also a codeword. Recall that, in **Subcase 2.2** we had established that if a non-codeword u has exactly two codeword neighbors, then they cannot be at distance exactly 2 from each other in the 6-cycle $\mathbf{G_6}[N(u)]$. Hence, $v_{\bar{1},2}$ is also a codeword. See Fig. 6(d) for reference. Therefore, $ch_1(v_{0,1}) \geq \frac{1}{5} + \frac{1}{5} + \frac{1}{6} = \frac{17}{30} = \frac{1}{2} + \frac{1}{15}$.

Note that, $v_{\bar{1},1}$ has (at least) two codeword neighbors: $v_{\bar{1},2}$ and $v_{\bar{1},0}$. Moreover, if $v_{\bar{2},1}$ is a non-codeword, then $v_{\bar{2},2}$ must be a codeword as $I(\{v_{0,0}, v_{0,1}\}) = I(\{v_{\bar{1},1}, v_{0,1}\})$. That means, either $v_{\bar{2},1}$ or $v_{\bar{2},2}$ must be a codeword. This will force $v_{\bar{1},1}$ to have at least three codeword neighbors, and thus $ch_1(v_{\bar{1},1}) \geq \frac{1}{2}$ due to **Case 1**. Hence, $v_{0,1}$ does not donate any charge to $v_{\bar{1},1}$ according to (R2).

Next, observe that, $v_{1,1}$ has two codeword neighbors $v_{1,0}$ and $v_{0,2}$. Due to **Case 2**, it is not possible to have $|I(v_{1,0})| = 2$ since $v_{1,0}$ and $v_{0,2}$ are at distance 2 in $\mathbf{G_6}[N(v_{1,1})]$. This will force $v_{1,1}$ to have at least three codeword neighbors, and thus $ch_1(v_{1,1}) \geq \frac{1}{2}$ due to **Case 1**. Hence, $v_{0,1}$ does not donate any charge to $v_{1,1}$ according to (R2).

That means, $v_{0,1}$ donates a charge of at least $\frac{1}{15}$ to $v_{0,0}$ according to (R2). $\circ$

If both $v_{1,1}$ and $v_{2,\bar{1}}$ are codewords, then $|I(v_{1,0})| \geq 3$ and $v_{1,0}$ donates at least $\frac{1}{4}$ to $v_{0,0}$ according to (R1).

If exactly one of $v_{1,1}$ and $v_{2,\bar{1}}$ are codewords, then $|I(v_{1,0})| \geq 2$ and $v_{1,0}$ donates at least $\frac{1}{5}$ to $v_{0,0}$ according to (R1). Moreover, one of $v_{0,1}$ and $v_{1,\bar{1}}$ donates $\frac{1}{15}$ to $v_{0,0}$ according to (R2) due to Claim 1. That means, $v_{1,0}, v_{0,1}, v_{1,\bar{1}}$ collectively donate a charge of at least $\frac{1}{5} + \frac{1}{15} = \frac{4}{15} > \frac{1}{4}$.

If both $v_{1,1}$ and $v_{2,\bar{1}}$ are non-codewords, then $|I(v_{1,0})| \geq 1$ and $v_{1,0}$ donates $\frac{1}{6}$ to $v_{0,0}$ according to (R1). Moreover, each of $v_{0,1}$ and $v_{1,\bar{1}}$ donates at least $\frac{1}{15}$ to $v_{0,0}$ according to (R2) due to Claim 1. That implies, $v_{1,0}, v_{0,1}, v_{1,\bar{1}}$ collectively donate a charge of at least $\frac{1}{6} + \frac{1}{15} + \frac{1}{15} = \frac{9}{30} > \frac{1}{4}$.

Overall, $v_{1,0}, v_{0,1}, v_{1,\bar{1}}$ collectively donate a charge of at least $\frac{1}{4}$ to $v_{0,0}$ according to (R1) and (R2). Similarly, $v_{\bar{1},0}, v_{\bar{1},1}, v_{0,\bar{1}}$ collectively donate a charge of at least $\frac{1}{4}$ to $v_{0,0}$ according to (R1) and (R2). This implies, $ch_2(v_{0,0}) \geq \frac{1}{4} + \frac{1}{4} = \frac{1}{2}$.

Case 3: $|\mathbf{I}(\mathbf{v_{0,0}})| \leq \mathbf{1}$. Note that, $I(v_{0,0})$ must be non-empty since C is a dominating set in particular. Thus, without loss of generality, let $v_{1,0}$ be the only codeword neighbor of $v_{0,0}$. Observe that, $I(\{v_{0,0}, v_{0,1}\}) = I(v_{0,1})$, which is not possible. $\square$

2.7   Proof of Theorem 2(c) (Lower Bound)

We omit the full proof due to space constraints and only present its outline. The proof is based on using the discharge function from Sect. 2.5 and showing that $ch_2(v) \geq \frac{4}{3} = (\frac{4}{7})/(1 - \frac{4}{7})$ for any non-codeword v and applying Lemma 1.

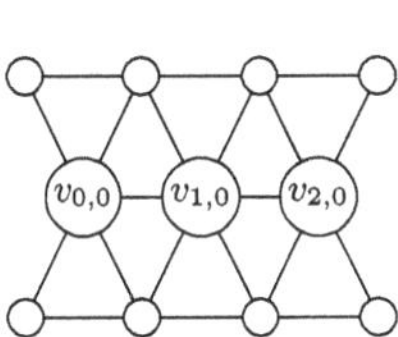

(a) Three non-codewords in a straight line.

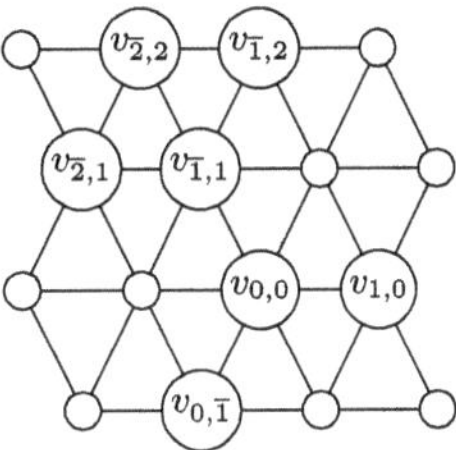

(b) Three non-codewords in a curve.

Fig. 7. Reference figures for the proof of Claim 2.

Claim 2. Let C be a $(1, \leq l)$-LDB code, $l \geq 3$, in the infinite triangular grid $\mathbf{G_6}$, then the following hold:

(a) There are no three distinct non-codewords u, v, w such that $N[w] \subseteq N[v] \cup N[u]$.
(b) There cannot exist three non-codewords, such as $v_{0,0}, v_{1,0}, v_{2,0}$, in a straight line as in Fig. 7(a).
(c) When we have the three non-codewords, such as $v_{0,\bar{1}}, v_{0,0}, v_{1,0}$, in a curve as in Fig. 7(b), then vertices $v_{\bar{1},1}, v_{\bar{2},1}, v_{\bar{1},2}$ are codewords. Moreover, if $l \geq 4$, then also vertex $v_{\bar{2},2}$ is a codeword.

Proof of Claim:

(a) If the three non-codewords u, v, w such that $N[w] \subseteq N[v] \cup N[u]$ exist, then $I(u, v) = I(w, v, u)$, a contradiction.
(b) When we have the three non-codewords in a straight line as in Fig. 7(a), we will have $N[v_{1,0}] \subseteq N[v_{0,0}] \cup N[v_{2,0}]$ which is not possible due to the part (a) of the claim.
(c) When we have the three non-codewords in a curve as in Fig. 7(b), the vertices $v_{\bar{2},1}, v_{\bar{1},1}$, and $v_{\bar{1},2}$ are codewords, as otherwise $I(v_{0,0}, v_{\bar{2},1}, v_{1,0}) = I(v_{\bar{2},1}, v_{1,0})$, $I(v_{0,0}, v_{\bar{1},1}, v_{1,0}) = I(v_{\bar{1},1}, v_{1,0})$, or $I(v_{0,0}, v_{\bar{1},2}, v_{0,\bar{1}}) = I(v_{\bar{1},2}, v_{0,\bar{1}})$ implying C is not a $(1, \leq 3)$-LDB code. Similarly, for $l \geq 4$, the vertex $v_{\bar{2},2}$ is codeword, as otherwise $I(v_{0,0}, v_{0,\bar{1}}, v_{1,0}, v_{\bar{2},2}) = I(v_{0,\bar{1}}, v_{1,0}, v_{\bar{2},2})$ implying C cannot be a $(1, \leq l)$-LDB code for $l \geq 4$.

248   S. S. Das et al.

By Claim 2(b), we have that for any $(1, \leq l)$-LDB code with $l \geq 3$, we have $|I(v)| \geq 3$ for any $v \in V(\mathbf{G_6}) \setminus C$. Hence, we divide the proof into four cases: $|I(v)| = 6, |I(v)| = 5, |I(v)| = 4$ and $|I(v)| = 3$. In each of these cases, we show that $ch_2(v) \geq \frac{4}{3}$, which will prove the desired lower bound.

We have included below, the proof in the case $|I(v)| = 5$. We choose without loss of generality $v = v_{0,0}$. Let $|I(v_{0,0})| = 5$ where $I(v_{0,0}) = \{v_{0,1}, v_{\bar{1},1}, v_{\bar{1},0}, v_{0,\bar{1}}, v_{1,\bar{1}}\}$. Notice that this is a general case since $\mathbf{G_6}$ is edge-transitive. Accordingly, we have assumed $v_{1,0}$ to be the only non-codeword neighbor of $v_{0,0}$.

Observe that, either of $v_{1,1}$ or $v_{2,\bar{1}}$ must be a codeword, or else we will have $I(\{v_{1,0}, v_{1,1}, v_{2,\bar{1}}\}) = I(\{v_{1,1}, v_{2,\bar{1}}\})$, a contradiction. Let us, thus, assume that $v_{1,1}$ is a codeword. This assumption is without loss of generality due to the reflectional symmetry (along X-axis) of the graph, and due to the positioning of the relevant codewords with respect to $v_{0,0}$.

Furthermore, consider the following four sets of vertices: $S_1 = \{v_{0,2}, v_{\bar{1},2}, v_{\bar{2},2}\}$, $S_2 = \{v_{\bar{2},2}, v_{\bar{2},1}, v_{\bar{2},0}\}$, $S_3 = \{v_{\bar{2},0}, v_{\bar{1},\bar{1}}, v_{0,\bar{2}}\}$ and $S_4 = \{v_{0,\bar{2}}, v_{1,\bar{2}}, v_{2,\bar{2}}\}$. Note that each of the above-listed sets of vertices must contain at least one codeword to avoid three non-codewords in a straight line due to Claim 2(b). Thus, observe that, we must have at least two such vertices from $\cup_{i=1}^{4} S_i$ as codewords, each of which is adjacent to at least one codeword neighbor of $v_{0,0}$ among $N(v_{0,0}) \setminus \{v_{1,\bar{1}}\}$.

Therefore, $v_{0,0}$ has exactly 5 codewords neighbors and among them at least two are adjacent to 3 or more codewords, two of them are adjacent to 2 or more codewords, and one of them is adjacent to one or more codewords. Hence, after applying Rule (R1), we have $ch_1(v_{0,0}) \geq \frac{1}{5} + \frac{2}{4} + \frac{2}{3} = \frac{41}{30} > \frac{4}{3}$. Therefore, in particular, $ch_2(v_{0,0}) \geq \frac{4}{3}$.

2.8   Proof of Theorem 2(d) (Lower Bound)

We again omit the full proof due to space constraints and only sketch its outline. Our goal is to show that $D_{1,l}^{B} \geq \frac{2}{3}$ for each $l \geq 4$. It is enough to show that $D_{1,4}^{B} \geq \frac{2}{3}$ since $D_{1,l}^{B} \geq D_{1,4}^{B}$ for each $l \geq 4$. In particular, we show that $ch_2(v) \geq 2 = (\frac{2}{3})/(1 - \frac{2}{3})$ for each non-codeword v and the result follows from Lemma 1 along with the discussion in Sect. 2.5. While calculating the values of $ch_2(v)$, we use Claims 2(b) and 2(c) extensively.

Similarly to the previous case, by Claim 2(b), we may consider separately the cases with $|I(v)| = 6, |I(v)| = 5, |I(v)| = 4$ and $|I(v)| = 3$ for some non-codeword v.

3   Conclusion

In this paper, we have given exact minimum densities $D_{1,2}^{A} = D_{1,2}^{B} = \frac{1}{3}$, $D_{1,l}^{A} = 1$ for each $l \geq 3$, $D_{1,3}^{B} = \frac{4}{7}$ and $D_{1,l}^{B} = \frac{2}{3}$, for all $l \geq 4$ in the infinite triangular grid. Together with known results for usual location-domination, these results cover every case for $(1, \leq l)$-LDA and -LDB codes. By [12], the result $D_{1,7}^{B} = \frac{2}{3}$, gives the maximum density of open irredundant sets as well as the minimum density of MILD codes in the infinite triangular grids. While we have considered only the case with $r = 1$, this is typical for the field and we have included the notation $r = 1$ since it has been the convention in the cases of LDA- and LDB-codes. Moreover, expanding the future research on the cases with $r > 1$ would be of interest.

We have presented constructions in each case which obtain the shown minimum densities. Our lower bound proofs are based on the discharging method, as is typical for these types of results. However, interestingly we have managed to implement two generic discharging rules which alone have yielded the exact lower bounds in all three cases. This is atypical since often different lower bound proofs based on the discharging method require varying numbers of different rules.

Additionally, software implementations [17, 19, 24, 25] exist to find the bounds using discharging and assignments for similar problems. To the best of our knowledge, no computer-aided approach has previously been applied to the $(r, \leq l)$-LDA or -LDB codes considered here. Moreover, none of the proofs in this article are computer-aided.

Acknowledgments. The research of Tuomo Lehtilä was in part supported by the Finnish Cultural Foundation and Academy of Finland grants 338797 and 358718. This work is also partially supported by the IFCAM project "Applications of graph homomorphisms" (MA/IFCAM/18/39).

Disclosure of Interests. The authors have no competing interests to declare that are relevant to the content of this article.

References

1. Bonnet, É., Foucaud, F., Lehtilä, T., Parreau, A.: Neighbourhood complexity of graphs of bounded twin-width. Eur. J. Comb. **115**, 103772 (2024)
2. Cohen, G.D., Honkala, I., Lobstein, A., Zémor, G.: Bounds for codes identifying vertices in the hexagonal grid. SIAM J. Discret. Math. **13**(4), 492–504 (2000)
3. Cranston, D.W., West, D.B.: An introduction to the discharging method via graph coloring. Discret. Math. **340**(4), 766–793 (2017)
4. Cukierman, A., Yu, G.: New bounds on the minimum density of an identifying code for the infinite hexagonal grid. Discret. Appl. Math. **161**(18), 2910–2924 (2013)
5. Honkala, I.: An optimal locating-dominating set in the infinite triangular grid. Discret. Math. **306**(21), 2670–2681 (2006)
6. Honkala, I., Laihonen, T.: On locating-dominating sets in infinite grids. Eur. J. Comb. **27**(2), 218–227 (2006)
7. Honkala, I., Laihonen, T., Ranto, S.: On locating-dominating codes in binary hamming spaces. Discrete Math. Theor. Comput. Sci. **6**(2) (2004)
8. Jean, D., Lobstein, A.: Watching systems, identifying, locating-dominating and discriminating codes in graphs. https://dragazo.github.io/bibdom/main.pdf
9. Jean, D.C., Seo, S.J.: Optimal error-detection system for identifying codes. Networks **85**(1), 61–75 (2025)

10. Karpovsky, M., Chakrabarty, K., Levitin, L.: On a new class of codes for identifying vertices in graphs. IEEE Trans. Inf. Theory **44**(2), 599–611 (1998). https://doi.org/10.1109/18.661507

11. Laihonen, T., Ranto, S.: Codes identifying sets of vertices. In: Boztaş, S., Shparlinski, I.E. (eds.) AAECC 2001. LNCS, vol. 2227, pp. 82–91. Springer, Heidelberg (2001). https://doi.org/10.1007/3-540-45624-4_9

12. Lehtilä, T.: Three ways to locate-dominate every vertex in a graph. In: Proceedings of Sixth Russian-Finnish Symposium on Discrete Mathematics, pp. 100–105 (2021)

13. Pelto, M.: On locating-dominating codes for locating large numbers of vertices in the infinite king grid. Australas. J. Comb. (2011)

14. Pelto, M.: On $(r, \leq 2)$-locating-dominating codes in the infinite king grid. Adv. Math. Commun. **6**(1), 27–38 (2012)

15. Pelto, M.: Optimal $(r, \leq 3)$-locating-dominating codes in the infinite king grid. Discret. Appl. Math. **161**(16–17), 2597–2603 (2013)

16. Rall, D.F., Slater, P.J.: On location-domination numbers for certain classes of graphs. Congr. Numer. **45**, 97–106 (1984)

17. Salo, V., Törmä, I.: Diddy: a python toolbox for infinite discrete dynamical systems. In: Manzoni, L., Mariot, L., Roy Chowdhury, D. (eds.) Cellular Automata and Discrete Complex Systems, pp. 33–47. Springer Nature Switzerland, Cham (2023). https://doi.org/10.1007/978-3-031-42250-8_3

18. Salo, V., Törmä, I.: Finding codes on infinite grids automatically. Fund. Inform. **191**(3–4), 331–349 (2024)

19. Salo, V., Törmä, I.: Griddy. http://github.com/ilkka-torma/griddy. Accessed 04 Apr 2026

20. Sampaio, R.M., Sobral, G.A., Wakabayashi, Y.: Density of identifying codes of hexagonal grids with finite number of rows. RAIRO Oper. Res. **58**(2), 1633–1651 (2024)

21. Slater, P.J.: Domination and location in acyclic graphs. Networks **17**(1), 55–64 (1987)

22. Slater, P.J.: Dominating and reference sets in a graph. J. Math. Phys. Sci. **22**(4), 445–455 (1988)

23. Slater, P.J.: Fault-tolerant locating-dominating sets. Discret. Math. **249**(1–3), 179–189 (2002)

24. Stolee, D.: Adage. http://github.com/derrickstolee/ADAGE. Accessed 04 Apr 2026

25. Stolee, D.: Automated discharging arguments for density problems in grids. arXiv:1409.5922 (2014)

26. West, D.B.: Introduction to Graph Theory (2nd Edition). Prentice Hall (2001)

Realizing Planar Linkages in Polygonal Domains

Thomas Depian[1]([✉])[iD], Carolina Haase[2][iD], Martin Nöllenburg[1][iD],
and André Schulz[3][iD]

[1] Algorithms and Complexity Group, TU Wien, Vienna, Austria
{tdepian,noellenburg}@ac.tuwien.ac.at
[2] Trier University, Trier, Germany
haasec@uni-trier.de
[3] FernUniversität in Hagen, Hagen, Germany
andre.schulz@fernuni-hagen.de

Abstract. A linkage $\mathcal{L}$ consists of a graph $G = (V, E)$ and an edge-length function ℓ. Deciding whether $\mathcal{L}$ can be realized as a planar straight-line embedding in $\mathbb{R}^2$ with edge length $\ell(e)$ for all $e \in E$ is $\exists\mathbb{R}$-complete [Abel et al., JoCG'25], even if $\ell \equiv 1$, but a considerable part of $\mathcal{L}$ is rigid. In this paper, we study the computational complexity of the realization question for structurally simpler, less rigid linkages inside an open polygonal domain P, where the placement of some vertices may be specified in the input. We show XP-membership and W[1]-hardness with respect to the size of G, even if $\ell \equiv 1$ and no vertex positions are prescribed. Furthermore, we consider the case where G is a path with prescribed start and end position and $\ell \equiv 1$. Despite the absence of any rigid components, we obtain NP-hardness in general, and provide a linear-time algorithm for arbitrary ℓ if G has only three edges and P is convex.

Keywords: Linkages · Polygonal Domain · Computational Complexity · Parameterized Complexity · Existential Theory of the Reals

1 Introduction

In this paper, we study *linkages* $\mathcal{L} = (G, \ell)$, which are graphs $G = (V, E)$ where every edge $e \in E$ has a predetermined length $\ell(e)$. Linkages can be used to model mechanical constructions, e.g., robot arms consisting of metal bars connected at joints, or folding proteins [13], and their study is well-established in computational geometry and mechanical engineering: famous results such as Kempe's universality theorem [22,23] date back to the 1870s. By now, there exists a plethora of work on linkages [1,18]. A realization of a linkage (an embedding of the vertices V such that edge lengths given by ℓ are realized) is called a *configuration*. A configuration in the plane is called *planar* if no two edges cross. In this work, we focus on planar configurations. Planarity is often a requirement given by the application. For example, bars of a mechanical framework might not have the option to overlap and hence have to stay disjoint. We refer to the book by Demaine and O'Rourke [13] for an overview of existing results and applications; both for planar but also for non-planar configurations.

PLANAR LINKAGE REALIZABILITY (PLR for short) is the problem that asks whether a linkage has a planar realization, i.e., deciding whether a linkage $\mathcal{L}$ admits a planar configuration. It is known to be complete for the *Existential Theory of the Reals* ($\exists\mathbb{R}$) (see also [1,2]), even if $\mathcal{L}$ is *unit-length*, i.e., all edge lengths are the same (also known as *matchstick graphs*). Known hardness constructions rely on many rigid components, whose embedding is fixed up to rigid motions [2,9,10,15]. This raises the question whether these hardness results carry over to more "flexible" linkages. The related *reconfiguration* problem, i.e., deciding whether there exists a continuous, edge-length-preserving, and sometimes also planar, motion from one configuration into another is PSPACE-hard [4,8,11,21,28]. This holds in many variants, but notably also if $\mathcal{L}$ is a *linear linkage* that is, G is a path [18,20] in the presence of obstacles. The problem remains NP-hard for obstacles made up of four axis-aligned segments [19]. Note that the above-mentioned reductions require non-unit edge lengths.

Motivated by the lack of analogous results for PLR, we investigate in this paper the realizability question for structurally simple linkages $\mathcal{L}$ in the presence of obstacles, modeled via an open polygonal domain P into which we must embed $\mathcal{L}$ in a planar way. This natural variant of the realization problem has, to the best of our knowledge, not been studied before. Note that related problems, e.g., finding a planar straight-line drawing of a graph inside a polygonal region with some prescribed vertex positions [24,25], do not have edge-length constraints.

Contributions. We show in Sect. 3 that PLR is in XP^1 with respect to the size of the graph (Theorem 1) by providing an $\exists\mathbb{R}$-formulation of our problem, which are known to be efficiently solvable for a bounded number of variables [17]. While PLR is $\exists\mathbb{R}$-complete even without polygonal domain P and, therefore, it is no surprise that it can be expressed as an $\exists\mathbb{R}$-formula, the following Sect. 4 underlines that this upper bound can be considered "best possible". More formally, we show that PLR is W[1]-hard when parameterized by the size of G even for unit-length linkages (Theorem 2). Apart from "justifying" Theorem 1, this also shows that no structural parameter of G, not even the number of rigid components, can lead to an FPT-algorithm for PLR.

We then focus in Sects. 5 and 6 on linear linkages, i.e., where G is a path. If the linkage is also unit-length, i.e., $\ell \equiv 1$, then PLR is equivalent to finding a placement for a single edge as we can place the entire linkage arbitrarily close to it. However, if we fix the position of the endpoints of the path, the problem is less trivial. Towards understanding the complexity of this special case of PLR, we show, on the one hand, that it is NP-hard in general (Theorem 3). On the other hand, if the polygonal domain P is convex and G has three edges, then we

---

[1] We assume familiarity with basic concepts of *parameterized complexity* [12].

can solve PLR in linear time, even for arbitrary ℓ (Theorem 4). The algorithm exploits the observation that three-edge linear linkages have only one degree of freedom, which allows us to examine a linear number of candidate configurations. Theorems 3 and 4 together suggest that the complexity of PLR in this surprisingly challenging setting arises from an interplay between the degree of freedom in $\mathcal{L}$ and the obstacles in P and we see these two results as a first step towards closing this complexity gap.

Full proofs of statements marked by ★ can be found in the full version [14].

2 Preliminaries

A *linkage* $\mathcal{L} = (G, \ell)$ consists of a simple, undirected, and connected graph G with vertex set $V(G)$ and edge set $E(G)$ and a function $\ell\colon E(G) \to \mathbb{R}$ assigning each edge $e \in E(G)$ a positive *length* $\ell(e)$. The linkage $\mathcal{L}$ is *linear* if G is a path and *unit-length* if $\ell \equiv 1$, abbreviated as $(G, \mathbb{1})$. A *configuration* Γ of a linkage $\mathcal{L}$ is a planar (i.e., crossing-free) straight-line drawing of G, such that we have $\|\Gamma(u) - \Gamma(v)\|_2 = \ell(uv)$ for every $uv \in E(G)$; we say Γ *respects* ℓ. Let P be an open polygonal domain, $P \subset \mathbb{R}^2$, i.e., a multiply connected region of $\mathbb{R}^2$ [26]. We say that Γ *lives* inside P if $\Gamma \subset P$ holds. In this paper, we allow the placement of some vertices $V' \subseteq V(G)$ to be *constrained* by a function δ and ask for the existence of a realization of $\mathcal{L}$ that *adheres* to δ within a polygonal domain P:

PLANAR LINKAGE REALIZABILITY (PLR)
Given: A linkage $\mathcal{L} = (G, \ell)$, an open polygonal domain $P \subset \mathbb{R}^2$, and a mapping $\delta\colon V' \to P$ for some $V' \subseteq V(G)$.
Question: Does there exist a planar configuration Γ of $\mathcal{L}$ that lives inside P such that $\Gamma(v) = \delta(v)$ for every $v \in V'$?

Let $\mathcal{I} = (\mathcal{L} = (G, \ell), P, V', \delta)$ be an instance of PLR. We denote with $n_G := |V(G)|$ and $n_P := |V(P)|$ the number of vertices of G and P, respectively and with $|\mathcal{I}|$ the size of the instance $\mathcal{I}$. We assume the *Real RAM* model of computation for our algorithmic results; see also [16,27]. We consider only rational edge lengths $\ell(\cdot)$ and polygonal domains P with rational vertex coordinates. This way, input numbers can be encoded with fixed precision (and bounded bit-complexity), as required by our hardness reductions.

3 LINKAGE REALIZABILITY is in XP

In this section, we show that PLR is in XP with respect to the number of vertices $n_G = |V(G)|$, even if $\mathcal{L}$ is not unit-length; Sect. 4 provides a complementing W[1]-hardness result. We devise an Existential Theory of the Reals ($\exists\mathbb{R}$) formulation φ using only $2n_G$ variables that expresses PLR. It is known that checking whether φ is feasible is in XP with respect to its number of variables [17]. Recall that deciding if a linkage admits a configuration is $\exists\mathbb{R}$-complete [2]. Therefore, it is not hard to see that we can express the corresponding problem as an $\exists\mathbb{R}$-formula.

However, to the best of our knowledge, the literature does not contain an explicit formula for PLR or a related problem that does not rely on some general position assumption that we cannot make. Since the running time of our XP-algorithm depends on the number of variables, (in)equalities and their degree in the $\exists\mathbb{R}$-formula, we sketch below an explicit formula $\varphi(\mathcal{I})$ for an instance $\mathcal{I}$ of PLR; details can be found in the full version [14].

A Sketch of $\varphi(\mathcal{I})$. We introduce two existentially-quantified variables x_v and y_v for every vertex $v \in V(G)$ to encode the position of the vertex v in a configuration Γ, i.e., $\Gamma(v) = (x_v, y_v)$. Moreover, $\varphi(\mathcal{I})$ contains four subformulas, ensuring that (1) Γ adheres to δ, (2) Γ respects the length constraint ℓ, (3) Γ is planar, and (4) Γ lives inside P, respectively. Subformulas (1) and (2) can be realized by forcing the value of some variables and by encoding the definition of the (squared) Euclidean distance, respectively. For (3), we can use the sign of the determinant $\Delta(p, q, r) = \begin{vmatrix} x_p & y_p & 1 \\ x_q & y_q & 1 \\ x_r & y_r & 1 \end{vmatrix}$ to check if the point r is to the left, to the right, or on the line $\overline{pq}$ through the points p and q [6]: If p, q are on different sides of $\overline{uv}$ and u, v are on different sides of $\overline{pq}$, then the edges $uv, pq \in E(G)$ cross. If three vertices are co-linear, the existence of a crossing depends on the concrete position of the vertices. Therefore, we test in this case if the x- and y-coordinate of a vertex, say p, is between those of u and v, respectively. The edges cross if this holds for at least one vertex. For (4), we first use $\Delta(\cdot)$ to force one vertex to lie inside P (which is already the case if there exists a vertex $v \in V'$). Afterwards, we ensure that the combined drawing of G and P is planar.

Theorem 1. PLANAR LINKAGE REALIZABILITY *is in* $\exists\mathbb{R}$ *and in* XP *with respect to the number* n_G *of vertices in* G.

Proof. Let $\mathcal{I}$ be an instance of PLR. Containment in $\exists\mathbb{R}$ follows from the existence of $\varphi(\mathcal{I})$. To show XP-membership, we analyze the formula. It has $2n_G$ variables and $\mathcal{O}(n_G^2 + n_G \cdot n_P)$ equalities and inequalities; note that $|E(G)| \in \mathcal{O}(n_G)$ holds as G can be assumed to be planar. It is known that feasibility of an $\exists\mathbb{R}$ formula on N variables and M (in)equalities can be checked in $L^{\mathcal{O}(1)}(M \cdot D)^{\mathcal{O}(N^2)}$ time, where D is the maximum degree of the (in)equalities and L denotes the bit-complexity of their coefficients [17]. In our case, L and D are polynomial in the size of $\mathcal{I}$. Thus, we can check whether $\varphi(\mathcal{I})$ is feasible in $|\mathcal{I}|^{\mathcal{O}(n_G^2)}$ time. $\square$

4   Unit-Length Linkage Realizability Is W[1]-Hard

We next complement Theorem 1 with a corresponding W[1]-hardness result. To show W[1]-hardness, we use a reduction from the W[1]-hard problem GRID TILING [12]. Here, given two integers k and m, as well as a collection S of size k^2 containing nonempty sets $S_{i,j} \subseteq [m] \times [m]$ with $1 \leq i, j \leq k$, the goal is to find for each of the sets a tuple $s_{i,j} \in S_{i,j}$, such that for two such tuples $(a, b) = s_{i,j}$ and $(a', b') = s_{i',j'}$ if $i = i' + 1$ and $j = j'$ ($j = j' + 1$ and $i = i'$), then $a = a'$ ($b = b'$); see Fig. 1c.

We construct a unit-length linkage $(G, \mathbb{1})$ by subdividing each edge of a $k \times k$ grid graph and adding to each vertex of the original grid two adjacent vertices to create an (equilateral) triangle. We refer to the vertices of the original grid graph as *grid vertices*, the vertices used to subdivide the edges as *subdivision vertices* and all other vertices as *triangle vertices*. We define a polygonal domain P' as a $k \times k$ grid, where every edge has width w. We refer to the square shaped regions in P' that correspond to the grid points as *grid squares*; see Fig. 1a. The goal is for each grid vertex of G to be placed in one of the grid squares and for no two grid vertices to be placed in the same grid square. In each grid square we define m^2 points on an $m \times m$ grid, such that every point represents a value $(a, b) \in [m] \times [m]$. We want each grid vertex to be placed on one of those points or in a small ε-region around it. If a grid vertex is placed in such an ε-region, we say it *uses* the corresponding point. For each grid square $R_{i,j}$ we need to ensure that only points (a, b) with $(a, b) \in S_{i,j}$ can be used. To this end, we add for each value $(a, b) \in S_{i,j}$ a so called triangle corridor to P'. Placing a grid vertex in an ε-region then corresponds to choosing a tuple in the GRID TILING instance.

Lemma 1 ($\bigstar$). *If a grid vertex lies inside a grid square $R_{i,j}$, then the adjacent triangle has to lie inside a triangle corridor and use a point $(a, b) \in S_{i,j}$.*

Note also that no two grid vertices can be in the same grid square, as their triangles would cross. Lastly, for our construction to work as intended, we have to choose the width w of the edges of P' such that if a point (a, b) is used by a grid vertex $g_{i,j}$ in $R_{i,j}$ then a grid vertex $g_{i,j+1}$ in $R_{i,j+1}$ can use all points $(a, b') \in S_{i,j+1}$, but no other points, and a grid vertex $g_{i+1,j}$ in $R_{i+1,j}$ can use all points $(b', a) \in S_{i+1,j}$, but no other points; see Fig. 1b.

Lemma 2 ($\bigstar$). *If a grid vertex $g_{i,j}$ uses (a, b) in $R_{i,j}$ then $g_{i,j+1}$ uses (a, b') in $R_{i,j+1}$ and $g_{i+1,j}$ uses (a', b) in $R_{i+1,j}$.*

The theorem follows from Lemmas 1 and 2.

Theorem 2. PLANAR LINKAGE REALIZABILITY *is* W[1]-*hard with respect to the number n_G of vertices of G.*

5   Realizing Linear Unit-Length Linkages Is NP-Hard

Theorem 2 rules out fixed parameter tractable algorithms for all common graph parameters. However, similar to previous reductions [2,10], the constructed graph G still contains rigid components. In this section, we consider linear linkages, i.e., where G is a path, which do not have any rigid structure. We show that PLR remains NP-hard even if $\mathcal{L}$ is a unit-length linear linkage where we constrain only the placement of the first and last vertex of G, i.e., its *endpoints*.

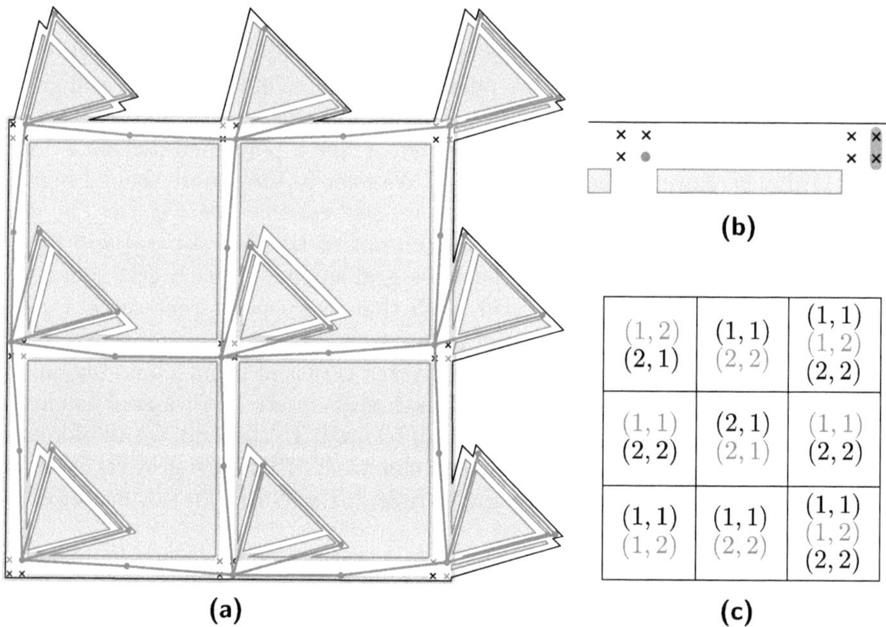

Fig. 1. **(a)** Construction for $k = 3$ and $m = 2$, P' highlighted in orange, grid squares marked in light blue. **(b)** If the blue point with tuple $(2, 2)$ is chosen, then to the right only points in the red area with tuples $(x, 2)$, $1 \leq x \leq m$ can be used. **(c)** GRID TILING instance of the construction in (a).(Color figures online).

Overview of the Reduction. We give a reduction from the NP-complete problem PLANAR MONOTONE 3-SAT [7]. An instance $\varphi = (\mathcal{X}, \mathcal{C})$ of this problem consists of N variables $\mathcal{X} = \{x_1, \ldots, x_N\}$ and M clauses $\mathcal{C} = \{c_1, \ldots, c_M\}$, partitioned into the positive clauses $\mathcal{C}^+$ containing only non-negated literals and negative clauses $\mathcal{C}^-$ containing only negated literals. Moreover, there must exist a planar rectilinear drawing $\mathcal{E}$ of the clause-variable incidence graph of φ where all variables are on a horizontal line, which separates $\mathcal{C}^+$ and $\mathcal{C}^-$; see Fig. 2a for an example. We can compute a drawing $\mathcal{E}$ with polynomial coordinates in polynomial time [7,9,10].

In our reduction, we create a unit-length linear linkage $\mathcal{L} = (G, \mathbb{1})$ of length n, where we specify n in the end. Conceptually, our construction will force a configuration Γ of $\mathcal{L}$ to visit each variable of φ and each clause it is contained in, set truth assignments for the former and verify their truth status for the latter components. The path that Γ follows will closely resemble the drawing $\mathcal{E}$ of φ, see also Fig. 2b, and we will replace each edge and vertex of $\mathcal{E}$ with a gadget, which is a part of the polygonal domain P. We highlight in each gadget dedicated regions in the plane, in the following called *areas*, where Γ must pass through. More concretely, once the configuration enters a gadget through an entry area, the construction ensures that it must leave the gadget at the respective exit

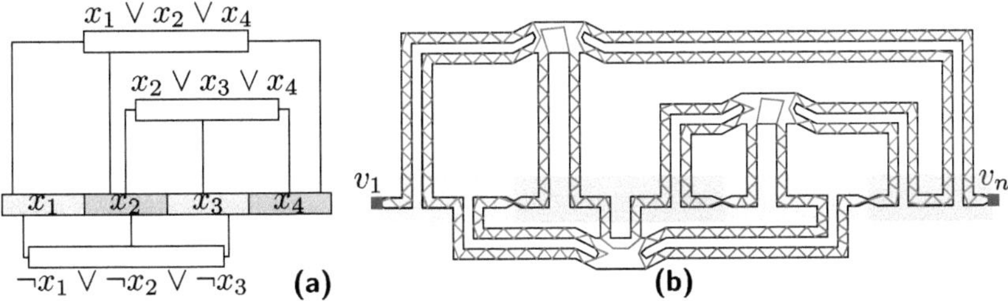

Fig. 2. A **(a)** formula φ and the **(b)** schematization of the constructed instance.

area. Areas are specified as quadrants of discs of radius ε, where ε is a small constant whose value we specify towards the end, and denoted as $\overset{\rightarrow}{\blacktriangle}_{i,t}(F)$ and $\blacktriangle^{\rightarrow}_{i,t}(F)$ for a gadget F related to a possible truth assignment $t \in \{0,1\}$ to the variable x_i, i.e., $x_i = t$, respectively. Furthermore, we consider for each of these *entrance* and *exit* areas the triangle that is inscribed in the same quadrant of a disc of radius $\varepsilon/2$, which we call the *start* and *end* area of a gadget and denote as $\overset{\rightarrow}{\blacktriangleright}_{i,t}(F)$, and $\blacktriangleright^{\rightarrow}_{i,t}(F)$, respectively. On a high level, to ensure that there exists a configuration Γ if φ is satisfiable, we want that, for a point p in the start area, there exists a configuration that starts at p and places a vertex somewhere in the respective end area. We create the polygonal region P by gluing the individual gadgets together at the correct areas.

In the following, we describe each of the gadgets on a high-level and refer to the full version [14] for the details. Note that all gadgets are agnostic to translations and rotations in the plane. Furthermore, to ease presentation, we will sometimes place vertices of $\mathcal{L}$ on the boundary of P. However, note that we can always slightly enlarge/shrink the gadgets to ensure that no vertex is forced to lie on the boundary of P.

Edge Gadget. There are three types of edge gadgets: tunnels, bends, and shifters. *Tunnels* will inhabit sawtooth-like shaped pairs of segments of height 0.6 and width 1.6. Any configuration Γ can embed at most two edges inside a tunnel enforced by the *obstacles*, i.e., holes in the polygonal domain, of the tunnel depicted in Fig. 3a. The edges *zig-zag* around the obstacles by alternating the placement of the vertices between a placement near the top and the bottom of the tunnel. This allows us to define two equivalence classes on the configurations depending on the side of the tunnel where they place the first vertex. They correspond to the truth assignment to a variable $x_i \in \mathcal{X}$, and we color areas and configurations from the classes in red (■) and green (■) in the figures, depending whether they correspond to $x_i = 0$ or $x_i = 1$, respectively. We place the first pair at the lower side of the tunnel and the second pair at the upper side of the tunnel as indicated in Fig. 3a.

Tunnels are accompanied by *bends*, which force the configuration Γ to perform a 90° turn and are depicted in Fig. 3b. Note that the obstacles force the green configuration to draw one edge (almost) horizontal and one (almost) ver-

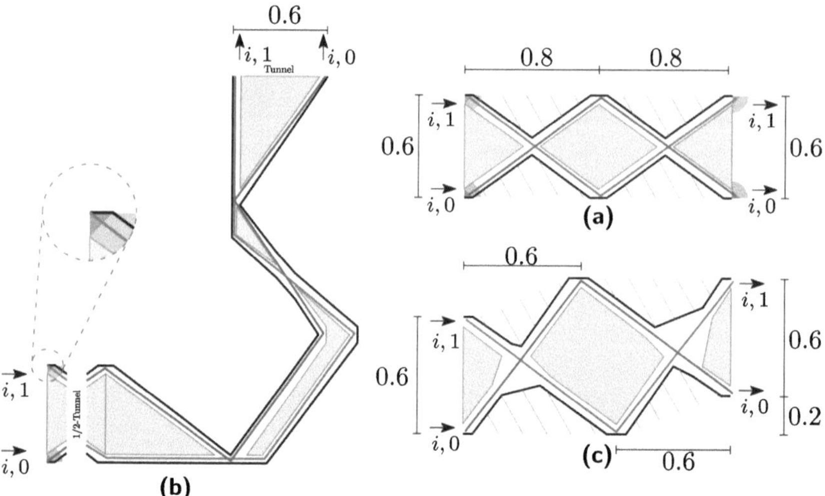

Fig. 3. A **(a)** tunnel, **(b)** bend, **(c)** and the main part of a shifter with configurations through them. We hatch in this and the following figures the outside of the polygonal domain if there is risk of confusion.

tical. Thus it performs, compared to the red configuration, a small detour to ensure that both configurations place the same number of vertices inside the gadget. This is crucial to ensure correctness of the reduction and is the main difficulty in constructing the gadget. Observe that the bend has at its start and end a height (or width) of 0.6, allowing us to attach tunnels on either of its ends. The entrance and exit areas of a bend are analogous to those of a tunnel.

Tunnels and bends can only start and end at specific coordinates due to their construction. With a *shifter*, we can shift tunnels up and down by 0.2 to give us more flexibility in later gadgets. Observe in Fig. 3a that inside a tunnel the distance of the endpoints of an edge of the linkage is approximately 0.8 and 0.6 in x- and y-direction, respectively. With the gadget from Fig. 3c, we can force an inverted behavior to move a configuration over the course of two edges up (or down) by 0.2. The shifter consists of the main part from Fig. 3c on whose two sides we attach a tunnel that help us establish correctness.

Clause Gadget. The main part of the clause gadget for the clause $x_i \vee x_j \vee x_k$ is depicted in Fig. 4 and has multiple obstacles that leave only seven narrow (possibly intersecting) *corridors* inside the gadget to limit how a configuration Γ can interact with it. In our reduction, we force the configuration to enter the main part three times, first via the entrance $\vec{\blacktriangle}_{i,t_i}$, then via $\vec{\blacktriangle}_{j,t_j}$, and finally via $\vec{\blacktriangle}_{k,t_k}$, for $t_i, t_j, t_k \in \{0, 1\}$. Observe that the distance between $\vec{\blacktriangle}_{i,t_i}$ and $\blacktriangle\vec{}_{i,t_i}$ can be spanned by a linkage of length two. The corridors leave little choice for Γ: If Γ enters the main part via $\vec{\blacktriangle}_{i,0}$, it is forced to leave it via $\blacktriangle\vec{}_{i,0}$, otherwise, i.e., if it enters the main part via $\vec{\blacktriangle}_{i,1}$, it is forced to leave it via $\blacktriangle\vec{}_{i,1}$. Note that placing a vertex in an area for x_j or x_k is impossible due to the unit-length requirement of the edges paired with the corridors. The same holds true for the entrance and exit areas corresponding to x_k. The three corridors in the middle constrain how a configuration can reach $\blacktriangle^{\rightarrow}_{j,t_j}$ from $^{\rightarrow}\blacktriangle_{j,t_j}$ using three edges. In particular, if Γ enters the main part via $^{\rightarrow}\blacktriangle_{j,0}$, the gadget contains two corridors, effectively giving the configuration the flexibility to lean more towards the left or right side of the main part; compare also Figs. 4a and b. Conversely, i.e., if Γ enters via $^{\rightarrow}\blacktriangle_{j,1}$, there is again only one corridor, giving the configuration little freedom in placing the remaining vertices.

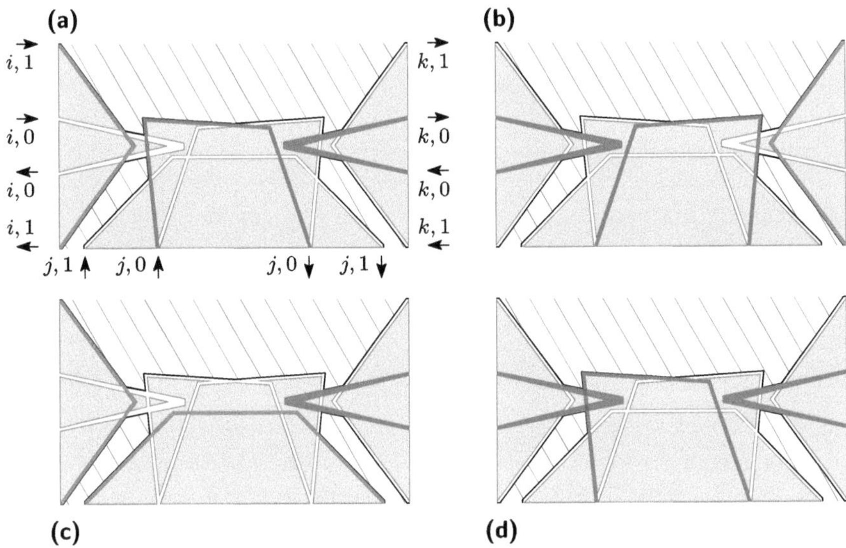

Fig. 4. (a)–(d) Different configurations Γ through the main part of the clause gadget. Dashed lines indicate different possibilities for the configuration to pass through the corridors.

The construction allows for the following crucial observation; compare also Figs. 4a–d: If a configuration Γ enters the main part via $^{\rightarrow}\blacktriangle_{i,0}$ *and* $^{\rightarrow}\blacktriangle_{k,0}$, a planar configuration that enters the main part via $^{\rightarrow}\blacktriangle_{j,0}$ becomes impossible. On the other hand, if Γ enters the main part via $^{\rightarrow}\blacktriangle_{i,1}$ *or* $^{\rightarrow}\blacktriangle_{k,1}$, it uses a corridor that does not intersect with the one(s) for $^{\rightarrow}\blacktriangle_{j,0}$, allowing a planar configuration even if Γ enters the main part via $^{\rightarrow}\blacktriangle_{j,0}$. The corridor connecting $^{\rightarrow}\blacktriangle_{j,1}$ with $\blacktriangle^{\rightarrow}_{j,1}$ can always be used.

Finally, we remark that for a suitable small constant ε it is not possible to enter the clause gadget at some entrance area assigned for one variable and leave it at an entrance/exit area assigned to a different variable. To see this recall, that we have seven possible "routes" in which the linkage is intended to pass through the gadget (two for x_i/x_k and three for x_j); compare this also to the seven corridors. We can find a constant α such that for every route all

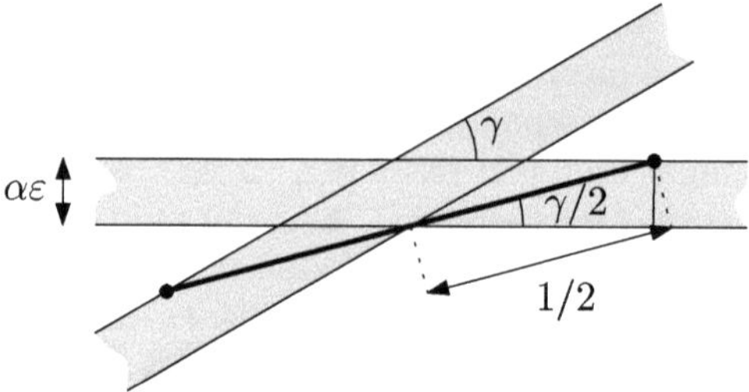

Fig. 5. A segment of length one can only have endpoints in two corridors if $\gamma \leq 2 \arcsin 2\alpha\varepsilon$.

possible actual configurations are contained in a polygonal corridor of width at most $\alpha\varepsilon$, which we use to refine the original corridors. The obstacles of P in the clause gadget are then defined by the points outside of the refined corridors. Let γ be the smallest "turning angle" for two intersecting corridors. Note that γ is independent from $\alpha\varepsilon$. In order to pass from one corridor to another, there has to be a segment of length one with endpoints in distinct corridors. A simple calculation shows that this is only possible if $\gamma \leq 2 \arcsin 2\alpha\varepsilon$; see Fig. 5. Thus, picking a rational value $\varepsilon \leq \frac{\alpha}{2} \sin \frac{\gamma}{2}$ ensures that we cannot deviate from the intended route through the clause gadget. Observation 1 summarizes this.

Observation 1. *Let Γ be a planar configuration of $\mathcal{L}$ that enters the main part C' of a clause gadget three times. For any variable x_i, truth assignment $t \in \{0,1\}$ to x_i, and vertex $v \in V(G)$, we have that $\Gamma(v) \in {}^{\rightarrow}\blacktriangle_{i,t}(C')$ implies that there is some $v' \in V(G)$ with $\Gamma(v') \in \blacktriangle_{i,t}^{\rightarrow}(C')$. Furthermore, there is some $v'' \in V(G)$ such that $\Gamma(v'') \in {}^{\rightarrow}\blacktriangle_{z,1}(C')$ for some $z \in \{i,j,k\}$.*

We attach on the left and right side of the main part two shifters, and at the bottom side two tunnels each to obtain the clause gadget and unify the entrance and exit areas.

Variable Gadget. The variable gadget for a variable $x_i \in \mathcal{X}$ consists of three main components with different roles: making Γ "set" the truth assignment $x_i = t$, propagating this to all relevant clauses and "resetting" Γ before entering the variable gadget of x_{i+1}, if it exists. The first component of the variable gadget is depicted in Fig. 6a. It consists of a triangular obstacle, forcing the configuration to place the next vertex either at the top or bottom end of the gadget, corresponding to setting the variable to *true* or *false*, respectively. The base of the first component has a height of 0.6, allowing us to attach a tunnel. To avoid an irrational coordinate for the tip of the obstacle, we reduce the height of the triangle slightly. Moreover, we force the linkage to place a vertex inside $\blacktriangle_{i,t}^{\rightarrow}$ for $t \in \{0,1\}$, effectively setting $x_i = t$. The third component of the variable gadget, depicted in Fig. 6b, uses an analogous idea to force Γ to approach the center of the gadget no matter if it passed through ${}^{\rightarrow}\blacktriangle_{i,t}$ for $t = 0$ or $t = 1$.

We combine different variable gadgets via entrance and exit areas, indicated for the ith variable as $\overrightarrow{\blacktriangleright}_i$ and $\blacktriangleright_i^{\rightarrow}$, and highlight two points s_i and t_i inside them that will act as a certificate necessary in the full proof. Note that in Fig. 6 the gadget for x_1 is closed around s_1; we do so likewise for t_N. Finally, the second component consists of multiple tunnels that connect the first component via the gadgets for clauses c_j with $x_i \in c_j$ to the third component of the gadget. For negative clauses, we attach to them half of a tunnel to toggle the configuration, i.e., its carried truth state, before visiting these clauses.

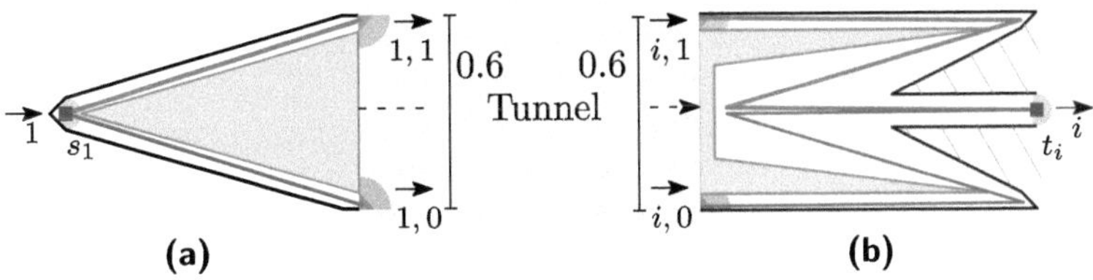

Fig. 6. The **(a)** first and **(b)** third component of the variable gadget for x_1 and x_i, respectively.

Complete Reduction. The placement of the obstacles in our gadgets ensures that any configuration follows pre-determined paths through the gadgets; see also the correctness arguments in the appendix.

We now combine the above-introduced gadgets by taking a suitable scaled planar rectilinear drawing $\mathcal{E}$ of the incidence graph of φ, replacing all components with the respective gadgets, and unifying all vertices in different gadgets that lie on the same point. To finish the construction of $(\mathcal{L}, P, V', \delta)$, we close the first and last variable gadget and set $V' = \{v_1, v_n\}$ and $\delta(v_1) = s_1$ and $\delta(v_n) = t_N$, where n denotes the number of vertices in G, which we specify next. We note that the obtained polygonal domain P contains (polynomially many) obstacles, i.e., holes. Let P be created using T tunnels, excluding those used in other gadgets such as the shifters, and B bends. The linear linkage $\mathcal{L}$ consists of $n = 2T + 7B + 4|\mathcal{X}| + 35|\mathcal{C}| + 2|\mathcal{C}^-| + 1$ vertices. When carefully analyzing our gadgets, we observe that any planar configuration Γ of $\mathcal{L}$ starting at s_1 must pass through every gadget exactly once. Otherwise it is too short to reach t_N. Using Observation 1, we conclude that for Γ to be planar, every clause must be satisfied, establishing NP-hardness of PLR:

Theorem 3 ($\star$). PLANAR LINKAGE REALIZABILITY *remains* NP-*hard even if* $\mathcal{L} = (G, \mathbb{1})$ *is a unit length linear linkage where we constrain the first and last vertex of* G.

6   Linear Linkages of Length Three in Convex Polygons

Theorem 3 raises the question in which settings we can solve PLR for linear linkages with constrained endpoints efficiently. Observe that Theorem 3 hinges on P being a polygonal domain with a polynomial number of holes. As our last result, we show that if P is convex and G consisting of three edges, we can solve PLR in linear time.

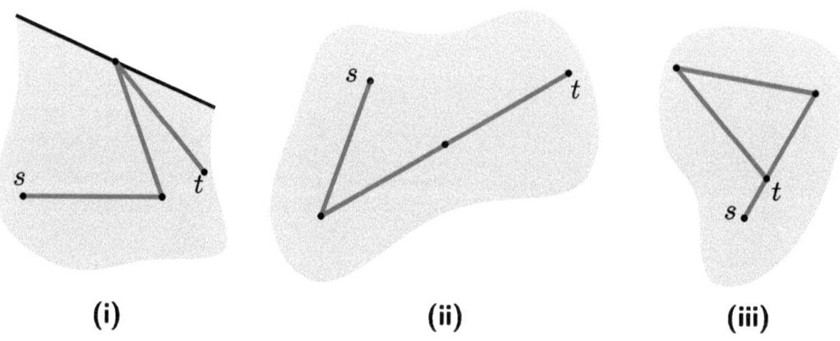

Fig. 7. Realizing a three-edge linkage. It suffices to check these types of configurations; see Theorem 4. The polygonal domain is hinted by the gray areas.

Theorem 4 (★). PLANAR LINKAGE REALIZABILITY *for three-edge linear linkages in a convex polygon P where we constrain the first and last vertex of G can be solved in $\mathcal{O}(n_P)$ time even for general linear linkages, i.e., for arbitrary ℓ.*

Proof Sketch. For this sketch, we allow degenerate configurations where a point might lie on the boundary of P or on an edge of the linkage. We discuss in the full proof how to deal with those.

Assume there exists a non-crossing configuration Γ of the linkage with vertices $v_1, \ldots, v_4$ in P. Let $s = \delta(v_1)$ and $t = \delta(v_4)$. Observe that any such configuration has at most one degree of freedom (1-dof). We now aim to reconfigure Γ until either (i) v_2 or v_3 hit the boundary of P, (ii) two incident edges become co-linear, or (iii) v_1 lies on $\Gamma(v_3 v_4)$ or v_4 lies on $\Gamma(v_1 v_2)$; see Fig. 7. If this is not possible, we have to be in case (i)–(iii) already. Observe that if the configuration space (allowing self-crossings) is not restricted by P, then we can realize the linkage such that it forms with the edge $\Gamma(s)\Gamma(t)$ a convex quadrilateral. This will either increase the angle at $\Gamma(v_2)$ or $\Gamma(v_3)$ (see [13, Lemma 5.3.2]) and hence leads to case (ii).

Consequently, if we start moving the linkage from configuration Γ we will either introduce a crossing (case (iii)), hit the boundary (and since P is convex this happens at some vertex, thus at v_2 or v_3, which is covered by case (i)), or two neighboring edges become co-linear (case (ii)). So it suffices to check all configurations for which one of the cases holds.

The number of configurations is in $\mathcal{O}(m_P)$, and in each case we have to solve an instance of PLR for a two-edge linkage, which takes constant time. Since $m_P \in \mathcal{O}(n_P)$, the theorem follows. $\qquad\square$

7 Concluding Remarks

We see this paper as a further step towards understanding the complexity of realizing linkages. Our paper shows that this problem is surprisingly hard in the presence of a polygonal domain even for relatively simple linkages. The W[1]-hardness from Theorem 2 underlines that, from a parameterized complexity perspective, we must parameterize the polygonal domain P, for example via its number of holes or concave corners. We see this as an interesting direction for future work. Our NP-hardness from Theorem 3 raises the question in which settings we can solve PLR for linear linkages where we constrain the first and last vertex of G in polynomial time. Theorem 4 already provides a partial answer to this question. However, Theorem 4 hinges on three-edge linkages having only one degree of freedom and an extension to four-edge linkages is highly non-trivial. Therefore, we see a generalization to arbitrary constant-size linear linkages (that stems from insights into the specific problem-variant and is thus more efficient as our XP-algorithm from Theorem 1) as a natural next step. The holes in the polygonal domain P were an important ingredient of the NP-hardness construction behind Theorem 3, and we cannot remove them without fundamentally changing the construction. Therefore, showing hardness for simple polygons P requires new approaches (as it has been the case for similar problems in polygonal domains, such as TWO-DIMENSIONAL PACKING [3]). Hence, we see establishing NP-hardness for simple/convex P and $\exists\mathbb{R}$-hardness for general P as interesting directions to pursue for linear linkages. Finally, extensions to other length-constrained drawings, such as dichotomous drawings [5], are also worth considering.

Acknowledgments. Thomas Depian and Martin Nällenburg acknowledge support from the Vienna Science and Technology Fund (WWTF) [10.47379/ICT22029]. This work started at the 19th European Research Week on Geometric Graphs (GGWeek) in Trier.

References

1. Abel, Z., Demaine, E.D., Demaine, M.L., Eisenstat, S., Lynch, J., Schardl, T.B.: Who needs crossings? Hardness of plane graph rigidity. In: Fekete, S.P., Lubiw, A. (eds.) Symposium on Computational Geometry (SoCG'16). LIPIcs, vol. 51, pp. 3:1–3:15. Schloss Dagstuhl – Leibniz-Zentrum für Informatik (2016). https://doi.org/10.4230/LIPICS.SOCG.2016.3
2. Abel, Z., Demaine, E.D., Demaine, M.L., Eisenstat, S., Lynch, J., Schardl, T.B.: Who needs crossings?: noncrossing linkages are universal, and deciding (global) rigidity is hard. J. Comput. Geometry **16** (2025). https://doi.org/10.20382/JOCG.V16I1A12

3. Abrahamsen, M., Miltzow, T., Seiferth, N.: Framework for $\exists \setminus$ -completeness of two-dimensional packing problems. TheoretiCS **3** (2024). https://doi.org/10.46298/THEORETICS.24.11

4. Alt, H., Knauer, C., Rote, G., Whitesides, S.: The complexity of (un)folding. In: Fortune, S. (ed.) Symposium on Computational Geometry (SoCG'03), pp. 164–170. ACM (2003). https://doi.org/10.1145/777792.777818

5. Angelini, P., et al.: Geometric realizations of dichotomous ordinal graphs. In: Aichholzer, O., Wang, H. (eds.) Symposium on Computational Geometry (SoCG'25). LIPIcs, vol. 332, pp. 9:1–9:16. Schloss Dagstuhl – Leibniz-Zentrum für Informatik (2025). https://doi.org/10.4230/LIPICS.SOCG.2025.9

6. de Berg, M., Cheong, O., van Kreveld, M., Overmars, M.: Computational Geometry: Algorithms and Applications, 3rd edn. Springer (2008). https://doi.org/10.1007/978-3-540-77974-2

7. de Berg, M., Khosravi, A.: Optimal binary space partitions for segments in the plane. Int. J. Comput. Geometry Appl. **22**(03), 187–205 (2012). https://doi.org/10.1142/s0218195912500045

8. Biedl, T., et al.: A note on reconfiguring tree linkages: trees can lock. Discret. Appl. Math. **117**(1–3), 293–297 (2002). https://doi.org/10.1016/s0166-218x(01)00229-3

9. Cabello, S., Demaine, E.D., Rote, G.: Planar embeddings of graphs with specified edge lengths. In: Liotta, G. (ed.) GD 2003. LNCS, vol. 2912, pp. 283–294. Springer, Heidelberg (2004). https://doi.org/10.1007/978-3-540-24595-7_26

10. Cabello, S., Demaine, E.D., Rote, G.: Planar embeddings of graphs with specified edge lengths. J. Graph Algorithms Appl. **11**(1), 259–276 (2007). https://doi.org/10.7155/jgaa.00145

11. Connelly, R., Demaine, E.D., Rote, G.: Straightening polygonal arcs and convexifying polygonal cycles. Discrete Comput. Geometry **30**(2), 205–239 (2003). https://doi.org/10.1007/S00454-003-0006-7

12. Cygan, M., et al.: Parameterized Algorithms. Springer, Cham (2015). https://doi.org/10.1007/978-3-319-21275-3

13. Demaine, E.D., O'Rourke, J.: Geometric Folding Algorithms: Linkages, Origami. Polyhedra. Cambridge University Press (2007). https://doi.org/10.1017/cbo9780511735172

14. Depian, T., Haase, C., Nöllenburg, M., Schulz, A.: Realizing planar linkages in polygonal domains (2026). https://doi.org/10.48550/ARXIV.2604.05786

15. Eades, P., Wormald, N.C.: Fixed edge-length graph drawing is NP-hard. Discret. Appl. Math. **28**(2), 111–134 (1990). https://doi.org/10.1016/0166-218x(90)90110-x

16. Erickson, J., van der Hoog, I., Miltzow, T.: Smoothing the gap between NP and ER. SIAM J. Comput. **53**(6), S20-102 (2024). https://doi.org/10.1137/20M1385287

17. Grigoriev, D., Jr., Vorobjov, N.: Solving systems of polynomial inequalities in subexponential time. J. Symbolic Comput. **5**(1–2), 37–64 (1988). https://doi.org/10.1016/s0747-7171(88)80005-1

18. Hopcroft, J., Joseph, D., Whitesides, S.: Movement problems for 2-dimensional linkages. SIAM J. Comput. **13**(3), 610–629 (1984). https://doi.org/10.1137/0213038

19. Hopcroft, J., Joseph, D., Whitcsides, S.: On the movement of robot arms in 2-dimensional bounded regions. SIAM J. Comput. **14**(2), 315–333 (1985). https://doi.org/10.1137/0214025

20. Joseph, D.A., Plantings, W.H.: On the complexity of reachability and motion planning questions (extended abstract). In: Symposium on Computational Geometry (SoCG), pp. 62–66. ACM (1985). https://doi.org/10.1145/323233.323242

21. Kantabutra, V.: Motions of a short-linked robot arm in a square. Discrete Comput. Geometry **7**(1), 69–76 (1992). https://doi.org/10.1007/bf02187825
22. Kapovich, M., Millson, J.J.: Universality theorems for configuration spaces of planar linkages. Topology **41**(6), 1051–1107 (2002). https://doi.org/10.1016/s0040-9383(01)00034-9
23. Kempe, A.B.: On a general method of describing plane curves of the nth degree by linkwork. Proc. Lond. Math. Soc. **1**(1), 213–216 (1875). https://doi.org/10.1112/plms/s1-7.1.213
24. Lubiw, A., Miltzow, T., Mondal, D.: The complexity of drawing a graph in a polygonal region. In: Biedl, T., Kerren, A. (eds.) GD 2018. LNCS, vol. 11282, pp. 387–401. Springer, Cham (2018). https://doi.org/10.1007/978-3-030-04414-5_28
25. Lubiw, A., Miltzow, T., Mondal, D.: The complexity of drawing a graph in a polygonal region. J. Graph Algorithms Appl. **26**(4), 421–446 (2022). https://doi.org/10.7155/jgaa.00602
26. Mitchell, J.S.B.: Geometric shortest paths and network optimization. In: Sack, J., Urrutia, J. (eds.) Handbook of Computational Geometry, chap. 15, pp. 633–701. Elsevier (2000). https://doi.org/10.1016/b978-044482537-7/50016-4
27. Preparata, F.P., Shamos, M.I.: Computational Geometry - An Introduction. Springer (1985). https://doi.org/10.1007/978-1-4612-1098-6
28. Whitesides, S., Pei, N.: On the reconfiguration of chains. In: Cai, J.-Y., Wong, C.K. (eds.) COCOON 1996. LNCS, vol. 1090, pp. 381–390. Springer, Heidelberg (1996). https://doi.org/10.1007/3-540-61332-3_172

Online Drone Coverage of Targets on a Line

Stefan Dobrev[1], Konstantinos Georgiou[2], Evangelos Kranakis[3](✉),
Danny Krizanc[4], Lata Narayanan[5], Jaroslav Opatrny[5], Denis Pankratov[5],
and Sunil Shende[6]

[1] Slovak Academy of Sciences, Bratislava, Slovakia
stefan.dobrev@savba.sk
[2] Department of Mathematics, Toronto Metropolitan University, Toronto, Canada
konstantinos@torontomu.ca
[3] School of Computer Science, Carleton University, Ottawa, Canada
evankranakis@gmail.com
[4] Department of Mathematics and Computer Science, Wesleyan University,
Middletown, CT, USA
dkrizanc@wesleyan.edu
[5] Department of Computer Science and Software Engineering, Concordia University,
Montreal, QC, Canada
lata@cs.concordia.ca, opatrny@cs.concordia.ca,
denis.pankratov@concordia.ca
[6] Department of Computer Science, Rutgers University, Camden, USA
sunil.shende@rutgers.edu

Abstract. We study a problem of online targets coverage by a drone or a sensor that is equipped with a camera or an antenna with fixed half-angle view α. The targets to be monitored appear at arbitrary positions on a line barrier in an online manner. When a new target appears, the drone has to move to a location that covers the newly arrived target, as well as already existing targets. The objective is to design a coverage algorithm that optimizes the total length of the drone's trajectory. Our results are reported in terms of an algorithm's competitive ratio, i.e., the worst-case ratio (over all inputs) of its cost to that of an optimal offline algorithm.

In terms of upper bounds, we present three online algorithms and prove bounds on their competitive ratios for every $\alpha \in [0, \pi/2]$. The best of them, called β-Hedge is significantly better than the other two for $\pi/6 < \alpha < \pi/3$. In particular, for $\alpha = \pi/4$, its worst case, β-Hedge has competitive ratio 1.25, while the other two have competitive ratio $\sqrt{2}$.

Finally, we prove a lower bound on the competitive ratio of online algorithms for a drone with half-angle $\alpha \in [0, \pi/4]$; this bound is a function of α that achieves its maximum value at $\alpha = \pi/4$ equal to $(1 + \sqrt{2})/2 \approx 1.207$.

S. Dobrev was supported in part by VEGA 2/0117/25, and K. Georgiou, E. Kranakis, L. Narayanan, D. Pankratov were supported in part by NSERC.

F. Foucaud and A. Parreau (Eds.): IWOCA 2026, LNCS 16587, pp. 266–280, 2026.
https://doi.org/10.1007/978-3-032-27732-9_19

Keywords: Competitive ratio · Online algorithm · Coverage algorithm · Drone

1   Introduction

We consider a problem of *path planning for online targets coverage* by a drone: given a sequence of target points that appear *online* in sequence on the ground, determine an optimal trajectory of the drone so that it can cover the collection of points seen so far. Each successive point, called a *request*, must be *serviced* by the drone by moving the drone into a position that covers the new request as well as all previous requests.

We study a simple model of a drone that can move and hover aerially at any given height above ground. The drone is equipped with a camera or scanning sensor that can see at any distance within its scanning sector on the ground below. We adopt a camera geometry that has been studied both experimentally and theoretically (see, for instance, [10]): the camera points straight at the ground and has a *fixed field-of-view* defined by the **half angle-of-view** α with respect to the vertical axis. We refer to the angle 2α as the **scanning angle** of the drone. The ground area scanned by the drone from a given height is defined by its *scanning cone* where the apex of the cone is the current position of the drone. This area is said to be *covered* or serviced by the drone.

We study the case where the targets/requests to be covered *all appear sequentially on a line*. Without loss of generality, we assume that the line is the x-axis. The drone is initially at position $(0, 0)$ and can move vertically and/or horizontally, to a position that covers all of the requests received so far. Since $\alpha \geq \pi/2$ would imply that at any height > 0, the drone covers the entire line, we assume that $\alpha < \pi/2$. Clearly, a drone at height h above the ground covers a segment on the line with length $2h \tan(\alpha)$.

Note that the drone can expand and/or translate its coverage by increasing its height and/or by moving parallel to the ground.

Displacing a drone from one position to another consumes energy; to simplify the model we will assume that the energy consumed by the drone is proportional to the total length of its trajectory (although horizontal and vertical energy consumption may differ), as has been used in the literature [17, 20]. The goal of our algorithms for drone coverage is to minimize the total sum of displacements of the drone needed to cover all the requests. If all the requests are known in advance, then the optimal (offline) algorithm would simply move directly to the closest point that covers all the points, however we study the online case.

1.1   Our Contributions

We prove upper and lower bounds for the competitive ratio of *online* algorithms for a single drone trajectory that covers target points arriving one at a time on a line.

We start in preliminaries with two simple and natural online algorithms. In the STRAIGHT-UP algorithm, in response to a new request, the drone simply moves straight up from its initial location exactly as much as needed to cover all existing points. In the GREEDY algorithm, in response to the new request, the drone moves to the *closest* point that would ensure coverage of all existing requests. We analyze the performance of these two algorithms and observe their weaknesses.

In Sect. 3 we define our main online algorithm, called β-HEDGE, that *hedges* between the two extreme angle displacements of STRAIGHT-UP and GREEDY algorithms, and travels at a fixed angle β to reach the closest feasible point that covers all requests. The angle β that we choose is a function of the scanning half-angle α of the drone. We derive an explicit formula that gives for each α the value of β that yields the best worst-case performance, and prove tight bound on the competitive ratio of β-HEDGE algorithm for every $\alpha \in [0, \pi/2]$. For $\pi/3 \leq \alpha < \pi/2$, it turns out that β-HEDGE is the same as STRAIGHT-UP and for $0 < \alpha \leq \pi/6$, β-HEDGE is the same as GREEDY. However, in the middle range $\pi/6 < \alpha < \pi/3$, β-HEDGE significantly outperforms both STRAIGHT-UP and GREEDY algorithms. The worst performance of β-HEDGE is at $\alpha = \pi/4$ with competitive ratio 1.25, while STRAIGHT-UP and GREEDY have competitive ratio $\sqrt{2}$ at this scanning angle.

Finally, in Sect. 4 we determine a lower bound on the competitive ratio of any online algorithm for $\alpha \in [0, \pi/4]$. The best lower bound is for $\alpha = \pi/4$, where the competitive ratio of any online algorithm is $\geq (1 + \sqrt{2})/2 \approx 1.207$. Figure 1 summarizes our results. Unfortunately, due to space constraints, we are unable to provide many of the intricate details in our proofs and explanations. These details can be found in the full version of the paper [5].

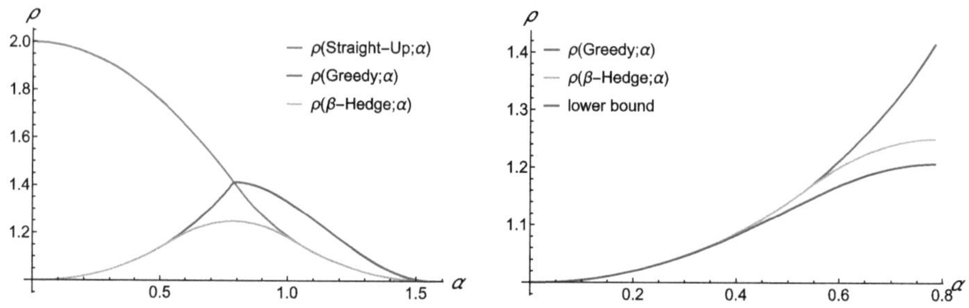

Fig. 1. This plot depicts competitive ratios of algorithms studied in this paper as functions of $\alpha \leq \pi/2$, as well as the lower bound (which holds for $\alpha \leq \pi/4$).

1.2 Related Work

Area coverage, barrier coverage, and target coverage are key application areas of sensor networks and have been extensively studied [8,12,18]. In area and barrier coverage [11,16], either an entire geometrical region's area or a specific barrier guarding the region is to be covered/monitored by a set of sensors. In target coverage [4,7,9,13,19], the goal is to cover a set of target points. There is also considerable research on coverage of points and regions by drones [14,15,20].

Some previous work has addressed questions about how single drones or swarms of drones can cover a specified area of interest using static path planning [1] or cover a set of moving targets [3].

Perhaps the theoretical work that comes closest to ours is the Convex Body Chasing by a guarding point [2,6]: a sequence of convex bodies is presented online and the goal is to move the guard with minimal total movement so that it always remains within the current convex body. As far as we are aware, there is no other work on *online* target coverage by sensors or drones. We give a fuller exposition of the related work in [5].

2   Preliminaries

In this section we present the necessary notation and terminology as well as two simple online algorithms.

2.1   Notation

We study the version of the online drone coverage problem where the requests appear on the x-axis in a two dimensional plane. The drone can move to any position in the upper part of the plane and the y coordinate of the drone specifies the height of the drone above the ground.

We will use uppercase letters for points in the plane, lowercase letters for coordinates, and boldface capital letters for sequences or sets of points. Given points $P_1 = (x_1, y_1)$ and $P_2 = (x_2, y_2)$, we use $P_1 P_2$ to denote the line segment between P_1 and P_2, and $|P_1 P_2| = \|P_1 - P_2\|_2 = \sqrt{(x_1 - x_2)^2 + (y_1 - y_2)^2}$ for the Euclidean distance between P_1 and P_2. The *coverage area* of a drone with scanning angle 2α, located at point $T = (t_x, t_y)$, is defined as the area of the triangle $\Delta P_l T P_r$ with apex T and its base points $P_l = (t_x - t_y \tan \alpha,\ 0)$ and $P_r = (t_x + t_y \tan \alpha,\ 0)$ on the x-axis.

We assume that the drone is initially located on the ground at point $(0,0)$. The input to the drone coverage problem is the scanning angle 2α of the drone, and a sequence of points $\mathbf{X} = (X_0 = (0,0), X_1, X_2, X_3 \ldots X_n)$ on the x-axis, where point $X_i = (x_i, 0)$ specifies the position of the i-th request. A solution to the drone coverage problem is a sequence $\mathbf{P} = (P_0 = (0,0), P_1, P_2, \ldots, P_n)$ of points, where P_i is the location of the drone providing the coverage of requests $X_0, X_1, \ldots, X_i$.

Let $l_i = \min(x_0, x_1, \ldots, x_i)$ and $r_i = \max(x_0, x_1, \ldots, x_i)$ be the extreme x-coordinates among $X_0, X_1, \ldots, X_i$, i.e., the interval spanned by $X_o, X_1, \ldots, X_i$ is $(l_i, 0)(r_i, 0)$ and hence the position P_i of the drone must be such that it covers this interval. Consequently, if request X_{i+1} is within the interval, then it is trivially covered by the drone from position P_i and the drone need not move.

We call such a request *redundant*. If request X_{i+1} lies outside interval $(l_i, 0)(r_i, 0)$ then the drone might need to move to position P_{i+1} different from P_i to cover the expanded interval spanned by the current requests. To provide a meaningful comparison between outputs of different inputs, we scale and possibly flip $\boldsymbol{X}$ so that $l_n = -1$ and $|l_n| \geq r_n$. Thus we assume in the rest of the paper that any given input sequence is *good*, i.e. without redundant requests, $X_0 = (0, 0)$, and scaled.

The cost of moving the drone from position P_i to P_{i+1} is assumed to be $|P_i P_{i+1}|$, and we define the *net cost* of an algorithm ALG with the drone positions $P_0, P_1, \ldots, P_n$ to be

$$\mathrm{ALG}(\boldsymbol{X}; \alpha) = \sum_{i=0}^{n-1} |P_i P_{i+1}|.$$

In the online problem, an adversary controls the sequence of requests while ALG must make local, irrevocable decisions about where to move the drone next to ensure coverage of all requests seen thus far. We use the standard measure of *competitive ratio* to measure the performance of ALG: it is the worst-case ratio (over the inputs) of the cost incurred by ALG divided by the corresponding cost incurred by an *optimal offline algorithm* OPT that knows the entire input instance in advance. Formally,

$$\rho(\mathrm{ALG}; \alpha) = \max \frac{\mathrm{ALG}(\boldsymbol{X}; \alpha)}{\mathrm{OPT}(\boldsymbol{X}; \alpha)}$$

Consider a target point X on the x-axis and a drone with scanning angle 2α that seeks to cover it. The *feasibility cone* of X, denoted $\mathrm{FC}(X)$, is defined as the locus of drone locations in the plane that cover X: it is the upward-facing cone with apex X whose flanking edges form angles $-\alpha$ and $+\alpha$ with respect to the y axis. In what follows, we will abuse notation by referring to the point $(x, 0)$ on the x-axis as the point X.

For sequence $\boldsymbol{X} = (X_0, X_1, \ldots X_n)$ of requests, the feasibility cone of $\boldsymbol{X}$, defined by $\mathrm{FC}(\boldsymbol{X}) = \bigcap_{0 \leq j \leq n} \mathrm{FC}(X_j)$, is the intersection of all the feasibility cones of requests in $\boldsymbol{X}$. Since X_n is non-redundant, X_n is one of the extreme requests. Let X_j be the other extreme request. Hence $\mathrm{FC}(\boldsymbol{X}) = \mathrm{FC}(X_j) \cap \mathrm{FC}(X_n)$. For any $0 \leq j \leq n$, we denote by T_j the apex of the feasibility cone $\mathrm{FC}((X_0, X_1, \ldots, X_j))$.

Lemma 1. *For any $0 \leq i \leq n$, consider the path $\mathcal{P}_i = T_0, T_1, \ldots, T_i$ that proceeds sequentially through the apexes of the feasibility cones of the first i subsequences of $\boldsymbol{X}$. Then the total length $\|\mathcal{P}_i\|$ of this path equals $|(x_i, 0)\, T_i|$, the distance between points $(x_i, 0)$ and T_i.*

Proof. The solid and the dotted blue lines in Fig. 2 give an intuition of the validity of the claim, a detailed proof is in [5]. $\square$

Lemma 2. *Consider the set of requests $\boldsymbol{X} = \{X_0, X_1, \ldots, X_n\}$ that span the interval $[-1, x_j]$ on the X-axis, where X_j is the rightmost request. Let $r = |x_j - x_0| \leq 1$. Then, the cost of the optimal offline algorithm is*

$$\text{OPT}(\boldsymbol{X}; \alpha) = \begin{cases} \frac{1}{2}\sqrt{(1+r)^2 \cot^2 \alpha + (1-r)^2} & \text{if } \alpha \leq \pi/4 \text{ or } r \geq \frac{\tan^2 \alpha - 1}{1 + \tan^2 \alpha}, \\ \cos \alpha & \text{otherwise.} \end{cases}$$

Proof. In the optimal offline algorithm the drone goes in a straight line from X_0 to the *point closest to it* on the feasibility cone $\text{FC}(\boldsymbol{X})$. There are two cases to consider: $\alpha \leq \pi/4$ and $\pi/4 < \alpha < \pi/2$. Details can be found in the full version of the paper [5]. $\qquad\square$

2.2   Two Simple Online Algorithms

Given a good instance $\boldsymbol{X} = (X_0, X_1, \ldots, X_n)$ of the Drone Coverage Problem, the execution of an online algorithm for this problem consists of $n + 1$ rounds. At round 0 the drone is at location $P_0 = X_0 = (0,0)$. At round i, $1 \leq i \leq n$, the algorithm calculates, using a specific strategy, the next position P_i of the drone based on the known requests $(X_0, X_1, \ldots, X_{i-1})$, the newly revealed request X_i, and the present position P_{i-1}. This increases the cost of the algorithm by $|P_i P_{i-1}|$. Clearly, several strategies can be considered. We first specify below two straight-forward strategies and determine their competitive ratios. These are used for comparison to our main β-HEDGE algorithm defined in Sect. 3.

The Straight-Up Algorithm. In round i of the STRAIGHT-UP algorithm, if the coverage area of the drone at its present location doesn't cover the next request X_i, the drone moves vertically, or we can say *straight up* from its present location P_{i-1} on the y axis until it reaches the feasibility cone of X_i. Thus, the x-coordinate of the drone remains fixed at 0, and the next position P_i of the drone is the smallest point on the intersection of the y-axis with the feasibility cone of sequence $(X_0, X_1, \ldots, X_i)$, see Fig. 2. We can see that the cost of the STRAIGHT-UP algorithm depends only on the value of the extreme x-coordinate among the request points in sequence $\boldsymbol{X}$. Since the input is assumed to be good, the extreme x coordinate is equal to -1, and the cost doesn't depend on r, the largest positive x coordinate in the input instance.

Theorem 1.

$$\rho(\text{STRAIGHT-UP}; \alpha) = \begin{cases} 2 \cos \alpha & \text{if } \alpha \leq \pi/4, \\ 1/\sin \alpha & \text{otherwise.} \end{cases} \qquad (1)$$

Proof. Observe in Fig. 2 that the cost of STRAIGHT-UP algorithm is the length of the red line equal to $\cot \alpha$, the cost of OPT is given in Lemma 2. The fraction of these two expressions gives the result. For details of calculations see the proof in [5]. $\qquad\square$

Theorem 1 shows that the competitive ratio of the STRAIGHT-UP algorithm is good for large values of α, but poor when α is small.

The Greedy Algorithm. During round i of the GREEDY algorithm, the drone moves from its present location P_{i-1} in a straight line to the *closest point* on the feasibility cone of the sequence $(X_0, X_1, \ldots, X_i)$ having apex T_i. When $\alpha \leq \pi/4$, the closest point remains the apex T_i of $FC(X_0, X_1, \ldots, X_i)$, but when $\alpha > \pi/4$, the closest point is on the side of the $FC(X_0, X_1, \ldots, X_i)$ cone, reached by traveling perpendicularly to the ray of the feasibility cone at T_i. Thus, when $\alpha \leq \pi/4$ the drone travels along the edge of the feasibility cone from $P_{i-1} = T_{i-1}$ to $P_i = T_i$, at angle α or $-\alpha$ to the vertical line. When $\alpha > \pi/4$ the drone travels from P_{i-1} at angle $\pi/2 - \alpha$ or $(\alpha - \pi/2)$ to P_i on the side of the feasibility cone with apex T_i.

Theorem 2.

$$\rho(\text{GREEDY}; \alpha) = \begin{cases} 1/\cos\alpha & \text{if } \alpha \leq \pi/4, \\ (1 + 2\cos^2\alpha)\sin\alpha & \text{otherwise.} \end{cases} \quad (2)$$

Proof. It is easy to see that the worst case ratio is when $r = 1$ and therefore we analyze this case below.

Case 1: $\alpha \leq \pi/4$. As discussed above, the greedy algorithm follows the path $X_0, T_1, T_2, \ldots, T_n$ where T_i is the tip of the feasibility cone of the first i requests. As shown in Lemma 1, the length of the path is equal to the distance between the leftmost request and T_n. Since we assume the input instance is good, the leftmost request is the point at $(-1, 0)$ and for $r = 1$ the rightmost request is at point $(1, 0)$. Thus, the length of the path is equal to $1/\sin\alpha$, and the cost of the OPT for $r = 1$ is equal to $\cot\alpha$. This gives the competitive ratio of $1/\cos\alpha$ in this case.

Case 2: $\alpha > \pi/4$. The cost of the optimal algorithm is the height h in the triangle X_0, X_i, T, with $X_0 = (0,0)$, $X_i = (-1, 0)$ and the cost of the GREEDY algorithm is $|X_0U| + |UT'|$. We have the following equations: $h = \cot\alpha$, $|X_0U| = h\sin\alpha$, $|UT| = \cos^2\alpha/\sin\alpha$, $|UT'|/|uT| = \cos(2\alpha - \pi/2) = \sin 2\alpha = 2\sin\alpha\cos\alpha$, and thus $|X_0U| + |UT'| = (1 + 2\cos^2\alpha)\cos\alpha$. Therefore, $\rho = (1 + 2\cos^2\alpha)\sin\alpha$. $\square$

The competitive ratio formulas above show that the STRAIGHT-UP algorithm performs better than the GREEDY algorithm when $\alpha > \pi/4$, and the GREEDY algorithm performs better than the STRAIGHT-UP algorithm when $\alpha < \pi/4$. However the competitive ratio for both algorithms is rather large for mid-range α values around $\pi/4$.

3 β-Hedge Algorithm

In the STRAIGHT-UP algorithm, the angle of displacement of the drone with respect to the y axis is always 0. In the GREEDY algorithm the angle of displacement of the drone with respect to the y axis is either α or $-\alpha$ when $\alpha \leq \pi/4$, and is either $\pi/2 - \alpha$ or $\alpha - \pi/2$ when $\alpha > \pi/4$. Algorithm STRAIGHT-UP is better than GREEDY when α is large, but GREEDY is better than STRAIGHT-UP when α is small. Both STRAIGHT-UP and GREEDY algorithms perform poorly when

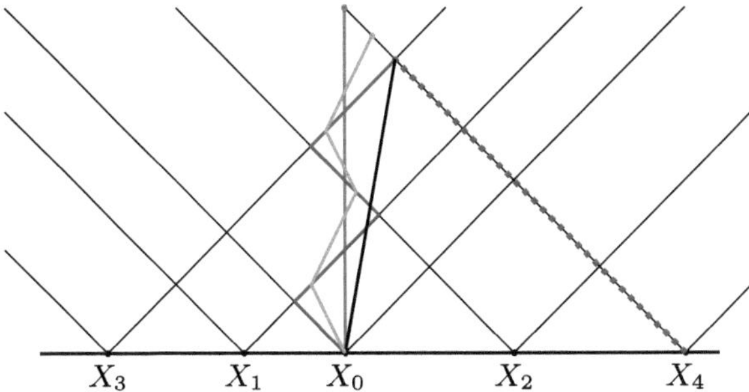

Fig. 2. Algorithm Trajectories for $\alpha = \pi/4$: OPT (in black), STRAIGHT-UP algorithm (in red), GREEDY algorithm (in solid blue, which is equivalent in distance to the dotted blue trajectory), and the β-HEDGE algorithm (in green). (Color figure online)

the scanning angle α is around $\alpha/4$. This motivates our β-HEDGE algorithm in which the angle β of the displacement of the drone, with respect to the perpendicular line $\beta \leq \alpha$, is a function of α that can be optimized.

In round i of the β-HEDGE algorithm, the drone travels from its current position P_{i-1} to point P_i on the side of the feasibility cone $\mathrm{FC}(X_0, X_1, \ldots, X_i)$ using either angle β or $-\beta$ with respect to the line perpendicular to the x-axis, see Fig. 3. Note that when $\beta = 0$, we obtain the STRAIGHT-UP algorithm, and when $\beta = \alpha$ and $\alpha \leq \pi/4$, we obtain the GREEDY algorithm. The optimized value of β is given in Theorem 3.

Lemma 3. *Given a good input instance with parameters α, r and an angle $0 \leq \beta < \alpha$, the distance travelled by the β-HEDGE algorithm with parameter β is equal to*

$$\beta\text{-HEDGE}\,(\alpha, r, \beta) = \frac{\tan \alpha + (1 + 2r)\tan \beta}{(\tan \alpha + \tan \beta)^2 \cos \beta}$$

Proof. Assume we are given a good input instance $X_0 = (0,0), X_1, \ldots, X_i$, where $X_j = (r,0)$ is the rightmost request. Observe in Fig. 3 that the total distance travelled by the β-HEDGE algorithm is equal to the sum of the distance traveled to the feasibility cone of points X_0 and X_j using angle β, i.e., from X_0 to point U, and the distance travelling from U to the feasibility cone of points X_i, X_j using angle $-\beta$, i.e., from U to T' in the figure. Clearly, this distance is equal to the length of the segment V, T'. Since the triangles $X_i V T'$ and $X_0 X_j U$ are similar, we have that $\frac{|VT''|}{1+z} = \frac{|X_0 U|}{r}$ where $z = |X_0 V|$. Since $\sin \beta = \frac{z}{2|X_0 U|}$, we have that $|VT'| = \frac{z(z+1)}{2r\sin \beta}$. We have the following equations: $\tan \beta = z/(2h')$ and $\tan \alpha = (2r - z)/2h'$, where h' is the height of U in the triangle $X_0 X_j U$. Solving the two equations we get $z = \frac{2r \tan \beta}{\tan \alpha + \tan \beta}$. Substituting the value of z into the expression for $|VT'|$ we get

$$|VT'| = \frac{2r \tan \beta}{2r \sin \beta(\tan \alpha + \tan \beta)} \cdot (1 + \frac{2r \tan \beta}{\tan \alpha + \tan \beta}) = \frac{\tan \alpha + (1 + 2r)\tan \beta}{(\tan \alpha + \tan \beta)^2 \cos \beta}$$

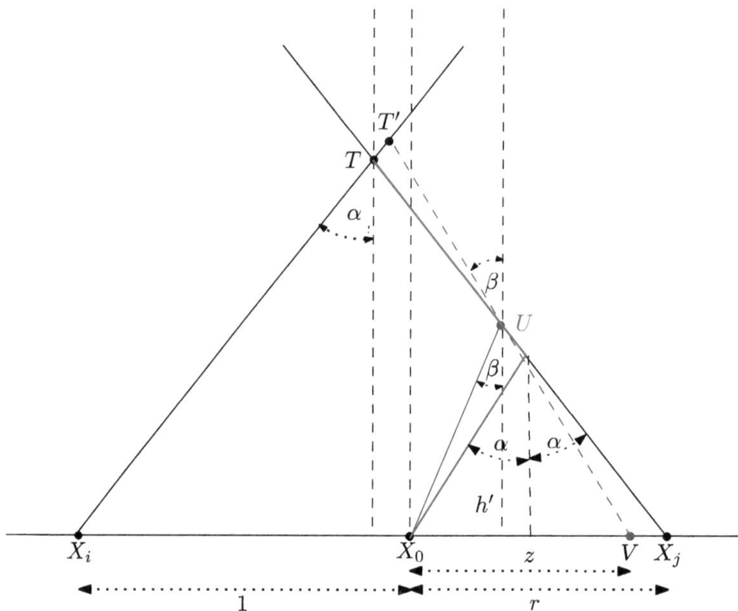

Fig. 3. The path X_0, U, T' is traveled by the β-HEDGE algorithm, while the length of the path traveled by the GREEDY corresponds to $|X_jT|$.

Obtaining an explicit expression for the optimal value of β for a given value of α and thereby of the competitive ratio of the algorithm, requires an extensive and detailed mathematical analysis that is presented fully in [5].

We define the function

$$f_1(\alpha, \beta, r) = \frac{2\sin(\alpha)}{\cos(\beta)(\tan(\alpha) + \tan(\beta))} \cdot \frac{1 + \frac{2\tan(\beta)}{\tan(\alpha)+\tan(\beta)}r}{\sqrt{1 + r^2 + 2\cos(2\alpha)r}} \tag{3}$$

over the domain $\{(\alpha, \beta, r) \in \mathbb{R}^3 : \alpha \in [0, \pi/2], \ \beta \in [0, \alpha], \ \text{and} \ r \in [0, 1]\}$.

We show that with an adversarial choice function $r_0(\alpha, \beta)$ and the specific function $f_1(\alpha, \beta, r)$, the following holds (details given in [5]):

Theorem 3. *For every* $\alpha \in [0, \pi/2]$, *the optimal angle* β *for the* β-HEDGE *algorithm is given by*

$$\beta_0 = \begin{cases} \alpha & \text{if } \alpha \in \left[0, \frac{\pi}{6}\right], \\ \frac{1}{2}\arccos\left(\frac{-2\cos(4\alpha)+\cos(6\alpha)+2}{3-2\cos(4\alpha)}\right) & \text{if } \alpha \in \left(\frac{\pi}{6}, \frac{\pi}{3}\right), \\ 0 & \text{if } \alpha \in \left[\frac{\pi}{3}, \frac{\pi}{2}\right], \end{cases}$$

where $\beta_0 = \beta_0(\alpha)$. *The choice of* β_0 *results in competitive ratio*

$$\rho(\beta\text{-HEDGE};\alpha) = f_1(\alpha, \beta_0, r_0(\alpha, \beta_0))$$

In particular, when $\alpha \le \pi/6$, the adversarial choice $r_0(\alpha, \beta_0) = 1$ and the competitive ratio is $\sec(\alpha)$, whereas when $\alpha \ge \pi/3$, $r_0(\alpha, \beta_0) = -\cos 2\alpha$ and the competitive ratio is $\csc(\alpha)$.

4   Lower Bound on the Competitive Ratio

In this section, we establish a lower bound (see Theorem 4 at the end of this section) on the competitive ratio of deterministic algorithms for $\alpha \in [0, \pi/4]$. As usual, we assume that α is fixed. For a given $s > 1$, the adversary defines the infinite sequence of requests $\boldsymbol{X}$, where $X_0 = (0,0)$ and $X_i = (x_i, 0)$ with $x_i = 2(-s)^{i-1}$ is the i-th request. Observe that the x_i form a geometric sequence with common ratio $(-s)$. Since an online algorithm does not know the length of the input, it must maintain good competitive ratio on every prefix $\boldsymbol{X}_i = (X_0, X_1 \ldots, X_i)$ of $\boldsymbol{X}$. To establish the lower bound we design an optimal online algorithm MAXHEDGE for $\boldsymbol{X}$, and show that any other algorithm that deviates from MAXHEDGE can be trapped by the adversary by terminating $\boldsymbol{X}$ at an appropriate time. Then we analyze the competitive ratio of MAXHEDGE to derive a closed-form formula for the lower bound. The proofs of lemmas and theorems are in [5].

Recall that T_i is the apex of the feasibility cone of $\boldsymbol{X}_i$. Observe that for the sequence $\boldsymbol{X}$ defined as above, we have the following:

$$T_1 = \left(1, \frac{1}{\tan(\alpha)}\right), T_i = \left((-s)^{i-2}(1-s), \frac{s^{i-2}(1+s)}{\tan(\alpha)}\right) \text{ for } i \ge 2.$$

Note that all T_i for even i lie on a straight line with $T_i = s^{i-2}T_2$. Similarly all T_i for odd i larger than 1 lie on a straight line, again with $T_i = s^{i-3}T_3$. Observe that these two lines are symmetric along the vertical axis, holding an angle $\beta = \arctan(\frac{s-1}{s+1} \tan \alpha)$ with it.

We also introduce the shorthand notation o_i to denote the cost of the optimal solution for $\boldsymbol{X}_i$. For $i \ge 2$ we have:

$$o_i = \mathrm{OPT}(\boldsymbol{X}_i; \alpha) = |X_0 T_i| = \sqrt{s^{-4+2i}\left(-4\,s + \frac{(1+s)^2}{\sin^2(\alpha)}\right)}.$$

Next, we are ready to define the algorithm MAXHEDGE. This algorithm has a parameter ρ, a desired/prescribed competitive ratio that MAXHEDGE tries to achieve on the input $\boldsymbol{X}$, regardless of when the infinite sequence is terminated. In essence, MAXHEDGE is doing maximal hedging (preparing for the next request on the other side of the y-axis) while ensuring that it does not exceed the desired competitive ratio ρ. Intuitively, no algorithm can achieve better performance than MAXHEDGE for this input sequence. More specifically, MAXHEDGE with parameter ρ is defined as follows:

-   Response to request X_i is a point $Z_i = T_i + z_i(T_{i+1} - T_i)$ for $z_i \ge 0$ such that $|Z_{i-1}Z_i| = \rho(o_i - o_{i-1})$,

- i.e. Z_i is the intersection of half-line $\overrightarrow{T_iT_{i+1}}$ with a circle centered at Z_{i-1} and radius $\rho(o_i - o_{i-1})$.
  - If such a point Z_i does not exist (i.e., the circle and the half-line do not intersect), we say that MaxHedge *fails* in round i.

If MaxHedge never fails, we say that it *succeeds*, the *domain* of MaxHedge is $[0, i-1]$ if MaxHedge fails in round i and it is $[0, \infty]$ otherwise. Note that the definition of MaxHedge inductively ensures that its competitive ratio is precisely ρ. Moreover, if MaxHedge succeeds, then it achieves a competitive ratio ρ on every prefix of $\boldsymbol{X}$. On the other hand, if MaxHedge fails, then it does not achieve a competitive ratio ρ on every prefix of $\boldsymbol{X}$. We claim that the optimal online algorithm for $\boldsymbol{X}$ is MaxHedge with the smallest value of ρ that makes it succeed.

Definition 1. *The execution of* MaxHedge *is valid, if for all rounds in the domain of* MaxHedge, Z_i *is on the same side of the y-axis as T_i.*

Note that when s is large enough and ρ is small enough, Z_i lies on the same side as T_i with respect to the y-axis (see Fig. 4a). For small values of s or for large values of ρ, the computed point Z_i may be too far towards T_{i+1} – it should not go beyond the vertical axis. This can be fixed by placing Z_i appropriately on the vertical axis in such a case. However, that would complicate the analysis of MaxHedge, and it happens only in situations that are irrelevant to our lower bound: a) a small value of s yields a small lower bound (Straight-Up is good enough for such cases), and b) we are looking at the smallest acceptable ρ for which MaxHedge succeeds. Thus, in the remainder of this section, we focus on valid executions of MaxHedge.

Consider now an arbitrary algorithm ALG to cover the requests. Let P_i for $i \geq 1$ be the response of ALG to X_i (by definition, $P_0 = X_0 = (0,0)$). Note that we do not insist on P_i's lying on the boundary of $\mathrm{FC}(\boldsymbol{X}_i)$. Let $c_i = \mathrm{ALG}(\boldsymbol{X}_i; \alpha) = \sum_{j=0}^{i-1} |P_j P_{j+1}|$ be the algorithm's cost for the first i requests. Let $\mathcal{S}(V, r)$ be the closed disk with its center at point V and with radius r.

The following terminology shall be used to compare the performance of ALG with MaxHedge:

- ALG is *cautious* in round i if $P_i \in \mathcal{S}(Z_{i-1}, |Z_{i-1}Z_i|)$;
- ALG is *aggressive* in round i if $P_i \notin \mathcal{S}(Z_{i-1}, |Z_{i-1}Z_i|)$;
- ALG is *always cautious* if it is cautious in all rounds in the domain of MaxHedge;
- ALG has *gone aggressive* in round i if it was cautious in all previous rounds, and is aggressive in round i.

First, we show that being cautious does not help:

Lemma 4. *Suppose that the execution of* MaxHedge *is valid for s and ρ, and* ALG *is cautious until and including round i. Then* $\mathrm{ALG}(\boldsymbol{X}_i; \alpha) \geq \mathrm{ALG}'(\boldsymbol{X}_i; \alpha)$ *where* ALG' *follows* MaxHedge *for the first $i-1$ steps and in the i-th step goes to P_i.*

The following lemma shows that MAXHEDGE is an optimal online algorithm for the request sequence $\mathbf{X}$.

Lemma 5. *If* MAXHEDGE *fails for a given choice of parameters s and ρ then the competitive ratio of any* ALG *is more than ρ.*

Next, we wish to compute the smallest value of ρ such that MAXHEDGE succeeds. Consider an execution of MAXHEDGE in an arbitrary round $i \geq 2$. Let $b_{i+1} = \rho(o_{i+1} - o_i)$ be its budget for responding to request X_{i+1}. Let $D_1^i = T_i + d_1(T_{i+1} - T_i) \in \overline{T_{i+1}T_i}$ be a point for which $|D_i^1 T_{i+1}| = b_{i+1}$ (note that due to regular scaling of T_i points, d_1 is the same for all $i > 2$). In other words, D_i^1 is the last point[1] on the halfline $\overrightarrow{T_{i+1}T_i}$ that is able reach (within the allowed budget b_{i+1}) the halfline $\overrightarrow{T_{i+1}T_{i+2}}$ (refer to Fig. 4b).

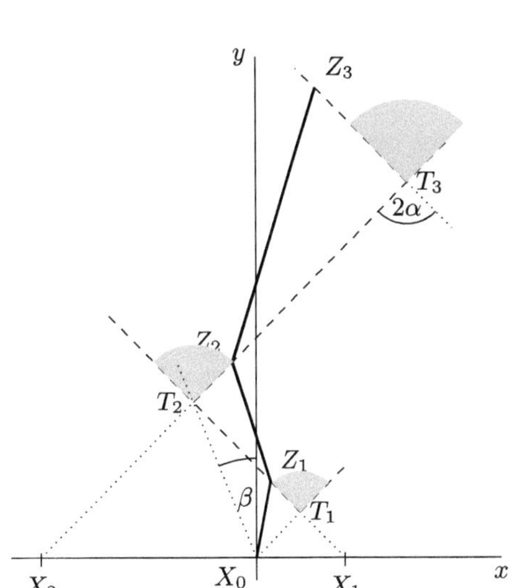

(4a) The requests X_i, feasibility cone tips T_i and MAXHEDGE's responses Z_i for $i = 1, 2, 3$ and $\alpha = \pi/4$.

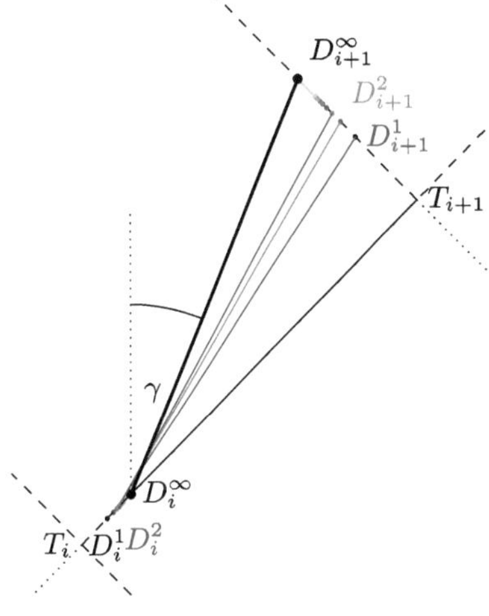

(4b) Progressively defining the dead zones: D_i^1 just barely reaches T_{i+1}, D_i^2 just reaches D_{i+1}^1 and in general D_i^{j+1} just reached D_{i+1}^j. $D_i^\infty = \lim_{j \to \infty} D_i^j$, if such limit exists.

Fig. 4. .

Clearly, if $z_i < d_1$ then MAXHEDGE fails in round $i+1$. Hence, the point set

$$\mathcal{D}_1 = \{T_i + r(T_{i+1} - T_i)|i > 1, r < d_1\}$$

represents *level-1 dead points* and should be avoided by MAXHEDGE.

---

[1] Note that for sufficiently large ρ we might have $d_1 \leq 0$, however we are interested in small ρ, for which $d_1 > 0$.

As the aim is for MAXHEDGE to succeed, $\mathcal{D}_1$ does not fully capture the points to be avoided. It is not sufficient from Z_i to reach $\overrightarrow{T_{i+1}T_{i+2}}$, it must reach outside of $\mathcal{D}_1$ on $\overrightarrow{T_{i+1}T_{i+2}}$. Let D_i^2 be the point of $\overrightarrow{T_iT_{i-1}}$ at a distance b_{i+1} from D_{i+1}^1. Again, we can express D_i^2 as $T_i + d_2(T_{i+1} - T_i)$. Since $\angle T_i T_{i+1} T_{i+2} \geq \pi/2$, if $z_i < d_2$ then Z_i can only reach level-1 dead points on $\overrightarrow{T_{i+1}T_{i+2}}$, hence it will fail in round $i + 2$.

The argument can be applied inductively, defining D_i^{j+1} as the first point on $\overrightarrow{T_iT_{i+1}}$ being able to reach a point outside $\mathcal{D}_j$ in $\overrightarrow{T_{i+1}T_{i+2}}$. From this, we can analogously define d_{j+1} and $\mathcal{D}_{j+1}$. Observe (basically from the definition of d_j) that if $z_i < d_j$ then MAXHEDGE fails in round at most $i + j$.

Lemma 6. *The sequence $\{d_i\}$ is a well defined and increasing sequence on the domain of* MAXHEDGE. *Conversely, if the sequence $\{d_i\}$ has a limit d_∞ and if $z_2 \geq d_\infty$ then* MAXHEDGE *succeeds.*

Recall that the adversarial sequence $\boldsymbol{X}$ defined in this section has a parameter s. In the full version of the paper (that appears in [5]), we show that there is an $s = s^*$, given by an explicit function of α, that establishes the following lower bound on the competitive ratio an arbitrary deterministic algorithm on the adversarial sequence $\boldsymbol{X}(s^*)$:

Theorem 4. *Fix $\alpha \in [0, \pi/4]$ and an arbitrary deterministic algorithm* ALG. *Then*

$$\rho(\mathrm{ALG}; \alpha) \geq \rho^* \ \text{where} \ \rho^* = \frac{s^*(1 + s^*)\sqrt{1 - \cos(4\alpha)}}{\sqrt{2(1 + (s^*)^4) - 4(s^*)^2\cos(4\alpha)}}.$$

In particular, for $\alpha = \pi/4$, no deterministic online algorithm can achieve a competitive ratio better than $\frac{1+\sqrt{2}}{2}$. $\qquad\square$

5   Conclusions

We considered an optimization problem for target coverage by a camera-equipped drone that can move to cover targets appearing online at arbitrary locations on a line. We designed path-planning algorithms which ensure coverage by a single mobile drone of new as well as old targets and gave upper and lower bounds on the optimal total movement of the drone. An interesting question for future research would be to study the drone coverage problem for camera-equipped swarms (drone networks) in more realistic geometric settings, e.g., in 3D space.

References

1. Bezas, K., Tsoumanis, G., Angelis, C.T., Oikonomou, K.: Coverage path planning and point-of-interest detection using autonomous drone swarms. Sensors **22**(19), (2022)
2. Bubeck, S., Klartag, B.A., Lee, Y.T., Li, Y., Sellke, M.: Chasing nested convex bodies nearly optimally. In: Proceedings of SODA, pp. 1496–1508 (2020)
3. Caillouet, C., Giroire, F., Razafindralambo, T.: Optimization of mobile sensor coverage with UAVs. In: IEEE INFOCOM 2018-IEEE Conference on Computer Communications Workshops, pp. 622–627. IEEE (2018)
4. Chen, Z., Gao, X., Wu, F., Chen, G.: A PTAS to minimize mobile sensor movement for target coverage problem. In: IEEE INFOCOM 2016 - The 35th Annual IEEE International Conference on Computer Communications, pp. 1–9 (2016). https://doi.org/10.1109/INFOCOM.2016.7524334
5. Dobrev, S., et al.: Online drone coverage of targets on a line (2026). https://arxiv.org/abs/2604.02491, arXiv:2604.02491
6. Friedman, J., Linial, N.: On convex body chasing. Discrete Comput. Geom. **9**(3), 293–321 (1993). https://doi.org/10.1007/BF02189324
7. Johnson, M.P., Bar-Noy, A.: Pan and scan: configuring cameras for coverage. In: 2011 Proceedings IEEE INFOCOM, pp. 1071–1079. IEEE (2011)
8. Junbin, L., Liu, M., Kui, X.: A survey of coverage problems in wireless sensor networks. Sens. Transducers **163**, 240–246 (2014)
9. Liao, Z., Wang, J., Zhang, S., Cao, J., Min, G.: Minimizing movement for target coverage and network connectivity in mobile sensor networks. IEEE Trans. Parall. Distrib. Syst. **26**(7), 1971–1983 (2015)
10. Longmore, S.N., et al.: Adapting astronomical source detection software to help detect animals in thermal images obtained by unmanned aerial systems. Int. J. Remote Sens. **38**(8-10), 2623–2638 (2017)
11. Ma, H., Yang, M., Li, D., Hong, Y., Chen, W.: Minimum camera barrier coverage in wireless camera sensor networks. In: 2012 Proceedings IEEE INFOCOM, pp. 217–225 (2012). https://doi.org/10.1109/INFCOM.2012.6195602
12. Meguerdichian, S., Koushanfar, F., Potkonjak, M., Srivastava, M.B.: Coverage problems in wireless ad-hoc sensor networks. In: Proceedings IEEE INFOCOM 2001 Conference on Computer Communications. 20th Annual Joint Conference of the IEEE Computer and Communications Society, vol. 3, pp. 1380–1387 (2001)
13. Munishwar, V.P., Abu-Ghazaleh, N.B.: Scalable target coverage in smart camera networks. In: Proceedings of the Fourth ACM/IEEE International Conference on Distributed Smart Cameras, ICDSC 2010, pp. 206–213, New York, NY, USA, ACM (2010)
14. Saeed, A., Abdelkader, A., Khan, M., Neishaboori, A., Harras, K.A., Mohamed, A.: On realistic target coverage by autonomous drones. ACM Trans. Sen. Netw. **15**(3), (2019). https://doi.org/10.1145/3325512
15. Savkin, A.V., Huang, H.: A method for optimized deployment of a network of surveillance aerial drones. IEEE Syst. J. **13**(4), 4474–4477 (2019)
16. Shih, K.P., Chou, C.M., Liu, I.H., Li, C.C.: On barrier coverage in wireless camera sensor networks. In: 2010 24th IEEE International Conference on Advanced Information Networking and Applications, pp. 873–879 (2010)
17. Shivgan, R., Dong, Z.: Energy-efficient drone coverage path planning using genetic algorithm. In: 2020 IEEE 21st International Conference on High Performance Switching and Routing (HPSR), pp. 1–6. IEEE (2020)

18. Wang, B.: Coverage problems in sensor networks: a survey. ACM Comput. Surv. **43**(4), (2011). https://doi.org/10.1145/1978802.1978811
19. Yao, P., Guo, L., Li, S., Peng, H.: Target coverage with minimum number of camera sensors. In: Proceedings of COCOA (LNCS), vol. 13135, pp. 12–24. Springer LNCS (2021)
20. Zorbas, D., Di Puglia, L., Pugliese, T.R., Guerriero, F.: Optimal drone placement and cost-efficient target coverage. J. Netw. Comput. Appl. **75**, 16–31 (2016)

Set Selection with Uncertain Weights: Non-Adaptive Queries and Thresholds

Christoph Dürr[1]([✉]) , Arturo Merino[2] , José A. Soto[3] , and José Verschae[4]

[1] Sorbonne Université, CNRS, Laboratoire d'informatique de Paris 6, LIP6, Paris, France
Christoph.Durr@lip6.fr
[2] Department of Computer Science, Universidad de Chile, Santiago, Chile
[3] Department of Mathematical Engineering and Center of Mathematical Modeling CNRS IRL 2807, Universidad de Chile, Santiago, Chile
[4] Institute for Mathematical and Computational Engineering, Faculty of Mathematics and School of Engineering, Pontificia Universidad Católica, Santiago, Chile

Abstract. For illustration, suppose we have to compute a shortest path in a graph between two vertices, but the edge weights are uncertain. All we know is that the weight w_e of every edge e belongs to some interval $[\ell_e, h_e]$. In this setting, for every fixed edge e, we can define two thresholds T_e^+, T_e^- such that e belongs to every shortest path if $w_e < T_e^+$ and to none if $w_e > T_e^-$. We study the problem of computing these thresholds for various set selection problems. In particular, we show that the problem is easy for the minimum spanning tree problem and the maximum weight matching problem on trees. In contrast, we show that the problem is NP-hard for the shortest path problem and for the minimum cost perfect matching problem.

Keywords: threshold · combinatorial optimization under uncertainty · uncertainty interval · NP-complete · minimum spanning tree · shortest path · matching

1 Introduction

We study set selection problems under uncertain weights. Without uncertainty, an instance of *(min-weight) set selection* consists of a *ground set E*, together with weights $w : E \to \mathbb{R}$, and an (implicitly defined) collection of *feasible sets* $\mathcal{F} \subseteq 2^E$. The goal is to find a feasible set $S \in \mathcal{F}$ that minimizes the total weight $w(S) := \sum_{e \in S} w(e)$.

C. Dürr—Part of the work was done while C.D. was affiliated with Universidad de Chile, CNRS, CMM, Santiago, Chile and while A.M. was affiliated with Saarland University, Computer Science department, Saarbrücken, Germany.

Set selection encodes many classical problems in combinatorial optimization. In particular, it encodes the problems of finding minimum spanning trees, shortest paths, and minimum-weight perfect matchings. Maximization problems can be modeled similarly by simply multiplying the weights by -1.

In an uncertain setting, we are given again a ground set E and an implicitly defined collection of feasible sets $\mathcal{F} \subseteq 2^E$. However, we do not have precise weights for each element $e \in E$. Instead, for each element $e \in E$, we know an interval $I_e := [\ell_e, h_e]$ in which the true weight w_e lies; i.e., $w_e \in I_e$ (see Fig. 1a). We call such an interval the *uncertainty interval* of e and denote the collection of intervals by $\mathcal{I} := \{I_e\}_{e \in E}$. Note that it is possible for these intervals to be singletons. In that case we call them *trivial*. Additionally, every weight function w that obeys $\mathcal{I}$ is called a *realization* of $\mathcal{I}$; i.e., $w \in \prod_{e \in E} I_e$ (see Fig. 1b and 1c). Naturally, realizations of $\mathcal{I}$ are the possible *true weights* of an uncertain instance.

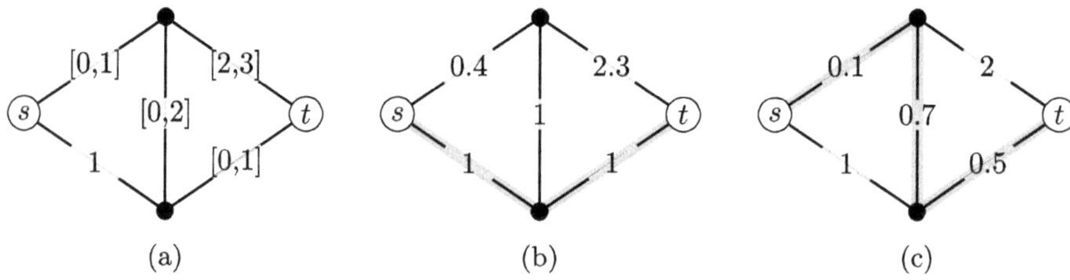

Fig. 1. (a) An instance of s–t paths with uncertain weights. (b)+(c) Two realizations of the instance in (a) with different optimal solutions marked in blue. (Color figure online)

Under uncertainty, we are again interested in minimizing the weight of the chosen solution. Note, however, that this could be realization dependent (see Figs. 1b and 1c); i.e., solutions could be optimal for some realizations and not for others. Thus, the best solution we could hope for is a feasible set that has the best weight independently of the realization. More formally, we say that a feasible set $S^* \in \mathcal{F}$ is *universally optimal* if we have $w(S^*) \leq w(S)$ for every feasible set $S \in \mathcal{F}$ and for every weight realization w of $\mathcal{I}$.

1.1   Minimum Cost Queries

Alas, not every uncertain set selection instance has universally optimal solutions (see, e.g., Fig. 1a). To counteract this, we consider the setting in which we can obtain more precise information about the weights at a cost. The reader can imagine that the original uncertainty intervals represent a rough estimate of the true weights and that we can spend additional resources to find out precise information about them.

The *(non-adaptive) query problem* models exactly the aforementioned situation: combinatorial optimization problems where the weights are uncertain, but can be queried at a cost. Querying an element reveals its true weight, replacing the possible weights it can take by a unique value. Here, our objective is to find the least costly set of queries that allows us to compute a universally optimal solution.

More specifically, we consider the setting where we can make exactly one (non-adaptive) set of queries; i.e., we can select a set $Q \subseteq E$ to be queried, and for every $e \in Q$ we will learn its true weight $w_e \in I_e$. A set of elements $Q \subseteq E$ is an *admissible query* if, after obtaining any possible answer to the queries in Q, there exists a universally optimal solution (see Fig. 2a for an illustration).[1] Note that admissible queries always exist, as querying every element is an admissible query. However, querying all weights is costly. Querying an element $e \in E$ comes at a cost c_e, which is known in advance. Thus, the objective is to find an admissible query Q that minimizes the total cost $c(Q)$.

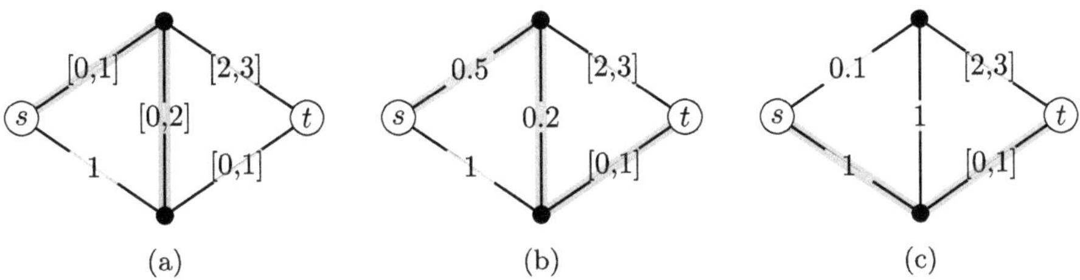

(a)   (b)   (c)

Fig. 2. (a) A minimum-sized admissible query marked in green. (b)+(c) Different true weights of the queried edges give rise to different universally optimal solutions (marked in blue). (Color figure online)

We are particularly interested in algorithms that not only compute a minimum-cost admissible query, but also compute a universally optimal solution (using the additionally queried information; see Figs. 2b and 2c). More concretely, these algorithms operate in the following stages.

1. Receive an instance $(E, \mathcal{F}, \mathcal{I})$.
2. *Query* a set of elements $Q \subseteq E$.
3. For the queried elements $e \in Q$, obtain the true weights $w_e \in I_e$.
4. Return a universally optimal solution $S^* \subseteq E$ for the intervals $\mathcal{I}'$ given by $\{w_e\}_{e \in Q} \cup \{I_e\}_{e \in E \setminus Q}$; i.e., $w(S^*) \leq w(S)$ for all feasible sets S and for all weight realizations $w \in \prod_{e \in Q} \{w_e\} \times \prod_{e \in E \setminus Q} I_e$.

We emphasize that it is *not* required that the algorithm knows the weight of the solution it outputs, only that it is *optimal*. In particular, the algorithm need not query all elements in the returned solution.

This model is a close variant of the *explorable uncertainty* model. The latter was introduced by Kahan in 1991 [Kah91], who studied the problem of identifying

---

[1] The standard term in the literature seems to be *feasible queries*. We use the term *admissible queries* to distinguish them from the *feasible sets* $S \in \mathcal{F}$.

the minimum value in a set of uncertain values using adaptive queries. Explorable uncertainty has attracted considerable attention since its introduction; see, e.g., the survey of Erlebach and Hoffmann [EH15].

Most of the work has focused on *adaptive* algorithms; i.e., algorithms that can query a few elements, learn their weights, decide whether to query more elements or compute an optimal solution, and repeat. We usually analyze these algorithms via competitive analysis; i.e., they compare against an adversary that knows the true weights beforehand. Some previous work also studied the question of not only identifying an optimal solution but also its value, which creates the need to query all the elements of the solution. Our understanding of the adaptive setting is quite good; almost optimal constant competitive algorithms are known for the problem of finding the k-th smallest element [Fed+00, GSS16], computing minimum spanning trees with uncertain edge weights [Erl+08, FMM17, MMS17], matroids [EH15], geometric problems [Bru+05], as well as for scheduling [LMS19, Gon+25] under slightly different models. Many of the results can even be obtained in a unified manner by using the technique of *witness sets* as introduced by Bruce et al. [Bru+05]. Furthermore, for other natural problems such as maximum weight matching, there are impossibility results showing that no constant-competitive algorithms exist [Mei18, Chapter 3], or [Sch25] for the knapsack problem. Algorithms with limited adaptivity were studied as well, see [Mei18, Chapter 4.5].

Surprisingly, we know very little about non-adaptive algorithms. They are used in settings with a huge fixed setup cost. Merino and Soto [MS19] studied the problem on matroids with uncertainty *sets* (i.e., not only intervals). They provide polynomial-time algorithms for finding minimum-cost admissible queries. Specializing their result for minimum spanning trees, gives implicitly an algorithm running in $\mathcal{O}(m^2\alpha(m, n))$ time; here, n is the number of vertices of the graph, m the number of edges, and α is the inverse Ackermann function. Despite that, their techniques do not seem to generalize easily to other problems. In particular, a natural direction is whether one can find minimum-cost admissible queries for s–t shortest paths or minimum-cost perfect matchings in polynomial time.

The shortest path problem has been studied in [Fed+07] in the non-adaptive setting. One is given a graph with edge-weight uncertainty intervals $[0, 1]$ or $\{0\}$, and the goal is to determine a minimum-cost edge set such that, after querying all edge weights from the set, one can determine an s–t shortest path and its length. The authors show that this problem is co-NP-complete, together with many more results in more general settings.

1.2   Thresholds of Inclusion and Exclusion

To better understand minimum admissible queries, we study a related problem on set selection under uncertainty. Note that even if an instance has no universally optimal solution, there are elements that belong to *all* optimal solutions and others that belong to *none* of them, for all realizations (see Fig. 3a). Identifying these key elements has proven useful for dealing with uncertainty; see, e.g., [Erl+08, MMS17, MS19].

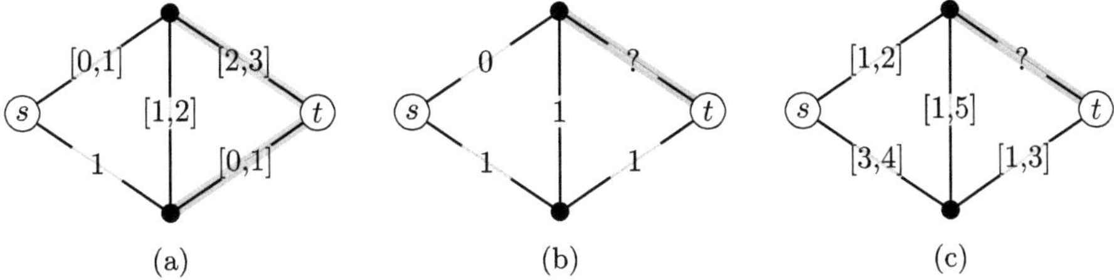

Fig. 3. (a) This instance has no universally optimal solution. (b) If the edge marked in green has weight less (resp. greater) than 2, then it is contained in *every* (resp. no) s–t shortest path. (c) If the green edge has weight smaller than 2 (resp. larger than 5), then it is included in all (resp. *no*) shortest s–t paths, regardless of the true weights.(Color figure online)

In the setting without uncertainty, these ideas are captured by the concept of *thresholds*. Here, we consider a fixed element $e \in E$ and are given all weights except the weight of e. Suppose that there are feasible sets with and without e: When is e contained in an optimal solution? If the weight of e is very small, say $w_e \approx -\infty$, then e must be included in all optimal solutions. Similarly, if the weight of e is very large, say $w_e \approx +\infty$, then every optimal solution must avoid e. The point where this behavior changes is called the *threshold* of e, denoted by T_e (see Fig. 3b). More formally, T_e is the unique number such that

- if $w_e > T_e$, then no optimal solution contains e,
- if $w_e < T_e$, then e belongs to every optimal solution,
- if $w_e = T_e$, there are optimal solutions with and without e.

Naturally, the threshold T_e is a function of the weights of all other elements w_{-e}. Furthermore, it is not hard to see that T_e is a piecewise linear function in each coordinate and can be easily computed whenever you can solve the corresponding min-weight set selection problem (see the full version for details).

In the uncertain setting, things are a bit more complicated, as we do not have a unique threshold that marks the two regimes (see Fig. 3c). For every item $e \in E$, we define the *threshold of inclusion* T_e^+ and the *threshold of exclusion* T_e^- as follows

$$T_e^+ := \inf\{T_e(w_{-e}) \mid w_{-e} \text{ realization of } \mathcal{I} \setminus \{I_e\}\}$$

$$T_e^- := \sup\{T_e(w_{-e}) \mid w_{-e} \text{ realization of } \mathcal{I} \setminus \{I_e\}\}.$$

In other words, T_e^+ is the minimum possible threshold, and T_e^- is the largest possible threshold. Note that $T_e^+ = T_e^- = \infty$ whenever e is in *every* feasible set; similarly, $T_e^+ = T_e^- = -\infty$ whenever e is in *no* feasible set. The interval $[T_e^+, T_e^-]$ indicates the possible behavior of e with respect to optimal solutions.

- If $h_e < T_e^+$, then e belongs to any optimal solution for every realization of $\mathcal{I} \setminus \{I_e\}$.

- If $\ell_e > T_e^-$, then e does not belong to any optimal solution for every realization of $\mathcal{I} \setminus \{I_e\}$.
- Otherwise, there are optimal solutions with and without e for specific realizations of $\mathcal{I} \setminus \{I_e\}$.

Also, the union of $T_e(w_{-e})$ over all realizations w_{-e} of $\mathcal{I} \setminus \{I_e\}$ is exactly the interval $[T_e^+, T_e^-]$.

The existence of thresholds is central to the isolation lemma, a technique from the 1980 s to guarantee uniqueness of optimal solutions; see, e.g., [Ta-15] and references therein. Despite its notoriety, to the best of our knowledge, the problem of computing thresholds under uncertain weights has not been studied prior to this work.

1.3   Our Contribution

Our main contribution is the introduction of thresholds under uncertainty and their use in analyzing minimum-cost admissible queries. As a first result, we show that computing thresholds and finding minimum-cost admissible queries are essentially equivalent.

- Given the thresholds of inclusion and exclusion for every element, we can compute a minimum-cost admissible query in linear time, and then compute a universally optimal solution by solving only one additional set selection problem (see Theorem 1 for a complete statement).
- Computing thresholds up to a constant additive error reduces to solving a logarithmic number of minimum-cost admissible query problems (see Theorem 2).

After showing this equivalence, we turn to computing thresholds in various settings.

- For the minimum spanning tree problem, we give an $\mathcal{O}(m\alpha(m, n) + n)$ algorithm for computing all $2m$ thresholds (see Theorem 5).
- For the maximum weight matching problem on trees, we give a linear-time algorithm that, given an element $e \in E$, computes its thresholds (see Theorem 10).

In other settings, we are able to show hardness. In particular, we show that computing the thresholds is NP-complete for shortest paths on DAGs (see Theorem 6) and for bipartite minimum-cost perfect matching (see Theorem 8).

As a consequence of the equivalence, we can translate the aforementioned results into the setting of minimum-cost admissible queries, obtaining the following results.

- An $\mathcal{O}(m\alpha(m, n) + n)$-time algorithm for finding minimum-cost admissible queries in the setting of minimum spanning trees, improving on the $\mathcal{O}(m^2\alpha(m, n))$ algorithm of Merino and Soto [MS19].

- A linear-time algorithm for minimum-cost admissible queries for maximum matching on trees.
- Deciding whether a given set is an admissible query for s–t shortest paths is NP-complete (on the full version).
- Deciding whether a given set is an admissible query for bipartite minimum-cost perfect matching is NP-complete (see Observation 9).

These last two points imply that there are no polynomial-time algorithms for finding minimum-cost admissible queries in the setting of shortest paths and bipartite minimum-cost perfect matching (unless $P = NP$).

Due to space constraints, many proofs and additional details are deferred to the full version.

2 Preliminaries

We begin by analyzing basic properties of the threshold functions for a fixed set selection problem, and refer to the full version of this paper for details. For this purpose, let $e \in E$ be an arbitrary fixed element. Let w_{-e} denote any realization of $\mathcal{I} \setminus \{I_e\}$, which we extend so that it assigns weight 0 to element e, regardless of whether 0 belongs to $[\ell_e, h_e]$ or not. By $\mathrm{OPT}^{+e}(w_{-e})$ we denote the weight of an optimal solution $S \in \mathcal{F}$ subject to $e \in S$. Similarly, we use $\mathrm{OPT}^{-e}(w_{-e})$ to denote the weight of an optimal solution subject to $e \notin S$. This allows us to reformulate the thresholds as

$$T_e^+ = \min_{w_{-e}} \left(\mathrm{OPT}^{-e}(w_{-e}) - \mathrm{OPT}^{+e}(w_{-e}) \right)$$

$$T_e^- = \max_{w_{-e}} \left(\mathrm{OPT}^{-e}(w_{-e}) - \mathrm{OPT}^{+e}(w_{-e}) \right).$$

Using a linearity argument, we can show that the maximizer/minimizer of the thresholds assigns to every element $f \neq e$ one of the extreme weights ℓ_f or h_f. For this reason, we call such a weight realization *extreme*. It follows that extreme realizations serve as witnesses for the decision problems $T_e^+ \leq y$ and $T_e^- \geq y$, which therefore belong to the class NP, provided that the underlying selection problem is polynomial-time solvable, in the sense that OPT^{+e} and OPT^{-e} can be computed in polynomial time. These thresholds form an interval $[T_e^+, T_e^-]$, which, when compared with $[\ell_e, h_e]$, provides crucial insight into the non-adaptive query problem. Namely, whenever $h_e \leq T_e^+$, it is safe to include e in an optimal solution, and whenever $T_e^- \leq \ell_e$ it is safe to exclude it. This allows us to state that the minimal admissible query set consists of all elements e for which the open intervals (T_e^+, T_e^-) and (ℓ_e, h_e) intersect, leading to the following algorithmic result.

Theorem 1. *Suppose all thresholds for $(E, \mathcal{F}, \mathcal{I})$ can be computed in time T_{THR} and that the optimization problem on $(E, \mathcal{F})$ can be solved in time T_{OPT}. We can find a minimum-cost admissible query set in time $\mathcal{O}(T_{THR} + |E|)$, and after the weights of the queries are revealed we can find a universally optimal solution in time $\mathcal{O}(T_{OPT} + |E|)$.*

The reverse direction is based on the following lemma. Given $e \in E$ and uncertainty intervals I_f for $f \in E - e$, we define $K = -\sum_{f \in E-e}(|\ell_f| + |h_f|)$, $I_e^\alpha = [K - 1, \alpha]$ and $\mathcal{I}^\alpha = \{I_f\}_{f \in E-e} \cup \{I_e^\alpha\}$.

Lemma 1. $T_e^+ \geq \alpha$ for $\mathcal{I}$ if and only if $E - e$ is an admissible query for $\mathcal{I}^\alpha$.

Combined with a binary search on α, this yields the converse reduction.

Theorem 2. *Suppose we can decide whether $E - e$ is an admissible query in time T, and that we can solve the corresponding optimization problem in time T_{OPT}. For every $e \in E$ and every additive precision $\delta > 0$, we can compute $a, b \in [-K, K]$ such that $T_e^+ \in [a, a + \delta]$ and $T_e^- \in [b, b + \delta]$ (or decide whether $T_e^-, T_e^+ = \pm\infty$). This algorithm runs in time $\mathcal{O}\big(T \log(\frac{2K}{\delta}) + T_{OPT}\big)$.*

The equivalence between threshold computation and minimum-cost admissible queries justifies focusing on threshold computation for different problems in the remainder of the paper. Section 3 gives an efficient algorithm for minimum spanning trees, Sects. 4 and 5 establish hardness for shortest paths and bipartite minimum-cost perfect matching, and Sect. 6 gives a linear-time algorithm for maximum-weight matching on trees.

3   Minimum Spanning Trees

In this section we consider the classical minimum spanning tree problem (MST). A *minimum spanning tree* of a graph $G = (V, E)$ with weights $w : E \to \mathbb{R}$ is a spanning tree T of G that minimizes $w(T)$. We show algorithms for computing the thresholds of inclusion and exclusion for MSTs. We assume that the edge $e = uv$ for which we want to compute the weights is not a *bridge*, also known as *cut-edge*, otherwise both thresholds are $+\infty$.

Given an undirected graph $G = (V, E)$ with weights $w \in \mathbb{R}^E$ and a path P, the *bottleneck of the path P* is given by $\max_{e \in P} w_e$. A *bottleneck path P* between $u, v \in V$ is a path of minimum bottleneck. The *bottleneck between u and v* is the bottleneck of a bottleneck u-v path. A classical result of Hu [Hu61] states that the paths in a minimum spanning tree are bottleneck paths.

Theorem 3 ([Hu61]). *Let $G = (V, E)$ be a graph with weights $w \in \mathbb{R}^E$ and let T be an MST of G. For every $u, v \in V$ the unique u-v path in T is a u-v bottleneck path.*

We consider a slight variation of the bottleneck problem. We say that an u-v path is *non-trivial* if it does not contain the edge uv. The non-trivial bottleneck between u and v, denoted by $b_w(u, v)$, is the minimum bottleneck of a non-trivial u-v path; i.e.,

$$b_w(u, v) = \min_{P \ u\text{-}v \ \text{nontrivial path}} \ \max_{e \in P} w_e.$$

In other words, the non-trivial bottleneck between u and v in G, is the bottleneck between u and v in $G - uv$.

We show that computing non-trivial bottlenecks is equivalent to computing thresholds.

Lemma 2. *Let $G = (V, E)$ be a graph with weights w. For every $uv \in E$, we have that $T_{uv}(w_{-uv}) = b_w(u, v)$.*

Combining Lemma 2 with the fact that $b_w(u, v)$ is increasing with respect to w, we obtain that for every $uv \in E$ it holds that $T_{uv}^+ = b_\ell(u, v)$ and $T_{uv}^- = b_h(u, v)$. Thus, the problem of computing the threshold for $e \in E$ is equivalent to computing $b_\ell(u, v)$ and $b_h(u, v)$.

We now provide a fast algorithm for computing all thresholds in the minimum spanning tree setting. Given a minimum spanning tree T and $e \in T$, we say that f is a *replacement* of e in T if $T + f - e$ is a minimum spanning tree of $G - e$. Replacements can be computed very efficiently by using the technique of path compression, as shown by Tarjan [Tar79].

Lemma 3 (Sect. 7 [Tar79]). *Given a weighted graph and a minimum spanning tree T, we can compute all $n - 1$ replacements of T in time $\mathcal{O}(m\alpha(m, n))$.*

We also make use of the computation of bottlenecks in a tree. This is usually a subroutine of minimum spanning tree verification algorithms, originally shown to run in linear time by Dixon, Rauch, and Tarjan [DRT92], with further simplifications by King [Kin97] and Hagerup [Hag10].

Lemma 4. *Given a tree T with weights on the edges and k pairs of vertices $(u_1, v_1), \ldots, (u_k, v_k)$, we can compute the bottlenecks in T of the k pairs in time $\mathcal{O}(n + k)$.*

The last ingredient we need, is Chazelle's algorithm for minimum spanning trees [Cha00].

Lemma 5. *Given a weighted graph, its minimum spanning tree can be computed in time $\mathcal{O}(m\alpha(m, n))$.*

We combine the aforementioned algorithms as follows.

Algorithm 4 *(Non-trivial bottlenecks).*
1. Compute a minimum spanning tree T via Chazelle's algorithm.
2. For every edge $e \in T$, compute its replacement $r_T(e)$ via path compression.
3. Using the tree-bottleneck algorithm, we compute $b_T(u, v)$ for $(u, v) \in E \setminus T$.
4. For $uv \in E \setminus T$, return $b_T(u, v)$. For $uv \in T$, return $w(r_T(e))$.

We can use Algorithm 4 using ℓ (resp. h) as weights to compute all thresholds T^+ (resp. T^-) for every $e \in E$.

Theorem 5. *For the minimum spanning tree problem, we can compute all thresholds of inclusion and exclusion in time $\mathcal{O}(m\alpha(m, n) + n)$.*

Proof. By Lemma 2 it suffices to compute non-trivial bottlenecks for every $e \in E$ with weights h and ℓ. It is clear that Algorithm 4 runs in the desired time, thus we only prove correctness.

Let T be the minimum spanning tree used by Algorithm 4. For $uv \notin T$, it is clear that T is a minimum spanning tree of $G - uv$. Thus, $b_T(u, v)$ is the non-trivial bottleneck for uv by Theorem 3. If $uv \in T$ and $r_T(uv) = f$, we have that $T' := T + f - uv$ is a minimum spanning tree of $G - uv$. By Theorem 3, the non-trivial bottleneck for uv is the bottleneck of the unique uv-path in T'. Note that this unique uv-path in T' is formed exactly by the edges of the fundamental cycle of f in T, excluding the edge uv. By the cycle property of minimum spanning trees, the weight of the non-tree edge f is greater than or equal to the weight of every tree edge on its fundamental cycle. Therefore, the maximum edge weight on this uv-path is exactly $w(f)$. Thus, returning $w(r_T(uv))$ correctly identifies the non-trivial bottleneck, concluding the proof. $\qquad\square$

4   Shortest Paths

In this section, we study the problem of computing a shortest path between two given vertices s and t with non-negative uncertain edge weights. In [Fed+07], this problem is studied in a slightly different model, where the algorithm aims to compute the length of the shortest path using as few queries as possible. In our setting, however, it is enough to output a path that is guaranteed to be a shortest s–t path, even though its exact length may be uncertain.

Consider an edge $e = (u, v)$. The threshold of exclusion T_e^- can be stated as the maximum, over all weight realizations w, of the difference between (i) the length ℓ of a shortest path from s to t that avoids e and (ii) the length of a shortest path P from s to t that is forced to use e.

Without loss of generality, for this maximum difference we can assume that all edges in P attain their lower limit and all edges not in P attain their upper limit. The idea is that increasing the weight of an edge not in P does not change the length of P, while the value of ℓ might increase (even though the corresponding shortest path might change). Similarly, if we decrease the weight of an edge in P by some amount $\delta > 0$, then the length of P decreases by δ, while ℓ can decrease by at most δ.

Theorem 6. *Computing the threshold of inclusion of an edge for the s–t shortest path problem is NP-hard.*

The proof is a reduction from 3-SAT, adapting a construction given in [GMO76]. It uses a layered graph: some layers correspond to clauses, others to variables, plus four additional layers for the vertices s, t, and the endpoints of e. Edges between adjacent layers have uncertainty intervals $[0, 1]$, while edges between literals and their occurrences in clauses have trivial uncertainty intervals $\{0\}$. The formal specification is postponed to the full version of this paper.

The idea is that an s–t path P^+ using edge e encodes one literal for each clause and one Boolean assignment for each variable. Let P^- be an arbitrary shortest path avoiding edge e. We first observe that the weight difference between P^- and P^+ is maximized if all edges in P^+ are realized at their lower limits and all edges not in P^- are realized at their upper limits. With these weights, the path P^- has positive total weight if and only if P^+ describes a Boolean assignment and, for each clause, selects a literal assigned to true.

For the threshold of exclusion, the reduction is much simpler and reduces from the Hamiltonian path problem.

Theorem 7. *Computing the threshold of exclusion of an edge for the s–t shortest path problem is NP-complete.*

5 Minimum Cost Perfect Matching

We complement the previous NP-hardness result by showing that computing the thresholds for matching is hard.

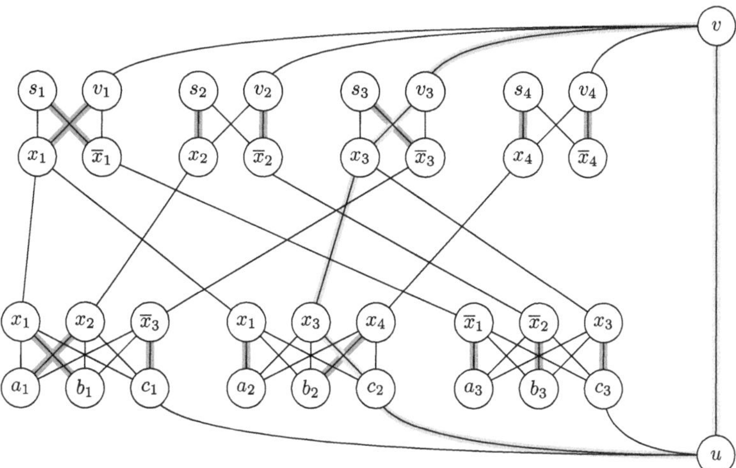

Fig. 4. The graph obtained by the reduction from the 3-SAT formula consisting of the clauses $C_1 = X_1 \vee X_2 \vee \overline{X_3}$, $C_2 = X_1 \vee X_3 \vee X_4$, $C_3 = \overline{X_1} \vee \overline{X_2} \vee X_3$. Edges in $M^+ \setminus M^-$ are represented by red lines, edges in $M^- \setminus M^+$ by blue lines, and edges in $M^+ \cap M^-$ by purple lines. For convenience we replaced vertices of the form c_{kj}, n_i, p_i by their corresponding literal.

Theorem 8. *Computing the threshold of exclusion of an edge in the minimum-cost perfect matching problem is NP-complete already for bipartite graphs.*

Proof. We reduce from 3-SAT, similarly to Theorem 6; see Fig. 4.

For every variable X_i, there is a *variable gadget* consisting of a complete bipartite graph, between vertices p_i, n_i on one side and vertices s_i, v_i on the other side. Vertex p_i corresponds to the variable X_i and n_i to its negation $\overline{X}_i$.

For every clause C_j there is a *clause gadget* in the form of a bipartite graph between vertices a_j, b_j, c_j on one side and vertices c_{1j}, c_{2j}, c_{3j} on the other side. Vertex c_{kj} corresponds to the k-th literal in C_j.

All edges inside these gadgets have uncertainty intervals of $[0, 1]$. In addition, there is a single edge between two vertices u and v, whose weight we do not specify, as we are only interested in its threshold. The construction is completed with the following additional edges, all with trivial uncertainty interval $\{0\}$. There is an edge from v to every vertex v_i, and from u to every vertex c_j. All vertices c_{kj} are connected to the corresponding literal vertex. Formally, if the k-th literal in C_j is X_i, then c_{kj} is connected to p_i; if this literal is $\overline{X}_i$, then c_{kj} is connected to n_i.

We claim that the threshold of exclusion of edge uv is 1 if the formula is satisfiable and 0 otherwise.

For a matching M and edge weights w, we use the notation $w(M) := \sum_{e \in M} w_e$. Recall that the threshold of exclusion $T^-(u, v)$ is defined as the difference $w(M^-) - w(M^+)$, maximized over all weight realizations w with $w_{uv} = 0$, where M^+ and M^- are minimum-cost matchings under the respective constraints $uv \in M^+$ and $uv \notin M^-$. Without loss of generality, for a fixed matching M^+ we can assume that $w_e = 0$ for all edges $e \in M^+$ and $w_e = 1$ for all other edges, since this maximizes the difference. Thus, the problem of maximizing the difference depends entirely on the choice of M^+.

The matching M^+ defines a boolean assignment to the variables, setting X_i true iff s_i is matched to p_i. If each vertex c_j is matched to a vertex which corresponds to a true valued literal, then we call M^+ *satisfying*.

In M^-, vertex u is matched to some vertex c_j and v to some vertex v_i. In addition M^- contains an edge with one endpoint in the j-th clause gadget and one endpoint in the i-th variable gadget. This edge corresponds to a literal z (either X_i or $\overline{X_i}$). Note that M^+ and M^- agree in all other than these two gadgets by the choice of the edge weights.

We consider two cases. Case 1: c_j is matched by M^+ to the clause-vertex corresponding to z and s_i is matched to the variable-vertex corresponding to z. Then the cost of M^- is 1, because it needs to match s_i to the variable-vertex corresponding to $\overline{z}$, which has edge weight 1. The cost is not more, because M^- can match a_j, b_j as in M^+. This means that if M^+ is satisfying, then the cost of M^- is necessarily 1.

Case 2: c_j is matched by M^+ to the clause-vertex corresponding to z and s_i is matched to the variable-vertex corresponding to $\overline{z}$. Then the cost of M^- is 0, as it can match a_j, b_j and s_i as in M^-. This means that if M^+ is not satisfying then the cost of M^- is 0.

To conclude the proof we observe that M^+ can be satisfying only if the formula is satisfiable.

$\square$

Observation 9. *We actually prove that it is NP-hard to distinguish whether the threshold of exclusion is 1 or 0. Combining this with Lemma 1, we obtain that it is NP-hard to decide whether a set is an admissible query. Moreover, this gap can be made arbitrarily large (as a constant) by scaling the weights.*

The NP-hardness of computing the threshold of inclusion can be shown via a similar construction. Replace the edge (u, v) by the path $u - u' - v' - v$, where the edges (u, u') and (v', v) have both lower and upper weight 0. Then every perfect matching that uses (u, v) in the original graph corresponds to a perfect matching that avoids (u', v') in the new graph, and vice versa. Hence, computing the threshold of inclusion of (u', v') is equivalent to computing the threshold of exclusion of (u, v) in the original graph.

6 Max Weight Matching in Trees

We contrast the previous hardness result with a special matching problem for which the thresholds can be computed efficiently.

Theorem 10. *For the maximum weight matching problem on trees, the thresholds of inclusion and exclusion for a fixed edge e can be computed in linear time.*

The idea is that, for a fixed edge e, the symmetric difference of a maximum weight matching M^+ that includes e and a maximum weight matching M^- that avoids e consists of a single path in the tree, alternating between the two matchings. Maximizing/minimizing the weight difference between these matchings reduces to computing a maximum-weight path in the tree with carefully defined edge weights. This can be done bottom-up via dynamic programming on the tree, rooted at the edge e.

7 Open Questions

We finish the paper by pointing out some open questions.

Further exploration of the non-adaptive explorable uncertain setting seems like an interesting direction. In particular, it would be interesting to better understand the boundary between polynomial-time tractability and NP-hardness for computing thresholds and finding minimum-cost admissible queries. Settings not covered in this paper include matchings in planar graphs and the edit distance between two strings under uncertain edit costs. Another intriguing question is whether the tractability of the threshold of inclusion is related to that of the threshold of exclusion, in the sense that either both can be solved in polynomial time or neither can.

Outside the setting studied in this paper, it would be interesting to consider models that interpolate between fully adaptive and non-adaptive algorithms, e.g., models with a fixed number of query rounds (see [Mei18] for some results in this setting).

Acknowledgments. We are very grateful to Shahin Kamali, Marcos Kiwi, and Claire Mathieu for helpful discussions.

This work was partially supported by grants ANR-19-CE48-0016 and ANR-23-CE48-0010 from the French National Research Agency, Proyecto Fondo Basal FB210005 CMM, and Fondecyt 11251528, 1221460 and 1231669 (ANID).

References

[Hu61]   Hu, T.C.: The maximum capacity route problem. Oper. Res. **9**(6), 898–900 (1961)

[GMO76]   Gabow, H.N., Maheshwari, S.N., Osterweil, L.J.: On two problems in the generation of program test paths. IEEE Trans. Softw. Eng. **SE-2.3**, 227–231 (1976). https://doi.org/10.1109/TSE.1976.233819

[Tar79]   Tarjan, R.E.: Applications of path compression on balanced trees. J. Assoc. Comput. Mach. **26**(4), 690–715 (1979). https://doi.org/10.1145/322154.322161

[Kah91]   Kahan, S.: A model for data in motion. In: Proceedings of the Twenty-Third Annual ACM Symposium on Theory of Computing - STOC 1991. The Twenty-Third Annual ACM Symposium, New Orleans, Louisiana, United States: ACM Press, pp. 265–277 (1991). https://doi.org/10.1145/103418.103449

[DRT92]   Dixon, B., Rauch, M., Tarjan, R.E.: Verification and sensitivity analysis of minimum spanning trees in linear time. SIAM J. Comput. **21**(6), 1184–1192 (1992). https://doi.org/10.1137/0221070

[Kin97]   King, V.: A simpler minimum spanning tree verification algorithm. Algorithmica **18**(2), 263–270 (1997). https://doi.org/10.1007/BF02526037

[Cha00]   Chazelle, B.: A minimum spanning tree algorithm with inverse-Ackermann type complexity. J. ACM **47**(6), 1028–1047 (2000). https://doi.org/10.1145/355541.355562

[Fed+00]   Feder, T., Motwani, R., Panigrahy, R., Olston, C., Widom, J.: Computing the median with uncertainty. In: Proceedings of the Thirty-Second Annual ACM Symposium on Theory of Computing, pp. 602–607. ACM (2000) https://doi.org/10.1145/335305.335386

[Bru+05]   Bruce, R., Hoffmann, M., Krizanc, D., Raman, R.: Efficient update strategies for geometric computing with uncertainty. Theor. Comput. Syst. **38**(4), 411–423 (2005). https://doi.org/10.1007/s00224-004-1180-4

[Fed+07]   Feder, T., Motwani, R., O'Callaghan, L., Olston, C., Panigrahy, R.: Computing shortest paths with uncertainty. J. Algorithms **62**, 1–18 (2007). https://doi.org/10.1016/j.jalgor.2004.07.005

[Erl+08]   Erlebach, T., Hoffmann, M., Krizanc, D., Mihalák, M., Raman, R.: Computing minimum spanning trees with uncertainty. In: The International Symposium on Theoretical Aspects of Computer Science. IBFI Schloss Dagstuhl, pp. 277–288 (2008). https://doi.org/10.4230/LIPIcs.STACS.2008.1358

[Hag10]   Hagerup, T.: An even simpler linear-time algorithm for verifying minimum spanning trees. In: Graph-Theoretic Concepts in Computer Science, vol. 5911. LNCS, Springer, Berlin, pp. 178–189 (2010). https://doi.org/10.1007/978-3-642-11409-0_16

[EH15]   Erlebach, T., Hoffmann, M.: Query-competitive algorithms for computing with uncertainty. Bull. EATCS **2**, 116 (2015)

[Ta-15]   Ta-Shma, N.: A simple proof of the Isolation Lemma. Electron. Colloquium Comput. Complex. (2015)

[GSS16]   Gupta, M., Sabharwal, Y., Sen, S.: The update complexity of selection and related problems. Theor. Comput. Syst. **59**(1), 112–132 (2016)

[FMM17]   Focke, J., Megow, N., Meißner, J.: Minimum spanning tree under explorable uncertainty in theory and experiments. In: LIPIcs-Leibniz International Proceedings in Informatics, vol. 75. Schloss Dagstuhl-Leibniz-Zentrum fuer Informatik (2017)

[MMS17] Megow, N., Meißner, J., Skutella, M.: Randomization helps computing a minimum spanning tree under uncertainty. SIAM J. Comput. **46**(4), 1217–1240 (2017). https://doi.org/10.1137/16M1088375

[Mei18] Meißner, J.: Uncertainty exploration (algorithms, competitive analysis, and computational experiments). PhD thesis. TU Berlin, Germany (2018). https://doi.org/10.14279/DEPOSITONCE-7327

[LMS19] Levi, R., Magnanti, T., Shaposhnik, Y.: Scheduling with testing. Manage. Sci. **65**(2), 776–793 (2019). https://doi.org/10.1287/mnsc.2017.2973

[MS19] Merino, A.I., Soto, J.A.: The minimum cost query problem on matroids with uncertainty areas. In: The International Colloquium on Automata, Languages, and Programming (ICALP 2019). Ed. by Christel Baier, Ioannis Chatzigiannakis, Paola Flocchini, and Stefano Leonardi, vol. 132. Leibniz International Proceedings in Informatics (LIPIcs). ISSN: 1868-8969. Dagstuhl, Germany: Schloss Dagstuhl–Leibniz-Zentrum fuer Informatik, 83:1–83:14 (2019). https://doi.org/10.4230/LIPIcs.ICALP.2019.83

[Gon+25] Gong, M., Fan, J., Lin, G., Su, B., Su, Z., Zhang, X.: Multiprocessor scheduling with testing: improved online algorithms and numerical experiments. J. Scheduling **28**(5), 513–527 (2025)

[Sch25] Schlöter, J.: On the complexity of knapsack under explorable uncertainty: hardness and algorithms. Technical report, 2507.02657. arXiv (2025)

Removable Online Knapsack: Exploiting Recourse and Bounded Item Sizes

Dimitris Fotakis[1,2] , Laurent Gourvès[3] , Aris Pagourtzis[1,2] ,
and Panagiotis Patsilinakos[3(✉)]

[1] National Technical University of Athens, Athens, Greece
`fotakis@cs.ntua.gr`, `pagour@cs.ntua.gr`
[2] Archimedes, Athena Research Center, Marousi, Greece
[3] Université Paris-Dauphine, Université PSL, CNRS, LAMSADE, Paris, France
`laurent.gourves@dauphine.fr`, `panagiotis.patsilinakos@dauphine.psl.eu`

Abstract. In the online unweighted knapsack problem, items arrive sequentially and must be accepted or rejected immediately for a bounded-capacity knapsack, aiming to maximize the total size of accepted items. When decisions are irrevocable, no algorithm achieves a non-trivial competitive ratio. An essential relaxation that circumvents this impossibility is to allow previously accepted items to be removed from the knapsack (a.k.a. removal). In this work, we investigate possible further improvements on the competitive ratio with removal by bringing together two other previously proposed model extensions: knowledge of an upper bound on the item sizes (a.k.a. bound awareness) and ability to reinsert previously rejected items (a.k.a. recourse). Our contributions are twofold. First, we establish close upper and lower bounds on the competitive ratio under bound awareness, removal and recourse. We show that significant improvements on the best known competitive ratio are possible when the upper bound on the item sizes is not too restrictive and only few uses of recourse are available. Second, we introduce the notion of adaptive online algorithms which do not require to know any upper bound on the item sizes but still achieve performance guarantees comparable to the ones obtained under the bound awareness model.

Keywords: Unweighted Knapsack · Adaptive Online Algorithms · Removable Items · Recourse

1 Introduction

We study the online unweighted knapsack problem in which some items, to be placed in a knapsack of capacity 1, are disclosed over time. Each item o_i has a *size* $|o_i| \in (0,1]$. Starting from an initial empty set S, item o_i is revealed at round i during which it must be decided to pack o_i in S or not. The value of S, denoted by $v(S)$ and defined as $\sum_{o_i \in S} |o_i|$, should never exceed the capacity. The goal is to maximize $v(S)$ when the instance stops, given that the number of items is not known in advance.

F. Foucaud and A. Parreau (Eds.): IWOCA 2026, LNCS 16587, pp. 296–310, 2026.
https://doi.org/10.1007/978-3-032-27732-9_21

The competitive ratio (CR for short) ρ of a deterministic online algorithm is the worst case ratio between the value of its solution $\hat{S}$ and the value of an optimal solution S^*: $\rho := v(\hat{S})/v(S^*)$. Every CR of this article is in $[0,1]$. Unfortunately, no deterministic algorithm can exhibit a non-trivial CR (i.e., bounded away from zero by any constant)[1] for the online unweighted knapsack problem [18]. This is mostly due to the fact that decisions are irrevocable. An essential relaxation that circumvents this impossibility is to allow items to be removed from S after their insertion. Iwama and Taketomi showed that the online unweighted knapsack problem with removable items admits a ϕ^{-1}-competitive algorithm with $\phi^{-1} := \frac{\sqrt{5}-1}{2} \approx 0.618$, and that no deterministic online algorithm can achieve a better CR [16].

Refined models have been proposed for improving upon the ϕ^{-1}-competitive algorithm. One such model is *recourse*, in which a recourse operation allows one item that was previously rejected to be eventually packed. Concretely, an item may be discarded from the solution upon its arrival and temporarily stored in a buffer. Subsequently, this item may eventually be inserted into the solution, which requires a so-called recourse operation.

In contrast to the unlimited number of removals, the number of recourse operations is typically small, as such operations can be costly in practice. Böckenhauer *et al.* have demonstrated (among other results) that the online unweighted knapsack problem with at most $r \in \{1, 2, 3\}$ recourse operation(s) and removable[2] items admits a deterministic $\frac{r+1}{r+2}$-competitive algorithm, and that no algorithm can achieve better CR with r recourse operation(s) [6,7]. Consequently, a single recourse operation suffices to improve upon the ratio of ϕ^{-1}.

The second model assumes that item sizes are upper-bounded by a value strictly smaller than the knapsack capacity. Moreover, when dealing repeatedly with such bounded instances, this bound can be known in advance and with precision.

Accordingly, *bound awareness* refers to the ability to know from the outset an upper bound $u \in (0, 1]$ such that $|o_i| \le u$ holds for every item o_i. As u approaches 0, the instance becomes increasingly constrained; however, the potential errors of an algorithm have a diminishing impact on the CR. When items are removable and their sizes are upper bounded by u, Gourvès and Pagourtzis [11] proposed an algorithm parameterized by u whose CR $\rho(u)$ improves upon ϕ^{-1} whenever $u \le \phi^{-1}$, given that ϕ^{-1} is known to be the optimal CR (without recourse) when $u \in (\phi^{-1}, 1]$ [16]. Figure 1 comprises an illustration of $\rho(u)$ while its formal definition is given in Theorem 3.

In comparison with the $\frac{r+1}{r+2}$-competitive algorithm of Böckenhauer *et al.*, $\rho(u)$ exceeds $2/3$ for $u < 0.6$, exceeds $3/4$ for $u < 0.4$, and exceeds $4/5$ for $u < 5/17$.

---

[1] Suppose a first item o_1 is released. If o_1 is not packed, then the instance stops. Otherwise, a second item o_2 of size 1 arrives and the instance stops. Clearly, o_2 cannot be packed because of o_1. The CR of any deterministic algorithm for this instance is at most $|o_1| = \max(0/|o_1|, |o_1|/1)$ but $|o_1|$ can be arbitrarily close to 0.

[2] Recourse operations are useless without the removability property [7].

The approaches of [7,11] are complementary yet incomparable: the former considers recourse without bound awareness, while the latter assumes bound awareness but allows no recourse.

1.1   Contribution

Our objective in this article is to extend the results of [7,11].

To this end, we explore in Sect. 2 the possibility of combining recourse with bound awareness for the online unweighted knapsack problem with removable items. We establish both upper and lower bounds on the optimal CR, given an upper bound u on item sizes and a fixed limit on the number of recourse operations. Our bounds, established for many discrete values of u (typically $u = 1/k$ for integers k), increase as $u = 1/k$ decreases, as does the number of possible recourse operations (on the order of $k/2$ for the lower bound and of k for the upper bound). Our most significant results concern the most relevant cases in which u is large (so the instance is only weakly constrained) and the number of possible recourse operations is small.

Figure 1 provides an illustration where bullets (resp., triangles) correspond to lower (resp., upper) bounds on the optimal CR for a small number of recourse operations (this number is indicated in parentheses, to the right of the bullet/triangle). Concretely, having $\blacktriangle(r)$ at coordinates (x, y) means that for instances whose items have size at most x, every algorithm using at most r recourse operations has CR at most y. Having $\bullet(r)$ at coordinates (x, y) means that there exists a y-competitive algorithm using at most r recourse operations for instances whose items have size at most x.

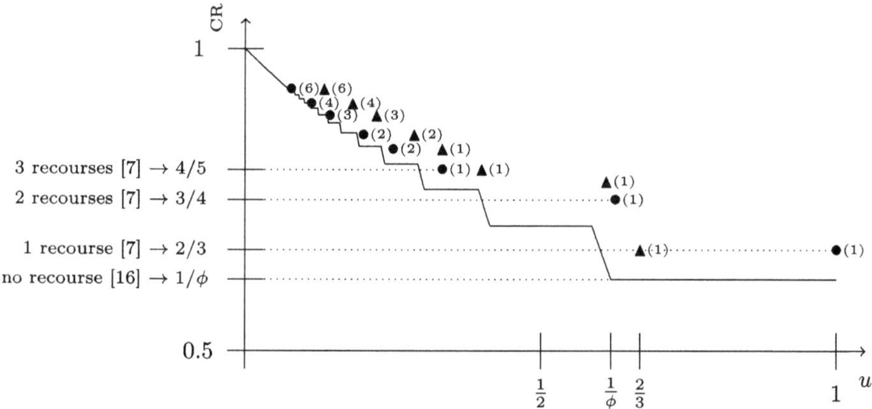

Fig. 1. Bounds on the CR as a function of u. The solid curve depicts $\rho(u)$, the CR of the bound aware algorithm of [11]. Bullets and triangles correspond to lower and upper bounds on the optimal CR when recourse is allowed (only drawn for a small number of recourse operations). The maximum number of recourse operations is indicated in parentheses. On the left are the CRs from [7,16], which are independent of u.

Our second approach, developed in Sect. 3, overcomes a drawback of bounded awareness. We continue to assume that item sizes are upper-bounded by a value u but u is *not* known in advance. Therefore, we propose an *adaptive* version of the parametric algorithm of [11]: its CR is approximately the same, but the parameter u is updated each time a (larger) new item is disclosed. Finally, an adaptive version of the algorithm that allows for one recourse operation is proposed, thereby improving the previously established competitive ratio.

Importantly, our adaptive algorithm yields a learning-augmented variant that can use an uncertain upper-bound prediction to match non-adaptive performance when correct, degrading by only a small factor in the worst case when incorrect.

All our bounds on CRs are for deterministic algorithms for the online unweighted knapsack problem with removable items, so we often omit this information. Due to space constraints, some technical parts are omitted and deferred to the full version.

1.2 Additional Related Work

We have already mentioned the most related to our work earlier results. Some additional related work is described below.

Regarding the possibility of item removal, an alternative model was proposed by Han *et al.* [12] where removal induces a cost either proportional to the item's size (by a value f known as the *buyback factor*) or equal to 1. The objective is to maximize the worth of packed items minus the cost of removed items, counting only those that were removed after been packed. Under this model, Han *et al.* studied the unweighted online knapsack problem [12], obtaining ratios that tend to $1/2$ when the buyback cost tends to 0. Babaioff *et al.* [2] considered, among others, the online weighted knapsack problem with a known upper bound $u \leq 1$ on the size of any item They gave a deterministic algorithm for the case $u < 1/2$ whose competitive ratio is $(1-2u)/(1+2f+2\sqrt{f(1+f)})$, which becomes $1-2u$ when $f = 0$. This is below the ratio ϕ^{-1} of [16] except for very small values of u, and is also inferior to the ratios obtained in [11] for every value of the item size bound u. However, the results of [2] hold for the weighted case as well; note that no deterministic algorithm can achieve any non-trivial CR for the online *weighted* knapsack, by removal only [17]. Another result of [2] is a randomized algorithm whose CR tends to $1/3$ as f goes to 0. Randomized algorithms for the online knapsack problem with (costless) removal can be found in [10, 13].

A variant to the idea of recourse was proposed by Böckenhauer *et al.* [3], where the possibility to reserve an item (i.e., postpone the decision about it) is studied. In that model items are non-removable and the decision on which to restore is taken at the end, assuming that the algorithm is informed once the instance ends and can take action afterwards. Assuming that reserving an item costs α times its value, the authors characterize the best competitive ratio for all possible values of $\alpha \in (0, 1)$. The best ratio in [3] is $1/2$, obtained for small values of α ($\leq 1/6$), and the performance deteriorates as α increases. Note that in [7] they achieve a $2/3$ CR with just one use of recourse and no information about the end of the instance; however, the two models are incomparable as in [3] there is no limit on the number of items that can be restored from the reserve, while in [7] there is no cost for reserving items in the recourse and bringing them into the solution. Another model supposes that the items can be placed in a buffer of size $K > 1$ before a selection of them is put in a knapsack of size 1 [14]; this model also assumes that the algorithm is informed when the input ends. Reservation with removability has been considered in [9].

Other proposed approaches for online knapsack include the possibility of *augmentation* of the knapsack size [17], the possibility of items being split and partially included in the knapsack [15], and the study of the *advice complexity* of the problem [4,8]. The work of Zhou *et al.* on the weighted knapsack problem also makes (restrictive) assumptions about both the size of objects and their weight-to-size ratio [20].

For more information about the online knapsack problem, the interested reader may refer to a recent survey by Böckenhauer *et al.* [5].

2   Algorithms with Recourse

2.1   Lower Bounds on the Best Competitive Ratio

We propose (and analyze the CR of) Algorithm 1 where for some integer $k \geq 3$ an upper bound on the item sizes of $u = 1/k$ is assumed and at most $\lfloor k/2 \rfloor$ recourse operations are allowed. Small and large items have size in $(0, \frac{1}{k+2}]$ and $(\frac{1}{k+2}, \frac{1}{k}]$, respectively. It follows that any feasible solution contains at most $k+1$ large items. Upon the arrival of item o_i, the high-level strategy of Algorithm 1 is to maintain a solution S according to the following rules: (i) do not modify S if $v(S) \geq \frac{k+1}{k+2}$; (ii) when $v(S) < \frac{k+1}{k+2}$, attempt to exploit o_i together with at most $\lfloor \frac{k}{2} \rfloor$ items from the buffer (via recourse operations) in order to construct a feasible solution of value at least $\frac{k+1}{k+2}$; (iii) if step (ii) fails, retain in the current solution the $\lceil \frac{k}{2} \rceil$ smallest large items, the $\lfloor \frac{k}{2} \rfloor$ largest large items, and all small items revealed so far.

Theorem 1. *For an integer $k \geq 3$, Algorithm 1 is $\frac{k+1}{k+2}$-competitive when $u \leq 1/k$. In addition, Algorithm 1 uses at most $\lfloor \frac{k}{2} \rfloor$ recourse operations.*

Proof (Sketch). The proof is by induction until it possibly happens that a feasible solution of value at least $\frac{k+1}{k+2}$ is built. As soon as the current solution has value at least $\frac{k+1}{k+2}$ (which occurs at line 4, or when the while loop at line 12 is executed,[3] or at line 16), the algorithm is clearly $\frac{k+1}{k+2}$-competitive because the optimal value is at most 1.

Consider the cases other than those of lines 4, 12, and 16. First, observe that if $v(S) < \frac{k+1}{k+2}$, then every small item disclosed so far belongs to S, and every item placed in the buffer is large. Indeed, an item is put into S if it fits (cf. line 7), which is always the case for small items because their size is at most $\frac{1}{k+2}$.

---

[3] The while loop always creates a feasible solution of value at least $1 - \frac{1}{k+2} = \frac{k+1}{k+2}$.

Algorithm 1 at most $\lfloor \frac{k}{2} \rfloor$ recourse operations, $u \le 1/k$, for any integer $k \ge 3$

1: $S \leftarrow \emptyset$ $\triangleright$ *Small and large items have size in* $(0, \frac{1}{k+2}]$ *and* $(\frac{1}{k+2}, \frac{1}{k}]$, *respectively*
2: **while** a new item o_i arrives **do**
3:     **if** $v(S) \ge \frac{k+1}{k+2}$ **then**
4:        Put o_i in the buffer
5:     **else**
6:        **if** $v(S \cup \{o_i\}) \le 1$ **then**
7:           $S \leftarrow S \cup \{o_i\}$
8:        **else**
9:           **if** $v(\{o \in S : |o| > \frac{1}{k+2}\} \cup \{o_i\}) \le 1$ **then**
10:             $S \leftarrow S \cup \{o_i\}$
11:             **while** $v(S) > 1$ **do**
12:                Remove a small item from S $\triangleright$ *chosen arbitrarily*
13:        **else**
14:           $\mathcal{T} \leftarrow$ collection of all sets of at most $k+1$ elements of $\{o_1, \ldots, o_i\}$
              which contain at most $\lfloor \frac{k}{2} \rfloor$ buffered items
15:           **if** $\exists T \in \mathcal{T}$ such that $1 \ge v(T) \ge \frac{k+1}{k+2}$ **then**
16:             $S \leftarrow T$
17:           **else**
18:             Keep in S the $\lceil \frac{k}{2} \rceil$ smallest large items seen so far, the $\lfloor \frac{k}{2} \rfloor$ largest
            large items seen so far, and every small item seen so far

Let OPT_i (resp., S_i) be an optimal solution (resp., the solution produced by Algorithm 1) for the item set $\{o_1, \ldots, o_i\}$. Let S_i^s and S_i^ℓ denote the restriction of S_i to small and large items of $\{o_1, \ldots, o_i\}$, respectively. Let Q_i be a subset of large items taken in $\{o_1, \ldots, o_i\}$ whose value is maximum under the constraint $v(Q_i) \le 1$. We claim that Algorithm 1 maintains a feasible solution which guarantees

$$v(S_j^\ell) \ge \frac{k+1}{k+2} v(Q_j) \tag{1}$$

for $j = 1..i-1$ until it possibly happens that $v(S_i) \ge \frac{k+1}{k+2}$. The base case $(j = 1)$ of the induction hypothesis (1) holds trivially: if o_1 is large then $S_1^\ell = Q_1 = \{o_1\}$, otherwise $S_1^\ell = Q_1 = \emptyset$, so $v(S_1^\ell) = v(Q_1)$. From now on, we assume that (1) holds for $j = i - 1$, right before o_i is disclosed. Our proof also relies on the following observation whose proof is deferred to the full version. $\quad\square$

Observation 1. $v(S_j^\ell) \ge \frac{k+1}{k+2} v(Q_j)$ *implies* $v(S_j) \ge \frac{k+1}{k+2} v(OPT_j)$.

At line 7 of Algorithm 1, o_i is put into the current solution because it fits. If o_i is small, then $S_i^\ell = S_{i-1}^\ell$, and $Q_i = Q_{i-1}$, so $v(S_i^\ell) = v(S_{i-1}^\ell) \ge \frac{k+1}{k+2} v(Q_{i-1}) = \frac{k+1}{k+2} v(Q_i)$. If o_i is large, then $S_i^\ell = S_{i-1}^\ell \cup \{o_i\}$, and $v(Q_i) \le v(Q_{i-1}) + |o_i|$, so $v(S_i^\ell) = v(S_{i-1}^\ell) + |o_i| \ge \frac{k+1}{k+2} v(Q_{i-1}) + |o_i| \ge \frac{k+1}{k+2} v(Q_i)$. In both cases, the insertion of o_i maintains the property of being $\frac{k+1}{k+2}$-competitive.

The last case to be analyzed is line 18. In this context we have $v(S_{i-1}) < \frac{k+1}{k+2}$, $v(S_{i-1} \cup \{o_i\}) > 1$ and $v(S_{i-1}^\ell \cup \{o_i\}) > 1$, implying that o_i is a large item.

Suppose that at some point i, $\{o_1, \ldots, o_i\}$ contains for the first time a set Y of $k+1$ large items such that $v(Y) \leq 1$. This means that $o_i \in Y$. Since Algorithm 1 maintains the $\lceil \frac{k}{2} \rceil$ smallest large items seen so far, it can exploit o_i and the $\lfloor \frac{k}{2} \rfloor$ recourse operations to produce a set Y' of $k+1$ large items such that $v(Y') \leq v(Y) \leq 1$. Since every large item has size at least $\frac{1}{k+2}$, it holds that $v(Y') \geq \frac{k+1}{k+2}$. Thus, the algorithm is able to produce a feasible solution of value at least $\frac{k+1}{k+2}$ as soon as a feasible set of $k+1$ large items exists.

It remains to consider the case where every set Y of large items such that $v(Y) \leq 1$ satisfies $|Y| \leq k$. This is true, in particular, for S_i^ℓ. Moreover, S_i^ℓ cannot contain less than k large items because the test of line 9 is false. Indeed, any set of k large items has size at most 1 because every item has size at most $1/k$. One cannot take a large item of the buffer, say $\hat{o} \notin S_i^\ell$, and have $v(S_i^\ell \cup \{\hat{o}\}) \leq 1$, because every set Y of large items such that $v(Y) \leq 1$ satisfies $|Y| \leq k$. This means that $v(S_i^\ell \cup \{\hat{o}\}) > 1 \Leftrightarrow v(S_i^\ell) > 1 - \frac{1}{k} = \frac{k-1}{k}$. If $v(Q_i) \leq \frac{k+1}{k+2}$, then $\frac{v(S_i^\ell)}{v(Q_i)} > \frac{k-1}{k} \cdot \frac{k+2}{k+1} \geq \frac{k+1}{k+2}$ when $k \geq 3$.

The final case is when $v(Q_i) > \frac{k+1}{k+2}$. By the definition of Q_i, $v(Q_i)$ is non-decreasing with i. Let us suppose that i is the first moment that $v(Q_i) > \frac{k+1}{k+2}$. This means that Q_i contains o_i. Let $\{o'_1, \ldots, o'_{\lceil \frac{k}{2} \rceil}\} \cup \{p_1, \ldots, p_{\lfloor \frac{k}{2} \rfloor}\}$ denote the k (large) items of Q_{i-1} where we assume w.l.o.g. that $|o'_1| \leq |o'_2| \leq \ldots \leq |o'_{\lceil \frac{k}{2} \rceil}| \leq |p_1| \leq |p_2| \leq \ldots \leq |p_{\lfloor \frac{k}{2} \rfloor}|$. We have $Q_i = Q_{i-1} \Delta \{o'_1, o_i\}$.

Since S_{i-1}^ℓ keeps the $\lfloor \frac{k}{2} \rfloor$ largest items seen so far, we can assume that $\{p_1, \ldots, p_{\lfloor \frac{k}{2} \rfloor}\} \subseteq S_{i-1}^\ell$. In other words, at most $\lfloor \frac{k}{2} \rfloor$ recourse operations are needed to produce Q_i from S_{i-1}^ℓ because $\{p_1, \ldots, p_{\lfloor \frac{k}{2} \rfloor}\}$ and o_i are accessible without any recourse. Thus, $Q_i \in \mathcal{T}$ at line 14 of Algorithm 1, meaning that the algorithm can produce a solution of value at least $\frac{k+1}{k+2}$.

Finally, Algorithm 1 is correct: it produces a feasible solution and uses at most $\lfloor k/2 \rfloor$ recourse operations (proof deferred to the full version). $\qquad \square$

Extending Theorem 1 to $k = 2$ yields a CR of 3/4 using at most one recourse operation when $u \leq 1/2$. In fact, a modified version of Algorithm 1, using four classes of items (small, medium, large and extra-large) gives the following stronger result.

Theorem 2. *There exists a deterministic $\frac{3}{4}$-competitive algorithm using at most one recourse operation when $u \leq 5/8$.*

Compared with [7], for identical competitive ratios, our results use fewer recourse operations, albeit under the assumption of bounded item sizes. Likewise, compared with [11], for identical bounds on item sizes, we achieve better competitive ratios thanks to the use of recourse.

2.2   Upper Bounds on the Best Competitive Ratio

It is known from [7, Thm. 1] that deterministic algorithms that use at most r recourse operations have a CR of at most $\frac{r+1}{r+2}$. This bound is established via instances in which the largest item has size $\frac{r+1}{r+2} + \varepsilon$. Consequently, both the size of the largest item and the upper bound on the CR increase with r. However, it is also known that the optimal CR increases as u decreases [11]. This motivates the need for novel upper bounds on the CR that reflect this dependence on u.

Lemma 1. *For every integer* $k \geq 3$*, there exists an upper bound of* $\frac{2k-1}{2k}$ *on the competitive ratio of deterministic algorithms with at most* $k - 2$ *recourse operations when* $\frac{1}{k} < u \leq 1$*.*

Proof. Take any online algorithm and suppose that an adversary reveals the following instance: $k - 1$ *large* items of size $\frac{1}{k} + \varepsilon$ followed by $2k$ *small* items of size $\frac{1}{2k}$, where $\frac{1}{2k} \gg \varepsilon > 0$. The instance can be interrupted at any moment if the algorithm's solution is (asymptotically) $(1 - \frac{1}{2k})$-approximate, or worse.

The first k items fit into the knapsack. If the algorithm discards any of them, then the CR is at most $1 - \frac{1}{2k-1}$ as $\varepsilon \to 0$. Indeed, the optimal solution has value $(k - 1)(\frac{1}{k} + \varepsilon) + \frac{1}{2k}$ whereas the algorithm's solution has value at most $(k-1)(\frac{1}{k}+\varepsilon)$. Henceforth, we assume that the first k items belong to the solution.

When the $(k+1)$-th item is revealed, it does not fit into the current solution, while the optimal value remains $(k - 1)(\frac{1}{k} + \varepsilon) + \frac{1}{2k} = 1 - \frac{1}{2k} + (k - 1)\varepsilon$. The optimal value increases to 1 only upon the arrival of the final item, when the optimum consists of $2k$ small items. At this point, the algorithm may (i) replace 1 large item with 1 small item, or (ii) buffer a small item, or (iii) replace 1 large item with 2 small items.

Action (i) decreases the value of the current solution by $\frac{1}{2k} + \varepsilon$, yielding a CR of $1 - \frac{1}{2k-1}$; hence, the adversary can terminate the instance. Actions (ii) and (iii) do not significantly affect the solution value, but action (iii) can be performed at most $k - 2$ times, as each use requires a recourse operation.

Since the instance contains $k - 1$ large items, the algorithm must either perform action (i), or reach the end of the instance, where the optimal value is 1 while the algorithm's solution has value at most $(k - 1)(\frac{1}{k} + \varepsilon) + \frac{1}{2k} = 1 - \frac{1}{2k} + (k - 1)\varepsilon$. In any case, the CR is at most $1 - \frac{1}{2k}$ as $\varepsilon \to 0$. $\square$

In the same spirit as Lemma 1, the following upper bounds can be derived.

Lemma 2. *For every integer* $k \geq 2$*, there exists an upper bound of* $\frac{2k}{2k+1}$ *on the CR of deterministic algorithms with at most* $k - 1$ *recourse operations when* $\frac{2}{2k+1} < u \leq 1$*.*

We now present a complementary upper bound on the CR for the case of a single recourse operation.

Lemma 3. *The competitive ratio of any algorithm with at most one recourse operation is upper-bounded by* $\sqrt{\frac{9-\sqrt{17}}{8}} \approx 0.78$ *when* $u \geq \frac{9-\sqrt{17}}{8} \approx 0.61$*.*

Let us summarize the bounds on the CR when a *single* recourse operation is possible (see Fig. 1). There exist lower bounds of $4/5$ when $u \leq 1/3$ (Theorem 1), $3/4$ when $u \leq 5/8$ (Theorem 2), and $2/3$ when $u \leq 1$ ([7, Thm. 2]). There exist upper bounds of $5/6$ when $u > 1/3$ (Lemma 1, $k = 3$), $4/5$ when $u > 2/5$ (Lemma 2, $k = 2$), $((9 - \sqrt{17})/8)^{1/2}$ when $u \geq (9 - \sqrt{17})/8$ (Lemma 3), and $2/3$ when $u > 2/3$ ([7, Thm. 1]).

In this section, we have proposed upper and lower bounds on the CR for certain combinations of upper bounds on object sizes, along with a limit on the number of allowable recourse operations. It would be interesting either to provide results for all possible combinations or to show that, for some combinations, recourse does not yield any significant improvement in the optimal CR. For example, what would be the best CR for all u with at most one recourse? Our results provide answers for $u \geq 1/3$, and we believe that a single recourse can only marginally improve the situation when u is very close to zero. Resolving this conjecture would be of great interest.

3 Adaptive Algorithms

3.1 An Adaptive Algorithm with Removal Only

Algorithms that are aware of an upper bound on item sizes and whose CR is lower bounded by $\frac{1}{\phi}$, but can exceed $\frac{1}{\phi}$ for most size upper bounds, exist [11]. Importantly, the smaller the size bound, the stronger the CR guarantees they provide (almost tight per upper bound). This raises the question:

Can we design algorithms whose competitive ratio (inversely) depends on the maximum item size, without any prior information?

Crucially, we want our adaptive algorithms' CR to still be lower bounded by $\frac{1}{\phi}$, in the worst case.

We begin by stating some useful definitions and theorems for the case of removable items without recourse. Consider an upper bound $u \in (0, 1]$ on the size of every item. Then, in [11] the following is shown:

Theorem 3 ([11]). *For all integers $k > 0$, there exists a γ_k-competitive algorithm for a known $u \in (0, \frac{1-\gamma_k}{\gamma_k^2}]$, where γ_k is a constant defined as $\gamma_k = \frac{k-2+\sqrt{k^2+4}}{2k}$. In addition, for all integers $k \geq 2$, the same algorithm is $\frac{1-u}{(k-1)u}$-competitive when $u \in (\frac{1-\gamma_k}{\gamma_k^2}, \frac{1}{k-1}]$.*

For all positive integers k, $\gamma_k \in (0, 1]$ increases with k, and $\gamma_1 = \frac{\sqrt{5}-1}{2} = \frac{1}{\phi}$. We will refer to this algorithm of [11] as "NONADAPTIVE" here.

NONADAPTIVE can be adapted to get improved guarantees (compared to $\frac{1}{\phi}$ [16]) without knowing the item size upper bound u in advance, at the cost of a small penalty in the CR. Our adaptive version, Algorithm 2 (a.k.a. ADAPTIVE), is identical to NONADAPTIVE with the addition of lines 3, 4 and 5. These lines are used to dynamically update the value of k whereas NONADAPTIVE directly uses a fixed value, optimal for the given item size upper bound u.

The adaptive version updates an over-estimation of the upper bound (setting $k' = \max\{k_{curr} - 2, 1\}$ in line 5) whenever a larger item arrives and continues running with the corresponding threshold. We show that its behavior matches that of NONADAPTIVE tuned from the start to the final over-estimation. Since the true bound is smaller, the corresponding approximation guarantee is preserved.

Informally, NONADAPTIVE defines a threshold $1 - \gamma_k$ and classifies items as small or non-small. When a new item does not fit into the knapsack and the guarantee is not met, items are removed based on the total sum of non-small item sizes. Specifically, when the sum of non-small item sizes is infeasible, it removes the largest item, otherwise it removes small items until it gets a feasible solution.

Definition 1 (k-small items). *An item o is said to be k-small if $|o| \leq 1 - \gamma_k$.*

The challenge is to identify the case where ADAPTIVE might remove an important item that NONADAPTIVE would not. In this case ADAPTIVE would delete its current largest item: For a current estimation of k, say k', the k'-non-small items seen-so-far do not fit in the knapsack, but for a later estimation, say k'', some of these items could be considered k''-small by NONADAPTIVE - allowing the rest to fit in the knapsack, not triggering the large item deletion. However, the over-estimation of the upper bound raises the non-small size threshold. Moreover, we target a lower CR which is easier to satisfy. Thus, as non-small items accumulate, the desired CR is achieved before they exceed the knapsack capacity.

For the following, all missing proofs of this section are included in the full version of the paper.

Observation 2. *Consider that the item upper bound is u. Then the optimal parameter k for NONADAPTIVE to guarantee the best possible approximation is:*

$$k^* = \max \left\{ k \in \mathbb{N} \mid \frac{2}{k - 1 + \sqrt{(k - 1)^2 + 4}} \geq u \right\}$$

Theorem 4. *For all positive integers k, Algorithm 2, a.k.a. ADAPTIVE, is $\gamma_{\hat k}$-competitive when $u \leq \dfrac{2}{k-1+\sqrt{(k-1)^2+4}}$ where $\hat k = \max\{k^* - 2, 1\}$.*

Proof. By Theorem 3 we have that NONADAPTIVE is guaranteed to achieve approximation $\gamma_{\hat k}$ whenever $u \leq \dfrac{2}{k-1+\sqrt{(k-1)^2+4}}$. That is since for any $k \geq 3$,

$$\frac{2}{k-1+\sqrt{(k-1)^2+4}} < \frac{1}{k-1} < \frac{1-\gamma_{k-1}}{\gamma_{k-1}} < \frac{1-\gamma_{\hat k}}{\gamma_{\hat k}}$$ (from properties in [11] and $\gamma_{\hat k} <$

γ_{k-1}). Finally for $k < 3$, $\hat k = 1$ and NONADAPTIVE gives CR γ_1 for any u.

Thus, it suffices to show that ADAPTIVE will behave exactly like NONADAPTIVE when run with parameter $\hat k$.

As ADAPTIVE runs, it computes the optimal parameter k_{curr} for NONADAPTIVE based on the largest item seen so far, using Observation 2, and sets $k' = \max\{k_{curr} - 2, 1\}$ (i.e., after an item of the maximum size u is revealed, we have $k' = \hat k$). The algorithm then updates the set B of non-small items according to $\hat k$ and follows the same steps as NONADAPTIVE with parameter k'.

Algorithm 2 : ADAPTIVE

1: $S \leftarrow \emptyset$, $MaxItem \leftarrow 0$, $B \leftarrow \emptyset$
2: **while** a new item o_i arrives **do**
3:     $MaxItem \leftarrow \max\{MaxItem, |o_i|\}$
4:     $k_{curr} \leftarrow \max\left\{k \in \mathbb{N} \mid \frac{2}{k-1+\sqrt{(k-1)^2+4}} \geq MaxItem\right\}$
        $\triangleright$ *Optimal k considering MaxItem as the upper bound*
5:     $k' \leftarrow \max\{k_{curr} - 2, 1\}$
6:     **if** $v(S) \geq \gamma_{k'}$ **then**
7:         Reject o_i $\triangleright$ *S is not changed afterwards*
8:     **else**
9:         $S \leftarrow S \cup \{o_i\}$
10:        **if** $v(S) > 1$ **then**
11:            Let $B = \{o \in S : |o| > 1 - \gamma_{k'}\}$ $\triangleright$ Reset B
12:            **if** B contains a subset $\hat{S}$ such that $1 \geq v(\hat{S}) \geq \gamma_{k'}$ **then**
13:                $S \leftarrow \hat{S}$ $\triangleright$ *S is not changed afterwards*
14:            **else if** $v(B) > 1$ **then**
15:                Remove the largest item of S
16:            **else**
17:                **while** $v(S) > 1$ **do**
18:                    Remove from S one element of $S \setminus B$ $\triangleright$ *chosen arbitrarily*

For the proof, let k'_i, for $i \in \mathbb{N}^*$, denote the value of k' after the ith iteration (i.e., after receiving item o_i and updating k'). We further denote by B_i and $B_{k,i}$ the sets B for ADAPTIVE and NONADAPTIVE respectively, after the ith iteration, where the k of $B_{k,i}$ is the parameter used in this instance of NONADAPTIVE.

Using the following claim, we show that for any k'_i, the set B of non-small items in ADAPTIVE coincides with the corresponding set in NONADAPTIVE run with parameter k'_i on the same input. That is, $B_i = B_{k'_i,i}$:

Claim 1. *If the largest item size seen so far satisfies* $MaxItem \leq \frac{2}{4-1+\sqrt{(4-1)^2+4}}$ *(i.e., $k \geq 4$), then* ADAPTIVE *has never removed any item in set B.*

We can now complete the proof of the theorem by induction on k'_i:
Base Case, $i = 1$: Since k'_1 is the first value for k' selected by ADAPTIVE, the first iteration is exactly the same as for NONADAPTIVE when running for k'_1.
Induction Step, $i > 1$: Suppose that the hypothesis is true for all k'_j, $1 \leq j < i$.
Case I, $k'_i = k'_{i-1}$: That is the simplest case. Since we have $B_{i-1} = B_{k'_{i-1},i-1}$ and k' doesn't change at step i, then ADAPTIVE and NONADAPTIVE behave exactly identical with regards to non-small items. That is, $B_{k'_i,i} = B_{k'_{i-1},i} = B_i$.

Case II, $k'_i \neq k'_{i-1}$: We identify the following sub-cases:
    Case II.a, $k'_i \neq 1$: We have $k_{curr} \geq 4$ since it must be $k' \geq 2$. By Claim 1, after iteration i, ADAPTIVE has either terminated with approximation at least $\gamma_{k'}$ or has not removed any item from B_i. Items in B_{i-1} are either reclassified

as small at the start of iteration i (and possibly deleted after) or remain in B_i. Thus, every item seen so far that is not k_i'-small remains in B_i.

We now consider NONADAPTIVE and the set $B_{k',i}$. Since all non-k_i'-small items fit in the knapsack, the condition to remove such items is not met, and $B_{k',i} = B_i$.

Case II.b, $k_i' = 1$: The case $k_{i-1}' = 1$ is covered by **Case I**, so we consider only $k_{i-1}' \neq 1$. By definition, this is the first iteration in which an item large enough to set $k' = 1$ appears. By induction, we have $B_{i-1} = B_{k_{i-1}',i-1}$. After item o_i arrives, ADAPTIVE updates k' to $k_i' = 1$ (equal to $\hat{k}$, as it will not change again) and resets B_i in Step 11 to include all non-1-small items seen so far, none of which have been deleted by Claim 1 and induction. It then handles item o_i.

Finally, it suffices to note that $B_{1,i-1}$ contains exactly the same items as the set B of ADAPTIVE immediately after the reset in Step 11. Since both algorithms then process the same item in iteration i, we obtain $B_{1,i} = B_i$.

This concludes the proof since the order of deletion of small items is arbitrary and therefore not important for the results guaranteed by NONADAPTIVE. $\square$

Learning-Augmented Variant. The adaptive algorithm also allows to create a *learning-augmented via predictions version* (cf. [1,19]).

The goal is to design algorithms that leverage predictive input, motivated by modern ML systems' ability to provide such insight. We are interested in two performance measures: (a) **Consistency**, the worst-case CR when predictions are perfect, and (b) **Robustness**, the worst-case CR when predictions are arbitrarily incorrect. In our setting, the prediction concerns the true item-size upper bound.

For our version, we identify the following key property:

Lemma 4. *Items of size up to* $\dfrac{2}{k-1+\sqrt{(k-1)^2+4}}$ *are* $\{k-c\}-small,\ for\ k > c \geq 3.$

This implies that the deletion order of previously seen items is irrelevant when targeting a $\gamma_{k'-3}$ approximation. This allows us to follow the non-adaptive strategy of [11] and, if the prediction is invalidated, switch to ADAPTIVE's strategy.

Specifically, the algorithm needs to follow the following basic steps:

1. For the prediction of the item-size upper bound u_p, let k_p be the optimal value of k, according to Observation 2.
2. If $k_p < 4$, then run ADAPTIVE until completion, else continue.
3. Start by running algorithm ADAPTIVE.
4. If at any step it holds that $k' \in \{k_p, k_{p-1}, k_{p-2}\}$, then keep the current knapsack state as is and start running NONADAPTIVE from there on.
5. If during step 4 an item o arrives such that $o > u_p$, then recalculate the new k_{curr} (as in step 6 of ADAPTIVE), keep the current knapsack state as is, and start running ADAPTIVE setting $k' = \min\{k_{p-3}, k_{curr-2}\}$.

Theorem 5. *There exists an algorithm that, given a prediction of the item-size upper bound, achieves* Consistency *equal to the* CR *of* NONADAPTIVE *from [11] (i.e., approximately* γ_{k^*}*) and* Robustness *of* γ_{k^*-3} *for all* $k^* \geq 4$*. For* $1 \leq k^* < 4$*, it guarantees both* Consistency *and* Robustness *of* $\gamma_{\hat{k}}$*, where* $\hat{k} = \max\{k^*-2, 1\}$*.*

3.2   Adaptive Algorithm with Removal and Recourse

In this section, we extend the previous result by turning our attention to adaptive algorithms (with unknown item size upper bound) that can also use a recourse0.

Algorithm 3 ADAPTIVEWITHRECOURSE

1: $S \leftarrow \emptyset$
2: $MaxItem \leftarrow 0$
3: $B \leftarrow \emptyset$
4: **while** a new item o_i arrives **do**
5:     $MaxItem \leftarrow \max\{MaxItem, |o_i|\}$
6:     $k_{curr} \leftarrow \max \left\{ k \in \mathbb{N} \mid \frac{2}{k-1+\sqrt{(k-1)^2+4}} \geq MaxItem \right\}$ $\triangleright$ Optimal k considering
    $MaxItem$ as the upper bound.
7:     $k' \leftarrow \max\{k_{curr} - 2, 1\}$
8:     $k'' \leftarrow \max\{k_{curr} - 1, 1\}$
9:     **if** $v(S) \geq \gamma_{k''}$ **then**
10:         Reject o_i $\triangleright$ *S is not changed afterwards*
11:     **else**
12:         $S \leftarrow S \cup \{o_i\}$
13:         **if** $v(S) > 1$ **then**
14:             Let $B = \{o \in S : |o| > 1 - \gamma_{k'}\}$ $\triangleright$ Reset B
15:             **if** S contains a subset $\hat{S}$ such that $1 \geq v(\hat{S}) \geq \gamma_{k''}$ **then**
16:                 $S \leftarrow \hat{S}$ $\triangleright$ *S is not changed afterwards*
17:                 **break**
18:             **if** $v(B) > 1$ **then**
19:                 Remove the largest item of S (send to the buffer)
20:             **else if** $|B| = 1 \wedge \exists\, o' \in Deleted\ Items$ s.t.: $S \bigcup o' \setminus B$ contains a
                subset $\hat{S}$ such that $1 \geq v(\hat{S}) \geq \gamma_{k''}$ **then**
21:                 Use the recourse to bring item o' back and set:
22:                 $S \leftarrow \hat{S}$ $\triangleright$ *S is not changed afterwards*
23:                 **break**
24:             **else**
25:                 **while** $v(S) > 1$ **do**
26:                     Remove from S the smallest element of $S \setminus B$ (send to the
                    buffer)

The goal is to incorporate the recourse into a new version of the adaptive algorithm so that we can create an equivalent of Theorem 4, that guarantees a CR even closer to the best possible that the non adaptive algorithm can achieve. Specifically, we aim for a CR of at least $\gamma_{\hat{k}+1}$ (in the general case). The main logic is as follows:

Maintain the same strategy as ADAPTIVE and keep considering items of size $1 - \gamma_{k'}$ to be non-small, where $k' = \max\{k_{curr-2}, 1\}$, but only terminate if the sum of collected items gives CR (at least) $\gamma_{k'+1}$. That is, run the algorithm as if we where trying to get CR $\gamma_{\hat{k}}$, but only accept the solution if a CR of $\gamma_{\hat{k}+1}$ is satisfied instead. While doing so, also keep track of the items of size larger than

$1 - \gamma_{k'+1}$, i.e., *"non-very-small items"*. I.e., the items we should have considered as "non-small" if we were aiming for CR of $\gamma_{\hat{k}+1}$, by the logic of NONADAPTIVE.

Unlike non-small items in the proof of Theorem 4, which always additively guarantee the desired CR before becoming infeasible, non-very-small items lack this property. As a result, the adaptive algorithm may remove items that it could later combine with new arrivals to achieve the target ratio. With careful analysis, we identify the unique case where this occurs and show that the algorithm can immediately detect and reverse it with a single recourse.

Refining the algorithm to delete small items in increasing order—without loss of generality, since arbitrary deletion sufficed before—we observe the following:

If any subset of the items seen so far is feasible and can yield an approximation of at least $\gamma_{k'+1}$ (for any current value of k'), then either there is a feasible subset among the items currently in the solution that satisfies this, or it can be achieved by a single swap between a rejected item and an item in the solution.

The technical analysis is deferred to the full version of the paper.

Theorem 6. *For all positive integers k, Algorithm 3 is $\gamma_{\tilde{k}}$-competitive when* $u \leq \dfrac{2}{k-1+\sqrt{(k-1)^2+4}}$, *where* $\tilde{k} = \max\{k^* - 1, 1\}$, *using at most one recourse.*

In Fig. 2, we show the performance of, ADAPTIVE and ADAPTIVEWITHRE-COURSE, compared to the upper-bound aware algorithm of [11].

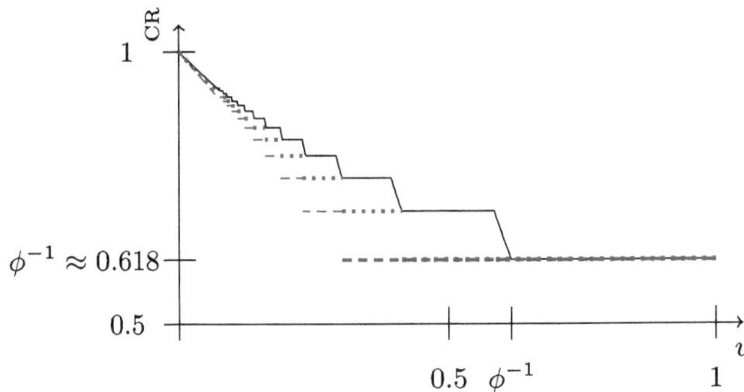

Fig. 2. The black continuous lines represent the CR of NONADAPTIVE, the red dashed the CR of ADAPTIVE and the blue dotted the CR of ADAPTIVEWITHRECOURSE. (Color figure online)

Acknowledgments. This work has been partially supported by project MIS 5154714 of the National Recovery and Resilience Plan Greece 2.0 funded by the European Union under the NextGenerationEU Program. Panagiotis Patsilinakos has received support under the program "Investissements d'Avenir" launched by the French Government, project PSL JRC 2024 - N 2024- 488. This work received the support of "Partenariat Hubert Curien France/Grèce – ARISTOTE".

References

1. ALPS: https://algorithms-with-predictions.github.io/
2. Babaioff, M., Hartline, J.D., Kleinberg, R.D.: Selling ad campaigns: online algorithms with cancellations. In: EC, pp. 61–70. ACM (2009)
3. Böckenhauer, H., Burjons, E., Hromkovic, J., Lotze, H., Rossmanith, P.: Online simple knapsack with reservation costs. In: STACS, pp. 16:1–16:18. LIPIcs, Schloss Dagstuhl - Leibniz-Zentrum für Informatik (2021)
4. Böckenhauer, H., Frei, F., Rossmanith, P.: Removable online knapsack and advice. In: STACS, pp. 18:1–18:17. LIPIcs, Schloss Dagstuhl - Leibniz-Zentrum für Informatik (2024)
5. Böckenhauer, H., Hromkovic, J., Komm, D., Rossmanith, P., Stocker, M.: A survey of online knapsack problems. Discret. Appl. Math. **378**, 492–507 (2026)
6. Böckenhauer, H., Klasing, R., Mömke, T., Rossmanith, P., Stocker, M., Wehner, D.: Online knapsack with removal and recourse. In: IWOCA, LNCS, vol. 13889, pp. 123–135. Springer (2023)
7. Böckenhauer, H., Klasing, R., Mömke, T., Rossmanith, P., Stocker, M., Wehner, D.: Online knapsack with removal and recourse. J. Comput. Syst. Sci. **155**, 103697 (2026)
8. Böckenhauer, H., Komm, D., Královic, R., Rossmanith, P.: The online knapsack problem: advice and randomization. Theor. Comput. Sci. **527**, 61–72 (2014)
9. Burjons, E., Gehnen, M., Lotze, H., Mock, D., Rossmanith, P.: The online simple knapsack problem with reservation and removability. In: MFCS, pp. 29:1–29:12. LIPIcs, Schloss Dagstuhl - Leibniz-Zentrum für Informatik (2023)
10. Cygan, M., Jez, L., Sgall, J.: Online knapsack revisited. Theor. Comput. Syst. **58**(1), 153–190 (2016)
11. Gourvès, L., Pagourtzis, A.: Removable online knapsack with bounded size items. In: SOFSEM, LNCS, vol. 14519, pp. 283–296. Springer (2024)
12. Han, X., Kawase, Y., Makino, K.: Online unweighted knapsack problem with removal cost. Algorithmica **70**(1), 76–91 (2014)
13. Han, X., Kawase, Y., Makino, K.: Randomized algorithms for online knapsack problems. Theor. Comput. Sci. **562**, 395–405 (2015)
14. Han, X., Kawase, Y., Makino, K., Yokomaku, H.: Online knapsack problems with a resource buffer. In: ISAAC, LIPIcs, vol. 149, pp. 28:1–28:14. Schloss Dagstuhl - Leibniz-Zentrum für Informatik (2019)
15. Han, X., Makino, K.: Online removable knapsack with limited cuts. Theor. Comput. Sci. **411**(44–46), 3956–3964 (2010)
16. Iwama, K., Taketomi, S.: Removable online knapsack problems. In: Widmayer, P., Eidenbenz, S., Triguero, F., Morales, R., Conejo, R., Hennessy, M. (eds.) ICALP 2002. LNCS, vol. 2380, pp. 293–305. Springer, Heidelberg (2002). https://doi.org/10.1007/3-540-45465-9_26
17. Iwama, K., Zhang, G.: Online knapsack with resource augmentation. Inf. Process. Lett. **110**(22), 1016–1020 (2010)
18. Marchetti-Spaccamela, A., Vercellis, C.: Stochastic on-line knapsack problems. Math. Program. **68**, 73–104 (1995)
19. Mitzenmacher, M., Vassilvitskii, S.: Algorithms with Predictions, p. 646–662. Cambridge University Press (2021)
20. Zhou, Y., Chakrabarty, D., Lukose, R.: Budget constrained bidding in keyword auctions and online knapsack problems. In: Papadimitriou, C., Zhang, S. (eds.) WINE 2008. LNCS, vol. 5385, pp. 566–576. Springer, Heidelberg (2008). https://doi.org/10.1007/978-3-540-92185-1_63

Layer-Based Width for **PAFP** on DAGs: A BFS-Width-2 Normal Form and Exact-Length Width-2 Tractability

Samuel German[(✉)]

University of California, San Diego, San Diego, USA
`sgerman@ucsd.edu`

Abstract. The Path Avoiding Forbidden Pairs (PAFP) problem asks whether, in a directed graph G with terminals s and t and a set $\mathcal{F}$ of forbidden vertex pairs, there is an s–t path that includes at most one endpoint from each pair. A standard width-based tractable regime for PAFP is obtained by adding the forbidden pairs as edges to form an undirected union/constraint graph and requiring that graph to have bounded treewidth or pathwidth.

We initiate the study of PAFP under a layer-based width measure. Let the BFS layer L_d be the set of vertices at directed shortest-path distance d from s, and define the BFS-width from s as $\max_d |L_d|$. We show that PAFP on DAGs remains NP-complete even when the union digraph $G \cup \mathcal{F}$—obtained by adding to G one arc per forbidden pair, oriented forward in a fixed deterministic topological order—is itself a DAG of BFS-width at most 2. We do this by giving a constructive polynomial-time transformation that converts any DAG PAFP instance into an equivalent DAG PAFP instance whose union digraph has BFS-width at most 2.

We also study *exact-length layers*, namely the sets of vertices reachable from s by a directed path of length exactly d. We show that PAFP, when the input graph is a DAG whose exact-length layers are of size at most 2, is in P; this width bound is tight, since a classical layered reduction already yields NP-completeness on DAGs of exact-length width 3. Previously studied polynomial-time cases of PAFP impose restrictions on $\mathcal{F}$ in order to achieve tractability, whereas our exact-length width-2 case combines unrestricted forbidden pairs with input graphs that may have exponentially many s–t paths.

Keywords: Forbidden pairs · Directed acyclic graphs · BFS layers · Width parameters · NP-completeness

1 Introduction

The *Path Avoiding Forbidden Pairs* problem (PAFP) asks, given a directed graph $G = (V, E)$, terminals $s, t \in V$, and forbidden vertex pairs $\mathcal{F} \subseteq \binom{V}{2}$,

© The Author(s), under exclusive license to Springer Nature Switzerland AG 2026
F. Foucaud and A. Parreau (Eds.): IWOCA 2026, LNCS 16587, pp. 311–326, 2026.
https://doi.org/10.1007/978-3-032-27732-9_22

whether there exists an s–t path that contains at most one endpoint from each pair in $\mathcal{F}$. PAFP is a natural abstraction for path selection under pairwise incompatibility constraints. It first arose in automatic software testing: a program is represented by a directed graph whose vertices represent code segments and whose edges represent the flow of control between code segments; an s–t path represents a complete entry-to-exit test execution of the program, and forbidden pairs encode mutually unexecutable branches. In this setting, PAFP was proposed as a means to make software testing more efficient by only considering paths through a program that contain at most one branch of each pair of mutually unexecutable branches [12]. The same abstraction appears in bioinformatics, where gene finding with RT-PCR evidence models candidate transcripts as s–t paths in a splicing graph whose vertices are non-overlapping DNA segments and whose edges indicate that one segment may immediately follow another in a transcript; an RT-PCR test selects two primer vertices u and v together with an observed product length ℓ, and only transcript paths whose u–v subpath has length ℓ explain the test. By setting lengths to an unattainable value and capturing combinations of segments that are incompatible with experimental evidence via forbidden pairs, PAFP is able to help gene-prediction algorithms rule out biologically implausible transcripts while incorporating RT-PCR data to improve sensitivity and reveal novel splicing variants [11].

In application settings of PAFP, the surrounding pipeline typically requires repeated path computations on large instances, so an exponential-time search would be rarely acceptable in practice. This difficulty is already present on directed acyclic graphs: PAFP is NP-complete even when the input digraph is acyclic [7]. Accordingly, positive results are valuable because they isolate structural assumptions under which exact computation becomes feasible, while hardness results show that some such extra structure is necessary and thereby motivate preprocessing, parameterization, or heuristics when no exploitable structure is available.

Union/Constraint Graph. A mainstream way to analyze constrained graph problems is to encode constraints as edges between vertices, the *primal graph* viewpoint in constraint satisfaction [13]. PAFP fits this unusually well: instances contain adjacency constraints (edges of G) and incompatibility constraints (pairs in $\mathcal{F}$) on the same vertex set. Bodlaender–Jansen–Kratsch exploit this interaction-graph view for forbidden-pairs path problems, obtaining algorithmic and kernel results under structural restrictions on an undirected union/constraint graph that contains the (undirected) edges of G together with an edge for each forbidden pair (e.g. treewidth-based parameters) [4].

For PAFP on DAGs it is common to orient forbidden pairs consistently with a DAG order (topological or reachability-based) and analyze the ensuing structure [9,10]. Accordingly, we study how certain width parameters affect PAFP on DAGs by considering the union graph obtained by adding the forbidden pairs as arcs to the input graph, where each forbidden pair is oriented according to a fixed topological order (see the union digraph in Sect. 2). The topological-order rule $G \mapsto \pi_G$ is fixed only to make the union digraph (and hence its BFS-width) uniquely defined. The particular topological-order rule is immaterial: all of our results hold for any fixed choice of $G \mapsto \pi_G$. In fact, our NP-completeness result holds when restricted to instances for which regardless of the topological order used to orient the forbidden pairs, the resulting union digraph is the same (Remark 2).

BFS-width. BFS layerings (grouping vertices by shortest-path distance from a root) are a standard structural tool in graph theory and graph algorithms, underlying, for example, layered path decompositions and separator theorems for minor-closed graph classes [2,5]. In particular, recent parameterized algorithms for length-constrained path problems on graphs reason explicitly about BFS distance layers along candidate paths, bounding the number of 'reused' layers as a function of the detour parameter [3]. A natural associated width statistic is the maximum size of any BFS distance layer. This statistic has been studied as a width parameter in the undirected setting, with emphasis on its polynomial-time computability [6]. Since PAFP is posed on directed graphs, in our work "BFS-width" and "BFS layer" denote the directed versions of these concepts unless stated otherwise, measured in terms of directed shortest-path distances (see Sect. 2 for a formal definition).

PAFP is an inherently directed s–t path problem, so distance layerings rooted at s are a natural structural lens; in particular any instance, if it has a safe s–t path, has a shortest one. Despite this, BFS-width has not been studied for PAFP. We initiate this direction by asking whether bounded BFS-width of the union digraph makes PAFP tractable, analogously to how the undirected variant of PAFP is fixed-parameter tractable parameterized by the treewidth of the undirected union/constraint graph [4].

There is a genuine positive baseline for this question: PAFP itself is indeed fixed-parameter tractable parameterized by the undirected BFS-width of the underlying undirected union/constraint graph[1]. This raises the possibility that for the union digraph which stays true to PAFP's directed nature, tractability can likewise be achieved by bounding its BFS-width.

Our Results. We refute this hope in the strongest possible form: *every* PAFP instance whose input graph is a DAG has an equivalent polynomial-time computable *normal form* whose union digraph is a DAG of BFS-width at most 2—

---

[1] Let U be the part reachable from s of the underlying undirected union/constraint graph, i.e., the graph obtained by adding an undirected edge uv for each forbidden pair $\{u, v\} \in \mathcal{F}$ to the underlying undirected graph of the digraph G. If $L_0, L_1, \ldots$ are the undirected BFS layers of U from s, then every edge of U has endpoints in the same or adjacent layers, and each vertex appears in at most two consecutive bags $B_d := L_d \cup L_{d+1}$. Hence $(B_d)_d$ is a path decomposition of width at most $2\,\mathrm{bfsw}_{\mathrm{und}}(U, s) - 1$, and therefore $\mathrm{tw}(U) \leq 2\,\mathrm{bfsw}_{\mathrm{und}}(U, s) - 1$. The bounded-treewidth MSO-on-structures argument used in Bodlaender–Jansen–Kratsch [4, Prop. 6] thus yields fixed-parameter tractability: their framework applies unchanged in the directed setting of PAFP via the standard mixed-graph encoding of directed edges by tail/head incidence relations [1, Sec. 3, Def. 3.1, pp. 314–315].

the smallest nontrivial value (Theorem 1). Hence PAFP remains NP-complete even in this ultra-thin, ladder-like regime where a shortest-path exploration has at most two frontier choices at every distance (Theorem 5). We also consider a second rooted layering notion on DAGs, namely exact-length layers $E_G(s, d)$, which record which vertices can occur at the dth position of a directed path from s. For this notion, we show that PAFP is polynomial-time solvable when $\mathrm{elw}(G, s) \leq 2$ (Theorem 7). This width bound is tight, as inspection of the classical layered reduction of Gabow et al. shows that PAFP remains NP-complete already on DAG instances of exact-length width 3 (Proposition 8). While there is substantial work on tractable special cases of PAFP, polynomial-time cases thereof rely on additional restrictions on the forbidden-pair set (see Table 1). By contrast, our exact-length width-2 tractability result simultaneously allows (i) arbitrary forbidden pairs and (ii) input DAGs with exponentially many s–t paths (Remark 3).

Table 1. Survey of tractable regimes for PAFP. Known polynomial-time results impose structure on the forbidden-pair set $\mathcal{F}$. By contrast, Theorem 7 allows for arbitrary $\mathcal{F}$, and the promise $\mathrm{elw}(G, s) \leq 2$ still permits exponentially many directed s–t paths.

Reference	Restriction	Arbitrary $\mathcal{F}$?
This paper	$\mathrm{elw}(G, s) \leq 2$ and G is a DAG; $\mathcal{F}$ can be any subset of $\binom{V}{2}$.	✓
Kováč (2013) [10]	For a fixed topological order $\prec$ of G, any two distinct forbidden pairs $\{u, v\}$ and $\{x, y\}$ with $u \prec v$, $x \prec y$, and $u \prec x$ satisfy $x \prec v \prec y$.	✗
Kolman– Pangrác (2009) [9]	*Hierarchical forbidden pairs*: With respect to the reachability order $\prec$ of G, no two oriented forbidden pairs (u, v) and (x, y) satisfy $u \prec x \prec v \prec y$.	✗
Bodlaender– Jansen– Kratsch (2013) [4]^a	Bounded treewidth of the underlying undirected union/constraint graph. This restricts $\mathcal{F}$ from making the union/constraint graph have large treewidth.	✗
Yinnone (1997) [14]	*Skew-symmetry*: The forbidden pairs are mutually disjoint, and for every $\{u, u'\}, \{v, v'\} \in \mathcal{F}$, the arc condition $(u, v) \in E \Rightarrow (v', u') \in E$ holds.	✗

^{a}This result was formulated in an undirected setting for PAFP; for the exact directed setting in this paper, combine [4, Prop. 6] with bounded-treewidth MSO model checking for structures with directed edge relations [1, Sec. 3, Def. 3.1, pp. 314–315].

2 Definitions

Arc Terminology. For a directed graph $G = (V, E)$ and arc $(u, v) \in E$, u is the *tail* and v the *head* of (u, v).

Problem Definition and Notation. A PAFP instance is a tuple $I = (G, s, t, \mathcal{F})$ where $G = (V, E)$ is a directed graph, $s \neq t \in V$, and $\mathcal{F} \subseteq \binom{V}{2}$. A directed s–t path P is *safe* if it contains at most one endpoint from each pair in $\mathcal{F}$; PAFP asks whether a safe s–t path exists. For $u, v \in V$, let $\mathrm{dist}_G(u, v)$ be the length of a shortest directed u–v path (or ∞ if none exists). Define the BFS layer $L_G(s, d) := \{v \mid \mathrm{dist}_G(s, v) = d\}$ and the BFS-width $\mathrm{bfsw}(G, s) := \max_{d \in \mathbb{Z}_{\geq 0}} |L_G(s, d)|$. Define the exact-length layer $E_G(s, d) := \{v \mid \exists \text{ an } s\text{–}v \text{ path of length exactly } d\}$, and the exact-length width $\mathrm{elw}(G, s) := \max_{d \in \mathbb{Z}_{\geq 0}} |E_G(s, d)|$.

σ-Oriented Forbidden-Pair Arcs. Let $G = (V, E)$ be a DAG, let $\mathcal{F} \subseteq \binom{V}{2}$, and let σ be a topological order of G. Define the σ-oriented forbidden-pair arcs $A_\sigma(\mathcal{F}) := \{(u, v) \mid \{u, v\} \in \mathcal{F}, \ \sigma(u) < \sigma(v)\}$.

Union Digraph. Fix a function $G \mapsto \pi_G$ that maps each DAG G to a topological ordering π_G of G. Let $I = (G = (V, E), s, t, \mathcal{F})$ be a PAFP instance where G is a DAG. Define the union digraph $G \cup \mathcal{F} := (V, \ E \cup A_{\pi_G}(\mathcal{F}))$.

Remark 1. The fixed rule $G \mapsto \pi_G$ is used only to make $G \cup \mathcal{F}$ uniquely defined on arbitrary DAG instances. Our results hold for any fixed topological-order rule. Moreover, we later note that our NP-completeness result holds when restricted to instances where the union digraph is independent of the chosen topological order (Remark 2).

3   BFS-Width-2 Normal Form

3.1   Intuition and Overview

The intuition behind our normal form is that we turn an arbitrary DAG PAFP instance into an "ultra-thin" ladder from a fresh start vertex s': we build a long directed path of new placeholder vertices (the *spine*) and then give each original reachable vertex a short detour (i.e., a directed path of length two via a fresh intermediate vertex) off a designated spine position (see Fig. 1 for an illustration). In this way, each original vertex is forced to appear at a controlled shortest-path distance from s'. These detours are placed at every other spine step so that each detour's intermediate vertex lies on an even BFS layer from s' and the corresponding original vertex lies on the next odd layer. This parity separation prevents multiple off-spine vertices from landing in the same BFS layer, keeping the frontier small.

(Distance Subtlety). The main subtlety for the BFS-width bound is preventing the *core union arcs* from creating unintended *distance-shortcuts*: once we consider the union digraph (i.e., the original arcs of G together with the π_G-oriented forbidden-pair arcs on the original vertices), either type of core arc could in principle combine with the spine and detours to reach some vertex earlier than its intended detour distance. We avoid this by using a topological order σ of the

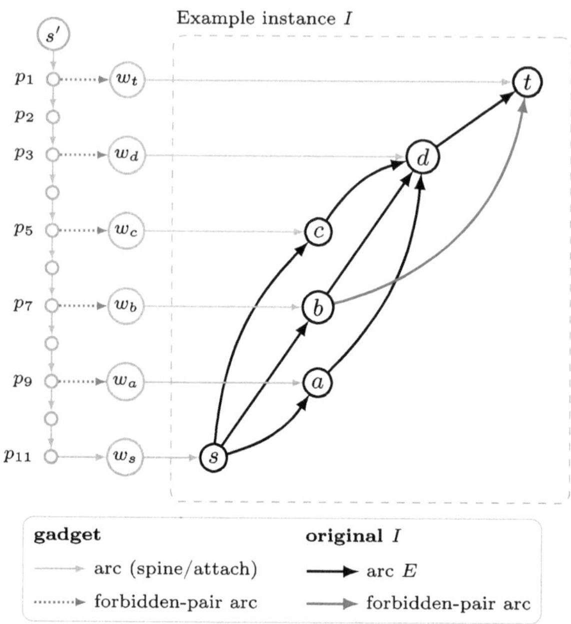

Fig. 1. The union digraph $G' \cup \mathcal{F}'$ of the normalized instance $I' = \mathrm{Normalize}(I)$, taking I to be the boxed structure. Note that although the union contains the forbidden-pair arc $b \to t$, the detour into t leaves the spine at p_1 while the detour into b leaves later (at p_7), so traversing $b \to t$ cannot create a shorter s'–t route than the direct detour into t.

reachable core as a *scheduling constraint* for the detours. Concretely, we assign vertices to detour slots in *reverse* topological order (so vertices later in σ are attached closer to s'). Think of the spine as a timeline: attaching a vertex closer to s' means it becomes reachable earlier. If a union arc $u \to v$ jumps forward by a positions in σ, then we "wait" and place the detour for u *after* the detour for v by at least a detour-slots along the spine. Thus v is scheduled a slots closer to s' than u, so traversing $u \to v$ can never beat the direct detour into v and cannot create a new shortest-path shortcut.

Equisatisfiability is enforced by "booby-trapping" each detour using its intermediate vertex: we add a forbidden pair that makes taking that detour unsafe, except for the unique detour that enters the original start s. Consequently, every safe s'–t path is forced to follow the spine into s and then continue inside the original instance. Finally, as a bookkeeping step, we insert the reachable-core forbidden-pair arcs oriented forward in the reachable core's topological order σ as ordinary arcs of the output graph G'. This ensures that forming the union digraph of the output instance introduces no additional arcs (so we may analyze BFS distances directly in G'), and it aligns all core arcs with the same order used in the detour scheduling. This step does not change satisfiability, since any path traversing such an inserted arc would contain both endpoints of that forbidden pair and hence be unsafe.

3.2   The BFS-Width-2 Normal Form Theorem

Theorem 1. *There exists a polynomial-time computable mapping* Normalize *such that for every* PAFP *instance* $I = (G, s, t, \mathcal{F})$ *where* G *is a DAG, the* PAFP *instance* $I' := \text{Normalize}(I) = (G', s', t, \mathcal{F}')$ *satisfies that* G' *is a DAG and* $\text{bfsw}(G' \cup \mathcal{F}', s') \leq 2$, *and* I *is a* YES-*instance* $\iff$ I' *is a* YES-*instance.*

Definition 1 (Normalization mapping Normalize**).** *Fix a* PAFP *instance* $I = (G = (V, E), s, t, \mathcal{F})$ *where* G *is a DAG. Define* $\text{Normalize}(I) = (G' = (V', E'), s', t, \mathcal{F}')$ *as follows.*

Reachable Core. *Let*

$$V_{\text{rch}} := \{v \in V \mid \text{dist}_G(s, v) < \infty\}, \qquad E_{\text{rch}} := E \cap (V_{\text{rch}} \times V_{\text{rch}}), \qquad (1)$$

and

$$\mathcal{F}_{\text{rch}} := \{\{u, v\} \in \mathcal{F} \mid u, v \in V_{\text{rch}}\}. \tag{2}$$

Let $G_{\text{rch}} := (V_{\text{rch}}, E_{\text{rch}})$. *Fix a total order* $\prec_{\text{enc}}$ *on vertex encodings, and let* σ *be the topological ordering of* G_{rch} *returned by Kahn's algorithm with* $\prec_{\text{enc}}$ *tie-breaking [8]. Set* $E_0 := E_{\text{rch}} \cup A_\sigma(\mathcal{F}_{\text{rch}})$, *where* $A_\sigma(\mathcal{F}_{\text{rch}})$ *is as defined in Sect. 2.*

Reverse-σ Enumeration. *Let* $q := |V_{\text{rch}}|$ *and fix an ordering* $(r_1, \ldots, r_q)$ *of* V_{rch} *such that* $\sigma(r_1) > \sigma(r_2) > \cdots > \sigma(r_q)$. *Note that* $q \geq 1$ *since* $s \in V_{\text{rch}}$. *If* $q = 1$, *then* $V_{\text{rch}} = \{s\}$ *and (since* $s \neq t$*) we have* $t \notin V_{\text{rch}}$, *so the instance is trivially* NO; *in this case, index ranges such as* $j = 1, \ldots, 2q - 2$ *are empty.*

Fresh Gadget Vertices. *Introduce pairwise distinct vertices*

$$s' \notin V, \qquad p_1, \ldots, p_{2q-1} \notin V \cup \{s'\}, \qquad w_1, \ldots, w_q \notin V \cup \{s'\} \cup \{p_1, \ldots, p_{2q-1}\}.$$

Let $P := \{p_1, \ldots, p_{2q-1}\}$ *and* $W := \{w_1, \ldots, w_q\}$.

Arcs. *Define*

$$E_{\text{spine}} := \{(s', p_1)\} \cup \{(p_j, p_{j+1}) \mid j = 1, \ldots, 2q - 2\}, \tag{3}$$

$$E_{\text{att}} := \{(p_{2i-1}, w_i) \mid i = 1, \ldots, q\} \cup \{(w_i, r_i) \mid i = 1, \ldots, q\}, \tag{4}$$

and set

$$V' := V_{\text{rch}} \cup \{t\} \cup \{s'\} \cup P \cup W, \qquad E' := E_0 \cup E_{\text{spine}} \cup E_{\text{att}}. \tag{5}$$

Forbidden Pairs. *Define*

$$\mathcal{F}' := \mathcal{F}_{\text{rch}} \cup \{\{p_{2i-1}, w_i\} \mid i \in \{1, \ldots, q\} \text{ and } r_i \neq s\}. \tag{6}$$

Finally, let $G' := (V', E')$ *and output* $\text{Normalize}(I) = (G', s', t, \mathcal{F}')$.

318   S. German

Proof (Proof of Theorem 1).

Standing notation for Normalize. Fix a PAFP instance $I = (G, s, t, \mathcal{F})$ where $G = (V, E)$ is a DAG, and let $I' = (G' = (V', E'), s', t, \mathcal{F}') := \text{Normalize}(I)$. Write $V_{\text{rch}}, E_{\text{rch}}, \mathcal{F}_{\text{rch}}, G_{\text{rch}}$, and σ for the reachable-core objects from Definition 1, where $E_0 := E_{\text{rch}} \cup A_\sigma(\mathcal{F}_{\text{rch}})$. Let $(r_1, \ldots, r_q)$ be the fixed reverse-σ listing of V_{rch} (with $q := |V_{\text{rch}}|$), and let $P = \{p_1, \ldots, p_{2q-1}\}$ and $W = \{w_1, \ldots, w_q\}$ denote the gadget vertices of G'. We also use E_{spine} and E_{att} for the spine and attachment arc sets, so that $E' = E_0 \cup E_{\text{spine}} \cup E_{\text{att}}$.

Output is a valid PAFP instance. By Definition 1, the output $I' = (G' = (V', E'), s', t, \mathcal{F}')$ satisfies $s', t \in V'$ with $s' \neq t$ (since $s' \notin V$ while $t \in V$), $E' \subseteq V' \times V'$, and $\mathcal{F}' \subseteq \binom{V'}{2}$. Hence I' is a valid PAFP instance.

Polynomial-Time Computability. The mapping of Definition 1 is computable in deterministic polynomial time. Indeed, V_{rch} is obtained by a single reachability search from s in G (e.g. by DFS), after which E_{rch} and $\mathcal{F}_{\text{rch}}$ are computed by one scan of E and $\mathcal{F}$. We next obtain σ by running Kahn's topological-sorting algorithm [8] on G_{rch}, maintaining the current zero-indegree vertices in a binary min-heap keyed by $\prec_{\text{enc}}$. Each vertex is inserted exactly once, namely when its indegree first becomes zero, and is extracted exactly once; furthermore, each arc is examined exactly once when its tail is extracted. Since each heap insertion and extraction costs $O(\log |V_{\text{rch}}|)$, this step takes $O(|E_{\text{rch}}| + |V_{\text{rch}}| \log |V_{\text{rch}}|)$ time. We then orient each pair in $\mathcal{F}_{\text{rch}}$ according to σ to obtain $A_\sigma(\mathcal{F}_{\text{rch}})$ and $E_0 = E_{\text{rch}} \cup A_\sigma(\mathcal{F}_{\text{rch}})$. Finally, we create $O(|V_{\text{rch}}|)$ fresh vertices and explicitly list the sets $E_{\text{spine}}, E_{\text{att}}$, and the additional $O(|V_{\text{rch}}|)$ "booby-trap" forbidden pairs, and assemble $(V', E', \mathcal{F}')$ by set unions. All of these operations run in time polynomial in $|V| + |E| + |\mathcal{F}|$ (in particular, $O((|V| + |E|) \log |V| + |\mathcal{F}|)$ suffices).

Claim 2 (Acyclicity and union-closure of the output). *In the standing notation above, the digraph $G' = (V', E')$ is a DAG. In particular, the topological ordering $\pi_{G'}$ is well-defined. Moreover, let $A_{\pi_{G'}}(\mathcal{F}')$ and $G' \cup \mathcal{F}' = (V', E' \cup A_{\pi_{G'}}(\mathcal{F}'))$ be defined as in Sect. 2. Then $A_{\pi_{G'}}(\mathcal{F}') \subseteq E'$ and $G' \cup \mathcal{F}' = (V', E')$.*

Proof. Acyclicity. We explicitly construct a topological ordering τ of G'. First, recall that σ is a topological ordering of $G_{\text{rch}} = (V_{\text{rch}}, E_{\text{rch}})$, hence $\sigma(u) < \sigma(v)$ for every $(u, v) \in E_{\text{rch}}$. Also, by definition of $A_\sigma(\mathcal{F}_{\text{rch}})$, every arc $(u, v) \in A_\sigma(\mathcal{F}_{\text{rch}})$ satisfies $\sigma(u) < \sigma(v)$. Therefore σ is a topological ordering of (V_{rch}, E_0), where $E_0 = E_{\text{rch}} \cup A_\sigma(\mathcal{F}_{\text{rch}})$.

Define $\tau_0 := s', p_1, p_2, \ldots, p_{2q-1}, w_1, w_2, \ldots, w_q, r_q, r_{q-1}, \ldots, r_1$ as a base sequence. If $t \notin V_{\text{rch}}$, set τ to be τ_0 with t appended at the end; otherwise (if $t \in V_{\text{rch}}$) set $\tau := \tau_0$. We verify that every arc $(x, y) \in E'$ goes forward in τ. Recall that $E' = E_0 \cup E_{\text{spine}} \cup E_{\text{att}}$.

<u>Spine arcs.</u> If $(x, y) \in E_{\text{spine}}$, then (x, y) is either (s', p_1) or (p_j, p_{j+1}) for some $j \in \{1, \ldots, 2q - 2\}$, and τ lists s' before p_1 and p_j before p_{j+1}.

<u>Attachment arcs.</u> If $(x, y) \in E_{\text{att}}$, then (x, y) is either (p_{2i-1}, w_i) or (w_i, r_i) for some $i \in \{1, \ldots, q\}$. In τ, all spine vertices appear before all w_i, and all w_i appear before all vertices of V_{rch}, hence before $r_i \in V_{\text{rch}}$.

<u>Core arcs.</u> If $(x, y) \in E_0$, then $x, y \in V_{\text{rch}}$ and $\sigma(x) < \sigma(y)$. Since $(r_1, \ldots, r_q)$ lists V_{rch} in *decreasing* σ-order, the sequence $(r_q, r_{q-1}, \ldots, r_1)$ inside τ lists V_{rch} in *increasing* σ-order; thus τ lists x before y inside the core block.

Thus every arc of E' goes forward in τ, so τ is a topological ordering of G'. Hence G' is a DAG.

Union-closure. We show $A_{\pi_{G'}}(\mathcal{F}') \subseteq E'$, i.e., for every forbidden pair $\{x, y\} \in \mathcal{F}'$, the arc oriented forward by $\pi_{G'}$ already lies in E'. By definition of $\mathcal{F}'$, there are two cases.

Case 1: $\{x, y\} \in \mathcal{F}_{\text{rch}}$. In the construction of $A_\sigma(\mathcal{F}_{\text{rch}})$, exactly one of (x, y) or (y, x) belongs to $A_\sigma(\mathcal{F}_{\text{rch}}) \subseteq E_0 \subseteq E'$; call this arc (a, b). Since (a, b) is an arc of G', every topological ordering of G' (in particular $\pi_{G'}$) satisfies $\pi_{G'}(a) < \pi_{G'}(b)$. Hence the oriented arc contributed by $\{x, y\}$ under $\pi_{G'}$ is exactly $(a, b) \in E'$.

Case 2: $\{x, y\} = \{p_{2i-1}, w_i\}$ *for some* $i \in \{1, \ldots, q\}$ with $r_i \neq s$. In this case $(p_{2i-1}, w_i) \in E_{\text{att}} \subseteq E'$, so every topological ordering of G' satisfies $\pi_{G'}(p_{2i-1}) < \pi_{G'}(w_i)$, and the oriented arc contributed by $\{x, y\}$ under $\pi_{G'}$ is $(p_{2i-1}, w_i) \in E'$.

In both cases, the arc contributed by $\{x, y\}$ under $\pi_{G'}$ lies in E', so $A_{\pi_{G'}}(\mathcal{F}') \subseteq E'$. Therefore $G' \cup \mathcal{F}' = (V', E' \cup A_{\pi_{G'}}(\mathcal{F}')) = (V', E')$. $\triangle$

Claim 3 (BFS-width-2 via a level function). *In the standing notation above, let $H := G' \cup \mathcal{F}'$ be the union digraph of the output instance. Then* $\text{bfsw}(H, s') \leq 2$.

Proof. By Claim 2, we have $H = (V', E')$.

Level Function. Define $\lambda : V' \to \mathbb{Z}_{\geq 0} \cup \{\infty\}$ by

$$\begin{aligned}
\lambda(s') &= 0, \quad \lambda(p_j) = j \ (1 \leq j \leq 2q - 1), \\
\lambda(w_i) &= 2i, \quad \lambda(r_i) = 2i + 1 \ (1 \leq i \leq q).
\end{aligned} \tag{7}$$

If $t \notin V_{\text{rch}}$, set $\lambda(t) = \infty$. If $t \notin V_{\text{rch}}$, then $\text{dist}_H(s', t) = \infty = \lambda(t)$. (In this case t is unreachable from s' in H and plays no role in BFS layers.) We first show that $\text{dist}_H(s', v) = \lambda(v)$ for every vertex $v \in V'$. Trivially, $\text{dist}_H(s', s') = 0 - \lambda(s')$.

(A) Spine Vertices $\{s'\} \cup P$. For each $j \in \{1, \ldots, 2q - 1\}$, the directed spine path $s' \to p_1 \to \cdots \to p_j$ has length j, hence $\text{dist}_H(s', p_j) \leq j$. Conversely, p_1 has the unique in-neighbor s', and for $j \geq 2$ the vertex p_j has unique in-neighbor p_{j-1}. By induction, any s'–p_j path has length at least j, so $\text{dist}_H(s', p_j) \geq j$. Thus $\text{dist}_H(s', p_j) = j = \lambda(p_j)$.

(B) Detour Vertices W. For each $i \in \{1, \ldots, q\}$, the path $s' \to p_1 \to \cdots \to p_{2i-1} \to w_i$ has length $2i$, hence $\text{dist}_H(s', w_i) \leq 2i$. Moreover, w_i has the unique in-neighbor p_{2i-1}, so any s'–w_i path must reach p_{2i-1} immediately before w_i.

Using (A), $\text{dist}_H(s', w_i) \geq \text{dist}_H(s', p_{2i-1}) + 1 = (2i - 1) + 1 = 2i$. Hence $\text{dist}_H(s', w_i) = 2i = \lambda(w_i)$.

(C) Core Vertices V_{rch}. Fix $i \in \{1, \ldots, q\}$, and consider $r_i \in V_{\text{rch}}$. The path $s' \to p_1 \to \cdots \to p_{2i-1} \to w_i \to r_i$ has length $2i + 1$, so $\text{dist}_H(s', r_i) \leq 2i + 1$. For the lower bound, let Q be any directed s'–r_i path and let r_k be the first vertex of V_{rch} appearing on Q. By construction, the only arcs with head in V_{rch} and tail in $V' \setminus V_{\text{rch}}$ are the attachment arcs $w_\ell \to r_\ell$; hence the predecessor of r_k on Q is exactly w_k. Therefore, $|Q| \geq \text{dist}_H(s', w_k) + 1 = 2k + 1$. After reaching r_k, the path cannot visit any gadget vertex because there is no arc from V_{rch} to $\{s'\} \cup P \cup W$ (indeed, $E_0 \subseteq V_{\text{rch}} \times V_{\text{rch}}$, while every arc of $E_{\text{spine}} \cup E_{\text{att}}$ has its tail in $\{s'\} \cup P \cup W$). Hence the suffix of Q from r_k to r_i lies in (V_{rch}, E_0). Now we argue that $k \geq i$. Recall that $(r_1, \ldots, r_q)$ lists V_{rch} in *reverse* σ-order, i.e., $\sigma(r_1) > \sigma(r_2) > \cdots > \sigma(r_q)$. Also recall that σ is a topological ordering of $(V_{\text{rch}}, E_{\text{rch}})$, and every arc in $A_\sigma(\mathcal{F}_{\text{rch}})$ is oriented forward with respect to σ by construction. Hence σ is a topological ordering of (V_{rch}, E_0), where $E_0 = E_{\text{rch}} \cup A_\sigma(\mathcal{F}_{\text{rch}})$. Now let $(r_a, r_b) \in E_0$. Since σ is topological for (V_{rch}, E_0), we have $\sigma(r_a) < \sigma(r_b)$. Because the list $(r_1, \ldots, r_q)$ is in decreasing σ-order, the inequality $\sigma(r_a) < \sigma(r_b)$ implies $a > b$. Therefore, if $r_{i_0} \to r_{i_1} \to \cdots \to r_{i_\ell}$ is any directed path in (V_{rch}, E_0), then $i_0 > i_1 > \cdots > i_\ell$. Thus Q reaching r_i from r_k is possible only if $k \geq i$. Consequently, $|Q| \geq 2k + 1 \geq 2i + 1 = \lambda(r_i)$ and $\text{dist}_H(s', r_i) \geq \lambda(r_i)$, and thus $\text{dist}_H(s', r_i) = \lambda(r_i)$.

BFS-Width Bound. For each depth $d \geq 0$, the BFS layer is

$$L_H(s', d) = \{v \in V' \mid \text{dist}_H(s', v) = d\} = \{v \in V' \mid \lambda(v) = d\}. \tag{8}$$

By definition of λ, for each d there is at most one spine vertex at depth d (namely p_d when $1 \leq d \leq 2q - 1$), and at most one non-spine vertex at depth d (namely $w_{d/2}$ if d is even and $d/2 \in [q]$, or $r_{(d-1)/2}$ if d is odd and $(d-1)/2 \in [q]$). Hence $|L_H(s', d)| \leq 2$ for all d, and therefore $\text{bfsw}(H, s') \leq 2$. Since $H = G' \cup \mathcal{F}'$, this proves the claim. $\triangle$

Claim 4 (Equisatisfiability of Normalize**).** *In the standing notation above, the original instance $I = (G, s, t, \mathcal{F})$ is a YES-instance if and only if the normalized instance $I' = (G', s', t, \mathcal{F}') = $ Normalize(I) is a YES-instance.*

Proof. **Trivial Unreachable Case.** If $t \notin V_{\text{rch}}$, then there is no directed s–t path in G, so I is a NO-instance. In this case, t has no incoming arc in G' (it is not among the r_i, and no gadget arc targets t), so t is unreachable from s' in G' and I' is also NO. Hence assume from now on that $t \in V_{\text{rch}}$.

($\Rightarrow$) Forward Direction. Let P be a safe directed s–t path in G. Every vertex on P lies in V_{rch}, and every arc of P lies in $E_{\text{rch}} \subseteq E_0 \subseteq E'$, so P is also a directed path in G'. Since V_{rch} consists of the vertices reachable from s, and thus there is a directed path from s to every vertex in $V_{\text{rch}} \setminus \{s\}$, the vertex s is first in the topological order σ. Hence in the reverse-σ enumeration we have $r_q = s$. Now define $P_{\text{pref}} := s' \to p_1 \to p_2 \to \cdots \to p_{2q-1} \to w_q \to r_q(= s)$, a (directed) gadget prefix which uses only arcs of $E_{\text{spine}} \cup E_{\text{att}} \subseteq E'$. Let Q be the directed s'–t path obtained by following P_{pref} and then appending the suffix of P from s to t (starting with the successor of s on P).

Safety. The only forbidden pairs in $\mathcal{F}'$ are: (i) the original pairs $\mathcal{F}_{\mathrm{rch}}$, whose endpoints lie in V_{rch}, and (ii) the booby-trap pairs $\{p_{2i-1}, w_i\}$ for indices i with $r_i \neq s$. The core vertices of Q are exactly the vertices of P, so Q is safe with respect to $\mathcal{F}_{\mathrm{rch}}$ because P is safe with respect to $\mathcal{F}$. Moreover, Q uses exactly one detour vertex, namely w_q, and the pair $\{p_{2q-1}, w_q\}$ is *not* in $\mathcal{F}'$ because $r_q = s$. Thus Q contains at most one endpoint of every booby-trap pair as well. Hence Q is safe, and therefore I' is a YES-instance.

($\Leftarrow$) **Reverse Direction.** Let Q be a safe directed s'–t path in G'. Since $t \in V_{\mathrm{rch}}$ and Q ends at t, the path Q must enter V_{rch}. Let r_k be the first vertex of V_{rch} that appears on Q. By the definition of $E' = E_0 \cup E_{\mathrm{spine}} \cup E_{\mathrm{att}}$, the only arcs with head in V_{rch} and tail outside V_{rch} are the attachment arcs $w_i \rightarrow r_i$. Therefore the predecessor of r_k on Q must be w_k. Also, the only arc entering w_k is $p_{2k-1} \rightarrow w_k$, so Q contains both p_{2k-1} and w_k. If $r_k \neq s$, then by definition of $\mathcal{F}'$ the booby-trap pair $\{p_{2k-1}, w_k\}$ belongs to $\mathcal{F}'$, contradicting that Q is safe. Hence necessarily $r_k = s$: every safe s'–t path enters the core at s. After reaching s, the path Q cannot return to gadget vertices, because there is no arc in E' from V_{rch} to $\{s'\} \cup P \cup W$ (indeed, $E_0 \subseteq V_{\mathrm{rch}} \times V_{\mathrm{rch}}$ and $E_{\mathrm{spine}} \cup E_{\mathrm{att}}$ has tail in $\{s'\} \cup P \cup W$). Thus the suffix of Q from s to t is a directed path entirely inside (V_{rch}, E_0). Finally, this suffix cannot use any arc from $A_\sigma(\mathcal{F}_{\mathrm{rch}}) \subseteq E_0$. Indeed, if it used an arc $(u, v) \in A_\sigma(\mathcal{F}_{\mathrm{rch}})$, then $\{u, v\} \in \mathcal{F}_{\mathrm{rch}} \subseteq \mathcal{F}'$, and the path would contain both endpoints u and v, contradicting safety. Therefore every arc on the suffix lies in $E_{\mathrm{rch}} \subseteq E$, so the suffix is a directed s–t path in G.

Safety in the Original Instance. Any forbidden pair $\{a, b\} \in \mathcal{F}$ whose endpoints both appear on this s–t path must satisfy $a, b \in V_{\mathrm{rch}}$, hence $\{a, b\} \in \mathcal{F}_{\mathrm{rch}}$. Since Q is safe with respect to $\mathcal{F}'$, it is safe with respect to $\mathcal{F}_{\mathrm{rch}}$, so the extracted s–t path in G is safe for $\mathcal{F}$ as well. Thus I is a YES-instance. $\triangle$

Taken together, this proves Theorem 1. $\square$

4   Two Layer-Based Width Complexity Results

4.1   Hardness for BFS Layering: NP-completeness at BFS-Width-2

Theorem 5 (*NP*-completeness on BFS-Width-2 Union DAGs). PAFP *is NP-complete even when restricted to instances* $I = (G = (V, E), s, t, \mathcal{F})$ *such that G is a DAG and* $\mathrm{bfsw}(G \cup \mathcal{F}, s) \leq 2$.

Proof. Membership in NP is immediate. For NP-hardness, reduce from PAFP on DAGs, which is NP-complete [7,9]. Given an arbitrary instance $I = (G, s, t, \mathcal{F})$ with G a DAG, output $I' := \mathrm{Normalize}(I)$. By Theorem 1, I is a YES-instance if and only if I' is a YES-instance, and the union digraph of I' is a DAG of BFS-width at most 2. Since Normalize is polynomial-time computable, this is a polynomial-time many-one reduction to the stated class. $\square$

Corollary 6 (*NP*-completeness on BFS-width-2 input DAGs). PAFP *is NP-complete when restricted to instances* $I = (G, s, t, \mathcal{F})$ *such that G is a DAG and* $\mathrm{bfsw}(G, s) \leq 2$.

Proof (Proof sketch). By Claim 2, Normalize produces PAFP instances whose union digraph and input graph are equal. Thus Theorem 1 implies that $\mathrm{bfsw}(G', s') \leq 2$ for the instance $\mathrm{Normalize}(I) = (G', s', t, \mathcal{F}')$, and that hard instances can be normalized in polynomial-time to equivalent instances whose input graph is a DAG of BFS-width at most 2. $\qquad\square$

Remark 2. The fact that Normalize produces PAFP instances whose input graph already embeds its forbidden pairs as arcs implies that for the instances to which it normalizes, every topological order of the input graph orients every forbidden pair the same way. Thus our NP-completeness result in fact holds when restricted to the order-independent setting in which regardless of the topological order used to orient the forbidden pairs, the resulting union digraph is the same.

4.2   Exact-Length Layering: a Polynomial-Time Case

Theorem 7 (PAFP on DAGs of exact-length width at most 2). *Let $I = (G = (V, E), s, t, \mathcal{F})$ be a PAFP instance where G is a DAG. Assume $\mathrm{elw}(G, s) \leq 2$. Then I can be decided in deterministic polynomial time.*

Remark 3 (the promise still allows many paths). The condition $\mathrm{elw}(G, s) \leq 2$ does *not* imply that G has few s–t paths. For example, let G be a properly layered DAG with layers $A_0 = \{s\}$, $A_d = \{u_d, v_d\}$ for $d = 1, \ldots, \ell - 1$, and $A_\ell = \{t\}$, and include all arcs from each layer to the next (i.e., every vertex in A_d has arcs to all vertices in A_{d+1}). Then $E_G(s, d) = A_d$ for all d, so $\mathrm{elw}(G, s) = 2$, but G has $2^{\ell-1} = 2^{\Theta(|V|)}$ distinct directed s–t paths. Thus Theorem 7 gives polynomial-time solvability even in instances with exponentially many candidate paths and completely arbitrary forbidden pairs.

Remark 4. For every DAG G, every $d \geq 0$, and every root s, one has $L_G(s, d) \subseteq E_G(s, d)$, and hence $\mathrm{bfsw}(G, s) \leq \mathrm{elw}(G, s)$. Thus the promise $\mathrm{elw}(G, s) \leq 2$ is stricter than $\mathrm{bfsw}(G, s) \leq 2$. Our results show that this stronger local promise yields a polynomial-time island, tight already at width 3, whereas BFS-width 2 does not suffice for tractability.

Proof (Proof of Theorem 7). Fix an instance $I = (G = (V, E), s, t, \mathcal{F})$ where G is a DAG and $\mathrm{elw}(G, s) \leq 2$. Let $n := |V|$. Since G is acyclic, every directed path is vertex-simple and hence has length at most $n - 1$. Therefore it suffices to consider lengths $\ell \in \{0, 1, \ldots, n - 1\}$.

Exact-Length Layers and their Computation. For each $d \in \{0, \ldots, n - 1\}$ define $D_d := E_G(s, d)$. We compute $D_0, \ldots, D_{n-1}$ by the DP

$$D_0 := \{s\}, \qquad D_d := \{v \in V \mid \exists (u, v) \in E \text{ with } u \in D_{d-1}\} \quad (d \geq 1). \qquad (9)$$

Correctness is immediate by induction on d (extend a length-$(d-1)$ path by one arc, and conversely take the last arc of a length-d path). Evaluating each D_d by scanning all arcs once takes $O(|E|)$ time per d, hence $O(n|E|)$ total time. By assumption $\mathrm{elw}(G, s) \leq 2$, we have $|D_d| \leq 2$ for all $d \geq 0$.

(Walks vs. Paths in a DAG.) We will use the following standard fact: in a DAG, every directed walk is vertex-simple, hence a directed path. Indeed, if a directed walk repeats a vertex v, then the subwalk between two occurrences of v forms a directed cycle, contradicting acyclicity. In particular, later any length-ℓ directed walk whose consecutive pairs are arcs is automatically a directed path.

Encoding Length-ℓ Safe Paths as 2-SAT. Fix $\ell \in \{0, \ldots, n-1\}$ with $t \in D_\ell$. *Idea.* Any directed s–t path of length exactly ℓ can be written as a sequence $s = X_0 \to X_1 \to \cdots \to X_\ell = t$, where X_d is the vertex used at *position* d along the path. Necessarily $X_d \in D_d$ for each d, and consecutive choices must satisfy $(X_d, X_{d+1}) \in E$. The forbidden pairs impose additional constraints of the form "it is not allowed that $X_i = a$ and $X_j = b$ simultaneously." Under the promise $|E_G(s,d)| \le 2$, each position d has at most two possible vertices, i.e. a Boolean choice. Moreover, each constraint forbids *one* joint assignment of *two* positions, so it is naturally expressible as a 2-CNF clause. Thus the fixed-length subproblem reduces to 2-SAT.

Variables (one Boolean Choice per Layer of Size 2). If $|D_d| = 2$, fix a deterministic ordering $D_d = \{v_d^0, v_d^1\}$ and introduce a Boolean variable x_d with the intended meaning: $x_d = \text{FALSE}$ chooses v_d^0 as the dth vertex on a length ℓ path, and $x_d = \text{TRUE}$ chooses v_d^1. If $|D_d| = 1$, write $D_d = \{v_d\}$; then position d is forced and we introduce no variable.

A Literal for "Position d Selects Vertex u." Define a literal (or constant) $\text{Sel}(d, u)$ for each $u \in D_d$ by

$$\text{Sel}(d, u) = \begin{cases} \neg x_d & \text{if } |D_d| = 2 \text{ and } u = v_d^0, \\ x_d & \text{if } |D_d| = 2 \text{ and } u = v_d^1, \\ \top & \text{if } |D_d| = 1 \text{ and } D_d = \{u\}. \end{cases} \qquad (10)$$

Thus $\text{Sel}(d, u)$ evaluates to TRUE exactly when the induced choice at position d is u (and it is identically true if the choice is forced).

Clauses. We add clauses of a single uniform form: to forbid the simultaneous event "$X_i = a$ and $X_j = b$," we add the clause $\neg\text{Sel}(i, a) \lor \neg\text{Sel}(j, b)$. We now list the constraints we need.

(i) Endpoint constraint at t. We enforce that position ℓ is t by adding the (unit) clause $\text{Sel}(\ell, t)$. (If $|D_\ell| = 1$ this is $\top$ and does nothing; otherwise it fixes x_ℓ appropriately.)

(ii) Adjacency constraints. For each $d \in \{0, \ldots, \ell-1\}$ and each $u \in D_d, v \in D_{d+1}$ with $(u, v) \notin E$, we forbid choosing u at position d and v at position $d + 1$ by adding $\neg\text{Sel}(d, u) \lor \neg\text{Sel}(d+1, v)$. Since $|D_d|, |D_{d+1}| \le 2$, there are at most 4 such checks per d.

(iii) Forbidden-pair constraints. For each forbidden pair $\{a, b\} \in \mathcal{F}$ and each pair of *distinct* positions $i \ne j$ with $a \in D_i$ and $b \in D_j$, we forbid simultaneously selecting a and b by adding $\neg\text{Sel}(i, a) \lor \neg\text{Sel}(j, b)$. (If a vertex is forced at some position, Sel becomes $\top$, and this correctly collapses to a unit clause forbidding the other endpoint at the other position.)

Let Φ_ℓ be the conjunction of all clauses above. The only Boolean variables occurring in Φ_ℓ are the variables x_d introduced for indices $d \in \{0,\ldots,\ell\}$ with $|D_d| = 2$. We substitute $\top$ and $\bot$ as Boolean constants and simplify: delete any clause containing $\top$, remove any occurrence of $\bot$ from a clause, and reject if some clause becomes empty. The resulting equivalent formula contains no constants and is a standard 2-SAT instance. For clarity, we work with the original, pre-simplified formula Φ_ℓ in the algorithm analysis.

Correctness for Fixed ℓ. We show that Φ_ℓ is satisfiable if and only if there exists a safe directed s–t path of length exactly ℓ.

$(\Rightarrow)$ *Satisfying Assignment $\Rightarrow$ Safe Path.* Given a satisfying assignment to the variables of Φ_ℓ, define X_d as follows: if $D_d = \{v_d\}$ then $X_d = v_d$, and if $D_d = \{v_d^0, v_d^1\}$ then $X_d = v_d^0$ when $x_d = \text{FALSE}$ and $X_d = v_d^1$ when $x_d = \text{TRUE}$. The endpoint clause enforces $X_\ell = t$, and since $D_0 = \{s\}$ we have $X_0 = s$.

By the adjacency clauses, for every $d \in \{0,\ldots,\ell-1\}$ we must have (X_d, X_{d+1}) in E; otherwise the unique clause forbidding that pair would be violated. Hence $s = X_0 \to X_1 \to \cdots \to X_\ell = t$ is a directed walk of length ℓ in G, and therefore (since G is a DAG) it is a directed path.

Finally, consider any forbidden pair $\{a, b\} \in \mathcal{F}$. If the path contained both a and b, then there would exist distinct positions $i \neq j$ such that $X_i = a$ and $X_j = b$, and the corresponding forbidden-pair clause would be violated. Thus the path is safe.

$(\Leftarrow)$ *Safe Path $\Rightarrow$ Satisfying Assignment.* Let $P := s = u_0 \to u_1 \to \cdots \to u_\ell = t$ be a safe directed s–t path of length ℓ. Then $u_d \in D_d$ for each d. For each d with $|D_d| = 2$ and $D_d = \{v_d^0, v_d^1\}$, assign x_d so that the induced choice X_d equals u_d. This assignment satisfies the endpoint clause (since $u_\ell = t$), satisfies every adjacency clause (since $(u_d, u_{d+1}) \in E$), and satisfies every forbidden-pair clause because P is safe. Hence Φ_ℓ is satisfiable.

Running Time. Let $m := |E|$ and $f := |\mathcal{F}|$. Computing $D_0, \ldots, D_{n-1}$ takes $O(nm)$ time. Now fix ℓ. Since $|D_d| \leq 2$, constraint **(ii)** inspects at most $|D_d| \cdot |D_{d+1}| \leq 4$ pairs per $d < \ell$, hence contributes $O(\ell)$ clauses. For **(iii)**, for each vertex v let $occ_\ell(v) := \{d \in \{0,\ldots,\ell\} \mid v \in D_d\}$; the sets $occ_\ell(v)$ can be built in $O(\ell)$ time by scanning $D_0, \ldots, D_\ell$, since $\sum_{d=0}^{\ell} |D_d| \leq 2(\ell+1)$. Each $\{a, b\} \in \mathcal{F}$ contributes at most $|occ_\ell(a)| \, |occ_\ell(b)| \leq (\ell+1)^2$ clauses, so $|\Phi_\ell| = O(\ell + f\ell^2)$. Building Φ_ℓ and solving it via 2-SAT both take $O(|\Phi_\ell|)$ time, hence $O(\ell + f\ell^2)$. Summing over $\ell \in \{0,\ldots,n-1\}$ yields $O(nm) + \sum_{\ell=0}^{n-1} O(\ell + f\ell^2) = O(nm + n^2 + fn^3)$, which is polynomial. $\qquad\square$

Proposition 8 (Exact-length width 3 is already hard). PAFP *is NP-complete on DAGs G with* $\mathrm{elw}(G, s) \leq 3$.

Proof. Membership in NP is immediate. For NP-hardness, we inspect the classical reduction of Gabow, Maheshwari, and Osterweil from 3-SAT to acyclic instances of the impossible-pairs constrained path problem (the problem now known as PAFP); see the proof of Lemma 2 and Fig. 2 in [7, pp. 229–230]. It therefore remains only to observe that their reduction graph has exact-length width at most 3. Given a 3-SAT instance $B = \bigwedge_{i=1}^{m}(p_{i1} \vee p_{i2} \vee p_{i3})$ the reduction constructs an acyclic layered digraph G with source s, sink t, and for each clause $i \in [m]$ exactly three vertices v_{i1}, v_{i2}, v_{i3}, one for each literal of the clause; arcs go only from s to the first clause-layer, between consecutive clause-layers (in fact as complete bipartite graphs), and from the last clause-layer to t [7, proof of Lemma 2, Fig. 2, pp. 229–230]. Hence every directed path from s advances exactly one layer per arc, so $E_G(s, 0) = \{s\}$, $E_G(s, i) = \{v_{i1}, v_{i2}, v_{i3}\}$ for $1 \le i \le m$, $E_G(s, m+1) = \{t\}$, and $E_G(s, i) = \emptyset$ for all other i. Therefore $\mathrm{elw}(G, s) \le 3$. $\square$

Disclosure of Interests. The author has no competing interests to declare.

References

1. Arnborg, S., Lagergren, J., Seese, D.: Easy problems for tree-decomposable graphs. J. Algorithms **12**(2), 308–340 (1991). https://doi.org/10.1016/0196-6774(91)90006-K
2. Bannister, M.J., Devanny, W.E., Dujmović, V., Eppstein, D., Wood, D.R.: Track layouts, layered path decompositions, and leveled planarity. Algorithmica **81**(4), 1561–1583 (2019). https://doi.org/10.1007/s00453-018-0487-5
3. Bezáková, I., Curticapean, R., Dell, H., Fomin, F.V.: Finding detours is fixed-parameter tractable. In: Chatzigiannakis, I., Indyk, P., Kuhn, F., Muscholl, A. (eds.) 44th International Colloquium on Automata, Languages, and Programming (ICALP 2017). Leibniz International Proceedings in Informatics (LIPIcs), vol. 80, pp. 54:1–54:14. Schloss Dagstuhl – Leibniz-Zentrum für Informatik, Dagstuhl, Germany (2017). https://doi.org/10.4230/LIPIcs.ICALP.2017.54, https://drops.dagstuhl.de/entities/document/10.4230/LIPIcs.ICALP.2017.54
4. Bodlaender, H.L., Jansen, B.M.P., Kratsch, S.: Kernel bounds for path and cycle problems. Theoret. Comput. Sci. **511**, 117–136 (2013). https://doi.org/10.1016/j.tcs.2012.09.006
5. Dujmović, V., Morin, P., Wood, D.R.: Layered separators in minor-closed graph classes with applications. J. Comb. Theory Ser. B **127**, 111–147 (2017). https://doi.org/10.1016/j.jctb.2017.05.006
6. Eppstein, D., Goodrich, M.T., Liu, S.A.: Bandwidth vs BFS width in matrix reordering, graph reconstruction, and graph drawing. In: Benoit, A., Kaplan, H., Wild, S., Herman, G. (eds.) 33rd Annual European Symposium on Algorithms (ESA 2025). Leibniz International Proceedings in Informatics (LIPIcs), vol. 351, pp. 69:1–69:17. Schloss Dagstuhl – Leibniz-Zentrum für Informatik, Dagstuhl, Germany (2025). https://doi.org/10.4230/LIPIcs.ESA.2025.69, https://drops.dagstuhl.de/entities/document/10.4230/LIPIcs.ESA.2025.69
7. Gabow, H.N., Maheshwari, S.N., Osterweil, L.J.: On two problems in the generation of program test paths. IEEE Trans. Software Eng. **2**(3), 227–231 (1976). https://doi.org/10.1109/TSE.1976.233819
8. Kahn, A.B.: Topological sorting of large networks. Commun. ACM **5**(11), 558–562 (1962). https://doi.org/10.1145/368996.369025

9. Kolman, P., Pangrác, O.: On the complexity of paths avoiding forbidden pairs. Discret. Appl. Math. **157**(13), 2871–2876 (2009). https://doi.org/10.1016/j.dam.2009.03.018

10. Kováč, J.: Complexity of the path avoiding forbidden pairs problem revisited. Discrete Appl. Math. **161**(10–11), 1506–1512 (2013). https://doi.org/10.1016/j.dam.2012.12.022, early version available as arXiv:1111.3996

11. Kovac, J., Vinař, T., Brejová, B.: Predicting gene structures from multiple RT-PCR tests. In: Algorithms in Bioinformatics (WABI 2009). Lecture Notes in Computer Science, vol. 5724, pp. 181–193. Springer (2009). https://doi.org/10.1007/978-3-642-04241-6_16

12. Krause, K.W., Smith, R.W., Goodwin, M.A.: Optimal software test planning through automated network analysis. In: Proceedings of the 1973 IEEE Symposium on Computer Software Reliability, pp. 18–22. IEEE (1973)

13. Samer, M., Szeider, S.: Constraint satisfaction with bounded treewidth revisited. J. Comput. Syst. Sci. **76**(2), 103–114 (2010). https://doi.org/10.1016/j.jcss.2009.04.003

14. Yinnone, H.: On paths avoiding forbidden pairs of vertices in a graph. Discrete Appl. Math. **74**(1), 85–92 (1997). https://doi.org/10.1016/S0166-218X(96)00017-0

Solid-Resolving Sets on Directed Graphs

Anni Hakanen[1,2]([✉])(iD) and P. D. Pavan[1](iD)

[1] Department of Mathematics and Statistics, University of Turku, Turku, Finland
ppadev@utu.fi
[2] Turku Collegium for Science, Medicine and Technology, University of Turku,
Turku, Finland
anehak@utu.fi

Abstract. A set $S \subseteq V(G)$ is a solid-resolving set of an undirected graph G if each vertex can be distinguished from any vertex subset by their distances from the set S. Solid-resolving sets are a generalization of resolving sets, which have been studied extensively in literature. We extend this concept to digraphs and oriented graphs. In this setting, we first define solid-resolving sets for directed graphs and then consider vertices that must be included in any solid-resolving set. We also study the solid-metric dimension, which is the minimum cardinality of a solid-resolving set for the given digraph. We determine the possible values that the solid-metric dimension can take for common graph classes such as tournaments, directed paths, oriented cycles and also for Cartesian products of oriented paths and cycles. Furthermore, we also show that two digraphs that differ by only one vertex or arc can have solid-metric dimensions that differ by an arbitrary k. This indicates that the solid-metric dimension is a global, rather than a local property.

Keywords: Digraph · Oriented graph · Resolving set · Metric dimension · Solid-resolving set · Solid-metric dimension · Forced vertex

1 Introduction

A resolving set is a vertex set with which we can distinguish all vertices of the graph from each other by looking at the distances from the vertices of the resolving set. More formally, we say that a vertex v *distinguishes* distinct vertices x and y if $d(v, x) \neq d(v, y)$, where $d(u, v)$ is the length of a shortest path from u to v. A set $S \subseteq V(G)$ is a *resolving set* of the graph G if for all distinct $u, v \in V(G)$ we have $d(s, u) \neq d(s, v)$ for some $s \in S$. The minimum cardinality of a resolving set of G is called the *metric dimension* of G.

Resolving sets and metric dimension were introduced for undirected graphs in the 1970s by Slater [17], and Harary and Melter [9], and have since been studied extensively. Over the years, the study has expanded to variations of the original concept: one can distinguish, for example, edges instead of vertices [11], use a different distance than the traditional graph distance [10, 15], or consider

© The Author(s), under exclusive license to Springer Nature Switzerland AG 2026
F. Foucaud and A. Parreau (Eds.): IWOCA 2026, LNCS 16587, pp. 327–340, 2026.
https://doi.org/10.1007/978-3-032-27732-9_23

more general graphs by including weights [6] or edge directions [5]. See the surveys [12,18] for more information on metric dimension and its variants.

Considering metric dimension in the context of directed graphs raises the question of how to interpret distance between vertices when there is no directed path from the start vertex to the target vertex, i.e. the target vertex is not reached from the start vertex. Metric dimension was first defined for digraphs by Chartrand, Raines, and Zhang [5], who dealt with this issue by avoiding it altogether! Their definition requires that all vertices of the resolving set are reachable from all vertices outside the resolving set. This leads to resolving sets, and consequently metric dimension being undefined for many digraphs. This definition of resolving sets in digraphs has its applications, however, in [7,13,14, 16,19], for example.

Another approach in answering the question was presented relatively recently in [1,2]: the distance from u to v is defined to be ∞ when v is not reached from u. In this case, the infinite distance can be used to distinguish vertices. If x is reachable from v but y is not, then we agree that $d(v,x) < d(v,y)$ and v distinguishes x and y. Using this interpretation of infinite distance allowed extending the definitions of resolving sets and metric dimension to all digraphs.

The *solid-metric dimension* was introduced for undirected graphs in [8]. Similar to classical resolving sets, solid-resolving sets can also distinguish distinct vertices from each other, however, they can also distinguish any single vertex from a set of vertices. Let the distance from a vertex to a set of vertices be defined as $d(s,Y) = \min_{y \in Y} d(s,y)$. Then for an undirected graph G, the set $S \subseteq V(G)$ is a solid-resolving set if for all $x \in V(G)$ and $Y \subseteq V(G)$, $Y \neq \emptyset$, $Y \neq \{x\}$, there exists $s \in S$ such that $d(s,x) \neq d(s,Y)$.

Before defining solid-resolving sets for directed graphs, we first present some basic definitions and notation. A *directed graph* (or *digraph*) D consists of vertices and arcs. We denote the vertex set of D by $V(D)$ and the arc set by $A(D)$. An arc is an ordered pair of vertices, and an arc from u to v is denoted by $\overrightarrow{uv}$ or (u,v). The digraphs we consider are simple, that is, without loops $\overrightarrow{uu}$. An *orientation* of an undirected graph G is a digraph where each edge of G is given a direction (and bidirected edges are forbidden). If necessary, say, if we need to make the distinction between an undirected graph G and its orientation, we may use the arrow notation $\overrightarrow{G}$, however, we usually denote digraphs without the arrow.

The *distance* from u to v is defined as the number of arcs in a shortest directed path from u to v, and it is denoted by $d_D(u,v)$ (where the subscript D may be omitted if the context is clear). If no such directed path exists, then v is not *reachable* from u and $d(u,v) = \infty$. The distance from a vertex u to a vertex set $X \subseteq V(D)$ is defined as $d(u,X) = \min_{x \in X} d(u,X)$. If no directed paths exist from u to any $x \in X$, then $d(u,X) = \infty$.

We now define solid-resolving sets for directed graphs as follows.

Definition 1. *Let D be a digraph. A set $S \subseteq V(D)$ is a solid-resolving set of D if for all vertices $v \in V(D)$ and nonempty subsets $X \subseteq V(D)$, there exists $w \in S$ such that $d(w,v) \neq d(w,X)$. The minimum cardinality of a solid-resolving set is called the solid-metric dimension of D, and is denoted by $\beta_S(D)$.*

In this work, we study solid-resolving sets and solid-metric dimensions of directed graphs. In Sect. 2, we generalize some basic results for solid-resolving sets from undirected graphs to directed graphs. Our exposition places special emphasis on forced vertices, that is, vertices which are in all solid-resolving sets. In Sect. 3, we consider directed graphs and orientations whose underlying graphs are paths, cycles or complete graphs, and bound their solid-metric dimensions and determine the exact solid-metric dimension in some cases. In Sect. 4, we prove that the solid-metric dimension of a directed graph can change by an arbitrary integer k when a vertex or an arc is removed. Finally, in Sect. 5, we consider the solid-metric dimension of Cartesian products of oriented paths and cycles.

2 Basic Results

We begin by generalizing some basic results on solid-resolving sets from [8] to digraphs. The following theorem gives us a useful characterization of solid-resolving sets, which we shall make use of extensively henceforth.

Theorem 1. *Let D be a nontrivial directed graph. A set $S \subseteq V(D)$ is a solid-resolving set of D if and only if for all distinct $u, v \in V(D)$, there is an element $s \in S$ such that $d(s, u) < d(s, v)$.*

Proof. Suppose first that S is a solid-resolving set. To prove the claim, let us assume on the contrary that there exist vertices $u, v \in V(D)$ such that $d(s, v) \leq d(s, u)$ for all $s \in S$. Let $X = \{u, v\}$. This implies that $d(s, X) = \min(d(s, u), d(s, v)) = d(s, v)$ for all $s \in S$. This contradicts the assumption that S is a solid-resolving set.

To prove the converse, let $v \in V(D)$ and let $X \subseteq V(D)$, with $X \neq \{v\}$. Let $u \in X$. By the hypothesis, there exists a vertex $s \in S$ such that $d(s, u) < d(s, v)$. But since $d(s, X) \leq d(s, u)$, we have $d(s, X) < d(s, v)$, and thus, $d(s, X) \neq d(s, v)$. Hence, S is a solid-resolving set.

In [8], it was noted that some vertices must belong to all solid-resolving sets, and thus they can be used to prove lower bounds on the solid-metric dimension. We make a similar observation in the context of directed graphs.

Definition 2. *A vertex $u \in V(D)$ is called a* forced vertex *of a solid-resolving set of D if every solid-resolving set of D contains u.*

The open in-neighborhood (resp. open out-neighborhood) of a vertex v is defined as $N_D^-(v) = \{u \in V(D) \mid \overrightarrow{uv} \in A(D)\}$ (resp., $N_D^+(v) = \{u \in V(D) \mid \overrightarrow{vu} \in A(D)\}$), and the closed in-neighborhood (closed out-neighborhood) as $N_D^-[v] = N_D^-(v) \cup \{v\}$ (resp., $N_D^+[v] = N_D^+(v) \cup \{v\}$). The in-degree $\deg_D^-(v)$ (resp., out-degree $\deg_D^+(v)$) of a vertex v is the cardinality of the open in-neighborhood $N_D^-(v)$ (resp., open out-neighborhood $N_D^+(v)$). Whenever the context is clear, we usually drop the subscript D for convenience. A vertex is called a *source*, if it has no in-neighbors, and a *sink*, if it has no out-neighbors.

The next theorem gives us a characterization of forced vertices.

Theorem 2. *A vertex $u \in V(D)$ is a forced vertex of a solid-resolving set of D if and only if there exists a vertex $v \in V(D) \setminus \{u\}$ such that $N^-(u) \subseteq N^-[v]$.*

Proof. First, suppose that u and v are distinct vertices in D with $N^-(u) \subseteq N^-[v]$. We will show that the set $S = V(D) \setminus \{u\}$ is not a solid-resolving set, which implies that every solid-resolving set of D must include u. Assume, to the contrary, that S is a solid-resolving set. By Theorem 1, there exists an element $s \in S$ such that $d(s, u) < d(s, v)$. Since $s \neq u$, any shortest path from s to u must pass through some $w \in N^-(u)$, so we can write $d(s, u) = d(s, w) + 1$. But since $w \in N^-(u) \subseteq N^-[v]$, we have $d(s, v) \leq d(s, w) + 1 = d(s, u)$, which is a contradiction. Thus, the set S and any of its subsets are not solid-resolving sets.

Conversely, assume that u is a forced vertex of a solid-resolving set of D, but $N^-(u) \not\subseteq N^-[v]$ for all $v \in V(D) \setminus \{u\}$. We will show that $S = V(D) \setminus \{u\}$ is a solid-resolving set, contradicting our assumption that u is forced. For any distinct $s \in S$ and $t \in V(D) \setminus \{s\}$ we trivially have $d(s, s) < d(s, t)$. Thus, it remains to distinguish u from any $s \in S$. Since $N^-(u) \not\subseteq N^-[s]$ for any $s \in S$, we can choose a vertex, say $t \in N^-(u) \setminus N^-[s]$. Then, $d(t, u) = 1$ and $d(t, s) \geq 2$, and thus $d(t, u) < d(t, s)$. Hence, by Theorem 1, S is a solid-resolving set, a contradiction.

Corollary 1. *A vertex $u \in V(D)$ with $\deg^-(u) \leq 1$ is a forced vertex of a solid-resolving set of D.*

In [8], the relationship between solid-resolving sets and the boundary of a graph was also explored. A definition of the boundary for digraphs was presented in [3]. However, we present a different definition for the boundary, which is more along the lines of the original definition for undirected graphs in [4]. This definition allows us to generalize results regarding solid-resolving sets and the boundary from the undirected case.

The *boundary* of a connected digraph D is the set $\partial(D) = \{v \in V(D) \mid \exists u \in V(D), \forall w \in N^+(v) : d(u, w) \leq d(u, v)\}$. Vertices in $\partial(D)$ are called *boundary vertices*. The boundary of a graph is always non-empty, for if the maximum distance between any pair of vertices in D is equal to $d(u, v)$, then u and v will always be boundary vertices. That is, $|\partial(D)| \geq 2$.

The *transpose* of a directed graph D is another directed graph D^T obtained by reversing all arcs, that is, $\overrightarrow{uv} \in A(D)$ if and only if $\overrightarrow{vu} \in A(D^T)$.

Theorem 3. *Let D be a finite nontrivial connected directed graph.*

1. *The boundary of D^T is a solid-resolving set of D.*
2. *Every forced vertex of a solid-resolving set of D is a boundary vertex of D^T.*
3. *There exist graphs that have solid-resolving sets of minimum cardinality that are not subsets of the boundary of the transpose.*

Proof (Omitted).

3   Graph Families

We utilize the characterizations of solid-resolving sets and forced vertices obtained in the previous section to study the solid-metric dimension of some graph families such as tournaments, paths and cycles in this section. We first turn our attention to the solid-metric dimension of tournaments. Let $\overrightarrow{K_n}$ be an orientation of the complete graph on n vertices.

Observation 1 *If* $\deg^-(u) \leq 2$ *for any vertex* $u \in V(\overrightarrow{K_n})$, *then* u *is a forced vertex of a solid-resolving set of* $\overrightarrow{K_n}$.

Proposition 1. *We have* $\beta_S(\overrightarrow{K_3}) = 3$, $\beta_S(\overrightarrow{K_4}) \geq 3$ *and* $\beta_S(\overrightarrow{K_5}) \geq 4$

Proof. No vertex of K_3 has degree greater than 2, and hence, by Observation 1, all vertices of $\overrightarrow{K_3}$ are forced. That is, $\beta_S(\overrightarrow{K_3}) = 3$.

Let us assume that there exists an orientation $\overrightarrow{K_4}$ of K_4 such that $\beta_S(\overrightarrow{K_4}) \leq 2$. Let $V(\overrightarrow{K_4}) = \{v_1, v_2, v_3, v_4\}$, and without loss of generality, let v_1 and v_2 be two vertices which are not in some solid-resolving set of $\overrightarrow{K_4}$. By Observation 1, we have that $3 \leq \deg^-(v_1), \deg^-(v_2) \leq 3$, that is $N^-(v_1) = \{v_2, v_3, v_4\}$ and $N^-(v_2) = \{v_1, v_3, v_4\}$. But this is a contradiction since $\overrightarrow{v_1 v_2}, \overrightarrow{v_2 v_1} \in A(\overrightarrow{K_4})$, while $\overrightarrow{K_4}$ is an orientation. Hence, $\beta_S(\overrightarrow{K_4}) \geq 3$.

Let us assume that there exists an orientation $\overrightarrow{K_5}$ of K_5 such that $\beta_S(\overrightarrow{K_5}) \leq 3$. Let $V(\overrightarrow{K_4}) = \{v_1, v_2, v_3, v_4, v_5\}$, and without loss of generality, let v_1 and v_2 be two vertices which are not in some solid-resolving set of $\overrightarrow{K_5}$. By Observation 1, we have that $3 \leq \deg^-(v_1), \deg^-(v_2) \leq 4$. If $\deg^-(v_1) = 4$, then $N^-(v_i) \subseteq N^-[v_1]$ for $2 \leq i \leq 5$. By Theorem 2, this implies that $\beta_S(\overrightarrow{K_5}) \geq 4$, a contradiction. Hence, $\deg^-(v_1) = \deg^-(v_2) = 3$. If $v_2 \in N^-(v_1)$, then without loss of generality, we must have $N^-(v_1) = \{v_2, v_3, v_4\}$ and $N^-(v_2) = \{v_3, v_4, v_5\}$. But now $N^-(v_1) \subset N^-[v_2]$ and by Theorem 2, v_1 is a forced vertex, a contradiction. Hence, $v_2 \notin N^-(v_1)$. Now if $v_1 \notin N^-(v_2)$, then $N^-(v_1) = N^-(v_2)$, which leads to a contradiction by Theorem 2 as before. Therefore, without loss of generality, we must have $N^-(v_1) = \{v_3, v_4, v_5\}$ and $N^-(v_2) = \{v_1, v_3, v_4\}$. But then $N^-(v_2) \subset N^-[v_1]$, which leads to a contradiction by Theorem 2. Hence, $\beta_S(\overrightarrow{K_5}) \geq 4$.

Proposition 2. *There exists an orientation* $\overrightarrow{K_n}$ *such that* $\beta_S(\overrightarrow{K_n}) = n$.

Proof. Let $\overrightarrow{K_n}$ be the oriented graph with $V(\overrightarrow{K_n}) = \{v_1, v_2, \ldots, v_n\}$ and $A(\overrightarrow{K_n}) = \{\overrightarrow{v_j v_i} : j > i\}$. Observe that $N^-(v_i) \subseteq N^-[v_{i+1}]$, for $i \in \{1, \ldots, n-1\}$. Hence, by Theorem 2, the vertices $\{v_1, \ldots, v_{n-1}\}$ are forced vertices of a solid-resolving set of $\overrightarrow{K_n}$. Moreover, since $\deg^-(v_n) = 0$, by Corollary 1, v_n is a forced vertex as well. Hence, we have $\beta_S(\overrightarrow{K_n}) = n$.

We next turn our attention to digraphs whose underlying undirected graphs are paths, and determine their solid-metric dimensions in the following theorem. We denote by $n_i^-(D)$ the number of vertices u with $\deg^-(u) = i$ in D.

Theorem 4. *Let $\overrightarrow{P_n}$ be a directed graph whose underlying graph is the path on n vertices P_n. Then $\beta_S(\overrightarrow{P_n}) = n - n_2^-(\overrightarrow{P_n})$.*

Proof. By Corollary 1, the vertices in the set $S = \{u \in V(\overrightarrow{P_n}) \mid \deg^-(u) \leq 1\}$ are forced vertices. Since, $|S| = n_1^-(\overrightarrow{P_n}) + n_0^-(\overrightarrow{P_n}) = n - n_2^-(\overrightarrow{P_n})$, it suffices to prove that S is a solid-resolving set.

For any distinct $s \in S$ and $t \in V(\overrightarrow{P_n}) \setminus \{s\}$ we trivially have $d(s,s) < d(s,t)$. Thus, it remains to distinguish any vertex $u \in V(\overrightarrow{P_n}) \setminus S$, that is, any vertex u with $\deg^-(u) = 2$, from every other vertex in $\overrightarrow{P_n}$. Let $N^-(u) = \{p_1, p_2\}$. For each $i \in \{1,2\}$, perform a depth-first search (DFS) that is restricted to unvisited in-neighbors of the vertices, starting from p_i (excluding u), and let p_i' be the first vertex encountered that lies in S. Such vertices can always be found since the graph is finite, and moreover, the initial and terminal vertices of the path are both in S. By construction, p_1' and p_2' lie on distinct unidirectional paths terminating at u. Now consider any vertex $x \neq u$. It follows that $d(p_1', u) > d(p_1', x) \implies d(p_2', u) < d(p_2', x)$ and symmetrically $d(p_2', u) > d(p_2', x) \implies d(p_1', u) < d(p_1', x)$. Hence, either p_1' or p_2' distinguishes u from every other vertex x, and S is a solid-resolving set.

Corollary 2. *Let $\overrightarrow{P_n}$ be a directed graph whose underlying graph is the path on n vertices P_n. Then, we have $2 \leq \beta_S(\overrightarrow{P_n}) \leq n$.*

Proof. By Theorem 4, we have $\beta_S(\overrightarrow{P_n}) = n - n_2^-(\overrightarrow{P_n})$. Hence, the extremal values of $\beta_S(\overrightarrow{P_n})$ occur when $n_2^-(\overrightarrow{P_n})$ is minimized or maximized. We can have $n_2^-(\overrightarrow{P_n}) = 0$ when $\overrightarrow{P_n}$ is a simple orientation of the undirected path from the initial vertex to the terminal vertex. We can have $n_2^-(\overrightarrow{P_n}) = n - 2$ (all vertices except the initial and terminal vertices) when all the edges are bidirected, with arcs in both directions.

Finally, we consider orientations of cycles. As with paths, we can determine their solid-metric dimensions by determining their forced vertices.

Theorem 5. *Let $\overrightarrow{C_n}$ be the n-edge oriented cycle. We have the following cases:*

- *If $n = 3$ or if $n = 4$ and $\overrightarrow{C_n}$ has only sources and sinks, then $\beta_S(\overrightarrow{C_n}) = n$.*
- *Otherwise, $\beta_S(\overrightarrow{C_n}) = n - n_2^-(\overrightarrow{C_n})$.*

Proof. By Corollary 1, the vertices in the set $S = \{u \in V(\overrightarrow{C_n}) \mid \deg^-(u) \leq 1\}$ are forced vertices. Since, $|S| = n_1^-(\overrightarrow{C_n}) + n_0^-(\overrightarrow{C_n}) = n - n_2^-(\overrightarrow{C_n})$, it suffices to prove that S is a solid-resolving set.

Observe that in $\overrightarrow{C_n}$, sources and sinks occur in pairs. If $\overrightarrow{C_n}$ does not have any sources or sinks, then $d^-(u) = 1$ for all $u \in V(\overrightarrow{C_n})$. Hence, all vertices of $\overrightarrow{C_n}$ are forced and the result is true.

Let us assume that $\overrightarrow{C_n}$ has at least one sink (or at least one source). For any distinct $s \in S$ and $t \in V(\overrightarrow{C_n}) \setminus \{s\}$ we trivially have $d(s,s) < d(s,t)$. Thus, it remains to distinguish any vertex $u \in V(\overrightarrow{C_n}) \setminus S$, that is, any vertex u with $\deg^-(u) = 2$, from every other vertex in $\overrightarrow{C_n}$. Let $N^-(u) = \{p_1, p_2\}$. Clearly, $p_1, p_2 \in S$. Now consider any vertex $x \neq u$. Depending on the distances between p_1, p_2 and x, we have the following possibilities:

- If $d(p_1, x) = 0$ and $d(p_2, x) = 1$, then $n = 3$ and we have either $N^-(u) \subseteq N^-[p_1]$ or $N^-(u) \subseteq N^-[p_2]$. In either case, by Theorem 2, u is also forced, and hence $\beta_S(\overrightarrow{C_3}) = 3$. We get a similar result if $d(p_1, x) = 1$ and $d(p_2, x) = 0$.
- If $d(p_1, x) = d(p_2, x) = 1$, then $n = 4$ and we have $N^-(u) = N^-(x)$. Hence, by Theorem 2, u and x are also forced, and therefore $\beta_S(\overrightarrow{C_4}) = 4$ when $\overrightarrow{C_4}$ has two sources and two sinks.
- If either $d(p_1, x) > 1$ or $d(p_2, x) > 1$, then we have that either $d(p_1, x) > d(p_1, u)$ or $d(p_2, x) > d(p_1, u)$. Hence, either p_1 or p_2 distinguishes u from every other vertex x, and S is a solid-resolving set.

Corollary 3. *Let $\overrightarrow{C_n}$ be the n-edge oriented cycle. If $n \geq 5$, then $\lceil \frac{n}{2} \rceil \leq \beta_S(\overrightarrow{C_n}) \leq n$.*

Proof. The upper bound can be obtained by considering the unidirectional oriented cycle (which has no sources or sinks). Since in this case, $n_2^-(\overrightarrow{C_n}) = 0$, we have from Theorem 5 that $\beta_S(\overrightarrow{C_n}) = n$.

To prove the lower bound, consider the oriented cycle $\overrightarrow{C_n}$ where consecutive arcs are oriented in opposite directions. In this graph, every vertex is either a source or a sink (except maybe one vertex), that is $n_2^-(\overrightarrow{C_n}) = \lfloor \frac{n}{2} \rfloor = n_0^-(\overrightarrow{C_n})$. Hence, from Theorem 5, we have $\beta_S(\overrightarrow{C_n}) = \lceil \frac{n}{2} \rceil$.

4   Vertex and Arc Deletions

In this section, we study the effect graph operations have on the solid-metric dimension. We begin by looking at vertex and arc deletion and prove that the solid-metric dimension can change drastically on performing such operations.

4.1   Vertex Deletion

We construct digraphs whose solid-metric dimension increases or decreases by an arbitrary integer k when one vertex is removed. This implies that the solid-metric dimension is not a local property but a global one.

Proposition 3. *For every integer $k \geq 1$, there exists a digraph D_k with $|V(D_k)| = 2k + 2$ and a vertex $u \in V(D_k)$ such that $\beta_S(D_k - u) - \beta_S(D_k) = k$.*

Proof. Let $k \geq 1$ and D_k be an oriented graph such that $V(D_k) = \{u, w\} \cup \{u_i, v_i \mid i \in [1, k+1]\}$ and $A(D_k) = \{\overrightarrow{uw}\} \cup \{\overrightarrow{uu_i}, \overrightarrow{wu_i}, \overrightarrow{wv_i}, \overrightarrow{v_iu_i} \mid i \in [1, k+1]\}$. The graph is illustrated in Fig. 1a. We will show that in this graph D_k, when the vertex u is removed, the solid-metric dimension increases by k.

We first determine the solid metric dimension of D_k. Observe first that since $\deg^-(v_i) = \deg^-(w) = 1$ and $\deg^-(u) = 0$, where $i \in [1, k+1]$, by Corollary 1, u, w and $v_i, i \in [1, k+1]$ are forced vertices. Therefore, $\beta_S(D_k) \geq k + 3$. We claim that $S = \{u, w\} \cup \{v_i \mid i \in [1, k+1]\}$ is a solid-resolving set and hence $\beta_S(D_k) = k + 3$. Observe that if s is a forced vertex, then $d(s, s) < d(s, t)$ for any other vertex $t \in V(D_k)$. Thus, by Theorem 1, to show that S is a solid resolving set, we only need to show that for all $u_i, t \in V(D_k), t \neq u_i$, there exists a vertex $s \in S$ such that $d(s, u_i) < d(s, t)$, where $i \in [1, k+1]$. Since $d(v_i, x) = 1 \iff x = u_i$, we have $d(v_i, u_i) < d(v_i, t)$, for $t \neq v_i$, and since $d(u, x) = 1 \iff x \neq v_i$, we have $d(u, u_i) < d(u, v_j)$, where $i, j \in [1, k+1]$. Hence, the claim is true.

Next, we determine the solid metric dimension of $D_k - u$. Observe that in $D_k - u$, since $\deg^-(v_i) = 1$ and $\deg^-(w) = 0$, where $i \in [1, k+1]$, by Corollary 1, w and $v_i, i \in [1, k+1]$ are forced vertices. Moreover, since $N^-_{D_k-u}(u_i) \subseteq N^-_{D_k-u}[v_i], i \in [1, k+1]$, by Theorem 2, $u_i, i \in [1, k+1]$ are also forced vertices in $D_k - u$. Hence, by definition, $\beta_S(D_k - u) = 2k + 3$. Thus, $\beta_S(D_k - u) - \beta_S(D_k) = k$.

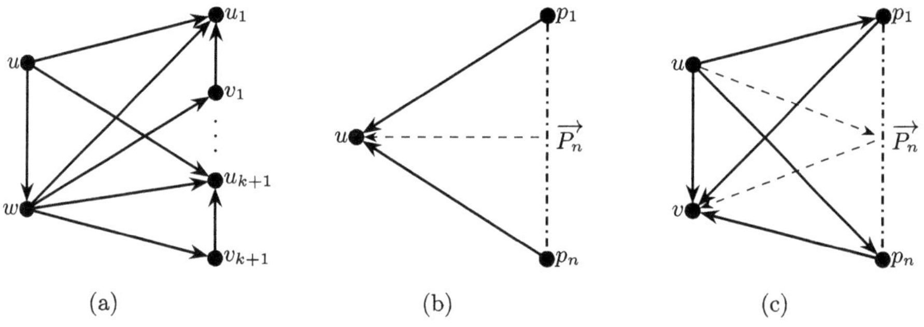

Fig. 1. Examples of graphs D_k where removing a vertex or an arc causes arbitrary changes in the solid-metric dimension: (a) $\beta_S(D_k) = \beta_S(D_k - u) - k$, (b) $\beta_S(D_k - u) = \beta_S(D_k) - k$, and (c) $\beta_S(D_k - \overrightarrow{uv}) = \beta_S(D_k) - k$.

Proposition 4. *For every integer $k \geq 1$, there exists a digraph D_k with $|V(D_k)| \geq \min\{5, k+3\}$ and a vertex $u \in V(D_k)$ such that $\beta_S(D_k) - \beta_S(D_k - u) = k$.*

Proof. Let $k \geq 1, n \geq \min\{4, k+2\}$ and let $\overrightarrow{P_n}$ be a directed graph whose underlying graph is the path on n vertices P_n. Let $n_2^-(\overrightarrow{P_n}) = k$ and let p_1 and p_n denote the initial and terminal vertices of $\overrightarrow{P_n}$. Construct the graph D_k by adding a new vertex u to $\overrightarrow{P_n}$ and arcs of the form $\overrightarrow{vu}$ for all $v \in V(\overrightarrow{P_n})$. The graph is illustrated in Fig. 1b. For every vertex $v \neq u$, we have $N^-(v) \subseteq N^-[u]$. Hence, by Theorem 2, all the vertices in $S = V(D_k) \setminus u$ are forced vertices. Since $n \geq \min\{4, k+2\}$, for any vertex $x \in V(D_k) \setminus \{u\}$, the structure of D_k ensures that either $d(p_1, u) < d(p_1, x)$ or $d(p_n, u) < d(p_n, x)$. Hence, S is a solid-resolving set, and $\beta_S(D_k) = |S| = n$. On the other hand, since $D_k - u \cong \overrightarrow{P_n}$, by Theorem 4, we have $\beta_S(D_k - u) = \beta_S(\overrightarrow{P_n}) = n - k$. Consequently, we have $\beta_S(D_k) - \beta_S(D_k - u) = k$.

4.2   Arc Deletion

Similarly to the previous subsection, in the following result, we construct a digraph whose solid-metric dimension decreases by an arbitrary integer k when one arc is removed.

Proposition 5. *For every integer $k \geq 1$, there exists a digraph D_k with $|V(D_k)| \geq \min\{6, k+4\}$ and an arc $\overrightarrow{uv} \in A(D_k)$ such that $\beta_S(D_k) - \beta_S(D_k - \overrightarrow{uv}) = k$.*

Proof. Let $k \geq 1$, $n \geq \min\{4, k+2\}$, and let $\overrightarrow{P_n}$ be a directed graph whose underlying graph is the path on n vertices P_n. Let $n_2^-(\overrightarrow{P_n}) = k$ and let p_1 and p_n denote the initial and terminal vertices of the path $\overrightarrow{P_n}$. Construct the graph D_k by adding new vertices u and v to $\overrightarrow{P_n}$ and arcs of the form $\overrightarrow{uw}$ for all $w \in V(\overrightarrow{P_n})$ and arcs of the form $\overrightarrow{wv}$ for all $w \in V(\overrightarrow{P_n}) \cup \{u\}$. The graph is illustrated in Fig. 1c. For every vertex $w \neq v$, we have $N^-(w) \subseteq N^-[v]$. Hence, by Theorem 2, all the vertices in $S = V(D_k) \setminus \{v\}$ are forced vertices. For any vertex $x \in V(D_k) \setminus \{v\}$, the structure of D_k ensures that either $d(p_1, v) < d(p_1, x)$ or $d(p_n, v) < d(p_n, x)$. Hence, S is a solid-resolving set, and $\beta_S(D_k) = |S| = n + 1$.

Next, consider the graph $D_k - \overrightarrow{uv}$. Observe that $\deg^-_{D_k - \overrightarrow{uv}}(w) = \deg^-_{\overrightarrow{P_n}}(w) + 1$, for $w \in V(\overrightarrow{P_n})$. Hence, vertices with in-degree 0 in $\overrightarrow{P_n}$ have in-degree 1 in $D_k - \overrightarrow{uv}$ and are still forced in $D_k - \overrightarrow{uv}$, by Corollary 1. On the other hand, if a vertex w has in-degree 1 in $\overrightarrow{P_n}$, with $\overrightarrow{w'w} \in A(\overrightarrow{P_n})$ for some vertex $w' \in V(\overrightarrow{P_n})$, then w will have in-degree 2 in $D_k - \overrightarrow{uv}$, but is still forced in $D_k - \overrightarrow{uv}$ by Theorem 2 as $N^-(w) \subseteq N^-[w']$. Since $\deg^-(u) = 0$ as well, if $S = \{w \in V(\overrightarrow{P_n}) \mid \deg^-(w) \leq 1\} \cup \{u\}$, then we have $\beta_S(D_k - \overrightarrow{uv}) \geq |S|$. The upper bound can be proved in a manner similar to that used in Theorem 4. Hence, we have $\beta_S(D_k - \overrightarrow{uv}) = \beta_S(\overrightarrow{P_n}) + 1 = n + 1 - k$. Consequently, we have $\beta_S(D_k) - \beta_S(D_k - \overrightarrow{uv}) = k$.

5   Cartesian Products of Orientations

In this section, we determine the solid-metric dimension of a Cartesian product consisting of oriented paths and cycles (with certain restrictions). As was seen in Sect. 3, the solid-metric dimension of oriented paths and cycles follows the number of forced vertices. We will prove that this is true also for their Cartesian products. To that end, we will first consider forced vertices in Cartesian products.

Let D_1 and D_2 be oriented graphs. Their *Cartesian product* $D_1 \square D_2$ is the digraph with

$$V(D_1 \square D_2) = V(D_1) \times V(D_2) = \{uv \mid u \in V(D_1), v \in V(D_2)\}$$
$$A(D_1 \square D_2) = \{(u_1 v_1, u_2 v_2) \mid (u_1, u_2) \in A(D_1) \text{ and } v_1 = v_2, \text{ or}$$
$$u_1 = u_2 \text{ and } (v_1, v_2) \in A(D_2)\}$$

where $\times$ is the Cartesian product of sets. The in-neighborhood of $uv \in V(D_1 \square D_2)$ can be written as $N^-(uv) = (N^-(u) \times \{v\}) \cup (\{u\} \times N^-(v))$.

5.1 Forced Vertices

We begin by focusing on oriented graphs that do not have forced vertices of in-degree more than 1. This is the type of oriented paths and cycles whose Cartesian products we will consider later in the following subsection.

Lemma 1. *If D_i, $i = 1, \ldots, k$, are orientations such that $\deg^-(u) \le 1$ for all forced vertices $u \in V(D_i)$, then $D_1 \square \cdots \square D_k$ is also such that $\deg^-(u) \le 1$ for all forced vertices $u \in V(D_1 \square \cdots \square D_k)$.*

Proof. Let $uv \in V(D_1 \square D_2)$ be a forced vertex. Due to Theorem 2, there exists $xy \in V(D_1 \square D_2)$ such that $xy \ne uv$ and $N^-(uv) \subseteq N^-[xy]$. Assume to the contrary that $\deg^-(uv) \ge 2$.

Suppose that $N^-(uv)$ is in one copy of D_1, that is, $N^-(uv) = N^-(u) \times \{v\}$. If $xy \in N^-(uv)$, then $N^-(u) \subseteq N^-[x]$, and u is a forced vertex of D_1 with in-degree at least 2, a contradiction. Thus, $xy \notin N^-(uv)$ and $N^-(uv) \subseteq N^-(xy)$. Let $av, a'v \in N^-(uv)$, $a \ne a'$. We have the following cases due to $av, a'v \in N^-(xy)$.

1. $a = x$ and $v \in N^-(y)$:
   (a) $a' = x$ and $v \in N^-(y)$: Now, $a = x = a'$, a contradiction.
   (b) $a' \in N^-(x)$ and $v = y$: Now, $xy = av \in N^-(uv)$, a contradiction.
2. $a \in N^-(x)$ and $v = y$:
   (a) $a' = x$ and $v \in N^-(y)$: Now, $xy = a'v \in N^-(uv)$, a contradiction.
   (b) $a' \in N^-(x)$ and $v = y$: Due to other cases being impossible, this implies that $N^-(u) \in N^-(x)$, and again u is a forced vertex in D_1 with in-degree at least 2, a contradiction.

Thus, $N^-(uv)$ cannot be in one copy of D_1 (or D_2 by similar arguments).

Suppose then that $N^-(uv)$ contains av and ub where $a \in N^-(u)$ and $b \in N^-(v)$. Notice that av and ub cannot be neighbors of each other, which implies that $xy \notin N^-(uv)$. We have the following cases due to $av, ub \in N^-(xy)$.

1. $a = x$ and $v \in N^-(y)$:
   (a) $u = x$ and $b \in N^-(y)$: Now, $u = x = a \in N^-(u)$, a contradiction.

(b) $u \in N^-(x)$ and $b = y$: Now, $u \in N^-(x) = N^-(a)$. However, we also have $a \in N^-(u)$ and thus a and u form a 2-cycle, which contradicts D_1 being an orientation.

2. $a \in N^-(x)$ and $v = y$:
   (a) $u = x$ and $b \in N^-(y)$: Now, $xy = uv$, a contradiction.
   (b) $u \in N^-(x)$ and $b = y$: Now, $v = y = b \in N^-(v)$, a contradiction.

The claim for $D_1 \square \cdots \square D_k$ follows by induction. Indeed, the above shows that the claim holds for the Cartesian product of two orientations $D_1 \square D_2$. Suppose that the claim holds for the Cartesian product $D = D_1 \square \cdots \square D_{k-1}$, that is, D is an orientation with no forced vertex of in-degree at least 2. Then, by the above arguments, the claim holds for $D \square D_k = D_1 \square \cdots \square D_k$.

Corollary 4. *Let D_i, $i = 1, \ldots, k$, be orientations such that $\deg^-(u) \leq 1$ for all forced vertices $u \in V(D_i)$. The vertex $u_1 \cdots u_k \in V(D_1 \square \cdots \square D_k)$ is a forced vertex in $D_1 \square \cdots \square D_k$ if and only if $\sum_{i=1}^{k} \deg^-(u_i) \leq 1$.*

Proof. The vertex $u_1 \cdots u_k \in V(D_1 \square \cdots \square D_k)$ is a forced vertex if and only if $\deg^-(u_1 \cdots u_k) \leq 1$ by Lemma 1 and Corollary 1. Since

$$N^-(u_1 \cdots u_k) = (N^-(u_1) \times \{u_2 \cdots u_k\}) \cup \cdots (\{u_1 \cdots u_{k-1}\} \times N^-(u_k)),$$

we have $\deg^-(u_1 \cdots u_k) \leq 1$ if and only if $\sum_{i=1}^{k} \deg^-(u_i) \leq 1$.

Corollary 5. *Let D_i, $i = 1, \ldots, k$, be orientations such that $\deg^-(u) \leq 1$ for all forced vertices $u \in V(D_i)$. The number of forced vertices in $D_1 \square \cdots \square D_k$ is*

$$\sum_{\mathbf{x} \in X} \prod_{i=1}^{k} n_{x_i}^-(D_i),$$

where $X = \{(x_1, \ldots, x_k) \in \{0,1\}^k \mid \sum_{i=1}^{k} x_i \leq 1\}$.

5.2 Cartesian Products of Oriented Paths and Cycles

In this section, we consider arbitrary Cartesian products of oriented paths and cycles. The paths we consider have at least two vertices and the cycles have at least five vertices. Additionally, the cycles are oriented such that they have at least two sources each. We will prove that the solid-metric dimension of the Cartesian product $D_1 \square \cdots \square D_k$ of such paths and cycles is equal to the number of forced vertices (see Corollary 5). The assumptions for the number of vertices in the orientations we consider enable us to use Corollaries 4 and 5.

The proof is largely based on which vertices are reached from which forced vertices. We denote the set of vertices that are reachable from v by $R(v)$. In a Cartesian product $D_1 \square \cdots \square D_k$, we can consider the reachability between vertices through the component graphs of the product. Denote $u = u_1 \cdots u_k$ and $v = v_1 \cdots v_k$, where $u_i, v_i \in V(D_i)$. There exists a directed path from v to u if and only if each u_i is reachable from each v_i. Thus, $R(v) = R(v_1) \times \cdots \times R(v_k)$. Moreover, much like in undirected Cartesian products we have

$$d(v, u) = \sum_{i=1}^{k} d(v_i, u_i),$$

where if $d(v_i, u_i) = \infty$ for some i, then we agree that $d(v, u) = \infty$.

Theorem 6. *Let D_i be an oriented path with at least two vertices or an oriented cycle with at least five vertices and two sources for $i \in \{1, \ldots, k\}$. We have*

$$\beta_S(D_1 \square \cdots \square D_k) = \sum_{\mathbf{x} \in X} \prod_{i=1}^{k} n_{x_i}^{-}(D_i),$$

where $X = \{(x_1, \ldots, x_k) \in \{0,1\}^k \mid \sum_{i=1}^{k} x_i \leq 1\}$

Proof. Due to Corollary 5, we only need to show that the forced vertices do indeed form a solid-resolving set of the Cartesian product. Denote the set of forced vertices by S. By Theorem 1, it suffices to prove that for all distinct vertices u and v, there exists an $s \in S$ such that $d(s, u) < d(s, v)$.

Let $u = u_1 \cdots u_k$ be an arbitrary vertex. If $u \in S$, then clearly for all $v \neq u$, we have $d(u, u) < d(u, v)$. Assume that $u \notin S$. Due to the assumptions on D_i, every vertex in D_i is reachable from some source in D_i. This implies that u is reachable from a source $x = x_1 \cdots x_k$ where u_i is reachable from x_i in each D_i. We assume to the contrary that there exists a vertex $v \neq u$ such that $d(s, v) \leq d(s, u)$ for all $s \in S$. In particular, v is reachable from every element in S that reach u.

Since the component graphs D_i do not have forced vertices with in-degree 2 or more, Corollary 4 implies that either at least one of the vertices u_i is a sink of in-degree 2 or at least two of the vertices u_i are of in-degree 1. We will prove that for such indices i either $v_i = u_i$ or there exists a source x_i in D_i such that $d(x_i, u_i) < d(x_i, v_i)$.

Assume u_i is of in-degree 1. Now $x^i = x_1 \cdots x_{i-1} u_i x_{i+1} \cdots x_k$ is a forced vertex and u is reachable from x^i. Moreover, for all t_i in D_i, u is reachable from $x_1 \cdots x_{i-1} t_i x_{i+1} \cdots x_k$ if and only if t_i is in the directed path from x_i to u_i (and then these are also forced vertices). Since $d(s, v) \leq d(s, u)$ for all $s \in S$, v must also be reachable from the aforementioned forced vertices. This implies that either $v_i = u_i$ or a directed path from x_i to v_i goes through u_i (in D_i), and thus $d(x_i, u_i) < d(x_i, v_i)$.

Assume then that, say, u_1 is a sink of in-degree 2. Now u is reachable from two sources: $x_\ell = x_1^\ell x_2 \cdots x_k$ and $x_r = x_1^r x_2 \cdots x_k$, where x_1^ℓ and x_1^r are the two sources in D_1 that reach u_1. Now v_1 must also be reached from both of these sources, which implies that v_1 is a sink of in-degree 2. Assume that $v_1 \neq u_1$. Then D_1 is a cycle with exactly two sources and u_1 and v_1 are the two sinks. However, since the cycle has at least five vertices, there exists a vertex t_1 of in-degree 1. Now, if u_1 is reachable from t_1, then v_1 is not. This implies that there is a forced vertex in the Cartesian product that reaches u but not v, a contradiction.

Thus, no in-degree 1 vertex in D_1 reaches u_1. Thus, the two sources are both in-neighbors of u_1 and we have $d(x_1^\ell, u_1) < d(x_1^\ell, v_1)$ or $d(x_1^r, u_1) < d(x_1^r, v_1)$. Therefore, either $v_1 = u_1$ or there exists a source x_1 in D_1 such that $d(x_1, u_1) < d(x_1, v_1)$.

Without loss of generality, we may assume that u_i are such that u_i are sinks of in-degree 2 when $1 \leq i \leq a$; of in-degree 1 when $a < i \geq b$, and sources x_i when $b < i \leq k$, where $0 \leq a \leq b \leq k$. By the considerations above, when $1 \leq i \leq b$, either $v_i = u_i$ or there exists a source x_i such that $d(x_i, u_i) < d(x_i, v_i)$. Let x_i be such sources if they exist. Now,

$$d(x, u) = \sum_{i=1}^{k} d(x_i, u_i) = \sum_{i=1}^{b} d(x_i, u_i) + \sum_{i=b+1}^{k} d(x_i, x_i) \leq \sum_{i=1}^{b} d(x_i, v_i) \leq d(x, v).$$

Equality in the above implies that $v_i = u_i$ for all $i \in \{1, \ldots, k\}$, and thus $v = u$, a contradiction. Therefore, we must have $d(x, u) < d(x, v)$, a contradiction to our assumption.

Acknowledgments. The authors were funded in part by the Research Council of Finland grant 358718. Pavan P D was funded in part by the Research Council of Finland grant 338797.

Disclosure of Interests. The authors have no competing interests to declare that are relevant to the content of this article.

References

1. Araújo, J., et al.: On finding the best and worst orientations for the metric dimension. Algorithmica **85**(10), 2962–3002 (2023). https://doi.org/10.1007/S00453-023-01132-0
2. Bensmail, J., Inerney, F.M., Nisse, N.: Metric dimension: from graphs to oriented graphs. Discret. Appl. Math. **323**, 28–42 (2022). https://doi.org/10.1016/j.dam.2020.09.013
3. Changat, M., Narasimha-Shenoi, P.G., Thottungal Joseph, M.S., Kumar, R.: Boundary vertices of cartesian product of directed graphs. Int. J. Appl. Comput. Math. **5**, 19 (2019). https://doi.org/10.1007/s40819-019-0604-4
4. Chartrand, G., Erwin, D., Johns, G.L., Zhang, P.: Boundary vertices in graphs. Discret. Math. **263**(1–3), 25–34 (2003). https://doi.org/10.1016/S0012-365X(02)00567-8
5. Chartrand, G., Raines, M., Zhang, P.: The directed distance dimension of oriented graphs. Mathematica Bohemica **125**(2), 155–168 (2000). http://eudml.org/doc/248478
6. Epstein, L., Levin, A., Woeginger, G.J.: The (weighted) metric dimension of graphs: hard and easy cases. Algorithmica **72**(4), 1130–1171 (2015). https://doi.org/10.1007/s00453-014-9896-2
7. Feng, M., Xu, M., Wang, K.: On the metric dimension of line graphs. Discret. Appl. Math. **161**(6), 802–805 (2013). https://doi.org/10.1016/j.dam.2012.10.018
8. Hakanen, A., Junnila, V., Laihonen, T.: The solid-metric dimension. Theoret. Comput. Sci. **806**, 156–170 (2020). https://doi.org/10.1016/j.tcs.2019.02.013

9. Harary, F., Melter, R.: On the metric dimension of a graph. Ars Combin. **2**, 191–195 (1976)
10. Jannesari, M., Omoomi, B.: The metric dimension of the lexicographic product of graphs. Discret. Math. **312**(22), 3349–3356 (2012). https://doi.org/10.1016/j.disc.2012.07.025
11. Kelenc, A., Tratnik, N., Yero, I.G.: Uniquely identifying the edges of a graph: the edge metric dimension. Discrete Appl. Math. **251**, 204–220 (2018). https://doi.org/10.1016/j.dam.2018.05.052
12. Kuziak, D., Yero, I.G.: Metric dimension related parameters in graphs: a survey on combinatorial, computational and applied results. ArXiV preprint abs/2107.04877 (2021). https://doi.org/10.48550/arXiv.2107.04877
13. Pancahayani, S., Simanjuntak, R.: Directed metric dimension of oriented graphs with cyclic covering. J. Comb. Math. Comb. Comput. **94** (2014)
14. Rajan, B., Rajasingh, I., Cynthia, J., Manuel, P.: Metric dimension of directed graphs. Int. J. Comput. Math. **91**, 1397–1406 (2014). https://doi.org/10.1080/00207160.2013.844335
15. Saenpholphat, V., Zhang, P.: Detour resolvability of graphs. Congr. Numer. **169**, 3–21 (2004)
16. Schmitz, Y., Wanke, E.: The directed metric dimension of directed co-graphs. ArXiV preprint abs/2306.08594 (2023). https://doi.org/10.48550/arXiv.2306.08594
17. Slater, P.J.: Leaves of trees. In: Proceedings of the Sixth Southeastern Conference on Combinatorics, Graph Theory, and Computing (Florida Atlantic Univ., Boca Raton, Fla., 1975), pp. 549–559. Congressus Numerantium, No. XIV. Utilitas Math., Winnipeg, Man. (1975)
18. Tillquist, R.C., Frongillo, R.M., Lladser, M.E.: Getting the lay of the land in discrete space: a survey of metric dimension and its applications. SIAM Rev. **65**(4), 919–962 (2023). https://doi.org/10.1137/21M1409512
19. Vetrík, T.: On the metric dimension of directed and undirected circulant graphs. Discussiones Mathematicae Graph Theory **40**(1), 67–76 (2020). https://doi.org/10.7151/dmgt.2110

Minimum Clique Bicoloring

Shunsuke Hamada$^{(\boxtimes)}$ [ID], Yuto Okada [ID], Hirotaka Ono [ID], and Yota Otachi [ID]

Nagoya University, Nagoya, Japan
`hamada.shunsuke.z9@s.mail.nagoya-u.ac.jp`, `research@yutookada.com`,
`{ono,otachi}@nagoya-u.jp`

Abstract. Clique coloring of a graph is a relaxation of proper vertex coloring that requires every maximal clique to contain at least two distinct colors. In this paper, we introduce and study an optimization variant called MINIMUM CLIQUE BICOLORING, which restricts the available colors to white and black and asks to minimize the number of black-colored vertices, and can be viewed as a coloring-type counterpart of the classical minimum clique transversal problem. We show that MINIMUM CLIQUE BICOLORING is Σ_2^P-hard even on weakly chordal graphs, and NP-complete on cubic planar graphs and on k-sun-free chordal graphs for every fixed $k \geq 4$. On split graphs, we further relate MINIMUM CLIQUE BICOLORING to the minimum clique transversal problem by showing that any polynomial-time algorithm for the latter yields one for MINIMUM CLIQUE BICOLORING. Regarding parameterized complexity, we also show that the problem is W[2]-hard when parameterized by the number of black vertices k, but fixed-parameter tractable (FPT) with respect to clique-width. Furthermore, we present a polynomial-time algorithm for MINIMUM CLIQUE BICOLORING on totally unimodular graphs, and, as consequences of our structural and parameterized results, we obtain polynomial-time solvability on several further well-studied graph classes. Finally, we give an exact exponential algorithm running in $O^*(2^n)$ time, which improves on a straightforward $O^*((2 \cdot 3^{1/3})^n)$ approach based on maximal-clique enumeration, where n is the number of vertices. These results nearly pinpoint the boundary between polynomial-time solvability and intractability of MINIMUM CLIQUE BICOLORING on standard graph classes; the remaining main gap concerns strongly chordal graphs.

Keywords: Clique coloring · Clique transversal · Chordal graphs

1 Introduction

Clique coloring assigns colors to the vertices of a graph so that no maximal clique is monochromatic: every maximal clique must contain at least two distinct colors. This notion relaxes the familiar concept of a proper vertex coloring, in which

Partially supported by JSPS KAKENHI Grant Numbers JP22H00513, JP24H00697, JP24K02898, JP25K03076, JP25K03077, by JST CRONOS, Grant Number JP MJCS24K2, and by JST SPRING, Grant Number JPMJSP2125.

F. Foucaud and A. Parreau (Eds.): IWOCA 2026, LNCS 16587, pp. 341–355, 2026.
https://doi.org/10.1007/978-3-032-27732-9_24

adjacent vertices receive different colors [2]. Consequently, every proper vertex coloring is a clique coloring, but the converse need not hold. For example, the triangle is 2-clique-colorable while its chromatic number is 3. In triangle-free graphs, the two notions coincide.

When an unlimited number of colors is available, both proper vertex coloring and clique coloring are trivial, and the combinatorial interest lies in minimizing the number of colors. The minimum number of colors that permits a proper vertex coloring of a given graph is its *chromatic number*, while the minimum number that permits a clique coloring is called the *clique chromatic number*. Because clique coloring is a weakening of the proper constraint, the clique chromatic number can be significantly smaller than the chromatic number for many graphs, which motivates both structural and algorithmic study.

Clique coloring is not only of theoretical interest but also arises in various applications, including redundancy and replica placement in data centers, shift scheduling, combinatorial test design, and key distribution in cryptographic systems (e.g., [7]). When edges encode pairwise failure correlations, a maximal clique represents a set of nodes likely to fail together, so requiring each maximal clique to contain at least one diversified node (for instance, holding a replica or running a different configuration) helps prevent domain-wide failures. Clique-coloring constraints naturally capture such requirements, and minimizing the number of diversified nodes is a practical optimization objective.

From a complexity-theoretic perspective, the decision problem "given a graph G and an integer k, does G admit a k-clique coloring?" is known to be hard in the second level of the polynomial hierarchy; in particular, the problem is Σ_2^P-complete, and this hardness holds even for $k = 2$ [29]. At the same time, theoretical and empirical results show that the clique chromatic number is small on many major graph classes. For instance, several well-studied subclasses of perfect graphs, including bipartite graphs, cographs, and various P_4-free classes, have clique chromatic number 2 or 3 [2,12,15,16,23].

There is also a close connection to hypergraph problems such as hypergraph coloring and transversal (hitting set) problems [5,26]. Viewing the maximal cliques of a graph as the hyperedges of a hypergraph, a clique coloring is exactly a vertex coloring of this hypergraph, and a *clique transversal* is simply a vertex set hitting all hyperedges. A practical distinction is that hyperedges are explicit in the hypergraph setting, whereas maximal cliques are only implicitly defined by the graph, which affects both algorithms and complexity. Nevertheless, the abundant work on 2-colorability and on minimum transversals in hypergraphs highlights the special role of threshold two and of covering all maximal cliques.

In this paper, we introduce and study the following natural optimization variant under the promise that the input graph is 2-clique-colorable. We restrict colors to two, white and black, and take white as the default color, that is, initially every vertex is white. Given a 2-clique-colorable graph G and a nonnegative integer k, we ask whether there exists a clique coloring of G in which at most k vertices are colored black. We refer to this decision problem as MINIMUM CLIQUE BICOLORING, or MCBC in short. This may be viewed as a coloring-type counterpart of the minimum clique transversal problem: instead of selecting a set that explicitly hits all maximal cliques, we require every maximal clique to contain both colors and minimize the number of vertices that deviate from the default color. This formulation models, for example, the task of minimizing the number of replicas to deploy in a data center application, where replication incurs costs. The problem is therefore a natural combinatorial optimization within the class of 2-clique-colorable graphs; our goal is to determine its computational complexity and to identify structural graph properties that admit efficient algorithms.

Related Work. Our study lies at the intersection of two classical lines of research: minimum transversals in hypergraphs, in particular the minimum maximal clique transversal problem, and the clique coloring problem. We first review related work on clique transversals and then on clique coloring.

The minimum transversal problem in hypergraphs is equivalent to the classical Minimum Hitting Set (or Minimum Set Cover) problem. The minimum clique transversal problem is the special case where the hyperedges are the maximal cliques of a graph. Since a general graph may have exponentially many maximal cliques and the problem is NP-hard in general, much of the algorithmic work has focused on restricted graph classes whose clique structure can be exploited. Even then, the problem remains computationally difficult: it is NP-complete already for chordal and even split graphs [9], as well as for several other natural classes with only polynomially many maximal cliques, including cocomparability graphs, planar graphs, and line graphs [24].

On the positive side, polynomial-time algorithms have been obtained for a number of "clique-friendly" classes. In the chordal world, this includes strongly chordal graphs and chordal graphs of bounded clique size, as well as some related subclasses [10,24]. There are also polynomial-time algorithms for certain non-chordal classes in which maximal cliques can still be handled efficiently, for example, comparability graphs [4], distance-hereditary graphs [28], Helly circular-arc graphs, and cographs [24]. Thus, chordal graphs and their close relatives remain among the structurally richest settings in which both maximal cliques and clique transversals are reasonably well understood.

Clique coloring originates from the work of Duffus et al. [16] on two-coloring maximal antichains, which, via comparability graphs, corresponds to forbidding monochromatic maximal cliques. While clique transversal problems aim to hit every maximal clique by selecting vertices, clique coloring seeks to avoid monochromatic maximal cliques by vertex colorings. These viewpoints impose different constraints but are closely related, and optimization variants of clique coloring often exhibit transversal-like behavior. Bacsó et al. [2] formally formulated clique coloring of a graph by introducing the clique-chromatic number and initiating its structural, algorithmic, and complexity-theoretic study. Unlike in classical hypergraph coloring, the hyperedges corresponding to maximal cliques are given only implicitly by the graph; thus clique coloring admits polynomial-time verification and belongs to NP on graph classes where all maximal cliques can be enumerated in polynomial time (such as chordal graphs), whereas the number of maximal cliques may be exponential in general, making even feasibility checking computationally difficult.

For example, Bacsó et al. [2] showed that verifying whether a given coloring is a clique coloring is coNP-complete and that 2-clique-coloring is NP-complete even for graphs of maximum degree three. Later, Marx [29] proved that for every fixed $k \geq 2$, the k-clique-coloring problem is Σ_2^P-complete, and for $k = 2$ this remains true even for weakly chordal graphs [17]. Further hardness and tractability results are known for restricted classes: 2-clique-coloring is polynomial-time solvable for planar graphs [27], but NP-complete for K_4-free perfect graphs [12] and for graphs of maximum degree three [27]. On the other hand, despite these hardness results, many important graph classes are known to be 2-clique-colorable. Many subclasses of perfect graphs are 2-clique-colorable, including strongly perfect and Meyniel [12], chordal and comparability [31], and claw-free perfect ones [2]. Beyond perfect graphs, every connected claw-free graph of maximum degree at most four, other than an odd hole, admits a 2-clique coloring [3]. For K_3-free graphs, clique coloring coincides with proper vertex coloring, since every maximal clique is an edge [2], so 2-clique-colorability is decidable in polynomial time.

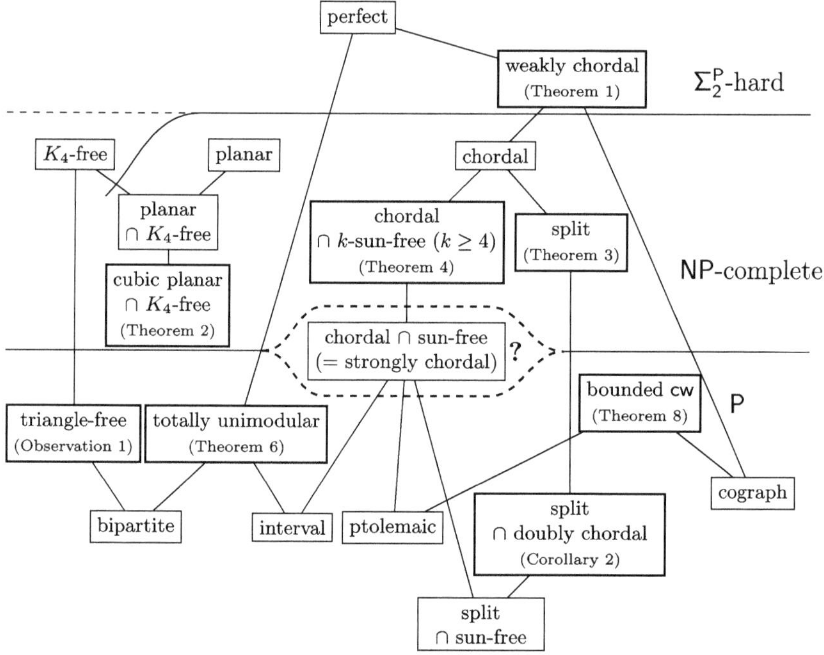

Fig. 1. Graph classes and summary of results.

Our Contributions. Figure 1 summarizes our results on the complexity of MCBC on various graph classes. Our main contributions are as follows.

- **High-level hardness.** We prove that MCBC is Σ_2^P-hard even on weakly chordal graphs (Theorem 1), a result that does not follow directly from the known Σ_2^P-completeness of deciding 2-clique-colorability on weakly chordal graphs, since our instances are promised to be 2-clique-colorable and the hardness lies in optimizing the number of black vertices. We also show that MCBC is NP-complete on cubic planar graphs (Theorem 2) and on k-sun-free chordal graphs for every fixed $k \geq 4$ (Theorem 4). In Fig. 1, we place K_4-free graphs in the NP region but outside the NP-complete area, since deciding 2-clique-colorability is NP-hard on this class.
- **Polynomial-time solvability.** We show that MCBC on totally unimodular graphs (Theorem 6) is polynomially solvable. For split graphs, we show that any polynomial-time algorithm for the minimum clique transversal problem can be turned into a polynomial-time algorithm for MCBC, so on split graphs the two problems have essentially the same algorithmic difficulty. Since MCBC is already hard on split graphs, this does not make the whole class tractable, but it immediately yields polynomial-time solvability on split graphs that are doubly chordal (Corollary 2).
- **FPT in clique-width.** We design an FPT algorithm parameterized by the clique-width k of the input graph (Theorem 8), running in $O^*(4^k)$ time, where k is the number of labels of the expression tree. As a consequence, MCBC is polynomial-time solvable on several bounded-clique-width classes, including Ptolemaic graphs and cographs.
- **W-hardness for the solution size.** When parameterized by the size of the black vertex set, MCBC is W[2]-hard even on split graphs (Theorem 3), so fixed-parameter tractability for this parameter alone is unlikely.
- **Exact exponential-time algorithm.** We give an exact $O(2^n)$-time algorithm for MCBC on n-vertex graphs, improving on the naive $O\big((2 \cdot 3^{1/3})^n\big)$ approach that enumerates all maximal cliques and colorings by using fast zeta transforms.

Taken together, these results nearly determine the boundary between NP-completeness and polynomial-time solvability for MCBC on the standard graph classes in Fig. 1. The only remaining standard class is that of strongly chordal graphs, for which the minimum clique transversal problem is polynomial-time solvable, so its complexity is a natural and interesting open question.

Paper organization. The remainder of this paper is organized as follows. Section 2 gives the formal problem definition and introduce basic terminology. Section 3 establishes several hardness results, including Σ_2^P-hardness, NP-completeness, and W[2]-hardness. Section 4 then turns to graph classes that admit polynomial-time algorithms. In Sect. 5, we present an exact exponential-time algorithm running in $O^*(2^n)$ time and show the tractability with respect to clique-width. Most of the proofs are omitted due to space limitations.

2   Preliminaries

We assume that the readers have basic knowledge about polynomial time hierarchy and parameterized complexity (see e.g. [1,18]). We follow the standard notations in graph theory (see e.g. [13]).

Let $G = (V, E)$ be a graph. For a graph H, we say that G is H-*free* if G does not contain H as an induced subgraph. A vertex set $C \subseteq V$ is called a *clique* of G if $E(G[C]) = \binom{C}{2}$. A clique C of G is called a *maximal clique* of G if there is no clique of G that is a proper super set of C. A vertex set $C \subseteq V$ is called an *independent set* of G if $E(G[C]) = \emptyset$.

Colorings. A *(proper) k-coloring* of G is a function $c\colon V \to \{0, \ldots, k-1\}$ such that every $\{u, v\} \in E$ satisfies $c(u) \neq c(v)$. A *k-clique-coloring* of G is a function $c\colon V \to \{0, \ldots, k-1\}$ such that every maximal clique C of G contains two vertices u, v such that $c(u) \neq c(v)$. When G admits a k-clique-coloring, we say that G is *k-clique-colorable*. For both of proper coloring and clique coloring, we may regard the codomain $\{0, \ldots, k-1\}$ as colors. Then, the problem we consider is formulated as follows.

Problem:	MINIMUM CLIQUE BICOLORING
Input:	A 2-clique-colorable graph $G = (V, E)$, an integer $k \geq 0$.
Question:	Is there a 2-clique-coloring $c\colon V \to \{0, 1\}$ of G that assigns the color 1 to at most k vertices?

Throughout this paper, we may refer to the two colors $0, 1$ as *white* and *black*, respectively. Since testing 2-clique-colorablity is already Σ_2^P-hard [29], we may treat MINIMUM CLIQUE BICOLORING as a promise problem[1] For this reason, we do not discuss completeness in some hardness results.

3   Hardness Results

3.1 Σ_2^P-Hardness for Weakly Chordal Graphs

In this section, we prove that MCBC is Σ_2^P-hard even on weakly chordal graphs. This may seem to follow from the known Σ_2^P-completeness of deciding 2-clique-colorability on weakly chordal graphs. However, MCBC is treated as a promise problem in which the input is assumed to be 2-clique-colorable, and the hardness here lies in optimizing the number of black vertices under this promise.

Theorem 1. MINIMUM CLIQUE BICOLORING *is* Σ_2^P-*hard, even if the input graph is a weakly chordal graph.*

---

[1] Formally, decision problems must assign YES or NO to every input, but we instead treat MCBC as a promise problem and assume that the input graph G is 2-clique-colorable, since recognizing this class may already be hard. For background on promise problems, see Goldreich [22].

Proof. Our reduction is inspired by the technique introduced by [17] to show the Σ_2^P-hardness of 2-CLIQUE-COLORING on weakly chordal graphs. We show the Σ_2^P-hardness by a reduction from MIN ONES QSAT$_2$.

Problem:	MIN ONES QSAT$_2$
Input:	A 3-DNF formula $\Phi(X,Y)$ (where $X = (x_1,\ldots,x_n), Y = (y_1,\ldots,y_m))$ and an integer $k \geq 0$.
Question:	Is there a truth assignment φ_X to the variables X such that φ_X assigns true to at most k variables in X and, for every assignment φ_Y to the variables Y, the formula $\Phi(X,Y)$ evaluates to true?

This problem is a variant of the well-known Σ_2^P-complete problem called QSAT$_2$ [34]. Since MIN ONES QSAT$_2$ clearly includes QSAT$_2$, it is Σ_2^P-hard.

Let $\langle \Phi(X,Y), k \rangle$ be an instance of MIN ONES QSAT$_2$. We may assume that no term in Φ contains contradicting literals, as such term will never be satisfied. Let $n = |X|, m = |Y|$ and q be the number of terms in Φ. For this instance, we construct an instance $\langle G, k' \rangle$ of MINIMUM CLIQUE BICOLORING.

To construct the graph G, we use a gadget NAS, which was used in [17]. The gadget NAS is the graph depicted in Fig. 2. For vertices u, v of G, by *attaching* NAS *to* u, v, we mean the operation of adding a copy of NAS to G and identifying u, v and u', v' respectively. We construct graph $G = (V, E)$ as follows. We first initialize V with $\emptyset$ and E with $\emptyset$, and then proceed as follows; see Fig. 3 for the whole image of G after the construction. Let $k' = 5n + 2m + q + k + 1$.

1. For each variable $x_i \in X$ (resp., $y_j \in Y$), we add vertices $x_i, \overline{x_i}$ to V (resp., $y_j, \overline{y_j}$ to V). For each term P_t of Φ, we add vertices p_t, p_t' to V.
2. For every $u, v \in \{x_1, \overline{x_1}, \ldots, x_n, \overline{x_n}, y_1, \overline{y_1}, \ldots, y_m, \overline{y_m}, z\}$, we add an edge $\{u, v\}$ if $\{u, v\}$ is not of the form $\{x_i, \overline{x_i}\}$ or $\{y_j, \overline{y_j}\}$.
3. For each p_t, we add an edge $\{p_t, z\}$ to E.
4. For each term P_t and each positive literal $w \in \{x_1, \ldots, x_n, y_1, \ldots, y_m\}$, we add an edge $\{p_t, w\}$ to E if $\overline{w}$ does not appear in P_t. For each term P_t and each negative literal $\overline{w} \in \{\overline{x_1}, \ldots, \overline{x_n}, \overline{y_1}, \ldots, \overline{y_m}\}$, we add an edge $\{p_t, \overline{w}\}$ to E if w does not appear in P_t.
5. For every $1 \leq i \leq n$, we attach NAS to x_i and $\overline{x_i}$.
6. For each vertex x_i, add a pendant neighbor x_i' adjacent to $\overline{x_i}$, and for each $v \in \{y_1, \overline{y_1}, \ldots, y_m, \overline{y_m}, z, p_1', \ldots, p_q'\}$, attach $k' + 1$ pendant neighbors to v.

This completes the construction. Graph G is 2-clique-colorable and weakly chordal. Furthermore, we can show the correctness of this reduction. $\square$

3.2 NP-Completeness for K_4-Free Cubic Planar Graphs

Before proceeding to the next result, we give the following observation.

Observation 1. *When the graph G is triangle-free,* MINIMUM CLIQUE BICOLORING *can be solved in linear time.*

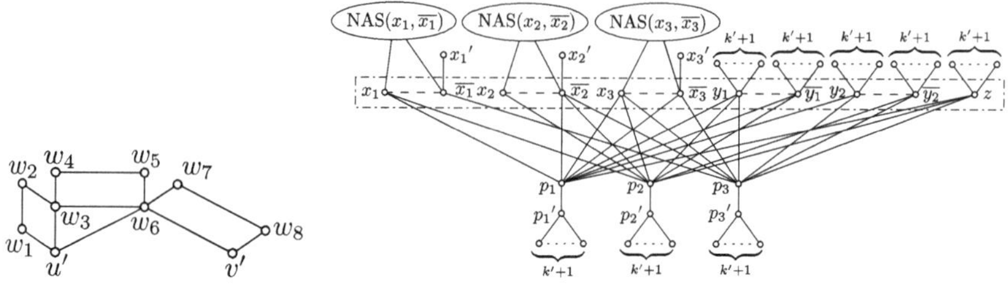

Fig. 2. NAS.   **Fig. 3.** The graph G we construct.

Recall that on triangle-free graphs, clique colorings coincide with proper colorings. Thus a triangle-free graph admits a 2-clique coloring iff it is bipartite, and when minimizing the number of black vertices one simply colors the smaller side of each bipartition black, which can be done in linear time. The next theorem shows that this approach cannot be extended to K_4-free graphs.

Theorem 2. *The* MINIMUM CLIQUE BICOLORING *problem on K_4-free cubic planar graphs is* NP-*complete.*

To show the above theorem, we first give two auxiliary lemmata. Let $\mathrm{vc}(G)$ denote the minimum size of a vertex cover of G and $\mathrm{mcbc}(G)$ denote the minimum number of black vertices in a 2-clique coloring of G.

Lemma 1. *Let $G = (V, E)$ be a graph, and let G' be the graph obtained by applying the following operations on G: 1. For each edge $e = \{u, v\}$ of G, subdivide it by inserting two vertices u_e and v_e. That is, replace $\{u, v\}$ with a path (u, u_e, v_e, v). 2. For each edge $e = \{u, v\}$ of G, add a vertex w_e and add the edges $\{u_e, w_e\}$ and $\{v_e, w_e\}$. Then, G' is 2-clique colorable, and $\mathrm{mcbc}(G') = \mathrm{vc}(G) + |E|$ holds.*

Lemma 2. *Let $G = (V, E)$ be a graph. For a vertex $v \in V$, let G' be the graph obtained by adding $\overline{P_2 \cup P_3}$ and connecting the degree-2 vertex to v. Then, $\mathrm{mcbc}(G') = \mathrm{mcbc}(G) + 2$ holds.*

Proof (Theorem 2). We show the theorem by a reduction from the VERTEX COVER problem. It is known that VERTEX COVER is in general NP-hard, and this holds even on cubic planar graphs [21]. Let $\langle G, k \rangle$ be an instance of VERTEX COVER with G being a cubic planar graph. We construct G' by applying the operation of Lemma 1 to each edge of G, and attach the gadget of Lemma 2 to every w_e. Finally, we set $k' = k + 3|E|$. This completes the construction. This graph is a cubic planar graph. The correctness of the reduction follows directly from Lemmas 1 an 2. Note that every connected cubic graph is K_4-free, except for K_4 itself. Hence, the graph G is a K_4-free cubic planar graph.   □

3.3 W[2]-Hardness for Split Graphs

Theorem 3. MINIMUM CLIQUE BICOLORING *is* W[2]*-hard when parameterized by* k*, even on split graphs.*

Proof. We show the W[2]-hardness by a parameterized reduction from SET COVER, which is defined as follows. This problem is known to be W[2]-complete when parameterized by d [14]. For simplicity, we exclude some trivial cases: we assume $d < m$ and $\bigcup_{S \in \mathcal{S}} S = U$.

Problem:	SET COVER
Input:	A set of integers $U = \{1, \dots, n\}$, a family of non-empty subsets $\mathcal{S} = \{S_1, \dots, S_m\} \subseteq 2^U$, and an integer $d \geq 1$.
Question:	Is there a family $\mathcal{T} \subseteq \mathcal{S}$ such that $|\mathcal{T}| \leq d$ and $\bigcup_{S \in \mathcal{T}} S = U$?

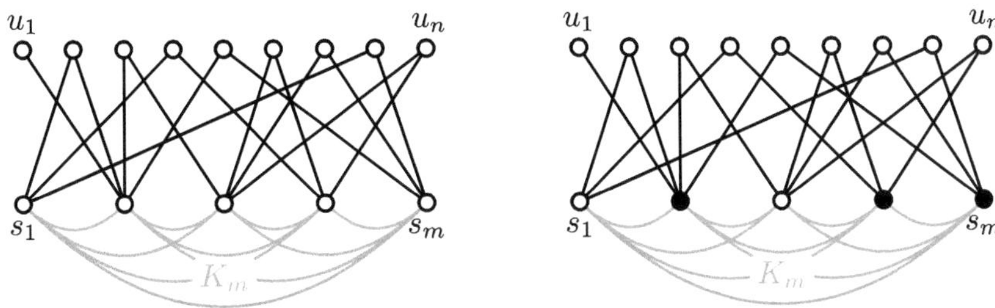

Fig. 4. The graph G (left) and its 2-clique-coloring c (right) we construct.

Construction. Let $\langle U, \mathcal{S}, d \rangle$ be an instance of SET COVER, $n = |U|$, and $m = |\mathcal{S}|$. For this instance, we construct an instance $\langle G = (V, E), k \rangle$ of MINIMUM CLIQUE BICOLORING; see Sect. 3.3. Let V be a set of $n+m$ vertices consisting of $u_1, \dots, u_n$ and $s_1, \dots, s_m$. Let $E = \{\{s_i, s_j\} \mid 1 \leq i < j \leq m\}$. Then, for every $1 \leq i \leq n$ and $1 \leq j \leq m$, we add $\{u_i, s_j\}$ to E if $i \in S_j$. Lastly, we let $k = d$. It is easy to see that graph G is a split graph, consisting of an independent set $\{u_1, \dots, u_n\}$ and a clique $\{s_1, \dots, s_m\}$. Note that this also implies the 2-clique-colorablity of G [31]. We can show the correctness of reduction (Fig. 4). $\qquad\square$

Corollary 1. MINIMUM CLIQUE BICOLORING *is* NP-*complete on split graphs.*

3.4 NP-Completeness on Recursively Chordal Graphs

A chordal graph that is k-sun-free for all $k \geq 4$ is called *recursively chordal*. In this section, we show that MCBC is NP-complete on recursively chordal graphs.

Reduction from Vertex Cover. Let (G, k) be an instance of *Vertex Cover*, where $G = (V, E)$ is an undirected graph. We assume that the vertex set and the edge set are ordered as $V = \{v_1, \ldots, v_n\}$ and $E = \{e_1, \ldots, e_m\}$. For an edge $\{v_p, v_q\} \in E$ with $p < q$, we sometimes write it as the ordered pair $(v_p, v_q) \in E$.

1. **Subdivision of Edges:** For each edge $e_i = (u, v) \in E$, we replace e_i by a path of length 3 (u, x_i, y_i, v). Let E' denote the resulting edge set. We also define the two layers $S_1 := \{x_i \mid i = 1, \ldots, m\}$, $S_2 := \{y_i \mid i = 1, \ldots, m\}$.
2. **Connections between S_1 and S_2:** For every pair of indices (i, l) with $l \geq i$, we add the edge between $x_i \in S_1$ and $y_l \in S_2$.
3. **Making $V \cup S_1$ and $V \cup S_2$ Cliques:** We add edges so that both $V \cup S_1$ and $V \cup S_2$ induce cliques.
4. **Adding Triangles:** For each edge $e = (u, v) \in E'$, add a new vertex t_e adjacent to u and v so that $\{u, v, t_e\}$ forms a triangle. Let $T := \{t_e \mid e \in E'\}$.
5. **Adding Stars:** For each $t_e \in T$, we create a star with $k + 4m$ leaves, whose center is a new vertex t'_e, and we connect t'_e to t_e. Let V_C be the set of all such centers and let V_L be the set of all leaves.

We denote by $\tilde{G}$ the graph obtained by the above construction.

Lemma 3. *Every maximal clique of $\tilde{G}$ is of one of the following types: (a) each edge of a star introduced in Step 5, (b) for each $e = (u, v) \in E'$, the triangle $\{u, v, t_e\}$, (c) for each $i = 1, 2, \ldots, m$, the set $V \cup \{x_1, \ldots, x_i\} \cup \{y_i, \ldots, y_m\}$.*

By using this lemma, we can show the following: The graph G has a vertex cover of size k (with $k < n$) if and only if $\tilde{G}$ admits a 2-clique-coloring with exactly $k + 4m$ black vertices. Also, the graph $\tilde{G}$ is recursively chordal. Consequently, the following holds.

Theorem 4. MINIMUM CLIQUE BICOLORING *is* NP-*complete even on recursively chordal graphs, i.e., chordal graphs that are k-sun-free for all $k \geq 4$.*

4 Polynomially Solvable Cases

4.1 Subclasses of Split Graphs

In this section, we study subclasses of split graphs for which the MINIMUM CLIQUE BICOLORING can be solved in polynomial time. We first observe the structure of optimal solutions on a split graph $G = (K \cup I, E)$.

Lemma 4. *Let B be an optimal solution of* MINIMUM CLIQUE BICOLORING *on a split graph $G = (K \cup I, E)$, and let $I_B = B \cap I$. Then, the following hold. (1) For every $x \in I_B$, the closed neighborhood $N[x]$ is the unique maximal clique containing x, and we have $B \cap N[x] = \{x\}$. (2) For any distinct vertices $x, y \in I_B$, if $\max(d(x), d(y)) \geq 2$ then $N(x) \cap N(y) = \emptyset$.*

A vertex of degree one is called a *pendant vertex*, and the edge incident to a pendant vertex is called a *pendant edge*. For each $v \in K$, define its pendant set by $P(v) := \{u \in I \mid N(u) = \{v\}\}$. Furthermore, for each $v \in K$ we define $c(v) := |P(v)|$. By using Lemma 4, we can show the following, which enables us to normalize the form of optimal solutions.

Lemma 5. *Let $G = (K \cup I, E)$ be a split graph in which every vertex $v \in K$ satisfies $c(v) \geq 1$. Then, the size of any 2-clique coloring B with the minimum number of black vertices is $\min\{c(v) \mid v \in K\} + |K| - 1$.*

By Lemma 5, for any split graph in which every vertex of K has pendant neighbors, an optimal solution can be obtained in linear time. Therefore, in what follows we consider split graphs that contain at least one vertex $v \in K$ with $c(v) = 0$. For such graphs we have the following lemma.

Lemma 6. *Let $G = (K \cup I, E)$ be a split graph having a vertex $v \in K$ with $c(v) = 0$. Define $K' = \{v \in K \mid c(v) > 0\}, I' = I \cap \bigcup_{v \in K'} N(v)$, and let $\bar{G} = G[(K \setminus K') \cup (I \setminus I')]$. Let $\bar{B}$ be an optimal solution of* MINIMUM CLIQUE BICOLORING *on $\bar{G}$ that does not contain any vertex in $I \setminus I'$. Then $B = \bar{B} \cup K'$ is an optimal solution of* MINIMUM CLIQUE BICOLORING *on G.*

By Lemmas 5 and 6, to solve MCBC on split graphs, it is sufficient to solve the problem on split graphs without pendant vertices.

Definition 1 (Minimum Clique Transversal (MCT)). *Let $G = (V, E)$ be a simple undirected graph. A vertex set $T \subseteq V$ is called a* clique transversal *of G if it intersects every maximal clique of G, that is, for the family $\mathcal{C}$ of maximal cliques of G we have $\forall C \in \mathcal{C} : T \cap C \neq \emptyset$. The* clique transversal number *of G is defined by $\tau(G) := \min\{|T| \mid T \in \mathcal{T}\}$, where $\mathcal{T} = \{T \subseteq V \mid \forall C \in \mathcal{C} : T \cap C \neq \emptyset\}$. The problem* MINIMUM CLIQUE TRANSVERSAL *takes a graph G as input and asks to compute $\tau(G)$ (and, if necessary, to output a clique transversal T with $|T| = \tau(G)$). Such a set T is called a* minimum clique transversal.

By definition, a set of black vertices B is a 2-clique coloring if and only if both B and $(K \cup I) \setminus B$ are clique transversals of G; the optimal value of MCBC on G is at least $\tau(G)$. Based on this observation we prove the following lemma.

Lemma 7. *Let $G = (K \cup I, E)$ be a split graph without pendant vertices. Then there exists a minimum clique transversal T of G with $T \cap I = \emptyset$, and any such T is also a minimum-black 2-clique coloring.*

By Lemma 7, in order to solve MCBC on a split graph without pendant vertices, it suffices to compute a minimum clique transversal.

Theorem 5. *If a minimum clique transversal of a split graph $G = (K \cup I, E)$ without pendant vertices can be found in polynomial time, then* MINIMUM CLIQUE BICOLORING *on G is solvable in polynomial time.*

The problem of computing a minimum clique transversal in split graphs is NP-complete, but for dually chordal graphs it is known to be solvable in linear time [8]. Graphs that are both dually chordal and chordal are known as doubly chordal graphs, so for doubly chordal split graphs, a minimum clique transversal can be computed in polynomial time. Note that doubly chordal graphs form a graph class that contains all strongly chordal graphs.

Corollary 2. MINIMUM CLIQUE BICOLORING *on doubly chordal split graphs can be solved in linear time.*

4.2   Totally Unimodular Graphs

In this section, we show that MINIMUM CLIQUE BICOLORING can be solved in polynomial time on totally unimodular graphs. This graph class is contained in the class of perfect graphs and includes several important subclasses such as bipartite graphs and interval graphs. The polynomial-time solvability is derived from their favorable linear programming properties [25] and that the number of maximal cliques is $O(n^2)$ [32].

Theorem 6. *When the graph G is totally unimodular,* MINIMUM CLIQUE BICOLORING *can be solved in polynomial time.*

5   Exact Exponential-Time and FPT Algorithms

Exact exponential-time algorithm. In this section, we give an exact exponential-time algorithm for MINIMUM CLIQUE BICOLORING. The key observation is that, after a singsle preprocessing step over all vertex subsets, the feasibility of a 2-clique-coloring for any fixed black set can be tested by two table lookups.

Preprocessing. We first enumerate all maximal cliques of G in $O^*(3^{n/3})$ time [33], and let $\mathcal{C} = \{C_1, \ldots, C_M\}$ be the resulting family. We then prepare an array A indexed by vertex subsets, setting $A(S) = 1$ if $S \in \mathcal{C}$ and $A(S) = 0$ otherwise, and compute $F(S) = \sum_{S' \subseteq S} A(S')$ for all $S \subseteq V$ by a subset zeta transform. Here $F(S)$ equals the number of maximal cliques contained in S, so S contains no maximal clique if and only if $F(S) = 0$. The transform takes $O(n2^n)$ time and $O(2^n)$ space [6, 19].

Scanning Black Sets and Running Time. For a candidate black set $B \subseteq V$ and the corresponding white set $W = V \setminus B$, the constraint that every maximal clique is non-monochromatic is equivalent to saying that no maximal clique is contained in B and none is contained in W, that is, $F(B) = 0$ and $F(W) = F(V \setminus B) = 0$. Thus we can test feasibility of B in constant time by two table lookups, and among all 2^n subsets B we keep a feasible one with minimum $|B|$. The total running time is $O^*(3^{n/3})$ for maximal-clique enumeration plus $O(n2^n)$ for the zeta transform and $O(2^n)$ for scanning, which is $O(n2^n)$ time and $O(2^n)$ space overall. In contrast, a naive algorithm that checks all maximal cliques separately for each B would take $O^*(3^{n/3}2^n)$ time, whereas the zeta transform reduces this per-set check to constant time and yields the $O^*(2^n)$ bound.

Theorem 7. MINIMUM CLIQUE BICOLORING *for a graph of order n can be solved in $O^*(2^n)$ time.*

FPT Algorithms. Since 2-clique-colorability is definable in MSO_1, MCBC can also be expressed in MSO_1. By the Courcelle-Makowsky-Rotics meta-theorem, this yields an FPT algorithm on a cw-expression tree [11], where cw denotes the clique-width of G. Although no FPT algorithm is known for constructing an optimal cw-expression, a polynomial-time algorithm computes a k-expression with $\mathsf{cw} \le k \le 8^{\mathsf{cw}} - 1$ [30], which implies the following.

Theorem 8. MINIMUM CLIQUE BICOLORING *is FPT by the clique-width of G.*

The meta-theoretic algorithm above satisfies the FPT criterion for MCBC, but its running time has a non-elementary dependence on the formula (a tower of exponentials) [20], and is thus impractical. Instead, we design a dynamic-programming algorithm running in $O^*(4^k)$ time, where k is the number of labels of the expression tree (details omitted).

References

1. Arora, S., Barak, B.: Computational Complexity - A Modern Approach. Cambridge University Press (2009). http://www.cambridge.org/catalogue/catalogue. asp?isbn=9780521424264
2. Bacsó, G., Gravier, S., Gyárfás, A., Preissmann, M., Sebö, A.: Coloring the maximal cliques of graphs. SIAM J. Discret. Math. **17**(3), 361–376 (2004). https://doi. org/10.1137/S0895480199359995
3. Bacsó, G., Tuza, Z.: Clique-transversal sets and weak 2-colorings in graphs of small maximum degree. Discret. Math. Theor. Comput. Sci. **11**(2), 15–24 (2009). https://doi.org/10.46298/DMTCS.453
4. Balachandran, V., Nagavamsi, P., Rangan, C.P.: Clique transversal and clique independence on comparability graphs. Inf. Process. Lett. **58**(4), 181–184 (1996). https://doi.org/10.1016/0020-0190(96)00054-3
5. Berge, C.: Hypergraphs - combinatorics of finite sets, North-Holland mathematical library, vol. 45. North-Holland (1989)
6. Björklund, A., Husfeldt, T., Kaski, P., Koivisto, M.: Fourier meets möbius: fast subset convolution. In: Johnson, D.S., Feige, U. (eds.) Proceedings of the 39th Annual ACM Symposium on Theory of Computing, San Diego, California, USA, June 11–13, 2007. pp. 67–74. ACM (2007). https://doi.org/10.1145/1250790.1250801
7. Bonomo, F., Mazzoleni, M.P., Stein, M.: Clique coloring b₁-epg graphs. Discret. Math. **340**(5), 1008–1011 (2017). https://doi.org/10.1016/J.DISC.2017.01.019
8. Brandstädt, A., Chepoi, V., Dragan, F.F.: Clique r-domination and clique r-packing problems on dually chordal graphs. SIAM J. Discret. Math. **10**(1), 109–127 (1997). https://doi.org/10.1137/S0895480194267853
9. Chang, G.J., Farber, M., Tuza, Z.: Algorithmic aspects of neighborhood numbers. SIAM J. Discret. Math. **6**(1), 24–29 (1993). https://doi.org/10.1137/0406002

10. Chang, M., Chen, Y., Chang, G.J., Yan, J.: Algorithmic aspects of the generalized clique-transversal problem on chordal graphs. Discret. Appl. Math. **66**(3), 189–203 (1996). https://doi.org/10.1016/0166-218X(95)00048-V
11. Courcelle, B., Makowsky, J.A., Rotics, U.: Linear time solvable optimization problems on graphs of bounded clique-width. Theory Comput. Syst. **33**(2), 125–150 (2000). https://doi.org/10.1007/S002249910009
12. Défossez, D.: Clique-coloring some classes of odd-hole-free graphs. J. Graph Theory **53**(3), 233–249 (2006). https://doi.org/10.1002/JGT.20177
13. Diestel, R.: Graph Theory, 6th Edition. Springer Nature (2025). https://doi.org/10.1007/978-3-662-70107-2
14. Downey, R.G., Fellows, M.R.: Fixed-parameter tractability and completeness I: basic results. SIAM J. Comput. **24**(4), 873–921 (1995). https://doi.org/10.1137/S0097539792228228
15. Duffus, D., Kierstead, H.A., Trotter, W.T.: Fibres and ordered set coloring. J. Comb. Theory A **58**(1), 158–164 (1991). https://doi.org/10.1016/0097-3165(91)90083-S
16. Duffus, D., Sands, B., Sauer, N., Woodrow, R.E.: Two-colouring all two-element maximal antichains. J. Comb. Theory A **57**(1), 109–116 (1991). https://doi.org/10.1016/0097-3165(91)90009-6
17. Filho, H.B.M., Machado, R.C.S., de Figueiredo, C.M.H.: Hierarchical complexity of 2-clique-colouring weakly chordal graphs and perfect graphs having cliques of size at least 3. Theor. Comput. Sci. **618**, 122–134 (2016). https://doi.org/10.1016/J.TCS.2016.01.027
18. Flum, J., Grohe, M.: Parameterized Complexity Theory. TTCSAES, Springer, Heidelberg (2006). https://doi.org/10.1007/3-540-29953-X
19. Fomin, F.V., Kratsch, D.: Exact Exponential Algorithms. TTCSAES, Springer, Heidelberg (2010). https://doi.org/10.1007/978-3-642-16533-7
20. Frick, M., Grohe, M.: The complexity of first-order and monadic second-order logic revisited. Ann. Pure Appl. Log. **130**(1–3), 3–31 (2004). https://doi.org/10.1016/J.APAL.2004.01.007
21. Garey, M.R., Johnson, D.S., Stockmeyer, L.J.: Some simplified np-complete graph problems. Theor. Comput. Sci. **1**(3), 237–267 (1976). https://doi.org/10.1016/0304-3975(76)90059-1
22. Goldreich, O.: On Promise Problems: A Survey. In: Goldreich, O., Rosenberg, A.L., Selman, A.L. (eds.) Theoretical Computer Science. LNCS, vol. 3895, pp. 254–290. Springer, Heidelberg (2006). https://doi.org/10.1007/11685654_12
23. Gravier, S., Hoàng, C.T., Maffray, F.: Coloring the hypergraph of maximal cliques of a graph with no long path. Discret. Math. **272**(2-3), 285–290 (2003). https://doi.org/10.1016/S0012-365X(03)00197-3
24. Guruswami, V., Rangan, C.P.: Algorithmic aspects of clique-transversal and clique-independent sets. Discret. Appl. Math. **100**(3), 183–202 (2000). https://doi.org/10.1016/S0166-218X(99)00159-6
25. Hoffman, A.J., Kruskal, J.B.: Integral Boundary Points of Convex Polyhedra. In: Jünger, M., et al. (eds.) 50 Years of Integer Programming 1958-2008, pp. 49–76. Springer, Heidelberg (2010). https://doi.org/10.1007/978-3-540-68279-0_3
26. Kang, D.Y., Kelly, T., Kühn, D., Methuku, A., Osthus, D.: Graph and hypergraph colouring via nibble methods: A survey. CoRR **abs/2106.13733** (2021). https://arxiv.org/abs/2106.13733
27. Kratochvíl, J., Tuza, Z.: On the complexity of bicoloring clique hypergraphs of graphs. J. Algorithms **45**(1), 40–54 (2002). https://doi.org/10.1016/S0196-6774(02)00221-3

28. Lee, C., Chang, M.: Distance-hereditary graphs are clique-perfect. Discret. Appl. Math. **154**(3), 525–536 (2006). https://doi.org/10.1016/J.DAM.2005.07.011
29. Marx, D.: Complexity of clique coloring and related problems. Theor. Comput. Sci. **412**(29), 3487–3500 (2011). https://doi.org/10.1016/J.TCS.2011.02.038
30. Oum, S.: Approximating rank-width and clique-width quickly. ACM Trans. Algorithms **5**(1), 10:1–10:20 (2008). https://doi.org/10.1145/1435375.1435385
31. Poon, H.: Coloring Clique Hypergraphs. Master's thesis, West Virginia University (2000)
32. Schrijver, A.: Theory of linear and integer programming. Wiley-Interscience series in discrete mathematics and optimization, Wiley (1999)
33. Tomita, E., Tanaka, A., Takahashi, H.: The worst-case time complexity for generating all maximal cliques and computational experiments. Theor. Comput. Sci. **363**(1), 28–42 (2006). https://doi.org/10.1016/J.TCS.2006.06.015
34. Wrathall, C.: Complete sets and the polynomial-time hierarchy. Theor. Comput. Sci. **3**(1), 23–33 (1976). https://doi.org/10.1016/0304-3975(76)90062-1

On the $(\leq p)$-Inversion Diameter of Oriented Graphs

Frédéric Havet[1] , Clément Rambaud[1] , and Caroline Silva[1,2]($\boxtimes$)

[1] Université Côte d'Azur, CNRS, Inria, I3S, Sophia Antipolis, France
[2] Universidade Estadual de Campinas, Campinas, Brazil
caroline.silva@ic.unicamp.br

Abstract. In an oriented graph $\vec{G}$, the *inversion* of a subset X of vertices consists in reversing the orientation of all arcs with both endvertices in X. The $(\leq p)$-*inversion graph* of a labelled graph G, denoted by $\mathcal{I}^{\leq p}(G)$, is the graph whose vertices are the labelled orientations of G in which two labelled orientations $\vec{G}_1$ and $\vec{G}_2$ of G are adjacent if and only if there is a set X with $|X| \leq p$ whose inversion transforms $\vec{G}_1$ into $\vec{G}_2$. In this paper, we study the $(\leq p)$-*inversion diameter* of a graph, denoted by $\mathrm{id}^{\leq p}(G)$, which is the diameter of its $(\leq p)$-inversion graph. We show that there exists a smallest number Ψ_p with $\frac{1}{4}p - \frac{3}{2} \leq \Psi_p \leq \frac{1}{2}p^2$ such that $\mathrm{id}^{\leq p}(G) \leq \left\lceil \frac{|E(G)|}{\lfloor p/2 \rfloor} \right\rceil + \Psi_p$ for all graph G. We then establish better upper bounds for several families of graphs and in particular trees and planar graphs. Let us denote by $\mathrm{id}_{\mathcal{T}}^{\leq p}(n)$ (resp. $\mathrm{id}_{\mathcal{P}}^{\leq p}(n)$) the maximum $(\leq p)$-inversion diameter of a tree (resp. planar graph) of order n. For trees, we show $\mathrm{id}_{\mathcal{T}}^{\leq 3}(n) = \left\lceil \frac{n-1}{2} \right\rceil$, $\mathrm{id}_{\mathcal{T}}^{\leq 4}(n) = \frac{3}{8}n + \Theta(1)$, $\mathrm{id}_{\mathcal{T}}^{\leq 5}(n) = \frac{2}{7}n + \Theta(1)$, and $\mathrm{id}_{\mathcal{T}}^{\leq p}(n) \leq \frac{n-1}{p-c\sqrt{p}} + 2$ with $c = \sqrt{2 + \sqrt{2}}$ for all $p \geq 6$. For planar graphs, we prove $\mathrm{id}_{\mathcal{P}}^{\leq 3}(n) \leq \frac{11n}{6} - \frac{8}{3}$, $\mathrm{id}_{\mathcal{P}}^{\leq 4}(n) \leq \frac{4n}{3} + \frac{10}{3}$, and $\mathrm{id}_{\mathcal{P}}^{\leq p}(n) \leq \left\lceil \frac{3n-6}{\lfloor p/2 \rfloor} \right\rceil + 8\lfloor p/2 \rfloor - 8$ for all $p \geq 6$.

Keywords: inversion graph · diameter · orientation · reconfiguration

1 Introduction

Making a digraph acyclic by removing a minimum cardinality set of arcs is an important and heavily studied problem, known as the FEEDBACK ARC SET problem. A **feedback arc set** in a digraph is a set of arcs whose deletion results in an acyclic digraph. The **feedback-arc-set number** of a digraph D, denoted by $\mathrm{fas}(D)$, is the minimum size of a feedback arc set. Note that if F is a minimum feedback arc-set in a digraph $D = (V, A)$, then we also obtain an acyclic digraph from D by reversing the arcs of F, that is replacing each arc $uv \in F$ by the arc

This work was partially supported by the french Agence Nationale de la Recherche under contract Digraphs ANR-19-CE48-0013-01. Caroline Silva was supported by FAPESP Proc. 2023/16755-2.

vu. Computing fas(D) is one of the first problems shown to be NP-hard listed by Karp in [16]. It remains NP-complete in tournaments [1,10].

Belkhechine et al. in [8] introduced the more general concept of inversion. In an oriented graph $\overrightarrow{G}$, if X is a set of vertices of $\overrightarrow{G}$, the **inversion** of X consists in reversing the orientation of all arcs with both endvertices in X. In particular, the reversal of a single arc corresponds to the inversion of the set of its endvertices. Belkhechine et al. in [8] studied the **inversion number**, denoted by inv(D), which is the minimum number of inversions that transform $\overrightarrow{G}$ into a directed acyclic graph. In particular, they proved that for every fixed k, determining whether a given tournament T has inversion number at most k is polynomial-time solvable. In contrast, Bang-Jensen et al. [7] proved that deciding whether a given oriented graph has inversion number 1 is NP-complete. The maximum inv(n) of the inversion numbers of all oriented graphs of order n has also been investigated. Independently, Aubian et al. [4] and Alon et al. [2] proved $n - 2\sqrt{n \log n} \leq \text{inv}(n) \leq n - \lceil \log(n+1) \rceil$.

Making the connection between feedback arc set (inversion over sets of size 2) and inversion over sets of unbounded size, Yuster [23] studied ($\leq p$)-**inversions**, which are inversions over sets of size at most p. (Bang-Jensen et al. [6] also studied ($= p$)-inversions, which are inversions over sets of size exactly p.) Given an oriented graph $\overrightarrow{G}$, its ($\leq p$)-**inversion number**, denoted by $\text{inv}^{\leq p}(\overrightarrow{G})$, is the minimum number of ($\leq p$)-inversions rendering $\overrightarrow{G}$ acyclic. Note that $\text{inv}^{\leq 2}(\overrightarrow{G}) = \text{fas}(\overrightarrow{G})$. Yuster studied the maximum $\text{inv}^{\leq p}(n)$ of the ($\leq p$)-inversion numbers of all oriented graphs of order n. Results of Spencer [20,21] (later improved by de la Vega [12]) on feedback arc sets show $\text{inv}^{\leq 2}(n) = \frac{1}{2}\binom{n}{2} + \Theta(n^{3/2})$. Yuster [23] proved $\frac{1}{12}n^2 + o(n^2) \leq \text{inv}^{\leq 3}(n) \leq \frac{257}{2592}n^2 + o(n^2)$ and conjectured $\text{inv}^{\leq 3}(n) = \frac{1}{12}n^2 + o(n^2)$. He also proved that, for every fixed p,

$$\left(\frac{1}{2p(p-1)} + \delta_p \right) n^2 + o(n^2) \leq \text{inv}^{\leq p}(n) \leq \left(\frac{1}{2\lfloor p^2/2 \rfloor} - \epsilon_p \right) n^2 + o(n^2)$$

where $\epsilon_p > 0$ for all $p \geq 3$ and $\delta_p > 0$ for all $p \geq p_0$ for some p_0.

Rather than reducing to an acyclic digraph, we can use inversions to reduce to other types of digraphs. Duron et al. [11] studied the minimum number of inversions to make a digraph k-arc-strong or k-strong. More generally, Havet et al. [13] studied the maximum over all pairs of orientations of a graph of the minimum number of inversions required to transform one of the orientations into the other. Formally, for a graph G with vertices labelled $v_1, \ldots, v_n$, the **(labelled) inversion graph** of G, denoted by $\mathcal{I}(G)$, is a graph with vertex set the labelled orientations of G. Then, two labelled orientations $\overrightarrow{G}_1$ and $\overrightarrow{G}_2$ of G are adjacent if and only if there is an inversion X transforming $\overrightarrow{G}_1$ into $\overrightarrow{G}_2$. The **(labelled) inversion diameter** of G is the diameter of $\mathcal{I}(G)$, denoted by id(G). Havet et al. [13] determined the maximum inversion diameter over all graphs on n vertices.

Theorem 1. *Let G be a graph on n vertices. Then* id(G) $\leq n - 1$.

This bound is tight for complete graphs. For sparser graphs, Havet et al. [13] showed that much better bounds can be obtained. Some of their bounds were subsequently improved by Arana et al. [3].

a) $id(G) \leq 12$ for every planar graph G [13], and there are planar graphs with inversion diameter at least 6 [3];
b) $id(G) \leq 3$ for every planar graph G of girth at least 7 [3], and there are planar graphs of arbitrary large girth with inversion diameter 3 [13];
c) $id(G) \leq 2\Delta(G)-1$ for every graph G with at least one edge and $id(G) = \Delta(G)$ for every complete graph G [13];
d) $id(G) \leq 2t$ for every graph G of treewidth at most t [13] and for every positive integer t, there are graphs of treewidth t with inversion diameter $2t$ [22].

The **(labelled) $(\leq p)$-inversion graph** of G, denoted by $\mathcal{I}^{\leq p}(G)$, is the graph whose vertices are the labelled orientations of G in which two labelled orientations $\overrightarrow{G}_1$ and $\overrightarrow{G}_2$ of G are adjacent if and only if there is an $(\leq p)$-inversion transforming $\overrightarrow{G}_1$ into $\overrightarrow{G}_2$. The **(labelled) $(\leq p)$-inversion diameter** of G is the diameter of $\mathcal{I}^{\leq p}(G)$, denoted by $id^{\leq p}(G)$. The $(\leq p)$-**converse number** of a graph G, denoted by $conv^{\leq p}(G)$, is the minimum number of $(\leq p)$-inversions needed to transform an orientation of G into its converse. Note that this number does not depend on the orientation. An $(\leq p)$-inversion reverses at most $\binom{p}{2}$ arcs, and one can reverse precisely one arc by inverting the set of its endvertices. Thus,

$$\frac{|E(G)|}{\binom{p}{2}} \leq conv^{\leq p}(G) \leq id^{\leq p}(G) \leq |E(G)|. \tag{1}$$

For $p = 2$, the above equation yields the equality $conv^{\leq p}(G) = id^{\leq p}(G) = |E(G)|$. The left-hand side inequality of Eq. (1) is attained for very sparse graphs, for example the disjoint union of complete graphs of order p. In contrast, the right-hand side inequality is not tight when $p > 2$. In Sect. 3, using strong edge-colourings, we show the following upper bound.

Theorem 2. *Let G be a graph and $p \geq 2$ be an integer. Then $id^{\leq p}(G) \leq \left\lceil \frac{|E(G)|}{\lfloor p/2 \rfloor} \right\rceil + \frac{1}{2}p^2$.*

This upper bound is tight up to the additive term $\frac{1}{2}p^2$. Indeed, consider a matching graph M, that is, the disjoint union of K_2 and K_1. Then, an $(\leq p)$-inversion reverses at most $\lfloor p/2 \rfloor$ edges, so $conv^{\leq p}(M) = id^{\leq p}(M) = \left\lceil \frac{|E(M)|}{\lfloor p/2 \rfloor} \right\rceil$. In particular, for $p \in \{2,3\}$, we get $conv^{\leq p}(M) = id^{\leq p}(M) = |E(M)|$, so the right-hand side inequalities of Eq. 1 are tight.

Hence, a natural question is to determine the smallest number Ψ_p, (resp. Ψ'_p) such that $id^{\leq p}(G) \leq \left\lceil \frac{|E(G)|}{\lfloor p/2 \rfloor} \right\rceil + \Psi_p$ (resp. $conv^{\leq p}(G) \leq \left\lceil \frac{|E(G)|}{\lfloor p/2 \rfloor} \right\rceil + \Psi'_p$) for every graph G.

Observe that if $n \leq p$, then $id^{\leq p}(K_n) = id(K_n) = n - 1$ as proved in [13], so $\Psi_p \geq n - 1 - \left\lceil \frac{\binom{n}{2}}{\lfloor p/2 \rfloor} \right\rceil \geq n - 1 - \left\lceil \frac{n(n-1)}{p-1} \right\rceil$. For $n = \lceil p/2 \rceil$, we obtain $\Psi_p \geq \frac{1}{4}p - \frac{3}{2}$.

We believe that this lower bound is tighter than the upper bound.

Problem 1. Does there exist a constant C such that $\Psi_p \leq C \cdot p$ for every integer p greater than 1 ?

The upper bound of Theorem 2 is tight for matching graphs, which are very sparse. It is however far from being tight for not too sparse graphs. In Theorem 6, we prove the following upper bound.

$$\mathrm{id}^{\leq p}(G) \leq \frac{1}{p-1}|E(G)| + \frac{p-2}{p-1}(|V(G)| - 1). \tag{2}$$

Note that this bound is better than the one of Eq. (1) if $|E(G)| \geq |V(G)| - 1$, and better than the one of Theorem 2 if $|E(G)| \geq (p-2)(|V(G)| - 1)$ when p is odd and if $|E(G)| \geq p(|V(G)| - 1)$ when p is even.

We can get an even better upper bound for very dense graphs (graphs of order n with $\Omega(n^2)$ edges). Using a classical theorem by Kővári, Sós, and Turán [18], Yuster [23] implicitly used the following theorem, which intuitively, means that after $\frac{|E|}{\lceil p/2 \rceil \cdot \lfloor p/2 \rfloor}$ inversions of well-chosen sets of size p, we have reversed exactly the edges in E, up to $o(n^2)$ errors (See Bang-Jensen et al. [6] for an explicit proof).

Theorem 3. *Let p be an integer with $p \geq 2$. There exist constants α_p and ϵ_p with $\epsilon_p > 0$ such that the following holds. Let n be a positive integer, and let $F \subseteq \binom{[n]}{2}$. There exists an integer ℓ with $\ell \leq \frac{|F|}{\lceil p/2 \rceil \cdot \lfloor p/2 \rfloor}$, and a family $X_1, \ldots, X_\ell$ of $(= p)$-subsets of $[n]$ such that*

$$\left| F \Delta \binom{X_1}{2} \Delta \ldots \Delta \binom{X_\ell}{2} \right| \leq \alpha_p \cdot n^{2 - \epsilon_p}.$$

This theorem directly implies

$$\mathrm{id}^{\leq p}(G) \leq \frac{|E(G)|}{\lceil p/2 \rceil \cdot \lfloor p/2 \rfloor} + o(|V(G)|^2). \tag{3}$$

This bound is asymptotically tight. Indeed, consider two orientations of a complete bipartite graph B which are converse to each other. Every inversion reverses at most $\lceil p/2 \rceil \cdot \lfloor p/2 \rfloor$ edges. Since all edges of B need to be reversed, we get $\mathrm{id}^{\leq p}(B) \geq \frac{|E(B)|}{\lceil p/2 \rceil \cdot \lfloor p/2 \rfloor}$.

In Subsect. 3.2, we give upper bounds of the (≤ 3)-inversion diameter of a connected graph in terms of its triangle-transversal numbers and its triangle-edge-packing number and derive the following theorem.

Theorem 4. *If G is a connected graph, then $\mathrm{id}^{\leq 3}(G) \leq \left\lceil \frac{3|E(G)|}{4} \right\rceil$.*

Let $\mathcal{C}$ be a graph class. It is (a, b)-**sparse** if $|E(G)| \leq a|V(G)| + b$ for every $G \in \mathcal{C}$, and **sparse** if it is (a, b)-sparse for some pair (a, b). The ($\leq p$)-**inversion diameter function** of $\mathcal{C}$, denoted by $\mathrm{id}_{\mathcal{C}}^{\leq p}$, is the smallest function f such that

$\mathrm{id}^{\leq p}(G) \leq f(|V(G)|)$ for every $G \in \mathcal{C}$. Similarly, the $(\leq p)$-**converse function** of $\mathcal{C}$, denoted by $\mathrm{conv}_{\mathcal{C}}^{\leq p}$, is the smallest function g such that $\mathrm{conv}^{\leq p}(G) \leq g(|V(G)|)$ for every $G \in \mathcal{C}$. Clearly, $\mathrm{conv}_{\mathcal{C}}^{\leq p}(n) \leq \mathrm{id}_{\mathcal{C}}^{\leq p}(n)$ for every n. If $\mathcal{C}$ is (a,b)-sparse, then Theorem 2 and Eq. (2) yield the upper bound

$$\mathrm{conv}_{\mathcal{C}}^{\leq p}(n) \leq \mathrm{id}_{\mathcal{C}}^{\leq p}(n) \leq \min\left\{\left\lceil\frac{an+b}{\lfloor p/2\rfloor}\right\rceil + \frac{1}{2}p^2, \frac{a+p+2}{p-1}n + \frac{b-p+2}{p-1}\right\}.$$

In the remaining of the paper, we improve on this bound for several well-known sparse classes of graphs. In Sect. 4, we determine the $(\leq p)$-inversion diameter function and $(\leq p)$-converse function of the class $\mathcal{F}$ of forests up to an additive constant when $p \leq 5$ and gives an upper bound on those parameters which improves on the above upper bounds for any value of p greater than 4.

Theorem 5. *(i)* $\mathrm{id}_{\mathcal{F}}^{\leq 3}(n) = \mathrm{conv}_{\mathcal{F}}^{\leq 3}(n) = \left\lceil\frac{n-1}{2}\right\rceil$.
(ii) $\mathrm{id}_{\mathcal{F}}^{\leq 4}(n), \mathrm{conv}_{\mathcal{F}}^{\leq 4}(n) = \frac{3}{8}n + \Theta(1)$.
(iii) $\mathrm{id}_{\mathcal{F}}^{\leq 5}(n), \mathrm{conv}_{\mathcal{F}}^{\leq 5}(n) = \frac{2}{7}n + \Theta(1)$.
(iv) $\mathrm{conv}_{\mathcal{F}}^{\leq p}(n) \leq \mathrm{id}_{\mathcal{F}}^{\leq p}(n) \leq \frac{n-1}{p-c\sqrt{p}} + 2$ *with* $c = \sqrt{2+\sqrt{2}}$.

In Sect. 5, we show that if G is a k-degenerate graph, then $\mathrm{id}^{\leq p}(G) \leq |V(G)| - 1$, for any $p \geq k + 1$. We show that it is tight when $k = 2$ and $p = 3$. We then consider the class $\mathcal{P}$ of planar graphs and show $\frac{7n}{6} - 2 \leq \mathrm{conv}_{\mathcal{P}}^{\leq 3}(n) \leq \mathrm{id}_{\mathcal{P}}^{\leq 3}(n) \leq \frac{11n}{6} - \frac{8}{3}$, $\mathrm{id}_{\mathcal{P}}^{\leq 4}(n) \leq \frac{4n}{3} + \frac{10}{3}$, and $\mathrm{id}_{\mathcal{P}}^{\leq p}(n) \leq \left\lceil\frac{3n-6}{\lfloor p/2\rfloor}\right\rceil + 8\lfloor p/2\rfloor - 8$ for every $p \geq 2$. We also give the easy lower bounds $\mathrm{conv}_{\mathcal{P}}^{\leq 3}(n) \geq \frac{7n}{6} - 2$ and $\mathrm{conv}_{\mathcal{P}}^{\leq p}(n) \geq \frac{p-1}{(p-2)^2+1} \cdot (n-2)$. It would be interesting to close the gaps between those upper and lower bounds.

Due to space constraints, we are unable to include all the proofs in detail. In the remainder of the paper, we outline the most significant ones. Complete proofs are provided in the full version available in [14].

2   Notations, Definitions, and Preliminaires

Notation not given below is consistent with [5]. An $(\leq p)$-**set** is a set of cardinality at most p.

Let D be an oriented graph and $\mathcal{X}$ be a family of subsets of $V(D)$. We say that $\mathcal{X}$ is an $(\leq p)$-**family** if all members of $\mathcal{X}$ are $(\leq p)$-sets. We denote by $\mathrm{Inv}(D; \mathcal{X})$ the oriented graph obtained after inverting all sets of $\mathcal{X}$ one after another. Observe that this is independent of the order in which we invert those sets: $\mathrm{Inv}(D; \mathcal{X})$ is obtained from D by reversing exactly those arcs for which an odd number of members of $\mathcal{X}$ contain both endvertices. If $\mathcal{X} = \{X\}$ for a set $X \subseteq V(D)$, then we write $\mathrm{Inv}(D; X)$ for $\mathrm{Inv}(D; \mathcal{X})$.

Let $\overrightarrow{G}_1$ and $\overrightarrow{G}_2$ be two orientations of a graph G. If an edge e has the same orientation in $\overrightarrow{G}_1$ and $\overrightarrow{G}_2$, we say that $\overrightarrow{G}_1$ and $\overrightarrow{G}_2$ **agree** on e; otherwise we say that they **disagree** on e. We denote by $E_=$ the set of edges of G on which $\overrightarrow{G}_1$

and $\vec{G}_2$ agree and by $E_{\neq}$ the set of edges of G on which $\vec{G}_1$ and $\vec{G}_2$ disagree. We set $G_{\neq} = (V(G), E_{\neq})$.

Proposition 1. *Let G be a graph. If G is the union of a subgraph H and an induced subgraph I, then $\mathrm{id}^{\leq p}(G) \leq \mathrm{id}^{\leq p}(H) + \mathrm{id}^{\leq p}(I)$.*

Since a graph whose edge set is a matching of size at most $\lfloor p/2 \rfloor$ has ($\leq p$)-inversion diameter 1, Proposition 1 immediately yields the following.

Corollary 1. *Let $p \geq 2$ be an integer. Let G be a graph and let M be an induced matching of G of size at most $\lfloor p/2 \rfloor$. Then $\mathrm{id}^{\leq p}(G) \leq \mathrm{id}^{\leq p}(G \setminus M) + 1$.*

Corollary 2. *Let $p \geq 2$ be an integer and set $q = \lfloor p/2 \rfloor$. Let G be a graph and v a vertex of G.*

(i) $\mathrm{id}^{\leq p}(G) \leq \mathrm{id}^{\leq p}(G - v) + \left\lceil \frac{d(v)}{p-1} \right\rceil.$

(ii) If $d(v) \geq q$, then $\mathrm{id}^{\leq p}(G) \leq \mathrm{id}^{\leq p}(G - v) + \lfloor d(v)/q \rfloor.$

3 General Upper Bounds

Let us sketch the proof of Theorem 2 which states $\mathrm{id}^{\leq p}(G) \leq \left\lceil \frac{|E(G)|}{\lfloor p/2 \rfloor} \right\rceil + \frac{1}{2}p^2$.

The proof is by considering a minimal counterexample G. Using Corollaries 1 and 2, one proves that G has no induced matching of size $\lfloor p/2 \rfloor$ and $\Delta(G) \leq q = \lfloor p/2 \rfloor - 1$. Recall that an **induced matching** of G is a set of edges that do not share any endvertices and that are the only edges connecting any two endvertices of them.

A **strong edge-colouring** of G is an edge-colouring of G such that each colour class is an induced matching of G. The **strong chromatic index** of a graph G, denoted by $\chi_s(G)$, is the minimum integer k such that G admits a strong edge-colouring with k colours. It is well-known and easy to show that $\chi_s(G) \leq 2\Delta(G)^2$. One can show $\mathrm{id}^{\leq p}(G) \leq \frac{|E(G)|}{\lfloor p/2 \rfloor} + \chi_s'(G)$. Since $\Delta(G) \leq q = \lfloor p/2 \rfloor - 1$, we deduce $\mathrm{id}^{\leq p}(G) \leq \left\lceil \frac{|E(G)|}{\lfloor p/2 \rfloor} \right\rceil + \frac{1}{2}p^2$, a contradiction.

Observe that the additive constant of $\frac{1}{2}p^2$ in the upper bound of Theorem 2 can be slightly improved using the results of [9,19], and [15] on strong-edge colouring. For small values of p, we prove $\mathrm{id}^{\leq p}(G) \leq \frac{|E(G)|}{\lfloor p/2 \rfloor} + \Psi_p$ with $\Psi_p = 0$ if $p \leq 5$ and $\Psi_p = 1$ if $p \in \{6, 7, 8, 9\}$.

3.1 Better Bounds for Not Too Sparse Graphs

Corollary 2 (i) and an easy induction yield the following.

Theorem 6. *Let p be an integer greater than 2 and let G be a graph. Then*

$$\mathrm{id}^{\leq p}(G) \leq \frac{1}{p-1}|E(G)| + \frac{p-2}{p-1}(|V(G)| - 1).$$

3.2   Better Bound for Connected Graphs

A **triangle-transversal** in a graph G is a set F of edges such that $G \setminus F$ has no triangle. The **triangle-transversal number** of a graph G, denoted by $\tau_3(G)$, is the minimum size of a triangle-transversal. Since every graph G has a bipartite subgraph with at least $\lceil |E(G)|/2 \rceil$ edges, $\tau_3(G) \leq \lfloor |E(G)|/2 \rfloor$.

Lemma 1. *Let G be a connected graph. Then* $\mathrm{id}^{\leq 3}(G) \leq \left\lceil \frac{|E(G)| + \tau_3(G)}{2} \right\rceil$.

Proof. Let G be a graph and let F be a minimum triangle-transversal in G.

Let $\overrightarrow{G_1}$ and $\overrightarrow{G_2}$ be two orientations of G. Let $\overrightarrow{H_1}$ and $\overrightarrow{H_2}$ be the orientations of $H = G \setminus F$ which are restrictions of $\overrightarrow{G_1}$ and $\overrightarrow{G_2}$ respectively.

Set $t = \lceil |E(H)|/2 \rceil$. One can show that H is connected. Thus, by a result due to Kotzig [17], it can be edge-decomposed into t paths $P_1, \ldots, P_t$ of order at most 3. For $i \in [t]$, there is a set $X_i \subseteq V(P_i)$ whose inversion transforms $\overrightarrow{H_1}\langle V(P_i)\rangle$ into $\overrightarrow{H_2}\langle V(P_i)\rangle$. Then, inverting $(X_i)_{i \in [t]}$ transforms $\overrightarrow{H_1}$ into $\overrightarrow{H_2}$. Thus $\mathrm{Inv}(\overrightarrow{G_1}; (X_i)_{i \in [t]})$ and $\overrightarrow{G_2}$ disagree only on edges of F. Each disagreeing edge can be reversed by inverting the set of its end-vertices. Thus $\overrightarrow{G_1}$ can be transformed into $\overrightarrow{G_2}$ by inverting at most $t + |F|$ (≤ 3)-sets.

Therefore $\mathrm{id}^{\leq 3}(G) \leq t + |F| = \lceil |E(H)|/2 \rceil + |F| = \left\lceil \frac{|E(G)| - |F|}{2} \right\rceil + |F| = \left\lceil \frac{|E(G)| + \tau_3(G)}{2} \right\rceil$. $\qquad\square$

The fact that $\tau_3(G) \leq \lfloor |E(G)|/2 \rfloor$ and Lemma 1 immediately imply Theorem 4, which states that $\mathrm{id}^{\leq 3}(G) \leq \left\lceil \frac{3|E(G)|}{4} \right\rceil$ for every connected graph G.

4   Trees and Forests

The aim of this section is to prove Theorem 5. We shall need the following lemma which allows to concentrate on $\mathrm{conv}_{\not{\mathcal{F}}}^{\leq p}(n)$.

Lemma 2. *If there exist two constants α and β such that $\mathrm{conv}_{\not{\mathcal{F}}}^{\leq p}(n) \leq \alpha n + \beta$ for all nonnegative integer n, then $\mathrm{id}_{\not{\mathcal{F}}}^{\leq p}(n) \leq \alpha n + 2\beta$ for all nonnegative integer n.*

Proof. Let F be a forest of order n, and let $\overrightarrow{F_1}$ and $\overrightarrow{F_2}$ be two orientations of F. Consider the graph $H_{\neq}$ whose vertices are the connected components of $(V(F), E_{\neq})$, and in which two such connected components C and C' are adjacent if and only if there is an edge uv in $E(F)$ with $u \in C$ and $v \in C'$. Observe that $H_{\neq}$ is a minor of F, and so is a forest. In particular, $H_{\neq}$ is bipartite. Let (A_1', A_2') be a bipartition of $H_{\neq}$ and let (A_1, A_2) the partition of $V(F)$ where a vertex is contained in A_j if it is contained in a component of $H_{\neq}$ that belongs to A_j' for $j \in [2]$. For every edge uv in $E(F)$, we have $uv \in E_{\neq}$ if and only if $\{u, v\} \subseteq A_1$ or $\{u, v\} \subseteq A_2$. For $j \in [2]$, let $F_j = F\langle A_j\rangle$, and let $n_j = |A_j|$. Since $\mathrm{conv}_{\not{\mathcal{F}}}^{\leq p}(n_j) \leq \alpha n_j + \beta$, there is a family $(X_i)_{i \in I_j}$ of at most $\alpha n_j + \beta$

subsets of A_j of size at most p whose inversion reverses all edges of F_j. Observe that by the construction of A_1 and A_2, the inversion of an $X_i \subseteq A_j$ in F only reverses edges with both end-vertices in X_j. Thus $(X_i)_{i \in I_1 \cup I_2}$ is family of at most $\alpha n_1 + \beta + \alpha n_2 + \beta = \alpha n + 2\beta$ sets that transforms $\overrightarrow{F_1}$ into $\overrightarrow{F_2}$. Thus $\mathrm{id}^{\leq p}(F) \leq \alpha n + 2\beta$. $\qquad\square$

<u>*Case* $p = 3$.</u> Let us prove $\mathrm{conv}^{\leq 3}(T) = \mathrm{id}^{\leq 3}(T) = \lceil \frac{n-1}{2} \rceil$ for every tree of order n, which directly imply assertion (i) of Theorem 5.

Let $\overrightarrow{T}$ an orientation of T and $\overleftarrow{T}$ its converse. Then $|E_{\neq}| = |E(T)| = n - 1$. Observe that a (≤ 3)-inversion reverses at most two edges of T, so $\overrightarrow{T}$ and $\overleftarrow{T}$ are at distance at least $\lceil (n-1)/2 \rceil$ in $\mathcal{I}^{\leq 3}(T)$. Hence $\mathrm{id}^{\leq 3}(T) \geq \mathrm{conv}^{\leq 3}(T) \geq \lceil \frac{n-1}{2} \rceil$.

Now the tree T has no triangle, so $\tau_3(T) = \nu_3(T) = 0$. Hence, by Lemma 1, $\mathrm{id}^{\leq 3}(T) \leq \lceil \frac{|E(T)|}{2} \rceil = \lceil \frac{n-1}{2} \rceil$.

<u>*Case* $p = 4$.</u> Let us now prove the assertion (ii) of Theorem 5 which states $\mathrm{id}^{\leq 4}_{\mathcal{F}}(n), \mathrm{conv}^{\leq 4}_{\mathcal{F}}(n) = \frac{3}{8}n + \Theta(1)$.

To obtain the upper bound, we first prove that every tree T of order n can be decomposed in $\lceil 3(n-1)/8 \rceil$ edge-disjoint induced subgraphs of order at most 4, from which we derive $\mathrm{conv}^{\leq 4}(T) \leq \left\lceil \frac{3(n-1)}{8} \right\rceil$. Thus $\mathrm{conv}^{\leq 4}_n \leq \left\lceil \frac{3(n-1)}{8} \right\rceil$, and so by Lemma 2, $\mathrm{id}^{\leq 4}_{\mathcal{F}}(n) \leq \frac{3}{8}n + 1$.

The lower bound comes from the following construction: Let n be a positive integer. Set $s = \lfloor (n-1)/2 \rfloor$ and $\epsilon = n - 1 - 2s$. Let T be the tree with vertex set $\{x\} \cup \{y_i \mid i \in [s+\epsilon]\} \cup \{z_i \mid i \in [s]\}$ and edge set $\{xy_i \mid i \in [s+\epsilon]\} \cup \{y_i z_i \mid i \in [s]\}$. One shows that $\mathrm{conv}^{\leq 4}(T) \geq \lceil \frac{3n-4}{8} \rceil$ and so $\mathrm{conv}^{\leq 4}_{\mathcal{F}}(n) \geq \lceil \frac{3n-4}{8} \rceil$.

<u>*Case* $p = 5$.</u> Let us now prove the assertion (iii) of Theorem 5 which states $\mathrm{id}^{\leq 5}_{\mathcal{F}}(n), \mathrm{conv}^{\leq 5}_{\mathcal{F}}(n) = \frac{2}{7}n + \Theta(1)$.

To obtain the upper bound, we first prove that every tree T of order $n \geq 5$ contains either a good 5-set or a good 5-pair. A **good 5-set with root** t in T is a set X of five vertices such that $T\langle X \rangle$ is a tree and $T - (X \setminus \{t\})$ is a tree. A **good 5-pair with roots** t_1, t_2 in T is a pair (X_1, X_2) of (≤ 5)-sets such that $T\langle X_1 \rangle$ has four edges, $T\langle X_2 \rangle$ has three edges, $T\langle X_1 \cap X_2 \rangle$ has no edge, and $T - ((X_1 \cup X_2) \setminus \{t_1, t_2\})$ is a tree. We then deduce that $\mathrm{conv}^{\leq 5}(T) \leq \lceil \frac{2n-2}{7} \rceil$. Thus $\mathrm{conv}^{\leq 5}_{\mathcal{F}}(n) \leq \lceil \frac{2n-2}{7} \rceil$, and so by Lemma 2, $\mathrm{id}^{\leq 5}_{\mathcal{F}}(n) \leq \frac{2}{7}n + \frac{8}{7}$.

The lower bound comes from the following construction. Let $n = 7q + 1$, and let T be the tree defined by

$$V(T) = \{r\} \cup \bigcup_{j=1}^{q} \{x_j, y_j^1, y_j^2, z_j^1, z_j^2, z_j^3, z_j^4\},$$

$$E(T) = \bigcup_{j=1}^{q} \{rx_j, x_j y_j^1, x_j y_j^2, y_j^1 z_j^1, y_j^1 z_j^2, y_j^2 z_j^3, y_j^2 z_j^4\}.$$

For $j \in [q]$, we define the *jth branch* as the subtree B_j with vertex set $\{r\} \cup \{x_j, y_j^1, y_j^2, z_j^1, z_j^2, z_j^3, z_j^4\}$ and edge set $\{rx_j, x_j y_j^1, x_j y_j^2, y_j^1 z_j^1, y_j^1 z_j^2, y_j^2 z_j^3, y_j^2 z_j^4\}$.

Let $\overrightarrow{T}$ be an orientation of T and let $\overleftarrow{T}$ be its converse, and let $(X_i)_{i\in I}$ be a family of (≤ 5)-sets whose inversions transform $\overrightarrow{T}$ into $\overleftarrow{T}$. The **trace** $T_{i,j}$ of a set X_i on branch B_j is the set of edges of B_j with both end-vertices in X_i. We assign weights to the traces as follows.

- If $|T_{i,j}| = 0$, then $w(T_{i,j}) = 0$.
- If $|T_{i,j}| = 1$, then $w(T_{i,j}) = \begin{cases} 5/3, & \text{if } rw_j \notin T_{i,j}, \\ 5/4, & \text{if } rw_j \in T_{i,j}. \end{cases}$
- If $|T_{i,j}| = 2$, then $w(T_{i,j}) = \begin{cases} 10/3, & \text{if } rw_j \notin T_{i,j}, \\ 9/4, & \text{if } rw_j \in T_{i,j}. \end{cases}$
- If $|T_{i,j}| = 3$, then $w(T_{i,j}) = \begin{cases} 5, & \text{if } rw_j \notin T_{i,j}, \\ 7/2, & \text{if } rw_j \in T_{i,j}. \end{cases}$
- If $|T_{i,j}| = 4$, then $w(T_{i,j}) = 5$.

On the one hand, the weight of each X_i which is $w(X_i) = \sum_{j\in[q]} w(T_{i,j})$ is at most 5.

On the other hand, let us consider the weight of branch B_j which is $w(B_j) = \sum_{i\in I} w(T_{i,j})$. Observe that B_j has at most one trace with four edges and that if it has one such trace, then rw is not in a trace of size 2 or 3. Hence, one easily sees that $w(B_j) \geq 10$. Thus $5|I| \geq \sum_{i\in I} w(X_i) = \sum_{j\in[q]} w(B_j) \geq 10q$, so $|I| \geq 2q = 2\left\lfloor \frac{n-1}{7} \right\rfloor$. Hence $\mathrm{conv}^{\leq 5}(T) \geq 2\left\lfloor \frac{n-1}{7} \right\rfloor$. Thus $\mathrm{conv}_{\overrightarrow{\mathcal{F}}}^{\leq 5}(n) \geq \mathrm{conv}^{\leq 5}(T) \geq 2\left\lfloor \frac{n-1}{7} \right\rfloor$.

General case. Let us prove the assertion (iv) of Theorem 5 which states $\mathrm{conv}_{\overrightarrow{\mathcal{F}}}^{\leq p}(n) \leq \mathrm{id}_{\overrightarrow{\mathcal{F}}}^{\leq p}(n) \leq \frac{n-1}{p-c\sqrt{p}} + 2$ with $c = \sqrt{2+\sqrt{2}}$ for any $p \geq 4$. We need the following lemma.

Lemma 3. *Let $p \geq 4$ be an integer. Let T be a tree with at least p vertices rooted at a vertex r. Then, there exists a set $X \subseteq V(T)$ such that*

(i) $|X| \leq p$,

(ii) $T\langle X\rangle$ has at least $p - \sqrt{(2+\sqrt{2})p}$ edges, and

(iii) every non-trivial connected component in $T \setminus E(T\langle X\rangle)$ contains r. In particular, this implies that there is at most one non-trivial connected component in $T \setminus E(T\langle X\rangle)$.

Proof. The proof follows by induction on $p + |V(T)|$. If $|V(T)| = p$, then $X = V(T)$ is a set satisfying Properties (i) to (iii). We may thus assume that $|V(T)| > p$.

Let $r_1,\ldots,r_k$ be the neighbours of r in T and let T_i be the subtree rooted at r_i. Without loss of generality, we may assume that $|V(T_1)| \geq \ldots \geq |V(T_k)|$. If $|V(T) \setminus V(T_k)| \geq p$, then, by applying the induction to p, $T - V(T_k)$ and r, there is a set X that satisfies Properties (i) to (iii) in $T - V(T_k)$ which directly implies that X satisfies the Properties (i) to (iii) in T.

Henceforth we may assume $|V(T) \setminus V(T_i)| < p$ for every $1 \leq i \leq k$. So $k \geq 2$. Let $\alpha = \sum_{i=1}^{k-1} |V(T_i)|$. Note that $\alpha \geq \frac{|V(T)|-1}{2} \geq \frac{p}{2}$. Set $c = \sqrt{2+\sqrt{2}}$.

Suppose first that $|V(T_k)| \leq c\sqrt{p}$. Let $X = V(T)\setminus V(T_k)$. Note that X clearly satisfies Properties (i) and (iii). Moreover, since $T\langle X\rangle$ is connected and has $\alpha+1$

vertices, $|E(T\langle X\rangle)| = \alpha$. Thus, $|E(T\langle X\rangle)| = \alpha \geq p - |V(T_k)| \geq p - c\sqrt{p}$, and X also satisfies Property (ii).

Henceforth, we may assume $|V(T_k)| > c\sqrt{p}$. This implies that $p > (k-1)c\sqrt{p}$. We will now define a set $X' \subseteq V(T_k)$ satisfying the following properties.

(a) $|X'| \leq p - \alpha$,
(b) $T_k\langle X'\rangle$ has at least $p - \alpha - c\sqrt{p - \alpha}$ edges, and
(c) every non-trivial connected component in $T_k \setminus E(T_k\langle X'\rangle)$ contains r_k.

If $p - \alpha \leq 3$, then set $X' = \{r_k\}$. It is easy to check that Properties (a) to (c) hold. Otherwise, by the induction hypothesis applied to $p - \alpha$, T_k, and r_k, there exists $X' \subseteq V(T_k)$ satisfying Properties (a) to (c).

Let $X = X' \cup (\bigcup_{i=1}^{k-1} V(T_i))$. Note that $|X| = |X'| + \alpha \leq p - \alpha + \alpha \leq p$. So Property (i) is satisfied. Moreover, since r and r_k are adjacent, Property (iii) is also satisfied. It suffices to show Property (ii). Note that

$$|E(T\langle X\rangle)| \geq |E(T_k\langle X'\rangle)| + E(T\langle \bigcup_{i=1}^{k-1} V(T_i)\rangle)|$$

$$\geq p - \alpha - c\sqrt{p - \alpha} + \alpha - (k-1) = p - c\sqrt{p - \alpha} - (k-1)$$

Since $\alpha \geq \frac{p}{2}$ and $p > (k-1)c\sqrt{p}$, it follows that

$$|E(T\langle X\rangle)| \geq p - c\sqrt{\frac{p}{2}} - \frac{p}{c\sqrt{p}} = p - c\sqrt{p}\left(\frac{\sqrt{2}}{2} + \frac{1}{c^2}\right)$$

Since $c = \sqrt{2 + \sqrt{2}}$, it follows that $\frac{\sqrt{2}}{2} + \frac{1}{c^2} = \frac{\sqrt{2}}{2} + \frac{1}{2 + \sqrt{2}} = 1$. Thus, $|E(T\langle X\rangle)| \geq p - c\sqrt{p}$. This finishes the proof. $\qquad \square$

Theorem 7. *Let $p \geq 4$ be an integer. Then,* $\mathrm{conv}_{\mathcal{F}}^{\leq p}(n) \leq \left\lceil \frac{n-1}{p - c\sqrt{p}} \right\rceil$ *with $c = $* $\sqrt{2 + \sqrt{2}}$.

Proof. We prove the result by induction on n, the result holding trivially if $n \leq p$.

Assume now that $n \geq p$, and let T be a tree on n vertices. By Lemma 3, there exists a subset X of $V(T)$ which satisfies the properties (i)-(iii) of this lemma. Property (i) asserts that $|X| \leq p$, so it can be inverted. Now $\mathrm{Inv}(T; X)$ disagrees with the converse of T on $E(T) \setminus E(T\langle X\rangle)$. By Property (iii), this set of edges induces a forest with unique connected component T'. Hence $\mathrm{conv}^{\leq p}(T) \leq 1 + \mathrm{conv}^{\leq p}(T')$. But by Property (ii), $|E(T')| \leq |E(T)| - (p - c\sqrt{p})$. Thus, by the induction hypothesis, $\mathrm{conv}^{\leq p}(T') \leq \left\lceil \frac{n - 1 - (p - c\sqrt{p})}{p - c\sqrt{p}} \right\rceil = \left\lceil \frac{n-1}{p - c\sqrt{p}} \right\rceil - 1$. Hence $\mathrm{conv}^{\leq p}(T) \leq \left\lceil \frac{n-1}{p - c\sqrt{p}} \right\rceil$. $\qquad \square$

This theorem and Lemma 2 directly imply the assertion (iv) of Theorem 5.

5 k-Degenerate and Planar Graphs

A graph G is **k-degenerate** if it has a **k-degenerate ordering**, that is, an ordering $(v_1, \ldots, v_n)$ of $V(G)$ such that v_i has at most k neighbours in $\{v_{i+1}, \ldots, v_n\}$ for all $i \in [n-1]$. An easy induction using Corollary 2 along a k-degenerate ordering yields the following.

Corollary 3. *Let G be a k-degenerate graph. For any $p \geq k+1$, $\mathrm{id}^{\leq p}(G) \leq |V(G)| - 1$.*

This corollary is tight for $k = 2$ and $p = 3$ as for every positive integer n, there is a 2-degenerate graph G_n^2 of order n such that $\mathrm{id}^{\leq 3}(G_n^2) = n - 1$.

A planar graph of order n has at most $3n - 6$ edges. Thus, by Theorem 2, $\mathrm{id}_{\mathcal{P}}^{\leq p}(n) \leq \left\lceil \frac{3n-6}{\lfloor p/2 \rfloor} \right\rceil + \frac{1}{2}p^2$. Using a more careful analysis, one can slighty improve on it and show $\mathrm{id}_{\mathcal{P}}^{\leq p}(n) \leq \left\lceil \frac{3n-6}{\lfloor p/2 \rfloor} \right\rceil + 8\lfloor p/2 \rfloor - 8$. For small values of p, since every planar graph is 5-degenerate, better upper bounds are given by Eq. (2). We now show even stronger bounds.

Theorem 8. *If G is a planar graph of order n, then $\mathrm{id}^{\leq 3}(G) \leq \frac{11}{6}n - \frac{8}{3}$ and $\mathrm{id}^{\leq 5}(G) \leq \mathrm{id}^{\leq 4}(G) \leq \frac{4}{3}n + \frac{10}{3}$*

Proof. Let G be a planar graph and let $p \in \{3, 4\}$. Let $\overrightarrow{G}_1$ and $\overrightarrow{G}_2$ be two orientations of G. We apply the following procedure.

1. First, as long as there is a K_3 with its three edges in $E_{\neq}$, we invert its vertex set. This reverses its three edges (which are removed from $E_{\neq}$).
2. As long as there is a 4-cycle $(v_1, v_2, v_3, v_4, v_1)$ with all its edges in $E_{\neq}$, we invert the sets $\{v_1, v_2, v_3\}$ and $\{v_3, v_4, v_1\}$. This reverses the four edges of the 4-cycle (which are removed from $E_{\neq}$) and no other.
3. As long as there are two edges $xy, yz \in E_{\neq}$ such that $xz \notin E(G)$, then we invert $\{x, y, z\}$. This reverses the two edges xy, yz (which are removed from $E_{\neq}$) without adding any new edges in $E_{\neq}$.
3+. If $p = 4$, then as long as there are two edges $wx, yz \in E_{\neq}$ such that $wy, wz, xy, xz \notin E(G)$, then we invert $\{w, x, y, z\}$. This reverses the two edges wx, yz (which are removed from $E_{\neq}$) without adding any new edges in $E_{\neq}$.
4. Finally, we reverse the remaining edges of $E_{\neq}$ one by one.

At Step 1, three edges are reversed per inversions; at Step 2, 3, and 3+, two edges (in average) are reversed per inversions; at Step 4, one edge is reversed per inversion. For $i \in [4]$, let E_i be the set of edges reversed at Step i, and set $m_i = |E_i|$.

The number of inversions of our procedure is $N = m_1/3 + m_2/2 + m_3/2 + m_4$. We have $m_1 + m_2 + m_3 + m_4 = |E_{\neq}| \leq |E(G)| \leq 3n - 6$, because G is planar. Observe that after Step 1, the graph $(V(G), E_2 \cup E_3 \cup E_4)$ has no triangle and is planar. So it has at most $2n - 4$ edges. Hence $m_2 + m_3 + m_4 \leq 2n - 4$.

Consider now the graph $H = (V(G), E_4)$.

- It has no triangle, nor 4-cycle, because of Step 1 and 2.
- The closed neighbourhood in H of every vertex is a clique in G for otherwise Step 3 would apply. Thus, as G is planar and thus has no clique of size 5, we get that $\Delta(H) \leq 3$.
- It has no odd cycle, since every odd cycle of a planar graph contains two consecutive edges xy, yz such that $xz \notin E(G)$, which should have been reversed at Step 3.
- Let $C = (u_1, v_1, u_2, v_2, \ldots, u_k, v_k, u_1)$ be an even cycle. Because Step 3 did not apply, $(u_1, u_2, \ldots, u_k, u_1)$ and $(v_1, v_2, \ldots, v_k, v_1)$ are cycles in G. Moreover, since Step 1 did not apply, no edge of those cycles is in $E_2 \cup E_3 \cup E_4$. Without loss of generality, the cycle $C' = (u_1, u_2, \ldots, u_k, u_1)$ is outside C. Note that, in H, there is no path from C to a vertex outside C'. Indeed, such a path would go through an edge $u_i w$ with $u_i \in V(C')$ and w outside C'. But then $u_i v_i$, $u_i w$ are edges in E_4 and $v_i w$ is not an edge since u_i is inside C' and w is outside C'. This is impossible because such a pair of edges is reversed at Step 3.

Therefore, there is no path between two cycles in H. Thus every component J of H has at most one cycle and so $|E(J)| \leq |V(J)|$. Hence $m_4 = |E(H)| \leq |V(H)| = n$.

Now,

$$N = \frac{1}{3}m_1 + \frac{1}{2}(m_2 + m_3) + m_4$$
$$= \frac{1}{3}(m_1 + m_2 + m_3 + m_4) + \frac{1}{6}(m_2 + m_3 + m_4) + \frac{1}{2}m_4$$
$$\leq \frac{1}{3}(3n - 6) + \frac{1}{6}(2n - 4) + \frac{1}{2}n \leq \frac{11}{6}n - \frac{8}{3}$$

Thus $\mathrm{id}^{\leq 3}(G) \leq \frac{11}{6}n - \frac{8}{3}$.

Assume now that $p = 4$. Then Step 3+ is also performed, giving more structure on H. In fact, one can prove $m_4 = |E(H)| \leq 12$. With a similar computation as above, we derive $\mathrm{id}^{\leq 4}(G) \leq \frac{4}{3}n + \frac{10}{3}$. $\qquad\square$

Let us now establish some lower bounds on $\mathrm{conv}^{\leq p}(G)$ (and so $\mathrm{id}^{\leq p}(G)$ as well). Let G be a graph of order n with n_o vertices of odd degree. One can easily show that $\mathrm{conv}^{\leq 3}(G) \geq \frac{|E(G)|}{3} + \frac{n_o}{6}$. But, for any positive integer q, there exists a plane triangulation P on $n = 4q$ vertices in which all vertices have odd degree. Such a plane triangulation has $3n - 6$ edges, so by above equation we get $\mathrm{conv}_p^{\leq 3}(n) \geq \mathrm{conv}^{\leq 3}(T) \geq \frac{7n}{6} - 2$.

The **double wheel** of order n, denoted by DW_n is the graph obtained from a cycle of order $n - 2$ by adding two vertices adjacent to all vertices of the cycle. A double wheel is clearly planar. Moreover assigning a weight of $\frac{p-3}{(p-2)^2+1}$ to the edges of the cycle and of $\frac{1}{(p-2)^2+1}$ to the others, one easily checks that $\sum_{e \in E(DW_n\langle X\rangle)} w(e) \leq 1$ for every ($\leq p$)-set $X \subseteq V(DW_n)$ and $\sum_{e \in E(DW_n)} w(e) = \frac{p-1}{(p-2)^2+1} \cdot (n - 2)$. Thus $\mathrm{conv}_p^{\leq p}(n) \geq \mathrm{conv}^{(\leq p)}(DW_n) \geq \frac{p-1}{(p-2)^2+1} \cdot (n - 2)$.

References

1. Alon, N.: Ranking tournaments. SIAM J. Discret. Math. **20**(1), 137–142 (2006). https://doi.org/10.1137/050623905
2. Alon, N., Powierski, E., Savery, M., Scott, A., Wilmer, E.: Invertibility of digraphs and tournaments. SIAM J. Discrete Math. **38**(1), 327–347 (2024). https://doi.org/10.1137/23M1547135, arXiv:2212.11969
3. Arana, C., Bellitto, T., Buffière, H., Chuet, Q., Pierron, T., Reinald, A.: Inversion diameter and 2-edge-colored homomorphisms (2026). https://arxiv.org/abs/2602.24171
4. Aubian, G., et al.: Problems, proofs, and disproofs on the inversion number. arXiv preprint (2022). https://doi.org/10.48550/ARXIV.2212.09188, https://arxiv.org/abs/2212.09188, arXiv:2212.09188
5. Bang-Jensen, J., Gutin, G.Z.: Digraphs: Theory, Algorithms and Applications. Springer-Verlag, London (2009)
6. Bang-Jensen, J., Havet, F., Hörsch, F., Rambaud, C., Reinald, A., Silva, C.: Making an oriented graph acyclic using inversions of bounded or prescribed size. arXiv preprint arXiv:2511.22562 (2025). https://arxiv.org/pdf/2511.22562, arXiv:2511.22562
7. Bang-Jensen, J., da Silva, J.C.F., Havet, F.: On the inversion number of oriented graphs. Discrete Math. Theoret. Comput. Sci. **23**(2) (2021). https://doi.org/10.46298/DMTCS.7474, https://doi.org/10.46298/dmtcs.7474, arXiv:2105.04137
8. Belkhechine, H., Bouaziz, M., Boudabbous, I., Pouzet, M.: Inversion dans les tournois. Comptes Rendus Mathématique **348**(13-14), 703–707 (2010). arXiv:1007.2103
9. Bonamy, M., Perrett, T., Postle, L.: Colouring graphs with sparse neighbourhoods: Bounds and applications. J. Comb. Theory Ser. B **155**, 278–317 (2022)
10. Charbit, P., Thomassé, S., Yeo, A.: The minimum feedback arc set problem is NP-hard for tournaments. Comb. Probab. Comput. **16**(1), 1–4 (2007). https://doi.org/10.1017/S0963548306007887, https://doi.org/10.1017/S0963548306007887
11. Duron, J., Havet, F., Hörsch, F., Rambaud, C.: On the minimum number of inversions to make a digraph k-(arc-)strong. arXiv preprint, arXiv:2303.11719 (2023)
12. Fernandez de la Vega, W.: On the maximum cardinality of a consistent set of arcs in a random tournament. J. Comb. Theory Ser. B **35**(3), 328–332 (1983). https://doi.org/10.1016/0095-8956(83)90060-6, https://www.sciencedirect.com/science/article/pii/0095895683900606
13. Havet, F., Hörsch, F., Rambaud, C.: Diameter of the inversion graph. arXiv preprint (2024). https://arxiv.org/abs/2405.04119, arXiv:2405.04119
14. Havet, F., Rambaud, C., Silva, C.: On the $(\leq p)$-inversion diameter of oriented graphs (2026), https://arxiv.org/abs/2604.04633, arXiv:2604.04633
15. Hurley, E., de Joannis de Verclos, R., Kang, R.J.: An improved procedure for colouring graphs of bounded local density. Adv. Comb. (2022). https://doi.org/10.19086/aic.2022.7
16. Karp, R.M.: Reducibility among Combinatorial Problems, p. 85–103. Springer US (1972). https://doi.org/10.1007/978-1-4684-2001-2_9
17. Kotzig, A.: From the theory of finite regular graphs of degree three and four. Časopis Pestov. Mat **82**, 76–92 (1957)
18. Kóvari, T., Sós, V.T., Turán, P.: On a problem of K. Zarankiewicz. Colloquium Mathematicum **3**(1), 50–57 (1954). https://doi.org/10.4064/cm-3-1-50-57

19. Molloy, M., Reed, B.: A bound on the strong chromatic index of a graph. J. Comb. Theory Ser. B **69**(2), 103–109 (1997)
20. Spencer, J.: Optimal ranking of tournaments. Networks **1**(2), 135–138 (1971). https://doi.org/10.1002/net.3230010204, https://onlinelibrary.wiley.com/doi/abs/10.1002/net.3230010204
21. Spencer, J.: Optimally ranking unrankable tournaments. Periodica Mathematica Hungarica **11**(2), 131–144 (1980). https://doi.org/10.1007/bf02017965
22. Wang, Y., Wang, H., Yang, Y., Lu, M.: Inversion diameter and treewidth (2025). https://arxiv.org/abs/2407.15384
23. Yuster, R.: On tournament inversion. J. Graph Theory **110**(1), 82–91 (2025). https://doi.org/10.1002/jgt.23251, arXiv:2312.01910

Hardness of SetCover Reoptimization

Klaus Jansen[1(✉)] [iD], Tobias Mömke[2] [iD], and Björn Schumacher[1] [iD]

[1] Kiel University, Kiel, Germany
`{kj,bsch}@informatik.uni-kiel.de`
[2] University of Augsburg, Augsburg, Germany
`moemke@informatik.uni-augsburg.de`

Abstract. We study the hardness of reoptimization of the fundamental and hard to approximate SetCover problem. Reoptimization considers an instance together with a solution and a modified instance where the goal is to approximate the modified instance while utilizing the information gained from the solution to the related instance. We study four different types of reoptimization for (weighted) SetCover: adding a set, removing a set, adding an element to the universe, and removing an element from the universe. Adding (unweighted and weighted) and removing (unweighted) elements is known to be easier to approximate than the classic SetCover problem. We show that all other cases are essentially as hard to approximate as SetCover.

The reoptimization problem of adding and removing an element in the unweighted case is known to admit a PTAS. For these settings we show that there is no EPTAS under common hardness assumptions via a novel combination of the classic way to show that a reoptimization problem is NP-hard and the relation between EPTAS and FPT.

Keywords: Set cover · Reoptimization · Approximation

1 Introduction

We consider the SETCOVER Problem according to the following definition.

Definition 1 (wSETCOVER). *Given a universe U, a collection of sets $\mathcal{S} \subseteq \mathcal{P}(U)$ such that $\bigcup \mathcal{S} = U$, and a weight function $w : \mathcal{S} \to \mathbb{Q}_{\geq 0}$, find a set $S \subseteq \mathcal{S}$ covering the universe U with minimum weight, i.e. that minimizes $\sum_{s \in S} w(s)$. The unweighted version of the problem (SETCOVER) is defined in the same way but w is the constant one-function, i.e. the objective is to find a cover of U with as few sets from $\mathcal{S}$ as possible.*

The SETCOVER problem is a fundamental problem and part of Karp's 21 NP-complete problems. A simple algorithm (greedily choosing the set that covers the most uncovered elements) achieves an approximation ratio of $\mathcal{O}\left(\ln|U|\right)$ [10,11] and can be extended to the weighted case [4]. The exact approximation ratio for the greedy algorithm was analyzed in [18]. There are several inapproximability results for SETCOVER showing that the greedy algorithm is essentially

© The Author(s), under exclusive license to Springer Nature Switzerland AG 2026
F. Foucaud and A. Parreau (Eds.): IWOCA 2026, LNCS 16587, pp. 370–384, 2026.
https://doi.org/10.1007/978-3-032-27732-9_26

optimal [1,5,8,12,17]. The strongest known hardness result is due to Dinur and Steurer [5] and states that there is no polynomial time approximation algorithm for SETCOVER with approximation ratio $(1 - \epsilon) \ln|U|$ for any $\epsilon > 0$ unless P = NP. Regarding parameterized complexity, SETCOVER is known to be $W[2]$-complete [6] when parameterized by the number of sets in the solution but is fixed-parameter tractable when parameterized either by $|U|$ or $|\mathcal{S}|$.

As SETCOVER is not just a NP-hard problem but also hard to approximate, it is interesting to consider what additional information helps to obtain better approximation ratios. The extra information provided by reoptimization is a related instance and a solution with a certain quality. The instance to solve is obtained by applying a specified modification to the related instance. Typical modifications include adding or removing certain parts of the input (vertices, edges, items) or changing weights or profits.

Reoptimization of (W)SETCOVER was considered before in [2,14,20]. Biló, Widmayer, and Zych [2] considered the four following reoptimization settings for SETCOVER: adding a constant number of new elements to the U and add them to the sets in $\mathcal{S}$ in an arbitrary manner, removing a constant number of elements from some sets in $\mathcal{S}$, complete removal of a constant number of elements, i.e. from U and all sets in $\mathcal{S}$, and adding an element from U to a set of $\mathcal{S}$. They provide a general framework for reoptimization and show how it applies to these settings to obtain approximation algorithms. For the reoptimization variants which do not alter the universe U they provide approximation lower bounds: Adding an element from U to a set of $\mathcal{S}$ cannot be approximated better than $\ln|U|$ and for any constant $\epsilon > 0$ a $(1 + \rho - \epsilon)$-approximation[1] for removing a constant number of elements from some sets in $\mathcal{S}$ is not possible unless P = NP. Mikhailyuk [14] showed that there is a $(2 - \frac{1}{\ln|\mathcal{S}|+1})$-approximation algorithm for the reoptimization setting when up to $|\mathcal{S}|$ elements from U are added to or removed from a set in $\mathcal{S}$.

We study four different types of reoptimization for SETCOVER: adding a set (S^+), removing a set (S^-), adding an element to the universe (e^+), and removing an element from the universe (e^-). The precise definitions of these modifications are given in Sect. 2. The settings e^+ and e^- overlap with the settings studied in [2,20] but there are no approximation lower bounds given for these settings.

1.1   Related Work

In general reoptimization of NP-hard problems is NP-hard as described in [3]. The structure of the proof is roughly as follows: For every instance, start from some trivial (polynomial time solvable) instance and apply the local modification (and the reoptimization algorithm) until we obtain the original instance together with a solution. In the same way, strongly NP-hard problems remain strongly NP-hard when considering the reoptimization variant. This rules out FPTAS in many cases and the best to hope for is an EPTAS.

---

[1] ρ describes the quality of the given solution, e.g. $\rho = 1$ is optimal and $\rho = 2$ is a 2-approximation.

Table 1. Overview of known and new results. Known results have an accompanying citation and new results are marked by a reference to the corresponding proposition in the paper. Lower bounds refer to approximation lower bounds of polynomial time approximation algorithms. ρ is the quality of the provided solution.

	SETCOVER	wSETCOVER
S^+	$\frac{1}{2}\ln|U|$ lower bound (Corollary 1)	$\ln|U|$ lower bound (Corollary 4)
S^-	$\ln|U|$ lower bound (Corollary 2)	$\ln|U|$ lower bound (Corollary 6)
e^+	PTAS [2, Lemma 2]	$1 + \rho$ approximation [20, Conclusion 3]
	no EPTAS (Lemma 4)	$\frac{1}{2} + \rho$ lower bound (Lemma 7)
e^-	PTAS [2, Lemma 2]	$\ln|U|$ lower bound (Corollary 7)
	no EPTAS (Lemma 5)	

It is a well-known fact that a reoptimization problem where the optimum changes only by a constant admits a PTAS in many cases due to a simple case distinction: Either the optimum is small and we can find an optimal solution (for example by enumeration) in polynomial time or the optimum is large and the given optimal solution can easily be turned into a good-enough solution for the modified instance. See, for example, Lemma 2 in [20] or [2] for a general description of this result. If the considered problem is FPT parameterized by the size of the solution the first case can be sped up to obtain an EPTAS. For example, this is the case for the APX-complete problem VERTEXCOVER and the special case of SETCOVER where the pairwise intersections of sets in $\mathcal{S}$ is bounded by a constant Δ [16]. Interestingly, the even more specialized case of SETCOVER where the sizes of all sets are bounded by a constant Δ cannot be approximated within a factor of $\ln \Delta - \mathcal{O}(\ln \ln \Delta)$ unless $\mathsf{P} = \mathsf{NP}$ [19].

1.2   Results

An overview of known and our results can be found in Table 1. The approximation lower bounds all follow the same pattern: For a given SETCOVER instance we construct a reoptimization instance with a simple optimal solution such that the reoptimization algorithm has to effectively solve the original SETCOVER instance. Since we construct optimal solutions, every lower bound also holds for the weaker reoptimization setting where only an approximate solution of a certain quality is given.

The proofs to rule out an EPTAS are more involved. Similar to the approach described above to show NP-hardness of reoptimization problems, we start with an easy instance and use a presumed EPTAS at every step of the way to construct an FPT algorithm parameterized by the size of the solution. This is similar to the classic result that an EPTAS for an optimization problem implies an FPT algorithm parameterized by the size of the solution (cf. Theorem 1.32 in [9]). We show a general statement for this result in Lemma 3.

Outline of the Paper. We begin by introducing notation and define our reoptimization settings in Sect. 2. Then, we discuss our results for the unweighted and weighted case in Sects. 3 and 4 respectively. We conclude by discussing interesting questions for future work in Sect. 5.

2   Preliminaries

For any $n \in \mathbb{N}_0$ let $[n] := \{1, 2, \ldots, n\}$ and $[n]_0 := [n] \cup \{0\}$. With $\langle \cdot \rangle$ we denote the encoding length. For an instance $I \in \mathcal{I}$ of an optimization problem $\mathcal{I}$ we denote the set of valid solutions by $\mathrm{sol}(I)$ and the value of a solution $S \in \mathrm{sol}(I)$ by $\mathrm{val}(S)$. The optimal value of a solution for an instance I is denoted by $\mathrm{OPT}(I)$.

We define reoptimization problems in this paper as follows. Let $\mathcal{I}$ be an optimization problem with a minimization objective. The definition for a maximization objective is analogous. Let $\mathcal{M} \subseteq \mathcal{I} \times \mathcal{I}$ be a relation on the set of instance (we write $I \sim_{\mathcal{M}} I'$ if $(I, I') \in \mathcal{M}$), which formalizes the valid modifications. We write $(\mathcal{M}, \rho)$-$\mathcal{I}$ where $\mathcal{I}$ is the problem, $\mathcal{M}$ the modification, and ρ is the quality of the given solution. An instance of this problem is a triple (I, S, I') with $(I, I') \in \mathcal{M}$, $S \in \mathrm{sol}(I)$ and $\mathrm{val}(S) \leq \rho\mathrm{OPT}(I)$. If an optimal solution is given we use $\mathcal{M}$-$\mathcal{I}$ as a shorthand for $(\mathcal{M}, 1)$-$\mathcal{I}$.

Next, we precisely define the modifications we study for wSETCOVER. The definitions for SETCOVER are the same, only with the restriction of the weight function to the constant one-function. Let $(U, \mathcal{S}, w)$ be a wSETCOVER instance.

Adding a set (S^+) Given $s' \in \mathcal{P}(U) \setminus \mathcal{S}$ and a value $w_{s'} \in \mathbb{Q}_{\geq 0}$ the modified instance is $(U, \mathcal{S}' := \mathcal{S} \cup \{s'\}, w')$ where

$$
w' : \mathcal{S}' \to \mathbb{Q}_{\geq 0}, s \mapsto \begin{cases} w_{s'} & \text{if } s = s' \\ w(s) & \text{otherwise.} \end{cases}
$$

Removing a set (S^-) Given $s' \in \mathcal{S}$ the modified instance is $(U, \mathcal{S}' := \mathcal{S} \setminus \{s'\}, w_{|\mathcal{S}'})$.

Adding an element (e^+) Given $e \notin U$ and $S_e \subseteq \mathcal{S}$ the modified instance is $(U \cup \{e\}, \mathcal{S}' := (\mathcal{S} \setminus S_e) \cup \{s \cup \{e\} : s \in S_e\}, w')$ where $w' : \mathcal{S}' \to \mathbb{Q}_{\geq 0}, s \mapsto w(s \setminus \{e\})$.

Removing an element (e^-) Given $e \in U$ the modified instance is $(U \setminus \{e\}, \mathcal{S}' := \{s \setminus \{e\} : s \in \mathcal{S}\}, w')$ where

$$
w' : \mathcal{S}' \to \mathbb{Q}_{\geq 0}, s \mapsto \begin{cases} w(s) & \text{if } s \in \mathcal{S} \\ w(s \cup \{e\}) & \text{otherwise.} \end{cases}
$$

Note that we require that every instance of SETCOVER is feasible and thus both the initial and the modified instance have to be feasible, i.e. $\bigcup \mathcal{S} = U$. This is not a restriction as we can test $\bigcup \mathcal{S} = U$ in polynomial time.

3 Unweighted Set Cover

Changes that only change the optimum by a constant in the unweighted case include adding/removing a set with constant size and adding/removing a constant number of elements. Thus, all these cases admit a PTAS as described in Sect. 1.1.

3.1 S^+-SetCover

We write our inapproximability results as a function of the universe such that they are compatible with the inapproximability results for SetCover. Typically, f will be slight variation of $x \mapsto \ln x$, e.g. $x \mapsto \frac{1}{3} \ln x$. Note, that this notation still allows constant factor approximation by using a constant function.

To obtain a SetCover instance with an obvious optimal solution we will duplicate every element of the universe and for every element of the universe add a set to cover it and its duplicate. Thus, all the newly added sets form an optimal solution and adding a new set covering the duplicates exposes the original SetCover instance.

We always build our reoptimization instances in such a way that they have obvious optimal solutions and thus showing our claims for $\rho = 1$. But since an optimal solution is also an ρ-approximate solution for any $\rho > 1$ we show the claim for every $\rho \geq 1$.

Lemma 1. *Let $\rho \geq 1$. Let $f : \mathbb{N} \to \mathbb{Q}_{\geq 1}$. An $f(|U|)$-approximation algorithm for (S^+, ρ)-SetCover implies an $2f(2|U|)$-approximation algorithm for Set-Cover.*

Proof. Let $(U, \mathcal{S})$ be an SetCover instance. W.l.o.g. assume $|U| \geq 1$ and $\mathrm{OPT}((U, \mathcal{S})) < |U|$, i.e. there exists a set in $\mathcal{S}$ with size at least 2. Obtain U' by adding a copy of every element such that $|U'| = 2|U|$. For every $u \in U$ let u' be the copy added to U'. Next, define

$$\mathcal{S}' := \mathcal{S} \cup \{\{u, u'\} : u \in U\}.$$

The optimal solution to this instance of SetCover is $S^* := \mathcal{S}' \setminus \mathcal{S} = \{\{u, u'\} : u \in U\}$. The construction is depicted in Fig. 1.

Now, consider the (S^+, ρ)-SetCover instance with instance $(U', \mathcal{S}')$, (optimal) solution S^*, and modified instance $(U', \mathcal{S}' \cup \{U' \setminus U\})$. Assume a solution of the modified instance contains a set of the form $\{u, u'\}$ then there are effectively two cases: Either the solution only contains sets of this form or it contains the set $U' \setminus U$. In both cases we can obtain a solution with smaller or equal value that does not contain any sets of the form $\{u, u'\}$: For the first case notice that all elements from U can easily be covered by at most $|U| - 1$ sets and we additionally add the set $U' \setminus U$. For the second notice that $U' \setminus U$ is already in the solution meaning all elements from U' are already covered. Thus, we can simply exchange every set with the form $\{u, u'\}$ with a set from $\mathcal{S}$ covering u while not increasing the size of the solution. Therefore, we may assume

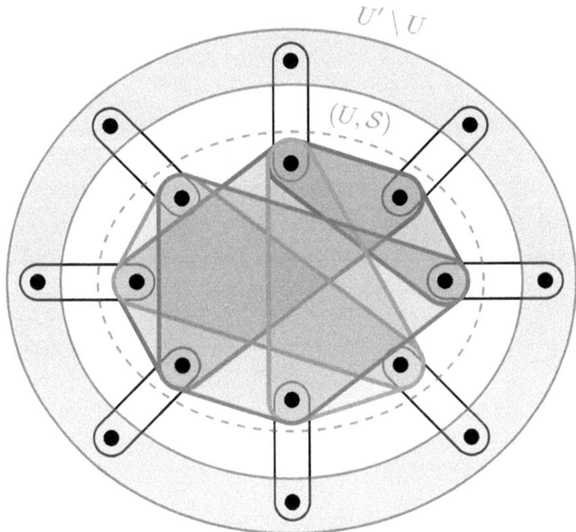

Fig. 1. Construction in Lemma 1. The colorful center is the original instance, the outer dots are the duplicates, and the black sets are the sets that cover every element together with its duplicate. The gray set is added as the local modification to cover all duplicates.

that no solution contains sets of the form $\{u, u'\}$. A direct consequence is that $\mathrm{OPT}((U', \mathcal{S}' \cup \{U' \setminus U\})) = \mathrm{OPT}((U, \mathcal{S})) + 1$.

We apply the presumed $f(|U|)$-approximation to the reoptimization instance and obtain a solution S such that $\{U' \setminus U\} \subseteq S \subseteq \mathcal{S} \cup \{U' \setminus U\}$ as discussed before. Notice that $S' := S \setminus \{U' \setminus U\}$ is a solution for the original instance. Hence, the following shows the claim

$$|S'| = |S| - 1 \le f(|U'|)\, \mathrm{OPT}((U', \mathcal{S}' \cup \{U' \setminus U\})) - 1$$
$$= f(2|U|)(\mathrm{OPT}((U, \mathcal{S})) + 1) - 1 \le 2f(2|U|)\, \mathrm{OPT}((U, \mathcal{S})). \qquad \square$$

Remark 1. The proof also shows that an exact algorithm for (S^+, ρ)-SETCOVER implies an exact algorithm for SETCOVER. This can be seen by analyzing the last inequality chain when $f = 1$.

For every $\alpha > 0$, Lemma 1 shows that an $\alpha \ln|U|$ approximation for (S^+, ρ)-SETCOVER implies a $2\alpha \ln(2|U|)$ approximation algorithm for SETCOVER. Let $\alpha \in (0, \frac{1}{2})$ and $|U| > \exp(\frac{2\alpha \ln 2}{(1-2\alpha)^2})$ which implies $2\alpha \ln 2 < (1 - 2\alpha)^2 \ln|U|$. Thus, we have

$$2\alpha \ln(2|U|) = 2\alpha \ln(2) + 2\alpha \ln|U| < (1-2\alpha)^2 \ln|U| + 2\alpha \ln|U| = (1 - 2\alpha + 4\alpha^2) \ln|U|$$

and $1 - 2\alpha + 4\alpha^2 \in (0, 1)$ for every $\alpha \in (0, \frac{1}{2})$. Combining this with the result that for every $\epsilon > 0$ there is polynomial time $(1 - \epsilon) \ln|U|$-approximation for SETCOVER unless $\mathsf{P} = \mathsf{NP}$ [5], yields the following corollary.

Corollary 1. *For any $\alpha \in (0, \frac{1}{2})$ and $\rho \ge 1$ there is no $\alpha \ln|U|$-approximation algorithm for (S^+, ρ)-SETCOVER unless $\mathsf{P} = \mathsf{NP}$.*

3.2 S^--SETCOVER

Lemma 2. *Let $\rho \geq 1$. Let $f : \mathbb{N} \to \mathbb{Q}_{\geq 1}$. An $f(|U|)$-approximation algorithm for (S^-, ρ)-SETCOVER implies an $f(|U|)$-approximation algorithm for SETCOVER.*

Proof. Given any SETCOVER instance. W.l.o.g. assume that there is no set covering the entire universe. We add a new set containing the entire universe to the sets. The newly added set is an optimal solution. Thus, we have constructed an (S^+, ρ)-SETCOVER instance (removing the newly added set) which has to exactly solve the original instance. Therefore, we obtain an $f(|U|)$-approximation for SETCOVER. $\qquad\qquad\square$

This results implies that the same inapproximability result that hold for SETCOVER also hold for S^--SETCOVER. Thus, the known $\mathcal{O}(\ln|U|)$-approximation is the best we can hope for under the assumption $\mathsf{P} \neq \mathsf{NP}$. Formally stated in the following corollary.

Corollary 2. *For any ϵ and $\rho \geq 1$ there is no $(1-\epsilon)\ln|U|$-approximation algorithm for (S^-, ρ)-SETCOVER unless $\mathsf{P} = \mathsf{NP}$.*

3.3 e^+-SETCOVER

The existence of a PTAS for e^+-SETCOVER was already discussed earlier. Here we focus on showing that the existence of an EPTAS is unlikely. First, we show the general framework and apply it afterwards to e^+-SETCOVER by giving an appropriate construction.

Similar to the way NP-hardness is shown for reoptimization problems (as described in Sect. 1.1) we can rule out the existence of an EPTAS under certain conditions. Lemma 3 shows how an EPTAS for the reoptimization problem can be used to build an FPTalgorithm for the original optimization problem.

Lemma 3 (EPTAS implies Parameterized Algorithm). *Let $\mathcal{I}$ be an optimization problem with integral solution values and $\mathcal{M} \subseteq \mathcal{I} \times \mathcal{I}$ a modification. Given an EPTAS for the reoptimization problem $\mathcal{M}\text{-}\mathcal{I}$, computable functions $f, f' : \mathbb{N} \to \mathbb{N}$, and for every $I \in \mathcal{I}$ we can find $I_0, I_1, \ldots, I_n \in \mathcal{I}$ (in $\mathrm{poly}(\langle I \rangle)$ time) such that for every $k \in \mathbb{N}$ we have*

- *if $\mathrm{OPT}(I) \leq k$ then $\mathrm{OPT}(I_i) \leq f(k)$ for all $i \in [n]_0$,*
- *we can find an optimal solution to I_0 or decide that $\mathrm{OPT}(I_0) > f(k)$ in time $f'(k)\,\mathrm{poly}(\langle I \rangle)$,*
- *$I_{i-1} \sim_{\mathcal{M}} I_i$ for all $i \in [n]$, and*
- *$\mathrm{OPT}(I_n) \leq f(k) \implies \mathrm{OPT}(I) \leq k$*

then $\mathcal{I}$ is fixed-parameter tractable parameterized by the solution size.

Proof. Suppose we have an EPTAS for $\mathcal{M}\text{-}\mathcal{I}$ with running time $g(\frac{1}{\epsilon})\,\mathrm{poly}(\langle I \rangle)$ for some computable function g. We construct an algorithm for $\mathcal{I}$ that given an instance $I \in \mathcal{I}$ and a $k \in \mathbb{N}$ either computes an optimal solution or decides that

$\mathrm{OPT}(I) > k$ with running time $f''(k)\,\mathrm{poly}(\langle I \rangle)$ for a computable function f''. This implies that $\mathcal{I}$ is fixed-parameter tractable parameterized by solution size.

Let $k \in \mathbb{N}$ and $I \in \mathcal{I}$ an arbitrary instance. First we calculate an optimal solution S_0 to I_0 or decide that $\mathrm{OPT}(I_0) > f(k)$ in time $f'(k)\,\mathrm{poly}(\langle I \rangle)$. If $\mathrm{OPT}(I_0) > f(k)$ we know that $\mathrm{OPT}(I) > k$ and we are done.

Let $\epsilon := \frac{1}{f(k)+1}$. Let $\mathcal{A}_\epsilon(I, S, I')$ be the solution that the EPTAS calculates for the instance I' given accuracy ϵ. Let $S_i := \mathcal{A}_\epsilon(I_{i-1}, S_{i-1}, I_i)$ for all $i \in [n]$. We have

$$|\mathrm{val}(S_i) - \mathrm{OPT}(I_i)| \leq \epsilon\mathrm{OPT}(I_i) = \frac{\mathrm{OPT}(I_i)}{f(k)+1} < 1$$

for any $i \in [n]$ assuming $\mathrm{OPT}(I) \leq k$. Thus, the EPTAS calculates an optimal solution in every step. Therefore, we can decide wether $\mathrm{OPT}(I) \leq k$ (due to $\mathrm{OPT}(I_n) \leq f(k) \implies \mathrm{OPT}(I) \leq k$), in time $f''(k)\,\mathrm{poly}(\langle I \rangle)$ where

$$f'' : \mathbb{N} \to \mathbb{N}, k \mapsto f'(k) + g(f(k) + 1)$$

which is a computable function. $\qquad\qquad\qquad\qquad\qquad\qquad\qquad\qquad\square$

Since Lemma 3 shows the existence of a fixed-parameter tractable algorithm for the original problem when the reoptimization problem admits an EPTAS, it allows to conditionally rule out an EPTAS when the problem is $W[t]$-hard for some $t \geq 1$ and the preconditions for Lemma 3 are fulfilled. We state this formally in the following corollary.

Corollary 3. *Let $t \in \mathbb{N}_{\geq 1}$. Given a $W[t]$-hard problem (parameterized by the solution size) with integral solution values, a reoptimization variant of the problem, and a construction that fulfills the preconditions of Lemma 3, then there is no EPTAS for the reoptimization problem, unless $W[t] = \mathsf{FPT}$.*

$W[t] \neq \mathsf{FPT}$ for any $t \geq 1$ is classic assumption in parameterized complexity. Note that $W[t] = \mathsf{FPT}$ implies that the ETH fails (cf. Theorem 29.4.1 in [7]).

Now we give the construction for e^+-SETCOVER to apply Corollary 3.

Lemma 4. *There is no EPTAS for e^+-SETCOVER, unless $\mathsf{FPT} = W[2]$.*

Proof. Let $I = (U, \mathcal{S})$ be an arbitrary SETCOVER instance and let $m := |\mathcal{S}|$ and $\mathcal{S} = \{s_1, \ldots, s_m\}$. We take $m+1$ fresh elements $e_1, \ldots, e_{m+1} \notin U$ and define the new instance $I' = (U', \mathcal{S}')$.

$$U' := U \sqcup \{e_1, \ldots, e_{m+1}\} \qquad \mathcal{S}' := \{s_i \sqcup \{e_i\} : i \in [m]\} \sqcup \{e_1, \ldots, e_{m+1}\}$$

We have $\mathrm{OPT}(I) + 1 = \mathrm{OPT}(I')$, as a solution has to always contain the set $\{e_1, \ldots, e_{m+1}\}$. Next consider the instance $I_0 = (U'', \mathcal{S}'')$ with

$$U'' := \{e_1, \ldots, e_{m+1}\} \qquad \mathcal{S}'' := \{\{e_i\} : i \in [m]\} \sqcup \{e_1, \ldots, e_{m+1}\}$$

for which $\{e_1, \ldots, e_{m+1}\}$ is the optimal solution. Now we define instances $I_1, \ldots, I_{|U|}$ by adding the elements of $|U|$ one by one. Due to the introduction of the elements e_i this construction guarantees that all sets are different at all points and we have $I_{|U|} = I'$. Further, we have $\text{OPT}(I_i) \le \text{OPT}(I) + 1$ for all $i \in [|U|]_0$.

This sequence of instances fulfills the conditions of Lemma 3 and thus there is no EPTAS for e^+-SETCOVER unless $\mathsf{FPT} = W[2]$ by Corollary 3 as SETCOVER is $W[2]$-complete. $\qquad\square$

3.4 e^--SETCOVER

Similar to the previous section on e^+-SETCOVER we only rule out an EPTAS for e^--SETCOVER under common assumptions. In this case we do not have a direct application of Corollary 3 but the proof structure remains similar. The challenge in this case is to find a bigger instance where optimum does not increase arbitrarily but still has an obvious optimal solution. To cope with this issue we instead show that a different but closely related problem is fixed-parameter tractable when an EPTAS exists. The concrete problem is DOMINATINGSET in unit disk graphs which has a PTAS [15] but is still $W[1]$-hard parameterized by the solution size [13]. In the proof we use the fact that a DOMINATINGSET instance can always be viewed as a SETCOVER instance and we use the PTAS to cope with the aforementioned issue. We start with an approximate solution to a DOMINATINGSET instance and add elements to make this solution an optimal solution. Next, we remove the added elements until we reach the original instance.

Lemma 5. *There is no EPTAS for e^--SETCOVER, unless $\mathsf{FPT} = W[1]$.*

Proof. Suppose we have an EPTAS for e^--SETCOVER.

Let $(G = (V, E), k)$ be a parameterized instance of DOMINATINGSET in unit disk graphs. W.l.o.g. assume that there are no vertices $u, v \in V$ with $N(u) \cup \{u\} = N(v) \cup \{v\}$. We apply the PTAS [15] as a 2-approximation and obtain a solution $S \subseteq V$. If $|S| > 2k$ we know that $\text{OPT} > k$ and there is no solution of size at most k. Otherwise, we build a SETCOVER Instance as follows. We set $U := V$ and

$$\mathcal{S} := \{N(v) \cup \{v\} : v \in V\}.$$

Note that due to our assumption above, the solutions to the DOMINATINGSET instance and the constructed SETCOVER instance are in a one-to-one relation that preserves the number of elements in a solution. Thus, it suffices to find a solution in the constructed instance with value at most k or decide that $\text{OPT}((U, \mathcal{S})) > k$. To do this we use the presumed EPTAS.

First we add $|S| \le 2k$ new elements to fix the solution S and make it the only optimal solution. For every $v \in S$ we add a v' to the universe and add it to $N(v) \cup \{v\} \in \mathcal{S}$. Call the resulting instance I'. The sets corresponding to the elements from S are the optimal solution for I' since we need to cover the newly added elements and the sets already cover the original instance. Now we remove the added elements one-by-one until we have a solution for $(U, \mathcal{S})$ as in the proof of Lemma 3. This works because the optimum in every step is bounded by $2k$ and we use the presumed EPTAS with accuracy $\frac{1}{2k+1}$.

Therefore, we showed that DOMINATINGSET in unit disk graphs is fixed-parameter tractable parameterized by solution size which implies $\mathsf{FPT} = W[1]$ since the problem is $W[1]$-hard [13]. $\square$

4   Weighted Set Cover

4.1 S^+-wSETCOVER

In the weighted case we can sharpen the results obtained in Lemma 1 by removing the factor of two before f which allows us to improve the inapproximability bound compared to Corollary 1.

Lemma 6. *Let $\rho \geq 1$. Let $f : \mathbb{N} \to \mathbb{Q}_{\geq 1}$. An $f(|U|)$-approximation algorithm for S^+-wSETCOVER implies an $f(2|U|)$-approximation algorithm for SETCOVER.*

Proof (Sketch). The proof is essentially the same as the proof of Lemma 1 where all sets have weight 1 except for the set $U' \setminus U$ which gets weight 0. Thus, the optimum of the original instance and reoptimization instance is the same yielding the sharper result. $\square$

Corollary 4. *For any $\epsilon \in (0,1)$ and $\rho \geq 1$ there is no $(1 - \epsilon)\ln|U|$-approximation algorithm for (S^+, ρ)-wSETCOVER unless $\mathsf{P} = \mathsf{NP}$.*

Proof (Sketch). Same arguments as used for Corollary 1. $\square$

4.2 S^--wSETCOVER

We obtain the following result as a straightforward corollary from Lemma 2.

Corollary 5. *Let $\rho \geq 1$. Let $f : \mathbb{N} \to \mathbb{Q}_{\geq 1}$. An $f(|U|)$-approximation algorithm for (S^-, ρ)-wSETCOVER implies an $f(|U|)$-approximation algorithm for SETCOVER.*

Corollary 6. *For any ϵ and $\rho \geq 1$ there is no $(1 - \epsilon)\ln|U|$-approximation algorithm for (S^-, ρ)-wSETCOVER unless $\mathsf{P} = \mathsf{NP}$.*

4.3 e^+-wSETCOVER

(e^+, ρ)-wSETCOVER can be approximated in polynomial with a ratio of $1 + \rho$. This follows from Conclusion 3 in [20] where they showed this for the more general reoptimization variant where a constant number of elements are added to $|U|$ and arbitrarily to sets in $\mathcal{S}$.

We show that an approximation algorithm for (e^+, ρ)-wSETCOVER with approximation ratio below $\frac{1}{2} + \rho$ implies a constant factor approximation for SETCOVER for every $\rho \geq 1$. The main idea is to add an extra set that covers the entire universe and is expensive but still an optimal solution. Then, we add an element to the instance but not to the expensive set. We show that any solution taking the expensive set weighs at least $\frac{1}{2} + \rho$ times the optimum and thus an approximation algorithm for (e^+, ρ)-wSETCOVER with ration smaller $\frac{1}{2} + \rho$ cannot choose this set. Furthermore, for the algorithm to achieve the approximation ratio of $\frac{1}{2} + \rho$, it has to find an approximate solution to the original SETCOVER instance within a ratio of 2ρ.

Lemma 7. *Let $\rho \geq 1$. (e^+, ρ)-WSETCOVER cannot be approximated within a factor smaller than $\frac{1}{2} + \rho$ in polynomial time, unless* $\mathsf{P} = \mathsf{NP}$.

Proof. Assume we have an approximation algorithm for (e^+, ρ)-WSETCOVER with approximation ratio below $\frac{1}{2} + \rho$.

Let $I = (U, \mathcal{S})$ be an instance of SETCOVER. W.l.o.g. assume that there are no singleton sets. If there is an element of the universe only covered by a singleton set we know that this singleton set is always in the solution and we can remove it and the corresponding element from the instance. Adding them back later to the instance and the singleton set to a solution only improves the approximation guarantee. If there are other singleton sets we may remove them because any solution containing singleton sets can easily be modified to only contain non-singleton sets and not increasing the size of the solution.

Next we add a singleton set for all elements and obtain the instance $I' = (U, \mathcal{S}')$ such that $\mathrm{OPT}(I') = \mathrm{OPT}(I)$ and $|\mathcal{S}'| \geq 2\mathrm{OPT}(I)$ as there is an optimal solution with no singleton sets and $|U| \geq \mathrm{OPT}(I)$. This will be important for the analysis later on.

We may assume that we know $\mathrm{OPT}(I')$ because we do the following construction for each value in $[|U|]$ and output the best valid solution we obtain. Thus, we only analyze the construction for the right guess yielding an upper bound on the output of the constructed algorithm.

We build a WSETCOVER instance I'' as follows. For every $s \in \mathcal{S}'$ we introduce a new element e_s to the universe. Call the new universe $U'' = U \cup \{e_S : S \in \mathcal{S}'\}$. The sets are

$$\mathcal{S}'' := \{s \cup \{e_s\} : s \in \mathcal{S}'\} \cup \{G, R\}$$

where $G := U''$ and $R := \{e_s : s \in \mathcal{S}'\}$ with weight function

$$w : \mathcal{S}'' \to \mathbb{Q}_{\geq 0}, s \mapsto \begin{cases} 2\rho\mathrm{OPT}((U, \mathcal{S}')) & \text{if } s = G \\ \mathrm{OPT}((U, \mathcal{S}')) & \text{if } s = R \\ 1 & \text{otherwise.} \end{cases}$$

The construction is depicted in Fig. 2. There are three candidates for an optimal solution of $I'' = (U'', \mathcal{S}'', w)$: First, just the set G with weight $2\rho\mathrm{OPT}((U, \mathcal{S}')) \geq 2\mathrm{OPT}((U, \mathcal{S}'))$, then the set R and an optimal solution for the instance $(U, \mathcal{S}')$ with weight $2\mathrm{OPT}((U, \mathcal{S}'))$, and all sets except for G and R with weight at least $2\mathrm{OPT}((U, \mathcal{S}'))$ due to the introduction of singleton sets above. Thus, G is a solution with $w(G) = 2\rho\mathrm{OPT}((U, \mathcal{S}')) \leq \rho\mathrm{OPT}(I'')$.

The local modification is the addition of new element e_{new} that is only added to the set R. Call this instance I^*. We apply our presumed algorithm on the instance $(I'', \{G\}, I^*)$ and obtain a solution S. Every solution for I^* has to include the modified version of R which means all elements e_s are covered. By combining an optimal solution for $(U, \mathcal{S}')$ and the modified version of R we obtain an optimal solution with weight $\mathrm{OPT}((U, \mathcal{S}')) + w(R) = 2\mathrm{OPT}((U, \mathcal{S}'))$. Since every solution has to contain the set R, a solution containing G has value at least $w(R) + w(G) = 1\mathrm{OPT}((U, \mathcal{S}')) + 2\rho\mathrm{OPT}((U, \mathcal{S}')) = (1 + 2\rho)\mathrm{OPT}((U, \mathcal{S}'))$.

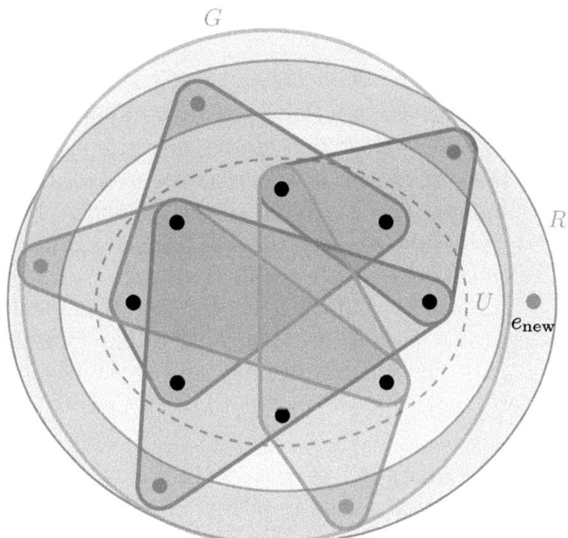

Fig. 2. Construction in Lemma 7. The black dots represent the original elements of the SETCOVER instance. The colorful sets are the sets of the original instance but each extended with a correspondingly colored element. The extra sets G and R are the brown and gray sets respectably. The added element is labeled e_{new}.

Our presumed algorithm has an approximation ratio of less than $\frac{1}{2} + \rho$ and thus the solution S cannot contain the set G. As the weight of every sets from $\mathcal{S}'$ is one, the number of sets from $\mathcal{S}'$ in the solution S is

$$\text{val}(S) - w(R) < \left(\frac{1}{2} + \rho\right) \text{OPT}(I^*) - w(R)$$

$$= (1 + 2\rho)\text{OPT}((U, \mathcal{S}')) - \text{OPT}((U, \mathcal{S}'))$$

$$= 2\rho\text{OPT}((U, \mathcal{S}')).$$

Thus, we have constructed a 2ρ-approximation for SETCOVER.   □

To improve the approximation lower bound of this lemma while using the same approach, we need a set of SETCOVER instances that cannot be approximated within a constant factor and that there is a $c > 2$ such that $|\mathcal{S}| \geq c\text{OPT}((U, \mathcal{S}))$ holds for every instance. Then, we can adjust the weights of G and R to $w(G) = c\rho\text{OPT}((U, \mathcal{S}'))$ and $w(R) = (c - 1)\text{OPT}((U, \mathcal{S}'))$ which allows us to rule out approximation ratios smaller than $\frac{c-1}{c} + \rho$ for (e^+, ρ)-wSETCOVER.

4.4 e^--WSETCOVER

Lemma 8. *Let $\rho \geq 1$. Let $f : \mathbb{N} \to \mathbb{Q}_{\geq 1}$. An $f(|U|)$-approximation algorithm for (e^-, ρ)-WSETCOVER yields an $f(|\bar{U}|)$-approximation algorithm for WSETCOVER.*

Proof. Given an instance $(U, \mathcal{S}, w)$ of WSETCOVER we construct an instance of e^--WSETCOVER as follows. Let $u' \notin U$ be a fresh element. The universe for the new instance is $U' := U \cup \{u'\}$, we keep the sets from $\mathcal{S}$ and add a single new set which contains all elements of U' with weight $W := (f(|U|)+1)\sum_{S \in \mathcal{S}} w(S)$. The newly added set is an optimal solution as it contains all elements from U and is the only sets that contains u'. Now using an $f(|U|)$-approximation algorithm on the constructed (e^-, ρ)-WSETCOVER instance solves the original instance with approximation factor $f(|U|)$ since $W > f(|U|)\mathrm{OPT}((U, \mathcal{S}, w))$. $\qquad\square$

Corollary 7. *For any ϵ and $\rho \geq 1$ there is no $(1 - \epsilon)\ln|U|$-approximation algorithm for (e^-, ρ)-WSETCOVER unless $\mathsf{P} = \mathsf{NP}$.*

Note that we construct an algorithm in the proof of Lemma 8 that evaluates $f(|U|)$ or at least needs to calculate an approximation of $f(|U|)$. If this is not possible in polynomial time we do not construct a polynomial time algorithm. This is not a problem for Corollary 7 since we only need to calculate the natural logarithm.

5 Conclusion

For all the reoptimization settings we discussed we essentially have matching approximation upper and lower bounds, except for (e^+, ρ)-WSETCOVER we where have an upper bound of $1 + \rho$ but only a lower bound of $\frac{1}{2} + \rho$. It is an interesting question what the best possible approximation ratio is for this problem. Since it seems promising to improve the construction in Lemma 7 it is probably not that close to $\frac{1}{2} + \rho$.

Other perspectives to consider include bounded set sizes and bounded element frequency which have better approximation guarantees but still strong hardness results. When the set sizes are bounded by a constant we obtain an EPTAS in cases we considered for SETCOVER as discussed in Sect. 1.1. It would be interesting to consider this restriction in the weighted case since most of our hardness construction use sets that are as big as or almost as big as U. For bounded set sizes as well as bounded frequencies the results in [2, 20] yield better approximation guarantees compared to the general case.

Acknowledgments. Klaus Jansen and Björn Schumacher were supported by the Deutsche Forschungsgemeinschaft (DFG, German Research Foundation) – Project number 528381760. We thank the anonymous referees for their insightful feedback and helpful comments.

Disclosure of Interests. The authors have no competing interests to declare that are relevant to the content of this article.

References

1. Alon, N., Moshkovitz, D., Safra, S.: Algorithmic construction of sets for K-Restrictions. ACM Trans. Algorithms **2**(2), 153–177 (2006). https://doi.org/10.1145/1150334.1150336
2. Bilò, D., Widmayer, P., Zych, A.: Reoptimization of weighted graph and covering problems. In: Bampis, E., Skutella, M. (eds.) WAOA 2008. LNCS, vol. 5426, pp. 201–213. Springer, Heidelberg (2009). https://doi.org/10.1007/978-3-540-93980-1_16
3. Böckenhauer, H.-J., Hromkovič, J., Mömke, T., Widmayer, P.: On the hardness of reoptimization. In: Geffert, V., Karhumäki, J., Bertoni, A., Preneel, B., Návrat, P., Bieliková, M. (eds.) SOFSEM 2008. LNCS, vol. 4910, pp. 50–65. Springer, Heidelberg (2008). https://doi.org/10.1007/978-3-540-77566-9_5
4. Chvátal, V.: A greedy heuristic for the set-covering problem. Math. Oper. Res. **4**(3), 233–235 (1979). https://doi.org/10.1287/MOOR.4.3.233
5. Dinur, I., Steurer, D.: Analytical approach to parallel repetition. In: Shmoys, D.B., Symposium on Theory of Computing, STOC 2014, New York, NY, USA, May 31 - June 03, 2014, pp. 624–633. ACM (2014). https://doi.org/10.1145/2591796.2591884
6. Downey, R.G., Fellows, M.R.: Fixed-Parameter tractability and completeness I: basic results. SIAM J. Comput. **24**(4), 873–921 (1995). https://doi.org/10.1137/S0097539792228228
7. Fundamentals of Parameterized Complexity. TCS, Springer, London (2013). https://doi.org/10.1007/978-1-4471-5559-1_35
8. Feige, U.: A threshold of ln n for approximating set cover. J. ACM **45**(4), 634–652 (1998). https://doi.org/10.1145/285055.285059
9. Parameterized Complexity Theory. TTCSAES, Springer, Heidelberg (2006). https://doi.org/10.1007/3-540-29953-X
10. Johnson, D.S.: Approximation algorithms for combinatorial problems. J. Comput. Syst. Sci. **9**(3), 256–278 (1974). https://doi.org/10.1016/S0022-0000(74)80044-9
11. Lovász, L.: On the ratio of optimal integral and fractional covers. Discret. Math. **13**(4), 383–390 (1975). https://doi.org/10.1016/0012-365X(75)90058-8
12. Lund, C., Yannakakis, M.: On the hardness of approximating minimization problems. J. ACM **41**(5), 960–981 (1994). https://doi.org/10.1145/185675.306789
13. Marx, D.: Parameterized complexity of independence and domination on geometric graphs. In: Bodlaender, H.L., Langston, M.A. (eds.) IWPEC 2006. LNCS, vol. 4169, pp. 154–165. Springer, Heidelberg (2006). https://doi.org/10.1007/11847250_14
14. Mikhailyuk, V.: Reoptimization of set covering problems. Cybern. Syst. Anal. **46**(6), 879–883 (2010)
15. Nieberg, T., Hurink, J.: A PTAS for the minimum dominating set problem in unit disk graphs. In: Erlebach, T., Persiano, G. (eds.) WAOA 2005. LNCS, vol. 3879, pp. 296–306. Springer, Heidelberg (2006). https://doi.org/10.1007/11671411_23
16. Raman, V., Saurabh, S.: Short cycles make W -hard problems hard: FPT algorithms for W -hard problems in graphs with no short cycles. Algorithmica **52**(2), 203–225 (2008). https://doi.org/10.1007/S00453-007-9148-9
17. Raz, R., Safra, S.: A sub-constant error-probability low-degree test, and a sub-constant error-probability PCP characterization of NP. In: Leighton, F.T., Shor, P.W., Proceedings of the Twenty-Ninth Annual ACM Symposium on the Theory of Computing, El Paso, Texas, USA, May 4–6, 1997, pp. 475–484. ACM (1997). https://doi.org/10.1145/258533.258641

18. Slavík, P.: A tight analysis of the greedy algorithm for set cover. J. Algorithms **25**(2), 237–254 (1997). https://doi.org/10.1006/JAGM.1997.0887
19. Trevisan, L.: Non-approximability results for optimization problems on bounded degree instances. In: Proceedings of the Thirty-Third Annual ACM Symposium on Theory of Computing. STOC '01, pp. 453–461. Association for Computing Machinery, Hersonissos, Greece (2001). https://doi.org/10.1145/380752.380839
20. Zych, A.: Reoptimization of NP-hard Problems, ETH Zurich, Zürich, Switzerland. PhD thesis (2012). https://doi.org/10.3929/ETHZ-A-007161496

Conflict-Free Cuts in Planar and 3-Degenerate Graphs with 1-Regular Conflicts

Subrahmanyam Kalyanasundaram[(✉)] and Subodh Kumar

Department of Computer Science and Engineering, IIT Hyderabad, Hyderabad, India
{subruk,cs23resch11009}@iith.ac.in

Abstract. A conflict-free cut F on a simple connected graph $G = (V, E)$ is defined as a set of edges $F \subseteq E$ such that $G - F$ is disconnected, and no two edges in F are conflicting. The notion of conflicting edges is represented using an associated conflict graph $\widehat{G} = (\widehat{V}, \widehat{E})$ where $\widehat{V} = E$. Deciding if a given planar graph G, with an associated conflict graph $\widehat{G}$, has a conflict-free cut is known to be NP-complete, when G has maximum degree four and $\widehat{G}$ is a line graph of G [Bonsma, JGT 2009].

In this paper, we prove the following for the case when $\widehat{G}$ is 1-regular.
- We completely resolve the complexity of the decision problem when G is planar. Towards this end, we show that (a) there always exists a conflict-free cut when the graph is planar and 4-regular unless it is the octahedron graph and (b) the decision problem is NP-complete, even in the case when G is planar with maximum degree 5.
- We also show that the decision problem is NP-complete when G is a 3-degenerate graph with maximum degree 5. This completely resolves the complexity status of the problem when G is 3-degenerate.
- We construct families of graphs with 1-regular conflict graphs that do not have a conflict-free cut.

Our results answer the questions posed in [Rauch, Rautenbach and Souza, IPL 2025].

Keywords: Conflict-free cut · NP-completeness · planar graphs · 3-degenerate graphs

1 Introduction

We study the conflict-free cut problem for planar and 3-degenerate graphs, when the conflict graph $\widehat{G}$ is 1-regular. The decision version of the problem is formally defined below:

Definition 1 (CF-cut problem). *Given a simple connected graph $G = (V, E)$ and a conflict graph $\widehat{G} = (\widehat{V}, \widehat{E})$ with $\widehat{V} = E$, the* Conflict-Free Cut (CF-cut) *problem is to decide if there is a set $F \subseteq E$ such that $G - F$ is disconnected and F is independent in $\widehat{G}$.*

© The Author(s), under exclusive license to Springer Nature Switzerland AG 2026
F. Foucaud and A. Parreau (Eds.): IWOCA 2026, LNCS 16587, pp. 385–399, 2026.
https://doi.org/10.1007/978-3-032-27732-9_27

The notion of conflicts on edges is represented using the conflict graph $\widehat{G} = (\widehat{V}, \widehat{E})$ where $\widehat{V} = E$. Edges e_i and e_j conflict in G if and only if there is an edge $\{e_i, e_j\} \in \widehat{E}$.

Several optimization problems have been studied under conflict constraints. The conflict constraints restrict the solution space by assigning conflicts which forbid items to be chosen simultaneously. Some examples include spanning tree problem [3,9], bin packing [11], knapsack [4,18], matching [2,9,13], and feedback vertex set [1]. Along the same lines, the CF-cut problem was introduced by Rauch, Rautenbach, and Souza in [19].

We obtain the matching cut problem as a special case of the CF-cut problem when the conflict graph $\widehat{G}$ is the line graph of G. The *line graph* of G is given by $L(G)$, where $V(L(G)) = E(G)$ and $\{e_i, e_j\} \in E(L(G))$ if and only if e_i and e_j share an endpoint in G. First discussed in [15], the matching cut problem is a well-studied problem in the literature. Matching cut is known to be NP-complete in general [7], and even in the case when G is planar [5]. The reader is referred to the following papers [6,12,14,16,17] for more results on matching cuts.

In this paper, we study the conflict-free cut problem in the case where the conflict graph $\widehat{G}$ is 1-regular. Apart from being a natural restriction, this case has been studied for other problems with conflict constraints. The conflict-free variant of maximum matching is NP-complete under the 1-regular constraint, but the spanning tree problem is polynomial time solvable [9]. For the matching cut problem on a graph, an edge e conflicts with as many as $(d_1 + d_2 - 2)$ edges where d_1 and d_2 are degrees of the end points of e. In other words, the degree of e in the conflict graph $\widehat{G}$ is $(d_1 + d_2 - 2)$. An immediate question is: what is the complexity of conflict-free cut when $\widehat{G}$ is much simpler, say, 1-regular?

In the paper that introduced the CF-cut problem, Rauch, Rautenbach and Souza [19] showed that the CF-cut problem is NP-complete even when the maximum degree of G is 5 with a 1-regular $\widehat{G}$. They also studied the complexity of the problem in the setting of parameterized complexity, showing that it is fixed parameter tractable with respect to the vertex cover number of G, and hard with respect to the size of the feedback vertex set of G, and the clique cover number of $\widehat{G}$. One natural direction here is the complexity of the CF-cut problem when G is planar, and has maximum degree 5 with 1-regular $\widehat{G}$. The authors of [19] also asked about the complexity of the problem when G is 3-degenerate with 1-regular $\widehat{G}$. In this paper, we show that both the problems are NP-complete.

When the conflict-graph $\widehat{G}$ is 1-regular, every edge in G conflicts with exactly one other edge. Proposition 4.1 in [19] notes the following when $\widehat{G}$ is 1-regular. If the edges incident with a vertex $v \in V(G)$ do not have conflicts among each other, then we have a CF-cut that separates v from the rest of the graph. Hence, if $|E(G)| < 2|V(G)|$, then there is a CF-cut in G.

Since we show NP-completeness when G is planar and has maximum degree 5, it is natural to study the complexity when G is planar and has maximum degree 4. For a graph G with maximum degree 4 that is not 4-regular, it follows that $|E(G)| < 2|V(G)|$ and hence there always exists a CF-cut, when the conflict graph $\widehat{G}$ is 1-regular. In this paper, we study the case when G is planar and 4-

regular, and show that it always contains a CF-cut unless it is the octahedron graph.

We present the following results in this paper.

1. Given a 4-regular planar graph G with a 1-regular $\widehat{G}$, there always exists a CF-cut unless it is the octahedron graph. This is shown in Theorem 2.
2. Given a planar graph G of maximum degree 5 with a 1-regular $\widehat{G}$, it is NP-complete to decide if G has a CF-cut. This is proved in Theorem 13.
   This result, along with the result in point 1, completely resolves the complexity of the problem when G is planar with 1-regular $\widehat{G}$.
3. Given a 3-degenerate graph $G = (V, E)$ of maximum degree 5 with a 1-regular $\widehat{G}$, it is NP-complete to decide if G has a CF-cut. This is proved in Theorem 21 and also answers a question by [19] (stated as Problem 4.3).
   It should be noted that when G is 3-degenerate with maximum degree 4, it must necessarily contain a CF-cut since $|E(G)| < 2|V(G)|$. Hence, our result completely settles the 3-degenerate case.
4. We present a family of uncuttable graphs G with 1-regular $\widehat{G}$ such that $|E(G)| = 2|V(G)|$. This is proved in Lemma 23.
   We also present another uncuttable family of graphs that are triangulations, one of which is the octahedron graph (depicted in Fig. 1). This is proved in Theorem 25. These families address questions posed by [19] (stated as Problems 4.1 and 4.2).

2   Preliminaries

Unless otherwise mentioned, we use G to denote a simple, connected graph. For a graph $G = (V, E)$ and two disjoint sets of vertices $A, B \subseteq V$, we use the notation $E(A, B)$ to describe the set of edges of G with one vertex in A and another in B. We define $M \subseteq E$ as a minimal edge cut of G if $V = A \cup B$ such that A and B induce connected subgraphs, and $M = E(A, B)$. We sometimes use the shorthand notation xy to denote an edge $\{x, y\}$. For a graph $G = (V, E)$, and a vertex set $S \subseteq V$, we use $G[S]$ to represent the induced graph on S. In this paper, we use the term cut, to refer only to an edge cut. We use standard graph-theoretic notation as presented in the book by Diestel [10]. We refer to a graph without any CF-cut as an *uncuttable graph*, or simply *uncuttable*. Sometimes, for a graph G with 1-regular $\widehat{G}$, we use the term: a graph with 1-regular conflicts. Due to space constraints, the proofs of some theorems/lemmas are omitted. These are marked with a $(\star)$ symbol.

We note some definitions that will be needed.

- A graph is k-*regular* if every vertex in the graph has degree k.
- A graph is said to be *planar* if it can be drawn in a plane with no two edges crossing each other. For a planar graph, such a drawing is called a *planar embedding*.

- An undirected graph G is *k-degenerate* if there is an ordering of the vertices of G such that each vertex x has at most k neighbors that precede x in the ordering.

  From the above definition, it follows that G is k-degenerate if and only if we can successively delete a vertex of degree at most k eventually resulting in an empty graph.
- Given a graph $G = (V, E)$, the *square graph of* G is defined as $G^2 = (V, \widetilde{E})$ as a graph on the same set of vertices and a superset $\widetilde{E} \supseteq E$ of edges. The set of edges $\widetilde{E}$ is all the pairs of vertices $x, y \in V$ such that either $xy \in E$ or there exists a vertex $w \in V$ such that both $xw, yw \in E$.

3   4-Regular Planar Graphs

In this section, we study the CF-cut problem when G is planar and 4-regular. As noted before, if $|E(G)| < 2|V(G)|$, there always exists a CF-cut when $\widehat{G}$ is 1-regular. This includes the case when G has maximum degree 4, but is not 4-regular. The complexity of the problem for the 4-regular case is not known. In this section, we show that when G is 4-regular and planar, there always exists a CF-cut except in the case when G is the octahedron graph (Fig. 1).

Theorem 2. *Given a 4-regular planar graph G with 1-regular conflict graph $\widehat{G}$, there always exists a CF-cut unless G is the octahedron graph.*

We give an overview of the proof. We first show that the planar embedding must contain faces that do not have conflicts between adjacent edges. If such a face is a triangle, we reason that there is always a CF-cut unless the graph is an octahedron. Otherwise, we can separate the face from the rest of the graph unless there are two conflicting edges that connect the face to the rest of the graph. We show that the graph contains a CF-cut in this case as well.

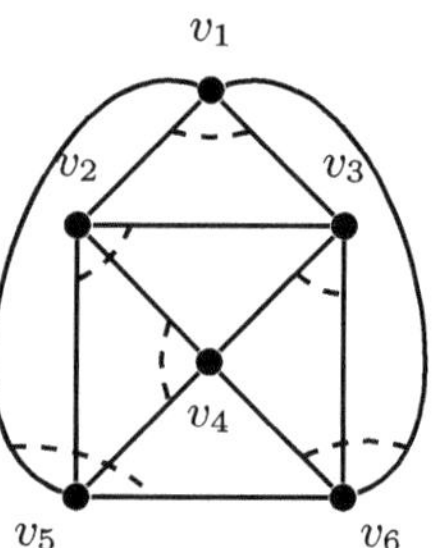

Fig. 1.   Uncuttable octahedron with 1-regular conflicts.

The proof is through detailed case analysis. We provide proof sketches in place of detailed proofs due to space constraints.

Throughout this section, we assume that the planar graph $G = (V, E)$ is given with an associated planar embedding. We may skip the explicit mention of the embedding. When a CF-cut contains only one vertex on one side of the cut, and all the remaining vertices on the other side, we refer to such a cut as a *trivial CF-cut*.

Observation 3. *In a 4-regular planar graph G with 1-regular $\widehat{G}$ without a trivial CF-cut, each vertex is incident with exactly two mutually conflicting edges.*

Proof. If the incident edges of any vertex do not have any conflicts with each other, we have a trivial CF-cut. Since we have $|E(G)| = 2|V(G)|$, each vertex must have exactly one conflict among its incident edges.   □

Observation 4. *For a 4-regular planar graph G with 1-regular $\widehat{G}$ to be uncuttable, a necessary condition is that all the conflicts must be between adjacent edges (edges incident with the same vertex).*

Proof. There are exactly $|E(G)|/2 = |V(G)|$ conflicts. If any conflict is between two non-adjacent edges, then at least one vertex has no conflict among the incident edges, resulting in a trivial cut.   □

For an uncuttable 4-regular planar graph $G = (V, E)$ with an associated planar embedding and a 1-regular $\widehat{G}$, we define two types of conflicts: (1) Type 1 (T_1) conflict when the pair of conflicting edges belongs to the same face, and (2) Type 2 (T_2) conflict when the pair of conflicting edges does not belong to the same face.

Observation 5. *For any 4-regular planar graph G with an associated planar embedding and a 1-regular $\widehat{G}$, there can be only two types of faces: one that has conflicting boundary edges and the other that does not.*

The following definition is useful for the remainder of this proof.

Definition 6 (CF face). *Given a planar graph G with planar embedding and 1-regular $\widehat{G}$, a face is said to be a conflict-free face or CF face, if no two boundary edges of the face are mutually conflicting. We explicitly refer to a CF face on three vertices as a CF triangle.*

Observation 7. *For $\ell \geq 3$, let $v_0 v_1 v_2 \ldots v_{\ell-1} v_0$ be a CF face in an uncuttable 4-regular planar graph G with an associated planar embedding and a 1-regular $\widehat{G}$. Then one of the following holds:*

- *For all $0 \leq i \leq \ell - 1$, the boundary edge $v_i v_{(i+1) \bmod \ell}$ conflicts with an edge incident with v_i (anticlockwise end point); or*
- *For all $0 \leq i \leq \ell - 1$, the boundary edge $v_i v_{(i+1) \bmod \ell}$ conflicts with an edge incident with v_{i+1} (clockwise end point).*

Proof. Since G is planar and 4-regular, as per Observation 3, it follows that every vertex has exactly one pair of incident edges conflicting. Since the face is a CF face, each boundary edge conflicts with one of its adjacent edges that are not part of the face boundary.   □

Lemma 8. *There always exist at least two CF faces in a 4-regular planar graph G with 1-regular $\widehat{G}$.*

Proof. We know that there are $2|V(G)|$ edges in a 4-regular graph G, and by Euler's theorem for a planar graph, there are $|V(G)| + 2$ faces. With 1-regular $\widehat{G}$, there are $|V(G)|$ conflicts in the graph G. Even when all conflicts are of T_1 type, i.e., two conflicting edges belong to a face, we have at least two faces without their boundary edges conflicting.   □

The proof of Theorem 2 relies on the above lemma. We consider various cases based on the CF face. The following four lemmas are also necessary ingredients of the proof of Theorem 2. Due to space constraints, we give overview of the proofs of these four lemmas below.

Lemma 9 ($\star$). *Suppose G is a 4-regular planar graph with 1-regular conflicts. If G is uncuttable and has a CF triangle, then G is the octahedron graph.*

Proof (Sketch of proof). Each of the boundary edges of the CF triangle must conflict with another edge. Using Observation 7, we arrive at four possibilities on how many of these conflicts can be T_1 and how many can be T_2. A detailed case analysis implies that the only case where G is uncuttable is when it is the octahedron graph. $\square$

Lemma 10 ($\star$). *For $\ell \geq 4$, let $F = v_0 v_1 v_2 \ldots v_{\ell-1} v_0$ be a CF face in an uncuttable 4-regular planar graph G with an associated planar embedding and a 1-regular $\widehat{G}$. Then there exist no edges whose one end point is at v_i and the other end point is at v_j for $|j - i| \bmod \ell \geq 2$.*

Proof (Sketch of proof). The edge $v_i v_j$ forms a cycle together with the path $v_i v_{i+1} \ldots v_{j-1} v_j$. Suppose the other neighbors of v_i and v_j are y and z respectively. The proof is divided into four cases depending on the position of the edges $v_i y$ and $v_j z$. A CF-cut is demonstrated in each of these cases. $\square$

Lemma 11 ($\star$). *Suppose G is a 4-regular planar graph with 1-regular conflicts and $v_0 v_1 v_2 \ldots v_{\ell-1} v_0$, $\ell \geq 4$, be one of its CF faces. Let v_i and v_j be two vertices of this CF face that are not adjacent on this face. If G has two conflicting edges, both incident with a common vertex, say x, and meeting the CF face at vertices v_i and v_j respectively, then there always exists a CF-cut.*

Proof (Sketch of proof). Let u and w be the other neighbors of x. The sequence $x v_i v_{i+1} \ldots v_{j-1} v_j x$ forms a cycle. The proof is divided into three cases based on how many of the edges xu and xw are inside the region bounded by the cycle. In each case, we demonstrate a CF-cut. $\square$

Lemma 12 ($\star$). *Suppose G is a 4-regular planar graph with 1-regular conflicts and $F = v_0 v_1 v_2 \ldots v_{\ell-1} v_0$, $\ell \geq 4$, be one of its CF faces. Suppose that for some $0 \leq i \leq \ell - 1$, the vertices v_i, v_{i+1}, together with a vertex x that is not part of F, form a triangle. If the edges $x v_i$ and $x v_{i+1}$ conflict, then there exists a CF-cut in G.*

Proof (Sketch of proof). We first classify the type of vertices that are possible in G. Note that this classification may neither be exhaustive nor be distinct.

- **Type A vertices:** The vertices of the face F.
- **Type X vertices:** The vertices like x that form a triangle with two adjacent vertices in F.
- **Type Y vertices:** These are the neighbors of type A vertices that are not of type X.

- **Type Z vertices:** These are the neighbors of type X vertices that are not of type A.

We further use types to refer to corresponding edges. For instance, an edge of type XZ is an edge whose one endpoint is a type X vertex and the other endpoint is a type Z vertex. The face F along with the type X vertices can be separated from the rest of the graph unless one of the following conditions holds:

- Two edges of type XZ mutually conflict.
- An edge of type XZ conflicts with an edge of type AY.
- There exist only vertices of type A and type X in G.

The rest of the proof demonstrates a CF-cut in each of the above cases.   □

We are now ready to prove Theorem 2 using the above lemmas.

Proof (Proof of Theorem 2*).* Suppose G is 4-regular and planar. By Lemma 8, it follows that there exist at least two CF faces in the embedding of G. Lemma 9 implies that if one of the CF faces is a triangle, then G has a CF-cut unless G is the octahedron graph.

The remaining cases are when the CF faces are not triangles. We argue that there exists a CF-cut in these cases. Let $F = v_0 v_1 \ldots v_{\ell-1} v_0$ be a CF face, where $\ell \geq 4$. We can separate the face F from the rest of the graph G, unless (i) all vertices of G are on the face F, or (ii) there exists a vertex x and edges $v_i x$ and $v_j x$ such that the edges $v_i x$ and $v_j x$ are conflicting with each other. In the former case, there exists a CF-cut because of Lemma 10. In the latter case, if v_i and v_j are not adjacent in F, then Lemma 11 implies that there is a CF-cut in G. If v_i and v_j are adjacent in F, Lemma 12 implies that there is a CF-cut in this case as well.   □

4   CF-Cut for Planar Graphs

In this section, we show that the CF-cut problem is NP-complete when G is planar, has maximum degree 5, and $\widehat{G}$ is 1-regular.

Theorem 13. *Given a planar graph G with maximum degree 5 and 1-regular conflict graph $\widehat{G}$, it is NP-complete to decide if there is a CF-cut in G.*

The NP-completeness of CF-cut follows from a reduction from the Clean 3-SAT problem, which was shown to be NP-complete [8].

Definition 14 (Clean 3-SAT). *A variant of 3-SAT in which each variable appears exactly thrice, at least once positive and at least once negative. Each clause in a clean 3-SAT formula can have two or three literals. A variable appears in a clause at most once. Moreover, by renaming a literal, we may assume that each variable in a given clean 3-SAT instance appears twice in positive form and once in negated form.*

It is straightforward to see that the CF-cut problem is in NP. We prove that Clean 3-SAT reduces to the CF-cut problem, where G is a planar graph with maximum degree 5 and a 1-regular conflict graph $\widehat{G}$. Our reduction is based on the reduction used by [19], but we use gadgets that are planar and uncuttable.

In the first step of this reduction, we reduce a given instance of clean 3-SAT to a multigraph G' with 1-regular conflict. We then construct a planar uncuttable gadget of maximum degree 5 with 1-regular conflict. Finally, we convert G' into a simple planar graph with maximum degree 5 and 1-regular conflicts using the uncuttable gadget. The whole reduction takes $O(mn)$ steps, where m is the number of clauses and n is the number of variables in the clean 3-SAT instance.

4.1   Reducing Clean 3-SAT to a Multigraph G' with 1-Regular Conflicts

Suppose we are given a clean 3-SAT formula $\mathcal{F}$ with n variables and m clauses. Let the variables be $(x_1, x_2, \ldots, x_n)$ and the clauses be $(C_1, C_2, \ldots, C_m)$. We use $\mathcal{F} = (x_1 \vee \bar{x}_3) \wedge (\bar{x}_2 \vee x_3) \wedge (\bar{x}_1 \vee x_2 \vee x_3) \wedge (x_1 \vee x_2)$ as an example for illustration. The variable gadget we use in this reduction is shown in Fig. 2.

To construct G' from $\mathcal{F}$, we start with two special vertices s and t. For every clause $C \in \mathcal{F}$, we add an $s - t$ path P_C of length $|C|$ in G'. In the path P_C, when seen from the vertex s, the jth edge represents the jth literal in C. Now, for every pair of distinct edges $e, f \in E(P_C)$, we add an edge e' parallel to e and an edge f' parallel to f, and we add a conflict between e' and f'.

Now we modify each path P_c as follows to implicitly create a variable gadget for each variable x_i in $\mathcal{F}$: we use $u_{i,k}v_{i,k}$, for $1 \leq i \leq n$ and $1 \leq k \leq 3$, to represent an edge $e \in E(P_C)$ corresponding to the kth occurrence of a variable x_i in some clause $C \in \mathcal{F}$. We emphasize that all three occurrences of x_i appear in different clauses $C \in \mathcal{F}$. Now, for every negative literal of x_i in $C \in \mathcal{F}$, we replace its corresponding edge $uv \in P_C$ with a pair of parallel edges, $u_{i,2}v_{i,2}$. Of these parallel edges, one conflicts with the edge $u_{i,1}v_{i,1}$ and another conflicts with $u_{i,3}v_{i,3}$. The graph G' is a planar multi-graph with maximum degree $4m$, as shown in Fig. 3. This completes the construction.

Lemma 15. *If there is a CF-cut in the multi-graph G' in the above construction then it is necessarily an $s - t$ cut.*

Proof. According to the construction of G', there is an $s - t$ path P_C of length $|C|$ for each clause $C \in \mathcal{F}$. Since each clause $C \in \mathcal{F}$ has two or three literals, each $s - t$ path has one or two intermediate vertices. We consider the case when C has three literals and P_C has two intermediate vertices. Let the intermediate vertices be u and v. In this case, edges su, uv, and vt will have two copies of parallel edges $(su)'$, $(uv)'$, and $(vt)'$ respectively. Of the two copies of $(su)'$, one conflicts with a copy of uv' and another with a copy of $(vt)'$. The remaining copy of $(uv)'$ conflicts with the remaining copy of $(vt)'$. For the sake of contradiction, let us assume that there is a CF-cut $M = E(A, B)$ such that both s and t belong to the same set, say B, and $u, v \in A$. Then we have two conflicting

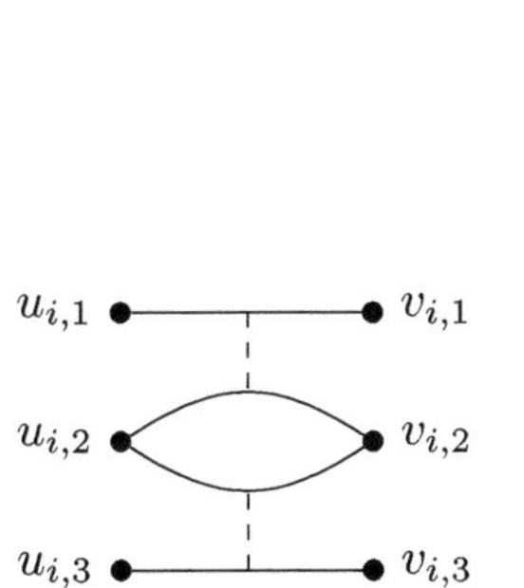

Fig. 2. The variable gadget representing the three literals of the variable x_i. The parallel edges $u_{i,2}v_{i,2}$ represent the negated literal and the edges $u_{i,1}v_{i,1}$, $u_{i,3}v_{i,3}$ represent the positive literals. Dashed edges represent conflicts.

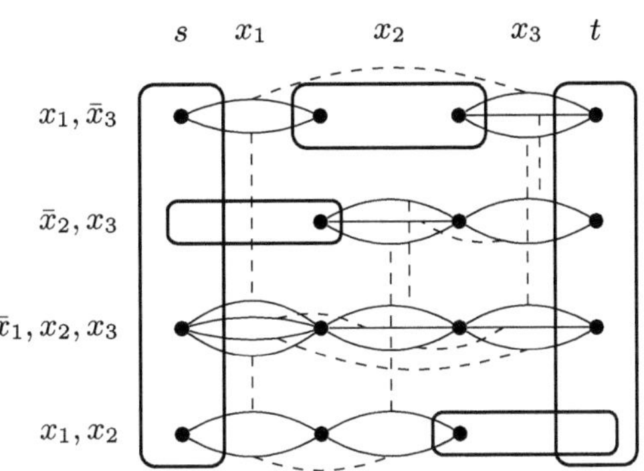

Fig. 3. Completed G' for $\mathcal{F} = (x_1 \vee \bar{x}_3) \wedge (\bar{x}_2 \vee x_3) \wedge (\bar{x}_1 \vee x_2 \vee x_3) \wedge (x_1 \vee x_2)$. Vertices inside overlapped rectangles are identified. Dashed edges represent conflicts.

edges $(su)', (vt)' \in M$, a contradiction. If $u \in A$ and $v \in B$, then there are two conflicting edges $(su)', (uv)' \in M$, again a contradiction. If $u \in B$ and $v \in A$, then we have two conflicting edges $(uv)'$ and $(vt)'$ in M, which is a contradiction. Therefore, all CF-cuts in G' are $s - t$ cuts. This completes the proof for the case when C has three literals. The case when C has two literals can be reasoned similarly. $\qquad\square$

Lemma 16. *The multi-graph G' constructed above has a CF-cut if and only if its corresponding clean 3-SAT formula $\mathcal{F}$ is satisfiable.*

Proof. For every edge $e \in P_C$, let $S(e)$ denote a multi-set of edges containing e and all its copy edges e'. The conflicts assigned in G' ensure the following: whenever $S(e)$ corresponding to a positive literal x_i becomes part of a CF-cut, the edges corresponding to the literal $\bar{x}_i$ do not, and vice versa. For one direction, let us assume that formula $\mathcal{F}$ is satisfiable and α is a satisfying assignment for $\mathcal{F}$. Each clause contains a true literal in α, and we select the edges corresponding to one of the true literals to be in the cut. This ensures that s and t are separated.

For the other direction, if M is a CF-cut in G', by Lemma 15, M must be an $s - t$ cut. Hence, for all clauses C in $\mathcal{F}$, M must contain the edges $S(e)$ for some $e \in P_C$. By setting the literals corresponding to the chosen edges to True, we recover a satisfying assignment to $\mathcal{F}$. The variable gadget ensures we cannot pick inconsistent assignments to any variable. $\qquad\square$

4.2   Constructing an Uncuttable Planar Gadget

This section explains how to obtain an uncuttable planar graph of maximum degree 5 and 1-regular conflicts from the square of an even cycle.

We will use this gadget to convert the multigraph G' obtained in the previous section into a simple planar graph with maximum degree 5.

Lemma 17. *Let H_{2n} be the square of a cycle graph C_{2n} $(n \geq 3)$ and $\{0, 1, 2, \ldots, 2n-1\}$ be its vertices when read clockwise. Suppose we assign a conflict between edges $\{i, (i+1)\} \bmod 2n$ and $\{i, (i+2)\} \bmod 2n$ for $0 \leq i \leq 2n-1$. The only CF-cut that H_{2n} possesses is the set of edges of the form $\{i, (i+1)\} \bmod 2n$ for $0 \leq i \leq 2n-1$.*

Proof. As we can observe from Fig. 4a, in the square of every cycle graph C_{2n} $(n \geq 3)$, there are only two types of edges: $\{i, (i+1)\} \bmod 2n$ and $\{i, (i+2)\} \bmod 2n$ for $0 \leq i \leq 2n-1$. We can easily verify from Fig. 4a that by deleting all the edges of the form $\{i, (i+1)\} \bmod 2n$ for $0 \leq i \leq 2n-1$, the graph becomes disconnected and forms two cycle graphs, each being C_n: one on all odd vertices and another on all even vertices. Moreover, the edges $\{i, (i+1)\} \bmod 2n$ for $0 \leq i \leq 2n-1$, that form the cut, are conflict-free. We refer to this cut as M.

To prove that M is the only CF-cut in H_{2n}, let us consider an arbitrary CF-cut, say M'. There must be some vertex j, $0 \leq j \leq 2n-1$, such that j and $(j+1) \bmod 2n$ are on opposite sides of the cut.

Let M' divide the vertices into parts A and B. We may assume that the vertex j belongs to A, and the vertex $(j+1) \bmod 2n$ belongs to B. Since $\{j, (j+1)\} \bmod 2n$ and $\{j, (j+2)\} \bmod 2n$ for are conflicting, the vertex $(j+2) \bmod 2n$ cannot belong to the set B. This implies that $(j+2) \bmod 2n$ must be in A. Thus the edge $\{(j+1), (j+2)\} \bmod 2n$ must belong to the cut M'. Therefore, if $\{i, (i+1)\} \bmod 2n$ belongs to the CF-cut M', then the edge $\{(i+1), (i+2)\} \bmod 2n$ also belong to M' for $0 \leq i \leq 2n-1$. This implies that M' contains all the edges in the cut M. For any edge outside M, it can be seen that there is an edge in M that conflicts with it. Since M' is conflict-free, this implies that $M' = M$. $\square$

Now we modify the graph in Fig. 4a so that it becomes uncuttable. Below we discuss such a construction.

Let H_{2n} be the square of C_{2n} $(n \geq 6)$ with 1-regular conflicts as shown in Fig. 4a. The vertices of even parity lie on the outer face of H_{2n}. We modify H_{2n} into an uncuttable planar graph $\widetilde{H}_{2n}$ as follows: we start with a vertex on the outer face of H_{2n}, say i and consider its four immediate successors $i+1, i+2, i+3, i+4$. Similarly, we consider another vertex on the outer face j such that $\{j, j+1, j+2, j+3, j+4\}$ is disjoint from $\{i, i+1, i+2, i+3, i+4\}$. All vertices are considered modulo $2n$. After this, we introduce four new vertices, namely, $i', (i+2)', j'$ and $(j+2)'$ that split the edges $\{i, i+2\}, \{i+2, i+4\}, \{j, j+2\}$ and $\{j+2, j+4\}$ respectively. We then add six new edges: $\{i+1, i'\}, \{i+3, (i+2)'\}, \{i', (i+2)'\}, \{j+1, j'\}, \{j+3, (j+2)'\}$ and $\{j', (j+2)'\}$. We establish a 1-regular conflict between these newly introduced 10 edges (four edges as a result of edge splits and six edges drawn separately as mentioned before) as depicted in Fig. 4b. This completes the construction of $\widetilde{H}_{2n}$.

We need the following lemma (proof of which is omitted due to space constraints) to complete the proof of Theorem 13.

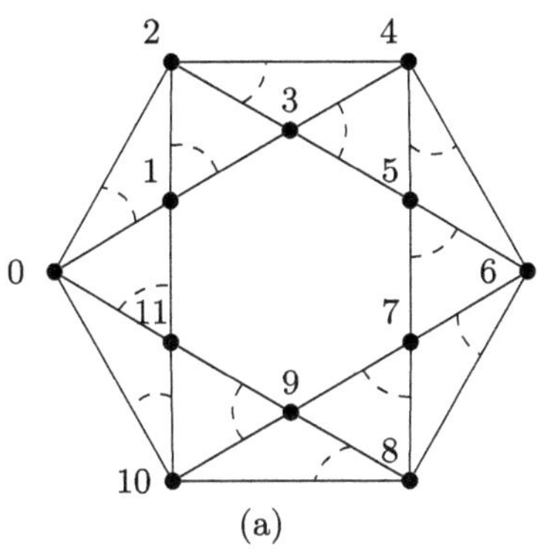

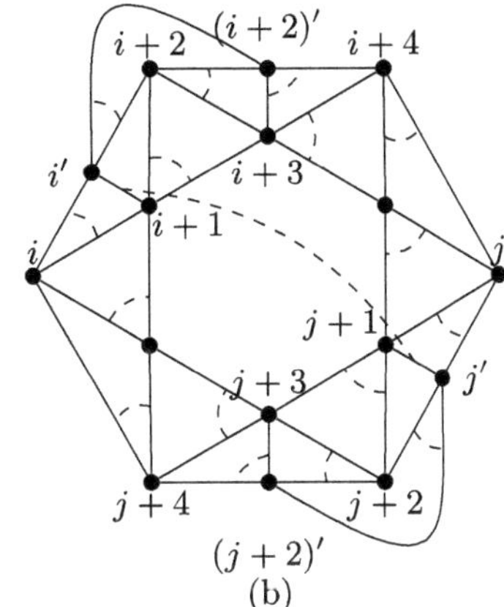

Fig. 4. (a) Square of C_{2n} ($n = 6$) with 1-regular conflict. Dashed edges represent conflicts. (b) Square of C_{2n} when modified into $\widetilde{H}_{2n}$.

Lemma 18 ($\star$). *For $n \geq 6$, the graph $\widetilde{H}_{2n}$ constructed above is an uncuttable planar graph of max degree 5.*

Proof (Proof of Theorem 13.) Recall that we have shown that the multigraph G' has an $s - t$ cut if and only if the formula $\mathcal{F}$ is satisfiable. We need to obtain a planar simple graph from G'. We replace each vertex of G' with a sufficiently large uncuttable planar graph $\widetilde{H}_{2t}$. Since the maximum degree of G' is $4m$ (where m is the number of clauses in $\mathcal{F}$), setting $t = 4m$ suffices. The adjacencies of a vertex $v \in G'$ are distributed among the vertices on the outer face of $\widetilde{H}_{2t}$ in such a way that each vertex on the outer face of $\widetilde{H}_{2t}$ is incident with at most one edge of v. Note that all the vertices on the outer face of $\widetilde{H}_{2t}$ are of degree 4. Thus we get an uncuttable, planar graph with 1-regular conflicts and maximum degree 5. The graph G' has at most $3n - 1$ vertices and at most $10n$ edges. The gadget $\widetilde{H}_{2t}$ can be constructed in polynomial time. This completes the reduction from a clean 3-SAT instance to an uncuttable, simple, planar graph. $\square$

4.3   NP-Completeness for Maximum Degree Beyond 5

Though it is not surprising, one may wonder if the NP-completeness result holds for planar graphs that have maximum degree larger than 5. We establish below that the NP-completeness result in Theorem 13 can be extended for planar G with maximum degree greater than 5 as well. We need the following lemma, the proof of which is straightforward.

Lemma 19 ($\star$). *Let H be a simple connected graph with 1-regular conflicts. We obtain a new graph H' with vertex set $V(H') = V(H) \cup \{u_1\}$ and edge set $E(H') = E(H) \cup \{u_1v_1, u_1v_2\}$, where v_1, v_2 are two vertices in H. Further we introduce conflicts between the new edges u_1v_1 and u_1v_2. Then H' is an uncuttable, 3-degenerate graph if and only if H is uncuttable and 3-degenerate.*

Theorem 20. *Let $D \geq 5$ be a positive integer. The CF-cut problem is NP-complete even when restricted to planar graphs G with maximum degree equal to D and 1-regular conflicts.*

Proof. The proof of Theorem 13 uses a reduction from Clean 3-SAT. The resulting graph G is a planar graph with maximum degree 5. For $D > 5$, we can use Lemma 19 to produce graphs that have maximum degree D as desired and are uncuttable if and only if G is uncuttable. By choosing v_1, v_2 from the same face, it can be ensured that the resulting graph remains planar. $\qquad\square$

5   CF-Cut on 3-Degenerate Graphs

In this section, we show that the CF-Cut problem is NP-complete when G is 3-degenerate and has maximum degree 5, with a 1-regular conflict graph $\widehat{G}$.

Theorem 21. *Given a 3-degenerate graph $G = (V, E)$ of maximum degree 5 and 1-regular conflict graph $\widehat{G}$, it is NP-complete to decide if a CF-cut exists in G.*

The proof starts with the graph G' derived in Sect. 4.1 but we need a different gadget as the goal is a 3-degenerate graph. We first observe that the graph H in Fig. 5(a) is 3-degenerate, has maximum degree 5 and is uncuttable for the assigned conflicts. Lemma 23 builds upon this graph to obtain a class of uncuttable 3-degenerate graphs.

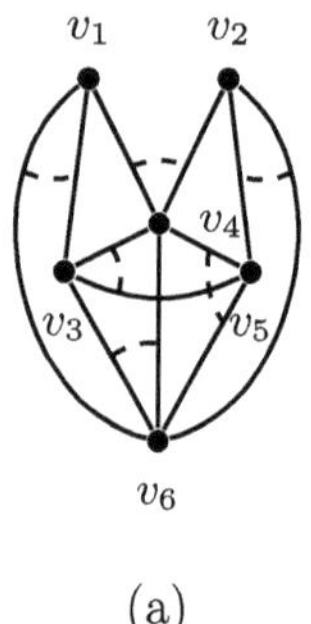

(a)

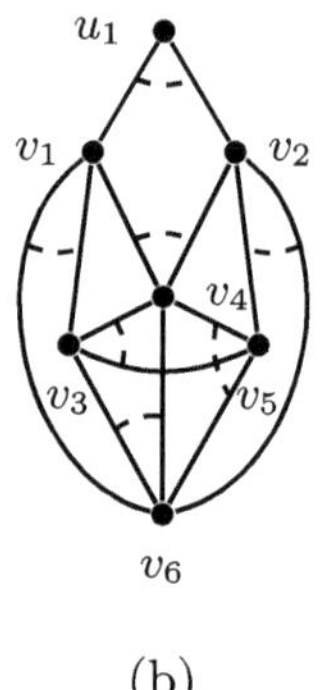

(b)

Fig. 5. (a) An uncuttable, 3-degenerate graph H with 1-regular conflicts, (b) An uncuttable, 3-degenerate graph H' derived from H after adding a vertex u_1 and a pair of conflicting edges u_1v_1 and u_1v_2 to H. Dashed edges represent conflicts.

Lemma 22. *The graph H with 1-regular conflicts as shown in Fig. 5(a), is an uncuttable, 3-degenerate graph.*

Proof. We can easily verify that the graph H is 3-degenerate. The sequence $v_1, v_2, v_3, v_4, v_5, v_6$ is a degeneracy ordering. To show that the graph with the shown conflicts is uncuttable, for the sake of contradiction, let us assume that $M = E(A, B)$ is a CF-cut of H.

We first note that the edge $v_1 v_6 \notin M$. Suppose not, say $v_1 v_6$ is part of the cut. Without loss of generality, we may assume that $v_1 \in A$ and $v_6 \in B$. This implies that $v_3 \in A$ and since $v_6 \in B$ and $v_3 \in A$, we can further infer that $v_4 \in B$. This implies that $v_5 \in B$. This makes two conflicting edges $v_3 v_4$ and $v_3 v_5 \in M$, a contradiction.

Next, we note that the edge $v_2 v_6 \notin M$. Suppose not, say $v_2 v_6$ is part of the cut. Without loss of generality, we may assume that $v_2 \in A$ and $v_6 \in B$. This implies that $v_5 \in A$. Since $v_6 \in B$ and $v_5 \in A$ we can infer $v_4 \in A$. Similarly, $v_6 \in B$ and $v_4 \in A$ together imply $v_3 \in B$. This makes two conflicting edges $v_3 v_4$ and $v_3 v_5 \in M$, a contradiction.

As explained above, we know that the edges $v_1 v_6, v_2 v_6$ are not in M. Without loss of generality, we assume that $v_1, v_2, v_6 \in A$. This successively implies that $v_4, v_5, v_3 \in A$. Hence all the six vertices are on the same side of the cut. This completes the proof. $\qquad\square$

We build on this graph H and repeatedly apply the construction in Lemma 19 to obtain uncuttable graphs of arbitrary size. This is summarized in the below lemma. The proof of this lemma is omitted due to space constraints.

Lemma 23 ($\star$). *There is a family of uncuttable graphs H_i with 1-regular conflicts that satisfy the following properties:*

- *$|V(H_i)| = 13 + 7i$, where 8 vertices are of degree 5 in H_i, $1 + 7i$ vertices are of degree 4, and 4 vertices are of degree 2.*
- *H_i has maximum degree 5 and $|E(H_i)| = 2|V(H_i)|$.*
- *Repeatedly deleting degree 2 vertices from H_{i+1} yields H_i for $i \geq 1$, and repeatedly deleting degree 2 vertices from H_1 yields H (as referred to in Lemma 22).*

Proof (Proof of Theorem 21). It is straightforward to see that the problem is in NP since we only need to verify that a given cut M separates the graph, and that M is conflict-free. We focus on showing the NP-hardness. We start with the reduction from a clean 3-SAT formula $\mathcal{F}$ stated in Sect. 4.1. By Lemma 16, it follows that the graph G' obtained as a result of the reduction has a CF-cut if and only if the given clean 3-SAT instance is satisfiable. We need to obtain a simple 3-degenerate graph from G' that has maximum degree 5. We replace the vertices of G' each with a sufficiently large uncuttable H_i, as mentioned in Lemma 23. The adjacencies of a vertex $v \in G'$ are distributed among the degree 4 and degree 2 vertices of H_i. Since the maximum degree of G' is $4m$ (where m is the number of clauses in $\mathcal{F}$), setting $i \geq 4m/7$ suffices.

The constructed graph is 3-degenerate. To observe this, note that by Lemma 23, the graph H_i contains 4 vertices of degree 2. So even with the additional adjacencies from G', these vertices have degree 3. We can remove the degree 2 vertices in H_i repeatedly and arrive at a copy of H disconnected from the rest of the graph. Since H is 3-degenerate, we can strip away the entire gadget. Once all the gadgets are removed, the resulting graph is empty. Thus, we get an uncuttable 3-degenerate graph G of maximum degree 5 such that G has a CF-cut if and only if the clean 3-SAT formula $\mathcal{F}$ is satisfiable. $\square$

6   An Uncuttable Class of Planar Graphs

In this section, we present a class of planar graphs with a 1-regular conflict assignment that is uncuttable. Although we have already presented a class of uncuttable planar graphs with 1-regular conflicts in Lemma 18, we present one more class of such graphs to answer the questions asked by [19] regarding the CF-cut on planar graphs. We first define prism graphs.

Definition 24 (Prism Graph). *The prism graph P_{2t} on $2t$ vertices contains two t-cycles $v_1, v_2, \ldots, v_t$ and $v_{t+1}, v_{t+2}, \ldots, v_{2t}$, with additional edges $v_i v_{t+i}$ for all $1 \leq i \leq t$.*

The main result of this section is stated below. Due to lack of space, we provide a sketch of the proof below and the full proof is omitted.

Theorem 25 ($\star$). *Let $G = P_{4t}^*$ be the planar dual of the prism graph on $4t$ vertices, where $t \geq 2$. There exists a 1-regular conflict graph $\widehat{G}$ for which the graph G does not have a conflict-free cut.*

Proof (Sketch of proof). We first note that every minimal cut in a planar graph is a cycle in its planar dual graph. We say that a cycle in a graph is *conflict-free* if it does not contain conflicting edges. We show that conflicts can be assigned for prism graphs P_{4t}, for $t \geq 2$, in such a way that there are no conflict-free cycles. We use two "building blocks": a graph H on 10 vertices and a graph H' on 6 vertices, both with associated conflicts. Depending on the value of t, the graphs P_{4t} are constructed using multiple copies of H and possibly a copy of H'.

7   Future Work

We conclude with some open directions:

1. The complexity of the CF-cut problem when G is 4-regular (not necessarily planar) with 1-regular $\widehat{G}$.
2. The complexity of the CF-cut problem for maximal planar graphs G with 1-regular $\widehat{G}$.
3. The complexity of counting CF-cuts for a given graph G with 1-regular $\widehat{G}$. It would also be an interesting direction for specific classes of G, such as when G is 4-regular and planar. There are existing results on the enumeration of matching cuts [14].
4. Taking a cue from the parameterized complexity results on matching cuts [12, 16], it would be interesting to study the complexity of CF-cuts with respect to different structural parameters of G.

References

1. Agrawal, A., Jain, P., Kanesh, L., Lokshtanov, D., Saurabh, S.: Conflict free feedback vertex set: a parameterized dichotomy. In: 43rd International Symposium on Mathematical Foundations of Computer Science, MFCS, Dagstuhl, Germany, pp. 53:1–53:15 (2018)
2. Agrawal, A., Jain, P., Kanesh, L., Saurabh, S.: Parameterized complexity of conflict-free matchings and paths. Algorithmica **82**(7), 1939–1965 (2020)
3. Barros, B.J.S., Ochi, L.S., Pinheiro, R.G.S., Souza, U.S.: On conflict-free spanning tree: mapping tractable and hard instances through the lenses of graph classes. Theor. Comput. Sci. **1031**, 115081 (2025)
4. Bettinelli, A., Cacchiani, V., Malaguti, E.: A branch-and-bound algorithm for the knapsack problem with conflict graph. INFORMS J. Comput. **29**(3), 457–473 (2017)
5. Bonsma, P.: The complexity of the matching-cut problem for planar graphs and other graph classes. J. Graph Theory **62**(2), 109–126 (2009)
6. Chen, C.Y., Hsieh, S.Y., Le, H.O., Le, V.B., Peng, S.L.: Matching cut in graphs with large minimum degree. Algorithmica **83**(5), 1238–1255 (2021)
7. Chvátal, V.: Recognizing decomposable graphs. J. Graph Theory **8**(1), 51–53 (1984)
8. Cygan, M., Marx, D., Pilipczuk, M., Pilipczuk, M.: Hitting forbidden subgraphs in graphs of bounded treewidth. Inf. Comput. **256**, 62–82 (2017)
9. Darmann, A., Pferschy, U., Schauer, J., Woeginger, G.J.: Paths, trees and matchings under disjunctive constraints. Disc. Appl. Math. **159**(16), 1726–1735 (2011)
10. Diestel, R.: Graph theory 6th ed. Graduate texts in mathematics, p. 173 (2024)
11. Epstein, L., Favrholdt, L.M., Levin, A.: Online variable-sized bin packing with conflicts. Disc. Optim. **8**(2), 333–343 (2011)
12. Feghali, C., Lucke, F., Paulusma, D., Ries, B.: Matching cuts in graphs of high girth and H-free graphs. Algorithmica **87**(8), 1199–1221 (2025)
13. Glock, S., Joos, F., Kim, J., Kühn, M., Lichev, L.: Conflict-free hypergraph matchings. J. Lond. Math. Soc. **109**(5), e12899 (2024)
14. Golovach, P.A., Komusiewicz, C., Kratsch, D., Le, V.B.: Refined notions of parameterized enumeration kernels with applications to matching cut enumeration. J. Comput. Syst. Sci. **123**(C), 76–102 (2022)
15. Graham, R.L.: Bounds on multiprocessing anomalies and related packing algorithms. In: Proceedings of the 16–18 May 1972, Spring Joint Computer Conference, pp. 205–217 (1971)
16. Komusiewicz, C., Kratsch, D., Le, V.B.: Matching cut: kernelization, single-exponential time FPT, and exact exponential algorithms. Disc. Appl. Math. **283**, 44–58 (2020)
17. Lucke, F., Paulusma, D., Ries, B.: Dichotomies for maximum matching cut: H-freeness, bounded diameter, bounded radius. Theor. Comput. Sci. **1017**, 114795 (2024)
18. Pferschy, U., Schauer, J.: Approximation of knapsack problems with conflict and forcing graphs. J. Comb. Optim. **33**(4), 1300–1323 (2017)
19. Rauch, J., Rautenbach, D., Souza, U.S.: On conflict-free cuts: algorithms and complexity. Inf. Process. Lett. **187**, 106503 (2025)

Tight Upper Bounds on Color Reversal by Local Inversions

Hitendra Kumar[1], Kumud Singh Porte[2(✉)] (iD), and R. B. Sandeep[2] (iD)

[1] Indian Institute of Science Education and Research, Pune, India
[2] Indian Institute of Technology Dharwad, Dharwad, India
`{cs24dp012,sandeprb}@iitdh.ac.in`

Abstract. We investigate the problem of reversing colors in bicolored graphs using the operation of *local inversion*. A bicoloration of a graph $G = (V, E)$ assigns each vertex a color from the set $\{-1, 1\}$. A local inversion at a vertex v reverses the colors of all neighbors of v and complements the subgraph induced by these neighbors, while leaving v and all other vertices unchanged. This operation was introduced by Sabidussi (Discrete Mathematics, 1987), who proved that any bicolored graph on n vertices without isolated vertices can be color-reversed using at most $6n + 3$ local inversions. He further showed that any bicolored graph can be transformed into any other bicolored graph on the same underlying graph using at most $9n$ local inversions. More recently, Porte et al. (CALDAM 2026) improved these bounds to $4n - 3$ for color reversal and to $\lfloor (11n - 3)/2 \rfloor$ for transforming one bicolored graph into another. In this paper, we further improve these results by showing that any bicolored graph on n vertices without isolated vertices can be color-reversed using at most $3n$ local inversions, and that this bound is tight for complete graphs on at least three vertices. Moreover, we show that any bicolored graph can be transformed into another bicolored graph on the same underlying graph using at most $5n - 3$ local inversions.

Keywords: local inversion · color reversal · bicolored graph · graph operations · tight bounds

1 Introduction

Local graph operations that change adjacency or vertex labels are fundamental tools in graph theory and combinatorics. One such operation, *local complementation*, modifies the adjacency among the neighbors of a vertex and has been widely studied [1, 4, 7–9, 11, 12, 18, 19, 21] due to its connections with matroids, vertex-minors [14], and quantum information theory [17]. When vertices carry additional information such as colors or signs, these operations naturally extend to transformations that affect both the graph structure and the vertex labels. In

Supported by ANRF MATRICS Grant MTR/2022/000692: "Algorithmic study on hereditary graph properties".

a two-colored graph, the act of reversing the colors of selected vertices is a classical operation already studied under the name *vertex switching* in signed graphs. Introduced by Harary [10] and later developed by Zaslavsky [20], this operation serves as a fundamental symmetry that preserves many global graph properties. In this work, we study a graph operation, termed *local inversion*, which simultaneously changes both adjacencies and vertex colors. Local complementation, vertex switching, and local inversion belong to a broad family of graph transformation problems that have been widely investigated in algorithmic and parameterized complexity [2,5,6]. Such problems typically ask whether a given graph can be transformed into one with a desired property using a bounded number of operations. These frameworks are useful for studying how graphs change under operations, what this reveals about their structure, and which transformations are computationally feasible. Vertex-minors are obtained using local complementation and vertex deletion, where a graph H is a vertex-minor of G if it is an induced subgraph of a graph locally equivalent to G. This concept has been widely studied due to its connections with rank-width, circle graphs, and structural graph theory. Since local inversion builds upon local complementation by additionally modifying vertex colors, it can be viewed as a natural extension of these local transformation frameworks. Therefore, the theory of vertex-minors provides useful background for understanding how sequences of local operations can produce global changes in graphs [13].

In this work, we consider simple, undirected graphs $G = (V, E)$ whose vertices are assigned binary colors via a mapping $\beta : V \to \{-1, +1\}$. Such a pair $B := (G, \beta)$ is called a *bicolored graph*. A basic operation, called a *local inversion* at a vertex $v \in V$, consists of applying local complementation (Sect. 3) at v to the underlying graph while simultaneously reversing the colors of all neighbors of v. Given a bicolored graph B, repeatedly applying local inversions according to a string $w \in V^*$ where V^* denotes the set of all finite sequences over V transforms a bicolored graph B into a new bicolored graph, denoted by B_w. From an algorithmic viewpoint, sequences of local inversions provide a framework for local-to-global transformations: simple local rules induce global effects whose complexity must be carefully controlled.

Our principal object of study is the *color reversal number*, denoted $cr(G)$. It is defined as the minimum integer ℓ such that, for every initial bicoloration β of G, there exists a string w with $|w| = \ell$ for which the resulting bicolored graph B_w is obtained from B by reversing the color of every vertex of G, while leaving the underlying graph unchanged. Observe that if there exists a string w such that, for one bicoloration β of G, the bicolored graph B_w is obtained from B by changing the color of every vertex of G, then the same string w has this property for every bicoloration of G. Consequently, $cr(G)$ depends only on the underlying graph G and not on the initial coloring.

The color-reversal problem under local inversions was first introduced and studied by Sabidussi [16]. His results were later utilized by Brijder and Hoogeboom [3] in their investigation of the algebraic structure induced by pivot and loop complementation on graphs and set systems. Sabidussi showed that for any bicolored graph with underlying graph G, where G has no isolated vertices, we have $cr(G) \leq 6n+3$, where $n = |V(G)|$. Recently, Porte et al. [15] improved this bound to $4n - 3$ using *parity-aware decompositions* (by tracking and controlling the parity (odd/even behavior) of color flips induced by local inversions) and repeated applications of constant-length strings. They used the perfect forest theorem and the partitioning of an odd tree to obtain such a decomposition. Decomposing this forest (or an odd-degree tree) into edges and small substructures lead each local operation to flip colors in a parity-controlled way. In this work, we substantially improve the bound by proving that $cr(G) \leq 3n$, for every graph G without isolated vertices, and we show that this bound is tight. Our proof relies on a simple strategy: we construct a spanning tree of G, extract vertex-disjoint stars in this tree, and apply carefully chosen local inversions to these stars. Porte et al. [15] also showed that for the complete graph K_n with $n > 2$ vertices, $cr(K_n) \leq 3n$ and asked whether $cr(K_n) = 3n$. In this work, we answer this question in the affirmative.

Sabidussi [16] also considered the more general transformation problem: given two bicolored graphs B and B' sharing the same underlying graph G, determine whether B can be transformed into B' by local inversions. He proved that such a transformation is always possible using at most $9n$ local inversions. Recently, Porte et al. [15] improved this bound to $\left\lfloor \frac{11n-3}{2} \right\rfloor$. By replacing their bound $4n-3$ with our $3n$ bound for $cr(G)$ in their proof and performing the resulting simple algebra, one can easily obtain a bound of $5n$. In this work, we adapt a similar method to that of Porte et al., but a careful case-by-case analysis allows us to improve the bound further to $5n - 3$.

2   Our Contribution

In this paper, we obtain a tight bound for the color-reversal problem under local inversions and improve the best known bounds for the more general transformation problem between bicolored graphs. Our main results are summarized as follows:

- For every graph G on n vertices with no isolated vertices,

$$cr(G) \leq 3n.$$

  This bound is obtained via a decomposition of a spanning tree into vertex-disjoint stars (see Theorem 1).
- **Exact value for complete graphs.** For every complete graph K_n with $n > 2$, we determine the exact value

$$cr(K_n) = 3n,$$

  answering an open question of Porte et al. [15] (see Theorem 2).

– **Improved transformation bound.** For any two bicolored graphs B and B' with the same underlying graph G on n vertices, B can be transformed into B' using at most

$$5n - 3$$

local inversions, improving the bound $\lfloor \frac{11n-3}{2} \rfloor$ of Porte et al. [15] (see Theorem 3).

3   Preliminaries

Unless stated otherwise, we follow the definitions and conventions introduced in [15]. In this work, we consider simple, undirected, and labelled graphs. For a graph G, $V(G)$ and $E(G)$ denote its vertex and edge sets, respectively. For any subset $S \subseteq V(G)$, we write $G[S]$ for the subgraph of G induced by S. The open neighborhood of a vertex v in G is denoted by $N_G(v)$. A path and star on t vertices are denoted by P_t and S_t, respectively.

A *bicoloration* of a graph $G = (V, E)$ is a mapping $\beta : V \to \{-1, 1\}$. A *bicolored graph* is a pair $B = (G, \beta)$, where G is a graph and β is a bicoloration of G.

Let $G = (V, E)$ be a graph and let $a \in V$. The *local complement* of G at a is the graph $G_a = (V, E_a)$ obtained by toggling adjacency among the neighbors of a. Formally, for two distinct vertices $x, y \in V$,

$$\{x, y\} \in E_a \iff \begin{cases} \{x, y\} \in E \text{ and } (x \notin N_G(a) \text{ or } y \notin N_G(a)), \\ \text{or} \\ \{x, y\} \notin E \text{ and } x, y \in N_G(a). \end{cases}$$

For a graph G, a *string* over $V(G)$ is a finite sequence of vertices from $V(G)$. The set of all such strings is denoted by $V(G)^*$, with ε representing the empty string. For $w \in V(G)^*$, the *length* of w, denoted by $|w|$, is the number of symbols in w. If a string $w \in V(G)^*$ occurs r number of times consecutively, we write it as w^r.

For any string $w \in V(G)^*$, the local complement of G with respect to w, denoted G_w, is defined inductively as follows:

– Base case: $G_\varepsilon = G$.
– Inductive case: if $w = w'a$ for some $a \in V(G)$, then $G_w = (G_{w'})_a$.

Let G be a graph and let $w, w' \in V(G)^*$ be two strings. We say that w and w' are *equivalent over* G if $G_w = G_{w'}$. In this case, we write $w \simeq w'$.

Given a bicoloration β, the bicoloration with respect to a vertex a, denoted β_a, is obtained by inverting the color of each neighbor of a:

$$\beta_a(x) = \begin{cases} -\beta(x) & \text{if } x \in N_G(a), \\ \beta(x) & \text{otherwise.} \end{cases}$$

More generally, for a string $w \in V(G)^*$, the bicoloration of G with respect to w, denoted β_w, is defined inductively as follows:

- Base case: $\beta_\varepsilon = \beta$.
- Inductive case: if $w = w'a$ for some $a \in V(G)$, then $\beta_w = (\beta_{w'})_a$.

Let $B = (G, \beta)$ be a bicolored graph and let $a \in V(G)$. The *local inversion* of B at a, denoted B_a, is defined as (G_a, β_a). More generally, for a string $w \in V(G)^*$, the local inversion of B with respect to w, denoted B_w, is defined inductively as follows:

- Base case: $B_\varepsilon = B$.
- Inductive case: if $w = w'a$ for some $a \in V(G)$, then $B_w = (B_{w'})_a$ (Fig. 1).

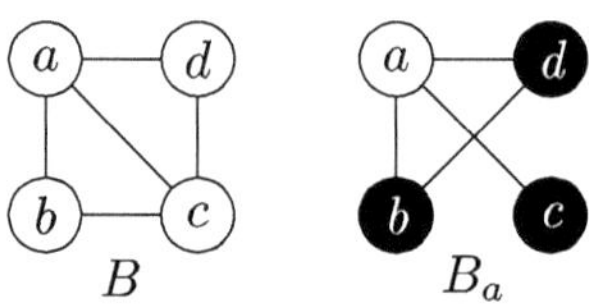

Fig. 1. Local inversion of a bicolored graph B with respect to vertex a.

Let $B = (G, \beta)$ be a bicolored graph and let $w, w' \in V(G)^*$ be two strings. We say that w and w' are *equivalent over* B if $B_w = B_{w'}$. In this case, we write $w \sim w'$.

Let $B = (G, \beta)$ be a bicolored graph, and let $A \subseteq V(G)$. Define the operator h_A by

$$\big(h_A(\beta)\big)(v) = \begin{cases} -\beta(v), & \text{if } v \in A, \\ \beta(v), & \text{if } v \notin A. \end{cases}$$

We write $B^A = (G, h_A(\beta))$ for the bicolored graph obtained from B by flipping the colors of all vertices in A and leaving all other colors unchanged. In particular, if $A = \{a\}$, we write B^a instead of $B^{\{a\}}$. We note that for a $w \in V(G)^*$, $w \simeq \varepsilon$ if and only if $B_w = B^A$ for some $A \subseteq V(G)$.

We recall a few known results, which will be used in subsequent sections. Proposition 1 states that for given graph G, a bicoloration β, and any vertex $v \in V(G)$, we have $aa \sim \varepsilon$. Proposition 2 states that any vertex of a bicolored graph can be color reversed in at 7 local inversions, without changing the underlying graph.

Proposition 1 ([16]). *For any bicolored graph $B = (G, \beta)$ and any vertex $a \in V(G)$, we have $B_{aa} = B$.*

Proposition 2 ([15]). *Let $B = (G, \beta)$ be a connected bicolored graph on $n \geq 3$ vertices. Let a be any vertex in G. Then there is a string w of length 7 such that $B_w = B^a$.*

Proposition 3 asserts that the colors of the two endvertices of any edge can be reversed using 6 local inversions without altering the graph, while Proposition 4 extends this result and states that the colors of all vertices of an induced triangle can be reversed using 9 local inversions, again preserving the underlying graph.

Proposition 3 ([16]). *Let $B = (G, \beta)$ be a bicolored graph. If $ab \in E(G)$ and $w = ababab$ (of length 6), then $B_w = B^{\{a,b\}}$.*

Proposition 4 ([16]). *Let $B = (G, \beta)$ be a bicolored graph. Let $a, b, c \in V(G)$ with a, b, c forming a triangle in G and $w = ababcbcac$ (of length 9). Then, $B_w = B^{\{a,b,c\}}$.*

Proposition 5 states that the end points of an induced P_3 can be color-reversed in 8 local inversions, without changing the underlying graph.

Proposition 5 ([15]). *Let $B = (G, \beta)$ be a bicolored graph, and let $a, b, c \in V(G)$ with $a, b \in N_G(c)$ and $ab \notin E(G)$. Let $w = cababab c$, a string of length 8. Then $B_w = B^{\{a,b\}}$.*

4   Achieving Tight Bound

In this section, we prove that $cr(G) \leq 3n$ for a graph without isolated vertices, and we show this bound is actually tight for the complete graph on $n > 2$ vertices.

Similar to Propositions 3 and 4, which handle an edge and an induced triangle respectively, Lemma 1 extends the color-reversal technique to induced paths of length two, by showing that, in a bicolored graph, the vertices of an induced P_3 can be color-reversed using 9 local inversions, while leaving the underlying graph unchanged.

Lemma 1. *Let $B = (G, \beta)$ be a graph such that $a, b, c \in V(G)$ with $b, c \in N_G(a)$ and $bc \notin E(G)$. Let $w = babcbcaca$ (of length 9). Then, $B_w = B^{\{a,b,c\}}$.*

Proof. Let $w' = ababcbcac$ (of length 9). We set $w'' = aw'a$. Since $b, c \in N_G(a)$ and $bc \notin E(G)$, the edge bc appears in G_a, so a, b, c form a triangle in G_a. In B_a, relative to B, the colors of the neighbors of a are flipped, while the colors of all other vertices remain unchanged. By Proposition 4, $(G_a)_{w'} = G_a$, and in $(B_a)_{w'}$ the colors of a, b and c are flipped (relative to B_a). Thus $B_{aw'}$ has underlying graph G_a, with the colors of vertex a and all neighbors of a except b and c are flipped, while the colors of the b, c and remaining vertices stay unchanged (relative to B). Finally, $B_{w''} = B_{aw'a}$ has underlying graph $G_{aa} = G$ (by Proposition 1), and the colors of a, b and c are flipped (relative to B) and all other vertices of G retain their original color. We notice that $w \sim w''$ (by Proposition 1). Therefore $B_w = B_{w''} = B^{\{a,b,c\}}$.

$\square$

Note that an induced P_3 is precisely a star S_3. Lemma 1 showed the existence of a string of length $9 = 3 \cdot 3$ such that S_3 can be color-reversed, while the underlying graph remains unchanged. We generalize the case $n = 3$ to arbitrary n using Lemmas 2 and 3.

Lemma 2. *Let $B = (G, \beta)$ be a bicolored graph and S_n be an induced subgraph of G, where S_n is a star on $n \geq 3$ vertices, with n odd. Then there exists a string w of length $3n$ such that $B_w = B^{V(S_n)}$.*

Proof. Let c_0 be the center of the star S_n, and let $c_1, c_2, \ldots, c_{n-1}$ be its leaves. Define the string

$$w' = c_1 c_0 c_1 c_2 c_1 c_2 c_0 c_2 c_0,$$

which has length 9. Since c_1, c_0, c_2 form an induced P_3, by Lemma 1, we have

$$B_{w'} = B^{\{c_0, c_1, c_2\}},$$

that is, the colors of c_0, c_1, c_2 are flipped, while all other vertex colors remain unchanged.

Applying a local inversion at c_0 to $B_{w'}$ flips the colors of c_1 and c_2 back to their original values, while flipping the colors of $c_0, c_3, c_4, \ldots, c_{n-1}$ with respect to B. Moreover, the underlying graph of $(B_{w'})_{c_0}$ is a clique on n vertices. Since n is odd, the clique $G_{w'c_0}[V \setminus \{c_0, c_1, c_2\}]$ contains a perfect matching of size $(n-3)/2$.

Define the string

$$w'' = (c_3 c_4)^3 (c_5 c_6)^3 \cdots (c_{n-2} c_{n-1})^3,$$

which has length $3(n-3)$. By repeated applications of Proposition 3, applying w'' to $(B_{w'})_{c_0}$ preserves the underlying graph and flips the colors of $c_3, c_4, \ldots, c_{n-1}$. Thus, in $((B_{w'})_{c_0})_{w''}$, all vertices except c_0 have their original colors with respect to B.

Applying a local inversion at c_0 restores the star S_n with center c_0 and flips the color of every vertex in the star. Let $w''' = w' c_0 w'' c_0$. Then

$$
\begin{aligned}
w''' &= w' c_0 w'' c_0 \\
&= (c_1 c_0 c_1 c_2 c_1 c_2 c_0 c_2 c_0)(c_0)((c_3 c_4)^3 (c_5 c_6)^3 \cdots (c_{n-2} c_{n-1})^3)(c_0) \\
&= (c_1 c_0 c_1 c_2 c_1 c_2 c_0 c_2)(c_0 c_0)((c_3 c_4)^3 (c_5 c_6)^3 \cdots (c_{n-2} c_{n-1})^3)(c_0) \\
&\sim (c_1 c_0 c_1 c_2 c_1 c_2 c_0 c_2)((c_3 c_4)^3 (c_5 c_6)^3 \cdots (c_{n-2} c_{n-1})^3)(c_0) \\
&\quad (\text{by } Proposition\ 1, c_0 c_0 \sim \varepsilon)
\end{aligned}
$$

By Proposition 3, we have $w'' \simeq \varepsilon$, and by Lemma 1, $w' \simeq \varepsilon$. Using cancellations permitted by Proposition 1, we have $w''' = w' c_0 w'' c_0 \simeq \varepsilon c_0 \varepsilon c_0 \simeq c_0 c_0 \simeq \varepsilon$, and therefore $B_{w'''} = B^{V(S_n)}$.

Define $w = (c_1 c_0 c_1 c_2 c_1 c_2 c_0 c_2)((c_3 c_4)^3 (c_5 c_6)^3 \cdots (c_{n-2} c_{n-1})^3)(c_0)$, which satisfies $w''' \sim w$. Therefore

$$B_w = B_{w'''} = B^{V(S_n)}$$

and also $w \simeq \varepsilon$. Now we compute the length of the string w

$$\begin{aligned}
|w| &= |(c_1 c_0 c_1 c_2 c_1 c_2 c_0 c_2)((c_3 c_4)^3 (c_5 c_6)^3 \cdots (c_{n-2} c_{n-1})^3)(c_0)| \\
&= |c_1 c_0 c_1 c_2 c_1 c_2 c_0 c_2| + |(c_3 c_4)^3 (c_5 c_6)^3 \cdots (c_{n-2} c_{n-1})^3| + |(c_0)| \\
&= 8 + 3(n-3) + 1 = 3n.
\end{aligned}$$

This completes the proof.

$\square$

Lemma 3. *Let $B = (G, \beta)$ be a bicolored graph and S_n be an induced subgraph of G, where S_n is a star on $n \geq 2$ vertices, with n even. Then there exists a string w of length $3n$ such that $B_w = B^{V(S_n)}$.*

Proof of Lemma 3). Let c_0 be the center of the star S_n, and let $c_1, c_2, \ldots, c_{n-1}$ be its leaves. Define the string

$$w' = c_1 c_0 c_1 c_0 c_1 c_0,$$

which has length 6. By Proposition 3, we obtain

$$B_{w'} = B^{\{c_0, c_1\}},$$

that is, the colors of c_0 and c_1 are flipped, while all other vertex colors remain unchanged.

Applying a local inversion at c_0 to $B_{w'}$ flips the color of c_1 back to its original value, while flipping the colors of $c_0, c_2, c_3, \ldots, c_{n-1}$ with respect to B. Moreover, the underlying graph of $(B_{w'})_{c_0}$ is a clique on n vertices. Since n is even, the clique $G_{w' c_0}[V \setminus \{c_0, c_1\}]$ contains a perfect matching of size $(n-2)/2$.

Let
$$w'' = (c_2 c_3)^3 (c_4 c_5)^3 \cdots (c_{n-2} c_{n-1})^3,$$

which has length $3(n-2)$. By repeated applications of the Proposition 3, applying w'' to $(B_{w'})_{c_0}$ preserves the underlying graph and flips the colors of $c_2, c_3, \ldots, c_{n-1}$. Thus, in $((B_{w'})_{c_0})_{w''}$, all vertices except c_0 have their original colors with respect to B, while the color of c_0 remains flipped.

Applying a local inversion at c_0 restores the star S_n with center c_0 and flips the color of every vertex. Let $w''' = w' c_0 w'' c_0$. Then

$$\begin{aligned}
w''' &= (c_1 c_0 c_1 c_0 c_1 c_0)(c_0)((c_2 c_3)^3 (c_4 c_5)^3 \cdots (c_{n-2} c_{n-1})^3)(c_0) \\
&= (c_1 c_0 c_1 c_0 c_1)(c_0 c_0)((c_2 c_3)^3 (c_4 c_5)^3 \cdots (c_{n-2} c_{n-1})^3)(c_0) \\
&\sim (c_1 c_0 c_1 c_0 c_1)((c_2 c_3)^3 (c_4 c_5)^3 \cdots (c_{n-2} c_{n-1})^3)(c_0) \\
&\quad (\text{by } Proposition\,1, c_0 c_0 \sim \varepsilon)
\end{aligned}$$

By Proposition 3, we have $w'' \simeq \varepsilon$, and $w' \simeq \varepsilon$. Using cancellations permitted by Proposition 1, we have $w''' = w' c_0 w'' c_0 \simeq \varepsilon c_0 \varepsilon c_0 \simeq c_0 c_0 \simeq \varepsilon$, and therefore $B_{w'''} = B^{V(S_n)}$.

We define $w = (c_1 c_0 c_1 c_0 c_1)((c_2 c_3)^3 (c_4 c_5)^3 \cdots (c_{n-2} c_{n-1})^3)(c_0)$, which satisfies $w''' \sim w$. Therefore

$$B_w = B_{w'''} = B^{V(S_n)}$$

and also $w \simeq \varepsilon$. Now we compute the length of the string w

$$
\begin{aligned}
|w| &= |(c_1 c_0 c_1 c_0 c_1)((c_2 c_3)^3 (c_4 c_5)^3 \cdots (c_{n-2} c_{n-1})^3)(c_0)| \\
&= |c_1 c_0 c_1 c_0 c_1| + |(c_2 c_3)^3 (c_4 c_5)^3 \cdots (c_{n-2} c_{n-1})^3| + |(c_0)| \\
&= 5 + 3(n-2) + 1 = 3n.
\end{aligned}
$$

This completes the proof.

Lemma 4. *Let T be a rooted tree with at least two vertices. Then T can be decomposed in polynomial time into vertex-disjoint stars, each of size at least two, such that every vertex in T belongs to some star.*

Proof. Let T be rooted at r. Choose any farthest leaf node, say x. Let $p(x)$ be the parent of x. Then $p(x)$ and all children of $p(x)$ form a star with $p(x)$ as the center. If the parent of $p(x)$ is the root, then check whether removing this star would leave r isolated. If so, then we make r a leaf of $p(x)$; else, we remove this star from T. T remains connected since we are removing the leaves-parent pair in a bottom-up manner. Let T' be the new tree obtained. We repeat the same process for T' and get another star and so on. Hence, we get a collection of vertex-disjoint stars in which each vertex is either a leaf or a center. This whole process takes polynomial time (Fig. 2). $\qquad\square$

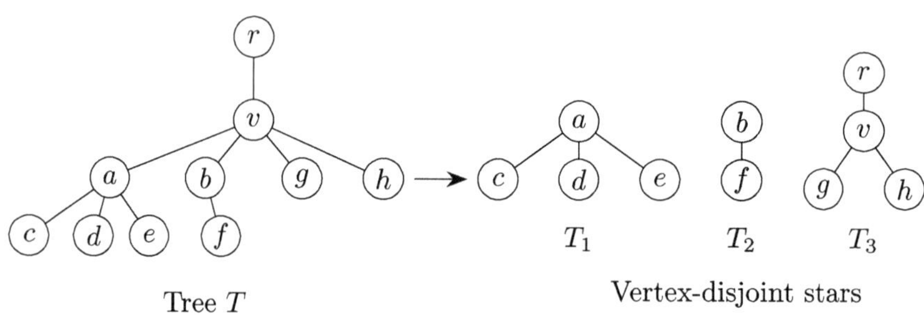

Tree T

Vertex-disjoint stars

Fig. 2. Decomposition of T into vertex-disjoint stars

Theorem 1. *Let $B = (G, \beta)$ be a connected bicolored graph on $n \geq 2$ vertices. Then $cr(G) \leq 3n$.*

Proof. Since G is connected, a spanning tree of G can be computed in polynomial time. Fix a rooted spanning tree T of G with vertex set $V(T) = V(G)$. By Lemma 4, the tree T can be decomposed in polynomial time into vertex-disjoint stars, each of size at least two, such that every vertex in T belongs to some star. Let

$$T = T_1 \cup T_2 \cup \cdots \cup T_t,$$

where each T_i is a star and $|V(T_i)| \geq 2$ for all $1 \leq i \leq t$.

We show that the colors of all vertices in each T_i can be reversed using at most $3|V(T_i)|$ local inversions, while preserving the underlying graph G. Summing over all stars then yields the desired bound.

Small Stars. If $|V(T_i)| = 2$, then T_i is an isolated edge. By Proposition 3, the colors of its two vertices can be reversed in $6 = 3|V(T_i)|$ local inversions.

If $|V(T_i)| = 3$, then T_i induces either a path P_3 or a triangle K_3 in G. In either case, the colors of all three vertices can be reversed in $9 = 3|V(T_i)|$ local inversions by Lemma 1 or Proposition 4, respectively.

Stars of Size at Least Four. Assume now that $|V(T_i)| \geq 4$. Let c_0 be the center of the star T_i. Consider the induced subgraph

$$H_i := G[V(T_i) \setminus \{c_0\}].$$

and compute a maximum matching M_i in H_i. We distinguish two cases.

Case 1: M_i is a Perfect Matching in H_i. Choose one edge $uv \in M_i$. Apply Proposition 3 to every edge of $M_i \setminus \{uv\}$. This flips the colors of all vertices in $V(T_i) \setminus \{c_0, u, v\}$.

The vertices u, v, and c_0 form a triangle in G. By Proposition 4, the colors of u, v, and c_0 can be reversed in 9 local inversions. Thus, all vertices of T_i have their colors reversed using

$$6(|M_i| - 1) + 9 = 6\left(\frac{|V(T_i)| - 3}{2}\right) + 9 = 3\,V(T_i)|$$

local inversions.

Case 2: M_i is not a Perfect Matching in H_i. Apply Proposition 3 to each edge of M_i, reversing the colors of all vertices in $V(M_i)$ using $6|M_i|$ local inversions. Let $R_i := V(T_i) \setminus V(M_i)$. Since M_i is maximal, the set $R_i \setminus \{c_0\}$ is independent in H_i. Therefore, the subgraph $G[R_i]$ is a star centered at c_0 with at least one leaf.

By Lemmas 2 and 3, the colors of all vertices in R_i can be reversed using $3|R_i|$ local inversions. Hence, the total number of local inversions used for T_i is

$$6|M_i| + 3|R_i| = 6|M_i| + 3(|V(T_i)| - 2|M_i|) = 3\,V(T_i)|.$$

Final Argument. In all cases, the colors of the vertices of each star T_i can be reversed using at most $3|V(T_i)|$ local inversions, without altering the underlying graph. Applying these sequences independently for $i = 1,\ldots,t$ yields a global color reversal of B using

$$\sum_{i=1}^{t} 3\,V(T_i)| = 3n$$

local inversions. This proves the theorem.

$\square$

Definition 1. *Let $B = (G,\beta)$ be a bicolored graph. A string $w \in V(G)^*$ is called a minimal string of B if there is no string $w' \in V(G)^*$ with $|w'| < |w|$ and $w \sim w'$.*

Lemma 5. *Let $B = (K_n, \beta)$ be a bicolored graph, where K_n is the complete graph on $n \geq 3$ vertices. If $w \in V(K_n)^*$ is a minimal string of length $3k$ with $k > 0$, then $G_w = K_n$ (a clique). Moreover, there exists a set $A \subseteq V(K_n)$ with $|A| \leq k$ such that $B_w = B^A$.*

Proof. We prove the lemma by induction on k.

Base Case ($k = 1$). Let $w = abc$ be a string of length 3. A direct computation shows that

$$B_{abc} = \begin{cases} B_c, & \text{if } a = b, \\ B^a, & \text{if } a \neq b \text{ and } a = c, \\ B_a, & \text{otherwise.} \end{cases}$$

If $a = b$, then $abc \sim c$; if $b = c$, then $abc \sim a$; and if $a \neq b$ and $a \neq c$, then $abc \sim a$. In each of these cases, abc is equivalent to a strictly shorter string, contradicting the minimality of w. Hence, if w is minimal string, we must have $a \neq b$ and $a = c$. In this case, $B_{abc} = B^a$, and therefore G_w is a clique. Moreover, $B_w = B^A$ with $A = \{a\}$ and $|A| = 1$.

Induction Hypothesis. Assume that the statement holds for all minimal strings of length $3k$, where $k \geq 1$.

Induction Step. Let w be a minimal string of length $3(k + 1)$. Write $w = w''\,a_{k+1}a_{k+2}a_{k+3}$, where $|w''| = 3k$.

Claim. w'' is a minimal string.

Proof of the Claim. Suppose that there exists a string z with $|z| < |w''|$ and $z \sim w''$. Then $|za_{k+1}a_{k+2}a_{k+3}| < |w|$ and

$$za_{k+1}a_{k+2}a_{k+3} \sim w''a_{k+1}a_{k+2}a_{k+3} = w,$$

which contradicts the minimality of w. Hence w'' is minimal. $\square$

By the induction hypothesis, $G_{w''}$ is a clique and $B_{w''} = B^A$ for some $A \subseteq V(K_n)$ with $|A| \leq k$.

Clique Preservation. If $a_{k+1} = a_{k+2}$, then by Proposition 1, $w \sim w''a_{k+3}$, which contradicts the minimality of w. Similarly, if $a_{k+2} = a_{k+3}$, then $w \sim$

$w''a_{k+1}$, again contradicting minimality of w. If $a_{k+1} \neq a_{k+2}$ and $a_{k+1} \neq a_{k+3}$, then $w \sim w''a_{k+1}$. Hence minimality forces

$$a_{k+1} \neq a_{k+2} \quad \text{and} \quad a_{k+1} = a_{k+3}.$$

Since $G_{w''}$ is a clique and the indices satisfy $a_{k+1} \neq a_{k+2}$ and $a_{k+1} = a_{k+3}$, the application of the string $a_{k+1}a_{k+2}a_{k+1}$ preserves the clique structure. Consequently, $G_w = K_n$.

Bounding the Number of Affected Vertices.

The string $a_{k+1}a_{k+2}a_{k+1}$ changes the color of exactly one vertex, namely a_{k+1}. Consequently, $B_w = B^{A'}$, where $A' \subseteq A \cup \{a_{k+1}\}$. Thus $|A'| \leq |A| + 1 \leq k + 1$, completing the induction.

$\square$

Theorem 2. *Let $B = (K_n, \beta)$ be a bicolored graph, where K_n is a complete graph on $n \geq 3$ vertices. Then $cr(K_n) = 3n$.*

Proof. Let $B = (K_n, \beta)$ be a bicolored graph. Porte et al. [15] showed that $cr(K_n) \leq 3n$ by giving a string w' of length $3n$ such that $B_{w'} = (K_n, -\beta)$.

Let w be a string of minimum length such that $B_w = (K_n, -\beta)$ and $|w| = p$. Clearly $p \neq 1$. Moreover, $p \neq 2$: if $w = ab$ with $a = b$, then $B_{ab} = B$, and if $a \neq b$, then B_{ab} is a star, contradicting the definition of w. Hence $p > 2$.

Let $w = x_1 x_2 \ldots x_{p-1} x_p$. Set $w_{[1,s]} = x_1 x_2 \ldots x_s$. Since w is a minimum length string such that $B_w = (K_n, -\beta)$, it is minimal. Furthermore, any $w_{[1,s]}$ is minimal; otherwise, replacing it by an equivalent shorter string would yield a shorter string equivalent to w, contradicting minimality,

Note that $p > 3n$ is not possible, as it is known that $cr(G) \leq 3n$. Hence, we assume that $p \leq 3n$. Suppose $p = 3\ell + r$ with $1 \leq \ell \leq n - 1$ and $r \in \{1, 2, 3\}$.

Case 1: $r = 1$. Let $w = w_{[1,3\ell]} x_{3\ell+1}$ for some ℓ. By Lemma 5, after exactly 3ℓ steps of the string w, we get the clique K_n. Applying local inversion on $x_{3\ell+1}$ gives us a star with center $x_{3\ell+1}$. Hence $r = 1$ is not possible.

Case 2: $r = 2$. Let $w = w_{[1,3\ell]} x_{3\ell+1} x_{3\ell+2}$ for some ℓ. By Lemma 5, after exactly 3ℓ steps of the string w, we get the clique K_n. Note that if $x_{3\ell+1} = x_{3\ell+2}$, then $w \sim w_{[1,3\ell]}$, which contradicts the minimality of w. Hence we assume that $x_{3\ell+1} \neq x_{3\ell+2}$. But in this case, B_w is a star with center $r_{3\ell+1}$. Hence $r = 2$ is not possible.

Case 3: $r = 3$. Let $w = w_{[1,3k]}$ for some k. By Lemma 5, after exactly $3k$ steps of the string w, we have the the clique and color of at most k vertices have changed (relative to B). Since we need to flip the color of n vertices, k must be at least n. This gives $p \geq 3n$ and we assumed $p \leq 3n$. Therefore $p = 3n$ and $cr(K_n) \geq 3n$.

$\square$

5   Transformations Between Bicolorings

Theorem 3 bounds the number of local inversions required to transform one bicolored graph into another bicolored graph of the same underlying structure.

Theorem 3. *Let G be a connected graph on $n \geq 2$ vertices. Let $B = (G, \beta)$ and $B' = (G, \beta')$ be two bicolored graphs on G. Then there exists a string w of length at most $5n - 3$ such that $B_w = B'$.*

Proof. Let $V_0 = \{v \in V(G) : \beta(v) = \beta'(v)\}$ and $V_1 = \{v \in V(G) : \beta(v) = -\beta'(v)\}$. Let $n := |V(G)|$, $n_0 := |V_0|$ and $n_1 := |V_1|$. Clearly $n = n_0 + n_1$. We distinguish two cases based on the value of n_1.

Case 1: *When $n_1 \leq \lfloor \frac{n}{2} \rfloor$.* In this case, we directly fix V_1 by applying Proposition 2 to each vertex of V_1, thereby ensuring that $\beta(v) = \beta'(v)$ for all $v \in V_1$. Since fixing the color of a single vertex can be done with a string of length 7, V_1 can be fixed by a string of length at most $7|V_1| = 7n_1 \leq 7 \lfloor \frac{n}{2} \rfloor$, which is at most $5n - 3$ for $n \geq 2$. This finishes the argument for this case.

Case 2: *When $n \geq n_1 \geq \lfloor \frac{n}{2} \rfloor + 1$.* If $n_1 = n$, then $\beta = -\beta'$. In this case, by Theorem 1, there is a string of length at most $3n \leq 5n - 3$ that transforms B to B'. So, from now on, we assume $n_1 \neq n$.

Now, to get B' from B, we consider two strategies and take the strategy for which the length of the string is minimum.

Strategy 1: Fix V_1.

We begin by examining the induced subgraph $G[V_1]$ and split the analysis into two cases, depending on whether $G[V_1]$ contains an edge.

i. $G[V_1]$ contains an edge, say ab. In this case, we use Proposition 3 to change the color of vertices a and b with a string of length 6. Applying Proposition 2 to each vertex of $V_1 \setminus \{a, b\}$ ensures $\beta(v) = \beta'(v)$ for all such $v \in V_1 \setminus \{a, b\}$, using a string of length $7|V_1 \setminus \{a, b\}|$. In this way, we have fixed V_1 with a string of length $7(n_1 - 2) + 6$.

ii. $G[V_1]$ contains no edge (i.e., $G[V_1]$ is independent). Since G is connected, $G[V_1]$ is independent and $n_1 > n_0$, there must exist a vertex $u \in V_0$ with $|N(u) \cap V_1| \geq 2$. Pick two arbitrary vertices $x, y \in N(u) \cap V_1$. Since $G[V_1]$ is independent, the vertices x, u, y form an induced P_3 with endpoints x and y. We use Proposition 5 to change the color of vertices x and y with a string of length 8. Applying Proposition 2 to each vertex of $V_1 \setminus \{x, y\}$ ensures $\beta(v) = \beta'(v)$ for all such $v \in V_1 \setminus \{x, y\}$, using a string of length $7|V_1 \setminus x, y|$. In this way, we have fixed V_1 with a string of length $7(n_1 - 2) + 8$.

In either subcase, we can obtain B' from B (by fixing V_1) using a string of length at most $7(n_1 - 2) + 8 = 7n_1 - 6 = 7n - 7n_0 - 6$.

Strategy 2: Flip V_0, then flip all of G.

We flip V_0 by applying Proposition 2 to each vertex of V_0, so that $\beta(v) = -\beta'(v)$ for all $v \in V_0$. Hence, V_0 can be flipped with a string of length $7|V_0|$. After this, we have $\beta(u) = -\beta'(u)$ for all $u \in V(G)$. By using Theorem 1, the color of the new graph can be fixed (i.e. $\beta(u) = \beta'(u) \ \forall u \in V(G)$) with a string of length at most $3n$. Hence, by this strategy, B can be transformed into B' using a string of length at most $7|V_0| + 3n \le 7n_0 + 3n$.

Thus, Strategy 1 uses length at most $p := 7n - 7n_0 - 6$, while Strategy 2 uses $q := 7n_0 + 3n$. For the given partition (n_0, n_1) we can get B' from B with a string of length at most $\min\{p, q\} \le \min\{\, 7n - 7n_0 - 6,\ 7n_0 + 3n \,\}$. Maximizing this upper bound over $0 \le n_0 \le \frac{n}{2}$ occurs when $7n - 7n_0 - 6 = 7n_0 + 3n$, which implies $2n - 7n_0 - 3 = 0$. Substituting back yields

$$\min\{p, q\} \le 5n - 3.$$

Hence, in both cases, we obtain $5n - 3$ as an upper bound. Therefore, there exists a string w of length at most $5n - 3$ such that $B_w = B'$.

$\square$

6 Concluding Remarks

We showed that any bicolored graph can be color-reversed in at most $3n$ local inversions. We also showed that the bound $3n$ is tight for the complete graphs. If there are two bicolorations of a graph G then we can obtain one from another in at most $5n - 3$ moves. Despite these results, several fundamental questions remain open.

- For any two bicolorations B and B' of G, can B' obtained from B in at most $3n$ local inversions?
- What are the graph classes for which $3n$ is tight and for which graph classes $cr(G) < 3n$?

References

1. Adcock, J.C., Morley-Short, S., Dahlberg, A., Silverstone, J.W.: Mapping graph state orbits under local complementation. Quantum **4**, 305 (2020). https://doi.org/10.22331/q-2020-08-07-305
2. Bodlaender, H.L., Heggernes, P., Lokshtanov, D.: Graph modification problems (DAGSTUHL seminar 14071). DAGSTUHL Reports **4**(2), 38–59 (2014). https://doi.org/10.4230/DAGREP.4.2.38

3. Brijder, R., Hoogeboom, H.J.: The group structure of pivot and loop complementation on graphs and set systems. Eur. J. Combin. **32**(8), 1353–1367 (2011). https://doi.org/10.1016/j.ejc.2011.03.002

4. Cabello, A., Parker, M.G., Scarpa, G., Severini, S.: Exclusivity structures and graph representatives of local complementation orbits. J. Math. Phys. **54**(7), 072202 (2013). https://doi.org/10.1063/1.4813438

5. Cai, L.: Fixed-parameter tractability of graph modification problems for hereditary properties. Inf. Process. Lett. **58**(4), 171–176 (1996). https://doi.org/10.1016/0020-0190(96)00050-6

6. Crespelle, C., Drange, P.G., Fomin, F.V., Golovach, P.A.: A survey of parameterized algorithms and the complexity of edge modification. Comput. Sci. Rev. **48**, 100556 (2023). https://doi.org/10.1016/j.cosrev.2023.100556

7. Danielsen, L.E., Parker, M.G.: Edge local complementation and equivalence of binary linear codes. Des. Codes Cryptogr. **49**(1–3), 161–170 (2008). https://doi.org/10.1007/s10623-008-9190-x

8. Ehrenfeucht, A., Harju, T., Rozenberg, G.: Transitivity of local complementation and switching on graphs. Discrete Math. **278**(1–3), 45–60 (2004). https://doi.org/10.1016/j.disc.2003.04.001

9. Hahn, F., Pappa, A., Eisert, J.: Quantum network routing and local complementation. npj Quantum Inf. **5**(1), 1–7 (2019). https://doi.org/10.1038/s41534-019-0191-6

10. Harary, F.: On the notion of balance of a signed graph. Michigan Math. J. **2**(2), 143–146 (1953). https://doi.org/10.1307/mmj/1028989917

11. Javelle, J., Mhalla, M., Perdrix, S.: On the minimum degree up to local complementation: bounds and complexity. In: Graph-Theoretic Concepts in Computer Science. Lecture Notes in Computer Science, vol. 7551, pp. 138–147. Springer (2012). https://doi.org/10.1007/978-3-642-34611-8_16

12. Joo, J., Feder, D.L.: Edge local complementation for logical cluster states. New J. Phys. **13**(6), 063025 (2011). https://doi.org/10.1088/1367-2630/13/6/063025

13. Kim, D., il Oum, S.: Vertex-minors of graphs: a survey. Discrete Appl. Math. **351**, 54–73 (2024). https://doi.org/10.1016/j.dam.2024.03.011, https://www.sciencedirect.com/science/article/pii/S0166218X24001136

14. il Oum, S.: Rank-width and vertex-minors. J. Comb. Theory Ser. B **95**(1), 79–100 (2005). https://doi.org/10.1016/J.JCTB.2005.03.003

15. Porte, K.S., Sandeep, R.B., Santra, K.: Improved upper bounds on color reversal by local inversions (2025). https://arxiv.org/abs/2510.00149

16. Sabidussi, G.: Color-reversal by local complementation. Discrete Math. **64**(1), 81–86 (1987). https://doi.org/10.1016/0012-365X(87)90240-8

17. Seidel, J.J.: A survey of two-graphs. In: Geometry and Combinatorics: Selected Works of J. J. Seidel, pp. 146–176. Academic Press (1991). https://doi.org/10.1016/B978-0-12-189420-7.50018-9

18. Traldi, L.: On the linear algebra of local complementation. Linear Algebra Appl. **436**(5), 1072–1089 (2012). https://doi.org/10.1016/j.laa.2011.06.048

19. Traldi, L.: Binary matroids and local complementation. Eur. J. Combin. **45**, 21–40 (2015). https://doi.org/10.1016/j.ejc.2014.10.001

20. Zaslavsky, T.: Signed graphs. Discrete Appl. Math. **4**(1), 47–74 (1982). https://doi.org/10.1016/0166-218X(82)90033-6

21. Zhang, J.: Local complementation rule for continuous-variable four-mode unweighted graph states. Phys. Rev. A **78**(3), 034301 (2008). https://doi.org/10.1103/PhysRevA.78.034301

Domination and Coverage Problems Under Vulnerability Constraints

Ioannis Lamprou, Nikolaos Lazaropoulos, Ioannis Sigalas,
Ioannis Vaxevanakis$^{(\boxtimes)}$, and Vassilis Zissimopoulos

Department of Informatics and Telecommunications, National and Kapodistrian
University of Athens, Athens, Greece
`vaxjohn@di.uoa.gr`

Abstract. In various domination and coverage problems, certain vertices or edges should not be dominated/covered and are designated as vulnerable. Motivated by this, we define the *k -Vertex Maximum Domination Ratio with Vulnerable Vertices* (*k-MaxDRVV*) problem, which extends the budgeted dominating set problem to include vulnerability constraints. We propose an approximation algorithm based on an unbudgeted variant of *k-MaxDRVV*, termed the *Maximum Domination Ratio with Vulnerable Vertices* (*DRVV*) problem. For bounded-degree graphs of order n, our algorithm provides an $O(k/n)$-approximation for the *k-MaxDRVV* problem. We also introduce the *Vertex Cover with Vulnerable Edges* (*VCVE*) problem, which can be naturally expressed as a special case of the Red-Blue Set Cover problem. We develop a 4-approximation algorithm for the *VCVE* problem.

Keywords: Approximation Algorithms · Coverage · Influence Maximization · Vulnerable · Domination

1 Introduction

Domination and Coverage problems have been studied extensively in the literature [7,8,20]. In this work, we focus on variants of these problems with constraints regarding vulnerable vertices or edges. These formulations are closely connected to the classical Red-Blue Set Cover problem [3,19] and are motivated by applications in influence maximization, where the diffusion cascade

This work was partially supported by the Special Account for Research Grants (ELKE) of the National and Kapodistrian University of Athens (NKUA).

I. Vaxevanakis—The research work was supported by the Hellenic Foundation for Research and Innovation (HFRI) under the 3rd Call for HFRI PhD Fellowships (Fellowship Number: 5245).

V. Zissimopoulos—The research work was partially supported by the Chinese Academy of Sciences President's International Fellowship Initiative (PIFI) under grant No. 2025PVA0150, and conducted during a visit to the Shenzhen Institutes of Advanced Technology (SIAT).

F. Foucaud and A. Parreau (Eds.): IWOCA 2026, LNCS 16587, pp. 415–429, 2026.
https://doi.org/10.1007/978-3-032-27732-9_29

model is typically used to capture how ideas or behaviors spread in networks while respecting vulnerable elements. In contrast to cascade-based formulations, our perspective abstracts the diffusion process into structural graph properties, aiming to maximize the number of dominated vertices or covered edges while controlling the number of vulnerable ones.

Influence maximization aims to select a set of seed vertices whose activation triggers the largest spread of ideas or behaviors throughout the network. There has been significant research [18,22] focusing on optimizing the influence on a specific target audience. Socially responsible influence maximization [4] further emphasizes minimizing influence on vulnerable vertices. A common metric to evaluate this trade-off is the Additively Smoothed Ratio (ASR), the fraction of expected influenced non-vulnerable vertices over vulnerable ones. We, on the other hand, focus on combinatorial formulations that represent vulnerability constraints directly on the graph.

Building on this perspective, we define the k-Vertex Maximum Domination Ratio with Vulnerable Vertices (k-$MaxDRVV$), which maximizes the ratio of dominated non-vulnerable over vulnerable vertices. This formulation offers a clear combinatorial abstraction of socially responsible influence maximization and provides a framework in which vulnerability constraints can be studied through domination and coverage problems. Vertex Cover with vulnerable edges ($VCVE$) is an example of coverage problem with vulnerable edges.

Beyond applications in social networks, the concept of Red-Blue vertices (or vulnerable and non-vulnerable vertices in our formulation) extends to a variety of other domains. For instance, data mining provides an important motivation, as illustrated by [3] in the context of detecting fraudulent claims within extensive datasets of legitimate Medicare records. Furthermore, the proposed framework may contribute to advancements in machine learning by enabling the prioritization of salient information and facilitating the enrichment of training datasets with data of higher relevance.

1.1 Related Work

The Dominating Set problem is well studied, and numerous variations have been introduced. One such budgeted variant is the k-Vertex Maximum Domination (k-$Max\ VD$) problem, defined as follows: Given a graph G, find a set with at most k vertices that maximizes the number of dominated vertices. Miyano and Ono [16] provide a $(1-\frac{1}{e})$-approximation algorithm. In the connected setting, the analogous problem is the budgeted connected dominating set problem ($BCDS$): Given a graph G, find a set with at most k vertices that induce a connected subgraph and maximize the number of dominated vertices. Khuller, Purohit and Sarpatwar [11] provide a $\frac{1}{13}(1-\frac{1}{e})$-approximation algorithm, which was gradually improved up to $\frac{1}{6}(1-\frac{1}{e})$ [12,14,15,21].

In the direction of Influence Maximization, Chen, Loukides, Pissis, and Chan [4] deal with the Independent Cascade Model under the ASR metric. Given a graph $G = (V, E)$ where vertices are partitioned into non-vulnerable V_N and vulnerable V_L, the ASR metric of a subset of vertices $S \subseteq V_N$ is

$ASR(S,c) = \frac{\sigma_N(S)+c}{\sigma_L(S)+c}$. The set S is called a seed-set and $\sigma_N(S)$ and $\sigma_L(S)$ are the expected number of influenced non-vulnerable vertices and the expected number of vulnerable vertices, respectively. A positive constant $c > 0$ is introduced primarily to allow ASR to be applied even when the seed set contains no vulnerable vertices. The choice of c influences the outcome, as discussed in [4]; specifically, increasing the value of c results in a larger output seed set. c is part of the input of any specific instance. The *ASR-Maximizing Seed-set problem* aims to identify a seed set containing no more than k vertices that maximizes the ASR metric. Note that ASR is a non-monotone, non-submodular, and non-supermodular function. The authors propose an algorithm for the problem, that guarantees $\mathbb{E}[ASR(S,c)] \geq \frac{\sigma_L(S^*)+c}{|V|+c}(1 - \frac{1}{e})ASR(S^*,c)$, where S^* is an optimal solution of size at most k with respect to ASR, and $\mathbb{E}[ASR(S,c)]$ denotes the expectation of ASR over every solution S constructed by their algorithm.

The Red-Blue Set Cover ($RBSC$) problem [3] is a natural generalization of the Set Cover problem with vulnerability constraints. In $RBSC$, the ground set is partitioned into "blue" elements, which must be covered, and "red" elements, whose coverage should be minimized. Formally, given a universe $U = R \cup B$ (with $R \cap B = \emptyset$), the goal is to select a subcollection of sets that covers all blue elements while covering as few red elements as possible. In [3,5], it is shown that it is quasi-NP-hard to approximate the Red-Blue Set Cover problem with ratio $2^{\log^\varepsilon |U|}$, for any $0 < \varepsilon < 1$, and Peleg [19] provided a $2\sqrt{|U| \cdot \log |B|}$-approximation algorithm for $RBSC$.

1.2   Contribution

This work investigates domination and coverage problems involving vulnerable vertices or edges. We define the *k-Vertex Maximum Domination Ratio with Vulnerable Vertices (k-MaxDRVV)* problem (Sect. 2). In order to provide an approximation algorithm for k-$MaxDRVV$, we define a version without the k restriction, named the *Maximum Domination Ratio with Vulnerable Vertices (DRVV)* problem (Subsect. 2.1). We prove that $DRVV$ is NP-hard by reducing its decision version from a special case of 3-$\mathcal{SAT}$. This implies hardness of the k-$MaxDRVV$. Next, we define the *Maximum Domination with a Bounded Number of Vulnerable Vertices (DVV)* problem (Subsect. 2.2) and provide the first $\frac{1}{\Delta+1} \cdot (1 - e^{-\frac{1}{\Delta}})$-approximation ratio, where Δ is the maximum degree of the graph. Using DVV, we propose an algorithm for $DRVV$ (Subsect. 2.3) and prove that it achieves the same approximation ratio as DVV. Then, we provide an approximation algorithm for k-$MaxDRVV$ (Subsect. 2.4) by restricting the solution of $DRVV$ to k vertices using the Maximum Coverage Problem. This algorithm guarantees a $\frac{\rho \cdot r}{d}$ approximation ratio for the k-$MaxDRVV$ problem, where ρ is the approximation ratio for $Max\ CP$, r is the approximation ratio for $DRVV$, $d = \lceil m/k \rceil$, $k \leq m < n$ and $m = |S|$ with S the solution provided by the algorithm $DRVV$. For the class of graphs with bounded maximum degree, the approximation ratio becomes $O(k/n)$. Also, in Sect. 3, we consider the Vertex Cover problem with vulnerability constraints and define the Vertex

Cover problem with vulnerable edges ($VCVE$). We propose an approximation algorithm that achieves a 4-approximation guarantee.

2 k-Vertex Maximum Domination Ratio with Vulnerable Vertices (k-$MaxDRVV$)

Inspired by the problem defined in [4], we propose a variant termed the k-Vertex Maximum Domination Ratio with Vulnerable Vertices (k-$MaxDRVV$). In contrast to [4], which uses the Independent Cascade Model to estimate expected influence, our work focuses on maximizing the domination ratio on a deterministic graph.

We define ASR as follows: Given a graph $G = (V, E)$ where vertices are partitioned into non-vulnerable V_N and vulnerable V_L, the ASR metric of a subset of vertices $S \subseteq V_N$ is $ASR(S, c) = \frac{\sigma_N(S) + c}{\sigma_L(S) + c}$, where c is a constant, $\sigma_N(S)$ is the count of non-vulnerable vertices in the closed neighborhood of S, and $\sigma_L(S)$ is the count of vulnerable vertices in the neighborhood of S. The closed neighborhood of S is the union of S and the set of adjacent vertices to S. With these in mind, we proceed to the problem definition:

Definition 1 ($k-MaxDRVV$). *Given a graph $G = (V, E)$ where vertices are partitioned into non-vulnerable V_N and vulnerable V_L, an integer $k \leq |V_N|$ and $c \in \mathbb{R}^+$ a constant, find a subset $S \subseteq V_N$ such that $ASR(S, c)$ is maximized and $|S| \leq k$.*

2.1   Maximum Domination Ratio with Vulnerable Vertices ($DRVV$)

In this subsection, we introduce the Maximum Domination Ratio with Vulnerable Vertices Problem, as an unbudgeted version of k-$MaxDRVV$. First, we prove that $DRVV$ is NP-hard by reducing its decision version to 3-$\mathcal{SAT}_{equal}$ [13]. Next, we define the Maximum Domination with a Bounded Number of Vulnerable Vertices, we prove its hardness and present an approximation algorithm for this problem. Finally, we use the result for DVV to obtain an approximation result for $DRVV$ (Fig. 1).

Definition 2 ($DRVV$). *Given a graph $G = (V, E)$ where vertices are partitioned into non-vulnerable V_N and vulnerable V_L, $c \in \mathbb{R}^+$ a constant, find a subset $S \subseteq V_N$ such that $ASR(S, c)$ is maximized.*

To prove that $DRVV$ is NP-hard, we define its decision version and reduce the NP-complete problem 3-$\mathcal{SAT}_{equal}$ to decision $DRVV$ [13].

Definition 3 (Decision $DRVV$). *Given a graph $G = (V, E)$ where vertices are partitioned into non-vulnerable V_N and vulnerable V_L and a rational number W and $c \in \mathbb{R}^+$ a constant, decide whether exists a subset of vertices $S \subseteq V_N$ such that it holds $ASR(S, c) \geq W$.*

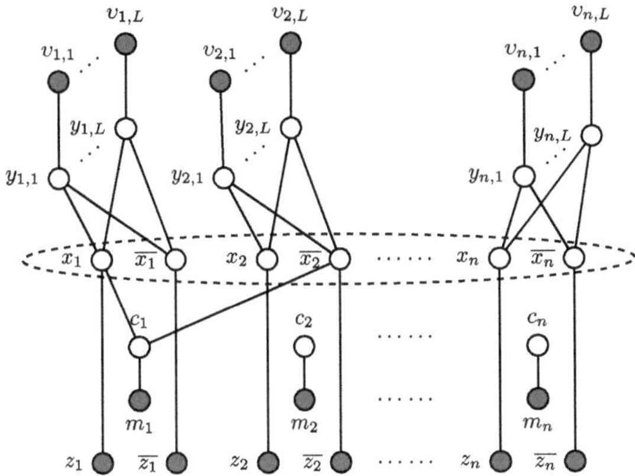

Fig. 1. The graph G constructed for the reduction. The dashed ellipsis is a clique. All leaves (red vertices) are vulnerable. (Color figure online)

Definition 4 (3-$\mathcal{SAT}_{equal}$ [13]). *Given a CNF formula ϕ with n variables and n clauses, where each clause is a disjunction of exactly 3 literals, decide whether ϕ is satisfiable.*

The Reduction. Given a 3-$\mathcal{SAT}_{equal}$ formula ϕ, we create a graph G. Let $x_1, \overline{x_1}, x_2, \overline{x_2}, \ldots, x_n, \overline{x_n}$ stand for the variables of ϕ and $c_1, c_2, \ldots, c_n$ for the clauses of ϕ (see Fig. 1). We construct the graph G in the following way: we place one vertex per literal $x_i, \overline{x_i}$ ($2n$ vertices in total), one vertex per clause c_i (n vertices) and a set of L vertices, where L is $poly(n)$ (definition of L follows later), for each variable (namely y_{ij} for $j = 1, \ldots, L$) summing up to $L{\cdot}n$ vertices. We call the two vertices $x_i, \overline{x_i}$ a *literal-pair* and each vertex c_i a *clause-vertex*. We add vulnerable vertices $z_i, \overline{z_i}$ connected to literals, vulnerable vertices m_i connected to clauses and each y_{ij} vertex is connected to a vulnerable vertex v_{ij}. Then, we connect each literal vertex to *all* the other literal vertices. Moreover, each literal-vertex is connected to all the corresponding clause-vertices where it appears in ϕ. Finally, x_i and $\overline{x_i}$ are connected to y_{ij} for all j. The constant c is part of the input and does not affect the proof. It is clear that the construction can be done in polynomial time. We define $\bar{c} = \left\lceil \max\{c, \frac{1}{c}\} \right\rceil$, $L = \lceil (1+\bar{c}+2n\bar{c})n \rceil$ and $\mathcal{S} = \{S \subseteq V : ASR(S, c) \geq ASR(S', c), \ \forall S' \subseteq V\}$ be the collection of all subsets of vertices V that maximize the function ASR.

Lemma 1. *If ϕ is satisfiable, there exists $S \subseteq V$ such that $ASR(S, c) \geq \frac{Ln+3n+c}{n+c}$.*

Lemma 2. *Let $S \in \mathcal{S}$ be a subset of vertices that maximize the function ASR. It holds $|S| = n$ and $S = \{s_1, s_2, \cdots, s_n\}$, where $s_i \in \{x_i, \overline{x_i}\}$ for all $i \in \{1, ..., n\}$.*

Lemma 3. *If there exists no satisfiable assignment for ϕ, then $\forall S \subseteq V$ it holds $ASR(S, c) < \frac{Ln+3n+c}{n+c}$.*

Theorem 1. *Decision DRVV is NP-Complete.*

Proof. The result stands by Lemma 1, Lemma 2 and Lemma 3. The detailed proof is deferred to the extended manuscript of this study.

Theorem 1 implies the hardness of $k\text{-}MaxDRVV$. Indeed, the existence of a polynomial-time algorithm for it would have led to a polynomial algorithm for $DRVV$.

Theorem 2. $k\text{-}MaxDRVV$ *is NP-Hard.*

2.2 Maximum Domination with Bounded Number of Vulnerable Vertices (DVV)

Following, we define the Maximum Domination with Bounded number of Vulnerable Vertices. The goal is to maximize the count of dominated non-vulnerable vertices while the vulnerable vertices are kept up to a bound.

Definition 5. (DVV). *Given a graph $G = (V, E)$ and an integer B, where vertices are partitioned into non-vulnerable V_N and vulnerable V_L, find a subset of vertices $S \subseteq V_N$ such that S dominates at most B vulnerable vertices ($\sigma_L(S) \leq B$, $0 \leq B \leq |V_L|$ integer) and maximizes the number of non-vulnerable vertices it dominates.*

DVV is NP-hard, and to prove this, we use a reduction from $k\text{-}Max\ VD$, which was proven to be NP-hard [16].

Definition 6 ($k-MaxVD$ *[16]*). *Given a graph $G(V, E)$ and an integer $k < |V|$ find a subset $S \subseteq V$ where $|S| \leq k$ such that the count of dominated vertices is maximized.*

Theorem 3. *The DVV is as hard as $k\text{-}Max\ VD$.*

Proof. Given an instance of $k\text{-}Max\ VD$ with a graph $G = (V, E)$, for every vertex v_i, add a vertex u_i and an edge $[v_i, u_i]$. Let V' be the set of new vertices and F the set of new edges. For the augmented graph $G' = (V \cup V', E \cup F)$, V is the set of non-vulnerable vertices and V' is the set of vulnerable vertices. Let S' be an optimal solution of DVV on G'. Since every vertex in G' has exactly one vulnerable neighbor by setting $B = k$, S' has at most k vertices and it dominates maximum count of non-vulnerable vertices. So, solution S' for DVV on G' is exactly an optimal solution S for $k\text{-}Max\ VD$ on G. □

To address the DVV problem, we formulate it as a Set Union Knapsack Problem ($SUKP$) [1]. This problem has a polynomial approximation algorithm with a constant approximation ratio of $(1 - e^{-\frac{1}{d}})$, when the number of items in which an element is present is bounded by a constant d.

Definition 7 (*SUKP*). *The problem consists of a set of elements $U = \{1, ..., n\}$ and a set of items $X = \{1, ..., m\}$. Each item $i \in X$ corresponds to a subset of elements $X_i \subseteq U$ with a nonnegative profit given by $p : X \to \mathbb{R}^+$. Each element has a nonnegative weight given by $w : U \to \mathbb{R}^+$. For a subset $A \subseteq X$, the weighted union of set A is $W(A) = \sum_{e \in (\cup_{i \in A} X_i)} w(e)$ and $P(A) = \sum_{i \in A} p(i)$. The goal is to find a subset of items $S \subseteq X$ such that $P(S)$ is maximized and $W(S) \leq B$, where B is a given budget. Let (X, p, w, B) denote a SUKP instance.*

The Algorithm. We transform each instance of DVV into an instance of $SUKP$ (see Algorithm 1). At first, all vertices that have no vulnerable neighbors are added to the solution. The remaining vertices are used to construct an instance of the $SUKP$. Each vulnerable vertex corresponds to an element, which is assigned a weight of 1. Each non-vulnerable vertex is represented as an item, including the vulnerable neighbors it dominates. The profit of each item is set equal to the number of non-vulnerable vertices it dominates. Based on this mapping, a $SUKP$ instance (X, p, w, B) is created and solved using a $SUKP$ algorithm provided by [1]. The solution obtained from $SUKP$ is then translated back into a feasible solution for the DVV problem. The complexity of this algorithm is dominated by the complexity of the $SUKP$ algorithm, which is $O(|V|^3)$. We prove that the approximation guarantee for DVV is $\frac{1}{\Delta+1}(1 - e^{-\frac{1}{\Delta}})$.

The Analysis. We proceed with some notation. Let S^* be an optimal solution of DVV, S^*_{SUKP} an optimal solution of the $SUKP$ problem, S_{SUKP} be a set of vertices that $SUKP$ algorithm returns, S_0 be the set of vertices with no vulnerable neighbors and S the set of vertices the algorithm returns as the solution of DVV. Also $P(S) = \sum_{v \in S} p(v)$ be the sum of profits of a set of vertices S, $\Delta_N = \max_{v \in V_N} |N(v) \cap V_N| \leq \Delta$, $\Delta_L = \max_{v \in V_L} |N(v) \cap V_N| \leq \Delta$

Algorithm 1: $DVV(G, B)$

    Input   : A graph $G = (V, E)$ and an integer $B \leq |V_L|$. The vertices V are
               partitioned into non-vulnerable V_N and vulnerable V_L
    Output: A subset of vertices $S \subseteq V_N$ such that S is a feasible solution

1 $S_0 \leftarrow \emptyset$
2 **foreach** $u \in V_L$ **do**
3     $w(u) \leftarrow 1$
4 **foreach** $v \in V_N$ **do**
5     $X_v \leftarrow N(v) \cap V_L$
6   **if** $X_v = \emptyset$ **then**
7        $S_0 \leftarrow S_0 \cup \{v\}$
8 **foreach** $v \in V_N$ **do**
9     $p(v) \leftarrow \sigma_N(v)$
10 $X \leftarrow V_N \setminus S_0$
11 $S_{SUKP} \leftarrow SUKP(X, p, w, B)$
12 $S \leftarrow S_0 \cup S_{SUKP}$
13 **return** S

---

and $N_A[u]$ is defined to be the closed neighborhood of vertex u in subset A. Recall that $\sigma_N(S)$ is the count of non-vulnerable vertices in the closed neighborhood of S (including S itself). For the analysis all non-vulnerable vertices are counted in the profit function $p(v) = \sigma_N(v)$, for every $v \in V_N$.

Lemma 4. *For any subset $A \subseteq V$ it holds: $\sigma_N(A) \leq P(A) \leq (\Delta_N + 1) \cdot \sigma_N(A)$.*

Proof. By Algorithm 1, every profit of a vertex is equal to the number of non-vulnerable neighbors. So, every non-vulnerable vertex adds one point to each neighboring vertex profit, including itself. So:

$$P(A) = \sum_{u \in A} p(u) = \sum_{u \in N[A] \cap V_N} |N_A[u]|$$

By definition of $N_A[u]$, for every $u \in N[A] \cap V_N$, $1 \leq |N_A[u]| \leq \Delta_N + 1$. So,

$$\sum_{u \in N[A] \cap V_N} 1 \leq \sum_{u \in N[A] \cap V_N} |N_A[u]| \leq \sum_{u \in N[A] \cap V_N} (\Delta_N + 1) \Rightarrow$$

$$\sigma_N(A) \leq P(A) \leq (\Delta_N + 1) \cdot \sigma_N(A)$$

$\square$

Theorem 4. *Algorithm 1 is a $\frac{1}{\Delta+1} \cdot (1 - e^{-\frac{1}{\Delta}})$ - approximation for DVV problem.*

Proof. There exists an optimal solution S^* such that $S_0 \subseteq S^*$. Let (X, p, w, B) an instance of the $SUKP$ problem that Algorithm 1 defines. By [1], S_{SUKP} is a feasible solution for $SUKP$ instance and $P(S_{SUKP}) \geq (1 - e^{-\frac{1}{\Delta_L}}) \cdot P(S^*_{SUKP})$. By definition, $S^* \setminus S_0$ is also a feasible solution for $SUKP$ instance and so $P(S^*_{SUKP}) \geq P(S^* \setminus S_0)$. We have:

$$P(S_{SUKP}) \geq (1 - e^{-\frac{1}{\Delta_L}}) \cdot P(S^*_{SUKP}) \geq (1 - e^{-\frac{1}{\Delta_L}}) \cdot P(S^* \setminus S_0) \Rightarrow$$

$$P(S_{SUKP}) + P(S_0) \geq (1 - e^{-\frac{1}{\Delta_L}}) \cdot (P(S^* \setminus S_0) + P(S_0)) \Rightarrow$$

$$P(S) \geq (1 - e^{-\frac{1}{\Delta_L}}) \cdot P(S^*).$$

By Lemma 4, $(\Delta_N + 1) \cdot \sigma_N(S) \geq P(S)$ and $P(S^*) \geq \sigma_N(S^*)$. So:

$$(\Delta_N + 1) \cdot \sigma_N(S) \geq P(S) \geq (1 - e^{-\frac{1}{\Delta_L}}) \cdot P(S^*) \geq (1 - e^{-\frac{1}{\Delta_L}}) \cdot \sigma_N(S^*) \Rightarrow$$

$$\sigma_N(S) \geq \frac{1}{\Delta_N + 1} \cdot (1 - e^{-\frac{1}{\Delta_L}}) \cdot \sigma_N(S^*) \geq \frac{1}{\Delta + 1} \cdot (1 - e^{-\frac{1}{\Delta}}) \cdot \sigma_N(S^*)$$

$\square$

2.3   Algorithm and Approximation for *DRVV*

In order to get an approximation algorithm for $DRVV$ (Algorithm 2), we are using Algorithm 1. We fix the number of vulnerable vertices (B) and we create an instance of DVV. Then we call the DVV algorithm, for B from 0 to $|V_L|$. The value giving the best ratio is chosen, with the corresponding solution.

Theorem 5. *Algorithm 2 is a* $\frac{1}{\Delta+1} \cdot (1 - e^{-\frac{1}{\Delta}})$ *-approximation for DRVV problem.*

Proof. Let S^*_{DRVV} be the optimal solution of $DRVV$ and $B^* = \sigma_L(S^*_{DRVV})$. Let S_{DVV} be the solution returned by Algorithm 1 and S^*_{DVV} be the optimal solution of DVV for the same instance G, B^*. If S_{DRVV} is the solution of Algorithm 2 we have the following. As described in Algorithm 2, the DVV algorithm is called for every value of B, including the parameter B^* corresponding to the optimal solution S^*_{DRVV}. So, it stands $\sigma_N(S^*_{DRVV}) = \sigma_N(S^*_{DVV})$ by the definition of DVV. The Algorithm 1 returns a set S_{DVV} such that $\sigma_N(S_{DVV}) \geq r \cdot \sigma_N(S^*_{DVV})$, (where $r \leq 1$ denotes the approximation ratio guaranteed by Algorithm 1), and the cost constraint is satisfied, i.e., $\sigma_L(S_{DVV}) \leq B^* = \sigma_L(S^*_{DRVV})$. We have:

$$ASR(S_{DRVV}, c) \geq ASR(S_{DVV}, c) = \frac{\sigma_N(S_{DVV}) + c}{\sigma_L(S_{DVV}) + c} \geq \frac{r \cdot \sigma_N(S^*_{DVV}) + c}{\sigma_L(S_{DVV}) + c} \Rightarrow$$

$$ASR(S_{DRVV}, c) \geq r \cdot \frac{\sigma_N(S^*_{DRVV}) + c}{\sigma_L(S^*_{DRVV}) + c} \geq r \cdot ASR(S^*_{DRVV}, c)$$

$\square$

Algorithm 2: $DRVV$ Algorithm

    Input   : A graph $G(V, E)$ where vertices are partitioned into vulnerable
                  V_L and non-vulnerable V_N and a positive constant c
    Output: A solution $S \subseteq V_N$
1 **foreach** $B \in \{0, 1, \cdots, |V_L|\}$ **do**
2    $|$ $S_B \leftarrow DVV(G, B)$
3 $S \leftarrow \underset{S_B \in \{S_0, S_1, \cdots, S_{|V_L|}\}}{\arg\max} (ASR(S_B, c))$
4 **return** S

Let's return back to the $k\text{-}MaxDRVV$ problem. Before proceeding, let's recall the definition of the $Max\ CP$ problem.

Definition 8 (*MaxCP*). *Given a universe* $U = \{e_1, e_2, \ldots, e_n\}$ *of n elements, a collection of subsets* $\mathcal{X} = \{X_1, X_2, \ldots, X_m\}$ *where each $X_i \subseteq U$, and an integer k, the goal is to select a subcollection $\mathcal{C} \subseteq \mathcal{X}$ such that: The collection has at most k subsets ($|\mathcal{C}| \leq k$) and the union of the sets in $\mathcal{C}$ maximizes the number of covered elements in U* $\left(\mathcal{C} = \arg\max_{\mathcal{A} \subseteq \mathcal{X},\ |\mathcal{A}| \leq k} \left| \bigcup_{X_j \in \mathcal{A}} X_j \right| \right)$.

2.4   Algorithm and Approximation for k-MaxDRVV

Given a solution S to the $DRVV$ problem, Algorithm 3 constructs a $Max\ CP$ instance such that, for every vertex $u \in S$, a subset of vertices $X_u = N[u] \setminus V_L$ is created, which includes all neighboring vertices of u that are not vulnerable, including u itself. The universe $U = N[S] \setminus V_L$ includes the close neighborhood of S excluding the vulnerable vertices. In this formulation, the vertices to be selected are restricted to S, whereas the dominated vertices correspond to the closed neighborhood of S. The algorithm then solves this $Max\ CP$ instance using the standard greedy algorithm [17] and returns the resulting subset. The $Max\ CP$ problem has an approximation ratio $(1 - 1/e)$ [9,17] and this bound cannot be improved unless $P = NP$ [6].

Let S be the output of $DRVV$ algorithm and S^* be the optimal solution of $DRVV$, S_k be the output of the k-$MaxDRVV$ algorithm, S_k^* be the optimal solution of k-$MaxDRVV$ and S_k' be the optimal solution for $Max\ CP$. We prove bicriteria Lemma 5 holds and then we convert this result to a true approximation.

Lemma 5. *Algorithm 2 is a (d, r) bicriteria algorithm for k-$MaxDRVV$ problem, where r is the approximation ratio for $DRVV$, $d = \lceil m/k \rceil$, $k \leq m < n$ and $m = |S|$ with S the solution provided by algorithm 2.*

Proof. Notice that S does not include vulnerable vertices. In the case where $m \leq k$ we have a feasible solution for k-$MaxDRVV$ problem with the approximation ratio of the $DRVV$ Algorithm 2. In the case where $m = n$, it means that we have an instance without vulnerable vertices and the approximation is that of $Max\ CP$. So, we focus in the case where $k \leq m < n$. By definition of S^*, it holds $ASR(S_k^*, c) \leq ASR(S^*, c)$ and $r \cdot ASR(S^*, c) \leq ASR(S, c)$. We have:

$$r \cdot ASR(S_k^*, c) \leq r \cdot ASR(S^*, c) \leq ASR(S, c) \Rightarrow r \cdot ASR(S_k^*, c) \leq ASR(S, c)$$

Algorithm 3: k-$Max\ DRVV$

    Input   : A graph $G = (V, E)$ where vertices are partitioned into non-vulnerable V_N and vulnerable V_L, an integer $k \leq |V|$ and a positive constant c

    Output: A subset S_k, such that $|S_k| \leq k$

1 $S \leftarrow DRVV(G)$

2 **if** $|S| \leq k$ **then**

3 | $S_k \leftarrow S$

4 **else**

5     $U \leftarrow N[S] \setminus V_L$

6   **foreach** $u \in S$ **do**

7 | $X_u = N[u] \setminus V_L$

8     $\mathcal{X} = \bigcup_{u \in S}\{X_u\}$

9     $C = Max\ CP(U, \mathcal{X}, k)$

10    $S_k \leftarrow \bigcup_{X_u \in C}\{u\}$

11 **return** S_k

Since $DRVV$ algorithm gives a subset S with ASR within r of the optimum and this subset has a size $m < n$, the lemma stands. $\square$

Converting the Bicriteria Approximation to a True Approximation.
We need to get a subset of S, of size at most k for a standard approximation algorithm.

Theorem 6. *Algorithm 3 is a $\frac{\rho \cdot r}{d}$ approximation for k-$MaxDRVV$ problem, where ρ is the approximation ratio for $Max\ CP$, r is the approximation ratio for $DRVV$, $d = \lceil m/k \rceil$, $k \leq m < n$ and $m = |S|$ with S the solution provided by algorithm 2.*

Proof. Let $m = |S| < n$ and $d = \lceil \frac{m}{k} \rceil$. By definition of σ_N, for any partition $\{S_1, S_2, \cdots, S_d\}$ of S, such that $|S_i| \leq k$ for all $i = \{1, \cdots, d\}$, and $|S_1| + |S_2| + \cdots + |S_d| = m$, it holds $\sigma_N(S) \leq \sigma_N(S_1) + \cdots \sigma_N(S_d)$. Clearly, there is a $S_{i*}, 1 \leq i^* \leq d$ such that $\sigma_N(S_1) + \cdots + \sigma_N(S_d) \leq d \cdot \sigma_N(S_{i*})$. We have:

$$\sigma_N(S) \leq d \cdot \sigma_N(S_{i*}) \Rightarrow \frac{\sigma_N(S) + c}{\sigma_L(S) + c} \leq \frac{d \cdot \sigma_N(S_{i*}) + c}{\sigma_L(S) + c} \Rightarrow$$

$$ASR(S, c) \leq \frac{d \cdot \sigma_N(S_{i*}) + c}{\sigma_L(S) + c} \tag{1}$$

By Lemma 5 it holds $r \cdot ASR(S_k^*, c) \leq ASR(S, c)$. By Eq. 1, we have:

$$r \cdot ASR(S_k^*, c) \leq \frac{d \cdot \sigma_N(S_{i*}) + c}{\sigma_L(S) + c} \leq d \cdot \frac{\sigma_N(S_{i*}) + c}{\sigma_L(S) + c} \tag{2}$$

By definition of S_k', it holds $\sigma_N(S_{i*}) \leq \sigma_N(S_k')$. Also, by [9,17] it holds $\rho \cdot \sigma_N(S_k') \leq \sigma_N(S_k)$. By Eq. 2, we have:

$$\frac{\rho \cdot r}{d} \cdot ASR(S_k^*, c) \leq \frac{\rho \cdot \sigma_N(S_{i*}) + \rho \cdot c}{\sigma_L(S) + c} \leq \frac{\rho \cdot \sigma_N(S_k') + c}{\sigma_L(S) + c} \leq \frac{\sigma_N(S_k) + c}{\sigma_L(S) + c} \tag{3}$$

Furthermore, by definition of σ_L it holds $\sigma_L(S_k) \leq \sigma_L(S)$ since $S_k \subseteq S$. So by Eq. 3, we have:

$$\frac{\rho \cdot r}{d} ASR(S_k^*, c) \leq \frac{\sigma_N(S_k) + c}{\sigma_L(S) + c} \leq \frac{\sigma_N(S_k) + c}{\sigma_L(S_k) + c} = ASR(S_k, c)$$

$\square$

Finally, under the current best known results for $Max\ CP$ and $DRVV$ we obtain for k-$MaxDRVV$ problem the approximation ratio $\frac{1}{\Delta+1} \cdot (1 - e^{-\frac{1}{\Delta}}) \cdot (1 - \frac{1}{e}) \cdot \frac{k}{n}$ for bounded degree graphs. We have the following corollary.

Corollary 1. *For the class of graphs with bounded maximum degree Δ, Algorithm 3 achieves an approximation ratio of $O(k/n)$.*

3 Vertex Cover with Vulnerable Edges ($VCVE$)

The well-known problem of vertex cover can be defined under the case of vulnerable edges. We seek a set of vertices covering all non-vulnerable edges, while covering minimum number of the vulnerable ones.

Definition 9 ($VCVE$). *Given a graph $G = (V, E)$ where edges are partitioned into non-vulnerable E_N and vulnerable E_L, find a set of vertices $S \subseteq V$ covering all E_N edges and the minimum number of E_L edges.*

The $VCVE$ problem is as hard as the Minimum Vertex Cover (Min-VC). Indeed, every instance $G = (V, E)$ of Min-VC can be transformed to an instance of $VCVE$ by adding a new vertex v and a new vulnerable edge (v, u) for each vertex u of G. Let G' be the created graph. Each solution for $VCVE$ on G' directly yields a Min-VC solution for G, and vice versa. We adopt the following notation: For any vertex $v \in V$, let $E(v)$ denote the set of edges incident to v. Furthermore, for a set of vertices $S \subseteq V$, $E_L(S)$ represents the set of vulnerable edges covered by S, defined as: $E_L(S) = \{(v, u) \in E_L : v \in S \vee u \in S\}$.

Algorithm 4: VCVE

Input   : A graph $G = (V, E)$ where edges are partitioned into
          non-vulnerable E_N and vulnerable E_L
Output: A feasible solution S for $VCVE$

1   $S \leftarrow \emptyset,\ S_0 \leftarrow \emptyset,\ V' \leftarrow V,\ E' \leftarrow E$
2   **foreach** $v \in V$ **do**
3   **if** $E(v) \cap E_L \neq \emptyset$ **then**
4      $w(v) \leftarrow |E(v) \cap E_L|$
5   **else**
6      $V' \leftarrow V' \setminus \{v\}$
7      $E' \leftarrow E' \setminus E(v)$
8      $S_0 \leftarrow S_0 \cup \{v\}$
9   $E' \leftarrow E' \setminus E_L$
10   **foreach** $v \in V'$ **do**
11   **if** $E(v) = \emptyset$ **then**
12      $V' \leftarrow V' \setminus \{v\}$
13   Let $G' = (V', E')$
14   $S_{WVC} \leftarrow WVC(G')$
15   $S \leftarrow S_0 \cup S_{WVC}$
16   **return** S

The Algorithm. In Algorithm 4, the Weighted Vertex Cover (WVC) [2] problem is used as a subroutine. In WVC, given an undirected graph $G = (V, E)$ with a positive weight function $w : V \to \mathbb{R}^+$ assigning a positive weight $w(v)$ to each vertex $v \in V$, the goal is to find a subset $C \subseteq V$ such that every edge $e \in E$ is incident to at least one vertex $v \in C$ and the total weight of the selected vertices, denoted as $W(C)$ is minimized. In the algorithm, we first select all vertices with no adjacent vulnerable edges and add them to the solution. These vertices are then removed from the initial graph. Each remaining vertex is assigned a weight corresponding to the number of vulnerable edges incident to it. Vulnerable edges and isolated vertices are removed, and the resulting graph G' is used in WVC algorithm to provide a feasible solution. The remaining graph G' might be disconnected. We apply the WVC algorithm for each of the components of G'.

Lemma 6. *For all $S \subseteq V$: $|E_L(S)| \leq W(S) \leq 2 \cdot |E_L(S)|$.*

Proof. By Algorithm 4, every weight of a vertex is equal to the number of vulnerable edges incident to it. So, every vulnerable edge adds one point to each neighboring vertex's weight. So:

$$W(S) = \sum_{v \in S} w(v) = \sum_{e \in E_L(S)} \delta_S(e)$$

where, for $e = (v, u)$, $\delta_S(e) = 2$ when $v, u \in S$; $\delta_S(e) = 0$ when $v, u \notin S$; and $\delta_S(e) = 1$ otherwise. By definition of $\delta_S(e)$, for every $e \in E_L(S)$, $1 \leq \delta_S(e) \leq 2$. So,

$$\sum_{e \in E_L(S)} 1 \leq \sum_{e \in E_L(S)} \delta_S(e) \leq \sum_{e \in E_L(S)} 2 \Rightarrow |E_L(S)| \leq W(S) \leq 2 \cdot |E_L(S)|$$

$\square$

Theorem 7. *Algorithm 4 provides a 4-approximation for VCVE.*

Proof. There exists an optimal solution S^* such that $S_0 \subseteq S^*$, where S_0 the subset of vertices with no vulnerable incident edges. Let $G' = (V', E')$ be the graph that Algorithm 4 defines and S^*_{WVC} the optimal solution for WVC. By [2,10] we obtain a feasible solution S_{WVC} for WVC, such that $2 \cdot W(S^*_{WVC}) \geq W(S_{WVC})$.

By definition, $S^* \setminus S_0$ is also a feasible solution for WVC on $G' = (V', E')$ and so $W(S^* \setminus S_0) \geq W(S^*_{WVC})$. We have:

$$2 \cdot W(S^* \setminus S_0) \geq W(S_{WVC}) \Rightarrow$$

$$2 \cdot (W(S^* \setminus S_0) + W(S_0)) \geq W(S_{WVC}) + W(S_0) \Rightarrow 2 \cdot W(S^*) \geq W(S)$$

By Lemma 6, $2 \cdot |E_L(S^*)| \geq W(S^*)$ and $W(S) \geq |E_L(S)|$. So:

$$4 \cdot |E_L(S^*)| \geq 2 \cdot W(S^*) \geq W(S) \geq |E_L(S)| \Rightarrow 4 \cdot |E_L(S^*)| \geq |E_L(S)| \square$$

4   Conclusion

This work addresses the challenge of socially responsible influence maximization by introducing new variants of domination and coverage problems. We defined the k-Vertex Maximum Domination Ratio with Vulnerable Vertices problem and provided an approximation algorithm. We also presented the Vertex Cover problem with vulnerable edges and proposed a 4-approximation algorithm.

A natural open question is how much we can improve upon the above results. It remains to be seen whether constant-factor approximation algorithms can be developed for specific classes of graphs. Beyond this, connected versions of these problems can also be examined from a vulnerability perspective.

Acknowledgements. The authors would like to thank the anonymous reviewers for their insightful comments and constructive suggestions, which significantly contributed to improving the quality and clarity of this paper.

References

1. Arulselvan, A.: A note on the set union knapsack problem. Discret. Appl. Math. **169**, 214–218 (2014)
2. Bar-Yehuda, R., Even, S.: A linear-time approximation algorithm for the weighted vertex cover problem. J. Algorithms **2**(2), 198–203 (1981)
3. Carr, R.D., Doddi, S., Konjevod, G., Marathe, M.: On the red-blue set cover problem. In: Proceedings of the Eleventh Annual ACM-SIAM Symposium on Discrete Algorithms, pp. 345–353. SODA '00, USA (2000)
4. Chen, H., Loukides, G., Pissis, S.P., Chan, H.: Influence maximization in the presence of vulnerable nodes: a ratio perspective. Theoret. Comput. Sci. **852**, 84–103 (2021)
5. Elkin, M., Peleg, D.: The hardness of approximating spanner problems. In: Proceedings of the 17th Annual Symposium on Theoretical Aspects of Computer Science, pp. 370–381. STACS '00, Springer-Verlag, Berlin, Heidelberg (2000)
6. Feige, U.: A threshold of ln n for approximating set cover. J. ACM (JACM) **45**(4), 634–652 (1998)
7. Haynes, T.W., Hedetniemi, S., Slater, P.: Fundamentals of domination in graphs, CRC Press (2013)
8. Haynes, T.W., Hedetniemi, S.T., Henning, M.A. (eds.): Topics in domination in graphs. DM, vol. 64. Springer, Cham (2020). https://doi.org/10.1007/978-3-030-51117-3
9. Hochbaum, D.S., Pathria, A.: Analysis of the greedy approach in problems of maximum k-coverage. Naval Res. Logistics (NRL) **45**(6), 615–627 (1998)
10. Karakostas, G.: A better approximation ratio for the vertex cover problem. ACM Trans. Algorithms (TALG) **5**(4), 1–8 (2009)
11. Khuller, S., Purohit, M., Sarpatwar, K.K.: Analyzing the optimal neighborhood: algorithms for budgeted and partial connected dominating set problems. SODA, pp. 1702–1713 (2013)
12. Khuller, S., Purohit, M., Sarpatwar, K.K.: Analyzing the optimal neighborhood: algorithms for partial and budgeted connected dominating set problems. SIAM J. Discret. Math. **34**(1), 251–270 (2020)

13. Lamprou, I., Martin, R., Schewe, S., Sigalas, I., Zissimopoulos, V.: Maximum rooted connected expansion. Theoret. Comput. Sci. **873**, 25–37 (2021)
14. Lamprou, I., Sigalas, I., Zissimopoulos, V.: Improved budgeted connected domination and budgeted edge-vertex domination. In: Gąsieniec, L., Klasing, R., Radzik, T. (eds.) IWOCA 2020. LNCS, vol. 12126, pp. 368–381. Springer, Cham (2020). https://doi.org/10.1007/978-3-030-48966-3_28
15. Lamprou, I., Sigalas, I., Zissimopoulos, V.: Improved budgeted connected domination and budgeted edge-vertex domination. Theoret. Comput. Sci. **858**, 1–12 (2021)
16. Miyano, E., Ono, H.: Maximum domination problem. In: Proceedings of the Seventeenth Computing on The Australasian Theory Symposium-Volume 119, pp. 55–62 (2011)
17. Nemhauser, G.L., Wolsey, L.A., Fisher, M.L.: An analysis of approximations for maximizing submodular set functions. Math. Program. **14**(1), 265–294 (1978)
18. Pasumarthi, R., Narayanam, R., Ravindran, B.: Near optimal strategies for targeted marketing in social networks. In: AAMAS, pp 1679–1680 (2015)
19. Peleg, D.: Approximation algorithms for the label-covermax and red-blue set cover problems. J. Discr. Algorithms **5**(1), 55–64 (2007)
20. Vazirani, V.V.: Approximation algorithms. Springer-Verlag, Berlin, Germany (2001)
21. Wang, Z., et al.: Approximation algorithm and applications for connected submodular function maximization problems. IEEE Trans. Netw. **33**(1), 241–254 (2025)
22. Wen, Y.T., Peng, W.C., Shuai, H.H.: Maximizing social influence on target users. In: Advances in Knowledge Discovery and Data Mining: 22nd Pacific-Asia Conference, PAKDD 2018, Melbourne, VIC, Australia, 3–6 June 2018, Proceedings, Part III 22, pp. 701–712. Springer (2018)

Fast Order Statistics with Group Inequality Testing

Adiesha Liyanage, Brendan Mumey[(✉)], and Braeden Sopp

Gianforte School of Computing, Montana State University,
Bozeman, MT 59717, USA
`{a.liyanaralalage,brendan.mumey}@montana.edu`,
`braeden.sopp@student.montana.edu`

Abstract. Suppose that a group test operation is available for checking order relations in a set, can this speed up problems like finding the minimum/maximum element, determining the rank of element, and computing order statistics? We consider a one-sided group inequality test to be available, where queries are of the form $u \leq_Q V$ or $V \leq_Q u$, and the answer is 'yes' if and only if there is some $v \in V$ such that $u \leq v$ or $v \leq u$, respectively. We restrict attention to total orders and focus on query-complexity; for min or max finding, we give a Las Vegas algorithm that makes $\mathcal{O}(\log^2 n)$ expected queries. We observe that rank determination can be solved with existing "defect-counting" algorithms, but also give a simple Monte Carlo approximation algorithm with expected query complexity $\tilde{\mathcal{O}}(\frac{1}{\delta^2} \log \frac{1}{\epsilon})$, where $1 - \epsilon$ is the probability that the algorithm succeeds and we allow a relative error of $1 \pm \delta$ for $\delta > 0$ in the estimated rank. We then give a Monte Carlo algorithm for approximate selection that has expected query complexity $\tilde{\mathcal{O}}(\frac{1}{\delta^3} \log \frac{1}{\epsilon \delta^2})$; it has probability at least $\frac{1}{2}$ to output an element x, and if so, x has the desired approximate rank with probability $1 - \epsilon$.

Keywords: Order statistics · Group inequality testing · Randomized algorithms

1 Introduction

In this work, we consider how a group inequality test can speed up order-finding problems, such as finding the minimum/maximum element of a totally-ordered set, determining the rank of an element, and selecting the element with a given rank. Given an order relation R on a set U of size n, we assume that an oracle is available that can answer one-to-many queries of the form $u \leq_Q V$ or $V \leq_Q u$, where the answer is yes if and only if there is some $v \in V$ that satisfies the inequality $u \leq v$ or $v \leq u$ respectively.

As a motivating example, imagine that we are surveying someone to determine their preferred choice among a set of options, but we can only ask yes or no questions. In this case the group test queries proposed are natural, i.e. do

F. Foucaud and A. Parreau (Eds.): IWOCA 2026, LNCS 16587, pp. 430–441, 2026.
https://doi.org/10.1007/978-3-032-27732-9_30

you prefer A versus B or C? Another example where such a test is available occurs in computational biology, where one seeks to identify consistent relationships among all parsimonious solutions for reconstructing tumor cell evolution from single cell sequence data [7]. In this case, the goal is to determine those order relationships among pairs of cells that hold true in all optimal phylogenetic trees on the cells. In [7], the authors first present a group test $u \leq_Q V$ based on integer-linear programming that can check whether cell u evolutionarily precedes some cell $v \in V$, then develop an algorithm based on the test to determine the complete relationship among all cells. In such contexts, if the goal is to find certain elements (e.g., the minimum or maximum), it is natural to ask whether group testing strategies can be effectively applied.

1.1   Related Work

Selection is a fundamental and well-studied problem in combinatorics and computer science; given an order relation R on a set U of size n, the task is to find the k-th smallest element. In the traditional comparison model, deterministic algorithms exist that can select the k-th element using at most $\mathcal{O}(n)$ comparisons [1], while a randomized selection algorithm exists that further lowers the number of comparisons to $n + \min(k, n-k) + o(n)$ in the average case [4]. In large-scale data applications, the space required for exact selection algorithms can be impractical, so approximate selection algorithms have been developed with a focus on space efficiency. Approximate selection algorithms are closely related to approximate quantile computation, as studied in [8]. Both aim to identify items whose ranks are near a target value. The key difference lies in their input specification: selection algorithms target an element with a specified integer rank, whereas quantile computation seeks an item whose rank corresponds to a given fraction (quantile) of the total input size. The quantile of an item x is defined as the fraction of items in the stream satisfying $x_i \leq x$. Approximate quantile computation algorithms such as the deterministic and randomized one pass algorithms presented in [8] achieve space complexity of $\mathcal{O}(\frac{1}{\epsilon} \log^2 \epsilon n)$. The current state-of-the-art algorithm for approximate quantiles computation in a streaming setting, such as the KLL sketch [6], achieves space complexity of $\mathcal{O}(\frac{1}{\delta} \log \log \frac{1}{\epsilon})$ for an additive error of δn with success probability $1 - \epsilon$.

Moreover, the group testing literature contains extensive work on estimating the number of defective items in a population [2,5,9–11]. In this setting, let X denote a set of elements and let $D \subset X$ be an unknown subset of defective elements. We are given access to an oracle that, on input a subset $T \subseteq X$, returns 'true' if and only if T contains at least one defective element. The objective is to estimate the cardinality $d = |D|$. Two main classes of algorithms have been studied in the literature: *non-adaptive* algorithms, in which all queries are fixed in advance, and *adaptive* algorithms, in which future queries can depend on the outcomes of previous queries. In [3], the authors proposed an adaptive randomized group testing algorithm that estimates the number of defective elements using a near-optimal number of queries. Specifically, their algorithm uses at most

$2 \log \log d + \mathcal{O}\left(\frac{1}{\delta^2} \log \frac{1}{\epsilon}\right)$ queries and outputs an estimate of d within a multiplicative factor of $1 \pm \delta$ with error probability at most ϵ. Furthermore, they established an information-theoretic lower bound showing that any adaptive algorithm must make at least $(1 - \epsilon) \log \log d - 1$ queries to estimate d for constant δ.

In this work, we propose adaptive algorithms for selection and rank determination, assuming an oracle is available that can answer group inequality tests. We focus on minimizing the query complexity.

1.2 Contributions

Our results can be summarized as follows:

- When considering min-finding, we introduce a Las Vegas algorithm to find the minimum element of a totally-ordered set U using an $\mathcal{O}(\log^2 n)$ expected group test queries where n is the size of U. The min-finding algorithm can also be adapted for max-finding (by just reversing the group tests).
- While we observe that rank-estimation can be solved using an existing defect-counting group testing algorithm, we also give a simple Monte Carlo algorithm to estimate the rank r of an element $x \in U$ using $\mathcal{O}(\frac{\log n}{\delta^2}(\log \log n + \log \frac{1}{\epsilon}))$ expected group test queries for any given ε probability of failure and approximation parameter $\delta \in (0, 1)$. The algorithm guarantees that $|rank(x) - r| < \delta \min(r, n - r)$ with probability $1 - \epsilon$.
- For approximate selection, we provide a Monte Carlo algorithm that uses $\mathcal{O}(\frac{1}{\delta} \log^2 \frac{n}{k} + \frac{1}{\delta^3} \log \frac{1}{\epsilon \delta^2})$ expected group test queries for a given $\delta \in (0, 1)$ and target rank k which returns an element $x \in U$ with probability $1/2$. When the algorithm returns an element, then with probability $1 - \epsilon$, we have $|\operatorname{rank}(x) - k| < \delta \min(k, n - k)$.

To the best of our knowledge, this work is the first to consider the problems of rank selection, min/max finding, and rank determination in a group testing framework.

2 Preliminaries

We assume that a total order $(S, \leq)$ exists, where S is a set of elements stored in an array or similar data structure. The algorithms make use of random sampling and partitioning in S, but we will not consider the time complexity of those operations, as we focus on the query-complexity, i.e., the number of group inequality tests performed. We assume that two group inequality test operations are available, $u \leq_Q V$ and $V \leq_Q u$, where the answer is yes if and only if there is some $v \in V$ that satisfies the inequality. Note that the algorithms do not have direct access to the total order; instead, they must rely on group inequality test operations to determine the ordering of the elements.

Definition 1. *For any subset $V \subseteq S$ and element $x \in S$, a left (right) group test operation is of the form $x \leq_Q V$ ($V \leq_Q x$), where the test returns true if and only if there is some $v \in V$ that satisfies $x \leq v$ ($v \leq x$).*

Definition 2. *The rank of an element $x \in S$ is its order position, i.e.,*

$$rank(x) = |\{y \in S : y \leq x\}|.$$

Definition 3. *The k-th order statistic (for $k \in [n]$) is the element $x \in S$ such that $rank(x) = k$. Given k, the problem of finding the k-th order statistic is called the* selection problem.

We first consider the min-finding problem, i.e., computing the 1-st order statistic, and develop an exact group testing algorithm for this problem. Let $\leq^{rev}$ be the reversed relation, i.e., $x \leq^{rev} y \Leftrightarrow y \leq x$ (clearly $\leq^{rev}$ is also a total order). Thus, max-finding for $\leq$ is the same problem as min-finding for $\leq^{rev}$. Furthermore, $x \leq_Q^{rev} V$ if and only if $V \leq_Q x$, and similarly, $V \leq_Q^{rev} x$ if and only if $x \leq_Q V$. This means that we can convert a min-finding algorithm to max-finding by simply reversing the direction of each group test call.

For the rank determination and selection problems, we consider approximate solutions, with the following problem definitions:

Definition 4. *A δ-approximation of $rank(x)$ (for $\delta \in (0, 1), x \in S$) is a value $r \in [n]$ such that $|\, rank(x) - r| \leq \delta \min(r, n - r)$. Given δ and x, the problem of finding a δ-approximation for $rank(x)$ is called the* approximate rank problem.

Likewise, it will be useful to assume that $k \leq n/2$ for selection, if this is not the case, then we can just select the $k' = (n - k)$-th element using $\leq^{rev}$ as the order, again using the same algorithm but reversing the direction of the group tests. We also remark that Definition 4 is stronger than the standard notion of a $1 \pm \delta$ approximation to $rank(x)$, since a δ-approximation is tighter when $r > n/2$.

Definition 5. *A δ-approximation for the k-th order statistic (for $\delta \in (0, 1), k \in [n]$) is an element $x \in S$ such that $|\, rank(x) - k| \leq \delta \min(k, n - k)$. Given δ and k, the problem of finding a δ-approximation for the k-th order statistic is called the* approximate selection problem.

3   Min-Finding

We first give a Las Vegas randomized algorithm to find the minimum element in S. The basic idea of the algorithm is to iteratively refine a candidate x for the minimum element (initially chosen at random). At each iteration, either x is determined to be the global minimum, or an element whose rank is at most that of the previous x. Furthermore, the rank of the next candidate found is expected to be half of the rank of the current x. The complete algorithm is given in Algorithm 1.

Lemma 1. *Let x be the current candidate, and let $Y = \{y \in S : y \leq x\}$. If $|Y| > 0$, then $\forall y \in Y$,*

$$\Pr[Swap(S, x) \text{ returns } y] = \frac{1}{|Y|}.$$

Algorithm 1. Min-finding algorithm

1: **function** MinFind(S)
2:     $x \leftarrow$ Randomly select an element from the set S.
3:     **while** $|S \setminus \{x\}| \geq 1$ **and** $S \setminus \{x\} \leq_Q x$ **do**
4:         $x \leftarrow$ Swap($S \setminus \{x\}, x$)
5:     **end while**
6:     **return** x.
7: **end function**
8: **function** Swap(A, x)
9:     **if** $A = \{a\}$ **then**
10:         **return** a
11:     **end if**
12:     $A_0, A_1 \leftarrow$ Partition A randomly into A_0, A_1 s.t. $||A_0| - |A_1|| \leq 1$
13:     **if** $A_0 \leq_Q x$ **then**
14:         **return** $Swap(A_0, x)$
15:     **else**
16:         **return** $Swap(A_1, x)$ $\triangleright$ A_1 must contain an element $\leq x$.
17:     **end if**
18: **end function**

Proof. The lemma follows from the random partitioning of elements in the Swap function.

Corollary 1. *The expected number of iterations needed by MinFind is* $\mathcal{O}(\log |S|)$.

Proof. Observe that rank(x) is initially uniformly distributed in $[1, n]$ and each iteration returns a new x whose rank is uniformly distributed in $[1, \text{rank}(x) - 1]$ by Lemma 1 (note rank(x) ≥ 2 if an iteration occurs). The loop ends when rank(x) $= 1$. Let $I(r)$ be the expected number of iterations needed assuming rank(x) $= r$. We have $I(1) = 0$ and for $r > 1$:

$$I(r) = 1 + \frac{1}{r-1} \sum_{i=1}^{r-1} I(i), \tag{1}$$

We can rewrite Eq. 1 as:

$$(r-1) \cdot I(r) = (r-1) + \sum_{i=1}^{r-1} I(i), \tag{2}$$

and so for $r > 2$, we have:

$$(r-2) \cdot I(r-1) = (r-2) + \sum_{i=1}^{r-2} I(i). \tag{3}$$

Subtracting Eqs. 3 from 2 yields

$$(r-1) \cdot I(r) - (r-2) \cdot I(r-1) = 1 + I(r-1),$$

and so

$$I(r) - I(r-1) = \frac{1}{r-1}.$$

From Eq. 1, $I(2) = 1$, so for $r \geq 2$, $I(r) = 1 + \frac{1}{2} + \frac{1}{3} + \ldots + \frac{1}{r-1} = \mathcal{O}(\log r)$. Note that the expected total number of iterations is $I(|S|+1) - 1$, since the first x sample does not require an iteration.

Lemma 2. *Swap does $\lceil \log_2(|A|) \rceil$ group tests.*

Proof. If $|A| > 1$, Swap makes one group inequality test and then recurses on either A_0 or A_1, which are of size at most $\frac{|A|+1}{2}$. Thus, the depth of recursion is $\lceil \log_2(|A|) \rceil$.

Theorem 1. *The min element of a set S of size n can be found with $\mathcal{O}(\log^2 n)$ expected group test queries.*

Proof. This follows directly from Corollary 1 and Lemma 2.

As previously mentioned, the min-finding algorithm can be converted to a max-finding algorithm by simply reversing the group test queries.

4   Rank Determination

We first observe that rank determination can be solved using group-testing defect counting; if we wish to determine the rank of an element $x \in S$, we can consider those elements less than or equal to x, i.e. $\{y \in S : y \leq x\}$ as being 'defective'. The group inequality test $V \leq_Q x$ is analogous to the standard group test to check if V contains any defective elements. The algorithm from [3] can thus find a $1 \pm \delta$ approximation to rank(x), using $2 \log \log n + \mathcal{O}\left(\frac{1}{\delta^2} \log \frac{1}{\epsilon}\right)$ group inequality queries, that succeeds with probability $1 - \epsilon$. Their algorithm is based on a multistage binary search. Below we describe a simple binary search approach that finds a δ-approximation with $\tilde{\mathcal{O}}(\frac{1}{\delta^2} \log \frac{1}{\epsilon})$ query complexity.

First, we introduce a useful randomized test routine that is also used later in our selection algorithm; specifically we would like to decide for any element x and some $r \in [n]$, whether rank$(x) \leq r$. We will assume $r \leq n/2$ (the case $r > n/2$ is handled by reversing the direction of the group tests and is discussed subsequently). We also assume that n is divisible by r; if not, add at most $r-1$ equivalent dummy elements to S that are greater than all other elements.

We first describe the test: Suppose $x \in S$ and S' is a random sample (with replacement) of S of size $N = n/r$. If rank$(x) \leq r - \delta r$,

$$\Pr[S' \leq_Q x] = 1 - \left(\frac{n - \text{rank}(x)}{n}\right)^N \leq 1 - \left(\frac{n - (r - \delta r)}{n}\right)^N \triangleq p_L.$$

Similarly, if rank$(x) \geq r + \delta r$,

$$\Pr[S' \leq_Q x] = 1 - \left(\frac{n - \text{rank}(x)}{n}\right)^N \geq 1 - \left(\frac{n - (r + \delta r)}{n}\right)^N \triangleq p_R.$$

Algorithm 2. Less-than rank test (for $r \leq \frac{n}{2}$)

1: **function** $\textsc{TestLE}(x, r, \delta, \epsilon)$
2:     $c \leftarrow 0$
3:     **for** $i = 1$ to N_t **do**
4:         $S' \leftarrow$ Randomly select n/r elements from S with replacement.
5:         **if** $S' \leq_Q x$ **then**
6:             $c \leftarrow c + 1$.
7:         **end if**
8:     **end for**
9:     **return** $(c \geq p^* \cdot N_t)$
10: **end function**

We have

$$p_R - p_L = (1 - p_L) - (1 - p_R)$$
$$= \left(\frac{n - r + \delta r}{n}\right)^N - \left(\frac{n - r - \delta r}{n}\right)^N$$
$$= \left(1 - \frac{1}{N} + \frac{\delta}{N}\right)^N - \left(1 - \frac{1}{N} - \frac{\delta}{N}\right)^N$$
$$= (\alpha + \beta)^N - (\alpha - \beta)^N \tag{4}$$
$$= \sum_{i=0}^{N} \binom{N}{i} \alpha^{N-i} (\beta^i - (-\beta)^i)$$
$$\geq N \alpha^{N-1} 2\beta$$
$$= 2 \left(1 - \frac{1}{N}\right)^{N-1} \delta$$
$$\geq 2 e^{-1} \delta, \tag{5}$$

where Eq. 4 uses the substitutions $\alpha = 1 - \frac{1}{N}$ and $\beta = \frac{\delta}{N}$, and Eq. 5 uses the inequality $(1 - \frac{1}{n})^{n-1} > e^{-1}$.

Suppose we do N_t random trials of this process (i.e., pick new random S' sets each trial) and let X be the number of positive responses. Let $p^* = \frac{p_L + p_R}{2}$. So, $p_R - p^* = p^* - p_L = \frac{1}{2}(p_R - p_L) \geq \delta/e$. If $\text{rank}(x) \leq r - \delta n$, by Hoeffding's inequality (since this sum has a binomial distribution),

$$\Pr[X \geq p^* N_t] = \Pr[X - p_L N_t \geq p^* N_t - p_L N_t]$$
$$\leq \Pr[X - p_L N_t \geq \delta/e \cdot N_t] \leq \exp(-2(\delta/e)^2 N_t)).$$

Similarly, if $\mathrm{rank}(x) \geq r + \delta n$,

$$
\begin{aligned}
\Pr[X \leq p^* N_t] &= \Pr[N_t - X \geq N_t - p^* N_t] \\
&= \Pr[(N_t - X) - (1 - p_R)N_t \geq N_t - p^* N_t - (1 - p_R)N_t] \\
&= \Pr[(N_t - X) - (1 - p_R)N_t \geq N_t(p_R - p^*)] \\
&\leq \Pr[(N_t - X) - (1 - p_R)N_t \geq \delta/e \cdot N_t] \leq \exp(-2(\delta/e)^2 N_t),
\end{aligned}
$$

where the last inequality again follows by Hoeffding's inequality on the sum of the negative trials. If $\epsilon > 0$ is the desired error rate, then we should set $\exp(-2(\delta/e)^2 N_t)) = \epsilon$, and take $N_t = \frac{\ln(1/\epsilon)}{2(\delta/e)^2} = \frac{e^2 \ln(1/\epsilon)}{2\delta^2}$. Algorithm 2 implements this strategy.

Using the observation that testing $\mathrm{rank}(x) \leq r$ for $\leq$ is equivalent to testing $\mathrm{rank}(x) \geq n - r$ for $\leq^{\mathrm{rev}}$, we can adapt the test to handle the case $r > n/2$ by replacing r with $n - r$ and reversing the direction of group tests. To summarize, we have a test $\mathsf{TestLE}(x, r, \delta, \epsilon)$ with the following property:

Lemma 3. $\mathsf{TestLE}(x, r, \delta, \epsilon)$ *makes* $\mathcal{O}(\frac{1}{\delta^2} \ln \frac{1}{\epsilon})$ *group test queries and is correct with probability* $1 - \epsilon$.

As mentioned, our rank determination strategy uses a binary-search approach. We assume that $\mathrm{rank}(x)$ has been found to lie in an interval $[a - \delta \min(a, n - a), b + \delta \min(b, n - b)]$ with high probability (initially $a = 1, b = n$). Let $m = \frac{a+b}{2}$ be the midpoint of the interval. We call $\mathsf{TestLE}(x, m, \delta, \epsilon/\lceil \log_2 n \rceil)$. If the call returns 'true', then we assume $\mathrm{rank}(x) \leq m + \delta \min(m, n - m)$ and set $b = m$, otherwise we assume $\mathrm{rank}(x) \geq m - \delta \min(m, n - m)$ and set $a = m$. Each call reduces the search range by a factor of 2, so $\lceil \log_2 n \rceil$ calls are needed to reduce the search range to a single value r, which is returned as the approximate rank. By the union bound, the probability that r is a δ-approximation to $\mathrm{rank}(x)$ is at least $1 - \epsilon$. The total number of group test queries is

$$
\begin{aligned}
N_t \cdot \lceil \log_2 n \rceil &= \frac{e^2}{2\delta^2} \ln(\lceil \log_2 n \rceil / \epsilon) \lceil \log_2 n \rceil \\
&= \frac{e^2}{2\delta^2} (\ln \lceil \log_2 n \rceil + \ln 1/\epsilon) \lceil \log_2 n \rceil \\
&= \mathcal{O}(\frac{\log n}{\delta^2} (\log \log n + \log 1/\epsilon)).
\end{aligned}
$$

Let this algorithm be denoted $\mathsf{ApxRank}(x, \delta, \epsilon)$. Thus we have,

Theorem 2. $\mathsf{ApxRank}(x, \delta, \epsilon)$ *makes* $\mathcal{O}(\frac{\log n}{\delta^2}(\log \log n + \log \frac{1}{\epsilon}))$ *group test queries and returns a rank* $r \in [n]$ *such that* $|\mathrm{rank}(x) - r| \leq \delta \min(r, n - r)$ *with probability* $1 - \epsilon$.

5   Selection

In this section, we describe a Monte Carlo randomized algorithm for the approximate selection problem. Our strategy uses two basic steps: the first step invokes a sampling procedure to generate a candidate x that is close to the desired rank with a certain probability (in this case x is said to be a good candidate). The second step uses a separate verification procedure to check that x meets the approximate rank requirements. These steps are repeated a fixed number of times so that we have at least a $\frac{1}{2}$ chance to find a good candidate. We will bound the query complexity and failure rates of both steps.

5.1   Step 1: Sampling Procedure

Let $k \leq n/2$; we will again assume that n is divisible by k; if not, we will add at most $k - 1$ equivalent dummy elements to S that are greater than all other elements. As before, the case $k \geq n/2$ can be handled by reversing the direction of the group tests in the algorithm and then selecting the $(n - k + 1)$-th element. We start by analyzing the chance that the minimum of a sample of size $N = n/k$ falls in the range $(k - \delta k, k + \delta k]$.

Lemma 4. *Let m be the minimum element of a randomly sampled (with replacement) subset from S of size $N = n/k$. Then,*

$$\Pr[k - \delta k < rank(m) \leq k + \delta k] \geq 2e^{-1}\delta.$$

Proof. Note that

$$\Pr[\mathrm{rank}(m) > k + \delta k] = \left(\frac{n - (k + \delta k)}{n}\right)^N$$

$$= \left(1 - \frac{1}{N} - \frac{\delta}{N}\right)^N.$$

Similarly,

$$\Pr[\mathrm{rank}(m) \leq k - \delta k] = 1 - \left(\frac{n - (k - \delta k)}{n}\right)^N$$

$$= 1 - \left(1 - \frac{1}{N} + \frac{\delta}{N}\right)^N. \tag{6}$$

Combining both bounds, we find

$$\Pr[k - \delta k < \mathrm{rank}(m) \leq k + \delta k] = 1 - \Pr[\mathrm{rank}(m) \leq k - \delta k] - \Pr[\mathrm{rank}(m) > k + \delta k]$$

$$= 1 - \left(1 - \frac{1}{N} - \frac{\delta}{N}\right)^N - \left(1 - \left(1 - \frac{1}{N} + \frac{\delta}{N}\right)^N\right)$$

$$= \left(1 - \frac{1}{N} + \frac{\delta}{N}\right)^N - \left(1 - \frac{1}{N} - \frac{\delta}{N}\right)^N$$

$$\geq 2e^{-1}\delta, \tag{7}$$

where Eq. 7 follows from the same bound derived in Eq. 5.

Algorithm 3 summarizes the sampling procedure. Let I be the number of iterations needed to first draw a δ-approximation to the k-th order statistic. From Lemma 4, we have $\mathbb{E}(I) \leq e/2\delta$. Furthermore, $\Pr[I \geq 2\,\mathbb{E}(I)] \leq \frac{1}{2}$, by Markov's inequality. Thus, if we draw e/δ samples, we will find a good sample with probability at least $\frac{1}{2}$. In fact, it will be useful to have a $\frac{1}{2}$ chance to get a $\delta/2$-approximation, so instead we will draw $2e/\delta$ samples.

Algorithm 3. Random sample - Step 1

1: **function** GETCANDIDATE(k, δ)
2:     $S' \leftarrow$ Randomly select $\frac{n}{k}$ elements from S with replacement.
3:     $x \leftarrow$ MinFind(S')
4:     **return** x
5: **end function**

5.2   Step 2: Checking a Sample

Next we need to check whether the sample x from Step 1 is a $\delta/2$-approximation to the k-th order statistic. We can use the rank comparison test, TestLE (Algorithm 2), from the previous section to check x: Specifically TestLE$(x, k + \frac{3}{4}\delta k, \delta', \epsilon')$ can check that $\mathrm{rank}(x) \leq k + \frac{1}{2}\delta k$ if we take $\delta' = \frac{\delta k/4}{k + \frac{3}{4}\delta k} = \frac{\delta/4}{1 + \frac{3}{4}\delta} > \delta/8$; this will be reliable with probability $1-\epsilon'$. Similarly, TestLE$(x, k - \frac{3}{4}\delta k, \delta'', \epsilon')$ can check that $\mathrm{rank}(x) \geq k - \frac{1}{2}\delta k$, if we take $\delta'' = \frac{\delta k/4}{k - \frac{3}{4}\delta k} = \frac{\delta/4}{1 - \frac{3}{4}\delta} > \delta/4$; this will again be reliable with probability $1 - \epsilon'$. If both of these inequality tests pass, then we accept x. We have:

Lemma 5. *Let x be a randomly sampled element from Step 1. If $|\mathrm{rank}(x) - k| \leq \frac{\delta}{2}\min(k, n-k)$, then x is accepted with probability $1 - 2\epsilon'$, and if $|\mathrm{rank}(x) - k| > \delta\min(k, n-k)$, then x is rejected with probability $1 - \epsilon'$.*

Proof. If $|\mathrm{rank}(x) - k| \leq \frac{\delta}{2}\min(k, n-k)$, then by Lemma 3, each inequality test is passed with probability $1 - \epsilon'$. By the union bound, both are passed with probability $1 - 2\epsilon'$. Suppose $|\mathrm{rank}(x) - k| > \delta\min(k, n-k)$. If $\mathrm{rank}(x) > k$ then TestLE$(x, k + \frac{3}{4}\delta k, \delta', \epsilon')$ returns false with probability $1 - \epsilon'$. Similarly, if $\mathrm{rank}(x) < k$, TestLE$(x, k - \frac{3}{4}\delta k, \delta'', \epsilon')$ will return true with probability $1 - \epsilon'$. Thus, x is correctly rejected with probability $1 - \epsilon'$.

If x is accepted, we report it as the selected element; if it is rejected, then another sample is drawn. The number of Step 2 checks performed is at most $2e/\delta$; any sample that should be rejected is rejected with probability $1 - \epsilon'$. If we draw a sufficiently good sample (a $\delta/2$-approximation) it is accepted with probability $1 - 2\epsilon'$. Thus, by the union bound, we should set $\epsilon' = \frac{1}{2e/\delta + 1} \cdot \epsilon$ to obtain an overall ϵ probability of failure (all bad elements are rejected and a good one is accepted, if drawn). We denote the full approximate selection algorithm as ApxSelect(k, δ, ϵ) (see Algorithm 4). We have,

---

Algorithm 4. Full selection algorithm

1: **function** APXSELECT(k, δ, ϵ)
2:     **for** $i = 1$ to $2e/\delta$ **do**
3:         $x \leftarrow$ GetCandidate($k, \delta/2$)
4:             **if** TestLE($x, k + \frac{3}{4}\delta k, \delta', \epsilon'$) **and not** TestLE($x, k - \frac{3}{4}\delta k, \delta'', \epsilon'$) **then**
5:                 **return** x
6:             **end if**
7:         **end for**
8:     **return** not found
9: **end function**

Theorem 3. *ApxSelect(k, δ, ϵ) makes an expected $\mathcal{O}(\frac{1}{\delta} \log^2 N + \frac{1}{\delta^3} \log \frac{1}{\epsilon\delta^2})$ group test queries and with probability at least $\frac{1}{2}$ returns an element $x \in S$ such that $|\,rank(x) - k| \leq \delta \min(k, n - k)$ with probability $1 - \epsilon$.*

Proof. As discussed above, if the algorithm outputs an element x, then it is a δ-approximation with probability $1 - \epsilon$, and there is at least a $1/2$ chance that this occurs. The number of iterations of the outer loop is at most $2e/\delta$. Each call to GetCandidate makes an expected $\mathcal{O}(\log^2 N)$ queries. Each call to TestLE makes at most

$$\frac{e^2 \ln(1/\epsilon')}{2(\delta/8)^2} = \mathcal{O}(\log(\frac{1}{\epsilon\delta^2})/\delta^2)$$

queries. The total expected number of queries is thus

$$\mathcal{O}(\frac{1}{\delta}(\log^2 N + \log(\frac{1}{\epsilon\delta^2})/\delta^2)).$$

Since this is a Monte Carlo algorithm, we can simply rerun it if we did not get an output; the expected number of times is at most two.

6   Discussion

In this work, we considered several classic problems on order-finding where a group inequality testing operation is available; we gave randomized algorithms for min-finding, rank determination, and selection. A natural question is whether group inequality testing might also be useful for similar problems on partially-ordered sets. For posets, there can be multiple minimum elements (up to the $width^1$ of the poset.). Algorithm 1 will find one, but there does not seem to be an efficient way to determine all minimum elements (or estimate the width of the poset). In addition, the rank of an element x of a poset is normally defined as the length of the longest left-extending chain from x, not the number of elements less than or equal to the element. Thus, the approach used in Algorithm 2 cannot be directly applied as the number of elements less than or equal to an element

---

[1] The size of a maximum antichain in the poset; i.e., the maximum number of mutually incomparable elements.

of rank r can vary between r and $wr - w + 1$, where w is the width of the poset. We note that the trivial lower bound of $\Omega(\log n)$ applies to each of the problems considered, so it is possible that non-trivial lowers bounds might be found, e.g. for min-finding with group testing. Furthermore, a more powerful group test of the form $U \leq_Q V$ may be considered, although it does not seem to benefit the algorithms presented. Another variation is a "noisy" version of the group test in which there is some probability to get the wrong answer from the oracle. In this case, the oracle can be resampled to decrease the error rate; the algorithms here can all be adapted to this setting, but we leave this as future work.

References

1. Blum, M., Floyd, R.W., Pratt, V., Rivest, R.L., Tarjan, R.E.: Time bounds for selection. J. Comput. Syst. Sci. **7**(4), 448–461 (1973)
2. Chen, C.L., Swallow, W.H.: Using group testing to estimate a proportion, and to test the binomial model. Biometrics, 1035–1046 (1990)
3. Falahatgar, M., Jafarpour, A., Orlitsky, A., Pichapati, V., Suresh, A.T.: Estimating the number of defectives with group testing. In: 2016 IEEE International Symposium on Information Theory (ISIT), pp. 1376–1380. IEEE (2016)
4. Floyd, R.W., Rivest, R.L.: Expected time bounds for selection. Commun. ACM **18**(3), 165–172 (1975)
5. Gastwirth, J.L., Hammick, P.A.: Estimation of the prevalence of a rare disease, preserving the anonymity of the subjects by group testing: application to estimating the prevalence of aids antibodies in blood donors. J. Stat. Plann. Inf. **22**(1), 15–27 (1989)
6. Karnin, Z., Lang, K., Liberty, E.: Optimal quantile approximation in streams. In: IEEE Symposium on Foundations of Computer Science (FOCS), pp. 71–78 (2016)
7. Liyanage, A., Burger, R., Shi, A., Sopp, B., Zhu, B., Mumey, B.: EssentCell: discovering essential evolutionary relations in noisy single-cell data (2025). https://doi.org/10.1101/2025.04.12.648524
8. Manku, G.S., Rajagopalan, S., Lindsay, B.: Approximate medians and other quantiles in one pass and with limited memory. ACM SIGMOD Rec. **27**(2), 426–435 (1998)
9. Swallow, W.H.: Group testing for estimating infection rates and probabilities of disease transmission (1985)
10. Thompson, K.H.: Estimation of the proportion of vectors in a natural population of insects. Biometrics **18**(4), 568–578 (1962)
11. Walter, S.D., Hildreth, S.W., Beaty, B.J.: Estimation of infection rates in populations of organisms using pools of variable size. Am. J. Epidemiol. **112**(1), 124–128 (1980)

An Algorithm for Monitoring Edge-Geodetic Sets in Chordal Graphs

Clara Marcille[1(✉)] and Nacim Oijid[2]

[1] Univ Lyon, EnsL, UCBL, CNRS, LIP, LYON Cedex 07 69342, France
clara.marcille@ens-lyon.fr

[2] Department of Mathematics and Mathematical Statistics, Umeå University, Umeå, Sweden
nacim.oijid@umu.se

Abstract. A *monitoring edge-geodetic set* (or meg-set for short) of a graph is a set of vertices M such that if any edge is removed, then the distance between some two vertices of M increases. This notion was introduced by Foucaud *et al.* in 2023 as a way to monitor networks for communication failures. As computing a minimum meg-set is hard in general, recent works aimed to find polynomial-time algorithms to compute minimum meg-sets when the input belongs to a restricted class of graphs. Most of these results are based on the property of some classes of graphs to admit a unique minimal meg-set, which is then easy to compute. In this work, we prove that chordal graphs also admit a unique minimal meg-set, answering a standing open question of Foucaud *et al.*

Keywords: Monitoring edge-geodetic · Polynomial-time algorithm · Chordal graphs

1 Introduction and Notations

In the area of network monitoring problems, graphs are used to emulate networks and their possible behaviors. For instance, there is a well-established way to represent a network by a graph where each terminal is represented as a vertex, such that two vertices are adjacent if the corresponding terminals can communicate directly. In this representation, deleting an edge can be interpreted as the failure of a communication, see for instance [1–3,5]. Monitoring edge-geodetic sets (or meg-sets for short) were introduced by Foucaud *et al.* in a recent paper [8], as a way to detect such failures. For a graph $G = (V, E)$, an edge $e \in E$ and two vertices $a, b \in V$, we say that $\{a, b\}$ *monitors* the edge e if it lies on all the shortest paths between a and b. Note that deleting e would increase the distance between a and b, which is detected. By extension, we say $M \subset V$ monitors e if there exist two vertices of M monitoring e. Finally, we say that M is a *meg-set*

N. Oijid—Partly supported by the Kempe Foundation Grant No. JCSMK24-515 (Sweden).

if it monitors all edges of E. As stated in [6], this can be understood as choosing a set of "probes" in the network which measure the distance to one another, and doing so, detecting that a failure in the network happened if one of the distance increases. As such measurements can be costly, we want to minimize the number of probes, that is, find the size of a minimum meg-set (w.r.t. size). This problem, known as MEG-SET, is hard in general [10], but can be solved in polynomial time on some classes of graphs using a characterization from [6].

A vertex $u \in V(G)$ is *mandatory* if it belongs to all meg-sets of G, and $Mand(G)$ denotes the set of mandatory vertices. In the same paper, the authors prove that computing the set of mandatory vertices can be done in polynomial time. This computation also provides a polynomial-time algorithm to compute the meg-sets of graphs which are *meg-minimal*, that is, with the property of admitting a unique minimal meg-set (w.r.t. inclusion), which is thus the set of mandatory vertices. These graph classes include namely interval graphs, cographs, block graphs, and well-partitioned chordal graphs [6]. In a more recent work [9], the authors prove that strongly chordal graphs are also meg-minimal. We refer the reader to the corresponding papers for a formal definition of these classes, and give an overview of the inclusion relations between these classes in Fig. 1. As all the above-mentioned graph classes (except cographs) are subclasses of chordal graphs, the next step is to inquire whether or not chordal graphs are meg-minimal. In particular, the best known algorithm on chordal graphs from [7] is an FPT algorithm parameterized by treewidth plus solution size.

In this work, we answer the question of the complexity of finding a minimum meg-set when the input graph is chordal by proving the following theorem:

Theorem 1. *Let G be a chordal graph. Then $Mand(G)$ is a meg-set of G.*

In other words, we prove that MEG-SET can be solved in polynomial time on chordal graphs. This result can be seen as a more fine-grained proof of Theorem 6.1 of [9], generalized to chordal graphs, thus requiring more work to handle the new possibilities. In particular, they use a strong elimination ordering for their induction, while we generalize to a perfect elimination ordering. In addition to that, we want to acknowledge that some technical lemmas are similar between the two proofs. For instance, Claim 6.3 of [9] is very close in spirit to Lemma 5.

We remind that if G is a graph, a vertex v of G is said to be *simplicial* if its neighborhood is a clique. Also a graph G is said to be chordal if it has a perfect elimination ordering, i.e., an ordering of its vertices $v_1, \ldots, v_n$ such that for any $1 \leq i \leq n-1$, $N(v_i) \cap \{v_{i+1}, v_{i+2}, \ldots, v_n\}$ is a clique, where $N(v_i)$ is the neighborhood of the vertex v_i.

Throughout this paper, we denote a path on the vertices $a_1, a_2, \ldots, a_k$ by the concatenation $a_1 a_2 \ldots a_k$. We also denote a cycle the same way, though repeating the first vertex at the end. For instance, the cycle of size three on a, b, c is denoted $abca$. Finally, to be coherent with previous works on meg-sets, we call an *induced 2-path* abc in a graph G a set of three vertices a, b, c of G such that abc is an induced P_3.

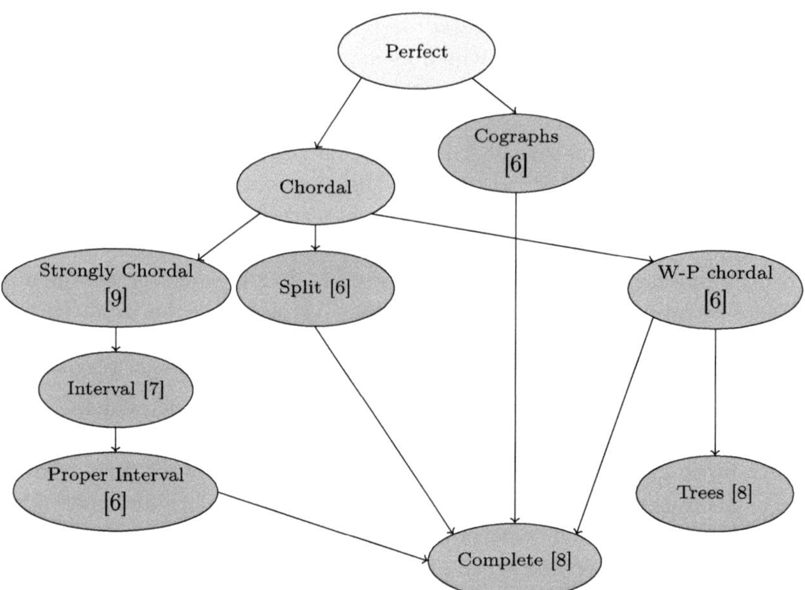

Fig. 1. A representation of the inclusion relations between the classes of graphs mentioned in the introduction. Our contribution is shown in blue, and the graph classes that are known not to be meg-minimal are in orange (perfect graphs). All the other depicted graph classes were previously known to be meg-minimal. (Color figure online)

2   Some General Lemmas

Theorem 2. (Foucaud *et al.* **[6, Theorem 4.1]).** *Let G be a graph. A vertex $v \in V(G)$ is mandatory if and only if there exists $w \in N(v)$ such that for any vertex $x \in N(v)$, the induced 2-path wvx is part of a 4-cycle.*

If it exists, such a vertex w is called a *support* of v.

Lemma 1. *Let G be a graph, and v be a simplicial vertex of G. Then $v \in Mand(G)$.*

Lemma 1 is a direct consequence of Theorem 2, but was already mentioned earlier in [8].

Lemma 2. *If $G = (V, E)$ is a chordal graph, and $a, b, c \in V$ are three vertices such that abc is an induced 2-path of G, which is in a cycle C. Then there is a chord in C incident to b, that is, an edge between b and some vertex of C different from a and c.*

Proof. By definition, C is a C_k for $k \geq 4$ of G. Thus, it is necessarily not induced (as G is chordal and hence admits no induced C_k). Hence, there is a chord in C. Either it is adjacent to b or it produced a smaller cycle C' containing $\{a, b, c\}$.

Up to applying the same argument to C', as abc is an induced P_3, a and c cannot be adjacent. Hence, when we cannot find such a C', we obtain a chord adjacent to b.

Lemma 3. (Foucaud *et al.* **[8, Lemma 2.3]).** *Let G be a graph with a cut-vertex G and $C_1, C_2, \ldots, C_k$ be the k components obtained when removing v from G. If $S_1, S_2, \ldots, S_k$ are meg-sets of the induced subgraphs $G[C_1 \cup \{v\}], G[C_2 \cup \{v\}], \ldots, G[C_k \cup \{v\}]$, then $S = (S_1 \cup \cdots \cup S_k) - v$ is an meg-set of G.*

3   Proof of the Main Theorem

We consider G a chordal graph which is a minimal counter-example (w.r.t. the number of vertices) to Theorem 1, and we prove that G cannot exist. To achieve this, let v be a simplicial vertex of G. By minimality of G, since $G - \{v\}$ is chordal, it is not a counter example. Hence, $M' = Mand(G - \{v\})$ is a meg-set of $G - \{v\}$. Let $W \subset V(G - \{v\})$ such that for all $w \in W$, w is mandatory in $G - \{v\}$ but not in G. Let $M = \{v\} \cup M' \setminus W$. We prove that M only contains mandatory vertices and that it is a meg-set of G, contradicting the fact that it is a counterexample.

3.1   Overview of the Proof

The proof of Theorem 1 considers a minimum counter-example G (w.r.t. the number of vertices) to its statement, and apply the well-known elimination rule of chordal graphs, that is, we consider a simplicial vertex v of G and remove it. This yields a chordal graph G' for which Theorem 1 holds. In Lemma 4, we prove that the mandatory vertices of $G - \{v\}$ keep monitoring in G the edges they monitor in $G - \{v\}$, that is, if an edge of $G - \{v\}$ is not monitored in G, then it is monitored by at least one vertex of $W = Mand(G - \{v\}) \cap (G - Mand(G))$. We then prove in Lemma 5 that vertices of W all belong to the neighborhood of v.

3.2   Preliminaries

Next lemma states that adding a simplicial vertex cannot create a shortcut.

Lemma 4. *If $a, b \in V(G) - \{v\}$ are two vertices such that $\{a, b\}$ monitors an edge e (in $G - \{v\}$), then $\{a, b\}$ monitors e in G.*

Proof. Let $e \in E(G)$ be an edge monitored by $\{a, b\}$ where $a, b \in V(G) - \{v\}$, then it means that e is on every shortest path between a and b. Let P be one of these shortest paths. Since v is simplicial, all its neighbors in P are adjacent. Thus, v has at most two neighbors in P and if so, they are consecutive on the path, otherwise, the path would have a chord. Thus, P is still a shortest path in G. Moreover, a shortest path between a and b in G cannot go through v, otherwise, it will contain v and two of its neighbors. Since $v \notin \{a, b\}$, it would contain an induced triangle, which cannot happen for a shortest path. Finally, e is in all the shortest path between a and b in G, and therefore is monitored by $\{a, b\}$.

Lemma 5. *If $u \notin N(v)$, then $u \in Mand(G - \{v\})$ if and only if $u \in Mand(G)$.*

Proof. Suppose first $u \in Mand(G)$. There exists a support vertex w of u. Hence, for any vertex $x \in V(G)$, such that wux is an induced 2-path, there exists v' such that $wuxv'w$ is a 4-cycle. Moreover, we must have $v' \neq v$, otherwise, since v is simplicial, wx would be an edge, and thus, wux would not be an induced 2-path. Since $v \notin N(u)$, we have $w \neq v$. Thus, for any vertex $x \in V(G) - \{v\}$, if wux is an induced 2-path, the same vertex v' ensures that $wuxv'w$ is a 4-cycle, and thus w is a support for u, and $u \in Mand(G - \{v\})$.

Reciprocally, suppose that $u \in Mand(G - \{v\})$, and let w be a support for u in $G - \{v\}$. Since $v \notin N(u)$, any induced 2-path wux of G is an induced 2-path of $G - \{v\}$. Thus, if v' is such that $wuxv'w$ is a 4 cycle in $G - \{v\}$, $wuxv'w$ is also a 4-cycle of G. Thus, u is mandatory in G.

Corollary 1. $W \subset N(v)$.

Recall that $M = \{v\} \cup Mand(G - \{v\}) \setminus W$.

Lemma 6. $M \subset Mand(G)$

Proof. First, as v is simplicial, by Lemma 1, we have $v \in Mand(G)$. Moreover, by Lemma 5, any vertex in $Mand(G - \{v\}) \setminus (N(v))$ is also mandatory in G. Finally, if $u \in (N(v) \cap M)$, then by hypothesis $u \in Mand(G - \{v\}) \setminus W$, and thus u is also mandatory in G. Thus, M only contains mandatory vertices, and $M \subset Mand(G)$.

Lemma 7. *G is 2-vertex connected.*

Proof. First, by minimality of G, G is connected, otherwise the union of the meg-sets of each of its connected components would be a meg-set of G, and would only contain mandatory vertices.

Suppose that G is not 2-vertex connected. Thus, there exists $u \in V(G)$ such that $G - \{u\}$ is disconnected. Let $C_1, \ldots, C_p$ be the connected components of $G - u$, and let $G_1, \ldots, G_p$ the subgraph induced by $V(C_1) \cup \{u\}, \ldots, V(C_p) \cup \{u\}$. By minimality of G, since $G_1, \ldots, G_p$ are chordal, $M_1 = Mand(G_1), \ldots, M_p = Mand(G_p)$ are meg-sets of $G_1, \ldots, G_p$. Let $M' = (M_1 \cup \cdots \cup M_p) - \{u\}$. By Lemma 3, M' is a meg-set of G. It remains to prove that M' only contain mandatory vertices.

If $v_i \in V(G_i) - \{u\}$ is mandatory, then there exists w_i such that any induced 2-path $w_i v_i x$ is in a 4-cycle. As $v_i \neq u$, any induced 2-path $w_i v_i x$ in G is an induced 2-path in G_i, and thus the 4-cycle in $G - \{v\}$ is also in G. which proves that v_i is mandatory in G.

Finally, we proved that if G is not 2-vertex connected, then either one of its 2-connected components is a smaller counter example, or G itself is not a counter example, which concludes the proof.

Lemma 8. *Let $G = (V, E)$ be a graph and $e = ab$ be an edge of G. Let $\{\alpha, \beta\}$ be two vertices such that $\{\alpha, \beta\}$ monitors e. Then, all the shortest paths from α to a contain e or all the shortest paths from β to a contain e.*

Proof. Since M is a meg-set of G, there must exist two vertices α, β monitoring e. That is, all shortest paths from α to β contain e. If one of α, β is a, then we are done. Otherwise, it must be that α is closer to one endpoint of e, say a than to b. Observe then that, since all shortest paths from α to β contain e they in particular contain a. Hence, since a is closer to α than b, a shortest path from α to β can be decomposed into a shortest path from α to a, the edge ab and a shortest path from b to β. In particular, $\{a, \beta\}$ monitors e.

Corollary 2. *Let $G = (V, E)$ be a graph, M a meg-set of G. If $a \in M$, then for any edge e incident to a, there exists $b \in M$ such that $\{a, b\}$ monitors e.*

3.3 Proof of the Main Result

In order to prove Theorem 1, we prove the following lemma that directly implies it. The proof of Lemma 9 is subdivided in four lemmas (Lemma 10 to 13).

Lemma 9. $Mand(G) = \{v\} \cup Mand(G - \{v\}) \setminus W$ *and is a meg-set of G.*

Recall that $W \subset N(v)$ is the set of vertices that are mandatory in $G - \{v\}$ but not in G. We prove that $M = \{v\} \cup Mand(G - \{v\}) \setminus W$ is a meg-set of G. Since M only contains mandatory vertices by Lemma 6, this is enough to yield Theorem 1. In what follows, we introduce notations that we use throughout Lemma 10 to Lemma 13. Let e be an edge of G. The proof will be a case distinction depending on where e lies in G. Finally, for a path $P = a_1 a_2 \ldots a_k$, we denote $|P|$ the length of P, which is the number of edges in P.

Proof. We treat cases based on the following disjunction:

- either $e \in E(G - \{v\})$, and
  - e is monitored (in $G - \{v\}$) by two vertices of $Mand(G - \{v\}) \setminus W$ (Lemma 10);
  - e is monitored (in $G - \{v\}$) by a vertex of $Mand(G - \{v\}) \setminus W$ and a vertex of W (Lemma 11) which are either non-adjacent (first case) or adjacent (second case);
  - e is monitored (in $G - \{v\}$) by two vertices of W (Lemma 12).
- or $e \notin E(G - \{v\})$, that is e incident to v (Lemma 13).

Since $Mand(G - \{v\})$ is a meg-set of $G - \{v\}$, then every $e \in E(G - \{v\})$ is monitored by two vertices of $Mand(G - \{v\})$, hence all cases are covered.

Lemma 10. *If $e \in E(G - \{v\})$, and e is monitored in $G - \{v\}$ by two vertices of $Mand(G - \{v\}) \setminus W$, then e is monitored by two vertices of $Mand(G)$.*

Proof. If $e \in E(G - \{v\})$ such that e is monitored by two vertices $\{a, b\}$ in $G - \{v\}$ that are in $Mand(G - \{v\}) \setminus W$, then by Lemma 4, $\{a, b\}$ monitors $e \in E(G)$.

Lemma 11. *If $e \in E(G - \{v\})$, and e is monitored by a vertex of $Mand(G - \{v\}) \setminus W$ and a vertex of W, then e is monitored by two vertices of $Mand(G)$.*

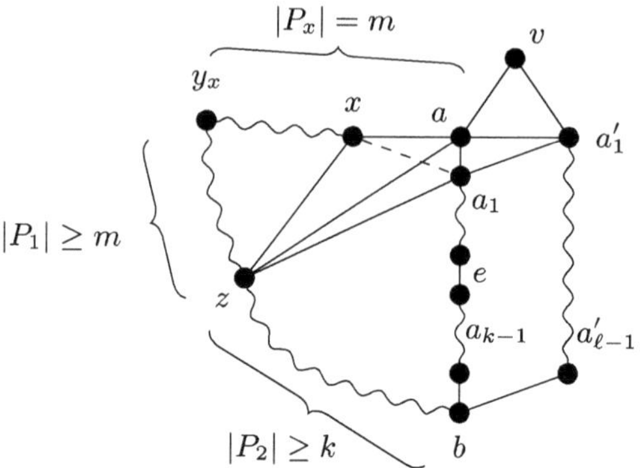

Fig. 2. Representation of the proof of Lemma 11, that is $e \in E(G - \{v\})$, $a \in W$, $b \notin W$, where a and b are not adjacent.

Proof. If $e \in E(G - \{v\})$ such that e is monitored by two vertices $\{a, b\}$ in $G - \{v\}$ with $a \in W$ and $b \in Mand(G - \{v\}) \setminus W$, we have the following two cases depending whether a and b are adjacent or not (Fig. 2).

- We first treat the sub-case where a and b are non-adjacent. Note that b cannot be adjacent to v since v is simplicial and adjacent to a. If all the shortest path from v to b start with the edge va, then e is on all the shortest path from v to b and thus it is monitored by $\{v, b\}$. Suppose this is not the case. Hence, there is a vertex $a' \in N(v)$ and a shortest path going from v to b that starts with the edge va'. Let $P = vaa_1 \ldots a_{k-1}b$ be a shortest path of length k from a to b going through a, and $P' = va'a'_1 \ldots a'_{\ell-1}b$ be a shortest path from v to b going through a' and avoiding e. Note that if such a path P' does not exist, then $\{v, b\}$ monitors e. Since P' is a shortest path, we have $|P'| \leq |P|$. Thus, $\ell \leq k$. Moreover, since $aa'a'_1 \ldots a'_{\ell-1}b$ is not a shortest path, from a to b, we have $k < \ell + 1$, hence $k = l$.
  Now, a cannot have any neighbor a'_i, otherwise $aa'_ia'_{i+1} \ldots a_{\ell-1}b$ would also be a shortest path from a to b, and a cannot have a neighbor a_i for $i \neq 1$ as P is chordless. Thus, by Lemma 2, applied to a_1aa' in the cycle $aa_1 \ldots c \ldots a'_1a'a$, where c is the first vertex common to P and P' other than v, we must have an edge a_1a'. Note that such a vertex c exists as $b \in P \cap P'$.
  We prove that under this conditions, either a is mandatory or we will find a vertex y such that $\{y, b\}$ monitors e. Let $x \in N(a)$ be a vertex. If $x \in N[v]$, since a' is a neighbor of a_1, $a_1axa'a_1$ is a 4-cycle, as v is simplicial. Otherwise, let y_x be a vertex such that $\{a, y_x\}$ monitors the edge ax in $G - v$. Note that this vertex exists by Lemma 8. Under this condition, y_x cannot be in $N(v)$, otherwise it would be adjacent to a, contradicting that all shortest paths from a to y_x goes through the edge ax. If $\{y_x, b\}$ monitors e, then we are done as y_x cannot be in $N(v)$, thus cannot be in W by Corollary 1.
  Otherwise, let P_x be a shortest path from a to y_x, and denote it by $P_x = axx_1 \ldots x_{m-1}y_x$. Let P_y be a shortest path from y_x to b that does not contain

e (possibly of length 0 if $y_x = b$). In particular P_y cannot contain a as all shortest paths from a to b contain e. Hence, there exists a cycle C containing only vertices from P_x, P_y and P. Note that by construction, this cycle has to contain a, x_1 and a_1 as P_y cannot contain a. If there are no chords adjacent to a in C, x_1a_1 is an edge by Lemma 2, and x_1xaa_1 is a 4-cycle. Otherwise there is a chord from a to some vertex z in C. Since both P and P_x are chordless and contain $\{a\}$, we necessarily have $z \in P_y$. Denote by P_1 the path from z to to y_x and by P_2 the path from z to b, such that $P_y = P_1P_2$. Since aP_1 is not a shortest path form a to y_x as it does not contain ax, we have $|P_1| \geq m$, and since P_2 is not a shortest path from a to b as it does not contains e, we have $|P_2| \geq k$. But, since P_y is a shortest path from y_x to b, we need to have $|P_1| + |P_2| \leq |P_x| + |P|$, i.e. $|P_1| + |P_2| \leq k + m$, which implies $|P_1| = m$ and $|P_2| = k$. Hence, a cannot be adjacent to any other vertex of P_1 (P_2 resp.) as otherwise, it would create a shortest path from a to y_x (to b resp.), that does not contain ax (e resp.). We denote c_1 the vertex common to P_1 and P_x that is the farthest away from y_x, so that $P_1 = y_x \ldots c_1P_1'$ and $P_x = y_x \ldots c_1P_x'$ for some paths P_1', P_x' which must be disjoint. Conversely, we denote c_2 the vertex common to P_2 and P that is the farthest away from b, so that $P_2 = P_2'c_2 \ldots b$ and $P = P''c_2 \ldots b$ for some paths P_2', P'' which must be disjoint. Thus, by Lemma 2, applied in the cycle obtained from az, P_1' and P_x'; or az, P_2' and P'', we have x and a_1 are both adjacent to z, ensuring that xaa_1zx is a 4-cycle. Finally, we proved that either one vertex y_x satisfy that $\{y_x, b\}$ monitors e, or all the induced 2-paths a_1ax with x a neighbor of a are in 4-cycles, and a_1 is a support vertex for a, contradicting that a is not mandatory and thus in W.

- We can now assume that a and b are adjacent, that is, $e = ab$. If $\{b, v\}$ monitors e, then we are done. Otherwise, if b is a support for a, then a is mandatory, which contradicts $a \in W$. Hence, there exists $a_1 \in N(a)$ such that baa_1 is an induced two path that is not is a 4-cycle. If a_1 is mandatory, as baa_1 is the only 2-path between b and a_1, then $\{b, a_1\}$ would monitor ab and we are done. Otherwise, by Lemma 8 there exist a_k such that $\{a, a_k\}$ monitors aa_1 in $G - \{v\}$. In particular $a_k \notin N(v)$, otherwise, since v is simplicial and adjacent to a, a_k would be adjacent to a. Therefore, by Lemma 5, $a_k \in M$. We denote by $P = aa_1a_2 \ldots a_k$ a shortest path from a to a_k. If all the shortest path from b to a_k goes through e, then $\{b, a_k\}$ monitors e and we are done. Otherwise there exist a shortest path $P' = b_\ell b_{\ell-1} \ldots b_1b$ of length ℓ from b to $b_\ell = a_k$ that does not go through e. Since it is a shortest path, we have $\ell \leq k+1$. Moreover since $abb_1 \ldots b_{\ell-1}a_k$ is not a shortest path, otherwise aa_1 would not be monitored by $\{a, a_k\}$, we have $\ell + 1 > k$. All together, we have $\ell \in \{k, k+1\}$. We also denote c the vertex common to P and P' which is farthest away from a_k, so that $P = aP_ac \ldots a_k$ and $P' = a_k \ldots cP_bb$ where P_a and P_b are disjoint.

If $\ell = k$: The vertex a cannot have any neighbor b_i in P' other than b, otherwise $ab_i \ldots b_k$ would be a shortest path from a to a_k that does not contain the edge aa_1. Hence, by Lemma 2 applied to a_1ab in the cycle aP_acP_bba, a_1

and b are adjacent, contradicting the fact that baa_1 is an induced 2-path. A representation of this configuration is shown in the left graph of Fig. 3.

If $\ell = k + 1$: The same argument as in the case $\ell = k$ works, except that now, the chord ab_1 can exist. Note that this is the only chord that can be attached to a without breaking the minimality of P, the minimality of P', or the fact that $\{a, a_k\}$ monitors aa_1. If ab_1 does not exist, then for the same reason as in the previous case, we would have the edge $a_1 b$ contradicting that $a_1 ab$ is an induced 2-path. Now if ab_1 exists, consider the cycle $aP_a cP_b a$. In this cycle, a cannot have any other chord adjacent to it, otherwise, it would contradict the fact that P or P' is a shortest path. Hence, by Lemma 2, there is an edge $a_1 b_1$. Hence, $baa_1 b_1 b$ is a 4-cycle, contradicting the fact that baa_1 is in no 4-cycle. A representation of this configuration is shown in the right graph of Fig. 3.

All together, b is a support for a in G, contradicting that a is not mandatory in G.

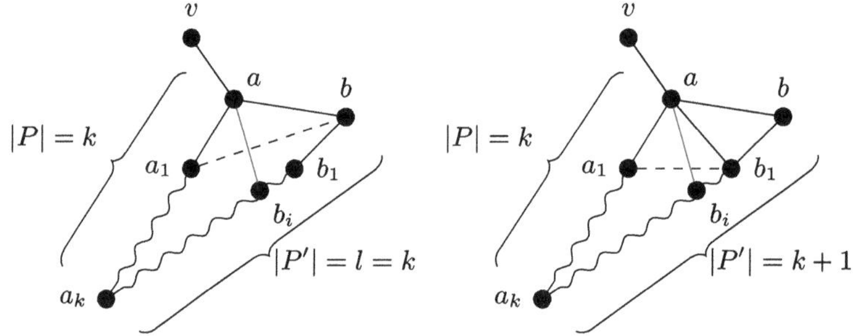

Fig. 3. Representation of the proof of Lemma 11, that is $e \in E(G - \{v\})$, where $a \in W$, $b \notin W$, a and b are adjacent.

Lemma 12. *If $e \in E(G - \{v\})$, and e is monitored by two vertices of W, then e is monitored by two vertices of $Mand(G)$.*

Proof. Suppose that $e \in E(G - \{v\})$ such that e is monitored by two vertices $\{a, b\}$ in $G - \{v\}$ with $a, b \in W$. Intuitively, we will have the same argument as in the case where $b \notin W$ and a and b are adjacent, but on the two extremities of e.

Since $W \subset N(v)$ and v is simplicial, necessarily, $e = ab$ at it is the only shortest path between a and b. Since a and b are in W, they are not mandatory. Thus, a is not a support for b neither is b for a. Let a_1 be a neighbor of a such that baa_1 is an induced 2-path that is not in a 4-cycle. Similarly, let b_1 be a neighbor of b such that abb_1 is an induced 2-path that is not in a 4-cycle. Note that this construction implies that $a_1 b$, ab_1, and $a_1 b_1$ are non-edges. Let a_n, b_m be two vertices such that $a_1 a$ is monitored by $\{a_n, a\}$ and $b_1 b$ is monitored by

$\{b_m, b\}$ in $G - \{v\}$. Let $P_a = aa_1a_2\ldots a_n$ be a shortest path from a to a_n and $P_b = bb_1\ldots b_m$ be a shortest path from b to b_m. Note that a_n and b_m cannot be in $N(v)$, otherwise, they would be adjacent to a and b, and thus they could not monitor aa_1 and bb_1. Hence, by Lemma 5, a_n and b_m are mandatory in G.

We first prove that P_a and P_b are disjoint. Suppose they are not and let $c = a_i = b_j$ be the closest vertex from a and b in which they intersect. Necessarily, we have $i \geq 2$ and $j \geq 2$ by construction of a_1 and b_1. Up to exchange P_a and P_b, suppose $i \leq j$. By application of Lemma 2, since ab_1 is a non-edge, there is a chord incident to b in the cycle $aa_1\ldots a_ib_{j-1}\ldots b_1ba$. Since P_b is a shortest path, this chord must be from b to some vertex a_ℓ for $2 \leq \ell \leq i$. Consider the path $ba_\ell a_{\ell+1}\ldots a_ib_{j+1}\ldots b_m$. As $i \leq j$, it is shorter that P_b and does not contain bb_1 contradicting that $\{b, b_m\}$ monitors bb_1. Thus P_a and P_b are disjoint.

We prove that $\{a_n b_m\}$ monitors e. Suppose by contradiction that this is not the case. Hence, there is a shortest path $P_c = a_nc_1,\ldots c_kb_m$ shorter or equal to $a_na_{n-1}\ldots a_1abb_1\ldots b_{m-1}b_m$. Hence $k+1 \leq (n-1)+(m-1)+3$, i.e. $k \leq n+m$. In this path, we denote P_a' (resp. P_b') the longest prefix (resp. suffix) of $a_nc_1,\ldots c_kb_m$ ending (resp. starting) with a vertex c_a of P_a (resp. c_b of P_b), so that we can define P_c' with $a_nc_1,\ldots c_kb_m = P_a'c_aP_c'c_bP_b'$. Note that $C = aa_1\ldots c_a P_c'c_b\ldots b_1ba$ is a cycle of length at least 4.

Since G is chordal and a_1b is a non-edge, by Lemma 2 any cycle containing a_1ab contains a chord incident to a. Similarly, any cycle containing abb_1 contains a chord incident to b. In particular, this is true for the cycle C. Since $aa_1\ldots a_n$ is a shortest path, their is no chord aa_i. Moreover, since $bb_1\ldots b_m$ is a shortest path, and ab_1 is a non-edge, for $1 \leq i \leq m$, there is no chord ab_i, as otherwise there would be a shortest path $bab_ib_{i+1}\ldots b_m$ that does not contain bb_1. Similarly, there are no chords ba_i nor bb_j. Hence, both a and b have a chord to some vertices c_i and c_j of P_c'.

We prove now that there exists $1 \leq i \leq k$ such that c_i is adjacent to a and to b. Let $1 \leq i, j \leq k$ such that a is adjacent to c_i, b is adjacent to c_j, $c_i, c_j \in P_c'$ and $|j-i|$ is minimized. Suppose by contradiction $i \neq j$, say $i < j$ without loss of generality, the case $i > j$ being similar. The cycle $ac_ic_{i+1}\ldots c_{j-1}c_jba$ has length at least 4, and contains no chord, as, for $i < \ell, \ell' < j$ there are no chord $c_\ell c_{\ell'}$ by minimality of the path $c_1\ldots c_m$, and no chord ac_ℓ nor bc_ℓ by minimality of $|j-i|$, which contradicts that G is chordal. Thus, there exists an integer i such that a and b are adjacent to c_i. Since $\{a, a_n\}$ monitors aa_1, we have $i+1 \geq n+1$, and since $\{b, b_m\}$ monitors bb_1, we have $k-i+2 \geq m+1$. We now prove that these two inequalities are strict. We prove it for the first one, the similar argument works for the second. Suppose $i = n$, and consider the cycle $aa_1\ldots c_a\ldots c_ia$. By minimality of the paths, a cannot be adjacent to any other c_j with $1 \leq j \leq i$, nor to any a_p for $1 \leq p \leq n$. Thus, by Lemma 2, a_1c_i is an edge. Thus $a_1abc_ia_1$ is a 4-cycle which is a contradiction with the definition of a_1. Thus, $i \geq n+1$. We prove similarly that $k - i + 2 \geq m + 2$

Combining the two inequalities, we have $k+2 \geq m+n+3$, i.e. $k \geq m+n+1$, which contradicts the fact that $a_nc_1\ldots_k b_m$ is a shortest path from a_n to b_m, as it is longer than $a_n a_{n-1} \ldots a_1 a b b_1 \ldots b_{m-1} b_m$ and thus, $\{a_n, b_m\}$ monitors e. A representation is shown in Fig. 4.

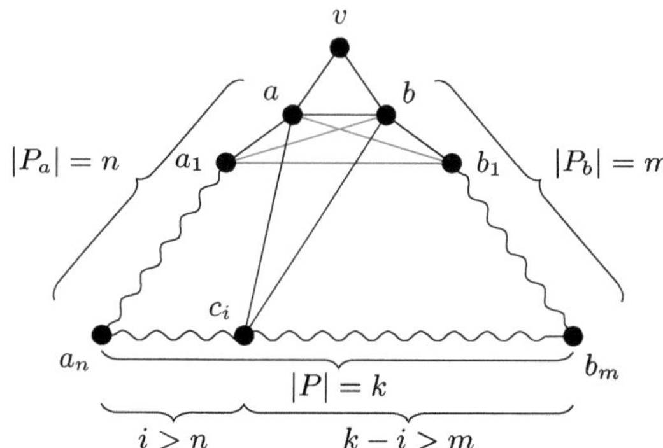

Fig. 4. A representation of the proof of Lemma 12, that is $e \in E(G - \{v\})$, where $a \in W$ and $b \in W$.

Lemma 13. *If $e \notin E(G - \{v\})$, then e is monitored by two vertices of $Mand(G)$.*

Proof. Assume $e = va \notin E(G - \{v\})$. Recall that v is simplicial, thus mandatory by Lemma 1. If all the vertices $a_1 \in N(a) \setminus N(v)$ have a neighbor w in $N(v) \setminus \{a\}$, by definition, v is a support for a, end thus, a is mandatory, and $\{a, v\}$ monitors e.

Otherwise, there exists a_1 in $N(a) \setminus N(v)$ that has no neighbor in $N(v) \setminus \{a\}$. Since a_1 is not incident to v, using the previous points of the proof, aa_1 is monitored by two vertices $\{b, a_n\} \in M$. Using Lemma 8, we can suppose that all the shortest path from a_n to a contains aa_1. We denote one of these paths by $aa_1 \ldots, a_n$ We prove that $\{a_n, v\}$ monitors e. Suppose that they don't. There exists another shortest path $va'a'_1 \ldots a'_{m-1} a_n$ that monitors e. By hypothesis, we have $a' \neq a$, otherwise, this path would also go through e. Since v is simplicial, aa' is an edge, and since a_1 has been chosen such that a_1 has no neighbors in $N(v)$, $a'a_1$ is not an edge.

As this path is a shortest path, we have $m \leq n$, otherwise it would be longer that $vaa_1 \ldots a_n$. Moreover, $a_n a'_{m-1} \ldots a'a$ cannot be a shortest path from a to a_n, by hypothesis on $\{a, a_n\}$. Thus $m + 1 > n$. This proves that $m = n$. Let us denote c the smallest value $1 \leq k \leq n$, such that $a_k = a'_k$ (that is, the first common vertex between $a'_1 \ldots a'_{n-1} a_n$ and $aa_1 \ldots, a_n$). Consider now the cycle $C = a_k a'_{k-1} \ldots a'_1 a'aa_1 \ldots a_k$. Since $a_1 a'$ is a non-edge, C has length at least 4, and there is a chord incident to a in that cycle by Lemma 2. This is a contradiction as, for $1 \leq i \leq n$, if this chord goes to a vertex a_i, in would contradict the minimality of $aa_1 \ldots a_n$, and if it goes to a vertex a'_i, the path $aa'_i a'_{i+1} \ldots a_n$ would be shorter or equal to $aa_1 \ldots a_n$ without containing aa_1. A representation of this configuration is shown in Fig. 5. This proves that $\{a_n, v\}$ monitors e.

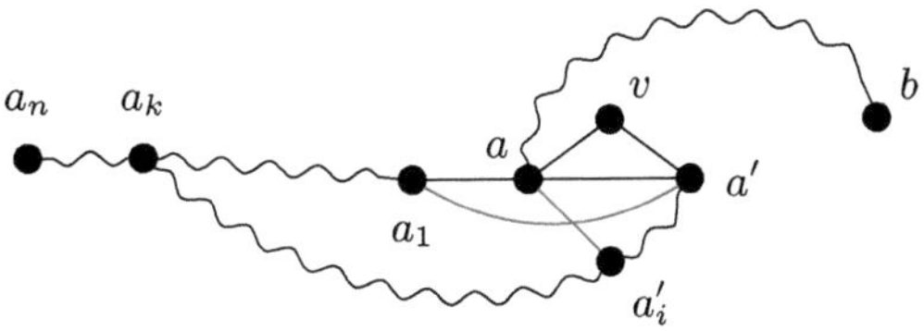

Fig. 5. A representation of the proof of Lemma 13, that is $e \notin E(G - \{v\})$, where $a \in W$ and $b = v$.

Finally, we have proved that any edge $e \in E(G)$ is monitored by M, hence M is a meg-set. Since all the meg-sets of a graph has to contain all its mandatory vertices, it proves that $mand(G) \subset M$. Overall, it proves that $M = Mand(G)$ is a meg-set of G.

4   An Efficient Algorithm to Compute $Mand(G)$

In this section, we provide an algorithm to computes $Mand(G)$ whose complexity time is $O\big(|V|(|V|+\Delta^2)\big)$ for any graph $G = (V, E)$. Using the structure provided in Lemma 9, this algorithm returns a minimum meg-set of G if G is chordal.

Theorem 3. *Let G be a graph. $Mand(G)$ can be computed in time $O\big(|V|(|V|+ \Delta^2)\big)$, where Δ is the maximum degree of G. In particular, if G is chordal, a minimum meg-set of G can be computed in time $O\big(|V|(|V|+\Delta^2)\big)$.*

Proof. Let $G = (V, E)$ be a graph. We suppose that the graph is connected, and given by its adjacency matrix A. We denote the vertices of G by $v_1, \ldots v_n$ We propose the following algorithm:

- For $1 \leq i \leq n$ and each vertex v_i we compute its set of neighbors $N(v_i)$, which can be done in $O(|V|)$ per vertex.
- Then we create a table N_2^i such that $N_2^i[j]$ contains the number of paths of length 2 between i and j. This can be done in time $O(\Delta^2)$ per vertex by initializing this matrix with zeros and then adding 1 in $N_2^i[j]$ for each integer $k \in N(v_j)$ such that $A[j, k] = 1$.
- We now compute M as a table of size n by induction. For each vertex, we will also keep in memory the list of its current support vertices. I.e. we start with $S(v_i) = N(v_i)$ for all $1 \leq i \leq n$.
- For $i = 1$ to n:
  - If $|S(v_i)| \geq 1$, set $M[i]$ to 1.
  - For each vertex $v_j \in N(v)$, for each vertex $v_k \in S(v_j)$, if $N_2^i[k] = 1$ and $A[i, k] = 0$, remove v_k for $S(v_j)$. Moreover, if $S(v_j) = \emptyset$, set $M[j]$ to 0. This step can be done in $O(\Delta^2)$. Note that we count the number of path of length 2 in G and not in G_i, thus, it is possible that during this step M is not a meg-set of G_i.
- Finally, returns the set of vertices with nonzero value in M.

We now prove that the set of vertices in M it the set of mandatory vertices of G. Let $v_i \in Mand(G)$. During step i, $M[i]$ is turned to 1. Now, if v_j is a support for v_i, for $1 \leq k \leq n$ during step k, if $v_k \in N(v_i)$, either $v_k \in N(v_j)$ or there are at least two paths of length 2 from v_j to v_k as v_j is a support vertex for v_i. Thus, v_j will not be removed from $S(v_i)$, and thus v_i will never be removed from M.

Reciprocally, suppose that $v_j \notin Mand(G)$. Let $v_k \in N(v_j)$ be a vertex. As v_k is not a support for v_j, there exist $v_i \in N(v_j) \setminus N(v_k)$ such that $v_j v_i v_k$ is an induced 2-path that is not in a 4-cycle. In particular, $v_j v_i v_k$ is the only path of length 2 between v_j and v_k. During step i, when v_j will be considered in $N(v_i)$, if v_k has not been removed yet form $S(v_j)$, we will have $N_2^i[k] = 1$ and $A[i, k] = 0$ by hypothesis. Thus, v_k will be removed from $S(v_j)$. Since this is true for all the vertices in $N(v_j)$, all the vertices of $S(v_j)$ will be removed at some points, and thus $M[j]$ will be set to 0 (or will not be set to 1 if v_j comes after all its neighbors).

Finally, at the end of the algorithm, all the vertices with nonzero values in M are mandatory, and the algorithm runs in $O\big(|V|(|V| + \Delta^2)\big)$.

5   Conclusion

The main contribution of this work is its answer to the open conjecture of [7] about chordal graphs. In addition, Theorem 3 provides a dedicated algorithm to compute the set of mandatory vertices of any graph G, and thus the minimum meg-set of a chordal graph. From our main result, as well as the structural insights gained through its proof, several research directions naturally emerge:

- A first, natural way to go beyond Theorem 1 is to try and extend it to other classes of graphs. In particular, with the notable exception of cycles, all current algorithms to solve MEG-SET in polynomial time are restricted to meg-minimal graph classes. This motivates a systematic study of the class of meg-minimal graphs. We proved that this class contains chordal graphs, and it is already known that it contains the class of cographs. It would be of interest to search for other structural characterizations of it. Another underlying question is to find other methods to compute minimum meg-sets of non meg-minimal graphs in polynomial time.
- An insight that can be gathered from our result, as well as the meg-minimality of cographs, is that the *meg* parameter seems to interact with density. This phenomenon was already observed in [6] for graphs with high girth. While some approaches already tackled graph sparsity parameters (see e.g. [7]), a parameterized study of MEG-SET by density parameters, like the independence number, is both relevant regarding our current knowledge, and well-fitting into some more recent framework (see e.g. [4]). Moreover, should this kind of parameterization exist, it would open a lot of possibilities to design algorithms to compute meg-sets in graphs of bounded tree-independence number. This is further supported by the fact that chordal graphs are the graphs of tree-independence number 1.

References

1. Bampas, E., Bilò, D., Drovandi, G., Gualà, L., Klasing, R., Proietti, G.: Network verification via routing table queries. J. Comput. Syst. Sci. **81**(1), 234–248 (2015)
2. Beerliova, Z., et al.: Network discovery and verification. IEEE J. Selected Areas Commun. **24**, 2168–2181 (2005). https://api.semanticscholar.org/CorpusID: 424573
3. Bejerano, Y., Rastogi, R.: Robust monitoring of link delays and faults in IP networks. In: IEEE INFOCOM 2003. Twenty-second Annual Joint Conference of the IEEE Computer and Communications Societies (IEEE Cat. No. 03CH37428). vol. 1, pp. 134–144. IEEE (2003)
4. Fomin, F.V., Golovach, P.A., Jedličková, N., Kratochvíl, J., Sagunov, D., Simonov, K.: Path cover, hamiltonicity, and independence number: an FPT perspective (2025). https://arxiv.org/abs/2403.05943
5. Foucaud, F., Kao, S.S., Klasing, R., Miller, M., Ryan, J.: Monitoring the edges of a graph using distances. Discret. Appl. Math. **319**, 424–438 (2022)
6. Foucaud, F., Marcille, C., Myint, Z.M., Sandeep, R., Sen, S., Taruni, S.: Bounds and extremal graphs for monitoring edge-geodetic sets in graphs. Discret. Appl. Math. **366**, 106–119 (2025)
7. Foucaud, F., Marcille, C., Sandeep, R., Sen, S., Taruni, S.: Algorithms and complexity for monitoring edge-geodetic sets in graphs. arXiv preprint arXiv:2409.19067 (2024)
8. Foucaud, F., Narayanan, K., Ramasubramony Sulochana, L.: Monitoring edge-geodetic sets in graphs. In: Algorithms and Discrete Applied Mathematics: 9th International Conference, CALDAM 2023, Gandhinagar, India, 9–11 February 2023, Proceedings, pp. 245–256 (2023)
9. Foucaud, F., Pandey, A., Paul, K.: Characterizing optimal monitoring edge-geodetic sets for some structured graph classes. Discret. Appl. Math. **389**, 92–105 (2026)
10. Haslegrave, J.: Monitoring edge-geodetic sets: hardness and graph products. Discret. Appl. Math. **340**, 79–84 (2023)

(Even Hole, Triangle)-Free Graphs Revisited

Beatriz Martins[(✉)] [iD] and Nicolas Trotignon [iD]

ENS de Lyon, CNRS, 69007 Lyon, France
`beatriz.martins@ens-lyon.fr`

Abstract. We revisit a classical paper about (even hole, triangle)-free graphs [Conforti, Cornuéjols, Kapoor and Vušković, Triangle-free graphs that are signable without even holes, *Journal of Graph Theory*, 34(3), 204–220, 2000]. In fact, the previous study describes a more general class, the so called triangle-free odd signable graphs, and we further generalise the class to the (theta, triangle, wac)-free graphs (not worth defining in an abstract).

We exhibit a stronger structure theorem, by precisely describing basic classes and separators named clique separator, 2-separator and P_3-separator. We prove that, under some additional conditions, the separators preserve the treewidth and several properties. Some consequences are a recognition algorithm with running time $O(|V(G)|^4|E(G)|)$, a proof that the treewidth of graphs in the class is at most 4 (improving a previous bound of 5), and a simple criterion to decide if a graph in the class is planar.

Keywords: Even-hole-free graphs · Triangle-free graphs · Treewidth · Algorithms · Structure

1 Introduction

We consider simple graphs (finite, undirected, with neither loops nor multiple edges). A graph G is H-free if G does not contain an induced subgraph isomorphic to the graph H. When L is a list of graphs, L-free means H-free for all H in L. A *hole* in a graph is a chordless cycle of length (number of edges) at least 4. A hole is even or odd according to the parity of its length.

Despite decades of work (see [11]) resulting in decomposition theorems [4,6] and polynomial time recognition algorithms [5,7], no structure theorem is known for even-hole-free graphs. A structure theorem should allow building all graphs in some class (and only them) starting from basic blocks by gluing previously constructed graphs along some simple operations. Also several related questions are still open, such as the existence of a polynomial time algorithm to colour them or to find a maximum independent set.

The present work proposes a full structural description of (even hole, triangle)-free graphs. A slightly more general class was studied by Conforti,

F. Foucaud and A. Parreau (Eds.): IWOCA 2026, LNCS 16587, pp. 456–469, 2026.
https://doi.org/10.1007/978-3-032-27732-9_32

Cornuéjols, Kapoor and Vušković in [2] that is the main inspiration of this work. Let us first explain why generalising the class is meaningful. Counting parities of cycles in proofs can be difficult. So Conforti et al. use a nice idea: taking advantage of some structures that enforce the presence of an even hole (let us call them *ehe* for Even Hole Enforcers). Let us give some example of ehes. By *ab-path* we mean a path from a vertex a to a vertex b in some graph. The *length* os a path is the number of its edges.

The first example of an ehe is the *theta*, that is a graph made of three chordless ab-paths of length at least 2, vertex-disjoint apart from a and b, and such that the only edges are the edges of the paths (note that since the paths are chordless, a and b are not adjacent and the union of any two paths of a theta forms a hole). By the pigeon hole principle, two among the three paths must have the same parity, and their union therefore induces an even hole, proving that thetas are indeed ehes. A theta such that a and b are the ends of the three paths is an *ab-theta*.

The second example is the *even wheel*. A *wheel* $W = (H, c)$ is a graph made of a hole H and a vertex c called the *centre* that has at least three neighbours in H (Fig. 1). It is *even* if c has an even number of neighbours in H. A *sector* of W is subpath of H of length at least 1 whose ends are adjacent to c and whose internal vertices are not. An even wheel is an ehe, because otherwise all sectors must have odd length since they form a hole with the centre. But then, the hole H is even (because it is made of an even number of sectors), a contradiction.

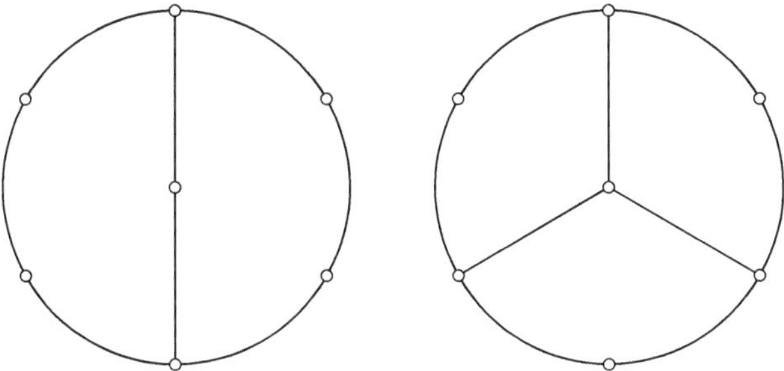

Fig. 1. A theta and a wheel

We now know that if a graph contains a theta or an even wheel, then it contains an even hole. Therefore, (theta, triangle, even wheel)-free graphs form a generalisation of (even hole, triangle)-free graphs, and this generalisation is precisely the one studied in [2] (under the name "odd signable triangle-free graphs"). But one may notice that the idea of enforcing even holes is not pushed as far as it can be since in proofs, the parity of the number of neighbours of the centre still has to be counted to find an even wheel.

We here propose a new even hole enforcer. A graph $W = (H, c, c')$ induced by a hole H together with two adjacent vertices c and c' that both have at least three

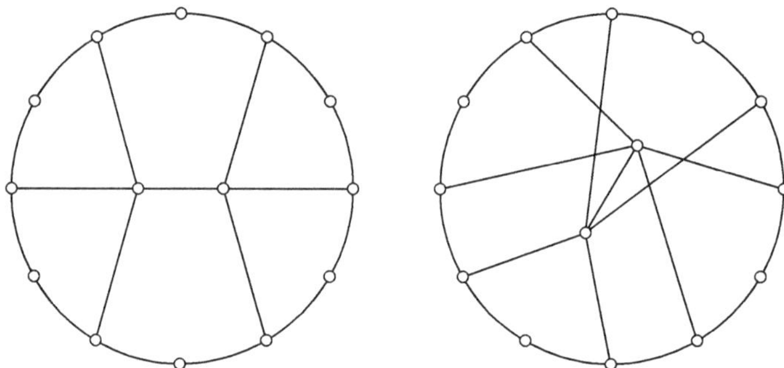

Fig. 2. Two examples of (theta, triangle)-free wacs

neighbours in H is called a *wac* (this stands for *Wheel with Adjacent Centres*), see Fig. 2 where planar and non-planar wacs are represented. We claim that every triangle-free wac $W = (H, c, c')$ contains an even wheel or a theta (and therefore an even hole). Otherwise c has an even number of neighbours in each sector S of (H, c') (or $(c'Sc', c)$ is an even wheel or $S \cup \{c'\}$ is a theta). Moreover, these neighbours are in the interior of S since G is triangle-free. Hence, the number of neighbours of c in H is the sum of even numbers, so (H, c) is an even wheel, a contradiction.

So, the class of (theta, triangle, wac)-free graphs is defined with no reference to parity, and it is a super-class of (theta, triangle, even wheel)-free graphs, that is in turn a super-class of (even hole, triangle)-free graphs. These inclusions are strict, as shown by an even hole and an even wheel. Here is our main result.

Theorem 1. *If G is a (theta, triangle, wac)-free graph, then either G is basic, or G has a clique separator, a proper 2-separator or a proper P_3-separator.*

Let us define all the concepts used in Theorem 1.

We view a *path in a graph* G as a sequence $P = v_1 \ldots v_k$ of vertices such that for all $i, j \in [k]$ (here the use of $[k]$ stands for $\{1, \ldots, k\}$), $v_i v_j \in E(G)$ if and only if $|j - i| = 1$. The vertices v_1 and v_k are the *ends* of P, and its other vertices are *internal*. Note that this is not standard, because this is usually called a chordless, or induced path, but since all the paths that we consider are induced, this is convenient. When a and b are vertices of some path P, we denote by aPb the subpath of P with ends a and b. A *hole* is defined as a path, with the additional conditions that $k \geq 4$ and $v_k v_1 \in E(G)$.

Since all the paper deals with induced subgraphs, we often make no difference between a set of vertices of some graph G (or path or hole) and the graph it induces. For instance, when P is a path, we may write $G \setminus P$ instead of $G[V(G) \setminus V(P)]$. We hope that this only make the notation less heavy and does not lead to any confusion.

A graph is *basic* if it is isomorphic to K_1, K_2, the cube or a daisy. Let us define all this. We denote by K_i the complete graph on i vertices. The *cube* is the graph represented on Fig. 3.

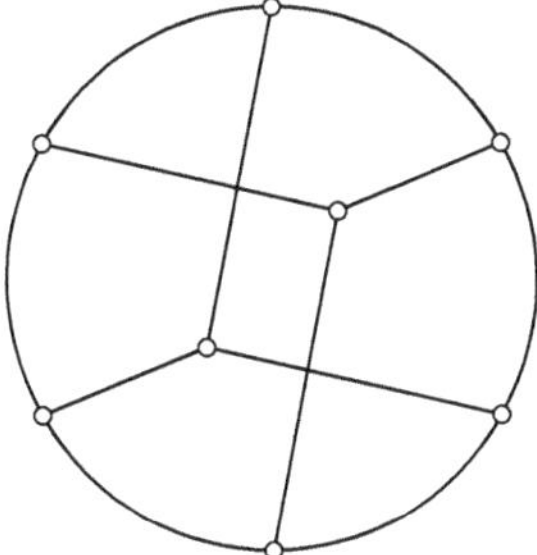

Fig. 3. The cube

A *petal* with respect to a hole $C = c_1 \ldots c_k c_1$ is a path $P = x \ldots y$ disjoint from C such that for some $i \in [k]$, x is adjacent to c_{i-1}, y is adjacent to c_{i+1}, c_i is adjacent to at least one internal vertex of P, no edge of $c_{i-1}xPyc_{i+1}$ has both its vertices adjacent to c_i, and there are no other edges between P and C (subscripts are modulo k). The petal is *centred at* c_i and c_i is the *centre* of the petal.

A *daisy* is a graph G formed of a hole C called the *base hole* together with petals with respect to C and such that no two petals have the same centre, the centres of the petals induce a (possibly empty) subpath of C or C itself, and there are no other edges than the edges of the petals, the edges of the hole and the edges between some petal and the hole. An example is represented in Fig. 4 (this graph is represented at the end of [2] as an example of a non-planar (even hole, triangle)-free graph). Observe that a daisy with no petal is a hole, and a daisy with one petal is a wheel.

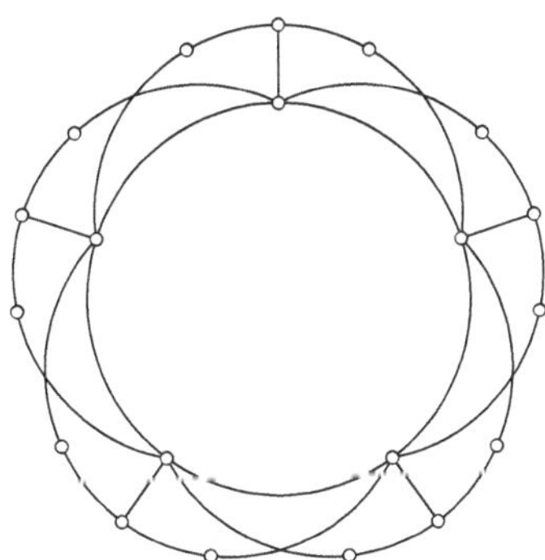

Fig. 4. The Vušković graph (the smallest non-planar daisy)

We call *separator* of G a set S of vertices of a graph G such that $G \setminus S$ has at least two connected components. An aXb-path in some graph G is a path from a vertex a to a vertex b whose internal vertices are all in some set $X \subseteq V(G)$.

A *clique separator* in a graph G is a (possibly empty) clique K such that $G \setminus K$ has at least two connected components. A graph is *atomic* if it has no clique separator. Note that basic graphs, thetas, triangle, wacs, even holes, even wheels and odd holes are atomic.

When a and b are non-adjacent vertices of some graph G such that $G \setminus \{a, b\}$ is not connected, we call $\{a, b\}$ a *2-separator* of G. A 2-separator $\{a, b\}$ of some graph G is *proper* if $G \setminus \{a, b\}$ has exactly two components X and Y and $G[X \cup \{a, b\}]$ (resp. $G[\{a, b\} \cup Y]$) contains an ab-path P (resp. Q) but is not equal to an ab-path, that is $V(P) \subsetneq X \cup \{a, b\}$ (resp. $V(Q) \subsetneq Y \cup \{a, b\}$).

When acb is a path in some graph G such that $G \setminus acb$ is not connected, acb is a *P_3-separator* of G. When acb is a P_3-separator, and X is a connected component of $G \setminus acb$, we say that X is *loose* if there exists an aXb-path with no internal vertex adjacent to c, and *tight* otherwise. A P_3-separator acb of some graph G is *proper* if $G \setminus acb$ has exactly two connected components X and Y, there exists an aXb-path and an aYb-path in G, c has neighbours in both X and Y, X is loose, Y is tight, and none of $G[X \cup \{a, b\}]$ and $G[Y \cup \{a, b\}]$ is an ab-path.

A (theta, triangle, wac)-free graph G with a proper 2-separator $\{a, b\}$ is represented on the left in Fig. 5. It is easy see that G is not basic, has no clique separator, no proper P_3-separator, and that $\{a, b\}$ is the unique proper 2-separator of G. On the right a similar example with a proper P_3-separator acb (or acb') is represented (observe that acb'' is not a proper P_3-separator because no component of $G \setminus acb''$ is loose).

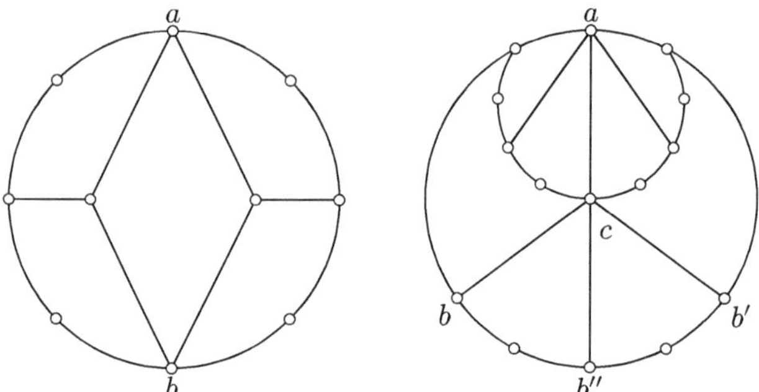

Fig. 5. Graphs with separators

In the rest of this extended abstract, we give applications of Theorem 1 (without proofs, see Sect. 2) and a self-contained proof Theorem 1 (see Sect. 3).

2   Applications

We insist that Theorem 1 is not only a decomposition theorem but a structure theorem for our class (and all its subclasses that we already mentioned). An evidence of its strength is the following corollary that is obtained by showing that our separators preserve the treewidth, which is of independent interest since they may serve to decompose other classes. Note that we improve a bound from [1], where it is proved that (theta, triangle, even wheel)-free graphs have treewidth at most 5. Also observe in contrast that the treewidth of (theta, triangle)-free and (even hole, K_4)-free graphs is unbounded as shown in [10].

Theorem 2. *Every (theta, triangle, wac)-free graph has treewidth at most 4 (and some have treewidth exactly 4).*

The following is also a corollary of Theorem 1.

Theorem 3. *There exists an algorithm that decides in time $O(|V(G)|^4|E(G)|)$ whether an input graph G is (theta, triangle, wac)-free (resp. (theta, triangle, even wheel)-free, (even hole, triangle)-free, bipartite (theta, wac)-free). Moreover, when it is, the algorithm outputs a tree decomposition of G of width $\mathrm{tw}(G)$ (in particular, of width at most 4).*

We also obtain the following (where a daisy is odd or even according to the parity of its base hole, and a daisy is *full* when all vertices of the base hole are centres of its petals).

Theorem 4. *A (theta, triangle, wac)-free graph is planar if and only if it contains no full odd daisy (in particular, all bipartite (theta, wac)-free graphs are planar).*

3   Proof of Theorem 1

Some results in this section are already known in which case we indicate a reference. But we include all proofs for the sake of completeness and also because it is sometimes not so easy to recover proofs from old papers because of a different focus, the precise class under consideration and changes in the notation.

Theorem 5 (See [3]). *If G is (theta, triangle, wheel)-free graph, then G is K_1, K_2, a hole or G has a clique separator.*

Proof. We may assume that G is an atomic graph. If G has no hole then, since G is triangle-free, G must be a forest. Hence G must be K_1 or K_2, since they are the only atomic forests.

Now, let C be a hole contained in G. We may assume that G is not a hole. Since G is atomic, there is a path $R = x \dots y$ in a component $Z \subseteq (G \setminus C)$ and there are non-adjacent vertices $u, v \in V(C)$ such that $xu, yv \in E(G)$. Assume that u, v and R are chosen to minimise the length of R.

Since G is triangle-free, $x \neq y$. Otherwise R is just a vertex and $C \cup R$ induces either a theta or a wheel. Furthermore, $N_C(x) = u$ (resp. $N_C(y) = v$), otherwise $V(C) \cup \{x\}$ (resp. $V(C) \cup \{y\}$) induces either a theta or a wheel.

If there is a vertex $w \in V(C) \setminus \{u, v\}$ such that $N_R(w) \neq \emptyset$, then w is adjacent to a vertex z in the interior of R. Notice that this contradicts the minimality of R

if w is a non-neighbour of either v or u. Hence, w must be adjacent to both u and v, which implies that at most two vertices of C can have neighbours in internal vertices of R. If only one vertex $w \in V(C) \setminus \{u, v\}$ has a neighbour in R, then $G[C \cup R]$ induces a wheel centred in w. So suppose that $w_1, w_2 \in V(C) \setminus \{u, v\}$ have neighbours in the interior of R say r_1 and r_2 respectively. Notice that in this case $C = uw_1vw_2u$. The path r_1Rr_2 has a smaller length than R and there are two non-adjacent vertices $w_1, w_2 \in V(C)$ such that $w_1r_1, w_2r_2 \in E(G)$, which contradicts the minimality of R. $\qquad\square$

Lemma 1 (Folk). *Suppose that G is a (theta, triangle)-free graph and H is a hole of G. If $v \in G \setminus H$ has at least two neighbours in H, then (H, v) is a wheel. In particular, no vertex in $G \setminus H$ has exactly two neighbours in H.*

Proof. If v has at least two neighbours in H and $V(H) \cup \{v\}$ is not a wheel, then $|N_H(v)| = 2$. So let u, w be the neighbours of v in H. Since G is triangle-free, $uw \notin E(G)$. Hence $V(H) \cup \{v\}$ induces an uw-theta. $\qquad\square$

Lemma 2 (see [9]). *If G is a (theta, triangle)-free graph that contains the cube, then G is isomorphic to the cube or G has a clique separator.*

Proof. We denote the vertices of a cube H contained in G with a bipartition $A = \{a_1, \ldots, a_4\}$ and $B = \{b_1, \ldots b_4\}$, in such a way that $a_ib_j \in E(H)$ if and only if $i \neq j$. If some vertex $v \in G \setminus H$ has at least two neighbours in A, say a_1 and a_2 up to symmetry, then $\{v, a_1, a_2, b_3, b_4\}$ contains a triangle or induces an a_1a_2-theta, a contradiction. Hence, v has at most one neighbour in A, and symmetrically in B. If v has two neighbours in H, they must therefore be a_1 and b_1 up to symmetry, so $\{v, a_1, a_2, a_3, b_1, b_2, b_3\}$ induces an a_1b_1-theta, a contradiction. Hence, a vertex $v \in G \setminus H$ has at most one neighbour in H.

We may assume that $G \setminus H$ has a connected component Z such that $N_H(Z)$ is not a clique. So, there exists a path $R = x \ldots y$ in Z and non-adjacent $u, v \in H$ such that $xu, yv \in E(G)$. We choose u, v and R subject to the minimality of R. This implies that if an internal vertex of R has a neighbour w in H, then w must be a common neighbour of u and v.

If $u \in A$ and $v \in B$, then up to symmetry, $u = a_1$ and $v = b_1$. Note that u and v have no common neighbours in H, so no internal vertex of R has a neighbour in H. Hence, R, a_1, a_2, a_3, b_1, b_2 and b_3 form an a_1b_1-theta, a contradiction. We may therefore assume that $u = a_1$ and $v = a_2$. No internal vertices of R is adjacent to a_3, a_4, b_1 and b_2 because they are not common neighbours of a_1 and a_2. Hence, R, a_1, a_2, a_3, a_4, b_1 and b_2 form an b_1b_2-theta, a contradiction. $\qquad\square$

Variants of the following are proved in [2, Theorem 2.5] for (theta, triangle, even wheel)-free graphs and in [8, Lemma 3.1].

Lemma 3. *Suppose that G is a (theta, triangle, wac, cube)-free graph and $W = (H, c)$ is a wheel of G. If v is a vertex of $G \setminus W$ then $N_W(v)$ is included in $\{c\}$ or in a sector of W.*

Proof. If v has at most one neighbour in H, then either $N_W(v) \subseteq \{c\}$, or $N_W(c)$ is included in some sector of W, or some sector of W, c and v form a theta or contain a triangle. Hence, by Lemma 1, we may assume that $|N_H(v)| \geq 3$. So $W' = (H, v)$ is a wheel, and $vc \notin E(G)$ for otherwise (H, c, v) is a wac. Suppose by contradiction that $N_W(v)$ is not included in some sector of W. So there exist distinct sectors $S = s \ldots s'$ and $T = t \ldots t'$ of W such that v has at least one neighbour in $S \setminus T$ and at least one neighbour in $T \setminus S$. Up to symmetry, we assume that s, s', t and t' appear in this order along H and that $s \neq t'$ (but possibly $s' = t$).

If $|N_S(v)| \geq 2$, then let x (resp. x') be the neighbour of v in S closest to s (resp. s') along S. Thus, $R = vx'Ss'csSxv$ is a hole in G. Moreover, $N_{T \setminus S}(v) \neq \emptyset$, so let y (resp. y') be the neighbour of v in T closest to t (resp. t') along T. Notice that $R' = vy'Tt'c$ is a chordless path in G and $G[R \cup R']$ is a vc-theta in G, unless $t = s'$ and $ty' \in E(G)$, in which case, R and y' form a vt-theta, a contradiction.

Thus, v can have at most one neighbour in each sector of W, and similarly c can have at most one neighbour in each sector of W'. Moreover, $N_H(c) \cap N_H(v) = \emptyset$, for otherwise either c has at least two neighbours in some sector of (H, v) or v has at least two neighbours in some sector of (H, c).

Hence, $|N_H(c)| = |N_H(v)|$ and the neighbours of c and v alternate in H. Consider three consecutive sectors of $W = (H, c)$: $Q_1 = q_1 \ldots q_2$, $Q_2 = q_2 \ldots q_3$ and $Q_3 = q_3 \ldots q_4$. Since the neighbours of c and v alternate, v has a unique neighbour v_i in each Q_i. If $v_3 q_1 \notin E(G)$, then $vv_2 Q_2 q_3$, $vv_3 Q_3 q_3$ and $vv_1 Q_1 q_1 c q_3$ form a theta, a contradiction. Hence, $v_3 q_1 \in E(G)$ (in particular $q_4 = q_1$ and W has three sectors). A symmetric argument shows that $v_i q_{i+1}$ and $q_i v_{i+1}$ are edges for all $i \in [3]$ (subscripts are modulo 3). Hence H and v form a cube, a contradiction. $\qquad\square$

Lemma 4. *Suppose that G is a (theta, triangle, wac, cube)-free graph and $W = (H, c)$ is a wheel of G. If Z is a connected induced subgraph of $G \setminus W$, then $N_H(Z)$ is included in some sector of W.*

Proof. Otherwise, there exist distinct sectors $S = s \ldots r$ and $T = t \ldots r'$ of W such that some vertex of $S \setminus T$ and some vertex of $T \setminus S$ both have neighbours in Z. So, Z contains a path $P = x \ldots y$ such that x has neighbours in $S \setminus T$ and y has neighbours in $T \setminus S$. We suppose that S, T and P are chosen subject to the minimality of P. By Lemma 3, P has length at least 1 (so $x \neq y$), $N_W(x) \subseteq S$ and $N_W(y) \subseteq T$ (in particular none of x and y is adjacent to c). By the minimality of P, internal vertices of P have no neighbours in H, except possibly the unique vertex in $S \cap T$ (if any). Note that internal vertices of P may be adjacent to c.

Let x' (resp. x'') be the neighbour of x in S closest to s (resp. to r) along S. Let y' (resp. y'') be the neighbour of y in T closest to t (resp. to r') along T. These names are given in such a way that s, x', x'', r, r', y'', y' and t appear in this order along H. Up to symmetry, we assume that $s \neq t$ (implying that $st \notin E(G)$ since G is triangle-free, but possibly $r = r'$).

Suppose that some internal vertex of P has a neighbour in H. Then, by the minimality of P, we must have $r = r'$ and r is the only vertex of H with

464   B. Martins and N. Trotignon

neighbours in the interior of P. Consider the hole $H' = x'(H \setminus r)y'yPxx'$. By Lemma 1, (H', c) is a wheel (because c has at least two neighbours in H': s and t). So, r has exactly one neighbour z in H' (that is in the interior of P), for otherwise by Lemma 1 (H', c, r) is a wac. Now, zrc, a shortest path from z to c in $zPxx'Ssc$, and a shortest path from z to c in $zPyy'Ttc$ form a zc-theta. Hence, no internal vertex of P has a neighbour in H.

If $x' = x''$, then $x'Ssc$, $x'Src$ and a shortest path from x' to c in $x'xPyy'Ttc$ form an $x'c$-theta, unless c has no neighbour in P, $r = r'$, $y' = y''$ and $y'r \in E(G)$. In particular, $y' = y''$ and a symmetric proof based on this fact shows that $x'r \in E(G)$. Hence, P and H form an $x'y'$-theta, a contradiction. Hence, $x' \neq x''$. Symmetrically, $y' \neq y''$.

Now, removing the interior of $x'Sx''$ and $y'Ty''$ in $P \cup H$ yields an xy-theta, unless $xy \in E(G)$. In this last case, by Lemma 1, x and y are both centres of a wheel with rim H, so (H, x, y) is a wac, a contradiction again. $\qquad\square$

The next lemma describes how a connected set of vertices may attach to the centre and the rim of a wheel. Before stating it, let us describe two examples represented in Fig. 6. In both examples, a path $P = x \ldots y$ attaches to a wheel $W = (H, c)$, and both examples turn out to be daisies. In the first one, the hole is $csSs'c$ and there are two petals: P and $H \setminus S$. In the second one, the hole is $csyt'Ss'c$ and there are three petals: the interior of $t'Ss$, $P \setminus y$ and $H \setminus S$. The next lemma shows that in some sense, these two examples are the only possibilities.

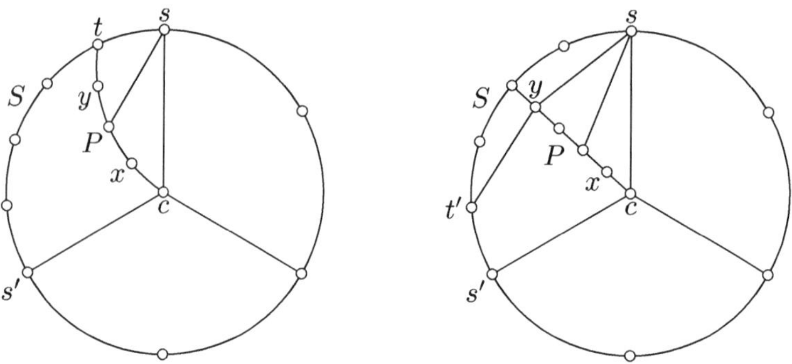

Fig. 6. Two examples of attachments to a wheel

Lemma 5. *Suppose that G is a (theta, triangle, wac, cube)-free graph and $W = (H, c)$ is a wheel of G. Let Z be a connected induced subgraph of $G \setminus W$ such that c has a neighbour in Z and $N_W(Z)$ is not a clique. Then Z contains a path $P = x \ldots y$ of length at least 1 and W has a sector $S = s \ldots s'$ such that $N_W(x) = \{c\}$, $N_W(y) \subseteq S$, y has at least one neighbour in the interior of S, s has at least one neighbour in the interior of P and no internal vertex of P has a neighbour in $W \setminus s$. Moreover, if y has a neighbour in S that is not adjacent to s, then $ys \in E(G)$. In all cases, W and P form a daisy with two or three petals.*

Proof. By Lemma 4, there exists a sector $S = s \ldots s'$ of W such that $N_H(Z) \subseteq S$. Suppose first that no internal vertex of S has a neighbour in Z. Then, since $N_W(Z)$ is not a clique, we have $N_W(Z) = \{s, c, s'\}$. Hence, H and a shortest path of Z from a neighbour of s to a neighbour of s' form an ss'-theta, a contradiction.

So let $P = x \ldots y$ be a shortest path in Z such that $xc \in E(G)$ and y has a neighbour in the interior of S. By Lemma 3, $x \neq y$. Let t (resp. t') be the neighbour of y in S that is closest to s (resp. s') along S. If both s and s' have neighbours in $P \setminus y$, then H and a shortest path of $P \setminus y$ from a neighbour of s to a neighbour of s' in $P \setminus y$ form an ss'-theta, a contradiction. Hence, up to symmetry, we may assume that s' has no neighbour in $P \setminus y$.

If s has no neighbour in the interior of P, then $S \cup P \cup \{c\}$ contains a cy-theta (if $t \neq t'$) or a ct-theta (if $t = t'$). So, s has at least one neighbour in the interior of P.

So, P has all the claimed properties. It remains to prove that if y has a neighbour in S that is not adjacent to s, then $ys \in E(G)$. Suppose not, that is $t's \notin E(G)$ and $t \neq s$. Hence, $ytSs$, $yt'Ss'cs$ and a shortest path from y to s in $P \cup \{s\}$ form a ys-theta if $t \neq t'$, or a ts-theta if $t = t'$, in both cases a contradiction.

The fact that W and P form a daisy with two or three petals has been explained in the paragraph before the statement. $\qquad\square$

Lemma 6. *If $\{a, b\}$ is a 2-separator of a theta-free graph G with no clique separator, then $G \setminus \{a, b\}$ has exactly two connected components X and Y, and G contains an aXb-path and an aYb-path.*

Proof. Let X and Y be connected components of $G \setminus \{a, b\}$. There exists an aXb-path (resp. an aYb-path) in G for otherwise a or b is a clique-separator of G, a contradiction. If $G \setminus \{a, b\}$ has a third component Z, then an aXb-path, an aYb-path and an aZb-path form an ab-theta, a contradiction. $\qquad\square$

Lemma 7. *If a theta-free graph G with no proper 2-separator contains a P_3-separator acb, then $G \setminus acb$ has exactly two connected components X and Y such that G contains an aXb-path and an aYb-path, and every aYb-path of G contains at least one internal vertex adjacent to c. Moreover, if none of $X \cup \{a, b\}$ and $Y \cup \{a, b\}$ induces an ab-path, then c has neighbours in both X and Y.*

Proof. Let X and Y be components of $G \setminus acb$. Since G is atomic, both a and b must have a neighbour in X for otherwise G has a clique separator (induced by some subset of acb). Hence $G[X \cup \{a, b\}]$ is connected and there exists an aXb-path in G. Similarly, there exists an aYb-path in G.

If $G \setminus acb$ has a third connected components Z, then an aXb-path, an aYb-path and an aZb-path would form a theta of G, a contradiction.

Now suppose that exists an aXb-path P and an aYb-path Q path that both have no internal vertex adjacent to c. Then, P, Q and acb form a theta, a contradiction. So, up to a swap of the names X and Y, every aYb-path has at least one internal vertex adjacent to c.

If c has no neighbours in X, then $\{a, b\}$ is a 2-separator of G, which is proper (because of the assumption that none of $X \cup \{a, b\}$ and $Y \cup \{a, b\}$ is an ab-path), a contradiction. $\qquad\square$

We need the following notation about daisies. Suppose that G is a daisy with notation as in the definition. We denote by Q_i the path of C from c_{i-1} to c_{i+1} that does not contain c_i. We denote by P_i the petal of the daisy centred at c_i (if any), and set $P_i = x_i \ldots y_i$ where x_i is adjacent to c_{i-1} and y_i is adjacent to c_{i+1}. We observe that P_i and Q_i form a hole H_i, and $W_i = (H_i, c_i)$ is a wheel. We observe that Q_i is a sector of W_i and the other sectors are subpaths of $c_{i-1}x_iP_iy_ic_{i+1}$ called the *external sectors of W_i*.

Lemma 8. *Suppose that G is an atomic (theta, triangle, wac, cube)-free graph and D is a daisy of G with at least two petals. If some connected induced set $Z \subseteq G \setminus D$ has neighbours in some petal of D, then G has a proper 2-separator or a proper P_3-separator.*

Proof. Since G is atomic, by Lemmas 6 and 7, to check that a 2-separator or a P_3-separator is proper, only the condition about $G[X \cup \{a, b\}]$ and $G[Y \cup \{a, b\}]$ not being an ab-path, and the condition about the loose component of a P_3-separator have to be checked.

Suppose that some vertex of Z has a neighbour in some petal P_i of D.

We claim that there exists an external sector $S = s \ldots s'$ of $W_i = (H_i, c_i)$ such that $N_D(Z)$ is included in $S \cup \{c_i\}$. Indeed, by Lemma 4, $N_{H_i}(D)$ is included in some sector S of W_i, and since some vertex of Z has neighbours in P_i, S must be an external sector of W_i. Let us check that $N_D(Z)$ is included in $S \cup \{c_i\}$. So suppose for a contradiction that Z contains a vertex with a neighbour $x \in D \setminus (S \cup \{c_i\})$. From the definition of S, x is not in W_i (in particular, $x \notin C$). So, $x \in P_j$ for some petal P_j, $j \neq i$. Now $Z \cup P_j$ is connected, so it contradicts Lemma 4 applied to W_i, because P_j has a neighbour in $C \setminus \{c_{i-1}, c_i, c_{i+1}\}$, so a neighbour in $H_i \setminus S$.

From the claim that we just proved, we see that $\{s, s'\}$ is a proper 2-separator of G (if c_i has no neighbour in Z) or $\{s, c_i, s'\}$ is a proper P_3-separator of G (the component containing Z is loose because of the path S). Note that the condition that none of $X \cup \{s, s'\}$ and $Y \cup \{s, s'\}$ is a path is satisfied. This is because of Z for the component that contains Z, and because of the second petal for the other side (this is the only place where we need the second petal). $\square$

End of the proof of Theorem 1

We may assume that G is atomic. By Lemma 2, we may assume that G is cube-free. By Theorem 5, we may assume that G contains a wheel. So G contains a daisy with at least one petal, and we consider such a daisy D of G that is maximal (in the sense of inclusion for the vertex-set). We may assume that $G \setminus D$ has a connected component Z for otherwise G is basic.

To describe D, we use notations as in the definition of a daisy. Since D has at least one petal, we assume up to symmetry that $c_1, \ldots, c_\ell$ where $1 \leq \ell \leq k$ are the centres of the petals of D.

Suppose first that some vertex of Z has a neighbour in some petal P_i of D. By Lemma 8, we may then assume that D has a unique petal, so it is the wheel $W_1 = (H_1, c_1)$. By Lemma 4, $N_{H_1}(Z)$ is included in some sector $S = s \ldots s'$

of W_1. If $N_D(Z) \subseteq S$, then $\{s, s'\}$ is a proper 2-separator of G. Otherwise, Z contains a neighbour of c_1, and by Lemma 5, either $N_D(Z)$ is a clique (so G has a clique separator, a contradiction), or D can be extended to a daisy with two or three petals, a contradiction to the maximality of D.

From here on, we may assume that no vertex of Z has a neighbour in some petal P_i of D. If no centre of a petal has a neighbour in Z, since G is atomic, then some vertex of C that is not the centre of a petal has a neighbour in Z (in particular, $\ell < k$). So, $\{c_{\ell+1}, c_k\}$ is a clique separator (a contradiction), or a proper 2-separator. Hence, we may assume that some centre c_i of some petal P_i has a neighbour in Z.

Let us apply Lemma 5 to Z and W_i. If $N_{W_i}(Z)$ is a clique, then so is $N_D(Z)$ (because $N_D(Z) = N_{W_i}(Z)$), and G has a clique separator, a contradiction. So, Z contains a path $P = x \ldots y$ of length at least 1, $xc_i \in E(G)$, and y has a neighbour in the interior of Q_i (recall that Q_i is the sector of W_i contained in C). Moreover, some end of Q_i, say c_{i+1} up to symmetry, has neighbours in the interior of P.

If the only neighbour of y in Q_i is c_{i+2}, then D has no petal centred at c_{i+1}, for otherwise, such a petal P_{i+1} together with P and $C \setminus c_{i+1}$ would form a theta from c_i to c_{i+2}. It follows that $i = \ell$, D has no petal centred at $c_{\ell+1}$ (in particular, $\ell < k$). Hence adding P to D yields a daisy that contradicts the maximality of D.

Hence, y has a neighbour in Q_i that is not adjacent to c_{i+1} (so it is not c_{i+2}). By Lemma 5, $yc_{i+1} \in E(G)$. We denote by Q_{i+1} the path $C \setminus c_{i+1}$. Let c_j be the neighbour of y in that is closest to c_i along the path Q_{i+1}. Note that $j \neq i$ since $x \neq y$. We now consider a daisy D' defined as follows. The hole is $C' = c_i c_{i+1} y c_j Q_{i+1} c_i$. The petals are $P \setminus y$, $c_{i+2} Q_{i+1} c_{j-1}$, and all petals of D centred at some vertex of $c_{j+1} Q_{i+1} c_i$ (in particular P_i). If all petals of D are petals of D', then D' contradicts the maximality of D. Hence, D has at least one petal R that is not a petal of D', and that is therefore centred at some vertex in $c_{i+1} Q_i c_j$. This petal R is therefore a connected subset of $G \setminus D'$ that has a neighbour in some petal of D'. Hence, by Lemma 8 applied to D', G has a proper 2-separator or a proper P_3-separator. This concludes the proof of Theorem 1.

4   Comments and Open Questions

The following is not needed but still striking.

Lemma 9. *Let G be a (theta, triangle, wac, cube)-free graph and (X, acb, Y) be a split with respect to a proper P_3-separator of G. Then, no internal vertex of any aXb-path is adjacent to c.*

Because of Lemma 9, it is tempting to define a loose component X of $G \setminus acb$ as one such that every aXb-path has no internal vertex adjacent to c. All our theorems would stay true and we would have a more precise conclusion of Theorem 1. However, by Theorem 6, it would be coNP-complete to decide if a graph contains a proper P_3-separator, so the theorem would be less practical.

Theorem 6. *Testing if a prescribed path acb of an input graph G is a path such that for every ab-path P in $G \setminus c$, no internal vertex of P is adjacent to c is a coNP-complete problem.*

We wonder whether proper 2-separators and proper P_3-separators preserve bounded cliquewidth. We propose the following conjecture.

Conjecture 1. There is a function f with the following property. Let $\mathcal{C}$ and $\mathcal{B}$ be two classes of graphs closed under taking induced subgraphs and such that every 2-connected graph from $\mathcal{C}$ is either in $\mathcal{B}$ or has a proper 2-separator or a proper P_3-separator. Moreover for some constant a, every graph in $\mathcal{B}$ has cliquewidth at most a. Then, every graph in $\mathcal{C}$ has cliquewidth at most $f(a)$.

We wonder whether it is possible to describe the most general (theta, triangle)-free graph with no separator as defined in the present work. We would need to combine in the description at least the (theta, triangle)-free layered wheels from [10] and the daisies.

Acknowledgments. This work is supported by Projet ANR GODASse, Projet-ANR-24-CE48-4377.

Disclosure of Interests. The authors have no competing interests to declare that are relevant to the content of this article.

References

1. Cameron, K., da Silva, M.V.G., Huang, S., Vušković, K.: Structure and algorithms for (cap, even hole)-free graphs. Discrete Math. **341**(2), 463–473 (2018)
2. Conforti, M., Cornuéjols, G., Kapoor, A., Vušković, K.: Triangle-free graphs that are signable without even holes. J. Graph Theor. **34**(3), 204–220 (2000)
3. Conforti, M., Cornuéjols, G., Kapoor, A., Vušković, K.: Universally signable graphs. Combinatorica **17**(1), 67–77 (1997)
4. Conforti, M., Cornuéjols, G., Kapoor, A., Vušković, K.: Even-hole-free graphs part I: decomposition theorem. J. Graph Theor. **39**(1), 6–49 (2002)
5. Conforti, M., Cornuéjols, G., Kapoor, A., Vušković, K.: Even-hole-free graphs Part II: recognition algorithm. J. Graph Theor. **40**, 238–266 (2002)
6. da Silva, M.V.G., Vušković, K.: Decomposition of even-hole-free graphs with star cutsets and 2-joins. J. Comb. Theor. Ser. B **103**(1), 144–183 (2013)
7. Lai, K.Y., Lu, H.I, Thorup, M.: Three-in-a-tree in near linear time. In: Makarychev, K., Makarychev, Y., Tulsiani, M., Kamath, G., Chuzhoy, J., (eds.) Proccedings of the 52nd Annual ACM SIGACT Symposium on Theory of Computing, STOC 2020, Chicago, IL, USA, June 22–26, 2020, pp. 1279–1292. ACM (2020)
8. Pilipczuk, M., Sintiari, N.L.D., Thomassé, S., Trotignon, N.: (Theta, triangle)-free and (even hole, K_4)-free graphs. part2: bounds on treewidth. J. Graph Theor. **97**(4), 624–641 (2021)
9. Radovanović, M., Vušković, K.: A class of three-colorable triangle-free graphs. J. Graph Theor. **72**(4), 430–439 (2013)

10. Sintiari, N.L.D., Trotignon, N.: (Theta, triangle)-free and (even hole, K_4)-free graphs - part 1: layered wheels. J. Graph Theor. **97**(4), 475–509 (2021). arXiv:1906.10998
11. Vušković, K.: Even-hole-free graphs: a survey. Appl. Anal. Discrete Math. **10**(2), 219–240 (2010)

One Sequence to Rule Them All: $\mathcal{O}(1)$-Time Parallel Generation of Mixed-Radix Gray Codes

Lucia Moura[1] , Prangya Parida[1], Brett Stevens[2] ,
and Aaron Williams[3]($\boxtimes$)

[1] University of Ottawa, Ottawa K1N 6N5, Canada
{lmoura,ppari017}@uottawa.ca
[2] Carleton University, Ottawa K1S 5B6, Canada
brett@math.carleton.ca
[3] Williams College, Williamstown, MA 01267, USA
aaron.williams@williams.edu

Abstract. In a mixed-radix word each digit has its own base. In the modular Gray code successive words differ in one digit by $+1$ mod its base. It can be constructed greedily: start at $00\cdots0$ then repeatedly create new words by cyclically incrementing the rightmost possible digit. For example, $M(b) = 00\overline{0}, 00\overline{1}, 0\overline{0}2, 01\overline{2}, 01\overline{0}, \overline{0}11, 11\overline{1}, 11\overline{2}, 1\overline{1}0, 10\overline{0}, 10\overline{1}, 102$ is the order for bases $b = 2, 2, 3$ where overlined digits cyclically increment to create the next word. Its change sequence (relative to $m_3\, m_2\, m_1$) is $\mathsf{ruler}(b) = 1, 1, 2, 1, 1, 3, 1, 1, 2, 1, 1$. This ruler sequence also guides lexicographic order and the reflected Gray code using ± 1 digit changes.

We generalize the greedy approach: start at word w, then repeatedly create new words that are greater than w by cyclically incrementing the rightmost possible digit. Pleasantly, we obtain a Gray code for the words greater than or equal to w. For example, when $b = 2, 2, 3$ and $w = 011$ the order $M_w(b) = 01\overline{1}, \overline{0}12, 11\overline{2}, 11\overline{0}, 1\overline{1}1, 10\overline{1}, 10\overline{2}, 100$ avoids the smaller words (i.e., 000, 001, 002, 010). While this order $M_w(b)$ is not a suffix of the full modular Gray code $M(b)$, its changes $1, 3, 1, 1, 2, 1, 1$ are a suffix of $\mathsf{ruler}(b)$. In other words, the words change but not the changes.

We generate our 'modded' modular Gray codes in worst-case $\mathcal{O}(1)$-time per word with $\mathcal{O}(n)$ additional memory. We do this by efficiently unranking the internal state (including focus pointers) of the classic loopless algorithm for ruler sequences. Warm starting this algorithm at any rank allows us to generate any suffix of the ruler sequence. Moreover, we can generate the modular and reflected Gray codes looplessly in parallel.

1 Introduction

Those in technical fields are familiar with binary (base-2), hexadecimal (base-16), and perhaps octal (base-8), while historians are often familiar with sexagesimal (base-60) bases[1]. In a *mixed-radix number* every digit has its own base, and a *mixed-radix word* is a mixed-radix number of a fixed length where leading

---

[1] Sexagesimal is often written with one or two glyphs per digit (i.e., $30\,31$ is two digits).

© The Author(s), under exclusive license to Springer Nature Switzerland AG 2026
F. Foucaud and A. Parreau (Eds.): IWOCA 2026, LNCS 16587, pp. 470–485, 2026.
https://doi.org/10.1007/978-3-032-27732-9_33

zeros are not suppressed. Let $\mathbf{W}(b)$ be the set of mixed-radix words with bases $b = b_1, b_2, \ldots, b_n$. For example, $\mathbf{W}(b) = \{00, 01, 02, 10, 11, 12\}$ for bases $b = 2, 3$. The number of such words is $|\mathbf{W}(b)| = b_1 \cdot b_2 \cdots b_n$ which we denote $b!$. We index the digits in a word as $w = w_1 w_2 \cdots w_n$ and let $m_i = b_i - 1$ be their maximums.

This paper is on efficient orders and generation algorithms of mixed-radix words. This section gives background on these topics and summarizes our results.

1.1 Lexicographic Order and Gray Codes

Lexicographic order applies our standard notion of alphabetic or numeric order to the digits from left-to-right. Similarly, if we consider the digits from right-to-left, then the order is *co-lexicographic order*. The orders are below for $b = 2, 3$:

$$L(b) = 00, 01, 02, 10, 11, 12 \qquad \lambda(b) = 00, 10, 01, 11, 02, 12 \qquad (1)$$

In lexicographic orders every digit can change between two successive words. For example, $0\, m_2\, m_3 \cdots m_n$ is followed by $1\, 0\, 0 \cdots 0$ in $L(b)$. In other words, the first digit is incremented and the others roll-over to 0 from their maximum.

A *Gray code* is an order in which successive objects differ in a 'small' manner; see the classic survey by Savage [32] and Mütze's updated dynamic survey [23]. Some Gray codes are difficult to construct. For example, Beckett Gray codes [7,34] and the doubly-adjacent Gray code of permutations [6] use challenging notions of a small change. The middle levels theorem instead uses a natural notion of a small change, but it took several decades to prove [21] and refine [14] into the current argument [22] from the 'book' [1]. In contrast, some Gray codes are simpler than lexicographic order. To illustrate this perspective we can compare $L(b)$ for bases $b = 2, 2, 2$ (i.e., binary strings of length $n = 3$) with the well-known *binary Gray code* [13] denoted as $G(b)$ below.

$$L(b) = 00\overline{0}, 00\overline{1}, 01\overline{0}, \overline{011}, 10\overline{0}, 10\overline{1}, 11\overline{0}, 111 \qquad (2)$$
$$G(b) = 00\overline{0}, 00\overline{1}, 01\overline{1}, \overline{0}10, 11\overline{0}, 1\overline{1}1, 10\overline{1}, 100 \qquad (3)$$
$$\mathsf{ruler}(b) = \quad 1, \ 2, \quad 1,3, \quad 1, \ 2, \quad 1 \qquad (4)$$

Here the overlines denote bits that are flipped to create the next word. Notice that when lexicographic order flips the rightmost i bits, the Gray code only flips the ith rightmost bit. Successive values of i create a mutual *change sequence* $\mathsf{ruler}(b) = 1, 2, 1, 3, 1, 2, 1$ known as the *binary ruler sequence* OEIS A001511 [24].

The binary Gray code is attributed to Frank Gray's patent at Bell Labs [13] despite an earlier patent by his coworker [36]. Similarly, plain changes of permutations from the 17th century [8] is better known as the Steinhaus-Johnson-Trotter algorithm for its 20th century rediscovers. Additional history [5,11,30] is found in the aforementioned surveys and in books by Ruskey [31] and Knuth [17].

Many classic Gray codes can be created greedily by choosing a start object and prioritizing the changes [40]. The order is then repeatedly extended to a new object by applying the highest priority change to the current object. For example, the binary Gray code starts at 0^n and then flips the rightmost bit. To clarify this idea, consider the start of (3): $G(b) = 000, 001, 011, 010$. Flipping the rightmost bit in the last word gives $01\overline{0} = 011$, which is already in $G(b)$. Similarly, flipping the middle bit recreates $0\overline{1}0 = 000$. But the third-highest priority change gives a new word $\overline{0}10 = 110$, so it is the next object in the order. Likewise, plain changes starts at $12\cdots n$ and then swaps the largest value [40]. The greedy approach has also led to deep theoretical results (see the *Permutation Languages* series of papers I–VII [15]– [4]) and answered open problems [19].

As discussed in [27], the greedy description of the binary Gray code can be generalized to mixed-radix words in several natural ways starting from 0^n.

– Reflected order $R(b)$ changes the rightmost digit by ± 1. [2]
– Modular order $M(b)$ changes the rightmost digit by $+1$ modulo its base.
– Max-right order $Q(b)$ sets the rightmost digit to the largest possible value.

In all three orders the index that changes is given by the mixed-radix ruler sequence $\mathsf{ruler}(b)$ defined in Sect. 2. Furthermore, $\mathsf{ruler}(b)$ gives the number of digits that change in lexicographic order. These points are illustrated below for bases $b = 2, 2, 3$ where overlines and underlines indicate how the digits change.

$$L(b) = 00\overline{0}, 00\overline{1}, 00\overline{2}, 01\overline{0}, 01\overline{1}, \overline{012}, 10\overline{0}, 10\overline{1}, 1\overline{02}, 11\overline{0}, 11\overline{1}, 112 \tag{5}$$

$$R(b) = 00\overline{0}, 00\overline{1}, 00\overline{2}, 01\underline{2}, 01\underline{1}, \overline{0}10, 11\overline{0}, 11\overline{1}, 1\underline{1}2, 10\underline{2}, 10\underline{1}, 100 \tag{6}$$

$$M(b) = 00\overline{0}, 00\overline{1}, 00\overline{2}, 01\overline{2}, 01\overline{0}, \overline{0}11, 11\overline{1}, 11\overline{2}, 1\overline{1}0, 10\overline{0}, 10\overline{1}, 102 \tag{7}$$

$$Q(b) = 00\overline{0}, 00\overline{2}, 00\overline{1}, 01\overline{1}, 01\overline{2}, \overline{0}10, 11\overline{0}, 11\overline{2}, 1\overline{1}1, 10\overline{1}, 10\overline{2}, 100 \tag{8}$$

$$\mathsf{ruler}(b) = \quad 1,\quad 1,\; 2,\quad 1,\quad 1,3,\quad\quad 1,\quad 1,\; 2,\quad 1,\; 1 \tag{9}$$

Note that these Gray codes have different properties. For example, $M(b)$ is *cyclic* for ternary words (i.e., $b = 3, 3, \ldots, 3$) meaning that a cyclic increment changes the last word of $M(b)$ into the first; the analogous property is not true for $R(b)$ and $Q(b)$. In fact, the cyclic property was the genesis of this project [25].

1.2   Combinatorial Generation

The goal of *combinatorial generation* is to efficiently generate every instance of a particular combinatorial object [31]. The best possible theoretical run-time is a *loopless algorithm* which generates each object in worst-case $\mathcal{O}(1)$-time. More specifically, the generation algorithm shares one object with the application and it always makes a worst-case $\mathcal{O}(1)$-time modification to create the 'next' object.

Loopless algorithms typically do not generate objects in lexicographic order as successive objects do not differ by a constant amount[3]. A loopless algorithm for $\mathbf{W}(b)$ was first published by Ehrlich [2,9]. He introduced the notion of focus pointers to compute successive values of $\mathsf{ruler}(b)$ in $\mathcal{O}(1)$-time to generate $R(b)$. Knuth [17] adapted this approach for $M(b)$ and it was later used for $Q(b)$ [27].

---

[2] The potential ambiguity of prioritizing $+1$ or -1 does not arise here [27] (or see [15]).
[3] The time required to modify an object depends on the underlying data structure [39].

Parallel generation algorithms have been studied for many years [18]. Some existing loopless algorithms can be parallelized to p processors for certain values of p. For example, we can generate the binary Gray code on $p = 2$ processors by hard-coding the first bit. However, the authors are not aware of existing research on loopless algorithms that can be split evenly across p processors for any p.

1.3   New Results

By focusing on the internal state of the loopless algorithms for $R(b)$, $M(b)$, and $Q(b)$ we are able to efficiently start them from any rank. This allows us to restart any long-running computation that is terminated unexpectedly. It also allows us to looplessly generate the orders in parallel for any number of processors p.

The relationship between the mixed-radix ruler sequence and its associated orders also leads to a new result blending lexicographic order and the modular Gray code. Let $\mathsf{rank}_b(w)$ be the *lexicographic rank* of $w \in \mathbf{W}(b)$. In other words, if $L(b) = w_0, w_1, \ldots, w_{b!-1}$, then $\mathsf{rank}_b(w) = r$ where $w_r = w$. For example,

$$L(2, 3, 2) = 000, 001, 010, 011, 020, 021, 100, 101, 110, 111, 120, 121 \qquad (10)$$
$$\mathsf{rank}_{2,3,2}(110) = 8. \qquad (11)$$

Let $\mathbf{W}_w(b) \subseteq \mathbf{W}(b)$ be the subset of words that are greater than or equal to w in terms of lexicographic order. We also let $\mathbf{W}_r(b) = \mathbf{W}_w(b)$ where $r = \mathsf{rank}_b(w)$ since the included words are precisely those with lexicographic rank at least r. For example, $\mathbf{W}_{110}(2, 3, 2) = \mathbf{W}_8(2, 3, 2) = \{110, 111, 120, 121\}$ by (10)–(11).

We present the first Gray code for $\mathbf{W}_w(b)$. Our *modified modular Gray code* $M_w(b)$ (or $M_r(b)$) generalizes the original modular Gray code $M(b)$ in two ways.

- Greedy algorithm. $M_w(b)$ starts at w and cyclically increments the rightmost digit that gives a new word greater than w (i.e., higher lexicographic rank).
- Change sequence. The change sequence $\mathsf{ruler}_r(b)$ is simply the suffix of the ruler sequence $\mathsf{ruler}(b)$ starting at rank $r = \mathsf{rank}_b(w)$.

We also looplessly generate these Gray codes, thereby strengthening the result for binary bases in [33]. Figure 2 illustrates several $M_r(b)$ with `Python` code.

1.4   Outline

We discuss ruler sequences and Gray codes followed by combinatorial generation.

- Sect. 2 discusses the ruler sequence $\mathsf{ruler}(b)$ and its properties.
- Sect. 3 describes Gray codes that use $\mathsf{ruler}(b)$ as a change sequence.
- Sect. 4 discusses suffixes of the ruler sequence $\mathsf{ruler}_r(b)$ and its properties.
- Sect. 5 introduces our new Gray codes $M_r(b)$ for $\mathbf{W}_r(w)$.
- Sect. 6 provides greedy interpretations of the new Gray codes.
- Sect. 7 (un)ranks the classic loopless algorithm for $\mathsf{ruler}(b)$, $R(b)$, and $M(b)$.
- Sect. 8 resumes and parallelizes the classic algorithms and generates $M_r(w)$.

Indexing Conventions. We informally mix our discussion of orders based on having the least-significant (and most rapidly changing) digit w_1 on the right (e.g., $L(w)$, $R(w)$, $M(w)$, $Q(w)$) or left (e.g., $\bot(w)$, $\mathcal{A}(w)$, $\mathbb{M}(w)$, $\mho(w)$). The distinction is mostly cosmetic for mixed-radix words as each digit is independently constrained by its base. (In general, the indexing $w_1 w_2 \cdots w_n$ aligns more naturally with bases $b_1, b_2, \ldots, b_n$, but the correct indexing for $L(n)$ is $w_n w_{n-1} \cdots w_1$.)
Pseudocode and Implementations. An implementation of all of our functions is available at [26] in `Python`. To match `Python`'s conventions we use 0-based indexing in our pseudocode and its discussion in Sect. 7. Also note that Fig. 2 reads and writes with `[::-1]` to display $M_r(b)$ instead of $\mathbb{M}_r(b)$

2 Mixed-Radix Ruler Sequences and Their Properties

Due to the cultural influences of America and the UK, students in Canada are often familiar with both meter sticks and yard sticks. Metric rulers have tick marks for each $\frac{1}{10}$th of an interval, while imperial rulers use $\frac{1}{2}$, $\frac{1}{4}$, $\frac{1}{8}$ and so on.

These differing tick patterns motivate the *mixed-radix ruler sequence*. It is defined recursively for bases $b = b_1, b_2, \ldots, b_n$ where $b' = b_1, b_2, \ldots, b_{n-1}$ if $n > 1$.

$$
\mathsf{ruler}(b) = \begin{cases}
\overbrace{1, 1, \ldots, 1}^{b_1 - 1 \text{ copies}} & \text{if } n = 1 \quad (12a) \\
\underbrace{\mathsf{ruler}(b'),\, n,\, \mathsf{ruler}(b'),\, n,\, \ldots,\, n,\, \mathsf{ruler}(b')}_{b_n - 1 \text{ copies of } n} & \text{if } n > 1 \quad (12b)
\end{cases}
$$

An example ruler sequence is shown below with several others in Fig. 1.

Example 1. $\mathsf{ruler}(b) = \mathsf{ruler}(b'), n, \mathsf{ruler}(b'), n, \mathsf{ruler}(b') = 1, 2, 1, 2, 1$ for $b = 2, 3$.

This section provides several properties of ruler sequences. In particular, the palindromic and length properties in Remarks 1–2 follow by induction on n.

Remark 1. Mixed-radix ruler sequences $\mathsf{ruler}(b)$ are palindromes.

Remark 2. Mixed-radix ruler sequences have $|\mathsf{ruler}(b)| = (b_1 \cdot b_2 \cdots b_n) - 1 = b! - 1$.

We use $[n]^k$ to denote sequences of length k whose entries are in $\{1, 2, \ldots, n\}$. For example, $(1, 3, 4, 4, 1) \in [4]^5$ and we often omit the parentheses.

Given sequence $S = s_1, s_2, \ldots, s_k$, a *subsequence* is any $s_i, s_{i+1}, \ldots, s_j$; it is a *prefix* if $i = 1$, a *suffix* if $j = k$, *non-empty* if $i \leq j$, and *strict* if $i \neq 1$ or $j \neq m$.

2.1 Closed and Cyclic Properties for Ruler Sequences

The following definitions and lemma show that ruler sequences do not have non-empty subsequences whose frequencies are multiples of their respective bases.

Definition 1. *Let $S = s_1, s_2, \ldots, s_k \in [n]^k$ be a non-empty sequence where f_i is its frequency of i. The sequence is b-closed relative to bases $b = b_1, b_2, \ldots, b_n$ if*

$$f_i \equiv 0 \bmod b_i \text{ for all } 1 \leq i \leq n. \tag{13}$$

That is, each i appears a multiple of b_i times. Otherwise, S is b-open.

Definition 2. *S is b-cyclic if it has a b-closed subsequence, or b-acyclic if not.*

Example 2. Consider $S = 1, 2, 1, 2, 1, 2$ relative to bases $b = b_1, b_2 = 2, 3$. Note that S is b-open because its frequency of 1s is $f_1 \bmod b_1 = 1$. However, its subsequence $2, 1, 2, 1, 2$ is b-closed. Therefore, S is b-cyclic rather than b-acyclic.

Lemma 1. *The mixed-radix ruler sequence $\mathsf{ruler}(b)$ is b-acyclic.*

Proof. We proceed by induction on the number of entries in the base n. The result holds for $n = 1$ by (12a). Now assume that the result is true for all bases with $n = k$ entries and consider some $b = b_1, b_2, \ldots, b_{k+1}$ with $b' = b_1, b_2, \ldots, b_k$. By induction the result holds for $\mathsf{ruler}(b')$. Therefore, any b-closed subsequence of $\mathsf{ruler}(b)$ contains at least one copy of $k+1$ by (12b). But $\mathsf{ruler}(b)$ only contains $b_{k+1} - 1$ copies of $k + 1$, so no such subsequence can exist.

2.2 Weak Left-to-Right Maxima for Ruler Sequences

Given sequence $S = s_1, \ldots, s_k$ the entry s_i is a *weak left-to-right maxima* if $i = 1$ or $s_i \geq \max(s_1, \ldots, s_{i-1})$. That is, s_i is greater than or equal to the previous entries. Let $\mathsf{ltr}(S)$ be the weak left-to-right maxima in S written as a word.

Example 3. If $S = \mathsf{ruler}(2, 3, 2) = 1, 2, 1, 2, 1, 3, 1, 2, 1, 2, 1$, then $\mathsf{ltr}(S) = 1223$.

Remark 3. If $S = \mathsf{ruler}(b)$ for bases $b = b_1, b_2, \ldots, b_n$, then its weak left-to-right maxima word is $\mathsf{ltr}(S) = 1^{f_1} 2^{f_2} \cdots n^{f_n}$ where $f_i = b_i - 1$ for all $1 \leq i \leq n$.

3 Ruler Change Sequences

Ruler sequences have the correct length to be change sequences of mixed-radix words by Remark 2. For example, Fig. 1 illustrates $\mathsf{ruler}(b)$ and lexicographic orders $L(b)$ for several bases b with ticks between consecutive words. More broadly, this section discusses several orders and their relation to Sect. 2.

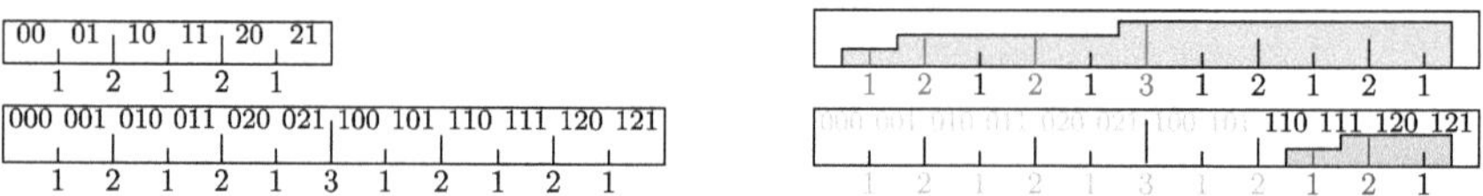

Fig. 1. Ruler sequences for $b = 2, 3, 2$ and $b' = 2, 3$ with lexicographic order (left); $\mathsf{ltr}(S) = 1223$ and $\mathsf{ltr}(S') = 12$ for $S = \mathsf{ruler}(b)$ and suffix $S' = \mathsf{ruler}_{110}(b)$ (right).

3.1   Modular Gray Code

Given a word $w = w_1 w_2 \cdots w_n \in \mathbf{W}(b)$ with bases $b = b_1, b_2, \ldots, b_n$ we let $\oplus_i(w)$ be the result of incrementing digit i modulo its base. That is,

$$\oplus_i (w) = w_1 w_2 \cdots w_{i-1} v_i w_{i+1} w_{i+2} \cdots w_n \tag{14}$$

where $v_i = 0$ if $w_i = b_i - 1$ or $v_i = w_i + 1$ otherwise. Given a word $w \in \mathbf{W}(b)$ and a sequence $S = s_1, s_2, \ldots, s_k \in [n]^k$ we let $w \oplus S$ be the result of repeatedly cyclically incrementing the digits given in the sequence. That is,

$$w \oplus S = w, \ \oplus_{s_1}(w), \ \oplus_{s_2}((\oplus_{s_1}(w))), \ldots \tag{15}$$

For example, $M(b)$ in (7) uses $\mathsf{ruler}(b)$ in (9). We now define the order in general.

Definition 3. *The* modular order *for bases b is $M(b) = 0^n \oplus \mathsf{ruler}(b)$. In other words, it starts at $w = 0^n$ and then repeatedly increments digit w_i cyclically modulo its base b_i, where i is the next value in the mixed-radix ruler sequence.*

Corollary 1. *$M(b)$ is a Gray code of $\mathbf{W}(b)$ using cyclic increments.*

$\square$

Proof. If a word $w = w_1 w_2 \cdots w_n \in \mathbf{W}(b)$ is repeated in $M(b)$, then some sublist cyclically increments each w_i a multiple of b_i times. But this implies that a subsequence of $\mathsf{ruler}(b)$ is b-closed which is impossible by Lemma 1.

By Corollary 1 we are justified in calling $M(b)$ the *modular Gray code*.

3.2   Reflected Gray Code and Other Gray Codes

The reflected Gray code $R(b)$ changes are also based on the ruler sequence. But each entry of $\mathsf{ruler}(b)$ indicates the index to increment *or* decrement. When generating $R(b)$ the previous words provide the missing information (i.e., only one of the changes gives a new word). Instead a direction for each index is typically maintained. Alternatively, a signed version of $\mathsf{ruler}(b)$ can be generated.

In the max-right Gray code $Q(b)$ the index of change is provided by $\mathsf{ruler}(b)$. The new value for the digit again follows from the previous words. Alternatively, one bit of information per index is sufficient for determining the value [28].

Many other Gray codes use ruler sequences. For example, permutation Gray codes often change the inversion vector via $\mathsf{ruler}(1, 2, \ldots, n)$ or ruler$(n, n-1, \ldots, 1)$ [12,16,35] while signed permutations add bases $2, 2, \ldots, 2$ [28,29]. Gray codes using ruler sequences are often *genlex* (*generalized lexicographic*) meaning that words with a given prefix are consecutive in the order [37,38]. The relationship between greedy and genlex Gray codes was investigated in Merino's thesis [10]. But the genlex property does not immediately lead to loopless algorithms [20].

3.3 Lexicographic Order

Consider a mixed-radix word $w = w_n w_{n-1} \cdots w_1 \in \mathbf{W}(b)$. We typically think of the next word in lexicographic order as incrementing w_i and setting $w_{i-1} \cdots w_1$ (which are at their maximum values) to 0. Equivalently, the digits $w_i w_{i-1} \cdots w_1$ are cyclically incremented. Indeed $L(b)$ is created by greedily cyclically incrementing the shortest suffix starting from 0^n, and the suffix lengths are ruler(b).

This perspective—cyclically increment suffixes of length ruler(b)—doesn't make it obvious that successive words are lexicographically larger. Section 4 clarifies this by generalizing the weak left-to-right maxima result from Remark 3.

4 Mixed-Radix Ruler Suffixes and Their Properties

If ruler$(b) = s_0, s_1, \ldots, s_k$, then ruler$_r(b) = s_r, s_{r+1}, \ldots, s_k$ is its suffix starting at the rth entry. Alternatively, we specify the suffix using the corresponding word in lexicographic order. That is, ruler$_w(b) = $ ruler$_r(b)$ if $r = $ rank$_b(w)$. More directly, the suffix is defined below for $w = w_1 w_2 \cdots w_n$ and $b = b_1, b_2, \ldots, b_n$ with $w' = w_1 w_2 \cdots w_{n-1}$ and $b' = b_1, b_2, \ldots, b_{n-1}$ if $n > 1$.

$$
\text{ruler}_w(b) = \begin{cases} \overbrace{1, 1, \ldots, 1}^{b_1 - 1 - w_1 \text{ copies}} & \text{if } n = 1 \quad (16a) \\ \underbrace{\text{ruler}_{w'}(b'), n, \text{ruler}(b'), n, \ldots, n, \text{ruler}(b')}_{b_n - 1 - w_n \text{ copies of } n} & \text{if } n > 1 \quad (16b) \end{cases}
$$

The full ruler sequence in (12) is obtained with $w = 0^n$ (or $r = 0$). Also note that only the first subsequence in (16b) is potentially a strict suffix of ruler(b').

Remark 4. Mixed-radix ruler suffixes have $|\text{ruler}_r(b)| = b! - 1 - r = |\mathbf{W}_r(b)| - 1$.

4.1 Weak Left-to-Right Maxima for Ruler Sequence Suffixes

Now we generalize Remark 3 to ruler sequence suffixes. In this generalization we refer to the corresponding word in lexicographic order, which is 0^n in Remark 3.

Example 4. Consider bases $b = 2, 3, 2$ and word $w = 010$ with $r = $ rank$_b(w) = 2$. Note that each digit in w is one increment from its maximum value. Hence, ltr$(S) = 123$ (i.e., values have frequency one) as $S = $ ruler$_r(b) = 1,2,1,3,1,2,1,2,1$.

Lemma 2. *Let $b = b_1, b_2, \ldots, b_n$ and $w = w_n w_{n-1} \cdots w_1 \in \mathbf{W}(b)$ with rank $r = \mathsf{rank}_b(w)$. If $S = \mathsf{ruler}_r(b)$, then $\mathsf{ltr}(S) = 1^{f_1} 2^{f_2} \cdots n^{f_n}$ with $f_i = b_i - 1 - w_i$.*

Proof. Consider $L(b)$ starting at w and two digit indices $i < j$. Note that w_i will be incremented to its maximum value $w_i = b_i - 1$ before w_j is incremented. Therefore, S has $f_i = b_i - 1 - w_i$ copies of i before j. Hence, $\mathsf{ltr}(S)$ has i^{f_i}.

5   New Modified Modular Gray Codes

We define our modified modular order analogously to $M(b)$ from Sect. 3.1.

Definition 4. *The* modified modular order *for bases b is $M_w(b) = w \oplus \mathsf{ruler}_w(b)$. In other words, it starts at w and then repeatedly increments digit w_i cyclically modulo its base b_i, where i is the next value in the mixed-radix ruler suffix.*

The modular Gray code is $M(b) = M_w(b)$ for $w = 0^n$. This is because $w = 0^n$ is the first word in $M(b)$, so applying $\mathsf{ruler}_w(b) = \mathsf{ruler}(b)$ gives the same order. We also start the new orders by rank. That is, $M_r(b) = M_w(b)$ for $r = \mathsf{rank}_b(w)$.

Example 5. Consider bases $b = b_1, b_2, b_3 = 4, 3, 5$ and word $w = w_3 w_2 w_1 = 312$ with lexicographic rank $r = \mathsf{rank}_b(w) = 42$. The ruler suffix is $S = \mathsf{ruler}_r(b) = 1, 2, 1, 1, 1, 3, 1, 1, 1, 2, 1, 1, 1, 2, 1, 1, 1$ and $\mathsf{ltr}(S) = 123$. Therefore, $M_r(b)$ is

$$31\overline{2}, 3\overline{1}3, 32\overline{3}, 32\overline{0}, 32\overline{1}, \overline{3}22, 42\overline{2}, 42\overline{3}, 42\overline{0}, 4\overline{2}1, 40\overline{1}, 40\overline{2}, 40\overline{3}, 4\overline{0}0, 41\overline{0}, 41\overline{1}, 41\overline{2}, 413$$

where the cyclic increments are overlined. Note that $32\overline{3} = 320$ is the first place that the least significant digit w_1 rolls over to 0. But the resulting word 320 is greater than the initial word 311 because an earlier transition $3\overline{1}3 = 323$ incremented a more significant digit w_2.

Now we prove our main mathematical result by generalizing Corollary 1.

Theorem 1. $M_w(b)$ *is a Gray code of $\mathbf{W}_w(b)$ using cyclic increments.*

Proof. Let $r = \mathsf{rank}(w)$. We know that $M_r(b)$ has $|\mathbf{W}_r(b)|$ words by Remark 4 and its words differ by a cyclic increment by Definition 4. So we can complete the proof by showing that it contains distinct members of $\mathbf{W}_r(b) = \mathbf{W}_w(b)$.

Let $S = \mathsf{ruler}_r(b)$. Since S is a subsequence of $\mathsf{ruler}(b)$, we know that it is b-acyclic by Lemma 1. Therefore, no words are repeated in $M_r(b)$.

In $M_r(b)$ the digit w_i rolls over from its maximum value to 0 for the first time after being cyclically incremented $b_i - w_i$ times. However, $\mathsf{ltr}(S) = 1^{f_1} 2^{f_2} \cdots n^{f_n}$ with $f_i = b_i - 1 - w_i$ by Lemma 2. Therefore, once i appears at least $f_i + 1$ times in S, a larger value $j > i$ appears at least once. Thus, any digit that is smaller than its initial value w_i is dwarfed by a more significant digit that is larger than its initial value w_j. In particular, the largest value in $\mathsf{ltr}(S)$ (i.e., the most significant digit that changes in $M_r(b)$ never rolls over. Therefore, the words in $M_r(b)$ are greater than or equal to w and hence in $\mathbf{W}_w(b)$.

6   Greedy Cyclic Increments

In this section we prove that our new Gray codes can instead be created greedily. We do this by arguing that ruler sequences and ruler suffixes are the longest and smallest sequences with the properties used in the proof of Theorem 1.

Lemma 3. *If $b = b_1, b_2, \ldots, b_n$, then any $S \in [n]^m$ with $m \geq b!$ is b-cyclic.*

Proof. There are $m + 1 \geq b! + 1$ words from $\mathbf{W}(b)$ in $0^n \oplus S$. But $|\mathbf{W}(b)| = b!$, so at least one word repeats. Hence, S has a b-closed subsequence.   □

Lemma 4. *Let $b = b_1, b_2, \ldots, b_n$ and $\mathsf{ruler}(b) = s_0, s_1, \ldots, s_{b!-1}$. If s_i is replaced by any x satisfying $1 \leq x < s_i$, then $S = s_0, s_1, \ldots, s_{i-1}, x$ is b-cyclic.*

Proof. By the choice of x we have that s_i is preceded by $\mathsf{ruler}(b')$ for the bases $b' = b_1, b_2, \ldots, b_x$. So the subsequence $\mathsf{ruler}(b'), x$ belongs to $[x]^m$ for $m = (b')!$. This subsequence is b'-cyclic by Lemma 3, so S is b-cyclic.   □

Algorithm 1 The classic loopless algorithm for $\mathsf{ruler}(b)$ and its associated reflected $R(b)$ and modular $M(b)$ Gray codes. The focus pointers can be controlled using the data structures for either Gray code (e.g., line 13 requires the reflected word R and its directions D while the equivalent line 14 requires the modular word M and its targets T. Arrays have length n (except for F of length $n + 1$) and use 0-based indices. We assume that the arrays are displayed from left-to-right, so the variable traces follow the co-orders. For example, column R in the variable trace is $\mathfrak{R}(b)$ rather than $R(b)$.

function `loopless`(B)				
1: R $\leftarrow [0, 0, ..., 0]$	▷ current word in $R(b)$
2: D $\leftarrow [1, 1, ..., 1]$	▷ directions for R
3: M $\leftarrow [0, 0, ..., 0]$	▷ current word in $M(b)$
4: T $\leftarrow [\mathsf{b_1}-1, \mathsf{b_2}-1, ..., \mathsf{b_n}-1]$	▷ targets for M
5: F $\leftarrow [0, 1, ..., n]$	▷ focus pointers $	F	= n+1$
6: **yield** R, M	▷ first word in $R(b)$, $M(b)$
7: **while** F[0] $<$ n **do**	▷ R[n], M[n] are invalid
8: j $\leftarrow$ F[0]	▷ index of change
9: b $\leftarrow$ B[j]	▷ base of change
10: F[0] $\leftarrow$ 0	▷ reset focus pointer
11: R[j] $\leftarrow$ R[j] + D[j]	▷ change digit by ± 1
12: M[j] $\leftarrow$ (M[j] + 1) mod b	▷ cyclic inc
13: limit $\leftarrow$ R[j] = 0 **or** R[j] = b−1	▷ test
14: limit $\leftarrow$ M[j] = T[j]	▷ equivalent test
15: **if** limit **then**	▷ R[j], M[j] at limit
16: D[j] $\leftarrow$ −D[j]	▷ direction for R[j]
17: T[j] $\leftarrow$ (T[j]−1) mod b	▷ M[j] target
18: F[j] $\leftarrow$ F[j + 1]	▷ inherit focus pointer
19: F[j + 1] $\leftarrow$ j + 1	▷ reset focus pointer
20: **yield** R, M	▷ next words in $R(b)$, $M(b)$

j	R	D	M	T	F
0		000	+++	000	221	0123
1	0	100	+++	100	221	0123
2	0	200	−++	200	121	1123
3	1	210	−++	210	121	0123
4	0	110	−++	010	121	0123
5	0	010	+++	110	021	1123
6	1	020	+−+	120	011	0223
7	0	120	+−+	220	011	0223
8	0	220	−−+	020	211	2123
9	2	221	−−−	021	210	0133
10	0	121	−−−	121	210	0133
11	0	021	+−−	221	110	1133
12	1	011	+−−	201	110	0133
13	0	111	+−−	001	110	0133
14	0	211	−−−	101	010	1133
15	1	201	−+−	111	000	0323
16	0	101	−+−	211	000	0323
17	0	001	++−	011	200	3123

Variable trace for bases B $= [3, 3, 2]$ at line 7; j is initially undefined.

Theorem 2. *The Gray code $M_w(b)$ is created greedily starting from w by cyclically incrementing the least significant digit giving a new word greater than w.*

Proof. Let $b = b_1, b_2, \ldots, b_n$ and $w = w_n w_{n-1} \cdots w_1$ with $r = \mathsf{rank}(b)$. Also let the ruler suffix that creates $M_w(b)$ be $\mathsf{ruler}_w(b) = \mathsf{ruler}_r(b) = s_0, s_1, \ldots, s_{b!-1-r}$.

If the statement is not true, then the greedy algorithm cyclically increments a less significant digit than the ruler suffix. More formally, there exists $S = s_0, s_1, \ldots, s_{i-1}, x$ with $0 \le x < s_i$ where $w \oplus S$ contains distinct words in $\mathbf{W}_r(b)$. We show that this is not possible by considering two cases.

Case 1: The last value in S is a weak left-to-right maxima. Since $\mathsf{ltr}(\mathsf{ruler}_w(b))$ includes x^{f_x} for $f_x = b_x - 1 - w_x$ by Lemma 2, we know that $\mathsf{ltr}(S)$ has x^{f_x+1}. But that rolls over digit w_x to 0 making the last word in $w \oplus S$ less than w.

Case 2: The last value in S is not a weak left-to-right maxima. In this case the impossibility follows from Lemma 4. $\qquad\square$

7   Ranking and Unranking the Classic Loopless Algorithm

A classic result in this area is that the ruler sequence $\mathsf{ruler}(b)$ can be generated in worst-case $\mathcal{O}(1)$-time per value [9]. This loopless algorithm appears in Algorithm 1 and also generates the reflected Gray code $R(b)$ and modular Gray code $M(b)$. For more details see Algorithm H and Exercise 77 in Sect. 7.2.1.1 of *The Art of Computer Programming* [17]. Small variations also generate other orders (e.g., $Q(b)$ in [28]) or subsets (e.g., linearly independent k-ary words in [3]).

To start Algorithm 1 at its rth iteration, we can set its variables to their respective values at this iteration. That is, we can *unrank* its internal state. Similarly, if Algorithm 1 unexpectedly terminates and we only have the most recently generated word in $R(b)$ or $M(b)$, then we can first *rank* the word relative to $R(b)$ or $M(b)$. We provide $\mathcal{O}(n)$-time `Python` code for these tasks in [26].

The most difficult and novel of these problems is unranking the focus pointers. A discussion appears in Sect. 7.1 along with pseudocode in Algorithm 2.

7.1   Focus Pointers

In Algorithm 1, F is initialized as an array with a sequence of integers $\mathsf{range}(0, n+1)$. The first index of F, $F[0]$ in each step points to the digit to be flipped (in case of binary) or altered in a "certain" way (in case of reflected or modular). Since $\lambda(b)$ shares the same underlying data structures with $\mathfrak{R}(b)$ and $\mathbb{M}(b)$, $F[0] = i$ means that the ith digit of the word is to be incremented while the first $i - 1$ digits are at their maximum corresponding to their bases. Before we discuss the pseudocode for providing the focus pointer corresponding to a given word in $\lambda(b)$ in Algorithm 2, we proceed with a few examples that are summarized in 17–19.

Example 6. $b = 3, 4, 5, 7, 2, 6$ and $w = \bar{1}12504$, the leftmost digit with a bar over it is given the highest priority to increment. Since each digit is strictly less than the maximum allowed digit corresponding to the base, $F = [0, 1, 2, 3, 4, 5, 6]$.

Example 7. $b = 3, 4, 5, 7, 2, 6$ and $w = 234304$. Since the digits at positions $0, 1, 2$ are at their maximum, the digit at position 3 is the top priority to increment. Thus, $F = [3, 1, 2, 3, 4, 5, 6]$.

Example 8. $b = 4, 3, 4, 5, 2, 6, 3, 2$ and $w = 32141001$. Since the digits at positions $0, 1, 3, 4, 7$ are the maximum corresponding to their bases, $F[0] = 2$ as the digit at position 2 has the highest priority to increment. Furthermore, $F(3) = 5$ since the digits at positions 3 and 4 are at their maximum while the digit at position 5 is not. Similarly, $F(7) = 8$. Thus, $F = [2, 1, 2, 5, 4, 5, 6, 8, 8]$.

Examples 6–8 are summarized below.

$$b = 345726 \qquad b = 345726 \qquad b = 43452632 \tag{17}$$
$$w = 112504 \qquad w = 234304 \qquad w = 32141001 \tag{18}$$
$$F = 0123456 \qquad F = 3123456 \qquad F = 212545688 \tag{19}$$

Algorithm 2 follows the ideas explained in Examples 6, 7, and 8 and returns the focus pointer corresponding to the word of r in $\lambda(b)$. In the algorithm, we

---

Algorithm 2 Unranking the focus pointers F at iteration r of Algorithm 1. Lists b, x, y are built based on runs of consecutive digits at their maximum values. For base B, word a, and its rank r in $\lambda(B)$, rankColex(B, a) and unrankColex(B, r) return the corresponding r and a, respectively.

```
function unrankFocus(B, r)          ▷ Bases B rank r      function unrankColex(B, r)
1:  a ← unrankColex(B, r)           ▷ Rank r word         1:  a = [0, 0, ..., 0]
2:  F ← [0, 1, ..., n]              ▷ Default focus pointers   2:  p ← [1, B[0], ···, B[0]·B[1]···B[n−2]]
3:  b ← []                          ▷ Indices in a at max value  3:  for i ← n−1, ..., 1, 0 do
4:  x ← []                          ▷ Start indices of runs  4:      b ← ⌊r / p[i]⌋
5:  y ← []                          ▷ Indices after each run  5:      1 ← b mod B[i]
6:  for i ← 0, 1, ..., n−1 do        ▷ Check all indices    6:      a[i] ← 1
7:      if a[i] = B[i]−1 then        ▷ Value at max?        7:  return a
8:          APPEND(b, i)             ▷ Add it index to list

9:  if |b| = 0 then                  ▷ No max values?
10:     return F                     ▷ Return default values
11: else                                                   function rankColex(B, a)
12:     z ← b[0]                     ▷ Start current run    1:  p ← [1, B[0], ..., B[0]·B[1]···B[n−2]]
13:     for j ← 1, 2, ..., |b|−1 do  ▷ Indices at max       2:  r ← 0            ▷ Initial rank
14:         if b[j] ≠ b[j − 1] + 1 then  ▷ End of run?      3:  for i ← n−1, ..., 1, 0 do
15:             APPEND(x, z)         ▷ Start of run         4:      r ← r + p[i] · a[i]  ▷ Update r
16:             APPEND(y, b[j − 1] + 1)  ▷ End of run       5:  return r
17:             z ← b[j]             ▷ Start new run

18:     APPEND(x, z)                 ▷ Start of last run
19:     APPEND(y, b[|b| − 1] + 1)    ▷ End of last run

20: for i ← 0, 1, ..., |x| − 1 do
21:     F[x[i]] ← y[i]               ▷ Update the focus pointer
22: return F
```

```python
def looplessMr(B, r):
    n = len(B)
    F = unrankFocus(B, r)
    M = unrankColex(B, r)
    T = [b-1-(M[i]==B[i]-1)
         for i,b in enumerate(B)]
    yield M
    while F[0] < n:
        r += 1
        j = F[0]
        b = B[j]
        F[0] = 0
        M[j] = (M[j]+1) % b
        if M[j] == T[j]:
            T[j] = (T[j]-1) % b
            F[j] = F[j+1]
            F[j+1] = j+1
        yield M

def unrankColex(B, r):
    n = len(B)
    prods = [1]
    p = 1
    for base in B[:-1]:
        p *= base
        prods.append(p)
    L = []
    for index in range(n-1,-1,-1):
        digit = r // prods[index]
        r -= digit * prods[index]
        L = [digit] + L
    return L

def unrankFocus(B, r):
    n = len(B)
    L = unrankColex(B, r)
    F = list(range(n + 1))
    lims = []
    for i in range(n):
        if L[i] == B[i] - 1:
            lims.append(i)
    if len(lims) == 0: return F
    x = []
    y = []
    z = lims[0]
    for j in range(1,len(lims)):
        if lims[j] != lims[j-1]+1:
            x.append(z)
            y.append(lims[j-1]+1)
            z = lims[j]
    x.append(z)
    y.append(lims[-1] + 1)
    for i in range(len(x)):
        F[x[i]] = y[i]
    return F

import sys
B = [int(w) for w in sys.argv[1][::-1]]
r = int(sys.argv[2])
for M in looplessMr(B, r):
    print(*M[::-1], sep="")
```

```
  0000
1 0001
2 0011 0010
1 0010 0011
2 0020 0021 0020
1 0021 0020 0021
3 0121 0120 0121 0100
1 0120 0121 0120 0101
2 0100 0101 0100 0111 0110
1 0101 0100 0101 0110 0111
2 0111 0110 0111 0120 0121 0120
1 0110 0111 0110 0121 0120 0121
4 1110 1111 1110 1121 1120 1121 1000
1 1111 1110 1111 1120 1121 1120 1001
2 1121 1120 1121 1100 1101 1100 1011 1010
1 1120 1121 1120 1101 1100 1101 1010 1011
2 1100 1101 1100 1111 1110 1111 1020 1021 1020
1 1101 1100 1101 1110 1111 1110 1021 1020 1021
3 1001 1000 1001 1010 1011 1010 1121 1120 1121
1 1000 1001 1000 1011 1010 1011 1120 1121 1120
2 1010 1011 1010 1021 1020 1021 1100 1101 1100
1 1011 1010 1011 1020 1021 1020 1101 1100 1101
2 1021 1020 1021 1000 1001 1000 1111 1110 1111
1 1020 1021 1020 1001 1000 1001 1110 1111 1110
4 2020 2021 2020 2001 2000 2001 2110 2111 2110
1 2021 2020 2021 2000 2001 2000 2111 2110 2111
2 2001 2000 2001 2010 2011 2010 2121 2120 2121
1 2000 2001 2000 2011 2010 2011 2120 2121 2120
2 2010 2011 2010 2021 2020 2021 2100 2101 2100
1 2011 2010 2011 2020 2021 2020 2101 2100 2101
3 2111 2110 2111 2120 2121 2120 2001 2000 2001
1 2110 2111 2110 2121 2120 2121 2000 2001 2000
2 2120 2121 2120 2101 2100 2101 2010 2011 2010
1 2121 2120 2121 2100 2101 2100 2011 2010 2011
2 2101 2100 2101 2110 2111 2110 2021 2020 2021
1 2100 2101 2100 2111 2110 2111 2020 2021 2020
```

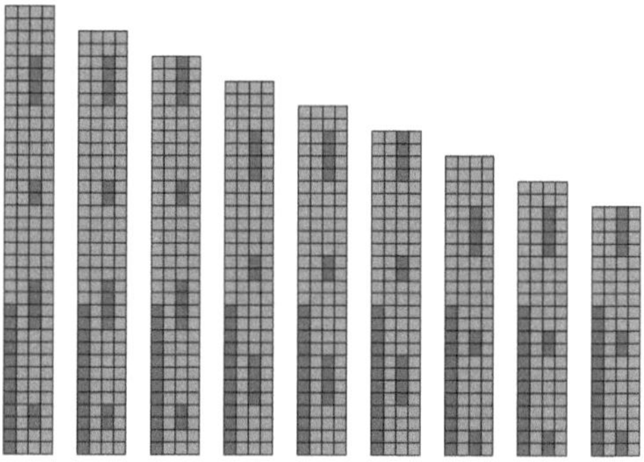

Fig. 2. Loopless generation of our modular Gray code $M_r(b)$ (left) for mixed-radix words $\mathbf{W}_r(b)$ of rank $\geq r$. The orders start at the smallest word w (i.e., $\mathsf{rank}_b(w) = r$), then new words greater than w cyclically increment the rightmost possible digit greedily. The index of change in $w_4 w_3 w_2 w_1$ is computed directly via the suffix $\mathsf{ruler}_r(b)$ of the mixed-radix $\mathsf{ruler}(b) = 1, 2, 1, 2, 1, 3, \ldots$. Output for $b = b_1, b_2, b_3, b_4 = 3, 2, 3, 2$ and selected r are provided (right). For example, `python3 looplessMr.py 3232 16` generates the rightmost column. Code is available [26].

consider three arrays b, x, and y such that b stores the indices of the word with maximum allowed digits corresponding to their bases. b may have some sequences of consecutive indices. Revisiting Example 8, we have b $= [0, 1, 3, 4, 7]$. Note that b in this case has some consecutive indices such as **01 34 7**. x then stores the minimum index in each subsequence, that is, x $= [0, 3, 7]$ while y stores 1 higher than the maximum index in each subsequence that is not in b, y $= [2, 5, 8]$. Then, we update the default focus pointer F by $F[0] = 2, F[3] = 5$, and $F[7] = 8$.

8 Loopless Generation Algorithms

This section discusses two additional loopless generation results.

8.1 Parallel Generation

Suppose that we want p processors to each looplessly generate roughly $m = \lceil \frac{b!}{p} \rceil$ distinct mixed-radix words from $\mathbf{W}(b)$. We can accomplish this with the results of Sect. 7. More specifically, we invoke Algorithm 1 in parallel at iterations $r = 0, m, 2m, \ldots, (p-1)m$ with each stopping after m iterations. Non-loopless generation of portions of the modular Gray code was recently considered in [3].

8.2 Modified Modular Gray Codes

We looplessly generate our modified modular Gray codes $M_r(b)$ in Fig. 2. The code in `looplessMr` initializes the focus pointers and the current word M to be the smallest in $\mathbf{W}_r(b)$ using `unrankFocus` and `unrankColex` from Algorithm 2. Finally, it sets the targets T (or sentinels [17]) for each digit w_i to be $b_i - 1$ or one less if it is already equal to this maximum value.

Appendices related to Sects. 7–8 are available in author versions of this paper.

Disclosure of Interests. The authors have no competing interests.

References

1. Aigner, M., Ziegler, G.M.: Proofs from the book. Berlin. Germany **1**(2), 7 (1999)
2. Bitner, J.R., Ehrlich, G., Reingold, E.M.: Efficient generation of the binary reflected gray code and its applications. Commun. ACM **19**(9), 517–521 (1976)
3. Bouyuklieva, S., Bouyukliev, I., Bakoev, V., Pashinska-Gadzheva, M.: Generating m-ARY gray codes and related algorithms. Algorithms **17**(7), 311 (2024)
4. Brenner, S., Cardinal, J., McConville, T., Merino, A., Mütze, T.: Combinatorial generation via permutation languages. VII. Supersolvable hyperplane arrangements. Eur. J. Combinatorics **135**, 104367 (2026)
5. Cohn, M.: Affine m-ARY Gray codes. Inf. Control **6**(1), 70–78 (1963)
6. Compton, R.C., Gill Williamson, S.: Doubly adjacent gray codes for the symmetric group. Linear Multilinear Algebra **35**(3–4), 237–293 (1993)

7. Cooke, M., North, C., Dewar, M., Stevens, B.: A note on Beckett-gray codes and the relationship of Gray codes to data structures. J. Comb. Math. Comb. Comput. **126**, 195–200 (2025)
8. Duckworth, R.: Tintinnalogia, or, the art of ringing (2010). Project Gutenberg
9. Ehrlich, G.: Loopless algorithms for generating permutations, combinations, and other combinatorial configurations. J. ACM **20**(3), 500–513 (1973)
10. Figueroa, A.I.M.: Combinatorial generation: greedy approaches and symmetry. Technische Universitaet, Berlin (Germany (2023)
11. Flores, I.: Reflected number systems. IRE Trans. Electron. Comput. **5**(2), 79–82 (1956)
12. Ganapathi, P., Chowdhury, R.: A unified framework to discover permutation generation algorithms. Comput. J. **66**(3), 603–614 (2023)
13. Gray, F.: Pulse code communication. United States Patent Number, p. 2632058 (1953)
14. Gregor, P., Mütze, T., Nummenpalo, J.: A short proof of the middle levels theorem. Discrete Anal. **8**, 1–12 (2018)
15. Hartung, E., Hoang, H., Mütze, T., Williams, A.: Combinatorial generation via permutation languages. I. Fundamentals. Trans. Am. Math. Soc. **375**(4), 2255–2291 (2022)
16. Holroyd, A.E., Ruskey, F., Williams, A.: Shorthand universal cycles for permutations. Algorithmica **64**(2), 215–245 (2012)
17. Knuth, D.E.: The art of computer programming, volumes 1–4B. Combinatorial Algorithms. Addison-Wesley Professional, Boxed Set (2022)
18. Lin, C.J.: A parallel algorithm for generating combinations. Comput. Math. Appl. **17**(12), 1523–1533 (1989)
19. Liu, B., Wong, D., Lam, C.T., Im, S.K.: Generating pivot gray codes for spanning trees of complete graphs in constant amortized time. In: Proceedings of the 2026 Annual ACM-SIAM Symposium on Discrete Algorithms (SODA), pp. 833–870. SIAM (2026)
20. Merino, A., Mutze, T., Williams, A.: All your bases are belong to us: listing all bases of a matroid by greedy exchanges. In: 11th International Conference on Fun with Algorithms (FUN 2022), vol. 226, p. 22. Schloss Dagstuhl-Leibniz-Zentrum für Informatik (2022)
21. Mütze, T.: Proof of the middle levels conjecture. Proc. Lond. Math. Soc. **112**(4), 677–713 (2016)
22. Mütze, T.: A book proof of the middle levels theorem. Combinatorica **44**(1), 205–208 (2024)
23. Mütze, T.: Combinatorial gray codes: an updated survey. Electron. J. combinatorics DS26 (2024)
24. OEIS foundation inc.: the on-line encyclopedia of integer sequences (2026). http://oeis.org
25. Parida, P.: Cover-free families on graphs and hypergraphs, University of Ottawa (2026)
26. Parida, P., Williams, A.: Mixed-radix ruler sequence gray codes (2026). https://github.com/ppari017/Mixed-radix-Ruler-Sequence-Gray-Codes
27. Qiu, Y., Sawada, J., Williams, A.: Maximize the rightmost digit: gray codes for restricted growth strings. In: International Conference and Workshops on Algorithms and Computation, pp. 296–311. Springer (2025)
28. Qiu, Y., Williams, A.: Generating signed permutations by twisting two-sided ribbons. In: Latin American Symposium on Theoretical Informatics, pp. 114–129. Springer (2024)

29. Qiu, Y.F.: Greedy and speedy: new gray code algorithms for generating signed permutations. Bachelor's thesis, Williams College (2024)
30. Rosenbaum, J.: Elementary problem E319. Am. Math. Mon. **45**(10), 694–696 (1938)
31. Ruskey, F.: Combinatorial generation. Working version (2003). 1j-CSC 425/520
32. Savage, C.: A survey of combinatorial gray codes. SIAM Rev. **39**(4), 605–629 (1997)
33. Sawada, J., Williams, A., Wong, D.: Flip-swap languages in binary reflected Gray code order. Theoret. Comput. Sci. **933**, 138–148 (2022)
34. Sawada, J., Wong, D.C.H.: A fast algorithm to generate Beckett-gray codes. Electron. Notes Discrete Math. **29**, 571–577 (2007)
35. Sedgewick, R.: Permutation generation methods. ACM Comput. Surv. (CSUR) **9**(2), 137–164 (1977)
36. Stibitz, G.: Binary counter. United States Patent Number, p. 2307868 (1943)
37. Walsh, T.: Generating gray codes in $O(1)$ worst-case time per word. In: International Conference on Discrete Mathematics and Theoretical Computer Science, pp. 73–88. Springer (2003)
38. Walsh, T.R.: A simple sequencing and ranking method that works on almost all Gray codes. Dept. Math. Comput. Sci. Université du Québeca Montréal Res. Rep. **243**, 53 (1995)
39. Williams, A.: $O(1)$-time unsorting by prefix-reversals in a boustrophedon linked list. In: International Conference on Fun with Algorithms, pp. 368–379. Springer (2010)
40. Williams, A.: The greedy gray code algorithm. In: Workshop on Algorithms and Data Structures, pp. 525–536. Springer (2013)

On the Complexity of Signed Domination

Sangam Balchandar Reddy[(✉)] [iD]

School of Computer and Information Sciences, University of Hyderabad,
Hyderabad, India
`21mcpc14@uohyd.ac.in`

Abstract. Given a graph $G = (V, E)$, a signed dominating function is a function $f : V \to \{-1, 1\}$ such that for every vertex $u \in V$, $\sum_{v \in N[u]} f(v) \geq 1$. The weight of f is defined as $\sum_{u \in V} f(u)$. The objective of the Signed Domination problem is to compute a signed dominating function f of minimum weight. The problem is known to be NP-complete even when restricted to bipartite, chordal, and planar graphs. In this paper, we extend the known complexity results for the Signed Domination problem. Since the problem is NP-complete on chordal graphs, we study its complexity on split graphs, a subclass of chordal graphs, and show that it remains NP-complete. Moreover, as the problem is W[2]-hard parameterized by weight, we investigate its parameterized complexity with respect to structural parameters. We prove that the problem is W[1]-hard when parameterized by feedback vertex set number (and hence by treewidth and clique-width). Motivated by this hardness result, we consider a more restrictive parameter, neighbourhood diversity, and present an FPT algorithm.

Keywords: Signed domination · NP-complete · W[1]-hard · FPT · Split graphs · Feedback vertex set number · Neighbourhood diversity

1 Introduction

The Dominating Set (DS) problem is one of the earliest problems shown to be NP-complete by Karp [9]. Given a graph $G = (V, E)$, a set $S \subseteq V$ is a dominating set if for each vertex $u \in V \setminus S$, $|N(u) \cap S| \geq 1$. The goal of the DS problem is to compute a dominating set of minimum size.

Since its inception, numerous variants of the DS problem have been introduced. One such variant is the Signed Domination (SD) problem. The concept of signed domination was introduced by Dunbar et al. [3] in 1995. Signed domination generalizes classical domination by allowing vertices to contribute either positively or negatively to the domination condition, thereby providing greater modeling flexibility. A signed dominating function on graph G is a function $f : V \to \{-1, 1\}$ such that for each vertex $u \in V$, $\sum_{v \in N[u]} f(v) \geq 1$. The weight of a signed dominating function f is $\sum_{u \in V} f(u)$. The minimum weight over all

F. Foucaud and A. Parreau (Eds.): IWOCA 2026, LNCS 16587, pp. 486–499, 2026.
https://doi.org/10.1007/978-3-032-27732-9_34

signed dominating functions for G is denoted by $\gamma_s(G)$. The objective of the SD problem is to compute $\gamma_s(G)$.

We formally define the decision version of the SD problem as follows.

SIGNED DOMINATION:

Input: An instance $I = (G, k)$, where $G = (V, E)$ is an undirected graph and an integer k.

Output: YES, if there exists a function $f : V \to \{-1, 1\}$ such that $\sum\limits_{u \in V} f(u) \le k$ and $\sum\limits_{v \in N[u]} f(v) \ge 1$ for each vertex $u \in V$; NO otherwise.

Hattingh et al. [7] proved that the SD problem is NP-complete on bipartite, and chordal graphs. They also showed that the problem admits a linear-time algorithm on trees. Damaschke [2] then showed that the problem remains NP-complete on planar graphs with maximum degree three. Further, Lee and Chang [12] proved that the problem is NP-complete even on doubly chordal graphs, planar bipartite graphs and chordal bipartite graphs. Later, Zheng et al. [16] showed that the problem is NP-complete on subcubic grid graphs. They also presented a kernel of size $\frac{k^2+k}{2}$ for general graphs, where k is the number of vertices with label 1. Recently, Lin and Poon [13] proved that the problem is W[2]-hard parameterized by weight. They also presented constant-factor approximation algorithms for subcubic graphs, graphs of maximum degree four and graphs of maximum degree five. In addition, they proved that the problem is FPT on subcubic graphs.

Several bounds have been established for the SD problem on various graph classes. In particular, Dunbar et al. [3] derived both upper and lower bounds on $\gamma_s(G)$. For r-regular graphs, Henning and Slater [8] showed that $\gamma_s(G) \ge \frac{n}{r+1}$ when r is even, and $\gamma_s(G) \ge \frac{2n}{r+1}$ when r is odd. Favaron [4] subsequently provided upper bounds, proving that $\gamma_s(G) \le n \cdot \frac{(r+1)^2}{r^2+4r-1}$ for odd r, and $\gamma_s(G) \le n \cdot \frac{r+1}{r+3}$ for even r. For general graphs, Zhang et al. [15] established the lower bound $\gamma_s(G) \ge 2 \left\lceil \frac{-1+\sqrt{1+8n}}{2} \right\rceil - n$.

Signed domination has applications in systems where elements exert both positive and negative influence, making it useful in real-world network analysis. For instance, in sensor networks, nodes can be in active $(+1)$ or inactive (-1) states, and the goal is to ensure that every node receives the positive support from its neighbours. This framework helps in designing efficient and fault-tolerant systems by minimizing resource usage while maintaining overall network stability. Similar interpretations arise in social and biological networks, where entities may promote or inhibit activity, and signed domination provides a mathematical tool to analyze and optimize such balanced interactions.

Our Results. In this paper, we investigate the SD problem in the realm of classical and parameterized complexity, and obtain the following results.

– We analyze the complexity of the problem on split graphs (a subclass of chordal graphs) and prove that it remains NP-complete.

- We then show that the problem is W[1]-hard parameterized by feedback vertex set number. From this result, it follows that the problem is W[1]-hard parameterized by treewidth or clique-width.
- Since the problem is W[1]-hard parameterized by clique-width, we further investigate its parameterized complexity for a larger parameter neighbourhood diversity, and present an FPT algorithm.

2   Preliminaries

Consider a graph $G = (V, E)$ with V as the vertex set and E as the edge set. For a vertex $u \in V$, the open neighbourhood of u is denoted by $N(u) = \{v : (u, v) \in E\}$ and the closed neighbourhood of u is denoted by $N[u] = N(u) \cup \{u\}$. The open neighbourhood of a set $T \subseteq V$ is denoted by $N(T) = (\bigcup_{u \in T} N(u)) \setminus T$ and the closed neighbourhood by $N[T] = N(T) \cup T$. The degree of a vertex u is represented by $d(u)$ and $d(u) = |N(u)|$. A graph G is r-regular if $d(u) = r$ for every vertex $u \in V$. A cubic graph is a 3-regular graph. A vertex of degree one is called a pendant vertex. A path of length l refers to a path with l edges and $l + 1$ vertices.

Let f be a signed dominating function. For a vertex $u \in V$, the value of $f(u)$ is also called the **label** of u. For a vertex u, **labelSum** is defined as the sum of the labels of all the vertices in its closed neighbourhood, i.e., $\sum_{v \in N[u]} f(v)$. For a set T, **weight** of T is the sum of the labels of all the vertices in the set, i.e., $\sum_{u \in T} f(u)$. Apart from this, standard notations as defined in West [14] are used.

Parameterized Complexity. A problem is said to be fixed-parameter tractable (FPT) with respect to a parameter k, if it can be solved by an algorithm with running time $\mathcal{O}(f(k) \cdot n^{\mathcal{O}(1)})$, where f is a computable function and n is the size of the input. The notation $\mathcal{O}^*(f(k))$ is used to suppress polynomial factors in n and to denote running times of this form. A problem is W[⋆]-hard with respect to a parameter if it is believed not to be fixed-parameter tractable. In parameterized complexity, there exists a hierarchy of complexity classes: FPT $\subseteq$ W[1] $\subseteq$ W[2] $\subseteq$ ... $\subseteq$ XP. For more information on *parameterized complexity*, the reader is referred to Cygan et al. [1].

We now define the class of graphs used in this paper.

Split Graphs. A graph $G = (V, E)$ is a split graph if the vertex set V can be partitioned into two sets V_1 and V_2 such that the subgraph induced by V_1 is a clique and the subgraph induced by V_2 is an independent set in G.

Next, we define the graph parameters used in this paper.

Feedback Vertex Set Number. For a graph $G = (V, E)$, the feedback vertex set number is the cardinality of a smallest vertex subset $F \subseteq V$ such that $G \setminus F$ is a disjoint union of trees.

Neighbourhood Diversity. Let $G = (V, E)$ be a graph. Two vertices u and v are of the same type if and only if $N(u) \setminus \{v\} = N(v) \setminus \{u\}$. The neighbourhood diversity of G is the smallest value of t for which there exists a partition of V into t sets $V_1, V_2, ..., V_t$ such that all the vertices in each set are of the same type.

A key result used in Sect. 4 of this paper is given as follows.

Lemma 1. *Let f be a signed dominating function. If $u \in V$ with $d(u) = 1$ and $v \in N(u)$, then $f(u) = f(v) = 1$.*

Proof. Consider a vertex $u \in V$ with $d(u) = 1$. Let v be the only neighbour of u. To ensure that u has a positive labelSum, neither u nor v can be assigned the label -1. If either vertex were labeled -1, the labelSum of u would be at most zero. Therefore, we must have $f(u) = f(v) = 1$. □

3   NP-Complete on Split Graphs

In this section, we prove that the SD problem is NP-complete on split graphs, a subclass of chordal graphs. We provide a polynomial-time reduction from the DS problem on cubic graphs to prove that the SD problem is NP-complete.

The complexity result related to the DS problem on cubic graphs is given as follows.

Theorem 1 ([10]). *DS problem is NP-complete even when restricted to cubic graphs.*

Given an instance $I = (G, k)$ of the DS problem where G is a cubic graph, we construct an instance $I' = (G', 2k)$ of the SD problem as follows.

- Two copies of the vertex set $V(G)$, denoted by A and X, are included in G'. A vertex $u_i \in V(G)$ is denoted by a_i in A and x_i in X. Thus, $A = \{a_1, a_2, ..., a_n\}$ and $X = \{x_1, x_2, ..., x_n\}$.
- Two additional sets B of size $4n$ and Y of size $2n$ are created. Let $B = \{b_1, b_2, ..., b_{4n}\}$ and $Y = \{y_1, y_2, ..., y_{2n}\}$.
- For each pair of vertices $a_i \in A$ and $x_j \in X$ with $i, j \in [n]$, an edge is added between a_i and x_j if and only if $u_i \in N_G(u_j)$. In addition, an edge is added between a_i and x_i for each $i \in [n]$.
- For each $i \in [n]$, the vertex $x_i \in X$ is made adjacent to four vertices $b_{4i-3}, b_{4i-2}, b_{4i-1}$ and b_{4i}.
- For each $i \in [2n]$, the vertex $y_i \in Y$ is made adjacent to both b_{2i-1} and b_{2i}.
- Finally, a clique is formed on the vertex set $A \cup B$.

This concludes the construction of G'. See Fig. 1 for an illustration.

Observe that G' is a split graph, where the subgraph induced by $A \cup B$ forms a clique and the subgraph induced by $X \cup Y$ forms an independent set.

Lemma 2.   *G has a dominating set of size at most k if and only if G' has a signed dominating function of weight at most $2k$.*

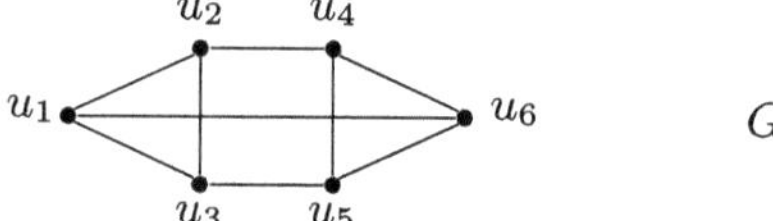

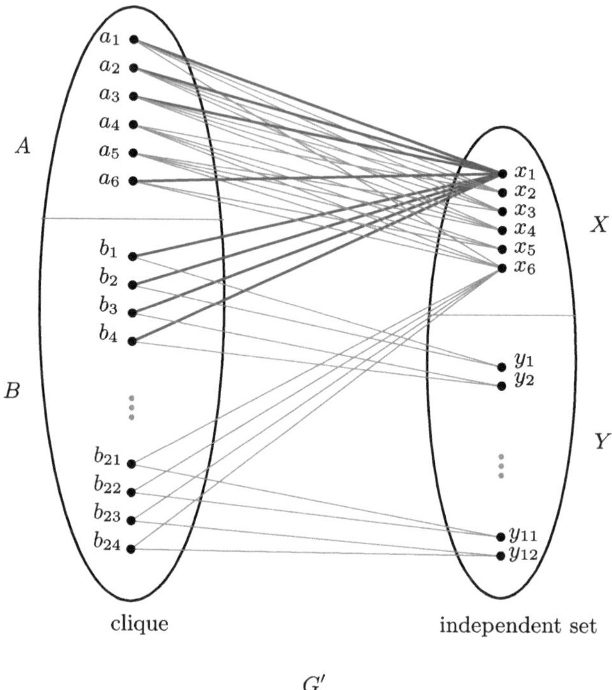

Fig. 1. Reduced instance of the SD problem constructed from an instance of the DS problem. Here, $k = 2$. Edges between the vertices of the clique $A \cup B$ are not shown in the figure. The edges incident on the vertex x_1 are colored in blue.

Proof. [$\Rightarrow$] Let S be a dominating set of size at most k. The corresponding signed dominating function f is given as follows.

$$f(w) = \begin{cases} -1, & \text{if } w \in \bigcup_{u_i \notin S} \{a_i\} \cup X \cup Y, \\ 1, & \text{if } w \in \bigcup_{u_i \in S} \{a_i\} \cup B \end{cases}$$

- Since G is cubic, each vertex $u \in X$ is adjacent to eight vertices in $A \cup B$. Vertex u has a labelSum of at least one as all four of its neighbours in B have label 1 and at least one of its neighbours in A has label 1.
- Each vertex in Y has a labelSum of at least one due to the positive labels for both of its neighbours in B.

- The weight of the clique $A \cup B$ is $3n + 2k$ which is at least $3n + 2$ (as $k \geq 1$). Therefore, each vertex in A has a labelSum of at least $3n - 2$, which is positive for all $n \geq 1$.
- Similarly, each vertex in B has a labelSum of at least $3n$, which is at least three for all $n \geq 1$.

The weight of A is at most $2k - n$, that of B is $4n$, that of X is $-n$ and that of Y is $-2n$. Since each vertex in G' has a positive labelSum, we conclude that f is a signed dominating function of weight at most $2k$.

[$\Leftarrow$] Let f be a signed dominating function of weight at most $2k$. The corresponding dominating set S is given as follows.

- For each $i \in [2n]$, vertex y_i is adjacent to two vertices in B, so its closed neighbourhood has size three. To ensure a positive labelSum for y_i, at most one of b_{2i-1}, b_{2i} and y_i can be labeled -1. Since y_i belongs to the independent set $X \cup Y$, we label y_i with -1 and both b_{2i-1} and b_{2i} with 1. Hence, the weight of B is $4n$ and the weight of Y is $-2n$.
- Since the weight of $B \cup Y$ is $2n$, the weight available for $A \cup X$ is at most $2k - 2n$. Therefore, at most k vertices in $A \cup X$ can be labeled 1, while the remaining vertices must be labeled -1.
- At this point, each vertex $u \in X$ has a labelSum of four from its neighbours in B. For u to have a positive labelSum, either u itself or at least one of its neighbours in A must be labeled 1. Since at most k vertices in $A \cup X$ can be labeled 1, we must label k vertices in A with label 1 so that each vertex in X can be adjacent to at least one vertex in A labeled 1.
- The k vertices in A that are labeled 1 will correspond to a k-vertex dominating set S for G.

$\square$

Hence, from Theorem 1 and Lemma 2, we arrive at the following theorem.

Theorem 2. *SD problem is NP-complete on split graphs.*

4 W[1]-Hard Parameterized by Feedback Vertex Set Number

In this section, we study the parameterized complexity of the SD problem parameterized by feedback vertex set number. We provide a parameterized reduction from a well known W[1]-hard problem, MULTIDIMENSIONAL RELAXED SUBSET SUM (MRSS). The MRSS problem is defined as follows.

> **Problem.** MULTIDIMENSIONAL RELAXED SUBSET SUM
> **Input.** An integer k, a set $S = \{s_1, s_2, ..., s_n\}$ of vectors with $s_i \in \mathbb{N}^k$ for every i with $1 \leq i \leq n$, a target vector $t \in \mathbb{N}^k$ and an integer m.
> **Parameter.** $k + m$
> **Output.** Does there exists a subset $S' \subseteq S$ with $|S'| \leq m$ such that $\sum_{s \in S'} s \geq t$?

The parameterized complexity of the MRSS problem has been established in the literature as follows.

Theorem 3 ([6]). *MRSS problem is W[1]-hard parameterized by $k + m$, even when all integers in the input are in unary.*

Construction. Consider an instance $I = (k, m, S, t)$ of the MRSS problem. Let $\max(s)$ denote the value of the largest coordinate of a vector s. We construct an instance $I' = (G, k')$ of the SD problem in the following way.

- k vertices $\{u_1, u_2, ..., u_k\}$ are created in G.
- A new set of vectors S' is constructed from S as follows:
  - For each vector $s \in S$, each coordinate of s is increased by 1 and the resulting vector is included in S'.
  - $\max(t)$ copies of the vector $(2, 2)$ are added to S'.
- The vector t' is obtained from t by adding m to each of its coordinates.
- For each $j \in [k]$, a set P_j is introduced with $\left(\sum\limits_{s_i \in S'} s_i(j) \right) - 2t'(j)$ vertices.
- For each $j \in [k]$, the vertex u_j is made adjacent to every vertex in the set P_j.
- For each $s_i \in S'$, $\max(s_i)$ paths of length two are created. For $l \in [\max(s_i)]$, the vertices of these paths are denoted by a_i^l, b_i^l and c_i^l, with b_i^l adjacent to both a_i^l and c_i^l. A vertex d_i is then created and made adjacent to every vertex in $\bigcup\limits_{l \in [\max(s_i)]} c_i^l$.
- Let $A_{s_i} = \bigcup\limits_{l \in [\max(s_i)]} a_i^l$, $B_{s_i} = \bigcup\limits_{l \in [\max(s_i)]} b_i^l$ and $C_{s_i} = \bigcup\limits_{l \in [\max(s_i)]} c_i^l$.
- For each $s_i \in S'$ and $j \in [k]$, the vertex u_j is made adjacent to exactly $s_i(j)$ vertices in A_{s_i} arbitrarily.

This concludes the construction of the reduced instance G. See Fig. 2 for an illustration.

The purpose of adding $\max(t)$ copies of the vector $(2, 2)$ to S' is to guarantee that, for each $j \in [k]$, the value of $\left(\sum\limits_{s_i \in S'} s_i(j) \right) - 2t'(j)$ is positive. This ensures that the set P_j, for each $j \in [k]$, contains at least one vertex.

We set $k' = \sum\limits_{j \in [k]} \left(\sum\limits_{s_i \in S'} s_i(j) - 2t'(j) \right) + k + \sum\limits_{s_i \in S'} (\max(s_i) - 1) + 2m$ and the set corresponding to feedback vertex set number is $\bigcup\limits_{j \in [k]} \{u_j\}$ of size k.

Lemma 3. *(k, m, S, t) is a yes instance of the MRSS problem if and only if (k, m, S', t') is a yes instance of the MRSS problem.*

Proof. By construction, the vectors in S' obtained from S have all coordinates at least 2. Consequently, the additional $\max(t)$ copies of vector $(2, 2)$ introduced in the construction of S' do not affect any feasible solution. Moreover, since each coordinate of the vectors in S is increased by one, the target vector t must be adjusted accordingly by increasing each of its coordinates by m, yielding t'. Therefore, we conclude that (k, m, S, t) is a yes instance of the MRSS problem if and only if (k, m, S', t') is a yes instance of the MRSS problem. $\square$

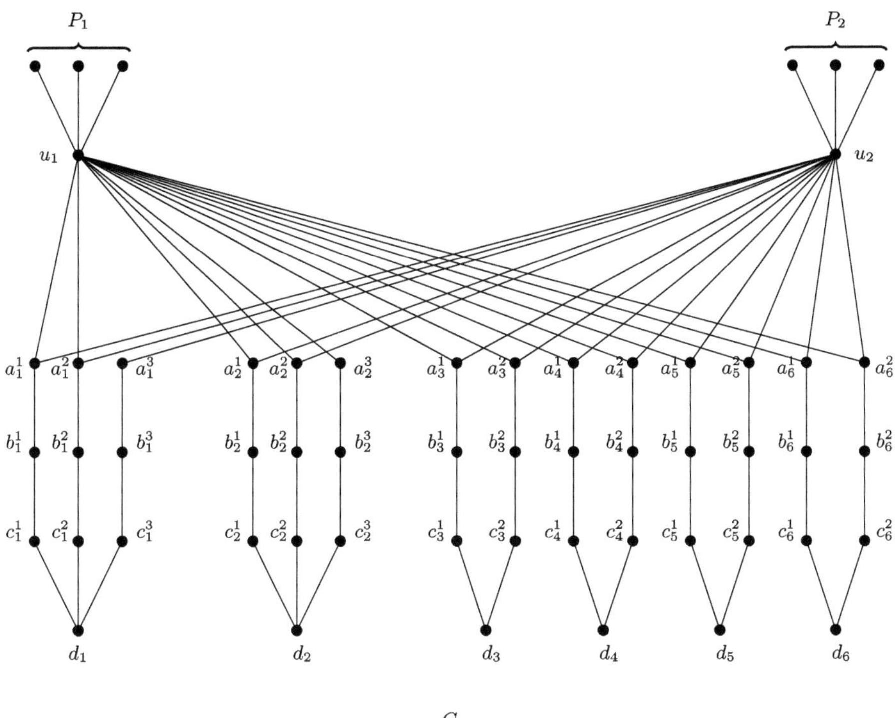

G

Fig. 2. Reduced instance of the SD problem constructed from the MRSS problem instance $S = \{(1,2),(2,1),(1,1)\}$, $t = (3,3)$, $k = 2$ and $m = 2$. Here, $S' = \{(2,3),(3,2),(2,2),(2,2),(2,2),(2,2)\}$ and $t' = (5,5)$.

From Lemma 3, it follows that (k, m, S, t) and (k, m, S', t') are equivalent instances of the MRSS problem. Therefore, for the rest of this section, we consider the instance (k, m, S', t') of the MRSS problem.

Lemma 4. *If (k, m, S', t') is a yes instance of the MRSS problem then there exists a signed dominating function f of weight at most k'.*

Proof. Let $R \subseteq S'$ such that $|R| \leq m$ and $\sum_{s_i \in R} s_i \geq t'$. The corresponding signed dominating function f is given as follows.

$$f(w) = \begin{cases} -1, & \text{if } w \in \bigcup_{s_i \in R} B_{s_i} \cup \bigcup_{s_i \notin R} (A_{s_i} \cup \{d_i\}), \\ 1, & \text{if } w \in \bigcup_{s_i \notin R} (B_{s_i} \cup C_{s_i}) \cup \bigcup_{s_i \in R} (A_{s_i} \cup C_{s_i} \cup \{d_i\}) \cup \bigcup_{j \in [k]} (P_j \cup \{u_j\}) \end{cases}$$

- For each $j \in [k]$, vertex u_j has label 1 and each vertex in P_j has label 1. This results in each vertex in P_j having a labelSum of exactly two.
- For each $j \in [k]$, the negative weight from the neighbours of u_j in $\bigcup_{s_i \in S} A_{s_i}$ is at most the positive weight from P_j. As u_j itself has label 1, the labelSum of u_j remains positive.

– For $s_i \notin S$, and for each $l \in [\max(s_i)]$, the vertex a_i^l has label -1, while b_i^l and c_i^l have label 1. The vertex d_i has label -1. Since all the neighbours of a_i^l, namely b_i^l and the vertices in $\bigcup_{j \in [k]} u_j$ have label 1, the labelSum of a_i^l is positive. The labelSum of b_i^l is one, as its neighbours a_i^l and c_i^l have labels -1 and 1, respectively. Similarly, the labelSum of c_i^l is one, since its neighbours b_i^l and d_i have labels 1 and -1, respectively. Finally, the labelSum of d_i is positive, as all of its neighbours $\bigcup_{l \in [\max(s_i)]} c_i^l$, have label 1.

– For $s_i \in S$, and for each $l \in [\max(s_i)]$, the vertices a_i^l and c_i^l have label 1, while b_i^l has label -1. The vertex d_i has label 1. Since all the neighbours of a_i^l in $\bigcup_{j \in [k]} u_j$ have label 1, the labelSum of a_i^l is positive. The labelSum of b_i^l is one, as both of its neighbours a_i^l and c_i^l have label 1. Similarly, the labelSum of c_i^l is one, since its neighbours b_i^l and d_i have labels -1 and 1, respectively. Finally, the labelSum of d_i is positive, as all of its neighbours $\bigcup_{l \in [\max(s_i)]} c_i^l$, have label 1.

The weight of $\bigcup_{j \in [k]} u_j$ is k, that of $\bigcup_{j \in [k]} P_j$ is $\sum_{j \in [k]} (\sum_{s_i \in S'} s_i(j) - 2t'(j))$ and that of $\bigcup_{i \in [n]} (A_{s_i} \cup B_{s_i} \cup C_{s_i} \cup \{d_i\})$ is $\max(s_i) - 1$ if $s_i \notin R$, while it is $\max(s_i) + 1$ if $s_i \in R$. Since each vertex in G has a positive labelSum, we conclude that f is a signed dominating function of weight at most $k' = \sum_{j \in [k]} (\sum_{s_i \in S'} s_i(j) - 2t'(j)) + k + \sum_{s_i \in S'} (\max(s_i) - 1) + 2m.$ $\qquad\square$

Lemma 5. *Let f be a minimum weighed signed dominating function. For any $s_i \in S'$, the weight of $A_{s_i} \cup B_{s_i} \cup C_{s_i} \cup \{d_i\}$ is*

1. $\max(s_i) - 1$ *in f, if for each $l \in [\max(s_i)]$, a_i^l is assigned label -1;*
2. $\max(s_i) + 1$ *in f, if for some $l \in [\max(s_i)]$, a_i^l is assigned label 1.*

Proof. We distinguish two cases according to the labels of the vertices in A_{s_i}.

1. Suppose $f(a_i^l) = -1$ for all $l \in [\max(s_i)]\}$. For b_i^l to have a positive labelSum, we must have $f(b_i^l) + f(c_i^l) \geq 2$ and hence $f(b_i^l) = f(c_i^l) = 1$. To obtain the minimum weight, the vertex d_i is assigned label -1. This yields a weight of $\max(s_i) - 1$ for $A_{s_i} \cup B_{s_i} \cup C_{s_i} \cup \{d_i\}$.
2. Now suppose that $f(a_i^l) = 1$ for some $l \in [\max(s_i)]$. In this case, to ensure that the vertex c_i^l has a positive labelSum, at most one of the vertices b_i^l, c_i^l, and d_i may be labeled -1, with the remaining two are assigned label 1. This increases the weight of $A_{s_i} \cup B_{s_i} \cup C_{s_i} \cup \{d_i\}$ by at least two compared to the previous case. Therefore, the weight of $A_{s_i} \cup B_{s_i} \cup C_{s_i} \cup \{d_i\}$ is at least $\max(s_i) + 1$.

$\qquad\square$

Lemma 6. *If there exists a signed dominating function f of weight at most k' then (k, m, S', t') is a yes instance of the MRSS problem.*

Proof. Let f be a signed dominating function of weight $k' = \sum\limits_{j \in [k]} (\sum\limits_{s_i \in S'} s_i(j) - 2t'(j)) + k + \sum\limits_{s_i \in S'} (\max(s_i) - 1) + 2m$.

- By Lemma 1, for each $j \in [k]$, we fix the label of u_j as 1 and assign label 1 to every vertex in P_j.
- After assigning labels to the vertices in $\bigcup\limits_{j \in [k]} (\{u_j\} \cup P_j)$, the vertex u_j has labelSum $\sum\limits_{s_i \in S'} s_i(j) - 2t'(j)$. For u_j to have a positive labelSum, we must assign label 1 to some of its neighbours in $\bigcup\limits_{s_i \in S'} A_{s_i}$. The labelSum of its neighbours in $\bigcup\limits_{s_i \in S'} A_{s_i}$ must be at least $2t'(j) - \sum\limits_{s_i \in S'} s_i(j)$.
- At this point, the remaining available weight for the set $\bigcup\limits_{s_i \in S'} (A_{s_i} \cup B_{s_i} \cup C_{s_i} \cup \{d_i\})$ is at most $\sum\limits_{s_i \in S'} (\max(s_i) - 1) + 2m$. By Lemma 5, for any $s_i \in S'$, if all the vertices in A_{s_i} are labeled -1, then the minimum weight of $A_{s_i} \cup B_{s_i} \cup C_{s_i} \cup \{d_i\}$ is $\max(s_i) - 1$. If instead, a subset of vertices in A_{s_i} are labeled 1, then the minimum weight increases to $\max(s_i) + 1$, requiring an additional weight of two for each such set.
- Based on the remaining weight $\sum\limits_{s_i \in S'} (\max(s_i) - 1) + 2m$ for $\bigcup\limits_{s_i \in S} (A_{s_i} \cup B_{s_i} \cup C_{s_i} \cup \{d_i\})$, we can afford to assign label 1 to all vertices in A_{s_i} for at most m such sets.
- Hence, if there exist at most m sets from $\bigcup\limits_{s_i \in S'} A_{s_i}$ such that assigning label 1 to every vertex in each A_{s_i} results in a positive labelSum for every vertex in $\bigcup\limits_{j \in [k]} u_j$, then there exist at most m vectors in S' whose sum is at least t'.

$\square$

From Theorem 3 and Lemmas 4 and 6, we obtain the following theorem.

Theorem 4. SIGNED DOMINATION *problem is W[1]-hard parameterized by feedback vertex set number.*

The parameters treewidth and clique-width are bounded by a function of the parameter feedback vertex set number. With this direct relation between the parameters, we have the following theorem.

Theorem 5. SIGNED DOMINATION *problem is W[1]-hard parameterized by treewidth or clique-width.*

5 FPT Parameterized by Neighbourhood Diversity

From Theorem 5, we have that the SD problem is W[1]-hard parameterized by clique-width. In this section, we investigate its parameterized complexity with respect to more restrictive parameter, neighbourhood diversity, and present an FPT algorithm.

For a graph $G = (V, E)$, Lampis [11] showed that the neighbourhood diversity can be computed in polynomial-time. For the rest of this section, we assume that a partition of V into t sets $V_1, V_2, ..., V_t$ is given to us. By the definition of neighbourhood diversity, each set V_i in the partition is either a clique or an independent set.

To establish fixed-parameter tractability, we transform the problem into an instance of Integer Linear Programming (ILP) problem, which is known to be FPT in the number of variables. The existing result related to the ILP problem is given as follows.

Theorem 6 ([5]). *The p-variable Integer Linear Programming problem can be solved using $\mathcal{O}(p^{2.5p+o(p)} \cdot L \cdot \log(MN))$ arithmetic operations and space polynomial in L, where L is the number of bits in the input, N is the maximum absolute value any variable can take, and M is an upper bound on the absolute value of the minimum taken by the objective function.*

We now present the details for constructing an ILP formulation for the SD problem parameterized by neighborhood diversity.

- Let $V_1, V_2, ..., V_t$ denote a partition of the vertex set V into corresponding sets. We use C to denote the union of cliques and I to denote the union of independent sets among $V_1, V_2, ..., V_t$.
- Let f be a minimum weighted signed dominating function. For each partition V_i, we define two boolean variables a_i and b_i, where $a_i = 1$ indicates that there exists a vertex in V_i with label -1 under f and $b_i = 1$ indicates that there exists a vertex in V_i with label 1 under f.
- We guess, for each partition V_i, whether there exists a vertex with label -1. If such a vertex exists, we set $a_i = 1$. Similarly, we guess whether there exists a vertex with label 1. If such a vertex exists, we set $b_i = 1$. If $a_i = 0$ and $b_i = 0$ for any partition, we reject the guess.
- For each partition V_i, we define a variable x_i representing the sum of the labels of all vertices in the partition. For some partitions, the values of x_i do not form a continuous range. To model the corresponding lower and upper bounds as a continuous range in the ILP formulation, we introduce an auxiliary variable y_i for such partitions.

The ILP formulation for the SD problem is given as follows.

$$\text{Minimize} \quad \sum_{i:a_i=1,b_i=1,|V_i|\equiv 0 \ (\mathrm{mod}\ 2)} 2y_i + \sum_{i:a_i=1,b_i=1,|V_i|\equiv 1 \ (\mathrm{mod}\ 2)} (2y_i+1)$$

Subject to

1. $x_i + A \geq 1$ for each $V_i \in C$,
2. $-1 + A \geq 1$ for each $V_i \in I$ with $a_i = 1$ and $b_i = 0$,
3. $1 + A \geq 1$ for each $V_i \in I$ with $a_i = 0$ and $b_i = 1$,
4. $\left\lfloor \frac{-|V_i|}{2} \right\rfloor + 1 \leq y_i \leq \left\lfloor \frac{|V_i|}{2} \right\rfloor - 1$, for each V_i with $a_i = 1$ and $b_i = 1$,

$$\text{Where} \quad A = -\sum_{j:V_j \in N(V_i),a_j=1,b_j=0} |V_j| + \sum_{j:V_j \in N(V_i),a_j=0,b_j=1} |V_j| + \sum_{j:V_j \in N(V_i),a_j=1,b_j=1,|V_j|\equiv 0 \ (\mathrm{mod}\ 2)} 2y_j + \sum_{j:V_j \in N(V_i),a_j=1,b_j=1,|V_j|\equiv 1 \ (\mathrm{mod}\ 2)} (2y_j+1)$$

We now discuss the constraints given in the ILP formulation.

1. For each partition V_i that is a clique, the sum of the labels of vertices in V_i together with the sum of the labels of vertices in its neighbouring partitions must be at least one.

2. For each partition V_i that is an independent set, if $a_i = 1$ and $b_i = 0$, then there exists a vertex in V_i with label -1 and no vertex with label 1. To ensure a positive labelSum for all the vertices in V_i, -1 plus the sum of the labels of the vertices in its neighboring partitions must be at least one.

3. For each partition V_i that is an independent set, if $a_i = 0$ and $b_i = 1$, then there exists a vertex in V_i with label 1 and no vertex with label -1. To guarantee a positive labelSum for all the vertices in V_i, the value 1 plus the sum of the labels of the vertices in its neighbouring partitions must be at least one.

4. For each partition V_i with $a_i = 1$ and $b_i = 1$, the value of x_i is minimum if there are $|V_i| - 1$ vertices with label -1 and one vertex with label 1. Therefore, the lower bound for x_i is $-|V_i| + 2$. The value of x_i is maximum if there are $|V_i| - 1$ vertices with label 1 and one vertex with label -1. Hence, the upper bound for x_i is $|V_i| - 2$.

    If $|V_i|$ is even then x_i can only take even values between $-|V_i|+2$ and $|V_i|-2$. Thus, we write $x_i = 2y_i$, where $y_i \in \left[\left\lfloor \frac{-|V_i|}{2} \right\rfloor + 1, \left\lfloor \frac{|V_i|}{2} \right\rfloor - 1 \right]$. Whereas if $|V_i|$ is odd then x_i can only take odd values between $-|V_i| + 2$ and $|V_i| - 2$. Thus, we write $x_i = 2y_i + 1$, where $y_i \in \left[\left\lfloor \frac{-|V_i|}{2} \right\rfloor + 1, \left\lfloor \frac{|V_i|}{2} \right\rfloor - 1 \right]$.

    Next, we describe how labels are assigned to the vertices in each partition V_i, based on the guesses for a_i and b_i.

- If $a_i = 0$ and $b_i = 1$, we assign label 1 to every vertex in V_i.
- If $a_i = 1$ and $b_i = 0$, we assign label -1 to every vertex in V_i.

For all other partitions, labels are assigned based on the outcome of the ILP formulation.

- If $a_i = 1$ and $b_i = 1$, we assign labels $\{-1, 1\}$ to the vertices in V_i such that the weight of V_i is exactly x_i and there must exist two vertices u and v in V_i such that $f(u) = -1$ and $f(v) = 1$.

In this way, we label all vertices of the partitions $V_1, V_2, ..., V_t$.

In our ILP formulation, we have t variables. The values of all the variables and the objective function are upper bounded by n. By Theorem 6, the problem can thus be solved in $\mathcal{O}^*(4^t \cdot t^{\mathcal{O}(t)})$ time.

Hence, we arrive at the following theorem.

Theorem 7. *SD problem is FPT parameterized by neighbourhood diversity.*

6   Conclusion

In this paper, we studied the complexity aspects of the SD problem. We proved that the problem remains NP-complete even when restricted to split graphs. We then showed that the problem is W[1]-hard parameterized by feedback vertex set number. Finally, we presented an FPT algorithm parameterized by neighbourhood diversity.
We conclude the paper with the following open questions on the SD problem.

1. What is the complexity of the problem on interval graphs, strongly chordal graphs, and block graphs?
2. What is the parameterized complexity of the problem parameterized by twin cover number, distance to disjoint paths, distance to cluster, and distance to co-cluster?
3. Does there exist polynomial kernels for the problem parameterized by vertex cover number, distance to clique, and max-leaf number?

References

1. Cygan, M., Fomin, F.V., Kowalik, L, Lokshtanov, D., Marx, D., Pilipczuk, M., Pilipczuk, M., Saurabh, S.: Lower bounds for kernelization. In: Parameterized Algorithms, pp. 523–555. Springer, Cham (2015). https://doi.org/10.1007/978-3-319-21275-3_15
2. Damaschke, P.: Minus domination in small-degree graphs. Discret. Appl. Math. **108**(1–2), 53–64 (2001)
3. Dunbar, J., Hedetniemi, S., Henning, M., Slater, P.: Signed domination in graphs. graph theory, combinatorics, and applications. John Wiley & Sons, Inc. 1, 311–322 (1995)
4. Favaron, O.: Signed domination in regular graphs. Discret. Math. **158**(1–3), 287–293 (1996)
5. Fellows, M.R., Lokshtanov, D., Misra, N., Rosamond, F.A., Saurabh, S.: Graph Layout Problems Parameterized by Vertex Cover. In: Hong, S.-H., Nagamochi, H., Fukunaga, T. (eds.) ISAAC 2008. LNCS, vol. 5369, pp. 294–305. Springer, Heidelberg (2008). https://doi.org/10.1007/978-3-540-92182-0_28

6. Ganian, R., Klute, F., Ordyniak, S.: On structural parameterizations of the bounded-degree vertex deletion problem. Algorithmica **83**(1), 297–336 (2021)
7. Hattingh, J.H., Henning, M.A., Slater, P.J.: The algorithmic complexity of signed domination in graphs. Australas. J. Comb. **12**, 101–112 (1995)
8. Henning, M.A., Slater, P.J.: Inequalities relating domination parameters in cubic graphs. Discret. Math. **158**(1–3), 87–98 (1996)
9. Karp, R.M.: Reducibility among Combinatorial Problems. Springer, US, Boston, MA (1972)
10. Kikuno, T., Yoshida, N., Kakuda, Y.: The np-completeness of the dominating set problem in cubic planer graphs. IEICE Trans. **6**, 443–444 (1980)
11. Lampis, M.: Algorithmic meta-theorems for restrictions of treewidth. Algorithmica **64**, 37–64 (2012)
12. Lee, C.M., Chang, M.S.: Variations of y-dominating functions on graphs. Discret. Math. **308**(18), 4185–4204 (2008)
13. Lin, J.-Y., Poon, S.-H.: Algorithms and Hardness for Signed Domination. In: Jain, R., Jain, S., Stephan, F. (eds.) TAMC 2015. LNCS, vol. 9076, pp. 453–464. Springer, Cham (2015). https://doi.org/10.1007/978-3-319-17142-5_38
14. West, D.B.: Introduction to Graph Theory, 2nd edn. Pearson, Chennai (2015)
15. Zhang, Z., Xu, B., Li, Y., Liu, L.: A note on the lower bounds of signed domination number of a graph. Discret. Math. **195**(1–3), 295–298 (1999)
16. Zheng, Y., Wang, J., Feng, Q.: Kernelization and Lower Bounds of the Signed Domination Problem. In: Fellows, M., Tan, X., Zhu, B. (eds.) AAIM/FAW -2013. LNCS, vol. 7924, pp. 261–271. Springer, Heidelberg (2013). https://doi.org/10.1007/978-3-642-38756-2_27

Improved Bounds on Proper Conflict-Free Coloring of Graphs

Ankit Sharma[1(✉)], Kaustav Paul[2], and Arti Pandey[1]

[1] Department of Mathematics, Indian Institute of Technology Ropar, Nangal Road,
Rupnagar 140001, Punjab, India
{ankit.23maz0013,arti}@iitrpr.ac.in
[2] Tel Aviv University, Tel Aviv, Israel

Abstract. Let $G = (V, E)$ be a simple connected graph with at least two vertices. A proper vertex coloring of G is called a *proper conflict-free coloring* if for every vertex $v \in V$, there exists a color that appears exactly once in the open neighborhood of v. The minimum number of colors required for such a coloring is called the *proper conflict-free chromatic number* of G and is denoted by $\chi_{pcf}(G)$.

Proper conflict-free coloring was introduced by Fabrici et al. Later, Caro et al. investigated how large $\chi_{pcf}(G)$ can be in comparison to the chromatic number $\chi(G)$. They showed that the ratio $\chi_{pcf}(G)/\chi(G)$ can be arbitrarily large for general graphs, while it remains bounded for certain restricted graph classes. They also conjectured that every connected graph G with maximum degree $\Delta \geq 3$ satisfies $\chi_{pcf}(G) \leq \Delta + 1$.

In this paper, we show that the ratio $\chi_{pcf}(G)/\chi(G)$ is bounded by a constant for AT-free graphs, P_4-sparse graphs, and split graphs. We further improve the upper bounds on $\chi_{pcf}(G)$ for several subclasses of AT-free graphs, including interval graphs, proper interval graphs, and permutation graphs. Additionally, we strengthen previously known bounds for minor-closed k-planar graphs and for K_t-minor-free graphs. In most of the cases, we prove that these bounds are tight. We also prove the validity of the $\Delta + 1$ conjecture for some of these graph classes.

Keywords: Proper Conflict-free Coloring · Conflict-free Coloring · Interval Graph · AT-free Graph · P_4-sparse Graph · Split Graph

1 Introduction

In this paper, we consider only finite, simple, and connected graphs with at least two vertices. For a graph $G = (V, E)$, V and E denote its vertex set and edge set, respectively. We denote the set $\{1, 2, ..., n\}$ by $[n]$. The open neighborhood of a vertex $v \in V$ for a graph $G = (V, E)$ is defined as $N(v) = \{u \in V : uv \in E\}$. The proper coloring of a graph $G = (V, E)$ refers to a coloring of the vertices of G such that no two adjacent vertices of G get the same color. The minimum number of colors required for proper coloring of a graph G is called the chromatic number of G and is denoted by $\chi(G)$. The square of G is the graph obtained by adding

© The Author(s), under exclusive license to Springer Nature Switzerland AG 2026
F. Foucaud and A. Parreau (Eds.): IWOCA 2026, LNCS 16587, pp. 500–514, 2026.
https://doi.org/10.1007/978-3-032-27732-9_35

an edge between every pair of vertices whose distance in G is at most 2. For other graph theoretic definitions and notations, we refer to [23]. The definitions required in the specific sections are mentioned in the beginning of the sections.

A *proper conflict-free coloring* (PCF coloring) of a graph G is a proper coloring such that every vertex has a color appearing exactly once in its open neighborhood. The minimum number of colors required for this coloring is called the *proper conflict-free chromatic number* of G and is denoted by $\chi_{pcf}(G)$. PCF m-coloring denotes a proper conflict-free coloring of a graph using m colors. Proper conflict-free coloring is motivated by applications in frequency assignment and wireless networks, where each vertex requires a uniquely identifiable signal among its neighbors while avoiding interference. It can be seen as the relaxed version of square coloring, where any two vertices at distance at most 2 receive distinct colors, equivalent to proper coloring of the square of the graph. See [17], for a survey on the distance-coloring of graphs. The graph hierarchy containing some important graph classes is shown in Fig. 1. A downward arrow from a graph class A to graph class B shows that $B \subseteq A$. The graph classes highlighted with blue are the graphs for which we give the bounds for $\chi_{pcf}(G)$.

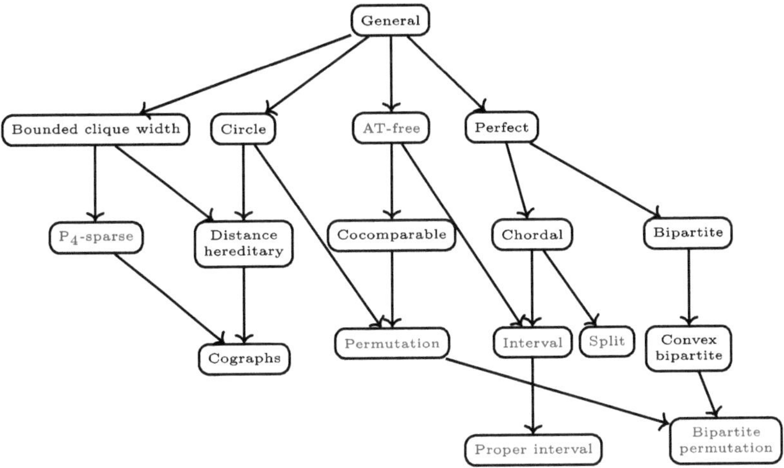

Fig. 1. Hierarchy of graph classes

The notion of a proper conflict-free coloring was introduced by Fabrici et al. in [12]. They showed that every planar graph admits a PCF 8-coloring and constructed a planar graph with no PCF 5-coloring. Later Caro, Petruševski, and Škrekovski [5] studied the notion of PCF coloring where they explored various combinatorial results. In particular, they showed $\chi_{pcf}(G) \leq 5\Delta(G)/2$ and characterize the equality. They also showed that the difference between $\chi_{pcf}(G)$ and the chromatic number $\chi(G)$ can be arbitrarily large; therefore, the ratio $\chi_{pcf}(G)/\chi(G)$ can also be arbitrarily large. In the same paper, they proposed several open questions, and in this paper we address the following ones:

Problem 1 (Caro et al. [5]). *Find other 'generic' graph families $\mathcal{G}$ for which there exists a constant $c = c(\mathcal{G})$ such that $\chi_{pcf}(G)/\chi(G) \leq c$ for every $G \in \mathcal{G}$.*

Caro et al. showed that the ratio $\chi_{pcf}(G)/\chi(G)$ is bounded by a constant for the claw-free graphs [5]. In this paper, we show that this ratio is bounded by a constant for AT-free graphs, P_4-sparse and split graphs. In particular, we show that $\chi_{pcf}(G) \leq \chi(G) + 3$ for AT-free graphs and $\chi_{pcf}(G) \leq \chi(G) + 1$ for split graphs. The class of Interval graphs, proper interval graphs, and permutation graphs are subclasses of AT-free graphs. Hence, the bound $\chi_{pcf}(G) \leq \chi(G) + 3$ holds for these graphs as well. We improve this bound for interval and proper interval graphs and show that $\chi_{pcf}(G) \leq \chi(G) + 2$ for interval graphs and $\chi_{pcf}(G) \leq \chi(G) + 1$ for proper interval graphs. In most cases, our bounds are tight. Some known results addressing this problem can be found in [16]. Caro et al. conjectured the following bound on $\chi_{pcf}(G)$ in terms of the maximum degree.

Conjecture 1 (Caro et al. [5]). *If G is a connected graph of maximum degree $\Delta \geq 3$, then $\chi_{pcf}(G) \leq \Delta + 1$.*

Caro et al. showed that $\chi_{pcf}(G) \leq 2\Delta + 1$ for claw-free and chordal graphs. A. Jiménez et al. improved the bound for claw-free graphs [16] and showed that $\chi_{pcf}(G) \leq \Delta + 6$. In this paper, we establish this conjecture for split graphs (which is a subclass of chordal graphs) and proper interval graphs. We also show that $\chi_{pcf}(G) \leq \Delta + \alpha$ for AT-free graphs and its subclasses, and P_4-sparse graphs, where α is a constant.

Some other relevant results on PCF coloring can be found in [2,6,8,18,19,22]. The main contributions of this paper are as follows:

- For split graphs, we prove that $\chi_{pcf}(G) \leq \chi(G) + 1$ and provide a necessary and sufficient condition for when $\chi_{pcf}(G) = \chi(G) + 1$. Using this characterization, we establish Conjecture 1 for split graphs and show that the ratio $\chi_{pcf}(G)/\chi(G)$ is bounded by a constant (Sect. 2).
- We improve existing upper bounds on the proper conflict-free chromatic number for minor-closed k-planar graphs (resp. K_t-minor free graphs) (Sect. 3).
- We derive new bounds for AT-free graphs and their subclasses, including permutation graphs, interval graphs, and proper interval graphs, showing that $\chi_{pcf}(G) \leq \chi(G) + 3$ in general and improving this bound for interval and proper interval graphs. Using the bound on $\chi_{pcf}(G)$ for proper interval graphs, we showed that Conjecture 1 is valid for the proper interval graphs (Sect. 4).
- We prove that $\chi_{pcf}(G) \leq \chi(G) + 2$ for P_4-sparse graphs and present a polynomial-time algorithm to compute $\chi_{pcf}(G)$ for this class (Sect. 5).
- We conclude with a discussion of the results and possible directions for future work (final section).

2   Split Graphs

Definition 1 (Split Graph). *A graph $G = (C \cup I, E)$ is called a split graph where the vertex set can be partitioned into two disjoint sets: a clique C (a* *complete subgraph where all vertices are adjacent) and an independent set I (a subgraph with no edges).*

In this section, we show that the ratio $\chi_{pcf}(G)/\chi(G)$ is bounded by a constant for the split graph. For this, we show that $\chi_{pcf}(G)$ is bounded by a linear function of $\chi(G)$. We take $C = \{c_1, c_2, ..., c_m\}$ and $I = \{i_1, i_2, ..., i_n\}$ where $|C| = m$ and $|I| = n$. In our proofs, we consider C as a maximal complete subgraph of G, *i.e.*, $\nexists\ v \in I$ such that $N(v) = C$. We first show that $\chi(G) \leq \chi_{pcf}(G) \leq \chi(G) + 1$. The proof is omitted due to space restrictions.

Theorem 1. *For a given split graph $G = (C \cup I, E)$, we have $\chi(G) \leq \chi_{pcf}(G) \leq \chi(G) + 1$.*

Next, we give a necessary and sufficient condition for $\chi_{pcf}(G) = |C| + 1 = \chi(G) + 1$. Let $X \subseteq C$. Define $I_X = \{i_j \in I : N(i_j) = X\}$. We denote $C - c_k$ as C^k where $c_k \in C$.

Theorem 2. *Let $G = (C \cup I, E)$ be a split graph such that $|C| \geq 3$. Then $\chi_{pcf}(G) = |C| + 1$ if and only if $|I_{C^k}| \geq 1$ for all $k \in [m]$ except possibly one, where $m = |C|$.*

Proof. Necessary part: Let $\chi_{pcf}(G) = |C| + 1$. Assume for the contradiction that $I_{C^k} = \emptyset$ for more than one vertex in C. Now we define a PCF coloring p of G using only m colors. Define $p : C \cup I \to [m]$ such that $p(c_j) = j$ for all $c_j \in C$. For every $i_j \in I$, each color is uniquely present in $N(i_j)$.

We now define p for the vertices of I so that for every $c_j \in C$, there is at least one color, uniquely present in $N(c_j)$. Let c_j and c_k be two vertices such that $I_{C^j} = \emptyset$ and $I_{C^k} = \emptyset$ (the existence is guaranteed by our assumption). First, we define the coloring p for the vertices of $N(c_j) \cap I$. Let $v \in N(c_j) \cap I$. There are two possible choices for v: $v \in N(c_k)$ or $v \notin N(c_k)$. If $v \in N(c_k)$, then there must exist a vertex $w \in C$ such that $w \notin N(v)$, otherwise C will not be the maximal clique. So in that case, color the vertex v with the color $p(w)$. In the other case, where $v \notin N(c_k)$, there exists a vertex $u \in C$ such that $u \notin N(v)$, otherwise $|I_{C^k}| \geq 1$. So, in this case, we give the color $p(u)$ to the vertex v. Thus, we have colored the vertices of $C \cup N(c_j)$ without using the color of c_k. Hence, $p(c_k)$ is a unique color in $N(c_j)$.

Now we define p for the vertices of $N(c_k) \cap I$ so that c_k has a unique color in $N(c_k)$. The vertices of $N(c_j) \cap N(c_k)$ got colored while coloring the neighborhood of c_j. Now, for the remaining vertices $v \in N(c_k)$, we have $v \notin N(c_j)$. As $I_{C^j} = \emptyset$, these vertices can be colored in a similar manner without using the color of c_j. So, $p(c_j)$ is the unique color in $N(c_k)$. Till now, we have not used the colors of c_j and c_k. For the remaining vertices of I, we color them with the color of c_j. So, for all other vertices of C, the color of c_k is unique as we have not used $p(c_k)$ in I. Thus, we can see that p is a PCF m-coloring of G, which is a contradiction. Hence, our assumption is wrong.

Sufficient part: Let $|I_{C^k}| \geq 1$ for all $c_k \in C$ except possibly one. Assume that there exists a PCF coloring $p : C \cup I \to [m]$. Now, we have two possible cases:

$|I_{C^k}| \geq 1$ for all $c_k \in C$ or $|I_{C^k}| \geq 1$ for all $c_k \in C$ except one. In the first case, we pick any arbitrary vertex $c_j \in C$ but in the latter case, we pick $c_j \in C$ such that $I_{C^j} = \emptyset$. Now, $|I_{C^k}| \geq 1$ for all $k \neq j$. Thus, every color except $p(c_j)$ will be present in $N(c_j)$ twice which contradicts our assumption that p is a PCF m-coloring of G. Hence, $\chi_{pcf}(G) = |C| + 1$. $\qquad\square$

In the next theorem, we give a necessary and sufficient condition for $\chi_{pcf}(G) = |C| + 1$, when $|C| = 2$. The proof is omitted due to space restrictions.

Theorem 3. *For a split graph $G = (C \cup I, E)$ with $|C| = 2$, $\chi_{pcf}(G) = |C| + 1$ if and only if $I \neq \emptyset$.*

Using the above theorems, we show that Conjecture 1 is true for split graphs.

Corollary 1. *For a given split graph $G = (C \cup I, E)$, we have $\chi_{pcf}(G) \leq \Delta + 1$.*

Proof. There are two cases here:

- Case 1: $I = \emptyset$. In this case, $\chi_{pcf}(G) = \chi(G)$ (using Theorem 2) and $\chi(G) \leq \Delta + 1$. This gives us that $\chi_{pcf}(G) \leq \Delta + 1$.
- Case 2: $I \neq \emptyset$. In this case, $\chi(G) \leq \Delta$ and thus, using Theorem 1, we have $\chi_{pcf}(G) \leq \Delta + 1$.

$\qquad\square$

3 K_t-Minor Free and Minor Closed k-Planar Graphs

In this section, we strengthen some known upper bounds on the proper conflict-free chromatic number for specific classes of graphs. We begin by recalling several definitions and known results that will be used throughout this section.

Definition 2 (Odd Tree (and Forest)). *An odd tree is a tree in which every vertex has odd degree. An odd forest is a forest whose connected components are odd trees.*

Definition 3 (Odd Forest Chromatic Number). *Let $G = (V, E)$ be a graph. The odd forest chromatic number of G, denoted by $\chi_{odF}(G)$, is the minimum integer t such that the vertex set V can be partitioned into t color classes $V_1, \ldots, V_t$ with the property that each induced subgraph $G[V_i]$ is an odd forest.*

Definition 4 (H-minor Free Graph). *A graph H is called a minor of a graph G if H can be obtained from G by a sequence of vertex deletions, edge deletions, and edge contractions. If G does not contain H as a minor, then G is said to be H-minor-free. A class of graphs $\mathcal{G}$ is said to be minor-closed if for every graph $G \in \mathcal{G}$, every minor of G also belongs to $\mathcal{G}$.*

Definition 5 (k-planar Graph). *Let $k \geq 0$ be an integer. A graph G is said to be k-planar if G admits a drawing in the plane such that each edge is crossed by at most k other edges.*

A class of graphs $\mathcal{G}$ is said to be *closed under pendant vertex addition* if for every graph $G \in \mathcal{G}$, the graph obtained from G by adding a new vertex and joining it by a single edge to an arbitrary vertex of G also belongs to $\mathcal{G}$.

By Theorem 1 of [13], every connected graph $G = (V, E)$ of even order admits a partition $\Omega = (V_1, \ldots, V_k)$ of V such that each induced subgraph $G[V_i]$ is an odd tree. Given such a partition Ω, we define an auxiliary graph $H_\Omega(G)$ as follows: $V(H_\Omega(G)) = \{v_1, \ldots, v_k\}$ and two vertices v_i and v_j are adjacent in $H_\Omega(G)$ if and only if there exists at least one edge of G with one endpoint in V_i and the other in V_j.

The relationship between $\chi_{odF}(G)$ and the chromatic number of $H_\Omega(G)$ is captured by the following theorem.

Theorem 4 ([1]). *Let $G = (V, E)$ be a connected graph of even order. Then $\chi_{odF}(G) \leq \chi(H_\Omega(G))$.*

Theorem 5. *Given a minor closed class of graphs $\mathcal{G}$, which is also closed under pendant vertex addition: $\chi_{pcf}(G) \leq 3\chi(\mathcal{G}) + 1$, for every $G \in \mathcal{G}$, where $\chi(\mathcal{G}) = max\{\chi(G) \mid G \in \mathcal{G}\}$.*

Proof. Let $G = (V, E)$ be a connected graph in $\mathcal{G}$. We split the proof in the two following cases:

Case 1: $|V|$ is even.

Let $(V_1, \ldots, V_k)$ be an optimal odd forest partition of G. Since every $G[V_i]$ is an odd forest, a proper conflict-free coloring $f_i : V_i \to \{3i - 2, 3i - 1, 3i\}$ can be defined for every $i \in [k]$. So, we define a coloring $f : V \to [3k]$ on G, as follows: $f(v) = f_i(v)$, if $v \in V_i$. It is easy to observe that f is also a conflict-free coloring. Hence, $\chi_{pcf}(G) \leq 3k = 3\chi_{odF}(G) \leq 3\chi(H_\Omega(G))$. Since $H_\Omega(G)$ is a minor of G,

$$\chi_{pcf}(G) \leq 3\chi(H_\Omega(G)) \leq 3\chi(\mathcal{G})$$

Case 2: $|V|$ is odd.

Pick any vertex $v \in V$ and add a pendant vertex u_v to v to form a new graph G'. Let $(V_1, \ldots, V_k)$ be an optimal odd forest partition of G'. Since every $G[V_i]$ is an odd forest, a proper conflict-free coloring $f_i : V_i \to \{3i - 2, 3i - 1, 3i\}$ can be defined for every $i \in [k]$. Without loss of generality, let $v, u_v \in V_1$. So, we define a coloring $f : V \to [3k]$ on G, as follows: $f(v) = f_i(v)$, if $v \in V_i$. Observe that f is a proper coloring of G and every vertex x in $V \setminus \{v\}$ has at least one unique color in its neighbourhood. Let v has no unique color in its neighbourhood (with respect to the coloring f). Pick an arbitrary neighbour $x_v \subset N_G(v)$, so we redefine the coloring f as $f' : V \to [3k + 1]$, which is defined as follows: $f'(x) = f(x)$ for every $x \in V \setminus \{x_v\}$ and $f'(x_v) = 3k + 1$. So, f' is a proper conflict-free coloring of G, hence $\chi_{pcf}(G) \leq 3k + 1 \leq 3\chi_{odF}(G') + 1$.

Note that $G' \in \mathcal{G}$ and $H_\Omega(G) \in \mathcal{G}$, as $\mathcal{G}$ is closed under pendant vertex addition and minor. Hence,

$$\chi_{pcf}(G) \leq 3\chi(H_\Omega(G')) + 1 \leq 3\chi(\mathcal{G}) + 1$$

$\square$

Now let $\mathcal{G}_t$ = class of all connected K_t-minor free graph and $\mathcal{G}_k$ = class of all connected minor closed k-planar graph; $t, k \in \mathbb{N}$. Note that both $\mathcal{G}_t$ and $\mathcal{G}_k$ are minor closed graph classes which are also closed under pendant vertex addition. For K_t-minor free graphs and k-planar graphs, the following upper bounds (mentioned in Theorem 6 and 7) on proper conflict-free chromatic number were given in [14].

Theorem 6 ([14]). *For every K_t-minor free graph G, $\chi_{pcf}(G) \leq 5(t-1)(t-2) - 1$.*

We improve this $O(t^2)$ bound to $O(t \log \log t)$ by using Theorem 5. Since $\chi(\mathcal{G}_t) = O(t \log \log t)$ [9], by Theorem 5 we state the following corollary:

Corollary 1. *For a K_t-minor free graph G, $\chi_{pcf}(G) = O(t \log \log t)$.*

Theorem 7. ([14]). *For every k-planar graph G, $\chi_{pcf}(G) \leq 60k + 59$.*

We could not improve this $O(k)$ bound for the class of all k-planar graphs, however we improve this bound for minor closed k-planar graphs. In [21] (Theorem 3), it was shown that for any k-planar graph $G = (V, E)$, $|E| \leq 4.108\sqrt{k}\,|V|$.

It follows that G contains a vertex of degree at most $\lfloor 8.216\sqrt{k} \rfloor$. A graph is said to be *d-degenerate* if every subgraph of G has a vertex of degree at most d. Consequently, every k-planar graph is $\lfloor 8.216\sqrt{k} \rfloor$-degenerate. It is well known (see [10]) that any d-degenerate graph admits a proper vertex coloring with at most $d+1$ colors. Therefore, $\chi(G) \leq \lfloor 8.216\sqrt{k} \rfloor + 1$ for every k-planar graph G. Hence, by Theorem 5, the following corollary follows:

Corollary 2. *For a minor closed k-planar graph G, $\chi_{pcf}(G) \leq \lfloor 24.648\sqrt{k} \rfloor + 4$.*

Hence in this section, we improve the upper bound for proper conflict-free chromatic number of minor closed k-planar graphs (resp. K_t-minor free graphs) from $O(k)$ to $O(\sqrt{k})$ (resp. $O(t^2)$ to $O(t \log \log t)$).

4   AT-Free Graphs and Its Sub-Classes

In this section, we establish a tight upper bound on the proper conflict-free chromatic number of AT-free graphs. Subsequently, we refine this bound for a smaller subclass of AT-free graphs, such as interval and proper interval graphs.

4.1   AT-Free Graphs

In this subsection, we show that $\chi_{pcf}(G) \leq \chi(G) + 3$ for AT-free graphs. We begin with the definition of AT-free graphs and their properties which will be useful later in this section.

Definition 6 (AT-free Graph). *An asteroidal triple (AT) in a graph $G = (V, E)$ is an independent set of three vertices $\{u, v, w\} \subseteq V$ such that between every pair of them there exists a path avoiding the neighborhood of the third. A graph is said to be AT-free if it contains no asteroidal triple.*

This family of graphs is a generalization of several well-studied graph families, including *interval graphs, cocomparability graphs*, and *permutation graphs*.

Definition 7 (Dominating Shortest Path). *Given a graph $G = (V, E)$, a pair of vertices (x, y) is called a dominating pair, if the vertex set of any path between x and y in G is a dominating set of G. A dominating shortest path (or DSP) is a shortest path connecting x and y in G.*

We use the following property of AT-free graphs in our result.

Theorem 8 ([7]). *A dominating pair exists in every AT-free graph, and it can be computed in linear time.*

Now, we are ready to state the main result of this section.

Theorem 9. *For any connected AT-free graph $G = (V, E)$, $\chi_{pcf}(G) \leq \chi(G) + 3$.*

Proof. Due to space constraints, the detailed proof is omitted. The proof idea is as follows. Let $P = (x_1, \ldots, x_l)$ be a dominating shortest path of G, and let f be an optimal proper coloring of G. Define p by keeping the colors of f on $V \setminus P$ and assigning three new colors to vertices of P cyclically. Then p is a proper coloring, as adjacent vertices on P receive different colors and vertices in $V \setminus P$ retain proper coloring.

Using the properties of the dominating shortest path P, we can show that p is a proper conflict-free coloring of G. Hence, p is a PCF coloring using at most $\chi(G) + 3$ colors. $\qquad\square$

Using Brook's inequality $(\chi(G) \leq \Delta + 1)$ for the proper coloring of G, the following bound can be obtained from Theorem 9 in terms of the maximum degree Δ of G.

Corollary 3. *For any connected AT-free graph $G = (V, E)$, $\chi_{pcf}(G) \leq \Delta + 4$.*

A *permutation graph* is the intersection graph of a set of line segments drawn between two parallel lines. Equivalently, it can be defined by a permutation π on n elements, where two vertices u and v are adjacent if and only if $u < v$ and $\pi(u) > \pi(v)$. As the permutation graphs are AT-free, we have the following result.

Corollary 4. *For any permutation graph G, $\chi_{pcf}(G) \leq \chi(G) + 3$.*

Since for a bipartite graph with at least two vertices, $\chi(G) = 2$, we have the following result for bipartite permutation graphs.

Corollary 5. *For any bipartite permutation graph G, $\chi_{pcf}(G) \leq 5$.*

Recently, A. Jiménez et al. have proved that for every permutation graph G, $\chi_{pcf}(G) \leq 3\chi(G)$ (see Theorem 1.4 in [16]). The bound in Corollary 4 (resp. corollary 5) is an improvement over this bound for all permutation graphs (resp. bipartite permutation) with $|V| \geq 2$.

Now we show that the bound for the bipartite permutation graph is tight. This shows that the bound for AT-free and permutation graphs is tight.

Lemma 1. *The bound for AT-free graphs (resp Permutation and Bipartite permutation graphs) is tight.*

Proof. Consider the bipartite permutation graph G as shown in Fig. 2. The figure itself provides a PCF coloring of G using 5 colors, hence $\chi_{pcf}(G) \leq 5$. One can also observe that at least 5 colors are required for any PCF coloring of G. The detailed proof is omitted due to space constraints.

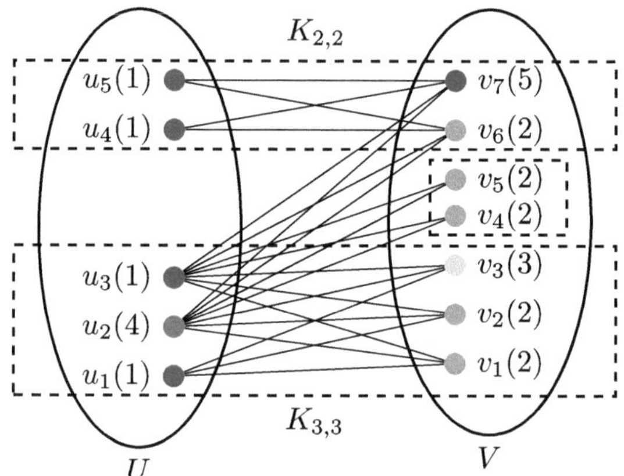

Fig. 2. Bipartite permutation graph $G = (U, V)$ with $\chi_{pcf}(G) = 5$.

$\square$

4.2   Interval Graph

Definition 8 (Interval Graph). *A graph $G = (V, E)$ is called an interval graph if each vertex can be represented by an interval on the real line, and two vertices are adjacent if and only if their intervals overlap.*

The class of interval graphs is a subclass of AT-free graphs. So, the bound in the above theorem holds for interval graphs as well. However, the interval graph has some important structural properties. One of the properties is *"Multi-chain ordering"*. Before defining this property, we present the necessary definitions.

Definition 9 (Chain Graph). *A bipartite graph $G = (U, V)$ is a chain graph if and only if for any two vertices $u_1, u_2 \in U$ either $N(u_1) \subseteq N(u_2)$ or $N(u_2) \subseteq N(u_1)$.*

Definition 10 (Multi-Chain Ordering [11]). *Let $G(V, E)$ be a graph, and $L_0, L_1, ..., L_p$ be a partition of the vertex set $V(G)$. We say that these layers form a multi-chain ordering of G if,*

- $|L_0| = 1;\ L_0 = \{v_0\}$

- $L_i = \{v \in V(G) : d(v, v_0) = i\}$, for all $i \in [1, p]$
- for every two consecutive layers L_i and L_{i+1}, where $i \in \{0, 1, ..., p-1\}$, the vertices in L_i and L_{i+1}, and the edges connecting these layers form a chain graph.

Theorem 10 ([11]). *Every connected interval graph admits a multi-chain ordering of G.*

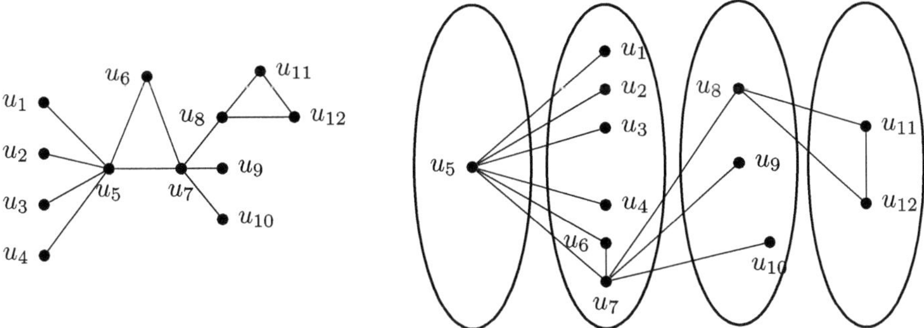

Fig. 3. An interval graph G (on the left) and a multi-chain ordering of G (on the right)

An example of an interval graph along with multi-chain ordering is given in Fig. 3. Using the multi-chain ordering of the interval graph, we find the better upper bound on $\chi_{pcf}(G)$. We begin with the following observations. The Observation 1 is same as the Observation 15 in [4].

Observation 1. *In every layer L_i, $i \in [0, p-1]$ there exist a vertex $v_{d_i} \in L_i$ such that $L_{i+1} \subseteq N(v_{d_i})$.*

Observation 2. *For $i \in [0, p-1]$, let $v_{d_i} \in L_i$ be such that $L_{i+1} \subseteq N(v_{d_i})$. Then $P = v_{d_0} v_{d_1} v_{d_2} ... v_{d_{p-1}}$ is a dominating path of the graph G with the property that it intersects with all cliques of maximum size in G.*

Using the above observations, we prove the following bound for interval graphs. The proof is omitted due to space constraints.

Theorem 11. *For a connected interval graph G, $\chi_{pcf}(G) \leq \chi(G) + 2$.*

The bound in terms of Δ is as follows:

Corollary 2. *For a connected interval graph G, $\chi_{pcf}(G) \leq \Delta + 3$.*

In the next subsection, we will improve this bound for proper interval graphs.

4.3   Proper Interval Graph

Definition 11 (Proper Interval Graph). *A graph $G = (V, E)$ is called a proper interval graph if there exists a family of closed intervals $\mathcal{I} = \{I_v : v \in V\}$ on the real line such that*

1. *for any two distinct vertices $u, v \in V$, $(u, v) \in E$ if and only if $I_u \cap I_v \neq \emptyset$, and*
2. *for any two distinct vertices $u, v \in V$, it is not the case that $I_u \subset I_v$ or $I_v \subset I_u$.*

In other words, proper interval graphs are precisely those interval graphs that admit an interval representation in which no interval is properly contained in another.

Definition 12 (Vertex Ordering Characterization [20]). *A graph $G = (V, E)$ is a proper interval graph if and only if there exists a linear ordering of its vertices $\sigma = (v_1, v_2, \ldots, v_n)$ such that for every triple of vertices v_i, v_j, v_k with $i < j < k$, if $(v_i, v_k) \in E$, then both $(v_i, v_j) \in E$ and $(v_j, v_k) \in E$. Such an ordering σ is referred to as a* proper interval ordering. *In this ordering, the neighbors of each vertex appear as a consecutive block.*

We use the above characterization of proper interval graphs to establish the following bound for $\chi_{pcf}(G)$ in terms of $\chi(G)$ (or $\omega(G)$). We also give the necessary and sufficient condition for $\chi_{pcf}(G) = \omega(G) + 1$. The proofs are omitted due to space constraints.

Theorem 12. *For every proper interval graph G, $\chi_{pcf}(G) \leq \omega(G) + 1$.*

Theorem 13. *For any proper interval graph G, $\chi_{pcf}(G) = \omega(G) + 1$, if and only if $\Delta(G) = 2\omega - 2$.*

Using the above two theorems, we have the following result.

Corollary 3. *For any proper interval graph G, we can find $\chi_{pcf}(G)$ in linear time.*

As $\chi(G) \leq \Delta$ for the interval graphs except the complete graph, using the above theorems, we have the following result.

Corollary 4. *For any proper interval graph G, $\chi_{pcf}(G) \leq \Delta + 1$.*

5 P_4-Sparse Graphs

In this section, we show that $\chi_{pcf}(G) \leq \chi(G) + 2$ for P_4-*sparse graphs* and give a linear-time algorithm for computing *PCF chromatic number* for P_4-sparse graphs.

Definition 13 (P_4-Sparse Graph). *A graph $G = (V, E)$ is said to be P_4-sparse if a subgraph induced on any 5 vertices of G contains at most one P_4 (path on 4 vertices).*

Definition 14 (Spider). *A spider is a graph $G = (V, E)$, where V admits a partition in three subsets S, C and R such that*

- *$C = \{c_1, \ldots, c_l\}$ ($l \geq 2$) is a clique.*
- *$S = \{s_1, \ldots, s_l\}$ is an independent set.*
- *Every vertex in R is adjacent to every vertex in C and nonadjacent to all vertex of S.*

A spider $G(S, C, R)$ is said to be a

- *thin spider if for every $i \in \{1, \ldots, l\}$, $N_C(s_i) = \{c_i\}$*
- *thick spider if for every $i \in \{1, \ldots, l\}$, $N_C(s_i) = C \setminus \{c_i\}$.*

An example of thick and thin spider can be seen in Fig. 4.

(a) Thin spider   (b) Thick spider

Fig. 4. Examples of spiders with spider partition (S, C, R)

More on P_4-sparse graphs and spider graphs can be found in [15]. The following theorem gives a characterization of P_4-sparse graphs.

Theorem 14 ([3]). *A graph G is said to be P_4-sparse if and only if one of the following conditions hold:*

- *G is a single vertex graph.*
- *G is a union of two P_4-sparse graphs G_1 and G_2.*
- *G is a join of two P_4-sparse graphs G_1 and G_2.*
- *G is a spider (thick or thin) that admits a spider partition (S, C, R) where $G[R]$ is a P_4-sparse graph or $R = \phi$.*

We first give bounds on $\chi_{pcf}(G)$ for P_4-sparse graphs; the proofs are omitted due to space constraints.

Theorem 15. *For any connected P_4-sparse graph G, we have $\chi(G) \leq \chi_{pcf}(G) \leq \chi(G) + 2$.*

The bound on $\chi_{pcf}(G)$ in terms of Δ for P_4-sparse graphs is as follows.

Corollary 5. *For any connected P_4-sparse graph G, we have $\chi_{pcf}(G) \leq \Delta + 3$.*

Now we give the algorithm to find the exact value of $\chi_{pcf}(G)$ in each case. We use the following two coloring parameters in the algorithm.

Definition 15 (PC1). *A proper coloring $f : V(G) \to [k]$ is called PC1 if there exists a color that appears only once in G. The smallest integer k for which such a coloring exists is denoted by $\chi'(G)$.*

Definition 16 (PCF1). *A proper conflict-free coloring $f : V(G) \to [k]$ is called PCF1 if there exists a color that appears only once in G. The smallest integer k for which such a coloring exists is denoted by $\chi'_{pcf}(G)$.*

Observation 3. *For any graph G, we have*

1. $\chi(G) \leq \chi'(G) \leq \chi(G) + 1$
2. $\chi_{pcf}(G) \leq \chi'_{pcf}(G) \leq \chi_{pcf}(G) + 1$

Due to space constraints, we omit the proofs of the following lemmas.

Lemma 2. *Let G be the disjoint union of two P_4-sparse graphs G_1 and G_2. Then*

1. $\chi(G) = max\{\chi(G_1), \chi(G_2)\}$.
2. $\chi'(G) = max\{\chi'(G_1), \chi'(G_2)\}$.
3. $\chi_{pcf}(G) = max\{\chi_{pcf}(G_1), \chi_{pcf}(G_2)\}$.
4. $\chi'_{pcf}(G) = max\{\chi'_{pcf}(G_1), \chi'_{pcf}(G_2)\}$.

Lemma 3. *Let G be the join of two P_4-sparse graphs G_1 and G_2. Then*

1. $\chi_{pcf}(G) = min\{\chi_{pcf}(G_1) + \chi_{pcf}(G_2),\ \chi'_{pcf}(G_1) + \chi(G_2),\ \chi(G_1) + \chi'_{pcf}(G_2),$
$$\chi'(G_1) + \chi'(G_2)\}.$$
2. $\chi'_{pcf}(G) = min\{\chi'_{pcf}(G_1) + \chi(G_2),\ \chi(G_1) + \chi'_{pcf}(G_2)\}$.
3. $\chi(G) = \chi(G_1) + \chi(G_2)$
4. $\chi'(G) = min\{\chi'(G_1) + \chi(G_2),\ \chi(G_1) + \chi'(G_2)\}$.

Lemma 4. *Let G be a spider (thick or thin) that admits a spider partition (S, C, R) with $R \neq \emptyset$ and $|C| \geq 2$. Then, $\chi_{pcf}(G) = \chi'_{pcf}(G) = \chi'(G) = \chi(G)$.*

Lemma 5. *Let G be a thin spider that admits a spider partition (S, C, R) with $R = \emptyset$ and $|C| \geq 3$. Then, $\chi_{pcf}(G) = \chi'_{pcf}(G) = \chi'(G) = \chi(G)$.*

Lemma 6. *Let G be a thick spider that admits a spider partition (S, C, R) with $R = \emptyset$ and $|C| \geq 3$. Then the following holds.*

1. $\chi'(G) = \chi(G) + 1$.
2. $\chi_{pcf}(G) = \chi'_{pcf}(G) = \chi'(G)$.

If G is a thick or thin spider with spider partition (S, C, R) such that $R = \emptyset$ and $|C| = 2$, then G is a path of length 4 (or P_4) and $\chi_{pcf}(G) = 3$. Using the above lemmas, and the fact that chromatic number in P_4-sparse graph can be found in polynomial time, we have the following theorem.

Theorem 16. *Given a P_4-sparse graph G, $\chi_{pcf}(G)$ can be computed in polynomial time.*

6 Conclusion

In this paper, we study proper conflict-free coloring for some restricted graph classes. We show that the ratio $\chi_{pcf}(G)/\chi(G)$ is bounded for AT-free graphs, P_4-sparse graphs, and split graphs. We also give upper bounds on $\chi_{pcf}(G)$ for several subclasses of AT-free graphs, including interval graphs, proper interval graphs, and permutation graphs. The bounds for AT-free graphs and proper interval graphs are tight while we do not know if the bound on interval graphs is tight or not. The characterization of graphs for which $\chi_{pcf}(G)$ achieves the given upper bound is open for AT-free and interval graphs. We also provide bounds in terms of the maximum degree Δ; for split graphs the conjectured bound $\chi_{pcf}(G) \leq \Delta + 1$ is verified, while for the other graph classes we obtain bounds that are close to $\Delta + 1$.

References

1. Aashtab, A., Akbari, S., Ghanbari, M., Shidani, A.: Vertex partitioning of graphs into odd induced subgraphs. Discuss. Math. Graph Theory **43**(2), 385–399 (2023)
2. Ahn, J., Im, S., Oum, S.-I.: The proper conflict-free k-coloring problem and the odd k-coloring problem are np-complete on bipartite graphs. Discret. Appl. Math. **377**, 10–17 (2025)
3. Bagan, G., Merouane, H.B., Haddad, M., Kheddouci, H.: On some domination colorings of graphs. Discret. Appl. Math. **230**, 34–50 (2017)
4. Bhyravarapu, S., Kalyanasundaram, S., Mathew, R.: Conflict-free coloring on subclasses of perfect graphs and bipartite graphs. Theoret. Comput. Sci. **1031**, 115080 (2025)
5. Caro, Y., Petruševski, M., Škrekovski, R.: Remarks on proper conflict-free colorings of graphs. Discret. Math. **346**(2), 113221 (2023)
6. Cho, E.-K., Choi, I., Kwon, H., Park, B.: Proper conflict-free coloring of sparse graphs. Discret. Appl. Math. **362**, 34–42 (2025)
7. Corneil, D.G., Olariu, S., Stewart, L.: Asteroidal triple-free graphs. SIAM J. Discret. Math. **10**(3), 399–430 (1997)
8. Cranston, D.W., Liu, C.-H.: Proper conflict-free coloring of graphs with large maximum degree. SIAM J. Discret. Math. **38**(4), 3004–3027 (2024)
9. Delcourt, M., Postle, L.: Reducing linear hadwiger's conjecture to coloring small graphs. CoRR, abs/2108.01633 (2021)
10. Diestel, R.: Graph Theory, volume 173 of Graduate Texts in Mathematics. Springer, 5th edn. (2017)
11. Enright, J., Stewart, L., Tardos, G.: On list coloring and list homomorphism of permutation and interval graphs. SIAM J. Discret. Math. **28**(4), 1675–1685 (2014)
12. Fabrici, I., Lužar, B., Rindošová, S., Soták, R.: Proper conflict-free and unique-maximum colorings of planar graphs with respect to neighborhoods. Discret. Appl. Math. **324**, 80–92 (2023)
13. Gutin, G.: Note on perfect forests. J. Graph Theory **82**(3), 233–235 (2016)
14. Hickingbotham, R.: Odd colourings, conflict-free colourings and strong colouring numbers. Australas. J. Comb. **87**, 160–164 (2023)
15. Jamison, B., Olariu, S.: A tree representation for p4-sparse graphs. Discret. Appl. Math. **35**(2), 115–129 (1992)

16. Jiménez, A., et al.: Boundedness for proper conflict-free and odd colorings. Discret. Math. **349**(2), 114730 (2026)
17. Kramer, F., Kramer, H.: A survey on the distance-colouring of graphs. Discret. Math. **308**(2–3), 422–426 (2008)
18. Liu, C.-H.: Proper conflict-free list-coloring, odd minors, subdivisions, and layered treewidth. Discret. Math. **347**(1), 113668 (2024)
19. Liu, C.-H., Reed, B.: Asymptotically optimal proper conflict-free coloring. Random Struct. Algorithms **66**(3), e21285 (2025)
20. Looges, P.J., Olariu, S.: Optimal greedy algorithms for indifference graphs. Comput. Math. Appl. **25**(7), 15–25 (1993)
21. Pach, J., Tóth, G.: Graphs drawn with few crossings per edge. Combinatorica **17**(3), 427–439 (1997)
22. Wang, Y., Wang, W., Liu, R.: Proper conflict-free 6-coloring of planar graphs without short cycles. Appl. Math. Comput. **499**, 129405 (2025)
23. West, D.B., et al.: Introduction to Graph Theory, vol. 2. Prentice hall Upper Saddle River (2001)

Cryptographic Applications
of Combinatorial Ranking for Integer
Compositions

Gustavo Zambonin[1]([✉])[iD], Larissa Gremelmaier Rosa[1][iD], Ricardo Custódio[1][iD],
and Daniel Panario[2][iD]

[1] Universidade Federal de Santa Catarina, Florianópolis, SC 88040-900, Brazil
{gustavo.zambonin,larissa.gremelmaier.rosa}@posgrad.ufsc.br,
ricardo.custodio@ufsc.br
[2] Carleton University, Ottawa, ON K1S 5B6, Canada
daniel@math.carleton.ca

Abstract. Bounded integer compositions are fundamental combinatorial objects employed in cryptography, yet they lack a unified theoretical treatment and efficient implementations for combinatorial (un)ranking. We propose generic algorithms that allow us to describe proposals found in the literature as specific parameter sets. Consequently, we are able to provide a single proof of correctness that applies simultaneously to all derived algorithms; characterize a number of ad hoc optimizations with respect to our generic framework; and discuss the storage complexity of pre-computed combinatorial count tables. Furthermore, drawing on Gaussian limit laws from analytic combinatorics, we propose performance and storage optimizations for the particular case of uniform random inputs to (un)ranking algorithms, a common setting in cryptographic applications. To validate our results, we give portable low-level implementations ready for use by algorithm designers, filling a practical gap in the literature.

Keywords: Uniform sampling · Gaussian law · Enumerative coding

1   Introduction

Combinatorial ranking provides a standard method for the uniform sampling and optimal compression of ordered structures, which are fundamental problems in cryptographic algorithms. In particular, ranking of (bounded) integer compositions is employed in (i) hash-based signatures, (ii) format-preserving encryption, and (iii) code-based cryptosystems. We expand on these use cases below, and argue how a unified approach joins the perspectives of the literature into a systematic framework of ranking compositions in applied cryptography.

In hash-based signatures like SLH-DSA (formerly SPHINCS$^+$) [25], mapping messages to bounded compositions is standard practice to improve performance and signature sizes [14]. While fundamental (un)ranking algorithms are

F. Foucaud and A. Parreau (Eds.): IWOCA 2026, LNCS 16587, pp. 515–530, 2026.
https://doi.org/10.1007/978-3-032-27732-9_36

frequently rediscovered [11,26,29], recent research has focused on optimizations like rejection sampling [12] and parameter tuning [14,27,35] to reduce overhead.

Format-preserving encryption (FPE) ensures plaintexts and ciphertexts maintain the same structural constraints, often via the *rank-then-encipher* paradigm introduced in [4]. In *sum-preserving* (SPE) schemes, messages are bounded compositions, which is relevant for thumbnail-preserving encryption [31], and statistical analysis of encrypted data [21].

Finally, a number of code-based cryptosystems require mapping binary strings to fixed-weight error vectors [16,19]. Although deterministic unranking has largely been superseded by randomized, constant-time alternatives [3] to mitigate side-channel risks, algorithmic improvements could reinstate combinatorial ranking as a competitive, deterministic solution.

Contributions. We present a comprehensive framework for the combinatorial ranking of bounded integer compositions. By parameterizing the problem into a strategy composed of three functions, focused on this particular structure, we are able to (i) recover classical algorithms and correctness proofs as specific instances of our framework; (ii) analyze common optimizations such as bisection search and partial sum re-usage; (iii) propose a probabilistic strategy that exploits the statistical structure of random compositions, suitable for cryptographic algorithms. We also measure the performance and storage requirements of several (un)ranking algorithms and pre-computed combinatorial count tables.

2   Preliminaries

Notation. We write $x \xleftarrow{r} X$ to mean x is sampled uniformly at random from X, $[x^j]f(x)$ to be the coefficient of x^j in the generating function f, $[d] = \{0, 1, \ldots, d\}$, and $\lg = \log_2$. For any statement P, $[[P]]$ is the Iverson bracket, where $[[P]] = 1$ if P is true and 0 otherwise. Let $()$ be the empty sequence. Given sequences $Y = (y_1, y_2, \ldots, y_\alpha)$ and $W = (w_1, w_2, \ldots, w_\beta)$ such that $|Y| = \alpha$ and $|W| = \beta$, let $y \oplus Y = (y, y_1, \ldots, y_\alpha)$, $Y \oplus y = (y_1, \ldots, y_\alpha, y)$, $Y \oplus W = (y_1, \ldots, y_\alpha, w_1, \ldots, w_\beta)$ and $Y^R = (y_\alpha, \ldots, y_1)$.

Combinatorial Ranking and Compositions. Let S be a finite set, $N = |S|$ and $s, t \in S$ such that $s \neq t$. A *ranking function* is a bijection $\mathbf{R} : S \to [N - 1]$ that defines a total order $\prec_{\mathrm{ord}}$ on the elements of S, i.e., $s \prec_{\mathrm{ord}} t$ if and only if $\mathbf{R}(s) < \mathbf{R}(t)$ [17, Section 2.1]. Given a ranking function, the *unranking function* is the inverse bijection $\mathbf{U} = \mathbf{R}^{-1}$, i.e., $\mathbf{U} : [N - 1] \to S$ such that $\mathbf{R}(s) = i$ if and only if $\mathbf{U}(i) = s$ for all $i \in [N - 1]$.

Consider $n, k \in \mathbb{Z}_{>0}$. A *weak k-composition* of n is a sequence $(a_1, \ldots, a_k)$ such that $\sum_{i=1}^{k} a_i = n$ and each part $a_i \in \mathbb{Z}_{\geq 0}$. An *$S$-bounded composition* with $S \subseteq \mathbb{Z}_{\geq 0}$ is a composition of n such that $a_i \in S$. If $S = [d]$ for $d \in \mathbb{Z}_{>0}$, we denote the compositions d-bounded [5]. We denote the set of d-bounded k-compositions of n by $\mathcal{C}_d(n, k)$, and its cardinality by $\mathfrak{C}_d(n, k)$, which is expressed by the generating function below.

Lemma 1 ([8, I.15]). $\mathfrak{C}_d(n, k) = [x^n] \left(\frac{1 - x^{d+1}}{1 - x} \right)^k$.

The polynomial interpretation yields a double combinatorial sum, deemed a "simple binomial convolution" [8, p. 43] and originally attributed to [22].

Lemma 2 ([1, **Remark 60**]). $\mathfrak{C}_d(n, k) = \sum_{i=0}^{k} (-1)^i \binom{k}{i} \binom{n - (d+1)i + k - 1}{k - 1}$.

We consider the upper limit of the summation to be $j = \min(k, \lfloor \frac{n}{d+1} \rfloor)$, which prevents unnecessary computations, and hereafter denote $\mathfrak{C}_d(n, k)_\alpha$ to be the α-th term of the right-hand side of Lemma 2, for $\alpha \in [j]$. The cardinality may be computed by two recursive formulas that generalize Pascal's and Vandermonde's identities, respectively.

Theorem 1 ([1, **Theorem 11**]). *If $k = 1$ and $n > d$, $\mathfrak{C}_d(n, k) = 0$. If $k = 1$ and $n \le d$, $\mathfrak{C}_d(n, k) = 1$. Otherwise, $\mathfrak{C}_d(n, k) = \sum_{i=0}^{d} \mathfrak{C}_d(n - i, k - 1)$.*

Theorem 2 ([6, **Table 1**]). *Let $k_1, k_2 \in \mathbb{Z}_{\ge 0}$ such that $k_1 + k_2 = k$. If $k = 1$ and $n > d$, $\mathfrak{C}_d(n, k) = 0$. If $k = 1$ and $n \le d$, $\mathfrak{C}_d(n, k) = 1$. Otherwise, $\mathfrak{C}_d(n, k) = \sum_{n_1=0}^{n} \mathfrak{C}_d(n_1, k_1) \cdot \mathfrak{C}_d(n - n_1, k_2)$.*

For further information on bounded integer compositions, we refer the reader to [6] for historical perspectives and applications to other disciplines.

Combinatorial Orders. We recall that for any set S of size N, there exist $N!$ total orders for S, corresponding to the number of distinct permutations of its elements. We give several descriptions of usual total orders found in the literature, following the "constructive" notation of [2, Section 14.1]. We use bold letters (e.g., $\mathbf{C}_d(n, k)$) to denote a permutation of $\mathcal{C}_d(n, k)$ related to a specific order.

The canonical total order is *lexicographic*, or "dictionary" order, which aligns naturally with the recursive structure of Theorem 1. A common variant, defined below, can be intuitively understood as "lexicographic read from right to left".

Definition 1. *The* colexicographic *order, or $\prec_{\mathsf{colex}}$, is defined as follows. If $k = 1$ and $n > d$, $\mathbf{C}_d(n, 1) = ()$. If $k = 1$ and $n \le d$, $\mathbf{C}_d(n, 1) = (n)$. Otherwise,*

$$\mathbf{C}_d(n, k) = \big(\mathbf{C}_d(n, k - 1) \oplus 0, \ \mathbf{C}_d(n - 1, k - 1) \oplus 1,$$

$$\dots, \ \mathbf{C}_d(n - d, k - 1) \oplus d \big).$$

We obtain the *reflected Gray code* [32, Theorem 1], or Eades-McKay [2, Section 6.5] order, by reversing the $\mathbf{C}_d(n - 1, k - 1)$ substructure at odd d. This order, which we denote by $\prec_{\mathsf{gray}}$, is *minimal-change*: two consecutive compositions differ only by two parts [2, Section 6.4].

We can also order compositions via the structure of Theorem 2, as proposed recently in [21], but first connected to cryptographic applications in [29]. Let $\mathbf{l} = (l_i)_{i=1}^{p}$, $\mathbf{r} = (r_j)_{j=1}^{q}$ be any two ordered lists of sequences with $p, q \in \mathbb{Z}_{\ge 0}$. The "product" operation is defined as $\mathbf{l} \odot \mathbf{r} = \oplus_{i=1}^{p} \left(\oplus_{j=1}^{q} (l_i \oplus r_j) \right)$, or $\mathbf{l} \odot \mathbf{r} = ()$ if $p = 0$ or $q = 0$. Explicitly, this produces the sequence $(l_1 \oplus r_1, \dots, l_1 \oplus r_q, \dots, l_p \oplus r_1, \dots, l_p \oplus r_q)$.

Definition 2. *The* recursive block *order, or $\prec_{\mathsf{rb}}$, is defined as follows. If $k = 1$ and $n > d$, $\mathbf{R}_d(n, 1) = ()$. If $k = 1$ and $n \leq d$, $\mathbf{R}_d(n, 1) = (n)$. Otherwise, for $k_l = \lfloor \frac{k}{2} \rfloor$ and $k_r = k - k_l$,*

$$\mathbf{R}_d(n, k) = \bigoplus_{n_l=0}^{\min(n, k_l \cdot d)} \left(\mathbf{R}_d(n_l, k_l) \odot \mathbf{R}_d(n - n_l, k_r) \right).$$

It is essentially a lexicographic ordering of the subproblems defined by Theorem 2 for k_1, k_2 split exactly at $\frac{k}{2}$. We note that $\prec_{\mathsf{rb}}$ is remarkably connected to the *boustrophedonic* order [20], hereafter denoted $\prec_{\mathsf{bous}}$. They differ only in how n_l and n_r are chosen; while for $\prec_{\mathsf{rb}}$, n_l grows monotonically from 0 to $\min(n, k_l \cdot d)$, for $\prec_{\mathsf{bous}}$ the iteration sequence is $0, n, 1, n - 1, \ldots$, giving rise to a more complex recursive case:

$$\mathbf{B}_d(n, k) = \bigoplus_{j=0}^{\lfloor \frac{n}{2} \rfloor} \left(\left(\mathbf{B}_d(j, k_l) \odot \mathbf{B}_d(n - j, k_r) \right) \oplus \tau \right),$$

where $\tau = \mathbf{B}_d(n - j, k_l) \odot \mathbf{B}_d(j, k_r)$ if $j < n - j$, and $()$ if $j = n - j$; see [9].

3   Generic (Un)ranking Algorithms

The design of (un)ranking algorithms is closely linked to the total order defined on the target combinatorial class. For instance, Theorem 1 implies that an unranking algorithm for $\prec_{\mathsf{colex}}$ must essentially determine

$$\min \left\{ p \in [\min(n', d)] : \sum_{j=0}^{p} \mathfrak{C}_d(n' - j, i) > r' \right\} \tag{1}$$

for every part $i \in (k - 1, \ldots, 1)$, while tracking the partial sum n' and remaining rank r'. (Of course, p is the i-th part of the resulting composition.) Historically, this iterative approach has dominated the literature, due to its intuitive definition and ease of verification [15, p. 358].

The generalized framework proposed in [20, 23] unified the derivation of (un)ranking algorithms for decomposable combinatorial structures, via the symbolic method of analytic combinatorics [8]. However, it prioritizes universality by relying on fixed orders ($\prec_{\mathsf{lex}}$ and $\prec_{\mathsf{bous}}$). As a result, research into alternative orders remains sparse, despite evidence that order choice significantly impacts the performance of enumeration and (un)ranking algorithms [20, Theorem 2].

Inspired by this generic framework, we take a complementary approach; rather than fixing the order to derive an algorithm for a generic class, we fix the combinatorial object (bounded compositions) and treat the *strategy* as a variable parameter. We let a strategy be a tuple of functions $\Sigma = (\mathsf{D}, \mathsf{P}, \mathsf{S})$, where the components are formally defined as follows.

Decomposition. The function $\mathsf{D} : \mathbb{Z}_{\geq 1} \to \mathbb{Z}_{\geq 0} \times \mathbb{Z}_{\geq 0}; \; k \mapsto (k_l, k_r)$ determines the recursive structure of the algorithms by mapping a problem size k to a pair of smaller sizes satisfying $k_l + k_r = k$. Examples are $\mathsf{D}_l(k) = (1, k - 1)$, $\mathsf{D}_r(k) = (k - 1, 1)$ and $\mathsf{D}_s(k) = (\lfloor \frac{k}{2} \rfloor, \lceil \frac{k}{2} \rceil)$.

Path. The function $\mathsf{P} : \mathbb{Z}_{\geq 0} \times \mathbb{Z}_{\geq 0} \times \mathbb{Z}_{\geq 0} \to \bigcup_{m=0}^{n+1}(\mathbb{Z}_{\geq 0})^m$; $(n, k_l, k_r) \mapsto \mathbf{w}$ generates the traversal schedule. It maps the current context to a sequence of values to visit. The output $\mathbf{w} = (w_1, w_2, \ldots)$ must be a permutation of the valid range $\{n_1 \in [n] : \mathcal{C}_d(n_1, k_l) > 0 \wedge \mathcal{C}_d(n - n_1, k_r) > 0\}$. Examples are $\mathsf{P}_{\mathsf{inc}}(n, k_l, k_r) = (w_1, \ldots, w_m)$, $\mathsf{P}_{\mathsf{dec}}(n, k_l, k_r) = (w_m, w_{m-1}, \ldots, w_1)$ and $\mathsf{P}_{\mathsf{alt}}(n, k_l, k_r) = (w_1, w_m, w_2, w_{m-1}, \ldots)$.

Remark 1. We restrict D to binary splits instead of a full m-ary decomposition $k \to (k_1, \ldots, k_m)$. The generic decomposition would require P to iterate over all m-compositions of n, increasing the size of the local search space to $\mathcal{O}(n^{m-1})$. Since any m-ary decomposition can be represented as a tree of binary splits with constant overhead, the binary restriction bounds the iteration size of P to $\mathcal{O}(n)$ at each step without loss of generality.

Split. The component S is a family of bijections indexed by dimensions $(u, v, w) \in \mathbb{Z}_{\geq 1}^2 \times \mathbb{Z}_{\geq 0}$. The function $\mathsf{S}^{(u,v,w)} : [uv - 1] \to [u - 1] \times [v - 1]$; $r \mapsto (r_l, r_r)$ decomposes a rank r to sub-ranks r_l, r_r satisfying $r_l + r_r = r$ based on the value w. Examples are $\mathsf{S}_{\mathsf{div}}^{(u,v,w)}(r) = (\lfloor \frac{r}{v} \rfloor, r \pmod{v})$, the usual specialization $\mathsf{S}_{\mathsf{id}}^{(1,v,w)}(r) = (0, r)$, and $\mathsf{S}_{\mathsf{g}}^{(1,v,w)}(r) = (0, r)$ if $w \equiv 0 \pmod 2$ and $(0, v - 1 - r)$ otherwise. Specific instantiations may discard certain parameters if they are not required by the algebraic structure of the split.

Recursive Construction. The ordered sequence of compositions $\mathcal{C}_d(n, k)$ generated by a strategy Σ, denoted $\mathbf{Z}_d(n, k)$, is defined recursively. We first extend the product operation $\odot$ to account for the operation of S. Let $\mathbf{L}$ and $\mathbf{R}$ be ordered lists of compositions with sizes u and v, respectively. The *permuted product* $\mathbf{L} \odot_{\mathsf{S}} \mathbf{R}$ is the list of size uv where the r-th element is $L_{r_l} \oplus R_{r_r}$, with $(r_l, r_r) = \mathsf{S}^{(u,v,w)}(r)$ for $r \in [uv - 1]$. For $(k_l, k_r) = \mathsf{D}(k)$, the generic sequence is

$$\mathbf{Z}_d(n, k) = \bigoplus_{w_l \in \mathsf{P}(n, k_l, k_r)} \left(\mathbf{Z}_d(w_l, k_l) \odot_{\mathsf{S}} \mathbf{Z}_d(n - w_l, k_r) \right).$$

Standard Instantiations. The strategy Σ can be instantiated to recover standard algorithms. We denote Σ_{ord} as the strategy "corresponding to $\prec_{\mathsf{ord}}$". The same convention is applied to (un)ranking functions. We give a number of strategies as follows: (i) $\Sigma_{\mathsf{lex}} = (\mathsf{D}_l, \mathsf{P}_{\mathsf{inc}}, \mathsf{S}_{\mathsf{id}})$; (ii) $\Sigma_{\mathsf{colex}} = (\mathsf{D}_r, \mathsf{P}_{\mathsf{inc}}, \mathsf{S}_{\mathsf{id}})$; (iii) $\Sigma_{\mathsf{gray}} = (\mathsf{D}_l, \mathsf{P}_{\mathsf{inc}}, \mathsf{S}_{\mathsf{g}})$; (iv) $\Sigma_{\mathsf{rb}} = (\mathsf{D}_s, \mathsf{P}_{\mathsf{inc}}, \mathsf{S}_{\mathsf{div}})$; and (v) $\Sigma_{\mathsf{bous}} = (\mathsf{D}_s, \mathsf{P}_{\mathsf{alt}}, \mathsf{S}_{\mathsf{div}})$.

Formalization. All components of the generic (un)ranking algorithms are mechanized in Lean 4 [24]. We prove their correctness via strong induction, and instantiate the Σ_{colex}, Σ_{rb} and Σ_{gray} strategies to show that the framework is sound. We also prove the auxiliary propositions, lemmas and theorems discussed in the following sections, which might be of independent interest. The complete artifact is available at [34].

We present the generic unranking and ranking algorithms for bounded integer compositions in Algorithm 1, respectively. We assume that the parameters

n, k, d are valid, such that $\mathfrak{C}_d(n, k) > 0$. Then, Theorem 3 states that the algorithms are mutual inverses, i.e., they establish a strict bijection between the set of combinatorial objects and their indices. In other words, composing the two operations yields the identity function on both domains, guaranteeing that the enumeration is exhaustive and collision-free.

$\mathbf{R}_\Sigma(\text{in: } n, k, d, z; \text{ out: } r)$.

1. If $k = 1$, set $r \leftarrow 0$ and exit. Otherwise, go to Step 2.
2. Set $(k_l, k_r) \leftarrow \mathsf{D}(k)$.
3. Set $w_l \leftarrow \sum_{i=1}^{k_l} z_i$.
4. Set $L \leftarrow (z_1, \ldots, z_{k_l})$.
5. Set $R \leftarrow (z_{k_l+1}, \ldots, z_k)$.
6. Set $\mathbf{w} \leftarrow \mathsf{P}(n, k_l, k_r)$.
7. Set $r \leftarrow 0$.
8. For each $v \in \mathbf{w}$:

   8A. If $v = w_l$, go to Step 9.
   8B. Set $r \leftarrow r + \mathfrak{C}_d(v, k_l) \cdot \mathfrak{C}_d(n - v, k_r)$.

9. Set $r_l \leftarrow \mathbf{R}_\Sigma(w_l, k_l, d, L)$.
10. Set $r_r \leftarrow \mathbf{R}_\Sigma(n - w_l, k_r, d, R)$.
11. Set $n_l \leftarrow \mathfrak{C}_d(w_l, k_l)$.
12. Set $n_r \leftarrow \mathfrak{C}_d(n - w_l, k_r)$.
13. Set $r \leftarrow r + (\mathsf{S}^{(n_l, n_r, w_l)})^{-1}(r_l, r_r)$.

$\mathbf{U}_\Sigma(\text{in: } n, k, d, r; \text{ out: } z)$.

1. If $k = 1$, set $z \leftarrow (n)$ and exit. Otherwise, go to Step 2.
2. Set $(k_l, k_r) \leftarrow \mathsf{D}(k)$.
3. Set $\mathbf{w} \leftarrow \mathsf{P}(n, k_l, k_r)$.
4. For each $w_l \in \mathbf{w}$:

   4A. Set $n_l \leftarrow \mathfrak{C}_d(w_l, k_l)$.
   4B. Set $n_r \leftarrow \mathfrak{C}_d(n - w_l, k_r)$.
   4C. Set $V \leftarrow n_l \cdot n_r$.
   4D. If $r < V$, go to Step 5.
   4E. Set $r \leftarrow r - V$.

5. Set $(r_l, r_r) \leftarrow \mathsf{S}^{(n_l, n_r, w_l)}(r)$.
6. Set $L \leftarrow \mathbf{U}_\Sigma(w_l, k_l, d, r_l)$.
7. Set $R \leftarrow \mathbf{U}_\Sigma(n - w_l, k_r, d, r_r)$.
8. Set $z \leftarrow L \oplus R$.

Algorithm 1. Generic ranking and unranking algorithms parameterized by a strategy $\Sigma = (\mathsf{D}, \mathsf{P}, \mathsf{S})$. We recall that $n, k, d \in \mathbb{Z}_{>0}$, $z \in \mathfrak{C}_d(n, k)$ and $r \in [\mathfrak{C}_d(n, k) - 1]$.

Theorem 3. *Let* $S = \mathbf{Z}_d(n, k)$. *Then, for any* $z \in S$ *and* $r \in [|S| - 1]$,

$$\mathbf{R}_\Sigma(n, k, d, \mathbf{U}_\Sigma(n, k, d, r)) = r \quad \text{and} \quad \mathbf{U}_\Sigma(n, k, d, \mathbf{R}_\Sigma(n, k, d, z)) = z.$$

We now discuss how specific choices for $\Sigma = (\mathsf{D}, \mathsf{P}, \mathsf{S})$ determine the efficiency of the resulting (un)ranking algorithms. We focus our analysis on D and P, as we observe that S performs only elementary arithmetic, with negligible cost compared to the search in P or how D controls the recursion structure.

3.1   Decomposition and Storage Requirements

The repeated evaluation of $\mathfrak{C}_d(n, k)$ for arbitrary n, k, d is not computationally trivial. In combinatorial literature, it is often assumed that needed counts are already pre-computed [23, Section 3.2.1]. On the other hand, cryptographic applications require careful management of the storage requirements associated with the usage of (un)ranking functions [21, 26, 36], particularly if arbitrary-precision integers are required to represent $\mathfrak{C}_d(n, k)$.

We provide a systematic view of this *tabulation* mechanism that complements the empirical evidence in the literature, and correlate it with the decomposition function D of a strategy. We define $\chi_{\mathsf{comb}}^{(d)}(n,k)$ as the $(k+1) \times (n+1)$ matrix with entries $(\chi_{\mathsf{comb}}^{(d)})_{i,j} = \mathfrak{C}_d(j,i)$ for $0 \le i \le k$ and $0 \le j \le n$. If $d = 1$, the entries are the binomial coefficients $\binom{n}{k}$, and we denote the table by $\chi_{\mathsf{bin}}(n,k)$. We drop the parameters from the notation if clear from the context.

Naturally, the choice of D controls which rows of this matrix are used by the (un)ranking algorithms. Consider $\mathsf{D}_l(k) = (1, k-1)$ and $\mathsf{D}_r(k) = (k-1, 1)$. In such "linear" orders, the recursion must solve problems of all sizes in the sequence $(1, \ldots, k)$. In other words, counts of the form $\mathfrak{C}_d(n', i)$ for all $1 \le i \le k-1$ and any partial sum $n' < n$ must be in χ_{comb}.

Remark 2. For the aforementioned "linear" orders, one can extend the table χ_{comb} to store cumulative partial sums of $\mathfrak{C}_d(n,k)$. However, this extension is only useful if P is specifically designed to utilize these aggregates. In the literature, this trade-off has been empirically explored in [26, Section 3.4]; we discuss the necessary interaction between storage and strategy in Sect. 3.2.

On the other hand, for $\mathsf{D}_s(k) = (\lfloor \frac{k}{2} \rfloor, \lceil \frac{k}{2} \rceil)$, only a sparse subset of counts is needed. While the reduced memory footprint of using $\prec_{\mathsf{rb}}$ (and therefore D_s) was empirically observed [21, Section 5], the authors did not provide a theoretical bound. We formalize this advantage by showing the exact set of "row" indices visited by (un)ranking algorithms whose decomposition strategy is D_s.

Definition 3. *Let $k \in \mathbb{Z}_{>0}$ and $\ell = \lfloor \lg k \rfloor$. The* iterated halving set *of k is $\mathsf{H}(k) = \bigcup_{p=1}^{\ell} \{\lfloor \frac{k}{2^p} \rfloor, \lceil \frac{k}{2^p} \rceil\}$, with $\mathsf{H}(1) = \varnothing$.*

It is straightforward to evaluate $\mathsf{H}(k)$ via the binary representation of k. Nonetheless, we provide a closed expression for its exact size, allowing precise estimates of the size of χ_{comb} as a function of parameters n, k, d for orders based on $\mathsf{D}_s(k)$. We let $\nu_2(k) = \max\{j \in \mathbb{Z}_{\ge 0} : 2^j \mid k\}$ be the 2-adic valuation of k.

Proposition 1. *For $k \in \mathbb{Z}_{>0}$ and $\ell = \lfloor \lg k \rfloor$, we have that*

$$|\mathsf{H}(k)| = 2\ell - \nu_2(k) - [\![2^\ell < k < 2^\ell + 2^{\ell-1}]\!].$$

Practical considerations on the storage requirements of the tabling mechanisms are further explored in Sect. 4.

3.2   Path and Algebraic Search

We discuss two optimization techniques for the case when P produces a monotonic sequence, for instance $\mathsf{P}_{\mathsf{inc}}(n, k_l, k_r) = (w_1, \ldots, w_m)$ which is featured in $\prec_{\mathsf{colex}}$. The search then reduces to finding an index in an ordered sequence.

The "bisect" Technique. We observe that if P is monotonic, Eq. (1) represents a search over a sorted sequence of partial sums, defined as

$$A(n', i, d) = \left(0, \mathfrak{C}_d(n', i), \ldots, \sum_{j=0}^{\min(n',d)} \mathfrak{C}_d(n' - j, i)\right) \tag{2}$$

for $1 \leq i \leq k-1$ and $n' \in [n]$. Consequently, the correct value w (or part p) can be found via binary search in $\log d$ instead of d steps. This approach is mentioned in [21, Section 4.4] and explored in [26, Section 3.2], who provide a closed-form identity to compute arbitrary terms of this sequence directly. We identified a substitution error in the proof [26, Section A.3], which actually yielded $\mathfrak{C}_d(n, k + 1)^{(l)}$. The formula is fixed here by incrementing k in the summation.

Lemma 3 ([26, Proposition 3]). *Let* $n, k, d \in \mathbb{Z}_{>0}$, $j = \min(k, \lfloor \frac{n}{d+1} \rfloor)$, $l \in [\min(n, d)]$ *and set* $u = n - (d + 1)i + k - 1$. *Then,*

$$\mathfrak{C}_d(n, k)^{(l)} = \sum_{i=0}^{j} (-1)^i \binom{k}{i} \left[\binom{u}{k} - \binom{u - l}{k}\right].$$

However, we identify redundant computations in the unranking algorithm proposed by the authors in [26]. Lemma 3 requires summing $\mathcal{O}(k)$ binomial terms; since the algorithm re-evaluates them independently at every step of the binary search, the computational cost to find a single part is $\mathcal{O}(k \log d)$. We argue that a direct bisection search on the sequence $A(n', i, d)$ is more efficient, as it can be constructed iteratively in $\mathcal{O}(d)$ steps.

Algorithms implementing this technique can make use of the tabling mechanism discussed in Remark 2. We define $\chi_{\mathsf{acc}}^{(d)}(n, k)$ to be a matrix holding accumulated sizes of composition sets, with entries $(\chi_{\mathsf{acc}}^{(d)})_{i,j} = A(j, i, d)$ for $0 \leq i \leq k$ and $0 \leq j \leq n$. As the size of each entry, itself a sequence, depends on the choice of parameters, we give a simple formula to calculate the number of entries of the entire matrix, such that it can be effectively packed.

Proposition 2. *The size of* $\chi_{\mathsf{acc}}^{(d)}(n, k)$ *is* $(k + 1)\left[\frac{d(2n-d+1)}{2} + 2(n + 1)\right]$.

We discuss the trade-offs of using χ_{acc} with respect to the performance of compatible unranking algorithms, and the general behavior of the "bisect" technique, in Sect. 4.

The "Partial Sum" Technique. If a binary search is not employed, the choice of $\mathsf{P}_{\mathsf{inc}}$ still allows for a different optimization. Instead of computing the combinatorial count $\mathfrak{C}_d(n' - p, i)$ from scratch at each step of Eq. (1), we can derive it from the previous term using an arithmetic ratio, which is a generalization of [10, Section 4.2]. A similar technique is found in [30, Section 1.13], in which the input to unranking is a number in $[0, 1)$ and no large integers are required.

Proposition 3. *Let* $n, k, d \in \mathbb{Z}_{>0}$ *and* $j \in [\min(i, \lfloor \frac{n'-p-1}{d+1} \rfloor)]$. *Then,*

$$\frac{\mathfrak{C}_d(n', i)_j}{\mathfrak{C}_d(n' - p, i)_j} = \prod_{g=1}^{p} \frac{(n' - g + 1) - (d + 1)j + i - 1}{(n' - g + 1) - (d + 1)j}.$$

We observe that Proposition 3 can be used to obtain $\mathfrak{C}_d(n' - p, i)$ from the previous term with only integer multiplications and exact divisions; it suffices to first set $p = 1$ in the product above, and then substitute n' by $n' - p$.

Corollary 1

$$\frac{\mathfrak{C}_d(n' - p, i)_j}{\mathfrak{C}_d(n' - p - 1, i)_j} = \frac{(n' - p) - (d + 1)j + i - 1}{(n' - p) - (d + 1)j}.$$

Thus, the summation of Lemma 2 must be evaluated only $k - 1$ times, once at the start of every recursion path, rather than at every step of $\mathsf{P}_{\mathsf{inc}}$.

We empirically measure the performance of both techniques in Sect. 4, comparing them against (un)ranking algorithms to other orders.

3.3   The "Spiral" Order for Cryptographic Applications

Cryptographic applications often require sampling combinatorial objects uniformly at random, and unranking algorithms are regularly employed for this purpose. A direct consequence of the change of variable lemma [18, p. 190] is that a bijection between two finite sets maps a uniform probability distribution on one set to another such distribution on the other. Thus, providing uniform random integers to unranking functions generates compositions drawn from the same distribution.

Combinatorial counts $\mathfrak{C}_d(n, k)$ used in unranking algorithms exhibit a strong concentration of measure under the uniform random input condition. As established in [8, Section VIII], the coefficients of powers of polynomials (such as the generating function in Lemma 1) converge asymptotically to a Gaussian limit. Particularly, it is shown in [28, Theorem 1.3] that $\mathfrak{C}_d(n, k)$ follows a normal distribution with an approximation error of $\mathcal{O}(n^{-\frac{1}{5}})$.

This behavior implies that for a random rank, the partial sums S_j calculated in the unranking "path" are also asymptotically Gaussian. Empirical data [28, Table 1] suggests the approximation error is small ($< 1\%$) even for modest $n \geq 80, k \geq 30$. As parameters n, k, d suitable for cryptographic applications are usually larger [21, Table 3], or bounded to a security level $\lambda \geq 128$ such that $\mathfrak{C}_d(n, k) \approx 2^\lambda$ (e.g., [14, Table 1] and [7, Table 5]), we argue that the Gaussian behavior can be exploited to accelerate unranking algorithms. Concretely, we propose a path function $\mathsf{P}_s = (\mu, \mu + 1, \mu - 1, \mu + 2, \mu - 2, \dots) \cap [d]$ for some expected value μ. We first properly characterize this quantity.

Statistics of Random Bounded Compositions. Let R be a discrete random variable that represents the total sum of a composition $z \xleftarrow{r} \mathcal{C}_d(n, k)^{(l)}$, where $\mathcal{C}_d(n, k)^{(l)} = \bigcup_{r=n-l}^{n} \mathcal{C}_d(r, k)$, generalizing the simpler setting of drawing from $\mathcal{C}_d(n, k)$. We recall that the p-th raw moment of R is $\mathbb{E}[R^p] = \frac{\mu_p}{\mu_0}$, with $\mu_p(n, k, d, l) = \sum_{r=n-l}^{n} r^p \cdot \mathfrak{C}_d(r, k)$. (We write μ_p whenever clear from the context.) For $p = 0$, the r^p term vanishes and $\mu_0 = \mathfrak{C}_d(n, k)^{(l)}$, calculated via Lemma 3. We provide a closed-form generalization of this result which might be of independent interest. First, we prove an auxiliary lemma on weighted sums of binomial coefficients.

Proposition 4. *Let* $s, c, y, l \in \mathbb{Z}_{\geq 0}$ *and* $W(s, c, y, l) = \sum_{r=y-l}^{y} r^s \binom{r}{c}$. *Then,*

$$W(s, c, y, l) = \begin{cases} \binom{y+1}{c+1} - \binom{y-l}{c+1} & \textit{if } s = 0, \\ c \cdot W(s-1, c, y, l) + (c+1) \cdot W(s-1, c+1, y, l) & \textit{otherwise.} \end{cases}$$

Now, we calculate the exact p-th raw moment.

Theorem 4. *Let* $n, k, d \in \mathbb{Z}_{>0}$ *as per Lemma 2,* $j = \min(k, \lfloor \frac{n}{d+1} \rfloor)$, $l \in [d]$ *and set* $u_i = (d+1)i - k + 1$ *for convenience. Then, for any* $p \in \mathbb{Z}_{\geq 0}$,

$$\mu_p(n, k, d, l) = \sum_{i=0}^{j} (-1)^i \binom{k}{i} \left[\sum_{s=0}^{p} \binom{p}{s} u_i^{p-s} W(s, k-1, n-u_i, l) \right].$$

To define the expected mean and variance of R in terms of μ_p, the following corollaries are useful. We let $W_0(k') = W(0, k', n - u_i, l)$ for readability.

Corollary 2

$$\mu_1(n, k, d, l) = \sum_{i=0}^{j}(-1)^i \binom{k}{i} \left[(d+1)i \cdot W_0(k-1) + k \cdot W_0(k) \right].$$

Corollary 3

$$\mu_2(n, k, d, l) = \sum_{i=0}^{j}(-1)^i \binom{k}{i} (\alpha_i + \beta_i + \gamma_i)$$

for $\alpha_i = ((d+1)i)^2 \cdot W_0(k-1)$, $\beta_i = k(2(d+1)i+1) \cdot W_0(k)$ *and* $\gamma_i = k(k+1) \cdot W_0(k+1)$.

The first and second raw moments of R are respectively $\mathbb{E}[R] = \frac{\mu_1}{\mu_0}$ and $\mathbb{E}[R^2] = \frac{\mu_2}{\mu_0}$. Consequently, the second central moment is $\mathbb{V}(R) = \mathbb{E}[R^2] - (\mathbb{E}[R])^2$ [18, p. 244]. We are now able to provide formulas for the first raw and second central moments of the partial sums of a random bounded composition from the accumulated set.

Proposition 5. *For* $k > 1$, *let* $X_1, \ldots, X_k$ *be discrete random variables that represent the* i-th *part of a composition* $z \xleftarrow{r} C_d(n, k)^{(l)}$ *for* $1 \leq i \leq k$, *and* $S_j = \sum_{i=1}^{j} X_i$ *be the* j-th *partial sum, with* $S_k = R$. *Then,*

$$\mathbb{E}[S_j] = j \cdot \frac{\mathbb{E}[R]}{k} \quad \textit{and} \quad \mathbb{V}(S_j) = \frac{j(k-j)}{k-1} \cdot \mathbb{V}(X_1) + \frac{j(j-1)}{k(k-1)} \mathbb{V}(R)$$

with $\mathbb{V}(X_1) = \frac{1}{\mu_0(n,k,d,l)} \left(\sum_{y=0}^{\min(n,d)} y^2 \mu_0(n-y, k-1, d, l) \right) - \left(\frac{\mathbb{E}[R]}{k} \right)^2$ *and the moments of* R *as given above.*

For the simpler case of $l = 0$, we have $\mathbb{V}(R) = 0$ and $\mathbb{E}[S_j] = j \cdot \frac{n}{k}$. During the computation of P_s, the remaining sum n' across i remaining parts shares the identical distribution as S_i due to the exchangeability of parts. Thus, at any recursive step, the expected value of the next part is $\frac{n'}{i}$, the "center of the spiral". We are also able to define a suitable tabling mechanism. For $\max(0, n-(k-i)d) \leq n' \leq \min(n, i \cdot d)$, $1 \leq i \leq k$, and $c \in \mathbb{Z}_{>0}$, the *probabilistic table*

$$\chi^{(d)}_{\mathsf{comb}\text{-}\sigma}(n, k) = \left\{ (n', i) \in \chi^{(d)}_{\mathsf{comb}}(n, k) : |n' - \mathbb{E}[S_i]| \leq c \cdot \sqrt{\mathbb{V}(S_i)} \right\}$$

is a truncation of χ_{comb} which covers the high-probability "band" delimited by Proposition 5. (Without loss of generality, we define $\chi^{(d)}_{\mathsf{acc}\text{-}\sigma}(n, k)$ analogously.) Relying on the asymptotic Gaussian convergence of these compositions, selecting $c = 4$ covers roughly 99.993% of the probability mass while storing only a fraction[1] of the original matrix. An unranking algorithm employing this table will rarely fall back to direct computation (e.g., via Lemma 2) if the required combinatorial count is not found.

4   Experiments

Next, we compare the performance and storage requirements of several algorithms and techniques previously discussed. To reproduce the experiments, we provide source code and accompanying documentation [33]. We choose parameter sets based on the TSL construction in [14, Table 1], which may be employed in SLH-DSA [25]. For the $\lambda = 128$ security level, equivalent to $\mathfrak{C}_d(n, k) \approx 2^\lambda$, we select $k \in \{25, 67, 136\}$ and calculate n, d according to [14, Section 4.2]. We denote by ps the "partial sums" technique, and by bis the "bisect" technique. Table 1 shows the results of our evaluations.

Experimental Setup. All benchmarks are run on an AMD Ryzen™ 7 9700X @ 3.8 GHz in a GNU/Linux environment (kernel 6.19.11) and gcc 15.2.1. Executables are compiled with the $\mathtt{-O3}$ $\mathtt{-march=native}$ $\mathtt{-mtune=native}$ set of optimization flags. We employ $\mathtt{\_BitInt(N)}$, a C23 type for (un)signed integers of bit length $\mathtt{N}$ [13, Section 6.2.5]. We set $\mathtt{N} = 256$ for the experiments discussed hereafter, and remark that the choice of $\mathtt{N}$ directly impacts storage requirements and performance of (un)ranking algorithms. We do not show $\prec_{\mathsf{gray}}$, $\prec_{\mathsf{bous}}$ or lexicographic variants due to their equivalent (or worse) performance compared to other orders. All tables are implemented as *ragged arrays* with appropriate metadata, packing the data structure optimally.

Performance Discussion. The performance of unranking algorithms is highly sensitive to the chosen parameter sets and the availability of pre-computed tables. In the absence of such mechanisms, $\prec_{\mathsf{rb}}$ is the most efficient strategy for larger k,

---

[1] The $\sqrt{\mathbb{V}(S_j)}$ term traces an elliptic segment (characteristic of a Brownian bridge) whose integral yields a total area of $\mathcal{O}(k\sqrt{k})$.

Table 1. Performance and behavior of selected unranking algorithms and tabling mechanisms. Table size is measured in 256-bit integers ($\times 10^3$) and performance is measured in CPU cycles ($\times 10^3$). Results are the average of 2^{10} calls with random seeds.

	Tabling mechanism			Unranking algorithm performance				
	Type	Size	Hits (%)	colex	ps	bis	rb	spiral
$C_{86}(384, 25)$	$-$	0.00	0.00	1141.03	178.78	9000.06	1613.85	1422.24
	χ_{bin}	10.34	100.00	74.60	102.53	382.53	221.10	92.33
	$\chi_{comb\text{-}2}$	3.25	97.63	10.79	103.41	383.40	49.92	13.78
	$\chi_{comb\text{-}4}$	5.55	99.99	9.13	103.84	381.19	50.13	12.73
	χ_{comb}	10.01	100.00	9.18	103.56	381.71	52.51	11.96
	$\chi_{acc\text{-}2}$	233.82	89.04	9.09	105.80	65.75	47.45	11.82
	$\chi_{acc\text{-}4}$	417.29	97.67	9.09	103.33	33.68	47.33	11.87
	χ_{acc}	753.48	100.00	9.12	103.82	7.46	47.53	11.98
$C_9(66, 67)$	$-$	0.00	0.00	490.13	282.91	3536.53	354.40	564.50
	χ_{bin}	6.83	100.00	37.24	41.93	171.70	77.20	42.58
	$\chi_{comb\text{-}2}$	1.29	97.36	6.12	45.59	171.56	24.34	8.18
	$\chi_{comb\text{-}4}$	2.28	100.00	5.72	42.25	172.74	25.25	7.20
	χ_{comb}	4.56	100.00	5.46	44.78	176.40	23.85	6.89
	$\chi_{acc\text{-}2}$	10.83	77.35	5.37	42.34	47.01	23.82	6.96
	$\chi_{acc\text{-}4}$	19.52	81.13	5.06	41.92	41.33	25.06	6.95
	χ_{acc}	46.36	100.00	5.11	42.34	6.45	24.13	6.90
$C_4(39, 136)$	$-$	0.00	0.00	932.53	759.97	5010.33	264.35	929.96
	χ_{bin}	14.80	100.00	52.92	53.76	206.63	68.89	57.28
	$\chi_{comb\text{-}2}$	1.65	96.25	7.59	54.16	206.42	34.37	10.68
	$\chi_{comb\text{-}4}$	2.88	99.96	5.52	54.46	202.73	35.03	8.91
	χ_{comb}	5.48	100.00	5.21	54.32	206.02	32.66	8.68
	$\chi_{acc\text{-}2}$	6.39	58.24	5.20	55.30	97.19	32.32	8.67
	$\chi_{acc\text{-}4}$	11.28	61.33	5.15	54.33	89.26	32.49	9.24
	χ_{acc}	31.28	100.00	5.15	54.73	5.93	32.90	8.64

particularly against the canonical $\prec_{colex}$. On the other hand, the ad hoc optimization ps drastically accelerates $\prec_{colex}$ for small k regimes and monotonic strategies. While $\prec_{spiral}$ offers a theoretically optimal search path, its practical advantages are restricted by the choice of parameters. In isolated synthetic evaluations with $C_8(200, 50)$ and $C_{42}(525, 25)$, performance is improved by up to 8% in the absence of tabling mechanisms.

However, parameters derived from known cryptographic algorithms are usually optimized for bandwidth and security bounds rather than statistical density. In these specific configurations, the variance of the partial sums remains sufficiently large such that $\prec_{spiral}$ does not yield a decisive computational advantage over other strategies previously discussed. Nonetheless, we emphasize that $\prec_{spiral}$ can be applied to future cryptographic primitives where larger, statistically dense parameter regimes are required.

Storage Discussion. The introduction of χ_{comb} essentially balances performance across the selected algorithms, except for ps, which does not benefit from any pre-computed tables other than χ_{bin}. If ample memory is available, bis paired with χ_{acc} is usually the fastest solution. Table 1 also shows how many wide integers are required to allocate several types of tabling mechanisms for D_l or a similar "linear" decomposition. Neither χ_{bin} nor χ_{acc} offer effective trade-offs; notably, χ_{acc} consumes 3–42$\times$ more storage than χ_{comb} for negligible speed gains ($< 3\times$). It also grows quickly even for moderate parameters and can only be used by a particular class of strategies.

Consequently, we identify χ_{comb} as the superior general-purpose choice. For constrained environments, probabilistic alternatives provide excellent value: setting $c = 4$ halves storage with only a 2–6% performance penalty. Finally, D_s remains highly efficient: as $|\mathsf{H}(k)| \approx \lg k$ by Proposition 1, it makes $\prec_{\mathsf{rb}}$ competitive against $\prec_{\mathsf{colex}}$ or $\prec_{\mathsf{spiral}}$, even when the recursion tree is unbalanced. For instance, the χ_{comb} table size is reduced from 5480 to only 400 integers if $\prec_{\mathsf{rb}}$ is employed for $\mathcal{C}_4(39, 136)$.

Constant-Time Implementations. We observe that while the generic framework does not eliminate side-channel vulnerabilities, it isolates the sources of information leakage. The side-channel resistance of any instantiated algorithm reduces to the choice of strategy $\Sigma = (\mathsf{D}, \mathsf{P}, \mathsf{S})$. Specifically, D operates only on public parameters, and S performs basic arithmetic on the rank r, which can be secured using standard constant-time arithmetic primitives. The attack surface is thus restricted to P; since it dictates the recursion path, it introduces data-dependent branching (loop terminations derived from z or r) and memory accesses (table lookups for $\mathfrak{C}_d(n, k)$). Consequently, producing a data-oblivious (un)ranking algorithm does not require restructuring the combinatorial logic, and reduces to the design of a constant-time evaluator for P.

5   Conclusion

We argue that our approach allows us to unify distinct algorithms found in the literature under a single algebraic definition, and study orders from a structural viewpoint. Furthermore, the choice of strategies is more suitable to finding new perspectives on optimizing (un)ranking algorithms for specialized applications. We provide an example of this methodology by proposing the "spiral" order, based on the uniform random input condition imposed by cryptographic algorithms. We hope this framework sheds light on the deep connections of the combinatorial theory underpinning the discussed cryptographic applications.

Future Work. We suggest the characterization of other ad hoc optimizations, such as pre-computing "pivots" to optimize the unranking algorithm, with respect to our strategy framework. Additionally, the statistics can be extended to analyze the storage requirements of the binomial table χ_{bin} under the uniform random input condition; preliminary evidence shows that similar high-probability patterns arise when Lemma 2 is used to evaluate $\mathfrak{C}_d(n, k)$. The formal verification of our generic implementations and the exploration of constant-time (un)ranking algorithms for cryptography are also highly interesting avenues of research.

Acknowledgments. We are grateful to the anonymous reviewers for their helpful comments and suggestions. G. Zambonin was partially supported by the Fundacao de Amparo a Pesquisa e Inovacao do Estado de Santa Catarina (FAPESC), Edital 62/2024, and by the National Operator of the Civil Registry of Natural Persons (ON-RCPN) of Brazil. L. G. Rosa was partially supported by the Fundacao de Amparo a Pesquisa e Inovacao do Estado de Santa Catarina (FAPESC), Edital 61/2024. D. Panario was partially funded by the Natural Sciences and Engineering Research Council of Canada (NSERC), reference number RGPIN-2024-05341.

Disclosure of Interests. The authors have no competing interests to declare that are relevant to the content of this article.

References

1. André, D.: Mémoire sur les combinaisons régulières et leurs applications. Annales scientifiques de l'École Normale Supérieure **5**, 155–198 (1876). https://doi.org/10.24033/asens.136. 2e série
2. Arndt, J.: Matters Computational: Ideas, Algorithms, Source Code, 1st edn. Springer, Heidelberg (2011)
3. Barenghi, A., Pelosi, G.: Constant weight strings in constant time: a building block for code- based post-quantum cryptosystems. In: CF 2020: Proceedings of the 17th ACM International Conference on Computing Frontiers, pp. 132–141 (2020). https://doi.org/10.1145/3387902.3392630
4. Bellare, M., Ristenpart, T., Rogaway, P., Stegers, T.: Format-preserving encryption. In: Jacobson, M.J., Rijmen, V., Safavi-Naini, R. (eds.) SAC 2009. LNCS, vol. 5867, pp. 295–312. Springer, Heidelberg (2009). https://doi.org/10.1007/978-3-642-05445-7_19
5. Eger, S.: Stirling's approximation for central extended binomial coefficients. J. Integer Sequences **16**(13.1.3) (2013)
6. Fahssi, N.E.: Polynomial triangles revisited. arXiv:1202.0228v7 (2012). https://doi.org/10.48550/arXiv.1202.0228
7. Filho, D.L.G., Adj, G., Bettaieb, S., Budroni, A., Chávez-Saab, J., Rodríguez-Henríquez, F.: Sota voce: low-noise sampling of sparse fixed-weight vectors. IACR Trans. Cryptogr. Hardw. Embed. Syst. **2026**(1), 353–375 (2026). https://doi.org/10.46586/tches.v2026.i1.353-375
8. Flajolet, P., Sedgewick, R.: Analytic Combinatorics, 1st edn. Cambridge University Press (2009)
9. Flajolet, P., Zimmermann, P., Cutsem, B.V.: A calculus for the random generation of labelled combinatorial structures. Theoret. Comput. Sci. **132**(1), 1–35 (1994). https://doi.org/10.1016/0304-3975(94)90226-7
10. Genitrini, A., Pépin, M.: Lexicographic unranking of combinations revisited. Algorithms **14**(3), 97 (2021). https://doi.org/10.3390/a14030097
11. Honda, M., Kaji, Y.: Optimum fingerprinting function for Winternitz one-time signature. In: 2024 IEEE International Symposium on Information Theory (ISIT), pp. 2898–2902 (2024). https://doi.org/10.1109/ISIT57864.2024.10619351
12. Hülsing, A., Kudinov, M., Ronen, E., Yogev, E.: SPHINCS+C: compressing SPHINCS+ with (almost) no cost. In: 2023 IEEE Symposium on Security and Privacy (S&P), pp. 1435–1453 (2023). https://doi.org/10.1109/SP46215.2023.10179381
13. ISO/IEC JTC 1/SC 22: Information technology — Programming languages — C. Standard 9899:2024, International Organization for Standardization (2024)
14. Khovratovich, D., Kudinov, M., Wagner, B.: At the top of the hypercube – better size-time tradeoffs for hash- based signatures. In: Kalai, Y.T., Kamara, S.F. (eds.) Advances in Cryptology – CRYPTO 2025. LNCS, vol. 16005, pp. 93–123 (2025). https://doi.org/10.1007/978-3-032-01887-8_4

15. Knuth, D.E.: The Art of Computer Programming, Volume 4A: Combinatorial Algorithms, Part 1, 1st edn. Addison-Wesley Professional (2011)
16. Krausz, M., Land, G., Richter-Brockmann, J., Güneysu, T.: A holistic approach towards side-channel secure fixed-weight polynomial sampling. In: Boldyreva, A., Kolesnikov, V. (eds.) Public-Key Cryptography – PKC 2023. LNCS, vol. 13941, pp. 94–124 (2023). https://doi.org/10.1007/978-3-031-31371-4_4
17. Kreher, D.L., Stinson, D.R.: Combinatorial Algorithms: Generation, Enumeration, and Search, 1st edn. CRC Press (1998)
18. Loève, M.: Probability Theory I, Graduate Texts in Mathematics, vol. 45. Springer (1977). https://doi.org/10.1007/978-1-4684-9464-8
19. López-García, M., Farouk-Marei, D.G., Cantó-Navarro, E.: Converting fixed-length binary strings into constant weight words: application on post-quantum cryptography. IEEE Trans. Dependable Secure Comput. $\mathbf{22}$(3), 3063–3074 (2024). https://doi.org/10.1109/TDSC.2024.3524626
20. Martínez, C., Molinero, X.: A generic approach for the unranking of labeled combinatorial classes. Random Struct. Alg. $\mathbf{19}$(3–4), 472–497 (2001). https://doi.org/10.1002/rsa.10025
21. Miracle, S., Yilek, S.: New algorithms and analyses for sum-preserving encryption. In: Agrawal, S., Lin, D. (eds.) Advances in Cryptology – ASIACRYPT 2022. LNCS, vol. 13793, pp. 3–31 (2022). https://doi.org/10.1007/978-3-031-22969-5_1
22. de Moivre, A.: Miscellanea analytica de seriebus et quadraturis, 1st edn. J. Tonson & J. Watts (1730)
23. Molinero, X.: Ordered generation of classes of combinatorial structures. Ph.D. thesis, Universitat Politècnica de Catalunya (2005)
24. Moura, L., Ullrich, S.: The lean 4 theorem prover and programming language. In: Platzer, A., Sutcliffe, G. (eds.) CADE 2021. LNCS (LNAI), vol. 12699, pp. 625–635. Springer, Cham (2021). https://doi.org/10.1007/978-3-030-79876-5_37
25. NIST: Stateless Hash-Based Digital Signature Standard. Federal Information Processing Standards Publication 205. National Institute of Standards and Technology (2024). https://doi.org/10.6028/nist.fips.205
26. Perin, L.P., Zambonin, G., Custódio, R., Moura, L., Panario, D.: Improved constant-sum encodings for hash-based signatures. J. Cryptogr. Eng. $\mathbf{11}$(4), 329–351 (2021). https://doi.org/10.1007/s13389-021-00264-9
27. Qin, Z., Sun, S.: Extending the SPHINCS+ framework: varying the tree heights and chain lengths. Cryptology ePrint Archive, Paper 2025/2236 (2025). https://eprint.iacr.org/2025/2236
28. Ratsaby, J.: Estimate of the number of restricted integer-partitions. Appl. Anal. Discrete Math. $\mathbf{2}$(2), 222–233 (2008). https://doi.org/10.2298/AADM0802222R
29. Reyzin, L., Reyzin, N.: Better than BiBa: short one-time signatures with fast signing and verifying. In: Batten, L., Seberry, J. (eds.) ACISP 2002. LNCS, vol. 2384, pp. 144–153. Springer, Heidelberg (2002). https://doi.org/10.1007/3-540-45450-0_11
30. Stojmenović, I.: Generating all and random instances of a combinatorial object. In: Handbook of Applied Algorithms: Solving Scientific, Engineering and Practical Problems, 1st edn., pp. 1–38. Wiley (2007). https://doi.org/10.1002/9780470175668.ch1
31. Tajik, K., et al.: Balancing image privacy and usability with thumbnail-preserving encryption. In: Network and Distributed Systems Security (NDSS) Symposium 2019 (2019). https://doi.org/10.14722/ndss.2019.23432

32. Vajnovszki, V., Vernay, R.: Restricted compositions and permutations: from old to new Gray codes. Inf. Process. Lett. **111**(13), 650–655 (2011). https://doi.org/10.1016/j.ipl.2011.03.022
33. Zambonin, G.: `bic`. Version 0.2.0 (2026). https://doi.org/10.5281/zenodo.19476406
34. Zambonin, G.: `bic-proofs`. Version 0.2.0 (2026). https://doi.org/10.5281/zenodo.19476432
35. Zhang, K., Cui, H., Yu, Y.: Revisiting the constant-sum Winternitz one-time signature with applications to SPHINCS$^+$ and XMSS. In: Handschuh, H., Lysyanskaya, A. (eds.) Advances in Cryptology – CRYPTO 2023. LNCS, vol. 14085, pp. 455–483 (2023). https://doi.org/10.1007/978-3-031-38554-4_15
36. Zhang, Y., Zhou, W., Zhao, R., Zhang, X., Cao, X.: F-TPE: flexible thumbnail-preserving encryption based on multi-pixel sum-preserving encryption. IEEE Trans. Multimedia **25**, 5877–5891 (2022). https://doi.org/10.1109/tmm.2022.3200310

Minimizing Makespan in Sublinear Time via Weighted Random Sampling

Bin Fu[1], Yumei Huo[2], and Hairong Zhao[3(✉)]

[1] University of Texas Rio Grande Valley, Edinburg, TX 78539, USA
`bin.fu@utrgv.edu`
[2] College of Staten Island, CUNY, Staten Island, NY 10314, USA
`yumei.huo@csi.cuny.edu`
[3] Purdue University Northwest, Hammond, IN 46323, USA
`hairong@purdue.edu`

Abstract. We consider the classical makespan minimization scheduling problem where n jobs must be scheduled on m identical machines. Using weighted random sampling, we develop two sublinear time approximation schemes: one for the case where n is known and one for the case where n is unknown. Both algorithms not only give a $(1 + 3\epsilon)$-approximation to the optimal makespan but also generate a sketch schedule.

Our first algorithm, which targets the case where n is known and draws samples in a single round under weighted random sampling, has a running time of $\tilde{O}(\frac{m^5}{\epsilon^4}\sqrt{n} + A(\lceil \frac{m}{\epsilon} \rceil, \epsilon))$, where $A(\mathcal{N}, \alpha)$ is the time complexity of any $(1 + \alpha)$-approximation scheme for the makespan minimization of $\mathcal{N}$ jobs. The second algorithm addresses the case where n is unknown. It uses adaptive weighted random sampling, and runs in sublinear time $\tilde{O}\left(\frac{m^5}{\epsilon^4}\sqrt{n} + A(\lceil \frac{m}{\epsilon} \rceil, \epsilon)\right)$.

Keywords: Sublinear time algorithms · makespan minimization · weighted random sampling

1 Introduction

We study the classical makespan minimization scheduling problem on parallel machines, where there are n jobs to be scheduled on m identical parallel machines. The jobs are all available at time 0 and are labeled $1, 2, \ldots, n$. Each job j, $1 \leq j \leq n$, has a processing time p_j and can be scheduled on any machine. At any time, only one job can be scheduled on each machine and a job can not be preempted. Given a schedule S, let C_j be the completion time of job j in S, the makespan of the schedule S is $C_{max} = \max_{1 \leq j \leq n} C_j$. The objective is to find the minimum makespan among all schedules, denoted as $OPT(I)$, or simply OPT if I is clear from the context. This classical scheduling problem, also known as the load balancing problem, has many applications in the manufacturing and service industries. Additionally, it plays significant roles in computer science, including applications in client-server networks, database systems, and cloud computing.

F. Foucaud and A. Parreau (Eds.): IWOCA 2026, LNCS 16587, pp. 531–545, 2026.
https://doi.org/10.1007/978-3-032-27732-9_37

In this paper, we aim to design sublinear-time approximation algorithms for this classical scheduling problem. Sublinear-time algorithms are suitable for settings where the input is so large that scanning the entire dataset is impractical. The sublinear-time algorithms compute a solution by examining only a small fraction of the input. Such algorithms are typically randomized and return approximate rather than exact solutions. Developing sublinear-time algorithms not only accelerates computation but also highlights interesting aspects of computation, particularly the power of randomization.

Many sublinear time algorithms have been developed based on uniform random sampling. While uniform sampling is easy to implement, it has a drawback in scheduling problems: When the jobs' processing times vary significantly, failing to sample the largest processing time may result in a significant loss of accuracy. Therefore, in this paper, we develop sublinear time algorithms using weighted random sampling which is defined below.

Definition 1. *Given n elements, where each element j is associated with a weight w_j, and the total weight is denoted by $W = \sum_{j=1}^{n} w_j$, **Weighted Random Sampling** (WRS) is a technique in which the probability of selecting an element is proportional to its weight, i.e., $\frac{w_j}{W}$.*

1.1   Related Work

The makespan minimization problem has been known to be NP-hard for a long time. In the 1960s, Graham showed that the List Scheduling rule and the Longest Processing Time First rule give approximations of $2 - \frac{1}{m}$ and $\frac{4}{3} - \frac{1}{3m}$, respectively (see [14] and [15]). Later, many approximation schemes have been developed for both the case where the number of machines m is fixed ([18]) and the case where m is arbitrary ([17,19,20]).

Over the past few decades, sublinear time algorithms have been developed for various applications including algebra, graph theory, geometry [2,3,5–8,11–13]. Most of these algorithms are based on uniform random sampling while only a few have used weighted sampling. For instance, weighted sampling has been used to estimate sums [1,4,22] and for bin packing [1]. In the latter, Batu, Berenbrink, and Sohler developed asymptotic approximation scheme that runs in $\tilde{O}(\sqrt{n} \cdot poly(1/\epsilon) + g(1/\epsilon)$ using weighted sampling alone, where $g(x)$ is an exponential function of x and in $\tilde{O}(n^{1/3} \cdot poly(1/\epsilon) + g(1/\epsilon)$ time when both weighted sampling and uniform sampling are allowed. In addition to an approximate value, the algorithm can also output a constant-size "template" of a packing that can later be used to find a near-optimal packing in linear time.

For scheduling problems, the only existing sublinear-time algorithms are those from [9] and [10]. In [9], they consider makespan minimization with chain precedence constraints. In [10], they develop a sublinear-time algorithm for the classical parallel machine makespan minimization problem. Both works use uniform random sampling and assume constrained input processing times.

In this paper, we consider the classical makespan minimization problem where the processing times are arbitrary, and develop the sublinear time approximation schemes using weighted sampling instead of uniform sampling. A brief discussion in [1] suggests that their sketch method for bin packing could be applied to approximate makespan minimization problem, but no detailed analysis is presented. Moreover, the algorithm in [1] only works for the case that n is known. In this paper, we develop two algorithms, one for the case in which n is known and another for the case in which n is unknown, which is more involved. Furthermore, as in [1], which outputs a constant-size "template" of a packing that can later be used to find a near-optimal packing in linear time, our algorithms can also be adapted to output not only an approximate value, but also a sketch of scheduling that takes a sublinear space and later can be used to generate a near optimal schedule in linear time. Such output is essential for many applications that require not only an approximate optimal value, but also a schedule of assigning jobs to machines.

1.2 New Contribution

In this paper, we develop sublinear time algorithms for the makespan minimization problem using weighted random sampling where the weight of a job is simply its processing time. Note that the lower bound of $\Omega(\sqrt{n})$ [22] for computing an approximate sum of n positive numbers using weighted random sampling also applies to our problem.

Our sublinear algorithms begin by constructing a sketch of the input instance using $\tilde{O}(\sqrt{n})$ samples. When the number of jobs n is given, we draw the samples in a single round. With high probability, the samples will include all jobs with large processing time and some medium jobs, but no small jobs. To construct the sketch of the input, we include all the sampled large jobs, divide the medium jobs into groups and use a generalized birthday paradox argument to estimate the number of jobs in each group, and disregard the small jobs. When the number of jobs n is unknown, we employ adaptive sampling: we sample jobs in several rounds, increasing the number of samples geometrically until we can apply the birthday paradox argument to estimate the number of jobs in all groups.

The second step of our algorithms is to compute an approximation of the optimal makespan for the scheduling problem based on the sketch constructed from the first step. To this end, we may employ any existing approximation scheme for makespan minimization ([16, 19–21]) as a black-box procedure and apply it to the large jobs.

As we mentioned earlier, many applications require not only an approximation of the optimal makespan but also a corresponding schedule that assigns jobs to machines. To meet this requirement, we use the concept of "sketch schedule" and show that we can modify our algorithms to compute "sketch schedule" while keeping the running time still sublinear in the number of jobs. Moreover, we present how "sketch schedule" can be used to construct a concrete schedule for jobs in I when full job information is available.

Let $A(\mathcal{N}, \alpha)$ be the time complexity of any $(1+\alpha)$-approximation scheme for $\mathcal{N}$ jobs. Our main results can be summarized in the following two theorems.

Theorem 1. *using non-adaptive weighted sampling, there is a randomized algorithm for the makespan minimization problem that gives a $(1+\epsilon)$-approximation in $\tilde{O}(\frac{m^5}{\epsilon^4}\sqrt{n}+A(\lceil\frac{m}{\epsilon}\rceil,\epsilon))$ time. Furthermore, it can compute a sketch schedule $\tilde{S}$, represented using $O(\frac{m^2}{\epsilon^2}\ln(mn^2))$ space, which can be used to generate a schedule of the jobs subsequently with a makespan of at most $(1+3\epsilon)$ times the optimal.*

Theorem 2. *When the number of jobs is unknown, there is a randomized algorithm for the makespan scheduling problem that uses adaptive weighted random samplings, and provides a $(1+\epsilon)$-approximation in $\tilde{O}\left(\frac{m^5}{\epsilon^4}\sqrt{n}+A(\lceil\frac{m}{\epsilon}\rceil,\epsilon)\right)$ time.*

Furthermore, it can compute a sketch schedule, represented using $O(\frac{m^2}{\epsilon^2}(\log\frac{nm}{\epsilon}))$ space, which can be used later to generate a schedule with a makespan at most $(1+3\epsilon)$ times the optimal.

The paper is organized as follows. Section 2 introduces key definitions and presents the framework of our sublinear-time algorithms, which consist of constructing an input sketch and computing an approximation for the optimal makespan. Subsection 2.3 develops sublinear-time approximation schemes assuming the sketch is given, while Sect. 3 designs and analyzes the sketch-construction algorithms, distinguishing between the cases where the number of jobs is known and unknown. Section 4 discusses applying the sketching framework to deterministic approximation algorithms. Section 5 concludes the paper.

Due to space constraints, we omit some of the proofs in Sect. 3.

2 Definitions and the Framework of The Algorithms

2.1 Definitions

Given an instance of the problem $I = \{p_j, 1 \leq j \leq n\}$, our algorithms draw a sublinear size of samples, then estimate the original job set based on the samples, and finally compute the approximate value of the optimal makespan based on the estimated job set. To estimate instance I, we construct a "sketch" instance $\tilde{I}$, which can be stored in a space of size sublinear in n. Let p_{max} be the largest job processing time in the input. Since p_{max} is unknown, our algorithm will first obtain an upper bound of p_{max}, denoted as p'_{max}. For a parameter $\delta, 0 < \delta < 1$, which will be decided later, we partition the jobs into groups such that group k contains all the jobs with the processing times in the interval $I_k = (p'_{max}(1-\delta)^k, p'_{max}(1-\delta)^{k-1}], k \geq 1$. For convenience, we also use $j \in I_k$ interchangeably with $p_j \in I_k$. A "sketch" instance $\tilde{I}$ is formally defined as follows.

Definition 2. *Given an instance of the problem $I = \{p_j, 1 \leq j \leq n\}$ and p'_{max} such that $p_{max} \leq p'_{max}$, let n_k be the number of jobs $j \in I_k$ where $I_k = (p'_{max}(1-\delta)^k, p'_{max}(1-\delta)^{k-1}], k \geq 1$. An input sketch of I is denoted as $\tilde{I} = \{\langle\tilde{n}_k, \tilde{p}_k\rangle\}$, where $\langle\tilde{n}_k, \tilde{p}_k\rangle$ estimates the jobs in interval I_k: $\tilde{n}_k$ approximates the number of jobs n_k, and $\tilde{p}_k = p'_{max}(1-\delta)^{k-1}$ approximates their processing time.*

Next, we define an $(\alpha, \beta_1, \beta_2)$-sketch of I, which characterizes how well a sketch approximates I in terms of the number of jobs in each interval, and $OPT(I)$. Without loss of generality, the sketch contains only those intervals I_k for which $\tilde{n}_k > 0$.

Definition 3. *Given an instance I we say an input sketch $\tilde{I} = \{\langle \tilde{n}_k, \tilde{p}_k \rangle : 1 \leq k \leq t\}$ is an $(\alpha, \beta_1, \beta_2)$-sketch of I if the following conditions hold:*

1. *For every $k \in [1, t]$ with $\tilde{n}_k > 0$, $(1 - \alpha)\, n_k \leq \tilde{n}_k \leq (1 + \alpha)\, n_k$*
2. *$(1 - \beta_1)\, OPT(I) \leq OPT(\tilde{I}) \leq (1 + \beta_1)\, OPT(I)$.*
3. *The total processing time of the jobs from intervals I_k not included in the sketch, i.e., those with $\tilde{n}_k = 0$ or the intervals $I_k, k > t$, is at most $\beta_2 \cdot OPT(I)$.*

2.2   The Framework of the Sublinear Time Algorithms

Our sublinear time algorithms follow the same structure, which consists of two stages: compute a sketch and derive an approximation from the sketch instance.

Main Algorithm:
  Input: ϵ, m, $I = \{p_j : 1 \leq j \leq n\}$
  Output: an approximate value of $OPT(I)$

1. generate an $(\alpha, \beta_1, \beta_2)$-sketch instance for the original instance I, $\tilde{I} = \{\langle \tilde{n}_k, \tilde{p}_k \rangle, 1 \leq k \leq t\}$, where α, β_1 and β_2 all depend on ϵ
2. compute an approximate value of $OPT(I)$ based on $\tilde{I}$.

Proposition 1. *Assuming that (1) there exists an algorithm $\mathcal{A}_1(.)$ that computes an $(\alpha, \beta_1, \beta_2)$-sketch for any given instance I of n jobs in time $T_1(\alpha, \beta_1, \beta_2, n, m)$, and (2) there exists another algorithm $\mathcal{A}_2(.)$ that computes a $(1 + \gamma)$-approximation for a sketch instance $\tilde{I} = \{\langle \tilde{n}_k, \tilde{p}_k \rangle : 1 \leq k \leq t\}$, $\tilde{n} = \sum_{k=1}^{t} \tilde{n}_k$, in $T_2(\gamma, t, \tilde{n}, m)$ time. Then there exists an approximation algorithm for any instance I that returns a value X, such that $(1 - \beta_1)OPT(I) \leq X \leq (1 + \gamma)(1 + \beta_1)OPT(I)$ with running time $O(T_1(\alpha, \beta_1, \beta_2, n, m) + T_2(\gamma, t, (1 + \alpha)n, m))$.*

So far, there is no existing sublinear time algorithm to obtain an $(\alpha, \beta_1, \beta_2)$-sketch of I. in sublinear time. Therefore, one of the main tasks in this paper is to develop sublinear time algorithms for obtaining an $(\alpha, \beta_1, \beta_2)$-sketch for any instance I. The details of these algorithms will be presented in Sect. 3. For the second step, while no sublinear-time algorithm exists to approximate from the input sketch, we can adapt existing classical makespan approximation schemes to develop one, with details given in the next subsection.

2.3   Sublinear Time Approximation Schemes for a Sketch Instance

Let A be any approximation scheme for the classical parallel machine makespan minimization scheduling problem. We design an efficient meta-algorithm that uses A as a black box to obtain $(1 + \epsilon)$-approximation for a sketch instance. The main idea is to apply A to the largest jobs and use a batch processing method to assign the remaining jobs.

Meta-Algorithm:

Input: ϵ, m

$\quad\quad\tilde{I} = \{\langle \tilde{n}_k, \tilde{p}_k \rangle : 1 \leq k \leq t\}$, $\tilde{p}_k > \tilde{p}_{k+1}$ for $k < t$, and $P = \sum_{k=1}^{t} \tilde{n}_k \tilde{p}_k$.

$\quad\quad A$: any approximation scheme for classical makespan minimization problem

Output: a $(1 + \epsilon)$-approximation of $OPT(\tilde{I})$

1. Let $\delta = \frac{\epsilon}{3}$ and $h(m, \delta) = \lceil \frac{m}{\delta} \rceil$
2. Apply A to the $h(m, \delta)$ largest jobs to obtain a schedule S whose makespan is at most $(1 + \delta)$ times that of the optimal schedule for these jobs
3. Let T_0 be the makespan of S
4. Return $T = (1 + \delta) \max(T_0, P/m)$

Theorem 3. *Assume A is a $(1 + \alpha)$-approximation scheme for the makespan minimization problem of scheduling n jobs on m machines, with running time $A(n, m, \alpha)$. Given any sketch instance $\tilde{I} = \{\langle \tilde{n}_1, \tilde{p}_1 \rangle, \langle \tilde{n}_2, \tilde{p}_2 \rangle, \cdots, \langle \tilde{n}_t, \tilde{p}_t \rangle\}$, the Meta-Algorithm returns a $(1 + \epsilon)$-approximation in time $O(A(h(m, \delta), m, \delta))$, where $\delta = \frac{\epsilon}{3}$, and $h(m, \delta) = \lceil \frac{m}{\delta} \rceil$.*

Proof. The time complexity is straight forward. We just show that $T \leq (1 + \epsilon)OPT(\tilde{I})$ and there exists a feasible schedule whose makespan is at most T.

First, by assumption, when we apply A to the $h(m, \delta)$ largest jobs, we get $T_0 \leq (1 + \delta)OPT(\tilde{I})$. Since P/m is a lower bound of $OPT(\tilde{I})$, we have

$$T = (1 + \delta) \max(P/m, T_0) \leq (1 + \delta)^2 OPT(\tilde{I}) \leq (1 + \epsilon)OPT(\tilde{I}).$$

Next, we show that there exists a feasible schedule whose makespan is at most T. Let $\tilde{p}_{k_0}$ be the largest processing time of the remaining jobs. Then, the processing time of each of the $h(m, \delta)$ largest jobs is at least $\tilde{p}_{k_0}$, and thus the makespan of S is at least $T_0 \geq h(m, \delta)\tilde{p}_{k_0}/m$. We claim that we can schedule all the remaining jobs based on S by time T using List Scheduling.

We consider two cases: (1) $P/m < T_0$. In this case, $T = (1 + \delta)T_0$, and $T - T_0 = \delta T_0 \geq \delta \cdot h(m, \delta) \cdot \tilde{p}_{k_0}/m = \delta \cdot \lceil \frac{m}{\delta} \rceil \cdot \tilde{p}_{k_0}/m \geq \tilde{p}_{k_0}$. This means that every remaining job can fit in the interval between T_0 and T. If at some point a job cannot be scheduled with the completion time at or before T on any machine, then we must have that all machines are busy at time $T_0 > P/m$. But this is impossible, since the total processing time is P. (2) $P/m \geq T_0$. In this case, $T = (1 + \delta)P/m$. Then the interval between T and P/m has length $T - P/m = \delta P/m \geq \delta T_0 \geq \tilde{p}_{k_0}$. That is, every remaining job can fit in the interval between P/m and T. If at some point a job cannot be scheduled with the completion time at or before T, we get a contradiction as in the first case.

The above Meta-algorithm returns only an approximation of the optimal makespan. However, in many applications, one may also need a schedule that specifies how the jobs are assigned to machines. In the following, we first introduce the concept of "sketch schedule" and show that we can modify the Meta-algorithm to compute a "sketch schedule" while still keeping the overall running time sublinear. Then, we show how a "sketch schedule" can be used to construct a concrete schedule for jobs in I when full job information is available.

2.4 Sketch Schedule

Definition 4. *Given a sketch instance $\tilde{I} = \{\langle \tilde{n}_k, \tilde{p}_k \rangle, 1 \le k \le t\}$ for instance I, we denote a sketch schedule for I as $\tilde{S} = \{\langle \tilde{n}_{i,j}, \tilde{p}_j \rangle, 1 \le i \le m, 1 \le j \le t, \sum_i \tilde{n}_{i,j} = \tilde{n}_j\}$, where $\langle \tilde{n}_{i,j}, \tilde{p}_j \rangle$ represents that $\tilde{n}_{i,j}$ jobs with processing time $\tilde{p}_j$ are assigned to machine M_i.*

Next, we show that we can modify the Meta-algorithm to generate a sketch schedule with additional time.

Theorem 4. *Given a sketch instance $\tilde{I} = \{\langle \tilde{n}_k, \tilde{p}_k \rangle : 1 \le k \le t\}$. A sketch schedule can be computed by the Meta-algorithm in $O(t + m)$ additional time.*

Proof. Let S be the schedule of the $h(m, \delta)$ largest jobs returned by A. Let $\tilde{S}$ be the sketch schedule that describes how the largest jobs are assigned to the m machines. For the remaining jobs, let p_{k_0} be the largest processing time, and let $\tilde{n}'_k$ be the number of jobs with processing time $\tilde{p}_k$ for $k_0 \le k \le t$. We can compute the assignment of the remaining jobs as follows:

1. $i = 1$
2. While $k_0 \le t$
   (a) Compute the number of jobs with processing time $\tilde{p}_{k_0}$ that can be scheduled on machine i by time T, that is, $\tilde{n}''_{k_0} = \min\{\tilde{n}'_{k_0}, \left\lfloor \frac{T - T_i}{\tilde{p}_{k_0}} \right\rfloor\}$
   (b) Update $T_i = T_i + \tilde{n}''_{k_0} \tilde{p}_{k_0}$ and $\tilde{n}'_{k_0} = \tilde{n}'_{k_0} - \tilde{n}''_{k_0}$
   (c) If $\tilde{n}'_{k_0} = 0$, then $k_0 = k_0 + 1$, else $i = i + 1$

For the time complexity, note that in each iteration, either k_0 or i is increased by one until $k_0 = t$; and $i \le m$. Thus we need $O(m + t)$ steps.

Next, we show that the sketch schedule can be used to generate a schedule of all n jobs when all the information of all jobs are accessed.

Theorem 5. *Given an instance I, an $(\alpha, \beta_1, \beta_2)$-sketch $\tilde{I}$ and a sketch schedule $\tilde{S}$, we can generate a schedule S for instance I based on $\tilde{S}$ and the makespan of S is at most $((1 + \beta_1)(1 + \frac{\alpha}{1-\alpha}m) + \beta_2)OPT(I)$.*

Proof. Based on the sketch schedule $\tilde{S}$, we can construct a schedule S for the job instance I interval by interval. For each interval $I_k, \tilde{n}_k > 0, 1 \le k \le t$, by the definition of $(\alpha, \beta_1, \beta_2)$-sketch, we have $(1 - \alpha) n_k \le \tilde{n}_k \le (1 + \alpha) n_k$, that is,

$$|(n_k - \tilde{n}_k)| \le \alpha \cdot n_k \le \frac{\alpha}{1-\alpha} \tilde{n}_k.$$

To construct S, we try to schedule the jobs from I_k in the same way as in $\tilde{S}$. In the case of $n_k > \tilde{n}_k$, there will be at most $\frac{\alpha}{1-\alpha}\tilde{n}_k$ jobs left unscheduled. We schedule these jobs to machine i that has the largest number of jobs from I_k scheduled, that is, $\tilde{n}_{i,k}$ is the largest among all $1 \leq i \leq m$, which implies $\tilde{n}_{i,k} \geq \frac{\tilde{n}_k}{m}$. Combining all the intervals, it can be shown that the makespan of S is increased by at most $\frac{\alpha}{1-\alpha}m$ times the makespan of $\tilde{S}$, which in turn is at most $(1+\beta_1)OPT(I)$. For the intervals I_k with $\tilde{n}_k = 0$ or $k \geq t$, we can schedule the jobs from these intervals on any machine, by definition of $(\alpha, \beta_1, \beta_2)$-sketch, the makespan will increase by at most $\beta_2 OPT$. In summary, the makespan of the constructed schedule is at most $((1+\beta_1)(1+\frac{\alpha}{1-\alpha}m)+\beta_2)OPT(I)$.

To summarize, we have shown that, for a given instance I, if we can obtain a $(\alpha, \beta_1, \beta_2)$-sketch $\tilde{I} = \{\langle \tilde{n}_k, \tilde{p}_k \rangle : 1 \leq k \leq t\}$ where t is sublinear in n, then we can get both an approximate value of $OPT(I)$ and a sketch schedule in sublinear time. From now on, we focus on designing sublinear-time algorithms to compute an $(\alpha, \beta_1, \beta_2)$-sketch for I. The algorithms are presented in two subsections: one for the case where the number of jobs n is known, and one where n is unknown. Throughout, we use weighted random sampling with replacement, distinguishing samples from jobs since a job may be sampled multiple times.

3   Computing the Sketch Instance in Sublinear Time

3.1   Generating a Sketch When n Is Known

We present a sublinear-time Sketch-Algorithm for computing an $(\alpha, \beta_1, \beta_2)$-sketch with known n. The algorithm has two main steps. In the first step, an upper bound p'_{max} on p_{max} is obtained by drawing a constant number of samples. We then consider jobs/samples by their processing time intervals, $I_k = (\tilde{p}_{k+1}, \tilde{p}_k]$ where $\tilde{p}_k = p'_{max} \cdot (1 - \delta)^{k-1}$.

In the second step, we construct the sketch instance $\tilde{I}$ by drawing $\tilde{O}(\sqrt{n})$ samples. We retain only the intervals I_k with $k \leq h = O(\log n)$ and with sufficiently many samples, discarding all others. To estimate the number of jobs in each remaining interval I_k $(1 \leq k \leq h)$, we distinguish two cases based on whether I_k contains a job that is sampled $\Omega(\log n)$ times. Let H_1 and H_2 denote the sets of intervals in these two cases, respectively. For intervals in H_1, which contain large jobs, we can directly estimate the number of jobs by counting the number of distinct sampled jobs. For intervals in H_2, which contain medium jobs, we obtain $\tilde{n}_k$ using a generalized birthday paradox argument. The algorithm details follow. For convenience, we define:

x_j: the number of times job j is sampled.
X_k: the number of samples whose processing times are in I_k.

Sketch-Algorithm
Input:
    the number of machines, m
    an instance $I = \{p_i : 1 \leq i \leq n\}$

a fixed parameter $\gamma_0 \in (0, \frac{1}{12}]$ to control the failure rate
the approximation error $\delta \in (0, \frac{1}{2})$

Output: a sketch instance $\tilde{I} = \{\langle \tilde{n}_i, \tilde{p}_i \rangle : 1 \leq i \leq h\}$ where $h = \frac{1}{\delta} \ln \frac{n^2}{\delta}$

1. Obtain an upper bound of p_{max}, p'_{max}, and determine intervals
   a. draw $K_0 = \lceil \log \frac{1}{\gamma_0} \rceil$ samples
   b. let p' be the maximum processing time among the K_0 samples
   c. let $p'_{max} = 2np'$
      $\tilde{p}_k = p'_{max} \cdot (1 - \delta)^{k-1}$, and $I_k = (\tilde{p}_{k+1}, \tilde{p}_k]$ for $k \geq 1$

2. Obtain the sketch instance $\tilde{I}$
   a. let $h = \frac{1}{\delta} \ln(\frac{n^2}{\delta})$
   b. let $\tau(n) = \delta h \ln(16h(\log_{1+\delta} 3\sqrt{n})/\gamma_0)$
   c. draw $K = \frac{36m}{\delta^4} \sqrt{n} \cdot \tau(n)$ samples
   d. let $H = \{I_k : k \leq h, \text{ and } X_k \geq \frac{\delta}{mh} K\}$
   e. let $f(n) = 36 \ln(\frac{2n}{\gamma_0})$
   f. let $H_1 = \{I_k : I_k \in H, \text{ and } \exists j, \text{ such that } j \in I_k, \text{ and } x_j \geq f(n)\}$ and
      $H_2 = H \setminus H_1$
   g. let $\tilde{I}_1 = \tilde{I}_2 = \emptyset$
   h. for each interval $I_k \in H_1$
          let $\tilde{n}_k$ be the number of jobs (not samples) from I_k that are sampled
          let $\tilde{I}_1 = \tilde{I}_1 \cup \{\langle \tilde{n}_k, \tilde{p}_k \rangle\}$
   i. for each interval $I_k \in H_2$
      I. let $h_0 = 3\sqrt{n}$
      II. $\tilde{n}_k = 1$
      III. partition the X_k samples from I_k into groups of size h_0, $G_1, \cdots, G_u$
      IV. for $l = 1, l \leq h_0, l = l(1 + \delta)$
             let $g_{l,k}$ be the number of groups among the u groups
             $G_1, \cdots, G_u$
             where the first l samples in the group are all distinct
             if $g_{l,k} \leq \frac{1}{\sqrt{e}} u$
                 let $\tilde{n}_k = l^2$, go to step V
      V. $\tilde{I}_2 = \tilde{I}_2 \cup \{\langle \tilde{n}_k, \tilde{p}_k \rangle\}$

   j. $\tilde{I} = \tilde{I}_1 \cup \tilde{I}_2$

In the following five lemmas, we show that p'_{max} is a good upper bound for p_{max} which is the maximum processing time among all n jobs; $\tilde{n}_k$ is a good estimation for n_k, $1 \leq k \leq h$, for all the intervals in H_1 and H_2, respectively; and the total processing time of jobs in the intervals not included in the sketch instance $\tilde{I}$ is very small.

Lemma 1. $Pr(p_{max} < p'_{max}) \geq 1 - \gamma_0$.

Lemma 2. *The probability that there exists an interval I_k such that $\langle \tilde{n}_k, \tilde{p}_k \rangle \in \tilde{I}_1$ and $\tilde{n}_k \neq n_k$ is at most γ_0.*

Lemma 3. *The probability that there exists an interval I_k such that $\langle \tilde{n}_k, \tilde{p}_k \rangle \in \tilde{I}_2$ and $\tilde{n}_k \notin [(1+\delta)^{-2c} \cdot n_k, (1+\delta)^{2c+2} \cdot n_k])$, $c = 4$, is at most $\frac{\gamma_0}{8}$.*

Lemma 4. *The total processing time of jobs in the interval I_k with $k > h$, is at most $2\delta OPT$.*

Lemma 5. *With probability at least $1 - \gamma_0$, the total processing time of jobs from the intervals I_k, $1 \leq k \leq h$ and $X_k < \frac{\delta}{mh} \cdot K$, is at most $2\delta OPT$.*

From the above lemmas, we can get the following.

Theorem 6. *With probability at least $1 - \frac{17}{8}\gamma_0$,*

$$(1 - 2cm\delta)(1 - 4\delta)\,OPT(I) \leq OPT(\tilde{I}) \leq (1 + 6cm\delta)(1 + \delta)\,OPT(I),$$

where $c = 4$ and $\delta \leq \frac{1}{4c}$.

Moreover, with high probability, the Sketch-Algorithm computes an $(\alpha, \beta_1, \beta_2)$-sketch with

$$\alpha = (1+\delta)^{2c+2}, \quad \beta_1 = (1 + 6cm\delta)(1 + \delta), \quad \beta_2 = 4\delta.$$

The sketch size is $O(h) = O\left(\frac{1}{\delta} \ln \frac{n^2}{\delta}\right)$. The running time is dominated by Step i, which computes $\tilde{n}_k$ for each interval $I_k \in H_2$. For each I_k, this can be done in $O(X_k)$ using a hash table to detect collisions in each group of $l \leq h_0$ samples, where $l = 1, (1+\delta), (1+\delta)^2, \ldots$. In total, the algorithm takes

$$O(K) = O\left(\frac{36\,m}{\delta^4}\sqrt{n} \cdot \tau(n)\right) = \tilde{O}\left(\frac{m}{\delta^4}\sqrt{n}\right).$$

Combining with the approximation scheme from Sect. 2.3, we are ready to prove Theorem 1.

Proof of Theorem 1: Given instance I, and parameters ϵ and γ_0, we follow the Main Algorithm. We first invoke the Sketch-Algorithm to get the sketch instance $\tilde{I}$ and then apply the Meta-algorithm to $\tilde{I}$ with appropriate parameters. For example, let $\delta = \frac{\epsilon}{12cm}$, and $\alpha = \frac{\epsilon}{4}$, by Theorems 6 and Theorem 3, we can get an approximation of $(1 + 3\epsilon)$ with probability at least $(1 - 3\gamma_0)$. The time complexity is $\tilde{O}(\frac{m^5}{\epsilon^4}\sqrt{n})$ for the Sketch-Algorithm, and $A(\lceil \frac{m}{\epsilon} \rceil, \epsilon)$ for the approximation scheme. Furthermore, the Meta-algorithm can generate a sketch schedule of size $O(\frac{m^2}{\epsilon^2}(\log \frac{nm}{\epsilon}))$. $\qquad\square$

3.2 Generating a Sketch When n is Unknown

We develop a sublinear time algorithm, Adaptive-Sketch-Algorithm, to compute input sketch $\tilde{I}$ when the number of jobs, n, is unknown. Similar to the Sketch Algorithm in Sect. 3.1, the Adaptive-Sketch-Algorithm has two main steps. First, we determine the processing time intervals I_k to be included in the sketch instance $\tilde{I}$; then we estimate the number of the jobs in each of the intervals.

In the first step, because n is unknown, there is a slight difference in determining the intervals I_k. Instead of defining the intervals based on an upper bound of p_{max}, we use a large job processing time w_0, where the jobs with processing times greater than w_0 have a very small overall impact on the makespan.

In the second step, we first determine h, the upper bound on the number of intervals $\tilde{I}$. Again, there is a slight difference since n is unknown. we choose h in such a way that jobs falling outside these intervals contribute minimally to the makespan. We then draw the first round of samples and try to construct $\tilde{I}$. As before, we consider only the intervals I_k for $k \leq h$ that have sufficiently many samples.

These intervals are divided into two sets, H_1 and H_2. While we can use the same method to estimate the number of jobs in the intervals of H_1, there is a key difference in estimating the number of jobs in the intervals of H_2 due to the unknown value of n. For these intervals, we may not be able to obtain an estimate using the birthday paradox from the first round of sampling, as we did in the Sketch Algorithm. So we increase the sample size geometrically and sample again. In each round, we try to estimate the number of jobs for as many intervals as possible. This process is repeated until an estimate can be obtained for all intervals.

We now formally present the Adaptive-Sketch-Algorithm. As before, X_k denotes the number of samples from interval I_k, and x_j the number of times job j is sampled. These values are updated whenever new samples are drawn.

Adaptive-Sketch-Algorithm
> Input: the instance $I = \{p_i : 1 \leq i \leq n\}$
>> the number of machines m
>> the approximation error $\delta \in (0, \frac{1}{3}]$, and
>> the failure rate parameter $\gamma_0 \in (0, \frac{1}{12}]$.
>
> Output: a sketch instance $\tilde{I} = \{\langle \tilde{n}_i, \tilde{p}_i \rangle\}$

> Steps:

1. Determine the intervals I_k
   a. let d_0 be the smallest integer such that $\left(1 - \frac{\delta}{m}\right)^{\frac{m}{\delta} d_0} \leq \gamma_0$
   b. draw $K = \frac{50m}{\delta} d_0$ random samples, and let w_0 be the largest processing time among the samples.
   c. for all $k \geq 1$, let $I_k = (\tilde{p}_{k+1}, \tilde{p}_k]$ where $\tilde{p}_k = w_0 \cdot (1 - \delta)^{k-1}$

2. Obtain the sketch instance $\tilde{I}$

a. let h be the smallest integer such that $\sum_{k>h} X_k \le \frac{\delta K}{4m} = \frac{d_0}{4}$
b. let $K_0 = \frac{8}{\beta_0} \cdot \ln \frac{e}{\beta_0 \gamma_0}$ where $\beta_0 = \frac{\delta^3}{32mh}$
c. draw K_0 random samples
d. let H be the set of intervals $H = \{I_k : 1 \le k \le h, \text{and } X_k \ge \frac{\delta}{mh} \cdot K_0\}$
e. let $H_1 = \{I_k : I_k \in H, \text{ and } \exists j, j \in I_k \text{ and } x_j \ge 8\beta_0 K_0\}$, and $H_2 = H \setminus H_1$
f. let $\tilde{I}_1 = \tilde{I}_2 = \emptyset$
g. for each interval $I_k \in H_1$
     let $\tilde{n}_k$ be the number of sampled jobs from I_k
     $\tilde{I}_1 = \tilde{I}_1 \cup \{\langle \tilde{n}_k, \tilde{p}_k \rangle\}$
h. for each interval $I_k \in H_2$ let $gs_k = 1$
i. let $j = 1$
j. while there exists $I_k \in H_2$ that are not marked
   I. discard the current samples, draw $K_j = 2^j K_0$ new samples
   II. for each unmarked interval I_k
     i. find the largest integer t such that $X_k \ge l_t \cdot u_t$ where $l_t = (1+\delta)^t$,
       $u_t = \lceil \frac{3e}{\delta^2}(\ln \frac{1}{\gamma_t}) \rceil$ and $\gamma_t = \frac{\delta}{hl_t}\gamma_0$
     ii. partition the samples in I_k evenly into u_t groups $G_1, \cdots, G_{u_t}$
     iii. for $i = gs_k$, $i \le t$ and I_k still unmarked, $i = i+1$
       let $l_i = (1+\delta)^i$, $\gamma_i = \frac{\delta}{hl_i}\gamma_0$, and $u_i = \lceil \frac{3e}{\delta^2}(\ln \frac{1}{\gamma_i}) \rceil$
       let $g_{l_i,k}$ be the number of groups among $G_1, \cdots, G_{u_i}$
       where the first l_i samples in the group are all distinct
       if $g_{l_i,k} \le \frac{1}{\sqrt{e}} u_i$
         let $\tilde{n}_k = l_j^2$
         let $\tilde{I}_2 = \tilde{I}_2 \cup \{\langle \tilde{n}_k, \tilde{p}_k \rangle\}$
         mark I_k
     iv. if I_k is still unmarked, $gs_k = t$

   III. $j = j+1$

k. $\tilde{I} = \tilde{I}_1 \cup \tilde{I}_2$

First, we show that, with high probability, w_0 provides a good bound in the sense that the total processing time of all jobs with processing times greater than w_0 has small impact on estimating OPT.

Lemma 6. *Let w_δ be the smallest processing time among all jobs such that $\sum_{p_j > w_\delta} p_j \le \frac{\delta P}{m}$. Then, with probability at least $1 - \gamma_0$, $w_0 \ge w_\delta$, which implies $\sum_{p_j > w_0} p_j \le \delta OPT$.*

Then we show that the number of intervals in H is bounded by $h = O(\log mn)$. Furthermore, for all the intervals in H_1 and H_2, the algorithm computes a good estimate of the number of jobs in each interval of H_1 and H_2.

Lemma 7. *With probability at most γ_0, we have $h > \frac{1}{\delta} \ln \frac{8nm}{\delta(1-\delta)}$.*

Lemma 8. *With probability at least $(1 - 2\gamma_0)$, $\tilde{n}_k = n_k$ for all k, $I_k \in H_1$.*

Lemma 9. *The probability that there exists an interval I_k such that $\langle \tilde{n}_k, \tilde{p}_k \rangle \in \tilde{I}_2$ and $\tilde{n}_k \notin [(1+\delta)^{-2c} \cdot n_k, (1+\delta)^{2c} \cdot n_k)])$ where $c = 4$, is at most γ_0.*

Based on the Lemmas 7, 8, and 9, we can show that the sketch instance $\tilde{I}$ provide a good estimation of OPT.

Theorem 7. *With probability at least $1 - 7\gamma_0$,*

$$(1 - 2cm\delta)(1 - 4\delta)OPT(I) \leq OPT(\tilde{I}) \leq (1 + 6cm\delta)(1 + \delta)OPT(I) \ ,$$

where $c = 4$.

The running time of the adaptive sketch algorithm is roughly proportional to the number of samples.

Lemma 10. *With probability at least $1 - 2\gamma_0$, Adaptive-Sketch-Algorithm draws $O(\frac{1}{\delta^4} m \sqrt{n} \cdot \ln(mn/\delta\gamma_0))$ weighted random samples.*

Proof of Theorem 2: By Theorem 7, we can invoke the Adaptive-Sketch-Algorithm with appropriate δ, to get the $(1 + \epsilon)$-approximation ratio and sketch size with probability at least $1 - 7\gamma_0$. Furthermore, it can generate a sketch schedule whose size is $O(mh) = O(\frac{m^2}{\epsilon^2}(\log \frac{nm}{\epsilon}))$ which can be used to generate a schedule whose makespan is at most $(1 + 3\epsilon)OPT(I)$.

The running time of the algorithm consists of two parts. The first is for computing the sketch instance using Adaptive-Sketch-Algorithm. For each iteration, we draw K_j samples, and try to estimate n_k for $I_k \in H_2$. This takes $O(K_j)$ time. From Lemma 10, the $\sum K_j = O\left(\frac{m^5}{\epsilon^4}\sqrt{n}\log^2 n)\right)$. The second part is for finding approximation of the sketch instance, which follows from Theorem 3.

4 Discussion: Application to Deterministic Algorithms Design

We show that by incorporating the idea of sketch instances and existing deterministic approximation algorithms, we can obtain faster deterministic approximation algorithms. The only source of randomness in our algorithms is in the first step, where the input sketch is constructed. To obtain a deterministic algorithm, all we need to change is to construct the input sketch deterministically. Once we get the input sketch, we can apply the Meta-Algorithm to get an approximation or use the modified version to get a sketch schedule which then can be converted to a real schedule.

Theorem 8. *Assume that there is a $(1 + \alpha)$-approximation algorithm A for the makespan problem of n jobs on m machines, with running time $A(n, m, \alpha)$. Then for any $\epsilon \in (0, 1)$, there is a $O(A(h(m, \delta), m, \delta) + n)$ time $(1 + \epsilon)$-approximation scheme for an input of n jobs, where $\delta = \frac{\epsilon}{3}$, and $h(m, \delta) = \lceil \frac{m}{\delta} \rceil$. A schedule can be obtained in additional $O(\frac{1}{\epsilon}\log\frac{n}{\epsilon} + m)$ time.*

Note that, from the theorem, whenever $A(\frac{3m}{\epsilon}, m, \frac{m}{\epsilon}) = O(n)$, we have an approximation scheme that runs in linear time. For example, for the algorithm in [20], $A(\frac{3m}{\epsilon}, m, \frac{m}{\epsilon}) = 2^{O(\frac{1}{\epsilon}(\log\frac{1}{\epsilon})^4)} + \frac{m}{\epsilon}(\log\frac{m}{\epsilon})$ which is $O(n)$ when $m = O(n/\log n)$. Consequently, by invoking the algorithm of [20] only on a subset of jobs in the sketch instance, we obtain a deterministic approximation scheme with overall running time $O(n)$. This is asymptotically faster than applying the algorithm of [20] directly to the entire instance.

5   Conclusions

In this paper, we designed the sublinear time algorithms for the classical parallel machine makespan minimization scheduling problem. Using weighted random sampling, our algorithms return an approximate value in sublinear time. Furthermore, our algorithms can generate a sketch schedule which, if needed, can be used to produce a real schedule when the complete job information is obtained. It is worth noting that weighted random sampling is well-suited for many scheduling problems with arbitrary job processing times. We believe our approach can be extended to other scheduling problems as well.

References

1. Batu, T., Berenbrink, P., Sohler, C.: A sublinear-time approximation scheme for bin packing. Theor. Comput. Sci. **410**(47), 5082–5092 (2009). issn: 0304-3975
2. Behnezhad, S.: Time-optimal sublinear algorithms for matching and vertex cover. In: 2021 IEEE 62nd Annual Symposium on Foundations of Computer Science (FOCS), pp. 873–884 (2022)
3. Behnezhad, S., Roghani, M., Rubinstein, A.: Sublinear time algorithms and complexity of approximate maximum matching. In: Proceedings of the 55th Annual ACM Symposium on Theory of Computing. STOC 2023, pp. 267–280. Association for Computing Machinery, Orlando (2023). isbn: 9781450399135
4. Beretta, L., Tětek, J.: Better sum estimation via weighted sampling. In: Proceedings of the 2022 Annual ACM-SIAM Symposium on Discrete Algorithms (SODA), pp. 2303–2338 (2022)
5. Chazelle, B., Liu, D., Magen, A.: Sublinear geometric algorithms. SIAM J. Comput. **35**, 627–646 (2005)
6. Czumaj, A., Sohler, C.: Estimating the weight of metric minimum spanning trees in sublinear-time. In: Proceedings of the 36th ACM Symposium on Theory of Computing, pp. 175–183 (2004)
7. Feige, U.: On sumes of independent random variables with unbounded variance and estimating the average degree in a graph. SIAM J. Comput. **35**, 964–984 (2006)
8. Fu, B., Chen, Z.: Sublinear-time algorithms for width-bounded geometric separators and their applications to protein side-chain packing problems. J. Comb. Optim. **15**, 387–407 (2008)
9. Fu, B., Huo, Y., Zhao, H.: Sublinear algorithms for scheduling with chain precedence constraints. In: Proceedings of the 30th International Computing and Combinatorics Conference (COCOON 2024) (2024)

10. Fu, B., Huo, Y., Zhao, H.: Sublinear time approximation schemes for makespan minimization on parallel machines. Math. Methods Oper. Res. **101**(3), 507–528 (2025). issn: 1432-5217
11. Goldreich, O., Ron, D.: Approximating average parameters of graphs. Technical report 05-73 (2005)
12. Goldreich, O., Goldwasser, S., Ron, D.: Property testing and its connection to learning and approximation. J. ACM **45**(4), 653–750 (1998)
13. Goldreich, O., Ron, D.: On testing expansion in bounded-degree graphs. Electron. Colloquium Comput. Complex. TR00 (2000)
14. Graham, R.L.: Bounds for certain multiprocessing anomalies. Bell Syst. Tech. J. **45**, 1563–1581 (1966)
15. Graham, R.L.: Bounds on multiprocessing timing anomalies. J. SIAM Appl. Math. **17**, 416–429 (1969)
16. Hochbaum, D.S., Landy, D.: Scheduling semiconductor burn-in operations to minimize total flowtime. Oper. Res. **45**(6), 874–885 (1997)
17. Hochbaum, D.S., Shmoys, D.B.: Using dual approximation algorithms for scheduling problems theoretical and practical results. J. ACM **34**(1), 144–162 (1987)
18. Horowitz, E., Sahni, S.: Exact and approximate algorithms for scheduling nonidentical processors. J. ACM **23**, 317–327 (1976)
19. K Jansen. An EPTAS for scheduling jobs on uniform processors: using an MILP relaxation with a constant number of integral variables. SIAM J. Disc. Math. **24**(2), 457–485 (2010). https://doi.org/10.1137/090749451
20. Jansen, K., Klein, K.M., Verschae, J.: Closing the gap for makespan scheduling via sparsification techniques. Math. Oper. Res. **45**(4), 1371–1392 (2020). https://doi.org/10.1287/MOOR.2019.1036
21. Leung, J.Y.-T.: Bin packing with restricted piece sizes. Inf. Process. Lett. **31**(3), 145–149 (1989). https://doi.org/10.1016/0020-0190(89)90223-8
22. Motwani, R., Panigrahy, R., Xu, Y.: Estimating sum by weighted sampling. In: International Colloquium on Automata, Languages and Programming, (ICALP), July 2007, pp. 53–64 (2007)

Author Index

A

Abu-Khzam, Faisal N.   1
Adak, Rajat   16
Akhtar, Sheikh Shakil   32
Amouzandeh, Aflatoun   45
Antony, Dhanyamol   60
Aráujo, Júlio   73
Aráujo, Samuel N.   73

B

Bakal, Deepak M.   88
Bar-Noy, Amotz   103
Beaton, Iain   118
Beisegel, Jesse   131
Bodlaender, Hans L.   146
Borse, Y. M.   88
Breitkopf, Tom-Lukas   161

C

Cameron, Ben   118
Chakraborty, Dipayan   1
Chatterjee, Sobyasachi   175
Choudhury, Neelabjo Shubhashis   190
Cioni, Lapo   204
Custódio, Ricardo   515

D

D'Ascenzo, Andrea   220
Das, Soura Sena   236
Depian, Thomas   251
Dobrev, Stefan   266
Dürr, Christoph   281

F

Fotakis, Dimitris   296
Froese, Vincent   161
Fu, Bin   531

G

Georgiou, Konstantinos   266

G

German, Samuel   311
Gourvès, Laurent   296
Gremelmaier Rosa, Larissa   515
Gupta, Chetan   190
Gupta, Sushmita   175

H

Haase, Carolina   251
Hakanen, Anni   327
Hamada, Shunsuke   341
Havet, Frédéric   356
Herrmann, Anton   161
Huo, Yumei   531

I

Isenmann, Lucas   1
Italiano, Giuseppe F.   220

J

Jacob, Dalu   60
Jansen, Klaus   45, 370

K

Kalinichev, Igor   103
Kalyanasundaram, Subrahmanyam   385
Kanellopoulos, Sotiris   220
Köhler, Ekkehard   131
Kranakis, Evangelos   266
Krizanc, Danny   266
Kumar, Hitendra   400
Kumar, Subodh   385

L

Lamprou, Ioannis   415
Lazaropoulos, Nikolaos   415
Lehtilä, Tuomo   236
Liyanage, Adiesha   430

M

Mallem, Maher 146
Marcille, Clara 442
Marino, Andrea 204
Martins, Beatriz 456
Medeiros, Pedro P. 73
Merino, Arturo 281
Mömke, Tobias 370
Moura, Lucia 470
Mpanti, Anna 220
Mumey, Brendan 430

N

Narayanan, Lata 266
Nichterlein, André 161
Nisse, Nicolas 73
Nöllenburg, Martin 251

O

Oijid, Nacim 1, 442
Okada, Yuto 341
Ono, Hirotaka 341
Opatrny, Jaroslav 266
Otachi, Yota 341

P

Pagourtzis, Aris 220, 296
Panario, Daniel 515
Pandey, Arti 500
Pankratov, Denis 266
Parida, Prangya 470
Patsilinakos, Panagiotis 296
Paul, Kaustav 500
Pavan, P. D. 327
Peleg, David 103
Pergaminelis, Christos 220
Philip, Geevarghese 32
Pirotton, Lis 45

R

Rambaud, Clément 356
Rawitz, Dror 103
Reddy, Sangam Balchandar 486
Richer, Camille 161

S

Sandeep, R. B. 60, 400
Saurabh, Saket 175
Scheffler, Robert 131
Schoeters, Jason 204
Schulz, André 251
Schumacher, Björn 370
Seetharaman, Sanjay 175
Sen, Sagnik 236
Sharma, Ankit 500
Shende, Sunil 266
Sigalas, Ioannis 415
Silva, Caroline 73, 356
Singh Porte, Kumud 400
Sopp, Braeden 430
Soto, José A. 281
Stevens, Brett 470
Strehler, Martin 131
Sunil Chandran, L. 60

T

Tewari, Raghunath 190
Trotignon, Nicolas 456

U

Uno, Takeaki 204
Upasana, Anannya 175

V

van Stee, Rob 45
Vaxevanakis, Ioannis 415
Verma, Pragya 16
Verschae, José 281

W

Wambsganz, Corinna 45
Williams, Aaron 470

Z

Zambonin, Gustavo 515
Zhao, Hairong 531
Zissimopoulos, Vassilis 415

MIX
Papier aus verantwortungsvollen Quellen
Paper from responsible sources
FSC® C105338

If you have any concerns about our products,
you can contact us on
ProductSafety@springernature.com

In case Publisher is established outside the EU,
the EU authorized representative is:
**Springer Nature Customer Service Center GmbH
Europaplatz 3, 69115 Heidelberg, Germany**

Printed by Libri Plureos GmbH
in Hamburg, Germany